# LACRIMAL GLAND, TEAR FILM, AND DRY EYE SYNDROMES

## Basic Science and Clinical Relevance

# ADVANCES IN EXPERIMENTAL MEDICINE AND BIOLOGY

# LACRIMAL GLAND, TEAR FILM, AND DRY EYE SYNDROMES

Basic Science and Clinical Relevance

Edited by

## David A. Sullivan
Schepens Eye Research Institute
and Harvard Medical School
Boston, Massachusetts

Associate Editors

## B. Britt Bromberg
University of New Orleans, New Orleans, Louisiana

## Michele M. Cripps
Louisiana State University Medical Center, New Orleans, Louisiana

Assistant Editors

## Darlene A. Dartt
Schepens Eye Research Institute and Harvard Medical School, Boston, Massachusetts

## Donald L. MacKeen
Georgetown University Medical Center, Washington, D. C.

## Austin K. Mircheff
University of Southern California School of Medicine, Los Angeles, California

## Paul C. Montgomery
Wayne State University Medical School, Detroit, Michigan

## Kazuo Tsubota
Ichikawa General Hospital and Tokyo Dental College, Chiba, Japan

## Benjamin Walcott
State University of New York at Stony Brook, Stony Brook, New York

PLENUM PRESS • NEW YORK AND LONDON

Library of Congress Cataloging in Publication Data

Lacrimal gland, tear film, and dry eye syndromes: basic science and clinical relevance
/ edited by David A. Sullivan; associate editors, A. Britt Bromberg, Michele M. Cripps;
assistant editors, Darlene A. Dartt . . . [et al.].
    p.    cm.—(Advances in experimental medicine and biology; v. 350)
  "Proceedings of an International Conference on the Lacrimal Gland, Tear Film, and
Dry Eye Syndromes: Basic Science and Clinical Relevance, held November 14–17,
1992, in Southampton, Bermuda"—T.p. verso.
  Includes bibliographical references and index.
  ISBN 0-306-44676-6
  1. Lacrimal apparatus—Physiology—Congresses. 2. Tears—Congresses. 3. Dry eye
syndromes—Congresses. 4. Sjögren's syndrome—Congresses. I. Sullivan, David A. II.
International Conference on the Lacrimal Gland, Tear Film, and Dry Eye Syndromes:
Basic Science and Clinical Relevance (1992: Southampton, Bermuda Islands) III.
Series.
  DNLM: 1. Lacrimal Apparatus—congresses. 2. Tears—physiology—congresses. 3.
Dry Eye Syndromes—diagnosis—congresses. 4. Dry Eye Syndromes—therapy—
congresses. W1 AD559 v.350 1994 / WW 208 L1453 1994]
QP188.T4L44  1994
612.8′47—dc20
DNLM/DLC                                               94-2222
for Library of Congress                                    CIP

Proceedings of an International Conference on the Lacrimal Gland, Tear Film, and Dry Eye Syndromes:
Basic Science and Clinical Relevance, held November 14–17, 1992, in Southampton, Bermuda

ISBN  0-306-44676-6

©1994 Plenum Press, New York
A Division of Plenum Publishing Corporation
233 Spring Street, New York, N.Y. 10013

Printed in the United States of America

## DEDICATION

We dedicate this book to Drs. Mathea R. Allansmith, Stella Y. Botelho, Frank Flynn, Frank J. Holly, Mogens S. Norn and O. Paul van Bijsterveld for their pioneering efforts and outstanding achievements in basic and clinical research on the lacrimal gland, tear film and dry eye syndromes.

The Editors

# ACKNOWLEDGMENTS

We would like to express our sincere appreciation to the following companies, whose generous financial contributions significantly offset the educational expenses and publication costs associated with the International Conference on the Lacrimal Gland, Tear Film and Dry Eye Syndromes: Basic Science and Clinical Relevance.

## ALCON OPHTHALMIC
Alcon Laboratories, Inc., 6201 South Freeway, Fort Worth, Texas 76134-2099

## SANDOZ PHARMACEUTICAL CORPORATION
3525 Lake Trail, Kenner, Louisiana 70065

## LACRIMEDICS, INC.
190 North Arrowhead Avenue, Suite #B, Rialto, California 92376

## DR. MANN PHARMA
Brunsbütteler Damm 165, Postfach 20 04 56, D-W-1000 Berlin, Germany

## IOLAB CORPORATION
500 Iolab Drive, Claremont, CA 91711

## CIBA VISION OPHTHALMICS
11460 John's Creek Parkway, Duluth, Georgia 30136

## ALLERGAN, INC.
2525 Dupont Ave., Irvine, CA 92715

## ROSS LABORATORIES
625 Cleveland Avenue, Columbus, Ohio 43215

## SANTEN PHARMACEUTICAL COMPANY, LTD.
9-19, Shimoshinjo 3-chome, Higashi Yodogawa-ku, Osaka 533, Japan

# PREFACE

During the past decade a significant international research effort has been directed towards understanding the composition and regulation of the preocular tear film. This effort has been motivated by the recognition that the tear film plays an essential role in maintaining corneal and conjunctival integrity, protecting against microbial challenge and preserving visual acuity. In addition, research has been stimulated by the knowledge that alteration or deficiency of the tear film, which occurs in countless individuals throughout the world, may lead to desiccation of the ocular surface, ulceration and perforation of the cornea, an increased incidence of infectious disease, and potentially, pronounced visual disability and blindness.7

To promote further progress in this field of vision research, the International Conference on the Lacrimal Gland, Tear Film and Dry Eye Syndromes: Basic Science and Clinical Relevance was held in the Southampton Princess Resort in Bermuda from November 14 to 17, 1992. This meeting was designed to assess critically the current knowledge and 'state of the art' research on the structure and function of lacrimal tissue and tears in both health and disease. The goal of this conference was to provide an international exchange of information that would be of value to basic scientists involved in eye research, to physicians in the ophthalmological community, and to pharmaceutical companies with an interest in the treatment of lacrimal gland, tear film or ocular surface disorders (e.g. Sjögren's syndrome).

To help achieve this objective, over 180 scientists, physicians and industry representatives from Australia, Brazil, Canada, Denmark, England, Finland, France, Germany, Israel, India, Italy, Japan, The Netherlands, Poland, Scotland, Spain and the United States registered as active participants in this conference. In addition, this volume, which contains summaries of the conference's keynote, oral and poster presentations, was created to provide an educational foundation and scientific reference for tear film research.

The editors would like to thank the following individuals, whose scientific, administrative, or technical advice and/or assistance helped make the International Conference on the Lacrimal Gland, Tear Film, and Dry Eye Syndromes: Basis Science and Clinical Relevance, as well as this book, a reality: Mark B. Abelson, M.D., Mathea R. Allansmith, M.D., Gillian Alexander, Alan Bergl, Anthony J. Bron, M.D., Marshall G.

Doane, Ph.D., Mary Gallagher, Michelle George, Jeffrey P. Gilbard, M.D., Jack V. Greiner, Ph.D., O.D., D.O., Louane E. Hann, Zhiyan Huang, M.D., Jan Jorrin, Ilya Kagansky, Robin S. Kelleher, Ph.D., Aize Kijlstra, Ph.D., Kevin J. Klein, Ross William Lambert, Ph.D., Carole Lanigan, Kathleen Lavin, Myca Mooshian, J. Daniel Nelson, M.D., Stephen C. Pflugfelder, M.D., Miguel F. Refojo, Ph.D., Margaret Rocco, Bernard Rossignol, Ph.D., Elcio H. Sato, M.D., Kendyl Schaefer, Amy G. Sullivan, Benjamin D. Sullivan, Rose M. Sullivan, Marva Trott, Janice Ubels, John L. Ubels, Ph.D., and L. Alexandra Wickham.

David A. Sullivan

# CONTENTS

## LACRIMAL GLAND: SIGNAL TRANSDUCTION
## AND PROTEIN SECRETION

## LACRIMAL GLAND: IMMUNOLOGY

## MOLECULAR BIOLOGICAL APPROACHES TO THE
## STUDY OF THE LACRIMAL GLAND AND TEAR FILM

## TEAR FILM COMPOSITION AND
## BIOPHYSICAL PROPERTIES

# TEAR PROTEINS AND GROWTH FACTORS

## TEAR FILM: PHARMACOLOGICAL APPROACHES
## AND EFFECTS

## TEAR FILM AND CONTACT LENS INTERACTIONS

# ARTIFICIAL TEARS

# DRY EYE SYNDROMES: PATHOGENESIS AND DIAGNOSIS

### DRY EYE SYNDROMES:
### TREATMENT AND CLINICAL TRIALS

## SJÖGREN'S SYNDROME: PATHOGENESIS AND DIAGNOSIS

## SJÖGREN'S SYNDROME: TREATMENT

## SJÖGREN'S SYNDROME ASSOCIATIONS

# REGULATION OF TEAR SECRETION

Darlene A. Dartt

Schepens Eye Research Institute
Department of Ophthalmology
Harvard Medical School
20 Staniford Street
Boston, MA 02114

## INTRODUCTION

The tear film, which is the interface between the external environment and the ocular surface has several differing functions.[1] It forms a smooth refracting surface over the otherwise irregular corneal surface and lubricates the eyelids. Moreover, it maintains an optimal extracellular environment for the epithelial cells of the cornea and conjunctiva because the electrolyte composition, osmolarity, pH, $O_2$ and $CO_2$ levels, nutrient levels, and concentration of growth factors in the tears is regulated within narrow limits. Tears dilute and wash away noxious stimuli.[1,2] They also provide an antibacterial system for the ocular surface and serve as an entry pathway for polymorphonuclear leukocytes, in the case of injury to the ocular surface. As tears have many and varied functions, it is not surprising that they have a complex structure and are produced by several different sources.

The tear film consists of three layers.[3] The inner layer is a mucous layer that coats the cornea and conjunctiva. It was previously thought to be 1-μm thick,[4] but new evidence suggests that it may be far thicker.[5] The mucous layer consists of mucins, electrolytes, water, IgA, and enzymes. The middle layer is an aqueous layer that is about 7-μm thick.[3] This layer contains electrolytes, water, IgA, and proteins, many of which are antibacterial enzymes. Finally, the outer layer is a lipid layer about 0.1-μm thick, which floats on the aqueous layer.[3] The lipid layer contains a complex mixture of hydrocarbons, sterol esters, wax esters, triacyglycerols, free cholesterol, free fatty acids, and polar lipids.[6]

Each layer of the tear film is secreted by a different set of orbital glands (Fig. 1). The mucous layer is secreted by the goblet cells, which in humans occur singly as well as in glands of differing structure.[7] In addition, the stratified squamous epithelial cells of the cornea and conjunctiva may contribute to the mucous layer by secreting glycoproteins that may be mucins.[8,9] The aqueous layer is secreted by the main and accessory lacrimal glands, which secrete electrolytes, water, and proteins. In addition, the corneal epithelial cells contribute to the aqueous layer by secreting electrolytes and water into tears.[10] Finally, the conjunctiva could contribute to the tear film by two different mechanisms. Like the cornea, the conjunctival epithelial cells could secrete electrolytes and water into tears. Unlike the cornea, the conjunctival blood vessels could leak serum (electrolytes, water, and plasma proteins) into tears. The lipid layer is secreted by the meibomian glands that line the eyelid.

It has been suggested that there are two types of secretion from the orbital glands and secretory epithelia, basic secretion and reflex secretion. Jones[11] defined basic secretion as a constant, slow baseline secretion and reflex secretion as an increased rate of secretion caused by stimulation, for example neural stimulation. The main lacrimal gland has been classified as a reflex secretor and the other orbital glands as basic secretors. This classification implies that secretion only from the main lacrimal gland is regulated and secretion of only the aqueous layer of the tear film is regulated. The remainder of this paper focuses on evidence suggesting that secretion by all the orbital glands and secretory epithelia of the ocular surface is regulated, and therefore secretion of all three layers of the tear film is regulated. Thus each gland and each layer of the tear film would be able to respond to challenges and changes in the environment. This would contribute to the maintenance of the remarkably stable and constant tear film.

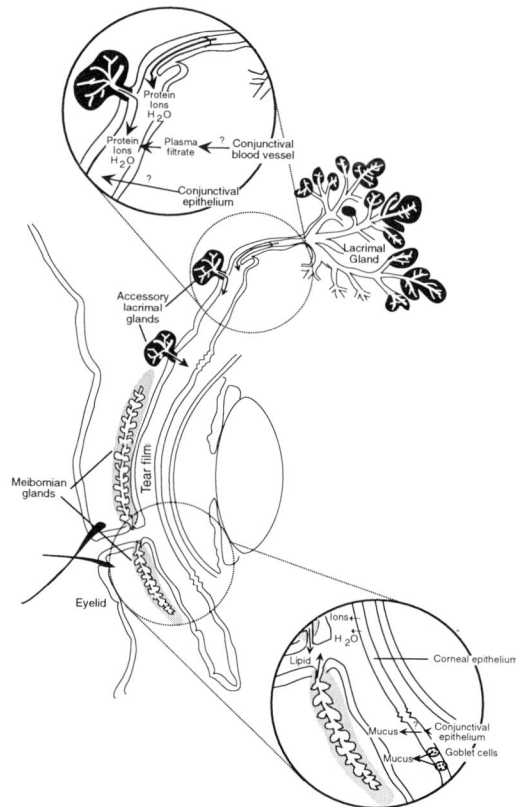

**Figure 1.** Schematic of the orbital glands and ocular epithelia that secrete the different layers of the tear film (reprinted with permission: D.A. Dartt, Physiology of tear production, *in*: "The Dry Eye: A Comprehensive Guide," M.A. Lemp and R. Marguardt, eds., Springer-Verlag, Berlin [1992]).

## REGULATION OF SECRETION OF THE MUCOUS LAYER

The goblet cells are the primary contributors to the mucous layer. Goblet cells are single cells or groups of cells located in the surface of the conjunctival epithelium and are connected to neighboring cells by tight junctions.[7] Goblet cells synthesize and secrete a heterogeneous group of O-linked glycoproteins called mucins.[12] The protein core of the mucin molecule is synthesized in the endoplasmic reticulum. and the complex branching chains of carbohydrates are added in the golgi apparatus and trans golgi network.[12] The synthesized mucins are stored in large secretory granules that fill a large portion of the goblet cell on its apical side. With stimulation, the mucin granule membranes fuse with the apical membrane of the cell or with other granule membranes in an explosive event that releases mucins onto the ocular surface. Goblet cells of the conjunctiva, like those of other epithelia, are not directly innervated.[7] However, the conjunctival stroma and the stratified squamous cells of the epithelium are innervated.[13] Sensory, sympathetic and parasympathetic nerves innervate the conjunctiva, and the loose stroma and widely spaced epithelial cells provide a pathway for diffusion of neurotransmitters to the goblet cells. In addition, paracrine and autocrine stimulation may be important mechanisms of stimulation.

Preliminary evidence suggests that goblet cells secrete mucin in response to neural stimulation.[14] In rats, a sensory stimulus, probably neural, from the cornea (a corneal debridement wound) caused goblet cell mucin secretion.[14] In addition, several neurotransmitters applied topically to the eye also caused mucin secretion. These neurotransmitters are epinephrine, phenylephrine, serotonin, dopamine and vasoactive intestinal peptide.[14] As an indication of the second messenger that these stimuli could be using, 1-isobutyl-3-methylxanthine (IBMX), a cAMP-dependent phosphodiesterase inhibitor that prevents breakdown of cAMP, stimulated goblet cell mucin secretion.

From these data it can be hypothesized that neural stimuli, such as reflexes from the cornea and conjunctiva, induce conjunctival goblet cell mucous secretion. Several different neurotransmitters appear to be involved, and cAMP may be the second messenger pathway used (Fig. 2). To summarize regulation of secretion of the mucous layer, secretion of goblet cell mucus to produce the mucous layer of the tear film appears to be under neural control and a cAMP-dependent signal transduction pathway may be involved.

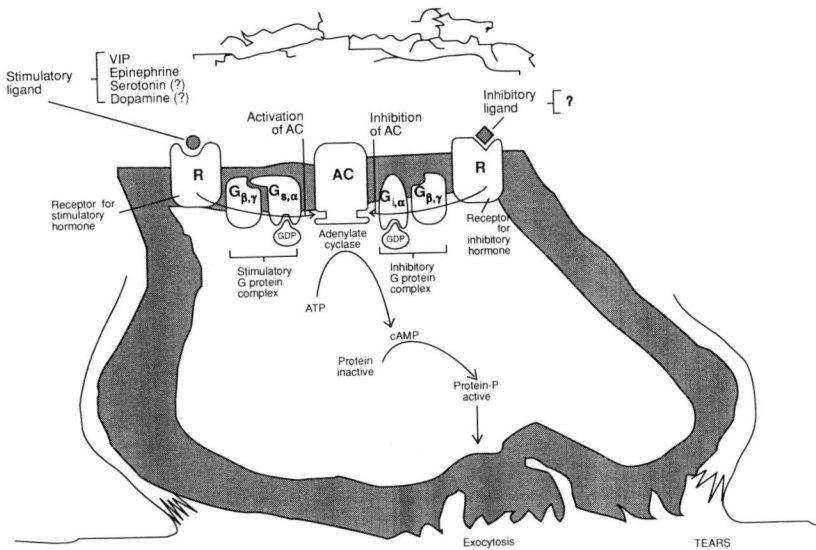

**Figure 2.** Schematic of the cAMP-dependent cellular signal transduction pathway used to stimulate goblet cell mucous secretion.

## REGULATION OF SECRETION OF THE AQUEOUS LAYER

### Secretion from the Main Lacrimal Gland

The main lacrimal gland is a compound tubular or a compound alveolar gland, depending on the species.[15] It consists of acini, ducts, nerves, myoepithelial cells, mast cells and plasma cells. About 80% of the gland is acini, which secrete electrolytes, water, and proteins to form primary fluid.[16] As the primary fluid moves along the duct system, duct cells modify the primary fluid by secreting or absorbing electrolytes. The final lacrimal gland fluid is then secreted onto the surface of the eye. It is well known that lacrimal gland fluid secretion is under neural control. Reflexes from the ocular surface and optic nerve, as well as from higher centers of the brain, stimulate lacrimal gland fluid secretion using parasympathetic and sympathetic efferent pathways. Parasympathetic and sympathetic nerves innervate the acinar cells, duct cells, and blood vessels of the lacrimal gland.[17] These nerves contain the neurotransmitters acetylcholine (ACh) and VIP in parasympathetic nerves, norepinephrine and perhaps NPY in sympathetic nerves, and substance P and perhaps calcitonin gene-related peptide (CGRP) in sensory nerves.[18,19] The proenkephalin family of peptides are also present in the lacrimal gland.[20]

The parasympathetic nerves provide the major short-term stimulatory regulation of lacrimal gland secretion. ACh stimulates both electrolyte and water, as well as protein, secretion by first activating muscarinic receptors on the plasma membrane of acinar and duct cells (Fig. 3). This interaction, via an unknown G protein, activates phospholipase C to breakdown phosphatidyl inositol bisphosphate into 1,4,5-inositol trisphosphate (1,4,5-IP$_3$) and diacylglycerol.[21-23] 1,4,5-IP$_3$ causes the release of intracellular $Ca^{2+}$. In addition, 1,3,4,5-IP$_4$ may also cause influx of $Ca^{2+}$, although this point remains controversial. $Ca^{2+}$, perhaps with calmodulin, activates $Ca^{2+}$/calmodulin protein kinases, which phosphorylate specific proteins to activate ion channels in the apical and basolateral membranes, causing electrolyte and water secretion, and to fuse secretory granule membranes with the apical plasma membrane, causing protein secretion. The diacylglycerol released with 1,4,5-IP$_3$ causes the translocation and activation of protein kinase C (PKC) from the cytosol to the plasma membrane where it phosphorylates specific proteins that cause secretion, similarly to those proteins phosphorylated by $Ca^{2+}$/calmodulin kinases.

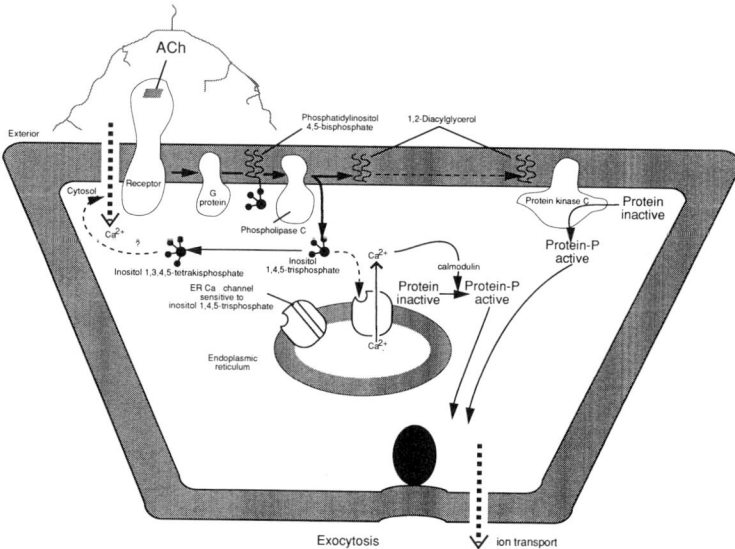

**Figure 3.** Schematic of 1,4,5 inositol trisphosphate/$Ca^{2+}$/diacylglycerol-dependent signal transduction pathway used to stimulate main lacrimal gland electrolyte water and protein secretion.

VIP is also released probably from parasympathetic nerves along with ACh and causes both electrolyte, water and protein secretion.[24] VIP activates specific receptors on the plasma membrane, which via a stimulatory G protein, activate adenylate cyclase (Fig. 4). Adenylate cyclase produces cAMP from ATP. cAMP, like $Ca^{2+}$/calmodulin and diacylglycerol, activates specific protein kinases, cAMP-dependent protein kinases, to phosphorylate proteins and cause secretion. β-Adrenergic agonists, adrenocrticotropic hormone (ACTH), and α-melanocyte stimulating hormone (α−MSH) also stimulate lacrimal gland secretion using the cAMP-dependent pathway.[25]

The sympathetic nerves, which release norepinephrine, also play a role in regulating lacrimal gland secretion. Norepinephrine, an α$_1$-adrenergic agonist in the lacrimal gland, causes protein secretion, but its effect on electrolyte and water secretion is unknown.[26,27] In vivo, norepinephrine is a potent vasoconstrictor, which would inhibit fluid secretion. Surprisingly, the mechanism by which α$_1$-adrenergic agonists stimulate lacrimal gland protein secretion is unknown, as they do not increase 1,4,5-IP$_3$ levels, $Ca^{2+}$ levels or cAMP levels, but they do translocate PKC.[27] α$_1$-Adrenergic agonists could stimulate protein secretion by activating phospholipase D and producing diacylglycerol that would in turn translocate PKC.

4

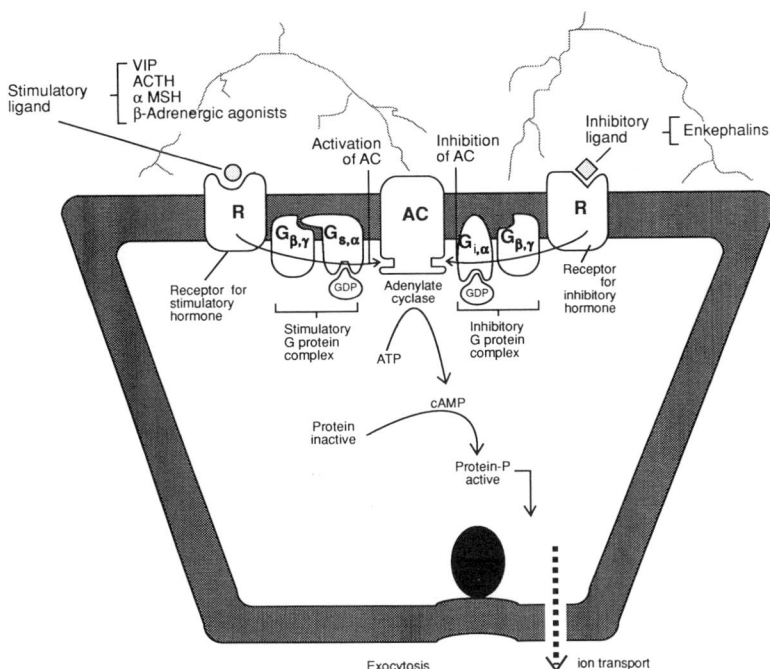

**Figure 4.** Schematic of cAMP-dependent signal transduction pathway used to stimulate or inhibit main lacrimal gland electrolyte water and protein secretion.

Main lacrimal gland secretion is regulated by neural stimuli that activate three different cellular signal transduction pathways: the muscarinic pathway which uses 1,4,5-IP$_3$, Ca$^{2+}$, and diacylglycerol; the $\alpha_1$-adrenergic pathway which uses an unknown mechanism, and the VIP pathway which uses cAMP. Activation of each of these pathways stimulates secretion.

Lacrimal gland secretion can be inhibited by the proenkephalin family of peptides, which are present in the lacrimal gland.[20] These peptides inhibit stimulated secretion by interacting with inhibitory G proteins that prevent activation of adenylate cyclase by stimulatory G proteins (Fig.4).[28]

### Accessory Lacrimal Gland Secretion

Accessory lacrimal glands are mini lacrimal glands embedded in the conjunctiva. These glands resemble the main lacrimal gland both structurally and histologically and secrete the same proteins.[29] There is no direct evidence for regulation of accessory lacrimal gland secretion, but there is indirect evidence. Furthermore, nerves have been identified in these glands, indicating for the first time that they appear to be innervated.[30] Indirect evidence for the stimulation of accessory lacrimal gland secretion was obtained using rabbits with experimental keratoconjunctivitis sicca.[31] In these rabbits, the lacrimal gland excretory duct was cauterized, and the harderian and nictitans glands removed. The accessory lacrimal glands were the primary tissues remaining that could contribute to the aqueous layer, although the corneal and the conjunctival epithelium were also possible sources. Agonists were applied topically to the ocular surface, and the volume of tears measured. VIP, glucagon, $\alpha-,\beta-$, and $\gamma-$MSH, which activate specific receptors; forskolin, which activates adenylate cyclase activity; 8-bromo cAMP, a permeable cAMP analog; and IBMX, a cAMP-phosphodiesterase inhibitor, each stimulated fluid secretion.[31] Pilocarpine, a cholinergic agonist, stimulated secretion, but by a sensory mechanism, not by activation of muscarinic receptors. This evidence suggests that there is neural regulation of accessory lacrimal gland secretion of electrolytes, water and protein and that agonists stimulate secretion by using the cAMP-dependent pathway (Fig. 5).

5

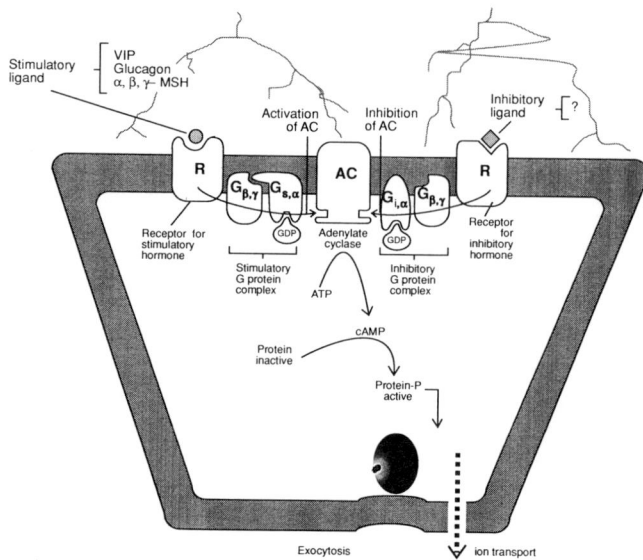

**Figure 5.** Schematic of cAMP-dependent signal transduction pathway used to stimulate accessory lacrimal gland electrolyte, water and protein secretion.

## Secretion by the Corneal Epithelium

Corneal epithelial cells are a third possible source of the aqueous layer of tears. The cornea is densely innervated with nerves containing classical neurotransmitters as well as neuropeptides.[32] Upon neural stimulation, the cornea secretes electrolytes and water into tears.[10] The effective stimuli include β–adrenergic agonists, dopamine, serotonin, dibutyryl cAMP a permeable cAMP analog and theophylline a phosphodiesterase inhibitor.[10] All these compounds stimulate secretion by increasing cellular cAMP levels. Thus, there is neural regulation of electrolyte and water secretion into the aqueous layer of tears by the corneal epithelium and the cAMP-dependent signal transduction pathway is employed (Fig. 6).

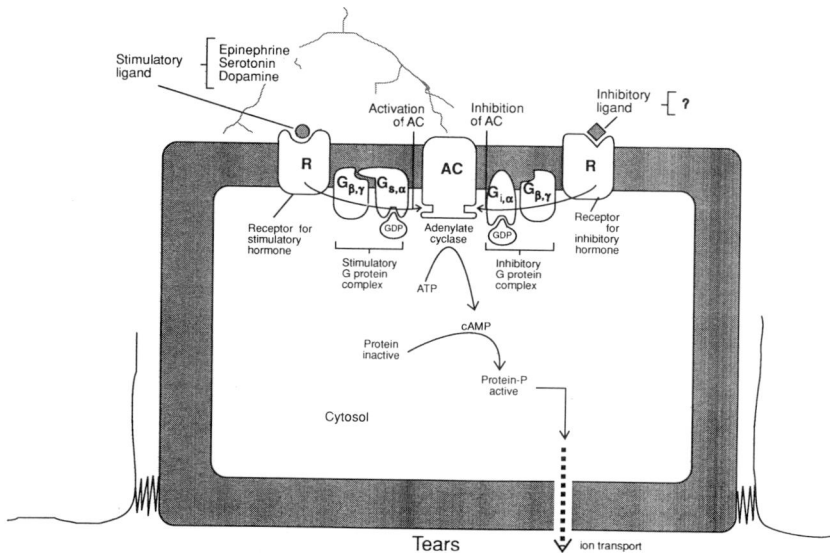

**Figure 6.** Schematic of cAMP-dependent signal transduction pathway used to stimulate corneal epithelial cell electrolyte and water secretion.

## Secretion by the Conjunctival Epithelium

Little is known about electrolyte and water secretion or absorption by the conjunctiva. The only study on conjunctival electrolyte and water transport found that the conjunctiva normally absorbs tears .[33] With stimulation, the conjunctival epithelium could secrete electrolytes and water into tears in a manner similar to that of the corneal epithelium. The stimulated conjunctival epithelium could produce a larger volume of tears than that of the cornea, as it is vascular and has a larger surface area. There is also the possibility of neural regulation, as the conjunctiva is innervated by sensory, parasympathetic and sympathetic nerves.[13]

The conjunctiva could also contribute to the aqueous layer of tears by a second mechanism, that is by an increase in the permeability of the conjunctival blood vessels. In the rabbit, this appears not to be the case.[34] Tears were collected from rabbits with experimental KCS, and the proteins separated by SDS-PAG electrophoresis. Topical application of agonists that increased tear secretion, VIP and IBMX, did not increase the concentration of plasma proteins in tears. Histamine, a compound known to increase conjunctival vascular permeability did increase the plasma proteins in tears.

To summarize the regulation of secretion of the aqueous layer, there are four possible sources for the aqueous layer, the main lacrimal gland, the accessory lacrimal glands, the corneal epithelium and the conjunctival epithelium. There is evidence that secretion by the main and accessory lacrimal glands and the corneal epithelium is regulated by nerves. Neurotransmitters that activate cAMP-dependent pathways stimulate secretion by these three tissues. In addition, neurotransmitters that active a $1,4,5\text{-}IP_3$, $Ca^{2+}$, diacylglycerol-dependent pathway, and an as yet unidentified pathway stimulate secretion by the main lacrimal gland. There is no evidence whether or not the conjunctival epithleium can contribute to the aqueous layer of tears. It is important to point out that the main lacrimal gland, the accessory lacrimal glands, and the corneal epithelium, although they may all be stimulated to secrete, do not secrete the same volume of fluid. In addition, there is a continuum of secretion from low to high. With low amounts of stimulation, all four tissues could secrete approximately the same volume of fluid, but, as the amount of stimulation increases, the main lacrimal gland could contribute a larger volume than the other tissues.

## REGULATION OF SECRETION OF THE LIPID LAYER

The meibomian glands are the major contributor to the lipid layer of the tear film. These glands lie in a row along the upper and lower lids, and their ducts open directly onto the inner margin of the lids in a perfect position for the lipids to spread over the aqueous layer. Although it has been shown that blinking fills and releases the meibomian gland fluid from the ducts, nothing is known about the stimulation of meibomian gland secretion. Meibomian glands are innervated as parasympathetic and sympathetic nerves as well as the neuropeptides substance P and CGRP are present.[35] Unlike the secretory fluid from the other orbital glands, meibomian gland fluid is secreted by a holocrine mechanism, that is, the entire cell including the stored lipids is released. With this type of secretion there are two possible levels at which meibomian gland secretion could be regulated. First, the rate of synthesis of lipids could be regulated; second, the rate of maturation and disintegration of cells could be regulated. These mechanisms are likely to be long-term mechanisms, compared to the short term regulation of secretion that has been described for the other layers of the tear film. Because the regulation may be long term, hormones rather than nerves may play the primary role in regulation, although nerves could play a secondary, modulatory role.

## SUMMARY

Although many questions remain about the regulation of secretion by the different layers of the tear film, several hypotheses can be suggested. We hypothesize first, that secretion of all layers of the tear film and all orbital glands and ocular epithelia that secrete tears is regulated; second, that neural regulation of secretion is of primary importance; and third, that the cAMP-dependent signal transduction pathway plays a pivotal role in this regulation and, except in the main lacrimal gland, $Ca^{2+}$ plays a secondary role. These hypotheses are suggested as the basis for further work and not as conclusions based on current knowledge.

## REFERENCES

1.    D.W. Lamberts, Physiology of the tear film, in: "The Cornea: Scientific Foundations and Clinical Practice," G. Smolin and R.A. Thoft, eds., Little, Brown and Company, Boston (1983).

2.    M.A. Lemp, Basic principles and classifications of dry eye disorders, in: "The Dry Eye: A Comprehensive Guide," M.A. Lemp and R. Marguardt, eds., Springer-Verlag, Berlin (1992).

3.    F.J. Holly and M.A. Lemp, Tear physiology and dry eye, Surv. Ophthalmol. 22:69 (1977).

4.    B.A. Nichols, M.L. Chiappino, C.R. Dawson, Demonstration of the mucous layer of the tear film by electron microscopy, Invest. Ophthalmol. Vis. Sci. 26:464 (1985).

5.    J.I. Prydal, F.W. Campbell, Study of the precorneal tear film thickness and structure by interferometry and confocal microscopy, Invest. Ophthalmol. Vis. Sci, 33:1996 (1992).

6.    J.M. Tiffany, Individual variations in human meibomian lipid composition, Exp. Eye Res. 27:289 (1978).

7.    S.V. Kessing, Mucous gland system of the conjunctiva, Acta Ophthalmol. 46(Suppl. 95):9 (1968).

8.    I.K. Gipson, M. Yankauckas, S.J. Spurr-Michaud, A.S. Tisdale, W Rinehart. Characteristics of a glycoprotein in the ocular surface glycocalyx. Invest. Ophthalmol. Vis. Sci. 33:218 (1992).

9.    J.V. Greiner, T.A. Weidman, D.R. Korb, M.R. Allansmith, Histochemical analysis of secretory vesicles in nongoblet conjunctival epithelial cells, Acta Ophthalmol. 63:89 (1985).

10.   S.D. Klyce and C.E. Crosson, Transport processes across the rabbit corneal epithelium: a reivew, Curr. Eye Res. 4:323 (1985).

11.   L.T. Jones, The lacrimal system and its treatment, Am. J. Ophthalmol. 62:47 (1966).

12.   M.R. Neutra and C.P. LeBlond. Synthesis of the carbohydrate of mucus in the golgi complex as shown by electron microscope radioautography of goblet cells from rats injected with glucose-$H_3$, J. Cell Biol. 30:119 (1966).

13.   G.L. Ruskell, Innervation of the conjunctiva. Trans. Ophthalmol. Soc. UK. 104:390 (1985).

14.   T. Kessler and D. Dartt, Neural stimulation of rat conjunctival goblet cell mucous secretion. This Volume--1993.

15.   B.B. Bromberg, M.H. Welch, R.W. Beuerman, S.-K. Chew, H.S. Thompson, D. Ramage, S. Githens, Histochemical distribution of carbonic anhydrase in rat and rabbit lacrimal gland. Invest. Ophthalmol. Vis. Sci. 34:339 (1993).

16.   S.Y. Botelho, Tears and the lacrimal gland, Sci. Am. 211:78 (1964).

17.   A. Ichikawa and Y. Nakajima, Electron microsope study on the lacrimal gland of the rat, Tohuku J. Exp. Med. 77:136 (1962).

18.   A. Nikinen, J.I. Lehtosalo, H. Uusital, A. Palkoma, P. Panula, The lacrimal glands of the rat and guinea pig are innervated by nerve fibers containing immunoreactivities for substance P and vasoactive intestinal peptide, Histochemistry. 81:23 (1984).

19.   R.M. Williams, J. Singh, K.A. Sharkey. Morphological differences in the innervation and mast cell density of the lacrimal gland in young and aged rats. This Volume--1993.

20.   J. Lehtosalo, H. Uusitalo, T. Mahrberg, P. Panula, A. Palkama, Nerve fibers showing immunoreactivities for proenkephalin A-derived peptides in the lacrimal glands of the guinea pig, Graefes Arch. Clin. Exp. Ophthalmol. 227:445 (1989).

21.  D.A. Dartt, D.M. Dicker, L.V. Ronco, I.M. Kjeldsen, R.R. Hodges, and S.A. Murphy. Lacrimal gland inositol trisophosphate isomer and tehakisphosphate production. *Am. J. Physiol.* 259:G274 (1990).

22.  O.H. Petersen. Cytoplasmic $Ca^{2+}$ signals, $Ca^{2+}$-dependent ion channels and tear formation, **This Volume--1993**.

23.  J.W. Putney, Jr., Inositol phosphates and calcium signalling in lacrimal acinar cells. **This Volume--1993**.

24.  D.A. Dartt, A.K. Baker, C. Vaillant, P.E. Rose, Vasoactive intestinal polypeptide stimulation of protein secretion from rat lacrimal gland acini, *Am. J. Physiol.* 247:G502 (1984).

25.  R. Jahn, U. Padel, P.-H. Porsch, H.-D. Soling, Adrenocorticotropic hormone and α-melanocyte stimulating hormone induce secretion and protein phosphorylation in the rat lacrimal gland by activation of a cAMP-dependent pathway, *Eur. J. Biochem.* 126:623 (1982).

26.  P. Maudiuit, G. Herman, and B. Rossignol, Protein secretion in lacrimal gland: $\alpha_1$-β synergism, *Am. J. Physiol.* 250:C704 (1985).

27.  R.R. Hodges, D.M. Dicker, P.E. Rose, and D.A. Dartt, $\alpha_1$-Adrenergic and cholinergic agonists use separate signal transduction pathways in lacrimal gland, *Am. J. Physiol.* 262:G1087 (1992).

28.  M.M. Cripps, and D.J. Bennett, Peptidergic stimulation and inhibition of lacrimal gland adenylate cyclase. *Invest. Ophthalmol. Vis. Sci.* 31:2145 (1990).

29.  T.E. Gillette, M.R. Allansmith, J.V. Greiner, M. Janersz, Histologic and immunohistologic comparison of main and accessory lacrimal tissue. *Am. J. Ophthalmol.* 89:724 (1980).

30.  P. Seifert and M. Spitznas, Light and electron microscopic morphology of accessory lacrimal glands, **This Volume--1993**.

31.  J.P. Gilbard, S.R. Rossi, K.G. Heyda, D.A. Dartt, Stimulation of tear secretion by topical agents that increase cyclic nucleotide levels. *Invest. Ophthalmol. Vis. Sci.* 31:1381 (1990).

32.  R.A. Stone, Y. Kuwayama, A.M. Laites, Regulatory peptides in the eye, *Experentia.* 43:791 (1987).

33.  D.M. Maurice, Electrical potential and ion transport across the conjunctiva. Exp. Eye Res. 15:527 (1973).

34.  D. Dartt, Y. Segal, S.R. Rossi, and J. P. Gilbard. Effect of topical stimuli on accessory lacrimal gland secretion and conjunctival blood vessel permeability in a rabbit model for keratoconjunctivitis sicca, *Invest. Ophthalmol. Vis. Sci.* 33 (ARVO Suppl.):1293 (1992).

35.  J. Luhtala, A. Palkawa, and H. Uusitalo, Calcitonin gene-related peptide immunoreactive nerve fibers in rat conjunctiva, *Invest. Ophthalmol. Vis. Sci.* 32:640 (1991).

# THE ANATOMY AND INNERVATION OF LACRIMAL GLANDS

Benjamin Walcott, Roger H. Cameron and Peter R. Brink

Departments of Neurobiology and Behavior and Physiology and Biophysics
SUNY at Stony Brook, Stony Brook, New York 11794

## INTRODUCTION

The comparative anatomic and physiological study of primary lacrimal glands raises a number of important questions the answers to which can provide insight into lacrimal gland function and dysfunction.

A typical mammalian lacrimal gland (fig 1) consists of lobes of secretory acini separated by connective tissue. These lobes are dominated by the tubular columnar secretory cells which form the acini which empty into secretory ducts leading to the main duct of the lobe and then gland. Associated with the acini in some areas are myoepithelial cells and small groups of lymphoid cells, mast cells and fibroblasts that are scattered in the interstitial spaces between the secretory tubules. This general pattern of organization is seen in most lacrimal glands. In the case of the main avian lacrimal glands (the harderian gland) (fig 2), however, the lobes have a cortex of secretory epithelial cells while the medulla is dominated by plasma cells of the immune system. Immunocytochemistry with antibodies specific for avian T-cells show positive staining cells only on the exterior of the gland and never in the medulla. Further, ultrastructural examination of the medulla reveals lymphoid cells all of which have extensive rough endoplasmic reticulum which is characteristic of plasma cells. This suggests that initiation of the immune response in the avian gland takes place in other lymphoid tissues such as gut associated lymphoid tissue (GALT). The difference in the numbers and types of lymphoid cells between mammals and birds is not due to an inflamatory process in individual birds but rather is probably due to a difference in the dominant immunoglobulin that is secreted. In the mammal, IgA is the dominant type that is actively transported across the secretory epithelium and secreted into the tears as a dimer coupled by secretory component. In the bird, the dominant immunoglobulin type is IgG that is presumed to passively diffuse

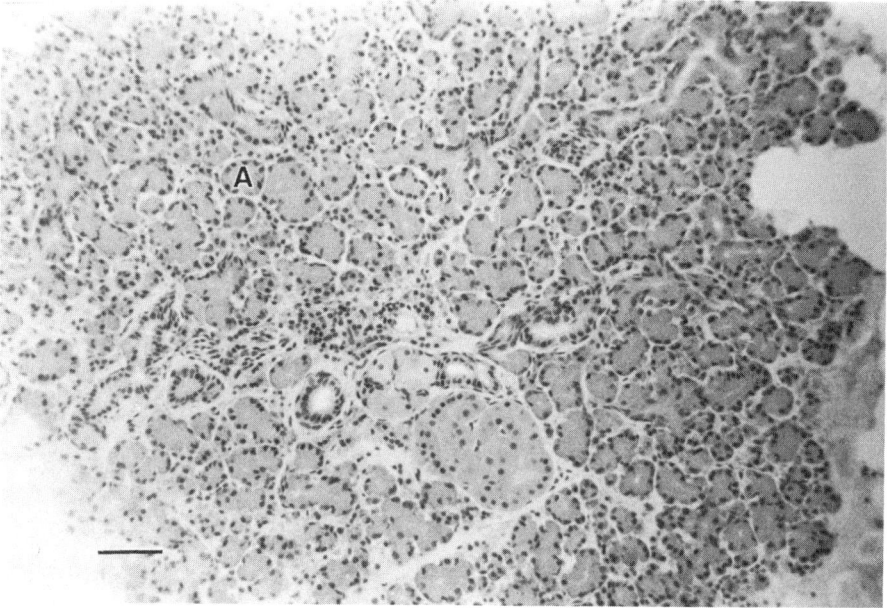

Figure 1. Light micrograph of goat lacrimal gland showing acini (A) with sparce accessory cells. Scale is 50 μm

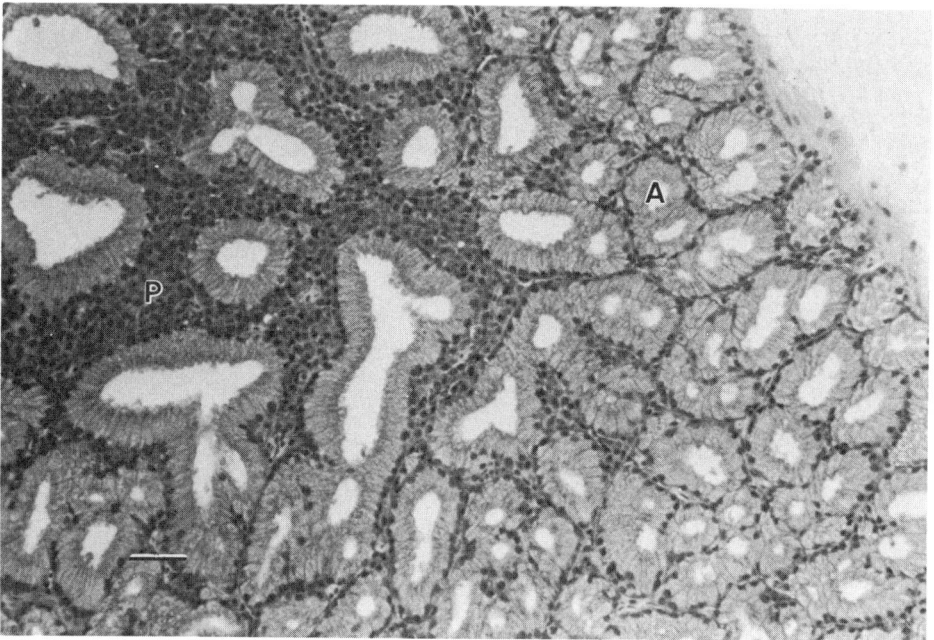

Figure 2. Light micrograph of chicken harderian gland showing the cortical secretory acini (a) and large number of plasma cells (p). Scale is 50 μm.

down its concentration gradient across the epithelium. Given that the concentration of immunoglobulins in the tears of mammals and birds is approximately the same[1,2], the large numbers of plasma cells in the bird are probably needed to generate the large concentration gradient necessary for significant diffusion across the secretory epithelium.

The secretory acini in mammal and avian lacrimal glands are similar, consisting of columnar secretory cells joined by junctional complexes with basal nuclei and apical secretory vesicles. The apical and lateral margins of the cells have many microvillar projections and folds providing an extensive cell surface (fig 3). In the bird, the junctional complex is very extensive (fig 4) and highly folded with large regions of gap junction. The anatomy strongly suggests that the cells are coupled. This is confirmed when dicarboxyflourescein is intracellularly injected into one cell resulting in spread of the dye into adjacent cells within 30 seconds (fig 5). This observation suggests that gap junctions could be important in the spread of excitation from one cell to the next in this tissue.

The innervation of lacrimal glands has been examined in a number of different species. In general, both parasympathetic (acetylcholine) and sympathetic (norepinepherine) fibers are present as well as fibers that contain neuropeptides such as Vasoactive Intestinal Polypeptide (VIP), Substance P (SP), Leu-Enkephaline (L-Enk), Neuropeptide Y (NPY) and Calcitonin gene-related peptide (CGRP). It seems likely that VIP co-exists with acetylcholine but the other peptides may be in different neurons. In most cases, a species will have both classical neurotransmitters and at least two of the peptides present. In the rat, for example, VIP, L-Enk and SP are present while in the chicken, VIP and SP are present (Fig 6, 7).The density of this innervation varies with the species and the region of the gland within a species. The chicken gland has a dense innervation pattern that is most pronounced in the medulla of the gland[3]. In the rat, the innervation is much less dense and is most abundant in the hilus of the gland where the main duct exits. In both species, however, at the light microscope level, the density of fibers does not seem to be sufficient for each acinar cell to be innervated. Ultrastructural examination of serial sections will be needed to effectively determine this issue. It is interesting that avian glands with the densest innervation of all lacrimal glands examined also have the largest number of plasma cells. The close anatomical relation in this gland between the nerve fibers with their vesicle filled varicosities and the plasma cells[1] suggests a functional relationship. Given that plasma cells are not electrically coupled, it is likely that a significant portion of the innervation may be needed to modulate the secretory immune response rather than that of the secretory epithelium where the cells are electrically coupled. Thus a mammalian gland with fewer plasma cells would require a less dense innervation.

The functional significance of the complex innervation with multiple transmitters/modulators is unclear. In the rat and rabbit, acetylcholine, norepinepherine and VIP all stimulate secretion of protein[4,5]. In salivary gland of the rat, SP also increases secretion[6]. One view of the function of the complex innervation is that this represents a necessary redundancy, a safety factor, in the control of tear production. Another view is that different mixes of neurotransmitters and neuropeptides will produce a different mix of proteins/fluids in the tears. There is evidence in the rabbit that suggests that norepinepherine, for example, preferentially stimulates peroxidase secretion when compared to acetylcholine[7].

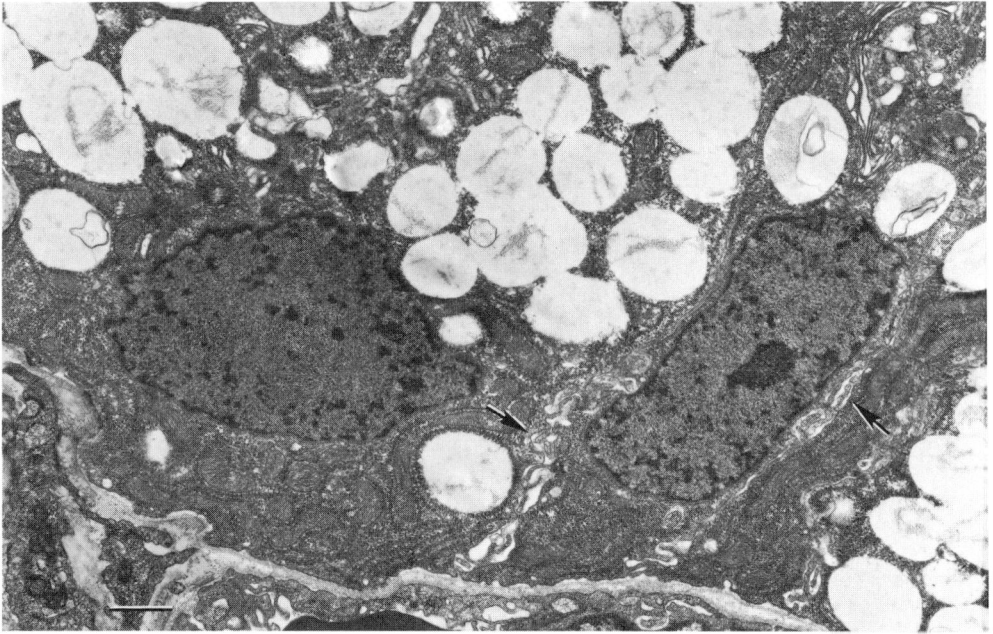

Figure 3. Electron micrograph of the base of chicken acinar cells showing extensive lateral membrane folding (arrows). Scale is 1 μm.

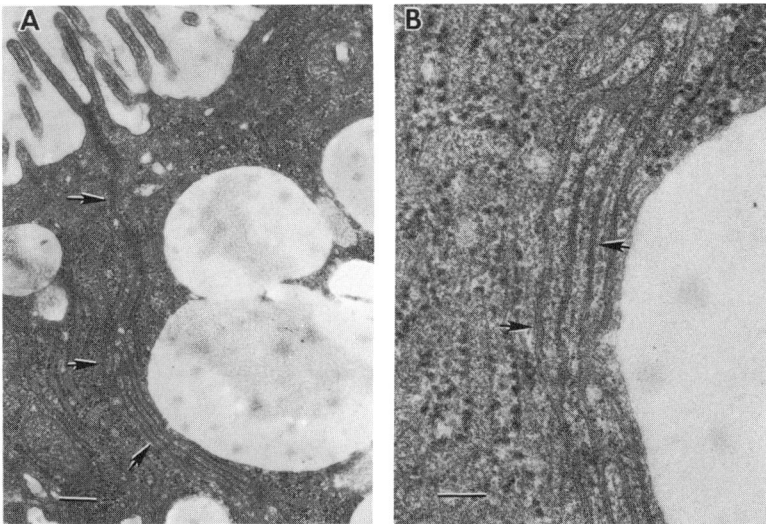

Figure 4. The junctional complex between chicken acinar cells is highly folded and extensive (arrows). Scale is 250 nm. B. At higher magnification, large areas of gap junction (arrows) are seen in the complex. Scale is 100 nm.

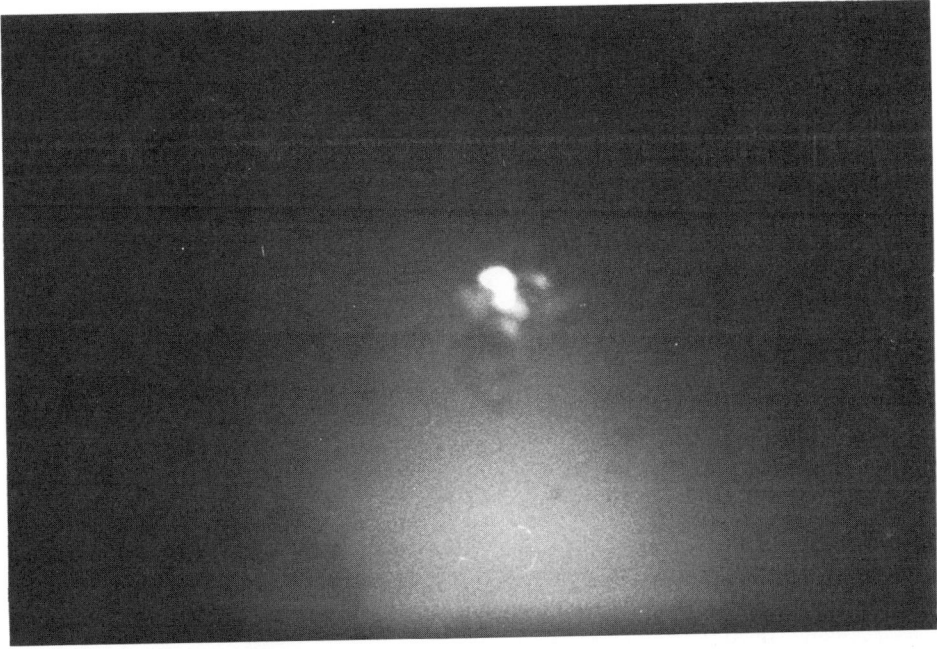

Figure 5. Carboxyflourescein was injected into one cell of a group of acinar cells and within 30 seconds spread to other cells.

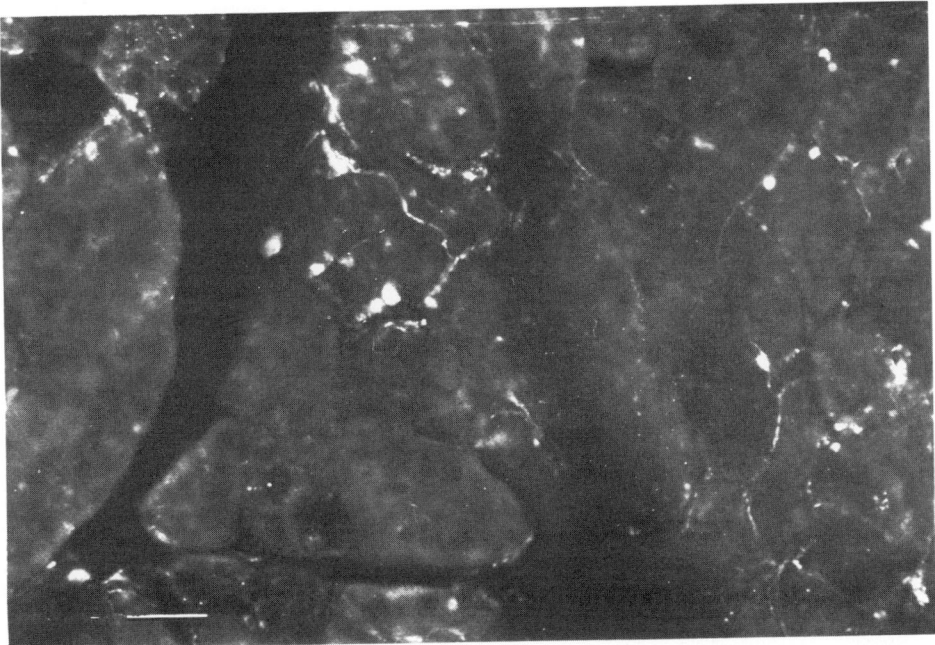

Figure 6. Rat extraorbital gland showing Leu-enkephaline-like immunoreactivity. Scale is 100 μm.

Differential secretion could be accomplished by stimulation of different vesicle pathways in cells or by activation of different populations of cells. In order to begin to distinguish between these two possibilities, we have started to examine the distribution of acetylcholine receptor in the avian and rat lacrimal gland.

In both the rat and the chicken gland, secretion stimulated by carbachol (an acetylcholine agonist) is blocked by atropine. This suggests the presence of muscarinic acetylcholine receptors. We have used a monoclonal antibody made to calf brain muscarinic actylcholine receptors[8] to examine at the light microscope level the distribution of the receptor in the two glands. In the rat, the anti-muscarinic acetylcholine receptor-like immunoreactivity (MAR-LI) is localized on the basal and lateral surfaces of the acini. The density appears to be

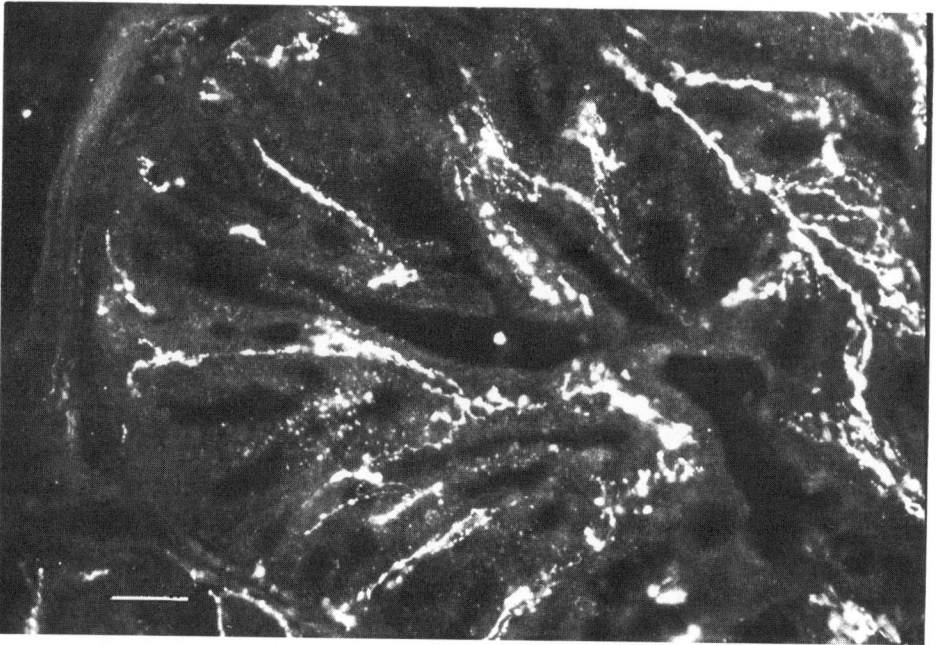

Figure 7. Substance P-like immunoreactivity in chicken harderian gland. Scale is 100 μm.

unveven both within an acinus and between them. However, in any section, there is MAR-LI associated with each acinus. In the chicken, the staining is most intense in the medullary regions. There is irregularly distributed MAR-LI on the basal regions of the acinar cells and to a much lesser extent on the lateral margins of the cells. The plasma cells also show extensive MAR-LI. We have isolated plasma cells from the chicken and shown that they stain both in vivo and in vitro. It is important to recognize that these experiments do not necessarily show active receptors. Funtional studies will be required to demonstrate that. One can conclude, however, that in the bird, both the acinar and plasma cell populations posess muscarinic actylcholine receptors.

# CONCLUSIONS

The structure and general organization of the acinar cells seems similar among lacrimal glands of different species. Acinar cells are coupled via apical gap junctions which, at least in the bird, are very extensive. The innervation of the glands is complex with both acetylcholine and norepinephrine present as well as at least two neuropeptides. Light microscopically, the density of the innervation does not seem to permit direct contact of neurons onto each of the acinar cells. The majority of the immunoglobulins that are found in tears are secreted from plasma cells within the lacrimal glands. In the bird, there are very large numbers of these plasma cells whose secretion of immunoglobulins (mainly IgG) is modulated by neurotransmitters.

## FUTURE ISSUES AND QUESTIONS

One issue is the relative role of the gap junctions between acinar cells and the direct innervation of these cells in the control of immunoglobulin secretion. The innervation of the gland does not appear to be sufficiently dense to permit direct innervation of each acinar cell but there is no data that indicates what faction of the cells are directly innervated. Therefore, structural analysis is needed for each neurotransmitter to determine the location of transmitter release sites in regions of glands. In addition, the analysis should also correlate the distribution of receptors for the transmitters on the target cells with this innervation pattern. If, as the data to date suggests, not all acinar cells are innervated, physiological studies will be needed to determine whether second message molecules move through the gap junctions between cells, thereby functionally coupling them. Further it is important to detect if that movement is regulated by the cell(s) and if some molecules move more readily than others. It is possible that different transmitters preferentially activate certain second message systems whose products move at different rates between cells. This process could effectively result in the neurotransmitters activating/affecting different populations of acinar cells thus producing a different secretory mix.

A second issue is the role of the complex innervation in the functioning of the gland. The neurotransmitters and neuropeptides present in the gland are known for a number of different species. What is not clear is their relative distribution within the gland. Detailed anatomical and immunocytochemical study of serial sections is needed to determine if individual acinar cells are innervated by neurons with each neurotransmitter. Further, virtually nothing is known about the distribution of receptors for these transmitters on cell surfaces. Do all cells have receptors for each of the neurotransmitters and are they uniformly distributed throughout the gland? In addition, functional studies are needed to resolve the role that each neurotransmitter plays in the secretion process.

A third issue concerns the role of the nervous system in the modulation of the activity of the immune system in the lacrimal gland. There are species differences in the operation of the secretory immune system and thus the modulatory role of the nervous system could be different. In the bird, for example, there are few if any T-cells present in the gland and programed B-cells are thought to migrate from other lymphoid areas such as GALT (gut associated lymphoid tissue). Further the dominant immunoglobulin is an IgG. In the rat, on the other hand, T-cells are present[8] and clonal expansion of B-cells can occur locally as evidenced by reaction centers seen in the glands. Here the dominant immunoglobulin is IgA. In the bird, the cholinergic system seems to stimulate an increase above a basal rate in IgG release from the plasma cells. This effect requires external calcium and depends on gated membrane calcium channels. The effect of other neurotransmitters is not known. In the rat lacrimal gland, there is little known about neuromodulation of the immune system. In other tissues, the rate of T-cell clonal expansion is modulated by adregergic transmitters[9]. Thus modulation of both T-cell and plasma cell activity needs to be examined.

## REFERENCES

1. J.M. Little, Y.M. Centifanto, and H.E. Kaufman, Immunoglobulins in human tears, *Amer. J. Ophthalmol.* 68: 898 (1969)
2. T. Baba, K. Matsumoto, T. Kajikawa, and M. Mitsui, Harderian gland dependency of immunoglobulin A production in the lacrimal fluid of chickens, *Immunology* 65:67 (1988)
3. B. Walcott, and J.R. McLean, Catecholamine containing neurons and lymphoid cells in a lacrimal gland of the pigeon, *Brain Res.* 328:129 (1985)
4. H.H. Stolze, and H.J. Sommer, Influence of secretogogues on volume and protein pattern in rabbit lacrimal fluid, *Current Eye Res.* 4:489 (1985)
5. D.A. Dartt, A.K. Baker, C. Vaillant and P.E. Rose, Vasoactive intestinal polypeptide stimulation of protein secretion from rat lacrimal gland acini, *Amer. J. Physiol.* 247:502 (1984)
6. T. Kudo, R. Inoki, T. Nishimoto, M. Akai, S. Shiosaka, and M. Tohyama, A possible role of substance P in the salivary secretion in rats, *Adv Exp. Med. Bio.* 156:681 (1983)
7. B.B. Bromberg, Autonomic control of lacrimal protein secretion, *Invest. Ophthalmol. Vis. Sci.* 20:110 (1981)
8. C. Andre, J.G. Guillet, J.-P. De Backer, P. Vanderheyden, J. Hoebeke, and A.D. Strosberg, Monoclonal antibodies against the native and denatured forms of muscarinic acetylcholine receptors, *EMBO J.* 3:17 (1984)
9. J. Pappo, J.L. Ebersole, and M.A. Taubman, Pheonotype of mononuclear leukocytes resident in rat major salivary and lacrimal glands, *Immunology* 64:295 (1988)

# LIGHT AND ELECTRON MICROSCOPIC MORPHOLOGY OF ACCESSORY LACRIMAL GLANDS *

Peter Seifert, Manfred Spitznas, Frank Koch,
and Andrea Cusumano

Alfried-Krupp-Laboratory, Department of Ophthalmology
Sigmund-Freud-Str. 25, D-53105 Bonn 1, Germany

The accessory lacrimal glands form up to 60 small nodules, which are located close to the conjunctival fornix and at the edge of the upper tarsus. In spite of the fact they are believed to play an important role in the pathogenesis of dry eye disease, they have escaped the attention of morphologists. Recently, we have developed a technique to biopsy those glands at the occasion of ptosis operations.

A total of 34 glands removed from human eyes were examined employing standard techniques for light and electron microscopy. In our material, the glands measured up to 1 mm in diameter, but most of them were considerably smaller.

The accessory lacrimal glands were surrounded by a thin connective tissue coat (Figure 1). On one side they exhibited a distinct hilus containing a single excretory duct and blood vessels. With the exception of one single specimen, the glands were not divided into typical lobules.

The excretory duct enters the gland in a zone of connective tissue. In semithin sections the epithelial lining stains only weakly. The epithelial cells are highly prismatic and occur in one or two layers (Figure 2). The free apical cell surfaces are characterized by microvilli and microplicae. The lateral surfaces show few but pronounced infoldings. In the excretory duct as well as in all further parts of the glandular epithelium the neighboring cells are joined by typical zonulae occludentes and, more basally, by zonulae adhaerentes and numerous desmosomes. The nuclei are oval in appearance and extend into the basal portion of the cells. The heterochromatin is sparsely developed, thus the nuclei appear rather light. The cytoplasm contains numerous mitochondria and few other organelles. Close to the free apical surface the cells contain sparse, relatively small secretory granules of different size and electron density (Figure 2).

Deeper in the gland, the excretory duct divides and subdivides at various angles into smaller intralobular ducts with narrower lumina (Figures 1,3). The epithelial cells resemble those of the excretory duct. However, they are isoprismatic, have a uniform height, form no more than one layer. In certain areas basal cells are present. The cells normally do not contain secretory granules and their nucleus is round. In circumscribed

---

* A more detailed original publication on this topic has been submitted to German J. Ophthalmol.

areas along the course of the ducts, the appearance of the epithelium changes and attains the characteristics of the secretory tubular epithelium described later. The intralobular ducts may terminate blindly with or without the formation of outpouchings. More importantly, they change into secretory tubules which also branch off laterally in great numbers (Figure 3).

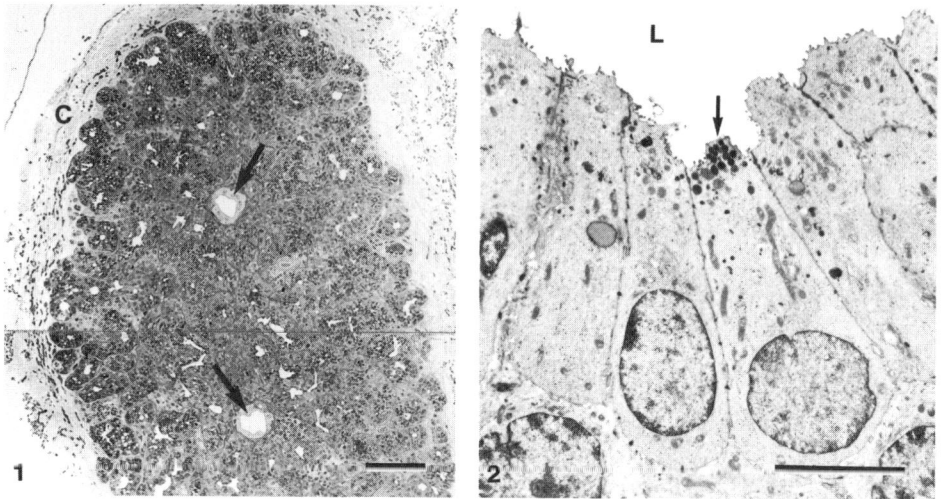

Figure 1. Accessory lacrimal gland not divided into lobules. Arrow = intralobular duct. C = connective tissue coat. Bar = 100 $\mu$m.
Figure 2. High prismatic epithelium of the excretory duct of an accessory lacrimal gland. Arrow = secretory granules. L = lumen of duct. Bar = 5 $\mu$m.

The epithelium of the tubules is composed of a monolayer of highly prismatic cells of different height, giving the lumen an uneven appearance (Figures 4,5). Depending upon their functional state, the cells have a very heterogenous appearance. The two extremes are the following: firstly, there are cells with a rather electron translucent cytoplasmic matrix. Their nucleus is round and located in the midportion of the cell. Often it appears rather light due to the sparse development of its heterochromatin. An abundance of mitochondria and rough-surface-endoplasmic reticulum is randomly distributed throughout the entire cell. A distinct Golgi apparatus is frequently visible. Secretory granules may be totally absent or occur in smaller numbers. If they are present, they are located mostly in the apical portion of the cell and their size and electron density varies. Only few apical microvilli are present. Secondly, there are cells with a rather electron dense ground plasm. Secretory granules are generally present. They may be only few in number, located in the apical portion of the cell, or they occupy almost the entire cell body. Various clusters of granules are characterized by different diameters (Figure 5). The granules are either uniformly electron dense or show different degrees of electron translucence, the latter being especially true for the big granules. In the lighter granules, the content has a finely granular appearance. It is conceivable that the different electron density of granules is due to the preparation technique. Depending upon the number of secretory granules the nucleus is either round and located in the middle of the cell or indented and displaced toward the base of the cell. In both locations the nucleus may show typical deep and narrow infoldings. The distribution of heterochromatin is similar to that in light tubular cells but the karyoplasm appears denser. The ergastoplasm is very well developed and

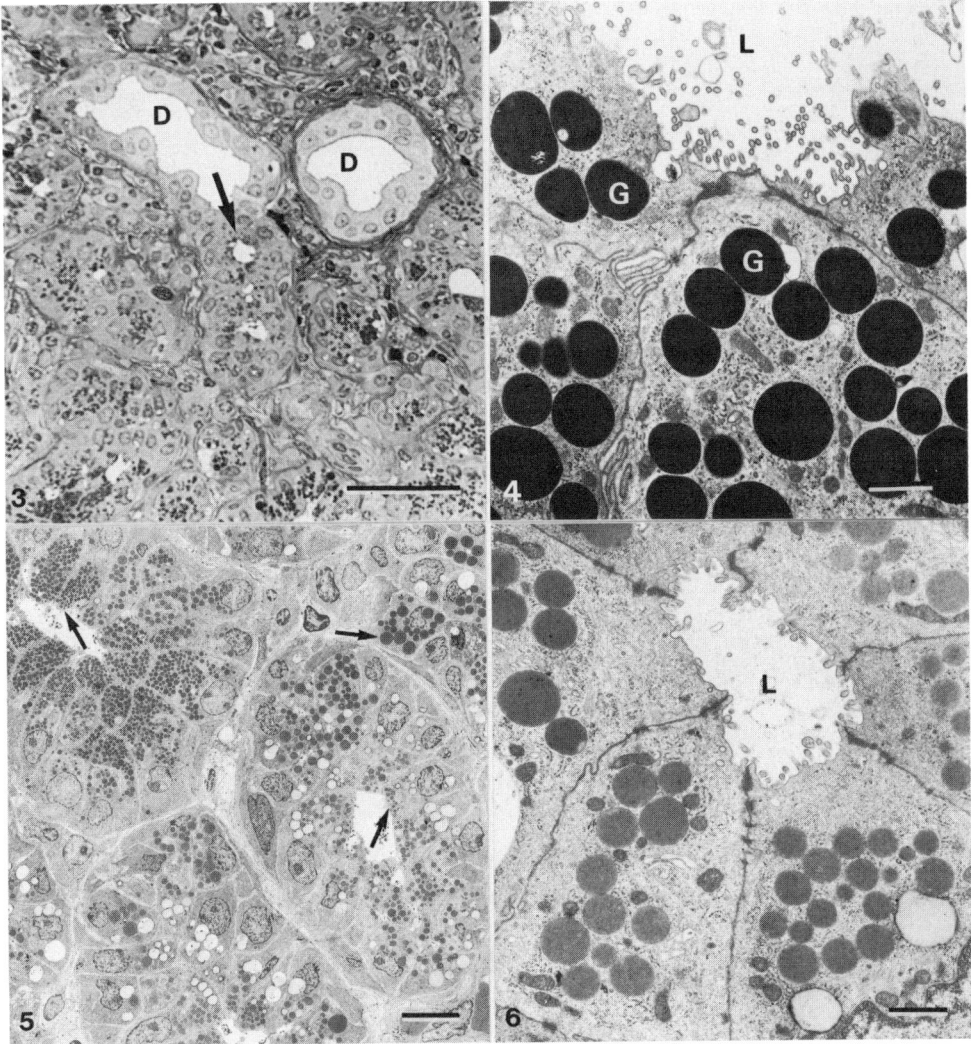

Figure 3.  Intralobular ducts (D) between tubules and end pieces. A tubule branches off
from an intralobular duct (arrow). Bar = 50 μm.

Figure 4.  Cells of a tubule. G = secretory granules. L = lumen of tubule. Bar = 1 μm.

Figure 5.  Tubules and end pieces contain granules in different size and amount (arrows).
Bar = 10 μm.

Figure 6.  A tube like end piece with narrow lumen (L). Bar = 1 μm.

located in the basal cell portion. The cytoplasmic islands between the secretory granules
contain numerous mitochondria and the Golgi apparatus. In these cells the apical surface
carries various amounts of microvilli. Moreover, intracellular canaliculi are present and
between all types of tubular cells  intercellular canaliculi  may be  present. Basal to the
occluding intercellular junctional complexes, the intercellular space is markedly widened
and shows microvilli that project into it from the lateral  surfaces of the neighbouring
cells. The basal plasmalemma is mostly smooth and runs parallel to the basement
membrane. At places, the cell expands with stump-like processes into the basement
membrane.

The tubules  terminate in  clusters of  secretory cells forming end pieces. As an exception, end pieces can also originate directly from intralobular ducts. The terminal clusters are connected to the tubular lumen by several secretory capillaries. Where a cluster of secretory cells is drained by a single capillary, it forms a tube-like structure (Figure 6). In such structures, the lumen of the secretory capillary can be considerably widened. Typical acini with a narrow intercalary duct were not observed in our material. The cellular composition of the end pieces is identical to that of the tubules. The end pieces and the tubules are covered with a loose net of processes from myoepithelial cells, characterized by bundles of myofilaments in the cytoplasm (Figure 7).

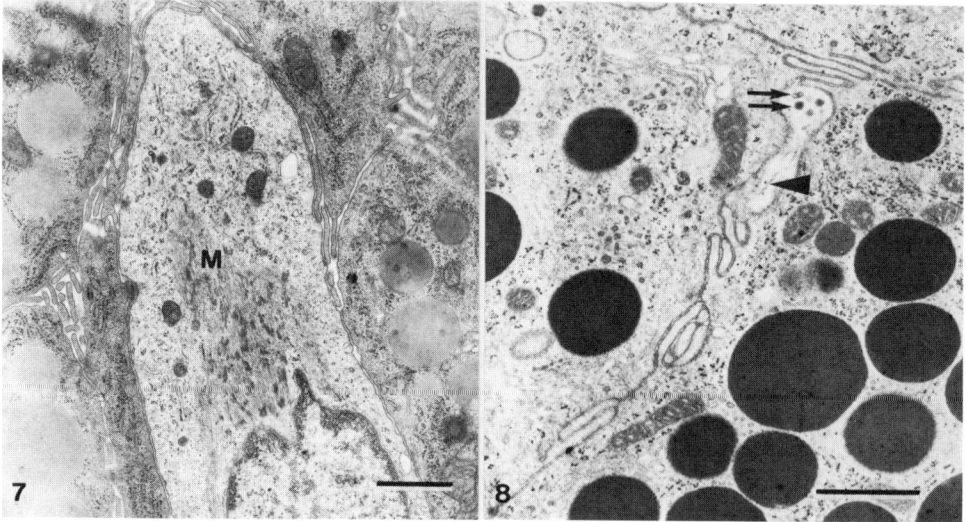

Figure 7.   Myoepithelial cell between secretory cells. M = myofilaments. Bar = 1 $\mu$m.
Figure 8.   A single axon between secretory cells. Arrows = clear and dense core vesicles.
            Arrowhead = neurotubule. Bar = 1 $\mu$m.

The myoepithelial cells are enclosed into the secretory epithelial basement membrane, which runs along their outer surface while their inner surface is in direct contact with the adjacent secretory epithelial cells.

The connective tissue between the glandular elements contains an abundance of collagen fibrils with a periodicity of 67 nm embedded into a clear matrix. Between the collagen fibrils, fibroblasts with long, slender prolongations and plasma cells are found as well as lymphocytes, which occasionally show distinct accumulations.

Blood vessels are numerous and consist often of capillaries whose lumen may be smaller than the diameter of a red blood cell. The endothelial lining has many fenestrations which are covered by a thin diaphragm.

Of particular interest is the finding of nerves. Between the glandular elements, non-myelinated nerve fibers consisting of a Schwann cell containing several axons can be identified. They are surrounded by a basement membrane. The axons contain mitochondria, microtubules, microfilaments and different types of vesicles, with various electron density. The clear granules measure up to 60 nm, the larger dense core vesicles up to 100 nm in diameter. The nerve fibers approach the secretory cells. Where they make contact, they become flattened, their basement membrane fuses with that of the epithelium, and several axons become naked. Infrequently, a single axon can be identified between secretory epithelial cells (Figure 8) as well as between cells of the intralobular duct. Special synaptic membrane structures were not identified. Additionally, morphological contacts between nerve fibers and plasma cells, fibroblasts or endothelial cells are present.

Because of the identity of the findings described in all specimens we feel safe in interpreting them as the regular structure of the accessory lacrimal glands.

## ACKNOWLEDGEMENT

This work was supported by "Aktion Kampf der Erblindung."

# IMMUNOHISTOCHEMISTRY AND PROTEIN SECRETION IN THE RAT LACRIMAL GLAND: A MORPHOPHYSIOLOGICAL STUDY

Jaipaul Singh[2], Ernest Adeghate[1], Shuna Burrows[2], Frank C. Howarth[2], and Ruth M. Williams[2]

[1]1st Department of Anatomy, Semmelweiss University Medical School Tuzolto, Budapest, Hungary
[2]Cell Communication Group, Department of Applied Biology, University of Central Lancashire, Preston PRI 2HE, England

## INTRODUCTION

The secretion of tears by the lacrimal glands is regulated by the parasympathetic and sympathetic nerves of the autonomic nervous system (Bothelho, 1964; Bromberg,1981). The autonomic nerves innervate the secretory acini, the duct and the blood vessels and they play important roles in the control of lacrimation and blood flow (Dartt,1989). The parasympathetic nerves form the majority of the structural and functional innervation and control mainly hypersecretion of tears. On the other hand the sympathetic innervation appears to be relatively sparse around the secretory acini but the main sites appear near the lacrimal blood vessels (Nikkinen et al,1985). Activation of both cholinergic and adrenergic nerves can elicit protein secretion from the lacrimal and these responses can be mimicked by the respective agonists (Dartt, 1989; Muller *et al*, 1985). Recent studies have shown that peptidergic nerves may also control lacrimal secretion and the putative neurotransmitter released by the nerve is vasoactive intestinal polypeptide (VIP) (Dartt *et al*, 1984; Hussain and Singh, 1988; Matsumoto *et al*, 1992). However, the involvement of other neurotransmitters and mast cell mediators such as substance P, dopamine, serotonin and histamine in the control of lacrimal gland function is less understood. Thus, this investigation was designed to undertake a morphophysiological study measuring protein secretion and employing immunohistochemistry to confirm the involvement of serotonergic, dopaminergic, peptidergic and histaminergic elements in the control of protein secretion in the rat lacrimal gland. In addition, the effects of acetylcholine (ACh) and noradrenaline (NA) were also studied for comparison.

## METHODS

All experiments were performed on the isolated lacrimal gland from adult rats. Immediately after killing the animal the glands were removed, placed in oxygenated

Krebs-Henseleit (K-H) solution and subsequently cut into small segments (5-10 mg). A total weight of about 200 - 250 mg was placed into a perspex flow chamber (2 ml volume) and continuously superfused at a constant rate of 1 ml min$^{-1}$ with a K-H solution comprising (mM); NaCl, 118; KCl, 3.7; CaCl$_2$, 2.56; NaHCO$_3$, 25; KH$_2$PO$_4$, 2.2; MgCl$_2$, 1.2 and glucose 10. The solution was kept at pH 7.4 and 37°C while being gassed with a mixture of 95% O$_2$: 5% CO$_2$. Total protein output in effluent samples were measured by an automated on-line colorimetric method (Hussain and Singh, 1988). During stimulation known concentrations of either ACh, NA, histamine, substance P, 5-hydroxytryptamine (5-HT) or dopamine were added to the perfusing medium. Bovine serum albumin (BSA) was used as standard. All values were expressed as micrograms of protein per millilitre effluent per 100 mg tissue above basal level.

Lacrimal glands from adult Sprague-Dawley rats were used for all immunohistochemical studies using established methods (Sharkey *et al*, 1990). Animals were anaesthetized with sodium pentobarbitone (60 mg Kg ip). They were then perfused via the ascending aorta with 200 ml phosphate buffered saline (PBS, pH 7.4) followed by 500 ml Zamboni's fixative. Both lacrimal glands were dissected and immersed fixed in the same fixative for 2 hours at 4°C. For the localization of histamine rats were perfused with 250 ml freshly prepared 1-ethyl-3-(3-dimethylaminopropyl) carbodamide instead of Zamboni's fixative. After fixation all tissues were washed in 0.1 M phosphate buffer containing 20% sucrose for 12 hours at 4°C. For the localization of tyrosine hydroxylase (TH), dopamine beta hydroxylase (DBH) 5-HT, histamine and substance P, frozen sections (10µm) were cut in a cryostat, thaw-mounted on to poly-D-lysine coated slides and processed for indirect immunofluorescence. Sections were incubated in a moist chamber at 4°C for 48 hours with the following rabbit polyclonal antisera against substance P (Instar; 1:1000), 5-HT (Instar; 1:800), histamine (Instar; 1:1000), TH (Pel-Freez; 1:100) and DBA; (Instar; 1:1000). After incubation sections were washed three times for 10 min in PBS and incubated with a fluorescein isothiocyanate conjugate swine anti-rabbit for 1 hour at 24°C in a moist chamber. Finally, the sections were washed three times for 10 min in PBS containing 0.1% Triton-X-100 and mounted in bicarbonate buffered glycerol (pH 8.6). Sections were subsequently examined using a Zeiss Axioplan fluorescent microscope. Photographs were taken using Kodak TMax (400 ASA) film for prints.

For the identification of acetylcholinesterase (ACHE), a modification of the method of Matsumoto *et al* (1992) was employed. Frozen sections were treated with H$_2$O$_2$ (1% for 60 min) to destroy endogenous peroxidase and then washed with three changes of 0.1M maleate buffer (pH 6.0) before treated with 10$^{-5}$ M tetraisopropyl pyrophosphoramide for 30 min at 24°C to inhibit pseudocholinesterase activity. Sections were then washed with five changes (2 min each) of 0.1 M maleate buffer (pH 6.0) and subsequently incubated for 30 min at 24°C in maleate buffer solution containing 24µM acetylcholine iodide, 30µM CuS04, 50µM C$_6$H$_5$O$_7$Na$_3$ and 5 µM KFe (N). Afterwards sections were washed with 5 changes of 50 mM Tris-HCl (pH 7.6) and incubated for 2½ min at 24°C in a solution of 40 mg DAB + 300 mg (NH) Ni(SO4) dissolved in 100 ml Tris buffer. A volume of 10µl of a 30% HO was added to the medium and the incubation continued for a further 8 min. Sections were washed with five changes of 50 mM Tris-HCl (pH 7.6), coversliped and subsequently examined under the microscope.

## RESULTS AND DISCUSSION

The mean (±SEM) basal protein output was $24.5\pm2.1$ µg ml$^{-1}$ (100 mg tissue)$^{-1}$, n=48. Figure 1 shows families of histograms of mean (± SE) protein output above basal levels for different concentrations of (a) ACh ($10^{-8}$-$10^{-5}$ M), (b) NA ($10^{-8}$ - $10^{-4}$ M) and (c) histamine ($10^{-6}$ - $10^{-3}$ M). These results show that the cholinergic and adrenergic agonists as well as the paracrine hormone, histamine can evoke marked increases in total protein output from the superfused rat lacrimal gland segments. Figure 2 shows the effects of different concentrations of (a) 5-HT ($10^{-7}$ - $10^{-5}$ M), (b) dopamine ($10^{-7}$ - $10^{-5}$ M) and (c) substance P ($10^{-8}$ - $10^{-6}$ M) on total protein output from superfused lacrimal segments. These results demonstrate that serotonergic, dopaminergic and peptidergic agonists have marked secretagogue effects on the rat lacrimal gland.

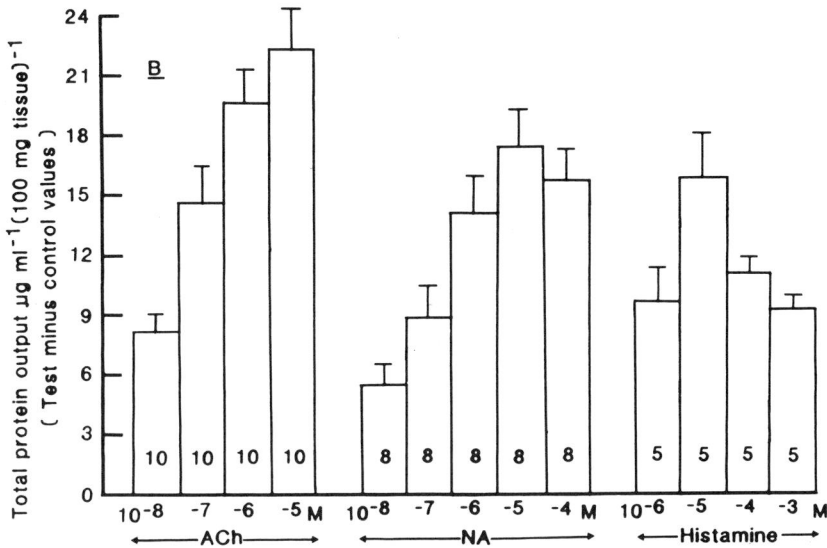

Figure 1. Families of histograms showing mean peak output in protein release above basal level from superfused lacrimal segments during stimulation with different concentrations of (a) ACh ($10^{-8}$ - $10^{-5}$M), (b) NA ($10^{-8}$ - $10^{-4}$M) and (c) histamine ($10^{-6}$ - $10^{-3}$ M). Each point is mean ±SE, n values are shown in each histogram.

Figure 3 shows the distribution pattern of (a) cholinergic and (b) adrenergic immunoreactive fibres in the rat lacrimal gland. The results show the lacrimal gland contains numerous thick and smooth AChE-positive nerve fibres which innervate the ducts, blood vessels and acini. In contrast, immunoreactive for the adrenergic marker, tyrosine hydroxylase was seen predominantly around blood vessels. In addition, mast cell for histamine were found predominantly near the blood vessels (Figure 3c).

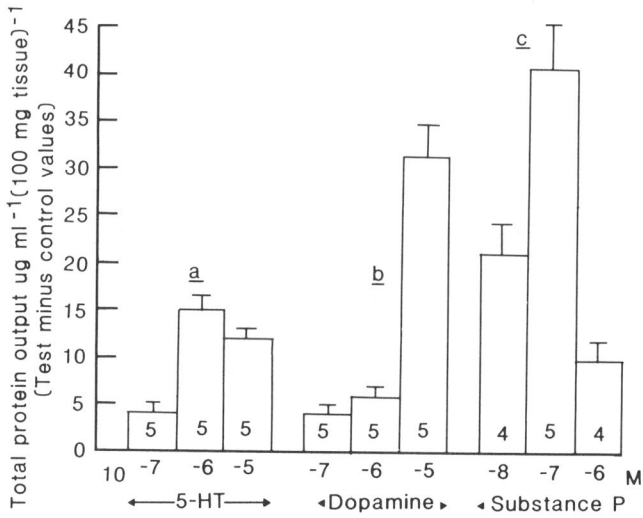

Figure 2. Families of histograms showing the effect of different concentrations of (a) 5-HT ($10^{-7}$ - $10^{-5}$ M), (b) dopamine ($10^{-7}$ - $10^{-5}$ M) and (c) substance P ($10^{-8}$ - $10^{-6}$ M) on mean peak total protein output from superfused rat lacrimal segments. Each point is mean ±SE, n values are shown in each histogram.

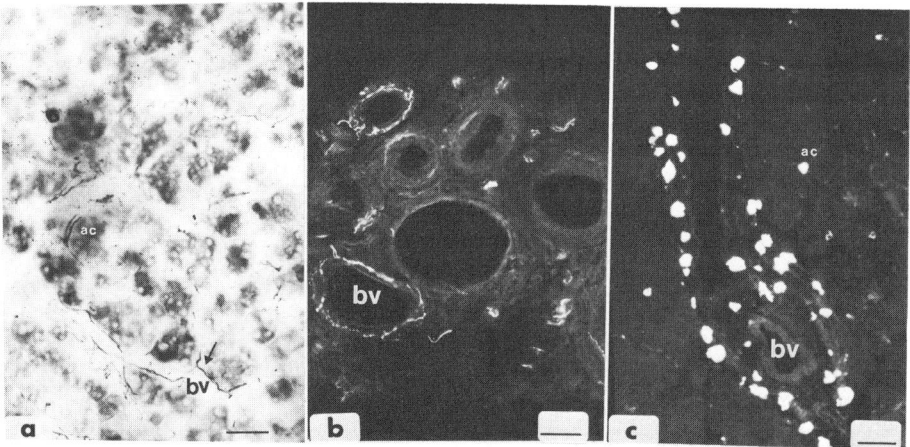

Figure 3. Light micrographs of rat lacrimal gland showing (a) AChE containing nerve fibres (arrow) (b) tyrosine hydroxylase (TH) immunoreactive fibres and (c) histamine immunoreactive mast cells. Note that the AChE positive fibres are found around ducts, acini (ac) and blood vessels (bv) whereas TH immunoreactive fibres are found in abundant quantities near the blood vessels (bv), scale bar = 50 μm.

The distribution pattern of immunoreactive fibres for (a) substance P, (b) dopamine beta hydroxylase (DBA), and (c) and (d) 5-HT is shown in Figure 4. The results show that substance P fibres are seen in the interacinar regions. DBA positive multipolar neurons were discerned in the periphery of the acinar cells. Similarly, both serotonin positive fibres and cells are seen in the periacinar areas.

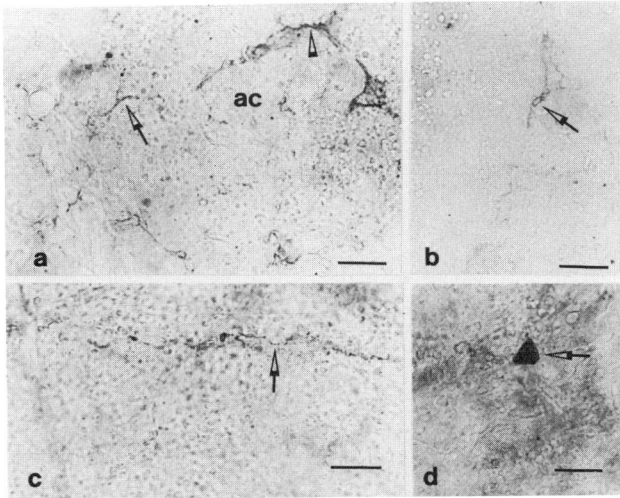

Figure 4. Light micrograph of the lacrimal showing (a) substance P positive neurons (arrow) and fibres (arrow head) in the interacinar regions. ac = acinus. (b ) DBA-immunopositive nerve (arrow) in the periacinar areas (c) serotonin positive nerve fibre (arrow) in the periacinar areas and (d) serotonin positive cell (arrow). Scale bar = 50 μm.

This study has demonstrated as before that cholinergic and adrenergic agonists can elicit marked increases in protein release from the rat lacrimal gland (Dartt, 1989). Our work has shown that such substances as 5-HT, dopamine, histamine and substance P can also stimulate protein secretion from the lacrimal gland. These observations suggest that the secretory acini may possess specific receptors for these aminergic, histaminergic and peptidergic secretagogues. The signal transduction mechanisms through which the classical neurotransmitters such as ACh and NA mediate their secretory responses are well documented (Dartt, 1989). However, the signal transduction mechanisms through which 5-HT, dopamine, histamine and substance P may act to elicit protein secretion are still unclear. Furthermore, since the lacrimal gland is known to secrete a large number of enzymes and different proteins then it is pertinent to measure specific enzymes rather than total protein output during applications of these aminergic, histaminergic and peptidergic secretagogues. This is important especially since different concentrations of a secretagogue may evoke both parallel and non-parallel secretions of protein and enzymes.

The present study has also shown the pattern of distribution of immunoreactive nerves, specific enzyme markers and mast cells for either ACh, NA, dopamine, substance P, histamine and 5-HT in the rat lacrimal gland. Like previous workers (Nikkinen et al, 1985; Matsumoto et al, 1992) this investigation has also demonstrated the presence of numerous  ACh positive nerves around the secretory acini, ducts and blood vessels. In contrast, NA immunoreactive nerves were distributed sparsely around ducts and acini but in abundant quantities near blood vessels. In addition, our immunohistochemical studies have shown that 5-HT, dopamine beta hydroxylase, substance P and histamine are present  in either nerves or cell bodies  in the rat lacrimal. Substance P, DBA and 5-HT positive nerves are observed in the peri-acinar areas. Similarly, 5-HT positive cells are  also seen around secretory acini. Furthermore, there was a high density of immunofluorescence for histamine  which is found in close apposition to the blood vessels. Previous studies have demonstrated the distribution of substance P around lacrimal blood vessels and it has been suggested that this neuropeptide may play a role in controlling lacrimal blood flow (Dartt, 1989; Matsumoto *et al*, 1992). However, there is little or no evidence for the involvement of serotonergic, dopaminergic and histaminergic control of lacrimal gland function. Our study is the first of its kind to show the distribution of nerves or cell bodies containing either DBA, 5-HT or histamine  in the rat lacrimal. Further experiments are required to elucidate precisely the relationship between the nerves and mast cell bodies. It may be possible that the peptidergic nerves may not only mediate vasodilation and protein secretion (Dartt,1989; Matsumoto *et al*, 1992), but also the release of mast cell mediators such as 5-HT and histamine.

## Acknowledgements

Supported by the Wellcome Trust and SERC.

## REFERENCES

Bothelho, S.Y.,  1964, Tears and the lacrimal gland, Sci. Amer., 211: 78.

Bromberg, B.B., 1981, Autonomic control of lacrimal protein secretion, Invest. Opthalmol. Vis. Sci., 20: 110.

Dartt, D.A., 1989, Signal transduction and control of lacrimal gland protein secretion: a review, Curr. Eye Res., 8: 619.

Dartt, D.A., Baker, A.K., Vaillant, C. and Rose, P.E., 1984, Vasoactive intestinal polypeptide stimulation of protein secretion from rat lacrimal acini, Am. J. Physiol., 247: G502.

Hussain, M. and Singh,.J., 1988, Is VIP the putative non-cholinergic non-adrenergic nerve controlling protein secretion in the rat lacrimal gland? Quart. J. Exp. Physiol., 73: 692.

Matsumoto, Y., Tanake, T., Ueda, S. and Kawata, M., 1992, Immunohistochemical and enzyme histochemical studies of peptidergic, aminergic and cholinergic innervation of the lacrimal gland of the monkeys (Macaca-fuscata), J. Auto. Nerv. Syst., 37: 207.

Muller, P., Chambaut-Guerin, A. and Rossignol, B., 1985, Comparitive effects of cytochalsin D on protein dischanrge induced by $\alpha$ and $\beta$ adrenergic and cholinergic agonist in the rat exorbital lacrimal glands, Biochim. Biophys. Acta., 844: 158.

Nikkinen, A., Unsitalo, H., Lehtosalo, J.A. and Palkama, A., 1985, Distribution of adrenergic nerves in the lacrimal glands of the guinea-pig and rat, Exp. Eye Res., 40: 751.

Sharkey, K.A, Mathison, R., Sharif, M.N. and Davison, J.S., 1985, The effect of streptozotosin induced diabetes on peptidergic innervation  and function of the rat parotid gland, J. Auto. Nerv., Syst., 27: 127.

# CARBONIC ANHYDRASE AND ACINAR CELL HETEROGENEITY IN RAT AND RABBIT LACRIMAL GLANDS

B. Britt Bromberg,[1,2] Cynthia W. Hanemann,[1] Mary H. Welch,[1] Roger W. Beuerman,[2] and Sherwood Githens[1]

[1]Department of Biological Sciences  [2]L. S. U. Eye Center
University of New Orleans       L. S. U. Medical Center
New Orleans, LA  70148         New Orleans, LA 70122

## INTRODUCTION

The principal function of the lacrimal gland is to provide an appropriate medium for the maintenance of the corneal epithelium. Insofar as the corneas of different species have unique requirements, we may expect that their lacrimal glands will have significant heterogeneity in form and function. The lacrimal contribution to tears is complex, and the protein secretory products of the lacrimal gland vary among different species.[1] Further, species specific heterogeneity is evident in the distinctly different organizations of the acini in rat and rabbit lacrimal glands.[2] For instance, rat lacrimal acini are spherical or oval in structure whereas the rabbit acini are elongate, tubular and branching. Both species secrete numerous proteins, but the rat lacrimal gland contains at least 2 exocytotic proteins, peroxidase (PX) and carbonic anhydrase (CA), that are not secreted by the rabbit. Species specific differences probably also exist in the manner in which the aqueous portion of the lacrimal fluid is generated.[3] The presence of a membrane-associated CA in the terminal acinar cells of rabbit but not rat lacrimal glands is consistent with this idea.[2] This CA isozyme may have a unique role in the unidirectional transport of water,[4] forming the aqueous component of the lacrimal fluid. The same task is likely achieved in a different manner in the non-terminal acinar cells of the rabbit lacrimal gland and in the rat lacrimal gland, which lacks altogether the acinar membrane-associated CA.[2] Even within the glands of a single species there is significant heterogeneity among the acini, reflecting some partitioning of functions among the principal secretory cells.[2,5] In this chapter, possible roles of CA are explored in both rat and rabbit lacrimal glands in view of acinar cell heterogeneity within the glands themselves and between glands of the two species.

## METHODS

Collagenase, hyaluronidase, Dulbbecco's modified Eagle's medium nutrient mixture

F-12 Ham (DME/F12), bovine serum albumin (BSA), calf serum (CS), carbamyl choline (CARB), phenylephrine HCl (PHEN), l-isoproterenol-d-bitartrate (ISOP), atropine sulfate (ATR), timolol maleate (TIM), o-phenylenediamine dihydrochloride (OPD), dexamethasone, and insulin/transferrin/selenium (ITS) were from Sigma Chemical Co, St. Louis MO. Matrigel, Nu-Serum V, and epidermal growth factor (EGF) were from Collaborative Research, Lexington, MA. HATF filters (25 mm) were from Millipore Corp., Bedford, MA.

Lacrimal glands were taken from male F344 rats (3-4 weeks) and male New Zealand albino rabbits (2-4 lb). Rats were killed by $CO_2$ inhalation and rabbits were killed by intravenous injection of sodium pentobarbital. All animals were treated in accordance with current NIH and USDA guidelines.

Isolated acini were prepared by mincing and digestion in collagenase (0.75 mg/ml) and hyaluronidase (1.1 mg/ml) in DME/F12 (manuscript in preparation). After washing sequentially in DME/F12 supplemented with 20% and 10% CS, the isolated acini were filtered through a 60 mesh stainless steel grid to remove large debris and lengths of intact ducts. The acini were washed in DME/F12 and pelleted through DME/F12 with 4% BSA to remove cellular debris. The viability of the cells constituting the acini was 96-98% as judged by trypan blue exclusion. Acini from rat lacrimal gland were prepared for in vitro protein secretory studies. Rabbit lacrimal acini, lacking histochemically demonstrable exocytotic CA activity,[2] were not examined for their ability to secrete CA. While both rat and rabbit lacrimal acini are mitotically active in cell culture (manuscript in preparation), growth of rabbit acini is much more robust; thus, only results of rabbit acinar cell culture are presented here.

The secretion of CA by rat lacrimal acini was determined in the incubation medium previously described.[6] Duplicate aliquots of the suspended acini were transferred to 5 ml siliconized Erlenmeyer flasks maintained at 37°C and gassed periodically with $O_2$. The acini were stimulated by the addition of 10 µM CARB, 100 µM PHEN, or 10 µM ISOP dissolved in incubation medium, or incubation medium alone (CTRL). Additional controls included a 5 min pre-incubation with 1 µM ATR or 1 µM TIM to inhibit responses to CARB and ISOP respectively. After 40 min, the acini were separated from the incubation medium by centrifugation and homogenized in 1 ml of distilled water. CA was assayed[7] in both the incubation medium and the acinar homogenate, and the percent of total CA secreted was calculated. Peroxidase (PX) secretion was similarly determined using OPD, as the indicator agent in 0.1 M citrate buffer, pH 5.

For cell culture, rabbit acini were plated either directly onto Matrigel coated HATF filters or suspended in ice cold, neutralized rat tail collagen and plated into 35 mm petri dishes (manuscript in preparation). In both cases, approximately $10^6$ cells were plated. Cultures were incubated at 37°C in a humidified atmosphere containing 5% $CO_2$. The cells were fed at 24 hrs and at 2-4 day intervals thereafter with medium composed of DME/F12 supplemented with 5% Nu-Serum V, 1 µM dexamethasone, ITS (5 µg/ml, 5 µg/ml, 5 ng/ml), and 10 ng/ml EGF. At intervals of up to two weeks, the acini were recovered from the cultures and stained for routine light microscopy and CA histochemistry.[2]

## RESULTS

### Rat Lacrimal Gland

Only about 10% of the acini contain histochemically detectable CA.[2] Within any positively stained acinus, as few as 1 cell (not shown) or as many as all of the cells were positive (Fig. 1). The apical cytoplasmic localization of CA activity suggested that the CA may be an exocytotic protein. This is supported by the presence of CA in the lumina of intercalated and intralobular ducts (Fig. 1) as well as in the interlobular and main lacrimal ducts.[2]

Isolated acini responded to autonomic stimulation by secreting PX and CA (Table 1). CARB, a muscarinic cholinergic agonist, and PHEN, an $\alpha_1$-adrenergic agonist stimulated PX secretion 3 times above control levels (p < 0.005) while ISOP, a $\beta$-adrenergic did not significantly stimulate PX secretion. For CA, stimulation was greater than 3 fold for CARB and ISOP (p < 0.005), but was not significant for PHEN. ATR inhibited CA secretion in response to CARB and TIM inhibited the response to ISOP to control levels (13.8% and 4.1% respectively).

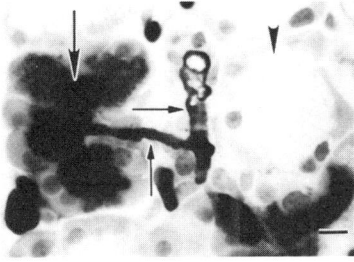

Figure 1. CA activity in rat lacrimal acinar cells (large arrow) and in associated ducts (small arrows). Unreactive acinus (arrowhead). Nuclei and basal cytoplasm counterstained with nuclear fast red. Bar = 15 µm.

Table 1. Secretion of PX and CA from isolated rat acini.

|  | % SECRETION | | | |
|---|---|---|---|---|
|  | PX* | p | CA* | p |
| CTRL | 6.3 ± 2.2 | .... | 12.9 ± 6.0 | .... |
| CARB | 19.6 ± 2.9 | <.005 | 45.3 ± 2.3 | <.002 |
| PHEN | 20.1 ± 2.0 | <.001 | 24.4 ± 6.3 | ns |
| ISOP | 15.9 ± 4.6 | ns | 44.4 ± 6.0 | <.005 |

*Mean±SEM; 3<N<7

## Rabbit Lacrimal Gland

Acini isolated from rabbit lacrimal glands retained the basolateral membrane-associated CA activity (Fig. 2A) characteristic of the terminal cells of the acini in the intact gland.[2] When placed in cell culture either within the collagen gel matrix or on coated filters, the cells proliferated. By day 4 of cell culture, lumina formed within the cell aggregates in collagen gel matrices (Fig. 2B). As with the intact acini, CA activity was associated with the basolateral membranes of the luminal cells and not the apical (luminal) cell membranes. The cells grown on HATF filters became stratified (Fig. 2C). Intense CA activity was associated with the plasma membranes except at the apical (free) and basal surfaces attached to the filters.

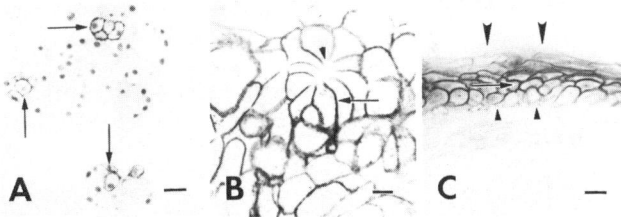

Figure 2. CA activity (arrows) in membranes of isolated rabbit lacrimal acini (A), in cultured acini within collagen gel matrix (B) and on coated filters (C). (A) Counterstained with nuclear fast red. (B) arrowhead indicates unreactive luminal (apical) cell membranes. (C) Small arrowheads indicate surface bound to filter; Large arrowheads indicate free surface. Bars = 20 µm in (A) and (C), 8 µm in (B).

## DISCUSSION

### Proposed CA Isoforms and Functions

In rat lacrimal gland, the histochemical localization of CA in the apical cytoplasm of the acinar cells and in ductal lumina suggests that this activity is due to the secretory isoform of CA designated as CA VI.[8] This is supported by the secretion of CA in response to autonomic agonists (Table 1). This tentative designation awaits confirmation by biochemical and immunological characterization and does not necessarily exclude the presence of other isoforms. In contrast, the principal CA in rabbit lacrimal acini is likely to be CA IV,[9] the membrane-associated isoform, though this also must be confirmed and does not exclude the presence of other isoforms.

The histoarchitectural and physiological differences between rat and rabbit lacrimal glands[2] suggest that the general function of the lacrimal secretions, viz., providing an appropriate medium to maintain the cornea, is achieved by somewhat different processes in these 2 species. CA isoforms have a variety of physiological functions.[8,10] All facilitate the maintenance of chemical equilibrium between dissolved $CO_2$, $H^+$ and $HCO_3^-$. Cytosolic CA's I, II and III, as well as CA IV in the plasma membrane, would ensure the rapid hydration of metabolic $CO_2$ and production of $H^+$ and $HCO_3^-$. These ions could be used in transepithelial transport processes which generate a flux of water into the acinar lumina.[3,10-12] However, by histochemical analysis, there is no detectable cytosolic CA activity in either rat or rabbit lacrimal acini.[2] Thus, the membrane-associated CA found in the terminal acinar cells of rabbit lacrimal gland may have a unique role in generating fluid flow by providing a catalytically mediated, vectorial transport of $H^+$ into the stromal space. The residual $HCO_3^-$ could be used basally in exchange for $Cl^-$ or, alternatively, could enter the acinar lumen through apically located selective channels.[12,13]

Rabbit lacrimal cells proliferated readily in cell culture, although the source of these cells, e.g., terminal acinar, non-terminal acinar or intercalated duct cells, currently is not known. While it is not yet proved that these cells are typical of acinar cells in vivo, at least one property, the presence of a membrane-associated CA, is consistent with idea that the cultured cells are similar to the cells forming the cul-de-sacs of the elongate acini. The ability to become confluent and the presence of a membrane-associated CA suggests that such cultures may be suitable for studying certain aspects of directional transport processes.

The absence of cytosolic CA activity in rat lacrimal glands is supported by lack the of immunohistochemically detectable CA's I and II.[14] This implies that either the functions provided by these enzymes in other cells are unnecessary in rat lacrimal acini, or that they are provided by other means, possibly by cytosolic CA III or mitochondrial CA V, though these were not detected by enzyme histochemistry.[2]

The apparent absence of cytosolic and membrane-associated CA in rat lacrimal gland suggests the uncatalyzed rate of $CO_2$ hydration is sufficient to take care of the demand. Perhaps $H^+$ and $HCO_3^-$ are less important, or are utilized differently, in generating the unidirectional flux of water necessary to create the required flow. It should be noted that a membrane-associated CA is present in the epithelial cells forming interlobular ducts in rat lacrimal gland, whereas it is not present in the rabbit interlobular ducts.[2]

A novel function for the CA secreted by the rat lacrimal gland is suggested by the presence of proton pumps localized in plasma membranes of the bovine corneal epithelium[15] and the requirement for $HCO_3^-$ in the medium which baths the cornea (Ubels et al, this volume). The direction in which $H^+$ is pumped by the corneal epithelial cells is not known with certainty. Whether the direction is into or out of the tear film, the presence of CA would facilitate achieving the chemical equilibrium necessary to prevent acidification or alkalinization of the corneal surface. In a survey of CA activity (CA I and II) in corneal epithelia, rabbit had the largest activity with humans having only about one third as much.[16] Though rat and bovine corneas were not studied, ovine corneal epithelium had the smallest amount of CA activity, 10% of that in rabbit. As a hypothesis, we suggest that rabbit corneas, with a very high level of endogenous CA activity, may not need exogenous CA provided by the lacrimal gland. Thus the rabbit lacrimal gland contains only the CA necessary for generating the unidirectional transport of water to develop tear flow. Other species, e.g., cat, sheep[16] and possibly rat and cow, may have very low endogenous levels of corneal CA, and therefore would require the presence of secretory CA (CA VI) provided by the lacrimal gland to ensure proper buffering of the tear film.

### Heterogeneity of Lacrimal Acinar Tissue

It is clear from both histochemical and physiological studies[2,5] that there is heterogeneity among the cells that comprise the acini in rat and rabbit lacrimal glands. In the rat lacrimal gland, most acinar cells synthesize and secrete PX, while only a minority of cells synthesize and secrete CA. Physiological heterogeneity is supported further by the order of potency in eliciting PX and CA secretion. For PX, the order of potency is CARB = PHEN >> ISOP. In contrast, the order for CA secretion is CARB = ISOP >> PHEN. If only the acinar cells that secrete CA contain β-adrenergic receptors, then this would explain the known weak protein secretory response of rat lacrimal gland to β-adrenergic stimulation.[5,17] In contrast to rabbit lacrimal gland, which secretes protein heartily upon β-adrenergic stimulation[18] and in which β-adrenergic receptors have been readily characterized by ligand methods,[19] direct assay for β-adrenergic receptors in rat lacrimal gland has not been reported, and they are difficult to detect (personal observation). Within rabbit lacrimal gland, heterogeneity among the acinar cells is indicated by the restriction of the membrane-associated CA to the terminal acinar cells.[2]

Many questions remain unresolved concerning the structural and functional differences in the lacrimal glands. Given the limited availability of human and other primate tissue, especially for physiological studies, exploration of the similarities and differences among a broad range of species may provide insight which can be more efficiently brought to bear on abnormal changes in human lacrimal gland function. It is unlikely that one or even a few non-human species will be sufficient to explore the full range of difference and similarities which must exist. Further, we must remain cognizant the lacrimal glands of each species is likely to be composed of a heterogenous population of secretory cells, and that biochemical and biophysical properties ascertained on cellular homogenates should be attributable to specific cell types within the gland, some of which may be may constitute only a small portion of the entire organ.

## ACKNOWLEDGMENTS

This work was supported in part by NIH EY0-4158.

## REFERENCES

1. L. Thorig, E.J. Van Agtmaal, E. Glasius, K.L. Tan, and N.J. Van Haeringen, Comparison of tears and lacrimal gland fluid in the rabbit and guinea pig, *Curr. Eye Res.* 4:913 (1985).
2. B.B. Bromberg, M.H. Welch, R.W. Beuerman, C. Sek-Jin, H.W. Thompson, D. Ramage, and S. Githens, Histochemical distribution of carbonic anhydrase in rat and rabbit lacrimal gland, *Invest. Ophthalmol. Vis. Sci.* 34:339 (1993).

3. A.K. Mircheff, Lacrimal fluid and electrolyte secretion: a review, *Curr. Eye Res.* 8:607 (1989).

4. T.H. Maren, Current status of membrane-bound carbonic anhydrase, *Ann. NY Acad. Sci.* 341:246 (1980).

5. B.B. Bromberg, M.M. Cripps, and M.H. Welch, Sympathomimetic protein secretion by young and aged lacrimal gland, *Curr. Eye Res.* 5:217 (1986).

6. M.M. Cripps, B.B. Bromberg, D.J. Bennett, and M.H. Welch, Structure and function of non-enzymatically dissociated lacrimal gland acini, *Curr. Eye Res.* 10:1075 (1991).

7. L.P. Brion, J.H. Schwartz, B.J. Zavilowitz, and G.J. Schwartz, Micro-method for the measurement of carbonic anhydrase activity in cellular homogenates, *Analyt. Biochem.* 175:289 (1988).

8. S.J. Dodgson, The carbonic anhydrases: Overview of their importance in cellular physiology and in molecular genetics, *in:* "The Carbonic Anhydrases," S.J. Dodgson, R.E. Tashian, G. Gros, and N.D. Carter, eds., Plenum Press, New York (1991).

9. D. Brown, X.L. Zhu, and W.S. Sly, Localization of membrane-associated carbonic anhydrase type IV in kidney epithelial cells, *Proc. Nat. Acad. Sci.* 87:7457 (1990).

10. T.H. Maren, The kinetics of $HCO_3^-$ synthesis related to fluid secretion, pH control, and $CO_2$ elimination, *Ann. Rev. Physiol.* 50:695 (1988).

11. A.K. Mircheff, C.E. Ingham, R.W. Lambert, K.L Hales, C.B. Hensley, and S.C. Yiu, $Na^+/H^+$ antiporter in lacrimal acinar cell basal-lateral membranes, *Invest. Ophthalmol. Vis. Sci.* 28:1726 (1987).

12. R.W. Lambert, M.E. Bradley, and A.K. Mircheff, $Cl^-$-$HCO_3^-$ antiport in rat lacrimal gland, *Am. J. Physiol.* 255:G367 (1988).

13. D.I. Cook and J.A. Young, Fluid and electrolyte secretion by salivary glands, in: "Handbook of Physiology Salivary, Gastric, Pancreatic, and Hepatobiliary Secretion," J.G. Forte, ed., American Physiological Society, Washington, DC (1989).

14. R.A. Hennigar, B.A. Schulte, and S.S. Spicer, Immunolocalization of carbonic anhydrase isozymes in rat and mouse salivary and exorbital lacrimal glands, *Anat. Rec.* 207:605 (1983).

15. V. Torres-Zamorano, V. Ganapathy, M. Sharawy, and P. Reinach, Evidence for an ATP-driven $H^+$-pump in the plasma membrane of the bovine corneal epithelium, *Exp. Eye Res.* 55:269 (1992).

16. W.C. Conroy, R.H. Buck, and T.H. Maren, The microchemical detection of carbonic anhydrase in corneal epithelia, *Exp. Eye Res.* 55:637 (1992).

17. Z.Y. Friedman, M. Lowe, and Z. Selinger, β-adrenergic receptor stimulated peroxidase secretion from rat lacrimal gland, *Biochim. Biophys. Acta* 675:40 (1981).

18. B.B. Bromberg, Autonomic control of lacrimal protein secretion, *Invest. Ophthalmol. Vis. Sci.* 20:110 (1981).

19. M.E. Bradley, R.W. Lambert, L.M. Lee, and A.K. Mircheff, Isolation and subcellular fractionation analysis of acini from rabbit lacrimal glands, *Invest. Ophthalmol. Vis. Sci.* 33:2951 (1992).

# BASEMENT MEMBRANE MODULATION OF STIMULATED

# SECRETION BY LACRIMAL ACINAR CELLS

Gordon W. Laurie, J. Douglas Glass, and Rebecca A. Ogle

Department of Anatomy and Cell Biology
University of Virginia
Charlottesville, VA 22908

## INTRODUCTION

Lacrimal acinar cells rest basally on a basement membrane and have a secretory granule-filled apical cytoplasm adjacent to a lumen into which tear proteins are released. An important array of proteins are produced. In man, tear proteins are estimated to number as many as sixty (Gachon et al, 1979) and include: *lysozyme*, which plays a prominent bacteriocidal role on the corneal surface; *lactoferrin*, which functions as both a bacteriocidal agent and as a potential inhibitor of complement activation; *secretory component*, which regulates the transcellular movement of IgA into acini lumen where it acts on the corneal surface to inhibit bacterial adhesion; *tear-specific prealbumin*, whose function is not known. In rats, *peroxidase* is a tear component which has served as a convenient marker in experimental studies. Tears not only have an important bacteriocidal role, they also keep the cornea clean and lubricated, and are important for the well-being of the corneal epithelium.

Tear protein secretion illustrates a number of age (McGill et al, 1984), gender and disease-specific alterations; contact lens wear can alter tear protein composition (Vinding et al, 1987). In ageing, tear secretion is reduced, a condition that may require topical artificial tear substitutes. Moreover, the second most common autoimmune disease is Sjögren's syndrome (Pflugfelder et al, 1991), a triad of diseases whose ocular component is dry eye - a condition affecting millions of people worldwide, particularly women (Kincaid, 1987).

The question of how basement membrane molecules may modulate tear protein secretion in development, ageing and disease is the interest of our laboratory. Surprisingly little is known on the molecular composition of lacrimal acinar basement membranes, nor on interactions of individual basement membrane molecules with lacrimal epithelia in vitro. This is unfortunate because epithelial cells of the mouse mammary gland, which are similar in many respects to lacrimal epithelia, are profoundly influenced in vitro by the presence of an underlying basement membrane (Li et al, 1987). Not only is the synthesis and secretion of milk proteins dramatically augmented, but also hormone responsiveness is restored. The molecular basis for this phenomenon remains largely unknown.

*Laminin* (800 kD), *entactin* (also known as nidogen, 158 kD), *collagen IV* (540 kD; Timpl, 1989) and heparan sulfate proteoglycan (now known as *perlecan*, 800 kD; Hassell et al, 1986) are all examples of basement membrane proteins with cell attachment activity, whereas *BM-40* (also known as osteonectin or SPARC; 40 kD) does not. Laminin also has EGF-like mitogenic activity (Panayotou et al, 1989). Entactin has EGF-like sequences (Mann et al, 1989) but these are not known to be functional. In addition, growth factor binding site(s) have been demonstrated on collagen IV for TGF-ß (Paralkar et al, 1991), and on perlecan for bFGF (Vlodofsky et al, 1991); BM-40, which binds $Ca^{+2}$, is thought to play a role in endothelial cell cycling (Funk and Sage, 1991). Since these molecules are large with multiple functional domains, proteolytic fragments and synthetic polypeptides have been useful in identifying active sites. This approach has identified several different attachment sites on laminin located on each of its three main constituent B1, B2 and A polypeptide chains which interact with distinct cell surface receptors (Beck et al, 1990). Collagen IV has at least one cell attachment site (Vandenberg et al, 1991), and entactin and heparan sulfate proteoglycan apparently one each. In addition, these molecules interact with one another through distinct binding sites (Laurie et al, 1986). Recent use of PCR technology has revealed novel homologous chains to laminin ('S-laminin', Hunter et al, 1989; 'merosin', Leivo et al, 1989) and collagen IV (alpha chains 3-5, Hudson et al, 1989; Barker et al, 1990) indicating that there exists a family of laminin and collagen IV molecules, each perhaps with different functions.

Although no information is available on whether any of the major component molecules are present in lacrimal basement membranes, their absence would be highly surprising since laminin, entactin, perlecan and collagen IV have been detected in almost all basement membranes examined (Laurie et al, 1983). A few functional studies, however, have been performed by plating isolated lacrimal acinar cells on a commercial crude extract of basement membrane known as 'Matrigel' (Kleinman et al, 1986; trade name for a urea extract of mouse basement membrane sold by Collaborative Research) whose composition includes all of the four main components listed above, and several growth factors (Taub et al, 1990; Mannuzza, 1992). In this way suitable lacrimal cell culture conditions have been gradually developed and several interesting observations have been made. Oliver et al (1987) compared the in vitro growth of isolated rat parotid, lacrimal and pancreatic acinar cells, and morphologically observed that cells attached via their basal surface and maintained a differentiated phenotype with peroxidase containing secretory granules. Sullivan's group (Hann et al, 1989) elegantly extended these studies by development of a serum-free medium wherein rat lacrimal acinar cells plated on Matrigel were monitored for secretory component production and MSH binding, and found to be functional up to 3 wks in culture. Recent advances in the composition of lacrimal culture media by Sullivan's group (Hann et al, 1991) included elimination of carbachol whose continual presence at $10^{-6}$ M was thought to be necessary for the well-being of the cells (Oliver et al, 1987). Also, the insulin level was increased. Under these conditions, both androgen amplified and constitutive secretion of secretory component by Matrigel-adherent lacrimal cells were found to be dramatically elevated over cells attached to fibronectin, collagen I or polylysine-coated plates. Similar, but less dramatic effects were observed for laminin (Hann et al, 1991). Taken together, the use of Matrigel made possible tremendous advances in lacrimal acinar cell culture conditions but little information was gained on the molecular mechanisms of action.

We therefore initiated studies using a combined molecular dissection/monoclonal antibody approach and as starting material a 10 mM EDTA extract ('BMS') of basement membrane which is similar in composition to Matrigel. These experiments have resulted in the identification of what appear to be a novel group of lower molecular weight polypeptides some of which have cell attachment activity, and at least one of which appears to be an important modulator of regulated tear secretion.


## METHODOLOGY

Since lacrimal acinar basement membranes are thin, making extraction difficult, we made use of the mouse EHS tumor from which gram amounts of basement membrane material can be readily obtained (Orkin et al, 1977). For this purpose, tumor matrix stored at -80°C was homogenized (4°C), then washed extensively in 150 mM NaCl, 50 mM Tris, pH 7.4 containing NEM and PMSF to remove serum proteins and cell debris. Matrix was extracted overnight in the same buffer which in addition, contained 10 mM EDTA (Paulsson et al, 1987); solubilized material was sterilized by dialysis versus PBS containing 0.5% chloroform, then dialyzed versus DMEM containing gentamicin (50 μg/ml). BMS was also divided into two peaks ('peak 1' and 'peak 2') by gel filtration; with each peak tested for activity. We used EDTA rather than urea for extraction because: (1) we obtain a similar spectrum of proteins, (2) EDTA is less harsh, therefore unique activities maybe revealed, and (3) EDTA serves as an effective inhibitor of metalloproteases. We isolated lacrimal acinar cells according to Oliver et al (1987), as modified by Hann et al (1989). Two changes were made in the protocol: (1) an initial cardiac perfusion with DMEM was performed to wash out glandular blood, and (2) cells were spun on a 10/30/60% Percoll step gradient giving rise to live cells at the 30/60% interface. Freshly isolated cells were then plated on BMS (EDTA extract of EHS tumor basement membrane) coated (37°C, 1 hr) at 0.6 mg/well in 48 well plates. 24 hr later floating cells were removed, fresh medium containing carbachol ($10^{-4}$ M) and VIP ($10^{-8}$ M) added and secreted peroxidase was assayed kinetically. Data was normalized to cellular DNA, compared to total cellular peroxidase and expressed as the mean ± the standard deviation. Secretory granules were isolated using isosmotic density gradient centrifugation (von Zastrow and Castle, 1987).

## RESULTS

Lacrimal acinar cells are polarized epithelial cells in continuous contact with basement membrane (Fig. 1a) - a thin, partially characterized extracellular matrix whose major component by mass is the cell adhesion protein laminin. Antilaminin antibodies immunostain lacrimal acinar basement membranes and freshly isolated lacrimal acinar cells adhere moderately well to laminin (Laurie et al, submitted), although not as well as Harderian acinar cells using the same isolation method (Laurie and Stone, submitted). Apical secretory granules maybe readily isolated (Fig. 1b).

To investigate how basement membrane may affect regulated tear protein secretion, lacrimal acinar cells were plated on BMS and incubated for 24 hr. Tear protein secretion by attached cells was monitored at various times after stimulation with carbachol ($10^{-4}$ M) and VIP ($10^{-8}$ M). In particular, we followed secretion of the tear for release. The effect, instead, was on the ability of cells to respond to secretagogue. It appeared therefore that: (1) BMS contained a factor(s) which is required for maintenance of stimulus-secretion coupling, (2) since BMS differed from the laminin/entactin preparation mainly in the presence of lower molecular weight peak 2 material, it was likely that the activity(s) resides in peak 2, and (3) regulated

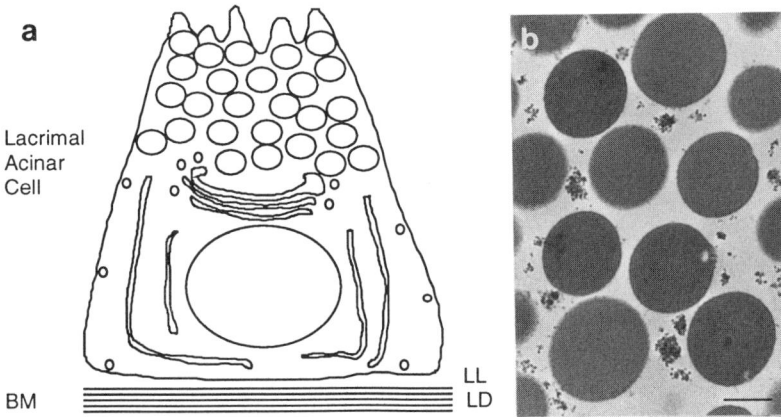

Figure 1. Intimate association with basement membrane. (a) Schematic diagram of lacrimal acinar cell attached basally to basement membrane (BM). In the past, basement membrane was described as consisting of a laminin lucida (LL) and lamina densa (LD), but tissue dehydration by freeze substitution has revealed the lamina lucida to be artifactual. The apical region of lacrimal acinar cells is packed with secretory granules. (b) Electron micrograph of secretory granules isolated from rat lacrimal acinar cells. Granules are very regular in size. Antibodies against granule content will be prepared for use in secretion studies. Bar = 0.4 μm.

protein peroxidase whose activity may be readily monitored spectrophotometrically using the method of Herzog and Miller (1976). We optimized cell number, concentration of secretogogue and validated the cellular DNA assay. BMS coating amount had little effect on constitutive (Fig. 2a) or regulated (Fig. 2b) peroxidase secretion. Subsequent experiments led to the interesting observation that attachment to BMS gives rise to higher levels of regulated peroxidase secretion versus attachment to laminin/entactin alone (Fig. 2b), an effect which was not due to differences in total peroxidase available secretion was not dependent on reaggregation of cells into acini-like structures, a process which requires about five days.

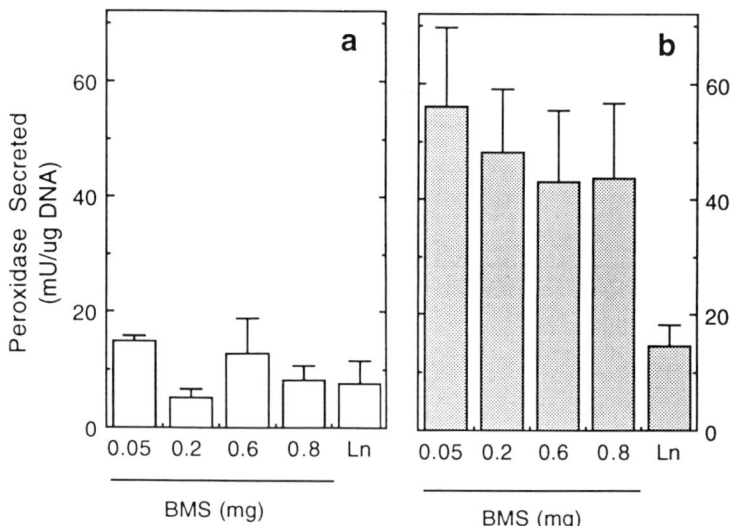

Figure 2. BMS coating concentration had little effect on constitutive (a) or regulated secretion (b). Regulated secretion by cells on 0.6 mg laminin (Ln) was dramatically less. Values represent peroxidase secreted in 100 min from one experiment. Regulated secretion was monitored after addition of carbachol/VIP at 0 min.

Since cells are in contact with BMS during the 24 hour period prior to stimulation, we asked whether the peak 2 activity(s) may be a cell adhesion protein(s). To test this possibility, attachment studies were carried out on peak 2 fractionated by gel filtration. Attachment activity was concentrated in several fractions which, via blot attachment analysis of DTT reduced peak 2, appeared to correspond to proteins of 25, 40 and 60 kD (Laurie et al, submitted). To pursue these activities, LOU/M rats were immunized over several months with peak 2, and an animal whose tail bleed stained lacrimal basement membranes and blocked attachment of HT1080 cells to peak 2 was chosen for fusion. Subsequent supernatants were screened for the ability to block cell adhesion and hybridomas were cloned. The antibodies do not appear to show identity with laminin, collagen IV, perlecan, vitronectin, BM-40 and fibronectin, nor do they inhibit cell adhesion to laminin, collagen IV or vitronectin. To then ask whether peak 2 adhesion protein(s) correspond to secretion activity(s), cells were plated on BMS which had been preincubated with each of the antibodies. One of the antibodies (3E12) dramatically suppressed regulated secretion to laminin-entactin levels without affecting total cellular peroxidase (Laurie et al, submitted). The nature and distribution of 3E12 antigen is currently under investigation.

# CONCLUSIONS

Basement membrane plays a key role in lacrimal acinar cell differentiation and physiology. In fact, the basement membrane and associated lacrimal acinar cell may be considered as a functional unit (Bissell et al, 1982). Since basement membrane is only partially characterized, molecular dissection of this relationship was initiated by testing gel filtration-separated fractions for affect on regulated or constitutive secretion and cell adhesion. This approach identified the importance of laminin and apparently novel peak 2 protein(s) in modulating stimulus-secretion coupling. Whether alterations in Sjögren's lacrimal basement membranes could lead to deficient tear secretion remains to be determined.

# REFERENCES

Barker, D.F., Hostikka, S.L., Zhou, J., Chow, L.T., Oliphant, A.R., Gerkin, S.C., Gregory, M.C.,Skolnick, M.H., Atkin, C.L., and Tryggvason, K., 1990, Identification of mutations in the Col4α5 collagen gene in Alport syndrome. *Science* 248:1224.

Beck, K., Hunter, I., and Engel, J., 1990, Structure and function of laminin: anatomy of a multidomain glycoprotein. *FASEB J.* 4:148.

Bissell, M.J., Hall, H.G., and Parry, G., 1982, How does the extracellular matrix direct gene expression? *J. Theor. Biol.* 99:31.

Funk, S.E., and Sage, H., 1991, The Ca+2-binding glycoprotein SPARC modulates cell cycle progression in bovine aortic endothelial cells. *Proc. Natl. Acad. Sci. USA* 88:2648.

Gachon, A.M., Verrelle, P., Betail, G., and Dastugue, B., 1979, Immunological and electrophoretic studies of human tear proteins. *Exp. Eye Res.* 29:539.

Hann, L.E., Kelleher, R.S., and Sullivan, D.A., 1991, Influence of culture conditions on the androgen control of secretory component production by acinar cells from the rat lacrimal gland. *Invest. Ophthal. Vis. Sci.* 32:2610.

Hann, L.E., Tatro, J.B., and Sullivan, D.A., 1989, Morphology and function of lacrimal gland acinar cells in primary culture. *Invest. Opthalmol. Vis. Sci.* 30:145.

Hassell, J.R., Kimura, J.H., and Hascall, V.C., 1986, Proteoglycan core protein families. *Ann. Rev. Biochem.* 55:539.

Herzog, V., Sies, H., and Miller, F., 1976, Exocytosis in secretory cells of rat lacrimal gland. *J. Cell Biol.* 70:692.

Hudson, B.G., Wieslander, J., Wisdom B.J., and Noelken, M.E., 1989, Goodpasture syndrome: Molecular architecture and function of basement membrane antigen. *Lab. Invest.* 61:256.

Hunter, D.D., Shah, V., Merlie, J.P., and Sanes, J.R., 1989, A laminin-like adhesive protein concentrated in the synaptic cleft of the neuromuscular junction. *Nature* 338:229.

Kincaid, M.C., The eye in Sjögren's syndrome, *in:* Sjögren's Syndrome. Clinical and Immunological Aspects, N. Talal, H.M. Moutsopoulos, and S.S.Kassan, eds., Springer-Verlag, Heidelberg (1987).

Kleinman, H.K., McGarvey, M.L., Hassell, J.R., Star, V.L., Cannon, F.B., Laurie, G.W., and Martin, G.R., 1986, Basement membrane complexes with biological activity. *Biochem.* 25:312.

Laurie, G.W., Bing, J.T., Kleinman, H.K., Hassell, J.R., Aumailley, M., Martin, G.R., and Feldmann, R.J. (1986) Localization of binding sites for laminin, heparan sulfate proteoglycan and fibronectin on basement membrane (type IV) collagen. J. Mol. Biol. 189: 205-216.

Laurie, G.W., Leblond, C.P., and Martin, G.R., 1983, Light microscopic immunolocalization of type IV collagen, laminin, heparan sulfate proteoglycan, and fibronectin in the basement membranes of a variety of rat organs. Am. J. Anat. 167:71.

Leivo, I., Engvall, E., Laurila, P., and Miettinen, M., 1989, Distribution of merosin, a laminin-related tissue-specific basement membrane protein, in human Schwann cell neoplasms. Lab. Invest. 61:426.

Li, M.L., Aggeler, J., Farson, D.A., Hatler, C., Hassell, J., and Bissell, M.J., 1987, Influence of a reconstituted basement membrane and its components on casein gene expression and secretion in mouse mammary epithelial cells. Proc. Natl. Acad. Sci. USA 84:136.

Mann, K., Deutzmann, R., Aumailley, M., Timpl, R., Raimondi, L., Yamada, Y., Pan, T., Conway, D., and Chu, M.-L., 1989, Amino acid sequence of mouse nidogen, a multidomain basement membrane protein with binding activity for laminin, collagen IV and cells. EMBO J. 8:65.

Mannuzza, F.J., 1992, Removal of soluble growth factors from matrigel basement membrane matrix and the demonstration and quantitation of insoluble, matrix-bound TGF-beta. Mol. Bio. Cell 3:226a.

McGill, J.I., Liakos, G.M., Goulding, N., and Seal, D.V., 1984, Normal tear protein levels and age-related changes. Br. J. Ophthal. 68:316.

Oliver, C., Waters, J.F., Tolbert, C.L., and Kleinman, H.K., 1987, Growth of exocrine acinar cells on a reconstituted basement membrane gel. In Vitro 23:465.

Orkin, R.W., Gehron, P., McGoodwin, E.B., Martin, G.R., Valentine, T., and Swarm, R., 1977, A murine tumor producing a matrix of basement membrane. J. Exp. Med. 145:205.

Panayotou, G., End, P., Aumailley, M., Timpl, R., and Engel, J., 1989, Domains of laminin with growth-factor activity. Cell 56:93.

Paralkar, V.M., Vukicevic, S., and Reddi, A.H., 1991, Transforming growth factor beta type-1 binds to collagen-IV of basement membrane matrix - Implications for development. Devel. Biol. 143:303.

Paulsson, M., Aumailley, M., Deutzman, R., Timpl, R., Beck, K., and Engel, J., 1987, Laminin-nidogen complex. Extraction with chelating agents and structural characterization. Eur. J. Biochem. 166:11.

Pfugfelder, C.C., Yen, M., and Atherton, S., 1991, Detection of Epstein-Barr virus antigens and receptor molecules on ocular surface and lacrimal gland epithelia. Invest. Ophthal. Vis. Sci. 32:807a.

Taub, M., Wang, Y., Szczesny, T.M., and Kleinman, H.K., 1990, Epidermal growth factor or transforming growth factor alpha is required for kidney tubulogenesis in matrigel cultures in serum-free medium. Proc. Natl. Acad. Sci. USA 87:4002.

Timpl, R., 1989, Structure and biological activity of basement membrane proteins. Eur. J. Biochem. 120:487.

Vandenberg, P., Kern, A., Ries, A., Luckenbill-Edds, L., Mann, K., and Kühn, K., 1991, Characterization of a type IV collagen major cell binding site with affinity for the $\alpha 1\beta 1$ and $\alpha 2\beta 1$ integrins. J. Cell Biol. 113:1475.

von Zastrow, M., and Castle, J.D., 1987, Protein sorting among two distinct export pathways occurs from the content of maturing exocrine storage granules. J. Cell Biol. 105:2675.

Vinding, T., Ericksen, J.S., and Nielsen, N.V., 1987, The concentration of lysozyme and secretory IgA in tears from healthy persons with and without contact lens use. Acta Ophthalmol. 65:23.

Vlodavsky, I., Fuks, Z., Ishaimichaeli, R, Bashkin, P., Levi, E., Korner, G., Barshavit, R., Klagsbrun, M., 1991, Extracellular matrix-resident basic fibroblast growth factor - Implication for the control of angiogenesis. J. Cell Biochem. 45:167.

# GALACTOSE- BINDING SITES IN THE ACINAR CELLS OF THE HUMAN ACCESSORY LACRIMAL GLAND

Winrich Breipohl[1], Manfred Spitznas[1], Fred Sinowatz[2], Oliver Leip[1], Wallid Naib-Majani[1], and Andrea Cusumano[1]

[1]Zentrum für Augenheilkunde, RFW-Universität, Bonn; [2]Institut für Veterinäranatomie, LM-Universität, München

## INTRODUCTION AND AIMS

Glycoproteins constitute a major component of human tears and play an important role in the maintenance of the precorneal tear film (Holly and Lemp 1977, Nichols et al. 1985).

Ahmed and Grierson (1989) have convincingly shown that human tears contain N-acetyl-galactosamine and galactose-N-acetylgalactosamine moieties, which can be identified by group III lectins (Goldstein and Poretz 1986). Versura et al. (1986) found altered tear concentrations of these glyconjugates in patients with dry eye syndrome. No reports have been published on lectin binding sites for N-acetyl-galactosamine moieties in the acinar cells of the accessory lacrimal gland.

Thus we investigated in the human accessory lacrimal gland the expression of glycoproteins binding lectins of group III (Goldstein and Poretz 1986) with different specificities to galactosamine and galactose sugar moieties.

## MATERIAL AND METHODS

Accessory lacrimal gland tissue became available from surgical material of normal patients (n= 6 - 10 for the different lectins) who underwent lid surgery for various reasons.

Samples were fixed in Bouins for 12 hours to 4 weeks at 4°C, embedded in Paraffin and serially sectioned at 6 μm. Sections were dewaxed in Xylene or Citric oil, rehydrated in decreasing series of methanol and ethanol respectively and 0.05 molar Tris or PBS at pH 6.8. This was followed by immersion in PBS and 1% Gelatine, and dark humid chamber incubation with 33 μg lectin (from Sigma, Munich) per ml over night at 4°C. After a short washing sections were inspected microscopically for overall staining. Renewed washing for another few hours allowed for removal of excessive unspecific labelling before coverslipping and storage at 4°C until final evaluation.

A panel of five FITC labeled group III lectins (Goldstein and Poretz 1986) which differ regarding their specific affinity to various mono- and disaccharide configurations of galactosamine and galactose sugar moieties were applied (Table 1).

Binding was classified as follows: No binding: -; moderate or regional binding only +/-; overall binding of acinar cells +. The labelling indices were set at 50%, 100% when all the samples showed +/- and + respectively. Other percentages were calculated accordingly. This form of evaluation does not intend to discriminate between weak and strong binding because patient related and tissue related specificities interfering with such differences cannot be excluded.

Negative controls consisted of tissue samples of mouse intestine with previously determined binding pattern to the above lectins (Döhrn et al. 1993). In positive controls the specificity of each lectin binding was determined by preincubation of the lectin with the respective sugar moiety in various molar concentrations (Table 1).

## RESULTS

Results are summarized in Table 1. In short: glycoconjugate binding of acinar cells was found for all five group III lectins reflecting the overall binding specificity to N-acetylgalactosamine and Galactose respectively. Labelling indices varied according to the differential binding capacity of the selected lectins. Monosaccharide binding was stronger than disaccharide binding and $\alpha$ and $\beta$ GalNAc was less effectively inhibiting than $\alpha$ and $\beta$Gal. DBA, the only lectin with a GalNAc$\alpha$1,3Gal-

**Table 1.** Survey on applied group III lectins, their carbohydrate specificities and labelling indices.

| LECTIN and PLANT ORIGIN | SUGAR SPECIFICITY | LABELLING INDEX |
|---|---|---|
| DBA (Dolichos biflorus) | N-Acetylgalactosamine (GalNAc$\alpha$1,3GalNAc > GalNAc$\alpha$1,3Gal) | 3% |
| GSA-I (Griffonia simplicifolia) | N-Acetylgalactosamine and Galactose ($\alpha$GalNAc > $\alpha$Gal ) | 56% |
| MPA (Maclura pomifera) | N-Acetylgalactosamine and Galactose (Gal$\beta$1,3GalNAc > $\alpha$Gal) | 56% |
| SBA (Glycine max) | N-Acetylgalactosamine and Galactose ($\alpha$ and $\beta$GalNAc > $\alpha$ and $\beta$Gal | 66% |
| PNA (Arachis hypogaea) | N-Acetylgalactosamine and Galactose ($\alpha$GalNAc > $\alpha$ and $\beta$Gal) | 78% |

LEGEND: $\alpha$ and $\beta$ indicate configuration of sugars.

ABBREVIATIONS USED: Gal: Galactose; GalNAc: N-Acetyl-galactosamine

NAc- and (less intensive) GalNAc1,3Galα disaccharide specificity led to a labelling index of only 3%. In contrast MPA, the only lectin with a GalNAcβGalNAc- and (less intensive) αGal-specificity led to a labelling of 56% of the acinar cells. A even higher labelling index was reached by SBA specific for the α and β form of both group III-monocsaccharides (GalNAc > Gal). PNA in comparison to SBA with a lack of binding to βGalNAC yielded the highest labelling index (78%). The method of deparaffinization had no influence on the results but with fixations exceeding a few weeks labelling intensities were reduced.

## DISCUSSION

This investigation is the first written report indicating a specific contribution of the accessory gland acinar cells to group III lectin binding glycoconjugates of human tears. αGal monosaccharide seems to be the main sugar moiety responsible for the observed results and labelling indices. Comparable investigations on the other tear glycoconjugate sources are missing as are studies differentiating to such extent between the various carbohydrate binding configurations involving N-Acetyl-Galactosamine and Galactose sugars. The results could help to identify a specific lectin binding pattern of the accessory lacrimal gland in comparison to the main lacrimal gland, conjunctival goblet and non-goblet cells. Overall the results are of special interest for our understanding of the importance and functional aspects of the accessory lacrimal gland. They could also help to differentiate between tear glycocon-jugate components from this organ and other sources under normal versus pathological conditions, e.g. the dry eye syndrome.

Using hexosamine as an indicator glycoprotein levels in the precorneal film have been stated to be normal in patients with Sjögren syndrome (Tabbara et al. 1978). This finding does not exclude, however, the possibility that the spectrum of tear glycoproteins has changed overall and that the individual sources for the tear glycoconjugates are differently affected. The origin of tear glycoconjugates has often been associated with the function of an appropriate number of conjunctival goblet cells and non-goblet epithelial cells (Greiner and Allansmith 1980, Kawano et al. 1984). Considerable controversy exists regarding the behaviour of tear glycoconjugate components in healthy probands versus patients with Keratoconjunctivitis Sicca (KCS) (Wright and Mackie 1977, Liotet et al. 1987). Goblet cell reduction and specific conjunctival glycoconjugate expression have been denied as likely reasons to explain the pathomechanisms involved in KCS and dry eye syndrome (Torök and Süveges 1982, Liotet et al. 1987, Johnson et al. 1990).

Both, the main and accessory lacrimal glands have been identified and assumed as additional sources of specific glycoconjugate origin for the tears (Versura et al. 1986, Ahmed and Grierson 1989, Goebbels 1990). Versura et al. (1986) have also shown that a specific group of glycoconjugates containing N-acetylgalactosamine and galactose-N-acetylgalactosamine sugar moieties are reduced while mannose moieties, present in the mucus-glycocalyx layer of conjunctival cells, but not conjunctival goblet cells (Kawano et al. 1984, Johnson et al. 1990), are increased in the tears of dry eye patients.

This investigation has revealed strong evidence in favour of an accessory lacrimal gland contribution to glycoconjugates with these carbohydrate moieties in the human tears. It has further shown that lectins with nominal identical carbohydrate specificities differ widely in the labelling intensity of the respective glycoconjugates. Partly these differences can be explained by their preferential sugar specificity (see Table 1). Other explanations develop from the specific site of glycoconjugate binding. It is known, e.g. that PNA binds especially to the preterminal N-acetyl-galactosamine. This specificity could help to explain the differences between PNA and SBA labelling indices, while βGalNAc inhibition experiments could not. It remains also to be seen whether in the accessory lacrimal gland, the 56% labelling index would change after neuraminidase exposure of the preterminal sugar residue.

The potential diagnostic value of a differential analysis of glycoprotein contributions to the tears could be tested e.g., in relation to therapeutic vitamin A treatments of drye eye syndrome (Sullivan et al. 1973). Further diagnostic implications develop from the following considerations. Mucin glycoprotein expression in conjunctival goblet cells (Versura et al. 1986), main lacrimal gland (Ahmed and Grierson 1989), and human accessory lacrimal gland (Breipohl et al. 1993), are assumed to help form an 0.02 to 0.04 μm thick hydrophobic/hydrophilic interface on the corneal surface. This way an intimate contact between the hydrophobic cornea and the hydrous tears are established, and normal break up times are achieved. The specific sialoglycoprotein contribution of the different sources may vary and it cannot be excluded that a reduced break up time could be caused by different pathways. Dry eyes can even be associated with an overproduction of specific glycoproteins (Versura et al. 1986, Goebbels 1990). Thus detailed studies with even more than the here applied glycoconjugate identifying lectins are required to differentiate between the specific behaviour of tear glycoconjugate producing sources (goblet cells, non-goblet epithelial cells, main and accessory lacrimal gland cells) in healthy probands and patients with KCS.

## SUMMARY

This investigation for the first time has collected evidence of a specific glycoconjugate contribution of the acinar cells from accessory lacrimal glands to human tears. Amongst group III lectin binding glycoconjugates, monosaccharides seem to be more prominent than disaccharides. $\alpha$ (and less obvious $\beta$) Galactose sugar moieties appear to be specifically important. A need for further differentiating investigations is outlined.

## REFERENCES

Ahmed, A., Grierson, I., 1989, Cellular carbohydrate components in human, rabbit and rat lacrimal gland. *Graefe's Arch Exp-Ophthalmol*, 227:78-87.

Breipohl. W., Sinowatz, F., Naib-Majani, W., Leip, O., Spitznas, M., 1993, Sialomucin expression in the human accessory lacrimal gland. *submitted*.

Doern, S., Breipohl, W., Lierse, W., Romaniuk, K., Young, W., 1993, Developmental changes in the distribution of cecal lectin binding sites of Balb-c mice. *Acta Anatomica*, in press.

Goebbels, M.J., 1990, Fluorophotometrie als objektive und quantitative Technik zur Bestimmung von Tränensekretion und Barrierefunktion des Hornhautepithels - neue Aspekte für Klinik und Pathophysiologie trockener Augen. *Thesis, Bonn*.

Goldstein, I.J., Poretz, R.D., 1986, Isolation, physicochemical characterization, and carbohydrate binding specificity of lectins, 35-250, in: Liener IE, Sharon NC, Goldstein ID, eds, "The Lectins: Properties, Functions and Applications in Biology and Medicine", Academic Press Inc., Orlando, Fla..

Greiner, J.V., Allansmith, M.R., 1981, Effect of contact lens wear on the conjunctival mucus system. *Ophtalmol*, 88:821-832.

Holly, F.J., Lemp, M.A., 1977, Tear physiology and dry eyes. *Surv Ophthalmol*, 22:69-87.

Johnson, W., Whitley, H.E,, McLaughlin, A., 1990, Effecfts of inflammation and aqueous tear film deficiency on conjunctival morphology and ocular mucus composition in cats. *Am J Vet Res*, 51:820-824.

Kawano, K., Uehara, F., Sameshima, M., Ohba, N., 1984, Application of lectins for detection of goblet cell carbohydrates of the human conjunctiva. *Exp Eye Res*, 38:439-447.

Liotet, S., Van Bijsterveld, O.P., Kogbe, O., Laroche, L., 1987, A new hypothesis on tear film stability. *Ophthalmol*, 195:119-124.

Nichols, B.A., Chiappino, M.L., Dawson, C.R., 1985, Demonstration of the mucous layer of the tear film by electron microscopy. *Invest ophthalmol Vis Sci*, 26:464-473.

Tabbara, K.F., Ostler, H.B., Daniels, T.E., Silvester, R.A., Greenspan, J.S., Talal, N., 1978, Sjögren's syndrome. *Trans Am Acad Ophthalmol Otolaryngol*, 18:121-124.

Sullivan, W.R., Culley, J.D., Dohlman, C.M., 1973, Return of goblet cells after vitamin A therapy in xerosis of the conjunctiva. *Am J Ophthalmol*, 75:720-725.

Törek, M., Süveges, I., 1982, Morphological changes in "Dry Eye Syndrome". *Graefe's Arch Clin, Exp Ophthalmol*, 219:24-28.

Versura, P., Maltarello, M.C., Cellini, M., Caramazza, R., Laschi, R.. 1986, Detection of mucus glycoconjugates in human conjunctiva by using the lectin-colloidal gold technique in TEM. II. A quantitative study in dry-eye patients. *Acta Ophthalmol*, 64:445-450.

Wright, P., Mackie, J.A., 1977, Mucus in the healthy and diseased eye. *Trans ophthal Soc*, 97:1-7.

# THOUGHTS ON THE DUCTULES OF THE AGING HUMAN LACRIMAL GLAND

Orkan George Stasior[1] and Janet L. Roen[2]

Albany[1] and New York[2], NY

As we age, there is a progressive decrease in secretion of tears from the lacrimal gland. The treatment of resultant dry eyes has been directed at the use of artificial tears and/or closing the lacrimal puncta.

As eye plastic and reconstructive surgeons, we have seen many patients with epiphora due to closure of the lacrimal puncta or canaliculi. In some patients, the narrowing and eventual closure of the canaliculi has been related to eye infections, inflammation or medications. We wondered, if the excretory canals could stenose due to medications, inflammation or infections, if the same process could also cause stenosis of the orifices of the tubules of the lacrimal glands? Could decreased lacrimal gland secretion be on an endocrine and anatomical basis?

A search of scientific literature on lacrimal gland anatomy and histopathology revealed little information about the status of tubules and their orifices in the aging lacrimal gland. Dry eye meetings that we attended focused mainly on the evaluation of the tear film, its composition and replacement. There were no papers dealing with revisiting the lacrimal gland to see if it could be stimulated medically or surgically to produce more tears.

We examined 32 lacrimal glands, removed at autopsy from persons aged 35 to 88 years. They were studied by light microscopy and characterized with regard to parenchyma, connective tissue, inflammation and ducts. Eight glands were found to be normal; 24 glands showed abnormalities (e.g. figures 1-3). The most common changes were chronic lymphocytic inflammation and periductular fibrosis. Stasis of tears, combined in five glands with enlarged efferent ducts, some massively ectatic, was noted in six patients in their seventh, eighth and ninth decades (Roen et al., 1985).

There were ductal abnormalities in 74% of the patients over fifty years of age. These consisted of epithelial changes in two, periductal fibrosis in 50% of patients over sixty years of age and enlarged or distended ductules in almost all patients over sixty years of age.

*Lacrimal Gland, Tear Film, and Dry Eye Syndromes*
Edited by D.A. Sullivan, Plenum Press, New York, 1994

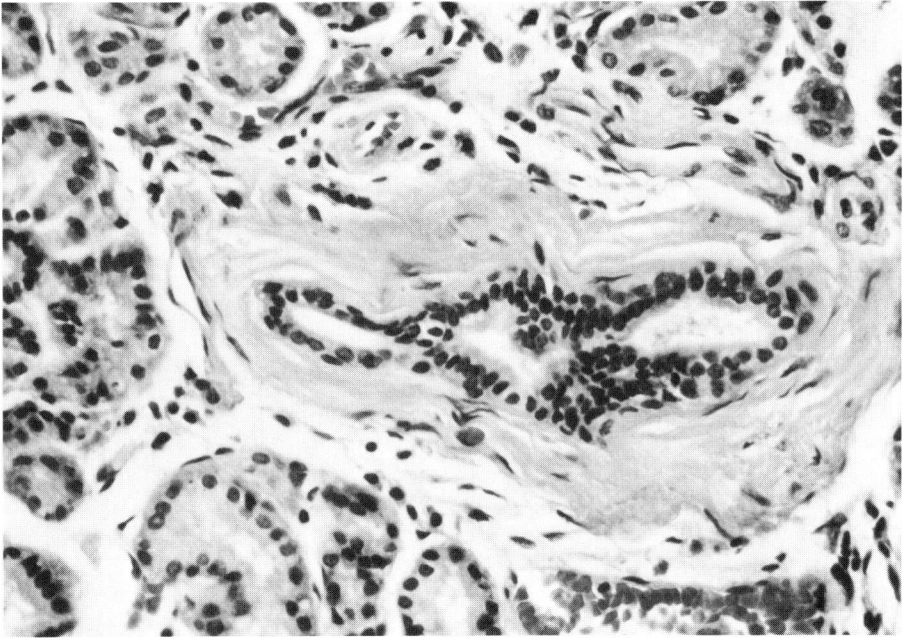

**Figure 1.** Periductal fibrosis.

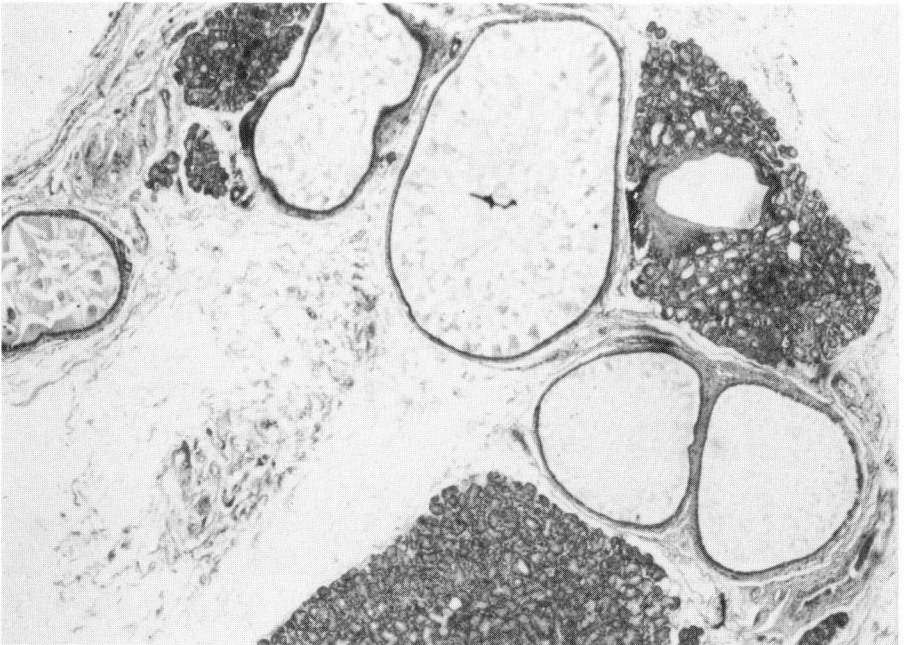

**Figure 2.** Distended ductules with stasis.

We first presented our findings at the Cambridge Dry Eye Meeting in 1984. Professor W.R. Lee at that meeting told me that his colleagues had just finished a similar study of 99 lacrimal glands and would soon be publishing their results (Damato et al., 1984).

Since then, Drs. Anthony Bron, Stephen Pflugfelder, Renee Kaswan, David Sullivan and others and, more recently, additional scientists at the International Conference on the Lacrimal Gland, Tear Film and Dry Eye symposium, have revisited the lacrimal gland. It now seems that it will soon be possible, through chemical means, to increase the secretions from the lacrimal gland.

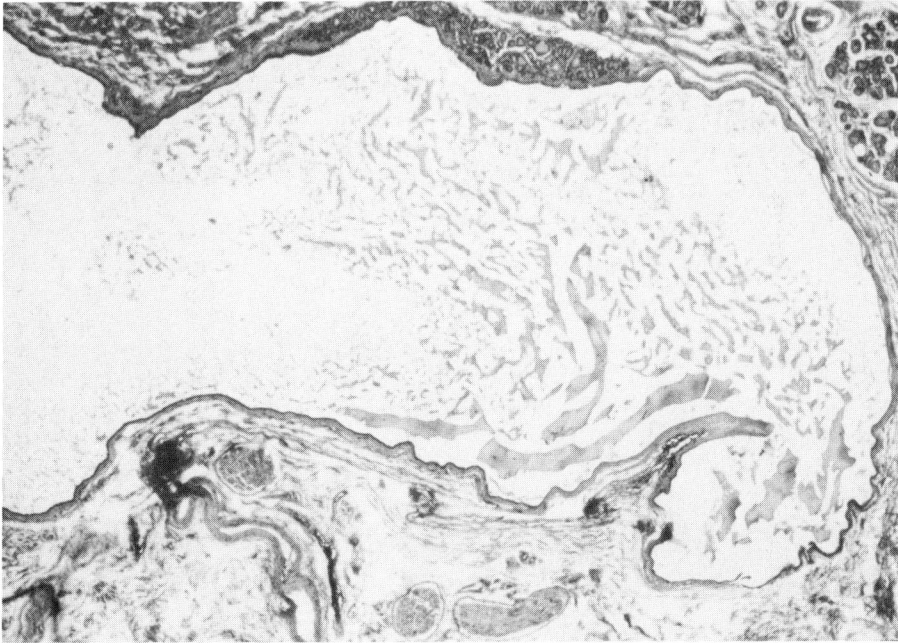

**Figure 3.** Huge distended ductule - "dacryops."

To continue successfully into a future of increased lacrimal gland secretion for our patients, we need more anatomical studies regarding the tubules of the lacrimal gland and their orifices. We should continue examining the openings of the tubules with the slit lamp in all of our patients. A non-invasive means of determining the status of the tubules and their orifices still has to be devised. Special probes for dilating the orifices and development of microsurgical lacrimal tubulotomy techniques, when distended tubules are present, may be used in the future. And even though we will soon be able to medically increase lacrimal gland secretions, we will still have to have functioning tubules to deliver the increased tear flow to the eyes.

# REFERENCES

Bron, A.J., 1986, Lacrimal streams: the demonstration of human lacrimal fluid secretion and the lacrimal ductules, Brit. J. Ophth. 70:241.

Damato, B.E., Allan, D., Murray, S.B., and Lee, R., 1984, Senile atrophy of the human lacrimal gland: the contribution of chronic inflammatory disease, Brit. J. Ophth. 68:674.

Kaswan, R., 1989, Cyclosporine drops: a potential breakthrough for dry eye, *in:* "Res. Prev. Blindness Writers Seminar."

Roen, J.L., Stasior, O.G., and Jakobiec, F.A, 1985, Aging changes in the human lacrimal gland: role of the ducts, CLAO J. 11(3):237.

Sullivan, D.A., and Sato, E.H, 1992, Potential therapeutic approach for the hormonal treatment of lacrimal gland dysfunction in Sjögren's syndrome, Clin. Immunol. Immunopath. 64(1):9.

# THE PARENCHYMA ACCOMPANYING MAJOR EXTRAGLANDULAR DUCTS WITHIN THE RAT LACRIMAL CORD

Mortimer Lorber

Department of Physiology & Biophysics
Georgetown University School of Medicine
Washington, DC 20007

## INTRODUCTION

While investigating regional differences in the duct-containing cord joining the rat exorbital lacrimal gland and conjunctiva (Fig. 1A), it was noted that some specimens also contained small masses of parenchyma (Fig. 1B). This study is concerned with the distribution and histological appearance of this cordal lacrimal tissue.

## MATERIALS AND METHODS

Under pentobarbital anesthesia, portions of 18 lacrimal cords were removed from 14 adult rats of both sexes and two strains (Sprague-Dawley and Wistar). Tissues were fixed in formalin, processed routinely, sectioned either transversely, longitudinally, or in both planes, and stained with hematoxylin and eosin.

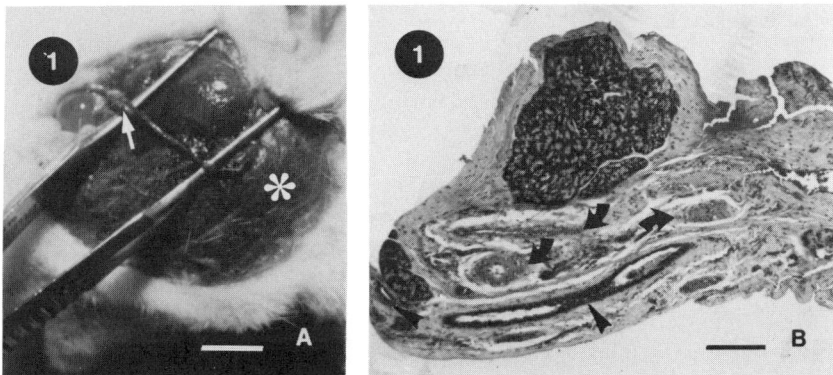

**Figure 1.** **A**. Following fascia removal, the lacrimal cord which extends between the exorbital gland (asterisk) and the upper lid rests on a forceps. It courses above the infraorbital gland (arrow). Bar = 5.5 mm. **B**. Longitudinal section of lacrimal cord near the exorbital gland contains a large and a small (left) mass of parenchyma. The former overlies two undulating arteries (curved arrows) and a similar duct (arrowheads). Bar = 167 μm.

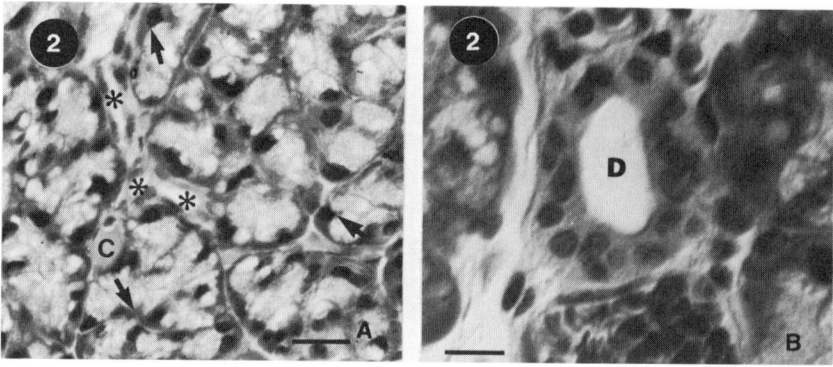

**Figure 2.** Cordal parenchyma. **A.** Numerous acini with myoepithelial cells (arrows) at the basal aspects of most acinar cells. A capillary (C) and a branching intercalated duct (asterisks) are evident. Bar = 25 μm. **B.** An interlobular duct (D) lies above an engorged capillary. Connective tissue separates both from the adjacent acini. Bar = 13 μm.

On the basis of anatomical position and their numbers of major ducts, eight regions of the lacrimal cord were identified (Lorber, 1993). Forty-five specimens were examined histologically and the presence or absence of parenchyma noted.

### RESULTS

Nine of the 18 lacrimal cords (50%) contained appreciable parenchyma in at least one region. Such tissue was noted in 13 of the 45 specimens (29%). It existed in five of the cords' eight regions, none occurring close to either lid or alongside the infraorbital gland. Lobules were, at times, present near the latter's anterior and posterior borders, being observed in one of four and two of six specimens respectively. They were most frequent in the center of the lacrimal cord where exocrine tissue occurred in five of eight specimens. Nearer the exorbital gland parenchyma was noted in three of twelve specimens. Next to that organ it was seen in two of seven tissue samples.

Cordal parenchyma resembles that of major lacrimal glands but lacks their largest ducts. Myoepithelial cells lie at the bases of the acinar cells. Intercalated ducts and small interlobular ducts are also present (Fig. 2).

This exocrine tissue appears as discrete masses of various sizes. Some extend for an appreciable distance along the length of the lacrimal cord (Fig. 3). At times, they border two sides of some of the major ducts (Fig. 4). In one specimen continuity of the

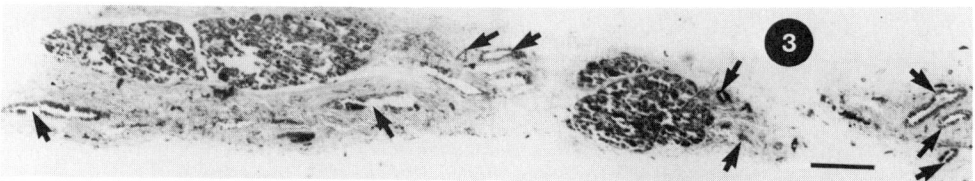

**Figure 3.** Montage of longitudinal section of the central region of a lacrimal cord with extensive areas of cordal parenchyma. Multiple regions of three major ducts (arrows) are present. Bar = 200 μm.

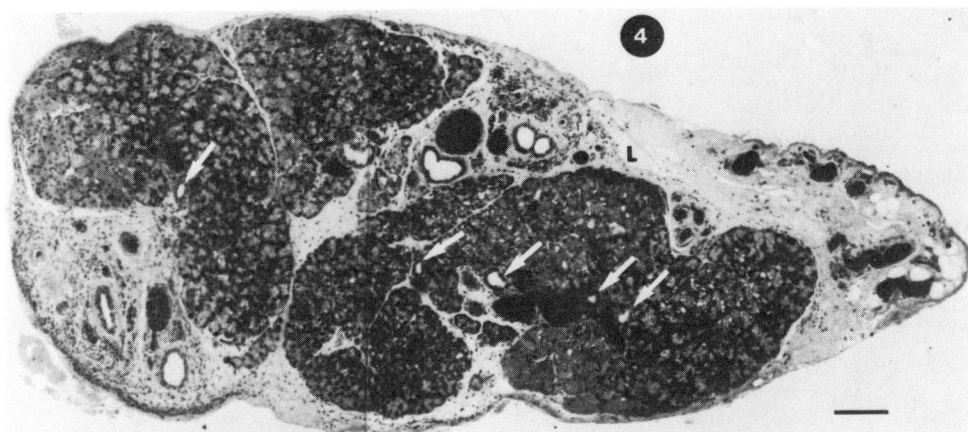

**Figure 4**. Montage of transected lacrimal cord near the exorbital lacrimal gland. Large lobulated parenchymal masses lie close to six major ducts, two at the left and four towards the top center. Right of the latter is a lymphatic (L). Five relatively large, mainly interlobular, ducts are evident (arrows). Many dark, engorged vessels are present intra- and extralobularly. Fat cells are at the right. Bar = 140 μm.

parenchyma between the lacrimal cord and the exorbital gland (Fig. 5) was seen.

Large parenchymal aggregates may thicken the cord (Fig. 4) which often bulges (Fig. 1). Many lacrimal cords, however, are of rather uniform diameter. The connective tissue overlying the masses may either form a capsule (Fig. 4) or be thick (Fig. 1B). The amount of connective tissue and the number of large diameter ducts present in a particular region would also affect cord thickness.

Cordal parenchyma may comprise one per cent of lacrimal parenchymal mass.

## DISCUSSION

In 1899, Loewenthal noted that the excretory duct of the rat infraorbital lacrimal gland was accompanied near the lid by a lobule of that organ. The latter may have been cordal lacrimal tissue.

The exorbital and infraorbital lacrimal glands of rodents, whose histology is essentially identical (Walker, 1958), have been considered to be two lobes of a single gland (Baquiche, 1959; Vianna et al., 1975; Sakai, 1989). Cordal parenchyma would represent additional lobules of that lacrimal organ because they, too, appear to provide serous secretion and so contribute to the welfare of the anterior eye region.

In man, lacrimal acini may be of neuroectodermal origin (Tripathi and Tripathi, 1990) or the parenchyma may derive from branchings of the main lacrimal duct that, in turn, arose from the conjunctiva (Mann, 1950). Perhaps in the rat the lacrimal cord did not provide too hospitable a microenvironment for prototypic cells or for pilot probes. The latter may have lacked the firmness or metabolic pathways that would have allowed them to extend far from their truncal duct of origin. Thus, unlike their counterparts for the major lacrimal glands, they were unable to ramify extensively.

Noting cordal parenchyma near the anterior and posterior borders of the infraorbital gland but not alongside the organ would be due to any in the latter location being viewed as part of the gland itself rather than being recognizable as a separate entity. Had entire lacrimal cords been examined, rather than merely portions of them, parenchyma might have been found in most, or perhaps all, specimens.

The current observations may have practicality. Parenchymal aggregates that drain along the lengths of major lacrimal ducts' extraglandular portions might affect analyses of secretion obtained by cannulation as accomplished by Alexander et al. (1972) and by Thörig et al. (1984). If the cordal parenchyma drains near the cannula tip, the secretion obtained would be unrepresentative of ductal fluid. To determine if that had been the case would require that each lacrimal cord whose duct had been cannulated would then

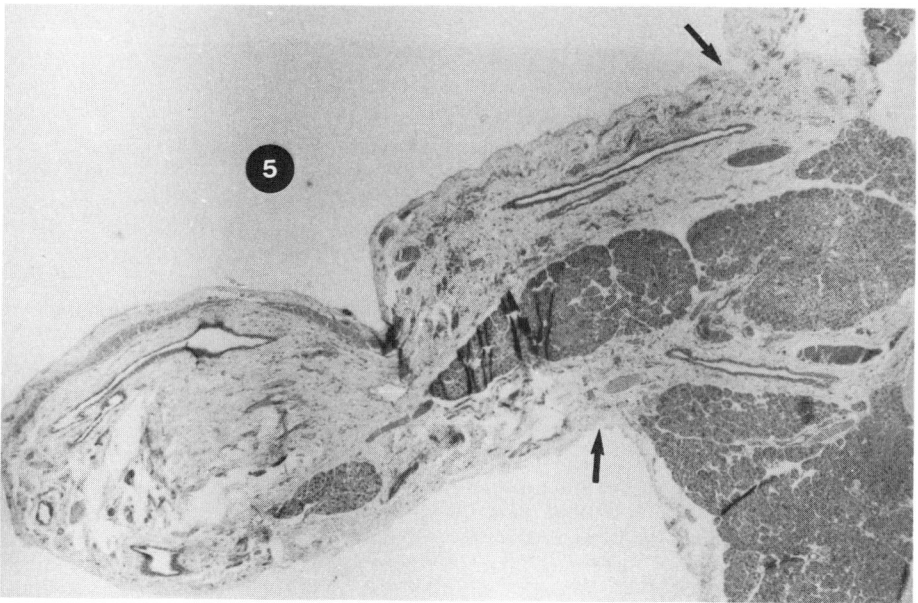

**Figure 5.** Continuity of parenchyma between the lacrimal cord (left and center) and the exorbital gland (right). Their junction is indicated by arrows. Three undulating major ducts extend through those regions. Two of them lie above the parenchymal masses. Bar = 230 μm.

have to be examined histologically. If on such inspection of multiple transections, parenchyma was noted in regions near where the cannula tip had lain, the purity of the "ductal" fluid would be in doubt. It would be advisable to discard its values and obtain additional samples of secretion from other cords in which parenchyma was subsequently shown to be absent on histological examination. Only in that way could the greatest accuracy regarding the chemical composition or rate of flow of pure ductal fluid be achieved.

## REFERENCES

Alexander, J. H., van Lennep, E. W., and Young, J. A., 1972, Water and electrolyte secretion by the exorbital lacrimal gland of the rat studied by micropuncture and catheterization techniques, Pflügers Arch. Ges. Physiol. 337:299.

Baquiche, M., 1959, Le dimorphisme sexuel de la glande de Loewenthal chez le rat albinos, Acta Anat. 36:247.

Loewenthal, N., 1899, A propos des glandes infraorbitaires, J. Anat. Physiol. Paris 35:130.

Lorber, M., 1993, Regional differences within the external "duct" of the rat exorbital lacrimal gland, Exp. Eye Res. In press.

Mann, I., 1950, "The Development of the Human Eye," Grune & Stratton, New York. p. 267.

Sakai, T., 1989, Major ocular glands (Harderian gland and lacrimal gland) of the musk shrew (Suncus murinus) with a review on the comparative anatomy and histology of the mammalian lacrimal glands, J. Morphol. 201:39.

Thörig, L., van Haeringen, N.J., and Wijngaards, G., 1984, Comparison of enzymes of tears, lacrimal gland fluid and lacrimal gland tissue in the rat, Exp. Eye Res. 38:605.

Tripathi, B.J., and Tripathi, R.C., 1990, Evidence for the neuroectodermal origin of the human lacrimal gland, Invest. Ophthalmol. Vis. Sci. 31:393.

Vianna, G.F., Cruz, A.R., and Azoubel, R., 1975, Allometric study of the lachrymal and Harderian glands of the rat during postnatal life, Acta Anat. 92:161.

Walker, R., 1958, Age changes in the rat's exorbital lacrimal gland, Anat. Rec. 132:49.

# PROTEIN SECRETION AND THE IDENTIFICATION OF NEURO-TRANSMITTERS IN THE ISOLATED PIG LACRIMAL GLAND

Jaipaul Singh[2], Ernest Adeghate[1], Shuna Burrows[2], Frank C. Howarth[2], and Tibor Donath[1]

[1] 1st Department of Anatomy, Semmelweiss University Medical School
Tuzolto, Budapest, Hungary
[2] Cell Communication Group, Department of Applied Biology,
University of Central Lancashire, Preston, PR1 2HE, England

## INTRODUCTION

The lacrimal gland is innervated with autonomic nerves which regulate protein secretion (Bothelho, 1964; Bromberg, 1981). Immunohistochemical studies have demonstrated the distribution of adrenergic, cholinergic and peptidergic nerves in the lacrimal of several animal species (Dartt et al, 1984; Ehinger, 1964; Lunberg et al, 1980). The known putative neurotransmitters released by these nerves include acetylcholine (ACh), noradrenaline (NA), vasoactive intestinal polypeptide (VIP), neuropeptide Y (NPY) and Substance P. Exogenous application of some of these putative neurotransmitters can stimulate protein secretion (Dartt et al, 1984; Hussain and Singh, 1988; Dartt, 1989) and exert vasodilatory effects in the lacrimal (Lunberg et al, 1980). However, studies involving protein secretion and the distribution of peptidergic and aminergic nerves in the pig lacrimal gland has long been neglected despite its similarity to human. This study investigates the effects of exogenous application of 5-hydroxytryptamine (5-HT), dopamine, VIP and NPY on total protein output and the distribution of nerve fibres containing neuropeptides including NPY, VIP and amines such as 5-HT and dopamine in the isolated pig lacrimal gland.

## METHODS

All experiments were performed on the isolated pig lacrimal gland obtained from the local abattoir. Immediately after killing the adult animal, the glands were removed, placed in an ice-cold oxygenated Krebs-Henseleit (K-H) solution and transported to the laboratory within 20 min. The lacrimal glands from 20 animals were cut into small segments (5-10 mg) and a total weight of about 250-350 mg was placed into a Perspex flow chamber (2 ml volume) and continuously superfused at a constant rate of 1 ml min$^{-1}$ with a K-H solution comprising (mM): NaCl, 118; KCl, 3.7; CaCl$_2$, 2.56; NaHCO$_3$, 25; KH$_2$PO$_4$, 2.2; MgCl$_2$, 1.2 and glucose 10. The solution was kept at pH 7.4 and 37° while being continuously gassed with a mixture of 95% O$_2$:5% CO$_2$.

*Lacrimal Gland, Tear Film, and Dry Eye Syndromes*
Edited by D.A. Sullivan, Plenum Press, New York, 1994

Total protein output in effluent samples were measured by an automated on-line colorimetric method (Hussain and Singh, 1988). During stimulation known concentrations of either VIP, NPY, 5-HT or dopamine were added to the perfusing medium. Bovine serum albumin (BSA) was used as standard. All value were expressed as micrograms of protein per millilitre effluent per 100 mg tissue above basal level.

Fresh pig lacrimal gland segments from 5 animals (10 glands) were immersion fixed for 48 hours in phosphate buffered (pH 7.4) para-formaldehyde (4%) and formaldehyde (0.5%) solution containing picric acid using established methods (Adeghate and Donath, 1990). The specimens were subsequently immersed overnight in 20% sucrose solution and were later sectioned into 20 μm thick slices using a cryostat. The sections were incubated for 48 hours at 4°C with either 1:1000 diluted

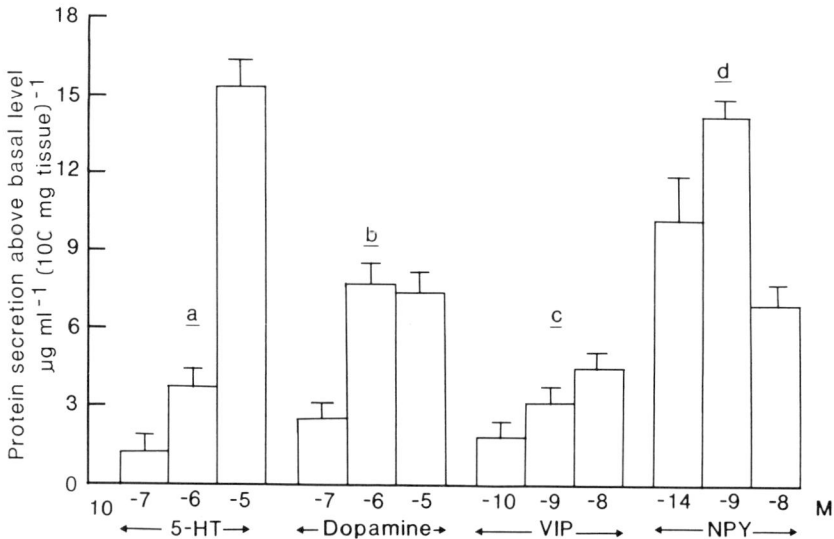

Figure 1. Families of histograms showing the effects of varying concentrations of (a) 5-HT ($10^{-7}$ - $10^5$M), (b) dopamine ($10^{-7}$-$10^{-5}$M), (c) VIP ($10^{-11}$ - $10^{-8}$M) and (d) NPY ($10^{-10}$ - $10^{-9}$M) on peak protein output above basal levels from superfused pig lacrimal segments. Each point is mean ± S.E.M (n=5).

rabbit anti-VIP (Amersham, England) or 1:1000 diluted rabbit anti-NPY (Amersham), or 1:500 diluted rabbit anti-DBH or with 1:1000 diluted rabbit anti serotonin sera. After several washings in phosphate buffered saline the sections were incubated for 1 hour at room temperature in biotinylated anti-rabbit immunoglobulin and later in Avidin-biotin-complex. Sites of immunoreactions were made visible by incubating the sections in diaminobenzidine solution (44 mg 300 ml$^{-1}$) of phosphate buffer (pH 7.4) containing 0.04 ml of $H_2O_2$ (0.03%) for 3 min. In all rinses phosphate buffered (pH7.4) saline containing 0.5% Triton was used. After staining the sections were dried, dehydrated and mounted in De-Pe-X. Anti sera for VIP, NPY, dopamine beta hydroxylase (DBA) and serotonin were raised by established methods (Adeghate and Donath, 1990). All data are expressed as mean ± SEM and were compared by Student's t test.

## RESULTS AND DISCUSSION

The mean ($\pm$ SEM) basal protein secretion in this study was 10.20 $\pm$0.33 $\mu$g ml$^{-1}$ (100 mg tissue)$^{-1}$, (n=60). Figure 1 shows families of histograms of mean ($\pm$ SE) data of protein release from superfused pig lacrimal segments during stimulation with (a) 5-HT ($10^{-7}$ - $10^{-5}$M), (b) dopamine ($10^{-7}$ - $10^{-5}$M), (c) VIP ($10^{-11}$ - $10^{-8}$M) and (d) NPY ($10^{-14}$ - $10^{-9}$ M). The results show that the aminergic and peptidergic neurotransmitters can exert marked secretagogue action on the pig lacrimal gland. The distribution of immunopositive nerves for NPY (1a), VIP (1b), dopamine beta hydroxylase (1c) and 5-HT (1d) in vibratome sectioned pig lacrimal gland is shown in Figure 2. Immunoreactive nerves for the aminergic and peptidergic neurotransmitters are distributed around the walls of the lacrimal ducts, the basolateral surfaces of the secretory acinar cells and in the interlobular areas.

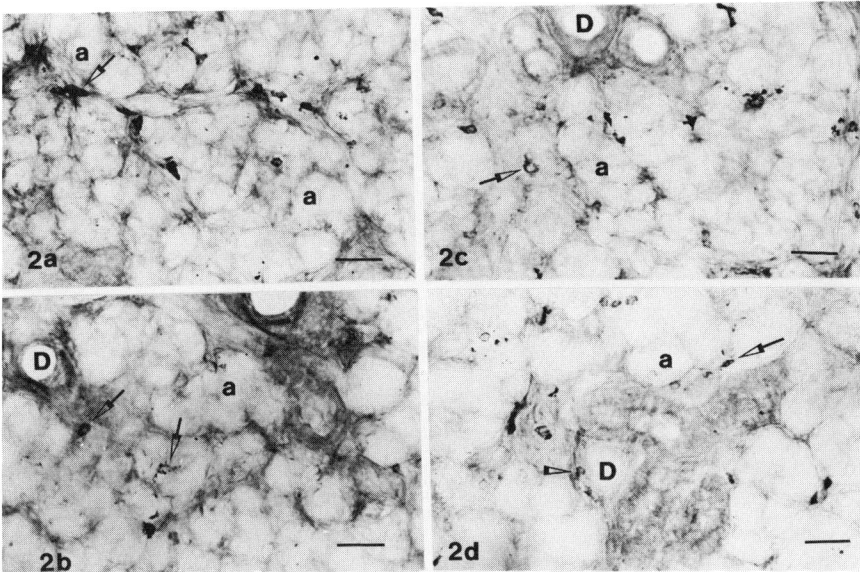

Figure 2. Light micrograph of vibratome sectioned pig lacrimal gland showing the distribution of immunoreactive nerves for (1a) neuropeptide Y (arrow) around the basolateral surfaces of lacrimal acini (a), (1b) VIP (arrows) in the interacinar spaces a = acinus, D = ducts. (1c) dopamine betahydroxylase (arrow) encircling the lacrimal acini (a), D = duct and (1d) serotonin (arrow) around the acini (a). These nerves (arrow head) are also found in the wall of the lacrimal duct. Scale bars = 50 $\mu$m.

Protein output from the lacrimal is secreted mainly by the acinar cells which form about 80% of the gland and several studies have demonstrated that cholinergic, peptidergic and adrenergic stimulation can elicit protein secretion in several animal species (Dartt, 1989). Unlike previous studies, this investigation employs the lacrimal gland of the pig to study mainly the secretagogue actions of some neuropeptides and biogenic amines and the distribution of the respective neurotransmitters in lacrimal nerves. Such studies on the pig have not yet been undertaken despite its similar anatomical topography to that in man. The results of our study have shown that exogenous application of either 5-HT, dopamine, VIP or NPY to pig lacrimal segments resulted in marked dose-dependent increases in total protein output. The results clearly demonstrate that the pig lacrimal acinar cells can secrete protein in a similar manner to other species previously studied (Bromberg, 1981; Dartt et al, 1984; Hussain and Singh, 1988) but the precise signal transduction mechanism controlling such secretion in the pig lacrimal still remains to be illucidated (Dartt, 1989).

Since the neuro-peptides and biogenic amines can exert marked secretory effects on the lacrimal then it was pertinent to ascertain whether intrinsic nerves may contain the respective neurotransmitters. Our immunohistochemical studies have shown that 5-HT, DBA, VIP and NPY are indeed present in intrinsic nerves of the pig lacrimal. These immunoreactive peptidergic and aminergic nerves are localized in the wall of the ducts, close to the basolateral surfaces surrounding the acinar cells and in the interlobular septa of the gland. The results show a similar distribution pattern and localization of the four different types of nerves. Apart from regulating lacrimal protein secretion, aminergic and peptidergic nerves may also regulate lacrimal blood flow and the rate of lacrimal fluid secretion thereby determining the quality of acinar and ductal secretions (Adeghate and Donath, 1990). Interestingly, NPY have been attributed to cause a sympathetic vasoconstriction and control blood vessels by interaction with noradrenaline while VIP have been shown to elicit lacrimal vasodilation (Lunberg et al, 1986; Dartt, 1989). However, further studies are required to determine precisely the physiological role of the aminergic and peptidergic neurotransmitters and their possible co-localization in the same nerves.

## Acknowledgements

Supported by the Wellcome Trust and British Council.

## REFERENCES

Adeghate, E. and Donath, T., 1990, Distribution of neuropeptide Y and vasoactive intestinal immunoreactive nerves in normal and transplanted pancreatic tissues, Peptides, 11: 1287.

Bothelho, S.Y., 1964, Tears and the lacrimal gland, Sci. Amer. 211: 78.

Bromberg, B.B., 1981, Autonomic control of lacrimal protein secretion, Invest. Opthalmol. Vis. Sci., 20: 110.

Dartt, D.A., Baker, A.K., Vaillant, C. and Rose, P.E., 1984, Vasoactive intestinal polypeptide stimulation of protein secretion from rat lacrimal acini, Am. J.Physiol., 247: G502.

Dartt, D.A., 1989. Signal transduction and control of lacrimal gland protein: A review, Curr. Eye Res., 8: 619.

Ehinger, B., 1964, Adrenergic nerves to the eye and its adnexa in rabbit and guinea pig, Acta Univ. Lundensis II, 20: 1.

Hussain, M and Singh, J., 1988, Is VIP the putative non - cholinergic, non - adrenergic neurotransmitter controlling protein secretion in rat lacrimal glands? Quart. J. Exp. Physiol., 73: 692.

Lunberg, J.M., Anggard, A., Fahrenkrug, J., Hokfelt, T and Mutt, V., 1980, Vasoactive intestinal polypeptide in cholinergic neurones of exocrine glands. Functional significance of co - existing transmitters for vasodilation and secretion, Proc. Natl. Acad. Sci., USA, 77: 1651.

# ANTI-MUSCARINIC ACETYLCHOLINE RECEPTOR-LIKE IMMUNOREACTIVITY IN LACRIMAL GLANDS

Benjamin Walcott, Roger Cameron, Elizabeth Grine, Elizabeth Roemer, Monica Pastor and Peter R. Brink

Departments of Neurobiology and Behavior, and Physiology and Biophysics
SUNY at Stony Brook, NY 11794

## INTRODUCTION

Lacrimal glands are extensively innervated by both the parasympathetic and sympathetic divisions of the autonomic nervous system. The neurotransmitters acetylcholine and norepinephrine are present in nerve fibers distributed among the secretory acini. Physiological studies using glands from a number of different species suggest that muscarinic acetylcholine receptors and both alpha and beta adrenergic receptors are present on the secretory acinar cells. There are no data, however, on the anatomic distribution of any neurotransmitter receptor in lacrimal glands. Given that the innervation density is sufficiently low so that it is unlikely that each acinar cell is directly innervated (see Walcott et al, this volume), it becomes even more important to determine the distribution of receptors in the glands in order to understand the control of lacrimal gland secretion.

## MATERIALS AND METHODS

Chickens and rats were sacrificed in accordance with the stated ARVO policies and glands (rat extraorbital and chicken Harderian) were immersion fixed in 4 % paraformaldehyde in phosphate buffer for 3 hours. The glands were washed in 25 % sucrose in the same buffer and stored in the sucrose solution at 4 degrees overnight. 14 - 20 μm thick cryostat sections of tissue were mounted on slides, dried and then processed for conventional immunocytochemistry. The primary antibody used was a mouse monoclonal antibody (1) made to calf brain muscarinic acetylcholine receptor. The antibody was diluted 1:200 with phosphate buffer containing 0.025% triton X-100 and applied to the sections overnight at room temperature. After washing with buffer for several minutes, an appropriate secondary antibody conjugated with FITC was applied. Controls where the primary antiserum was not applied were run in parallel with each batch of sections. After washing with buffer, the sections were coverslipped and examined in an epiflourescence microscope.

Isolated cells were prepared by mechanical maceration of the gland in RPMI medium. This process releases large clouds of cells into the medium, many of which are plasma cells. Large pieces of gland were discarded after mechanical agitation had dislodged most of the loose cells. The resulting cell suspension was washed several times in RPMI with gentle centrifugation between. Cells were then allowed to settle on coverslips, were fixed for a few minutes with 4 % para-formaldehyde and then stained as described above.

## RESULTS

The chicken gland consists of a cortex of secretory tubules with a medulla of tubules and plasma cells (see Walcott et al, this volume). The innervation is most dense in the medulla with many fine nerve fibers among the plasma cells (2). Sections of the gland exposed to the antibody showed prominent muscarinic acetylcholine receptor-like immunoreactivity (MARLI) in the medulla (fig 1). The antibody bound to cell surfaces

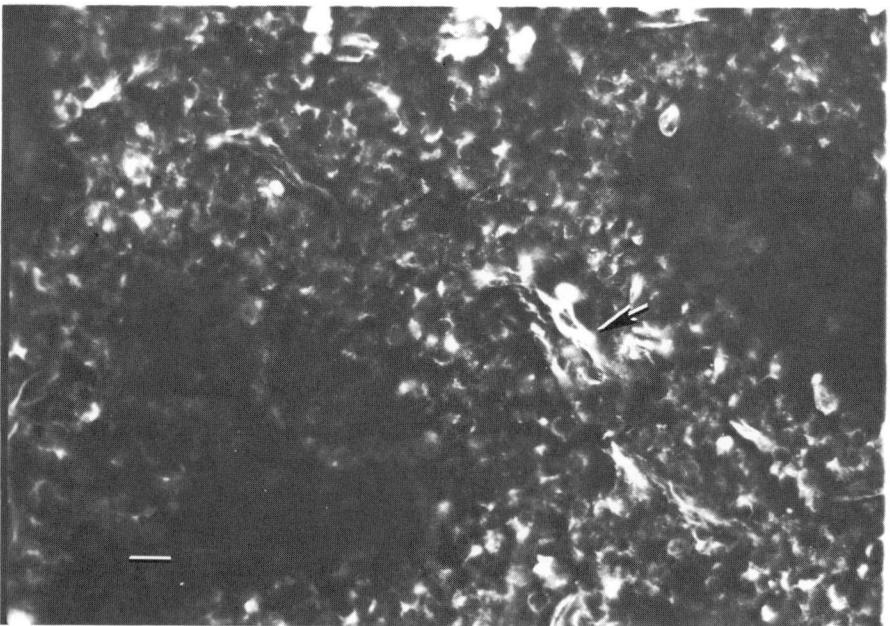

Figure 1. Section of chicken lacrimal gland showing extensive MARLI in the medullary region including the base of the acinar cells and the interstitial cells. Arrow points to nerve fibers. Scale is 50 µm.

particularly the plasma cells that occur in great numbers in this area of the gland. Acinar cells stained particularly at their basal pole and to a lesser extent on their lateral margins. The staining pattern was very irregular suggesting that the receptors were not uniformly distributed on the cell surfaces. Intracellular staining of cells was not observed. Certain fibers of large nerve bundles also showed MARLI, a pattern observed in other tissues (3). There was very little staining of the cortex of the gland which consists primarily of the secretory acini. Controls in which the primary antisera was not added did not show any positive staining.

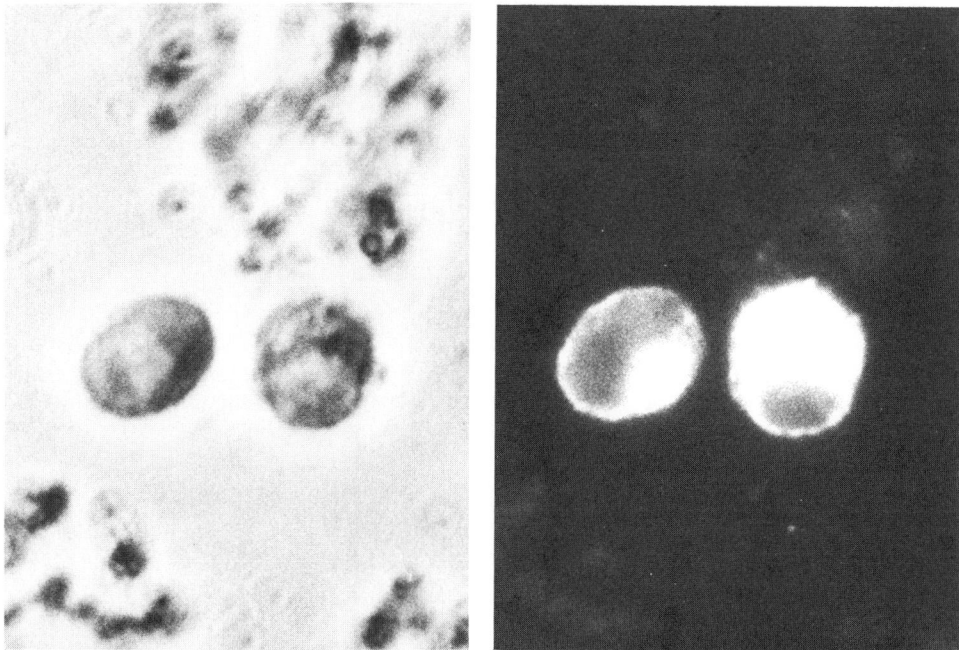

Figure 2. Isolated plasma cells imaged in phase (left) and the same cells (right) showing MARLI when illuminated in epiflourescence.

In order to determine if the plasma cells really had muscarinic receptors, we examined the immunoreactivity of isolated cells from the chicken gland. The technique for isolation was the same used for the patch-clamp analyses (4) except the cells were allowed to settle on coverlips and were then fixed. The results (Fig. 2) show a phase image of an isolated plasma cell on the left and on the right the MARLI of the same cell under epiflourescence illumination. Of the hundreds of cells identified as plasma cells using phase optics, all showed positive MARLI. Again, as in the sections, the staining is not evenly distributed over the cell surface.

To see if these results were applicable to other species, we examined the receptor distribution in rat exorbital lacrimal gland. Figure 3 shows positive MARLI associated with the secretory acini, particularly on the basal pole of the cells. Again, the staining is not uniform with an irregular distribution on each cell and acinus. However, it does appear that all acinar cells do have some positive immunoreactivity.

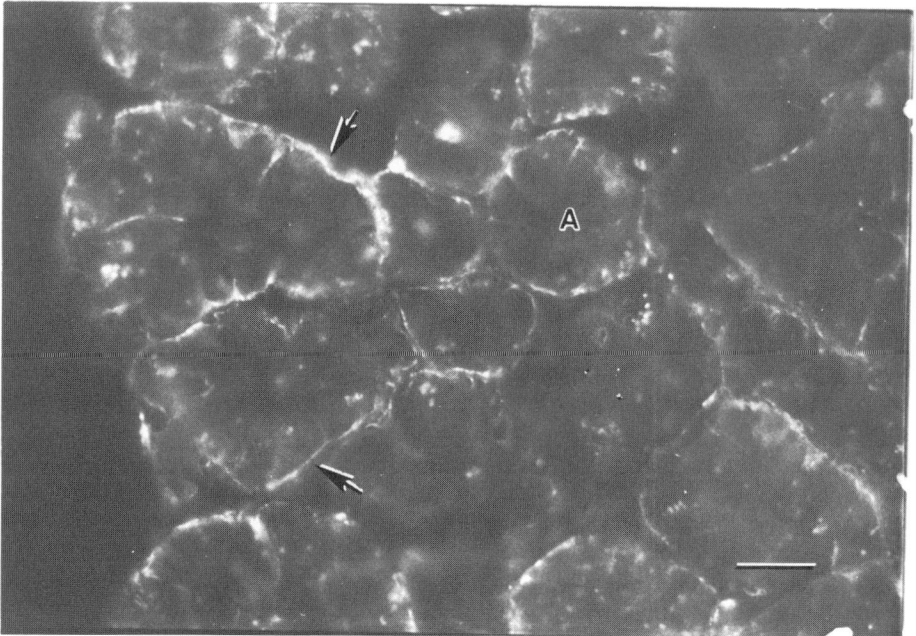

Figure 3. Anti-muscarinic acetylcholine receptor-like immunoreactivity in the rat exorbital lacrimal gland. Note the irregular staining of the base of the acinar cells (arrows). Scale is 100 μm.

## DISCUSSION

The observation of muscarinic acetylcholine receptor-like immunoreactivity in the chicken gland is consistent with the physiology of this system. It is known that both protein and immunoglobulin secretion is increased by carbachol, an acetylcholine agonist, and that this effect is blocked by atropine (6). Patch-clamp studies of isolated plasma cells show that carbachol increases the open time probability of the large maxi-K channels, an effect that is blocked by atropine (7). The higher density of the receptors in the medulla of the gland is also consistent with the distribution of the innervation. Acetylcholinesterase positive fibers are found predominantly in the medulla among the extensive numbers of plasma cells found there (8).The presence of MARLI in the rat extraorbital lacrimal gland is also consistent with the physiology of the gland which shows increased lacrimation with parasympathetic activity (9). It is interesting that the density of receptor appears to be higher than that of the innervation.

## ACKNOWLEDGMENT

This research was supported by NIH grant EY09604.

## REFERENCES

1. C. Andre, J.G. Guillet, P. DeBaker, P. Vanderheyden, J. Hoebeken and A.D. Strosberg, Monoclonal antibodies against muscarinic acetylcholine receptor recognize active and denatured forms, *EMBO Journal* 3:17 (1984)
2. B. Walcott, and J.R. McLean, Catecholamine-containing neurons and lymphoid cells in lacrimal gland in pigeon, *Brain Res.* 328:129 (1985)
3. B.A. Vogt, P.R. Crino, and E.L. Jensen, Multiple heteroreceptors on limbic thalamic axons: M2 acetylcholine, serotonin, B2-adrenoceptors, μ-opioid and neurotensin, *Synapse* 10:44 (1992)
4. P.R. Brink, E. Roemer, and B. Walcott, Maxi-K channels in plasma cells, *Pflugers Archiv.* 417:349 (1990)
5. P.R. Brink, B. Walcott, E. Roemer, R. Cameron, and M. Pastor The role of membrane channels in IgG secretion by plasma cells in the lacrimal gland of chicken, This volume (1993)
6. D.A. Dartt, A.K. Baker, C. Vaillant, and P.E. Rose, Vasoactive intestinal polypeptide stimulation of protein secretion from rat lacrimal gland acini, *Amer. J. Physiol.* 247:502 (1984)

# INNERVATION AND MAST CELLS OF THE RAT LACRIMAL GLAND: THE EFFECTS OF AGE

Ruth M. Williams,[1,2] Jaipaul Singh,[2] and Keith A. Sharkey [1]

[1]Department of Medical Physiology, University of Calgary, Calgary, Alberta, Canada
[2]Department of Applied Biology, University of Central Lancashire, Preston, U.K.

## INTRODUCTION

The secretion of tears by the lacrimal gland is primarily regulated by the autonomic nervous system.[1-3] The components of tears are regulated by nerves to reflect different requirements, but the exact nature of neurosecretory interactions in the lacrimal has only been partially elucidated. One reason for this is that the entire pattern of innervation and its relationship to the cells of the lacrimal has not been fully elaborated. In this summary, the innervation of the normal and aged rat lacrimal is described in the context of a brief review of lacrimal innervation. In addition, we describe the presence of mast cells in the gland. A full report of this work has been submitted for publication.[4]

### Autonomic Innervation of the Lacrimal Gland

The majority of the structural and functional innervation of the lacrimal gland is parasympathetic.[2,5,6] It has been demonstrated that lacrimal fluid and electrolyte secretion is principally regulated by parasympathetic nerves. Indeed, sympathetic stimulation inhibits cholinergic-stimulated lacrimal flow.[7] The sympathetic innervation is in part responsible for basal secretion by regulating blood flow to the gland and in some species has a direct action on the acini to cause protein secretion.[1-3] Bromberg demonstrated that adrenergic and cholinergic agonists interact in a synergistic manner to promote protein secretion in isolated lacrimal gland slices of the rabbit.[3] Thus protein secretory rate elicited by a low dose of the beta-adrenergic agonist isoproterenol, increased secretion by the muscarinic cholinergic agonist carbachol by more than the additive secretory effects of the two drugs given alone. Furthermore, it was shown that effects of carbachol could be attenuated by the addition of adrenergic antagonists as well as being completely blocked by atropine.[3] This data suggests that both divisions of the autonomic nervous system interact at the level of the acini and that lacrimal protein output is the product of both sympathetic and parasympathetic stimulation, whereas fluid and electrolyte output is under antagonistic control: parasympathetic stimulation and sympathetic inhibition.

*Lacrimal Gland, Tear Film, and Dry Eye Syndromes*
Edited by D.A. Sullivan, Plenum Press, New York, 1994

It is now recognized that numerous biologically active peptides present in nerves regulate secretory processes in glandular tissues and the lacrimal gland is no exception to this. Dartt and colleagues were the first to show that vasoactive intestinal polypeptide (VIP) caused a peroxidase secretion that was similar in magnitude to that elicited by carbachol.[8] In addition it was shown that VIP-immunoreactive (-IR) nerves were present in the gland, associated with acini and blood vessels. Physiologically, a non-adrenergic, non-cholinergic (NANC) lacrimal protein secretion was first shown by Hussain and Singh, who demonstrated that electrically field-stimulated segments of rat lacrimal were resistant (by about 50%) to combined autonomic blockade.[9] At the present time the mediator of the NANC protein secretion is not known, but preliminary experiments by us indicate that antagonism by a putative VIP antagonist eliminates some of the remaining response to field stimulation.[10]

Other investigators have also reported the occurrence of VIP-IR fibers in the lacrimal gland of various species including the cat, guinea pig, rat, monkey and human.[11-14] Generally, a dense VIP-IR plexus surrounds the glandular acini, and lower densities of VIP-IR are found surrounding ducts and blood vessels. Evidence suggests that the distribution of VIP-LI in the human lacrimal gland to be similar to that described in other mammalian species.[14] It seems likely that VIP-IR nerves in the lacrimal is mostly of parasympathetic origin, where it presumably coexists with acetylcholine.[11] In addition to VIP, substance P, calcitonin gene-related peptide (CGRP)- and neuropeptide Y (NPY)-IR fibers have been demonstrated in the lacrimal gland of the monkey[13] and substance P-IR fibers have been demonstrated in the rat and guinea pig[12], however, other neuropeptides have not been completely described in the rat lacrimal gland. Substance P- and CGRP-IR are generally sparse and most likely are of primary sensory origin (probably afferents from the trigeminal ganglion).[13] These fibers are found associated with ducts and blood vessels and in the interlobular connective tissue. Neuropeptide Y-IR in the periphery in most cases coexists in postganglionic sympathetic neurons with norepinephrine.[13] It seems that in the monkey this is also true since the distribution of the enzyme tyrosine hydroxylase (a marker for sympathetic nerves) and NPY-IR were very similar.[13]

## Neuroimmune Interactions in the Lacrimal Gland

Experimental evidence from many fields now points towards the existence of interactions and communication between the nervous and immune systems.[15] The lacrimal is a source of IgA derived from a resident population of B-lymphocytes. The IgA antibody receptor, a protein called secretory component (SC), is synthesized by acinar cells, and combines with IgA to form secretory IgA in tears. Recently, Sullivan and his colleagues have examined the effects of neural peptides and classical transmitter agonists on the secretion of SC from purified acinar cells.[16] VIP and isoproterenol increased the output SC in control and androgen-stimulated acini, whilst carbachol reduced it. Whether neurotransmitters in the lacrimal can modulate IgA secretion in the lacrimal is not known (though they do elsewhere), however, it has been shown that IgA content and distribution of IgA containing cells was not affected by denervation of the gland.[17]

Though not described in the rat, mast cells have been shown in the lacrimal gland.[18] In many organs, mast cells have been associated with nerves containing neuropeptides and there is evidence for modulation of mast cell secretion by neurotransmitters.[19,20] There are no previous reports describing the possible relationship between nerves and mast cells in the lacrimal gland.

We have investigated the presence and distribution of a number of putative neurotransmitters in the lacrimal gland in young and old rats. Aging is accompanied by a diminished tear production[21] and reduced responsiveness to autonomic stimulation,[22] however, there is little or no evidence as to whether there are any differences in the innervation pattern of the lacrimal gland with age. In addition, we examined the distribution of mast cells in the gland and their relationship to nerves.

## METHODS

Experiments were performed on Sprague-Dawley rats of 3-5, 14 and 24 months of age. Animals were anesthetized with sodium pentobarbitol (60mg/kg, i.p.) and perfused via the ascending aorta with phosphate-buffered saline followed by Zamboni's fixative (for peptides, tyrosine hydroxylase [TH] and serotonin [5-HT]) or freshly prepared 4% 1-ethyl-3(3-dimethylamionpropyl)carbodiimide (for histamine). To examine mast cells histologically, glands were immersed in a 1:1:8 mixture of 10% formaldehyde, glacial acetic acid and methanol. Frozen sections ($10\mu$m) were processed for indirect immunofluorescence using antibodies raised against VIP, NPY, CGRP, substance P, TH, 5-HT and histamine. Mast cells were stained histologically with toluidine blue in paraffin sections ($5\mu$m). Routine histology was examined using hematoxylin and eosin stained sections.

In 3 month old rats, the relationship between nerves and mast cells was studied. In order to do this sections were incubated with mixture of anti-5-HT and antibodies to either CGRP, NPY, substance P or VIP. In all cases the same secondary antibody was used to visualize the mixture.

## RESULTS

The glands from 3-5 month (regarded as normal, control rats) were composed of tubulo-acinar tissue arranged in discrete lobules separated by interlobular connective tissue. Interspersed in the glands were neurovascular bundles, consisting of venules, arterioles and nerve fibre bundles; large ducts were also observed in association with these bundles. In 14 month old animals, the morphology of the gland was similar for the most part. It was noted that the gland was stained less intensely with eosin and occasionally an inflammatory infiltrate was observed. By 24 months, all animals examined had morphological changes. These consisted of a luminal swelling with concomitant reduction in acinar cell volume, loss of discrete acinar cell morphology and a weakly eosinophilic cytoplasm. Furthermore, there was evidence of chronic inflammation with lymphocyte and macrophage infiltration into the glandular tissue and a marked thickening of the connective tissue surrounding the gland. In 24 month old rats there was a massive loss of acinar tissue.

### Mast Cell Histochemistry

In 3-5 month old rats, mast cells stained metachromatically with toluidine blue (Figure 1). They were principally localized around blood vessels, in the interlobular connective tissue and were abundant in the connective tissue sheath surrounding the gland. Occasionally, mast cells were observed in the acinar tissue. In 14 and 24 month old rats there was an increase in mast cells associated with the sites of inflammation. This was localized to a few sites in the 14 month old rats, but in the 24 month animals there was an extensive mast cell hyperplasia (Figure 1). Furthermore, mast cells in older rats were not confined to the connective tissue sites; they were distributed throughout the gland and were very numerous in the acinar tissue.

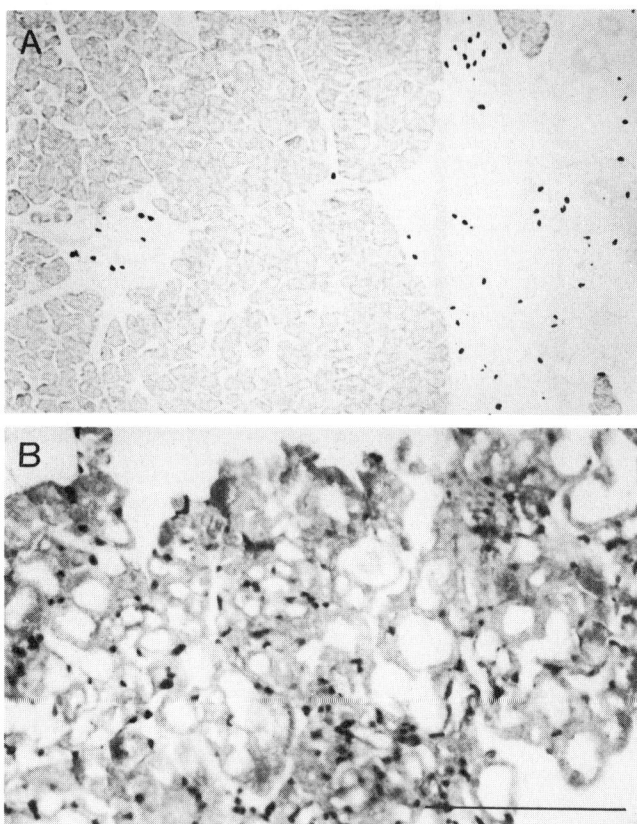

**Figure 1.** Micrographs of toluidine blue-stained mast cells in the rat lacrimal gland. **A.** Normal distribution of cells in a 3 month old animal. Note that mast cells are not present in the acinar tissue. **B.** Mast cells distributed throughout the gland in a 24 month old rat. Note that the morphology of the gland differs considerably from normal, with extensive luminal swelling in the acinar tissue. Scale bar: 200$\mu$m.

## Immunohistochemistry

The glands of the 3 different age groups were innervated by VIP-, NPY-, CGRP-, TH- and substance P-IR nerve fibres. VIP-IR nerves were distributed around acini, blood vessels and ducts (Figure 2). Substance P- and CGRP-IR nerves were sparse and found mostly associated with blood vessels, ducts and in the interlobular connective tissue. NPY- and TH-IR nerves were found around blood vessels and in fiber bundles close to blood vessels.

The 5 and 14 month rats had a similar pattern of innervation, however, by 24 months there was a reduction in the number and intensity of all IR nerves. The loss of nerves was particularly associated with damage to the gland (Figure 2).

Histamine- and 5-HT-IR was confined to cells localized around blood vessels and in the connective tissue (Figure 3). Because of their distribution and the fact that these chemicals are well described in mast cells, it is assumed that these label mast cells in the gland.

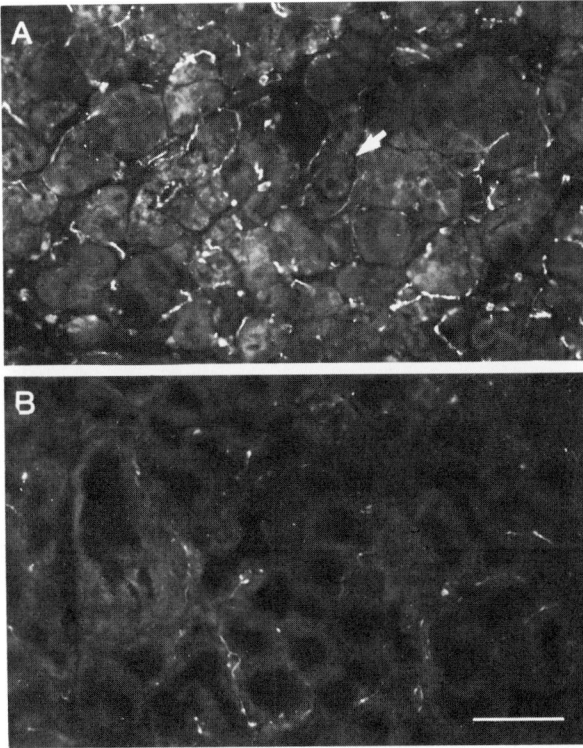

**Figure 2.** Fluorescence micrographs of VIP-IR in the rat lacrimal gland. **A.** VIP-IR from a 3 month old rat. VIP-IR is distributed in the acinar tissue and close to ducts (arrow). **B.** VIP-IR in a 24 month old rat. Note the loss of intensity and a reduced density of VIP-IR. Scale bar: 50μm.

### Nerve-Mast Cell Relationships

Since 5-HT-IR was localized only in mast cells, it was used as a marker in combination with antibodies against the peptides (which labelled only nerves), to examine nerve-mast cell relationships (in young rats only). Immunoreactive nerves to all of the four peptides examined were found in close apposition to mast cells. This was particularly striking for substance P- and CGRP-IR nerves, which though sparse, were almost invariably found close to mast cells (Figure 3).

## DISCUSSION

We have demonstrated that the rat lacrimal is innervated by a number of peptides that may represent putative neurotransmitters in the gland. Our findings confirm earlier observations regarding the localization of VIP and substance P in the rat[12] and extend them by the addition of other peptides (NPY and CGRP) previously described only in the monkey.[14] In the rat, VIP is quantitatively the predominant peptide in the gland and is presumably a marker of the parasympathetic innervation. Based on distribution, it is

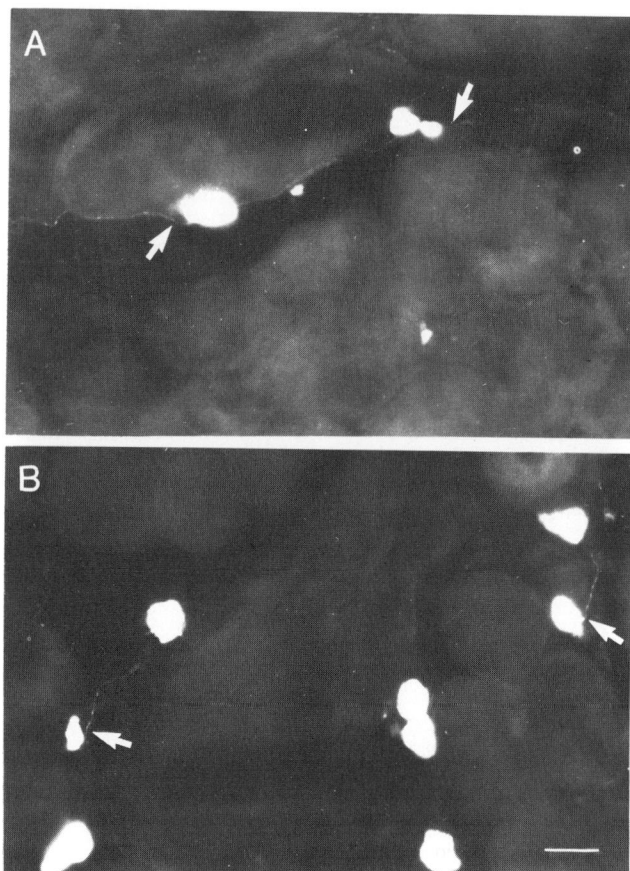

**Figure 3.** Fluorescence micrographs of 5-HT- and CGRP-IR in the rat lacrimal gland. 5-HT-IR is localized in mast cells and CGRP-IR in fine varicose fibers in close association with the mast cells (arrows). **A and B.** In both these examples, mast cells are situated in the interlobular connective tissue. Scale bar: 10$\mu$m.

probable that NPY is in sympathetic nerves in the gland and that substance P and CGRP are in primary afferents. The assumptions about the origins of peptides in the lacrimal have yet to be confirmed using immunohistochemistry in combination with retrograde tracing and/or nerve lesions, as has been described for the parotid gland.[23]

In all cases the extent of the innervation is reduced with age, however, this occurs in parallel with overt morphological changes in the tissue. Both of these changes could lead to the reduction in tear output associated with aging. From this study it is unclear whether the loss of IR nerves is secondary to the chronic inflammation seen in 24 month old rats or occurs as an independent effect of age. Some of the morphological changes in this study resemble keratoconjunctivitis sicca (KCS) in other animals.[24] Further studies that examine the rat as a model for aging and potentially for autoimmune KCS may be warranted.

The light microscopic observations of nerve-mast cells relations in the lacrimal are interesting. Though there is no conclusive evidence to describe a functional interaction between mast cells and nerves in the gland, we might speculate that these peptides have the ability to cause mast cell degranulation. If that is so, then the potential exists for mast cell mediators, including histamine and 5-HT, to modify lacrimal function. This might take the form of an increased vascular permeability and hyperemia (as in the skin) or through actions on the ducts, acini or other cell types in the gland (i.e. lymphocytes). Another possibility is that mast cells influence nerves and increase (or reduce) transmitter release. If this were the case, tear production may be affected in situations where mast cell degranulation occurs, for example in some allergic reactions. Further research to examine nerve-mast cell interactions is required to explore the full repertoire of neuroimmune interactions in the lacrimal gland.

## Acknowledgements

This work was supported by the Medical Research Council of Canada, the British Council and the Wellcome Trust. Ruth Williams is a Science and Engineering Research Council (UK) Student and Keith Sharkey is a Scholar, supported by the Alberta Heritage Foundation for Medical Research. We thank Winnie Ho for excellent technical assistance and Dr. N. Yanaihara, Shizuoka, Japan for generously providing the VIP antiserum used in this study.

## REFERENCES

1.  S.Y. Botelho, M. Hisadad, and N.Fuenmayor, Functional innervation of the lacrimal gland in the cat, Arch. Ophthal. 76:581 (1966).
2.  L.T. Jones. The lacrimal secretory system and its treatment, Am. J. Ophthalmol. 62:47 (1966).
3.  B.B. Bromberg, Autonomic control of protein secretion, Invest. Opthalmol. Vis. Sci. 20:110 (1981).
4.  R.M. Williams, J. Singh, and K.A. Sharkey, The innervation and mast cell content of the rat exorbital lacrimal gland: the effects of age. J. Auton. Nerv. Syst. submitted for publication.
5.  G.L. Ruskell, Changes in nerve terminals and acini of the lacrimal gland and changes in secretion induced by autonomic denervation. Z. Zellforsch. Mikrosk. Anat. 94:261 (1969).
6.  G.L. Ruskell, The distribution of autonomic post-ganglionic nerve fibres to the lacrimal gland in monkeys, J. Anat. 109:229 (1971).
7.  S.Y. Botelho, E.V. Martinez, C. Pholpramool, H.C. van Prooyen, J.T. Janssen, and A. De Palau, Modification of stimulated lacrimal gland flow by sympathetic nerve impulses in rabbit, Am. J. Physiol. 230:80 (1976).
8.  D.A. Dartt, A.K. Baker, C. Vaillant, and P.E. Rose, Vasoactive intestinal polypeptide stimulation of protein secretion from rat lacrimal gland acini, Am. J. Physiol. 10:G502 (1984).
9.  M. Hussain, and J. Singh, Is VIP the putative non-cholinergic, non-adrenergic neurotransmitter controlling protein secretion in the rat lacrimal glands?, Quart. J. Exp. Physiol. 73:135 (1988).
10. R.M. Williams, J. Singh, R.W. Lea, E. Adeghate, and K. Sharkey, Evidence for the involvement of vasoactive intestinal polypeptide and neuropeptide Y in the control of enzyme secretion in the isolated rat lacrimal gland, J. Physiol. 459:32P (1993).
11. J.M Lundberg, Evidence for co-existance of vasoactive intestinal polypeptide (VIP) and acetylcholine in neurons of cat exocrine glands, Acta Physiol. Scand. Suppl. 112:1-57 (1981).
12. A. Nikkinen, J.I. Lehtosalo, H. Uusitalo, A. Palkoma, and P. Panula, The lacrimal glands of the rat and guinea pig are innervated by nerve fibers containing immunoreactivities for substance P and vasoactive intestinal polypeptide, Histochemistry. 81:23 (1984).
13. Y. Matsumoto, T.Tanabe, S. Ueda, and M. Kawata, Immunohistochemical and enzymehistochemical studies of peptidergic, aminoergic and cholinergic innervation of the lacrimal gland of the monkey (Macaca Fuscata), J. Auton. Nerv. Syst. 37:207 (1991).
14. P.A. Sibony, B. Walcott, C. McKeon, and F.A. Jakobiec, Vasoactive intestinal polypeptide and the innervation of the human lacrimal gland, Acta. Ophthalmol. 106:1085 (1988).
15. E. Weihe, D. Norhr, S. Michel, S. Muller, H-J. Zentel, T. Fink, and J. Krekel, Molecular anatomy of the neuro-immune connection, Int. J. Neurosci. 59:1 (1991).
16. R.S. Kelleher, L.E. Hann, J.A. Edwards, and D.A. Sullivan, Endocrine, neural, and immune control of secretory component output by lacrimal gland acinar cells, J. Immunol. 146:3405 (1991).
17. D.A. Sullivan, L.E. Hann, C.H. Soo, L. Yee, J.A. Edwards, and M.R. Allansmith, Neural-immune

interrelationship: effect of optic, sympathetic, temporofacial, or sensory denervation of the secretory immune system of the lacrimal gland, Reg. Immunol. 3:204 (1991).

18. C.L.Martin, J. Munnell, R. Kaswan, Normal ultrastructure and histochemical characteristics of canine lacrimal glands, Am. J. Vet. Res. 49, 9:1566 (1988).

19. F. Shanahan,J.A. Denburg, J. Fox, J. Bienenstock, and D. Befus, mast cell heterogenity: effects of neuroenteric peptides on histamine release, J. Immunol. 135:1331 (1985).

20. P.J. Barnes, Regulatory peptides in the respiratory system, Experientia. 43: 832 (1987).

21. B.B. Bromberg, and M.H. Welch, Lacrimal protein secretion: Comparison of young and aged lacrimal gland, Exp. Eye Res. 40:313 (1985).

22. B.B. Bromberg, M.M. Cripps, and M.H. Welch, Sympathomimetic protein secretion by young and aged lacrimal gland, Curr. Eye Res. 5:217 (1986).

23. K.A. Sharkey, and D. Templeton, Substance P in the rat parotid gland: evidence for a dual origin from the otic and trigeminal ganglia, Brain Res. 304:392 (1984).

24. R.L. Kaswan, C.L. Martin, and W.L. Chapman, Keratoconjunctivitis sicca: Histopathological study of nicitating membranes and lacrimal glands from 28 dogs, Am. J. Vet. Res. 45:112 (1984).

# IMMUNOGOLD LOCALIZATION OF PROLACTIN IN ACINAR CELLS OF LACRIMAL GLAND

Richard L. Wood, Kyung-Ho Park, J. Peter Gierow and Austin K. Mircheff

Department of Anatomy and Cell Biology
Department of Physiology and Biophysics
and Department of Ophthalmology
University of Southern California
Los Angeles, CA

## INTRODUCTION

Prolactin (PRL) is known to be involved in osmoregulation and control of cellular differentiation in lower vertebrates (Horobbin, 1980). In mammals, PRL influences sodium and water transport in kidney (Loretz and Bern, 1982) and sweat glands (Robertson et al., 1989), and stimulates the secretion of both fluid and protein in the mammary gland (Olivier-Bousquet, 1978; Houdebine and Dijane, 1989). Although PRL is primarily a product of mammotrophes in the anterior pituitary, it is known to be elaborated by other cell types in the body, including lymphocytes (Buskila, et al., 1991; Lavelle, 1992). PRL receptors also have been found on a number of cells besides mammary secretory cells, again including lymphocytes, and these facts implicate the hormone in immunomodulation, as well as in its better known function in lactogenesis. In fact, Buskila and co-workers (1991) have proposed that PRL may serve as a master hormone controlling immune system function.

In previous studies, our laboratories have reported the presence of immunoreactivity for PRL in the exorbital lacrimal gland of rat (Mircheff et al., 1992). The immunoreactivity was localized to apical regions of acinar cells and was punctate, suggesting an association with secretory vesicles. The present study extends the earlier immunofluorescence observations to the electron microscopic level in order to localize the immunoreactivity more precisely. In addition, we present results of EM immunogold localization of PRL immunoreactivity to isolated acinar cells from the rabbit lacrimal gland, and compare those results with the findings from intact lacrimal glands of both animals.

## METHODS

Tissues obtained from anesthetized animals were prepared for morphological study

*Lacrimal Gland, Tear Film, and Dry Eye Syndromes*
Edited by D.A. Sullivan, Plenum Press, New York, 1994

in several different ways. For immunofluorescence, the tissues were either immersed directly in OCT and frozen in liquid nitrogen, or fixed 30-60 min in buffered paraformaldehyde prior to freezing. Cryostat sections were stained by conventional indirect immunolabeling, using rabbit anti-sheep prolactin, or rat anti-rabbit prolactin as primary antibodies. Controls consisted of: 1) substitution of non-immune IgG for primary antibody, 2) use of rabbit anti-sheep growth hormone in place of anti-PRL, and 3) pre-absorption of primary antibody with ovine prolactin.

The procedure for acinar cell isolation was a modification of that published by Hann et al., (1989). Briefly, it consisted of mincing glands in balanced salt solution, digestion with collagenase and hyaluronidase, and course filtration. Isolated cells were either used immediately, or maintained in Ham's/DME culture medium for 18-48 hr (Mircheff et al., 1992).

For electron microscopy, tissues were fixed in buffered paraformaldehyde containing 0.1% glutaraldehyde, and embedded in LR White. Thin sections were picked up on coated grids, and stained by flotation on droplets of fluid placed on parafilm. Sections were etched with saturated sodium metaperiodate prior to the immunostaining. In one experiment, tissues were fixed with the ethylacetimidate method (Geiger et al., 1981), which is reported to provide the improved morphology obtained with the presence of glutaraldehyde with minimal loss of antigenicity. After exposure to primary antibody, the signal was amplified by applying a biotinylated second antibody and streptavidin gold.

## RESULTS

In rat lacrimal gland, immunofluorescence denoting the presence of PRL-like reactivity was confined primarily to acinar cells, and was predominantly apical in location. At the EM level, immunogold localization occurred mostly over secretory vesicles of both the foamy and electron dense types. There was some reactivity over the Golgi region, the cytosol, and the nucleus.

In rabbit lacrimal gland, the distribution pattern of the immunoreactivity for PRL was similar to that for the rat lacrimal gland, except there was a more pronounced staining of secretory vesicles of moderate electron density, as compared to the more numerous secretory vesicles of low electron density. The pattern was virtually identical for isolated acinar cells, indicating that the isolation procedure did not cause a marked disruption of cellular morphology and function.

## DISCUSSION

These studies have confirmed that PRL-like immunoreactivity occurs in the acinar cells of mammalian lacrimal gland. Current evidence indicates that the PRL in lacrimal acinar cells has a dual origin. In previous work we have demonstrated the presence of PRL receptors in membrane fractions of acinar cells (Mircheff, 1992), and this suggests that the hormone is sequestered from the circulation. At the same time, we demonstrated by Northern blot hybridization that mRNA for PRL is present in homogenates of male but not female lacrimal glands, and this suggests that some PRL is synthesized endogenously, at least in males.

The function of PRL in lacrimal acinar cells can only be speculated upon at the present time. One obvious possibility is that PRL secreted into the tear fluid serves as a growth factor affecting the differentiation and maintenance of epithelium on the ocular surface and conjunctiva. PRL has been reported to be present in the tear fluid of humans (Frey, et al., 1986), but direct evidence is lacking for its function as a growth factor for epithelium at the ocular surface.

A second possible function of PRL in the lacrimal gland is that it may serve as a modulator of gene expression in the acinar cell. This function has been demonstrated in the mammary gland, where the transcription of mRNA for casein is modulated by PRL (Houdebine and Dijane, 1980). Our present data show that there is immunoreactivity in the nucleus as well as the cytosol of lacrimal acinar cells, and this is consistent with a regulatory function at the level of transcription.

Third, we hypothesize that PRL is released from the acinar cell both by regulated secretion at the apical surface, and by constitutive secretion at the basal-lateral surface. Release at the basal-lateral surface would implicate PRL in autocrine regulation of lacrimal fluid secretion, or as a potential co-stimulator of lymphocytes in immune responses, a paracrine function. We have no direct evidence at present to support this hypothesis, but we anticipate that it will be readily testable.

## ACKNOWLEDGEMENTS

Supported by NIH Grant EY04801, and by grants from the Wright Foundation, Fight-For-Sight Prevent Blindness, Inc., and the Sjögren's Syndrome Foundation, Inc.

## REFERENCES

Buskila, D., Sukanik, S., and Shoenfeld, Y., 1991, The possible role of prolactin in autoimmunity, *Amer. J. Reprod. Immunol.* 26:123.

Frey, W.H. II, Nelson, D., Frick, M.L., and Elde, R.P., 1986, Prolactin immunoreactivity in human tears and lacrimal gland: Possible implications for tear production, *in:* "The Preocular Tear Film in Health Disease, and Contact Lens Wear," Holly, F.J., ed., p. 798.

Geiger, B., Dutton, A.H., Tokuyasu, K.T., and Singer, S.J., 1981, Immunoelectron microscope studies of membrane-microfilament interactions: distributions of actinin, tropomyosin, and vinculin in intestinal epithelial brush-border and chicken gizzard smooth muscle cells, *J. Cell Biol.* 91:614.

Hann, L.E., Tatro, J.B., and Sullivan, D.A., 1989, Morphology and function of lacrimal gland acinar cells in primary culture, *Invest. Ophthalmol. Vis. Sci.* 30:145.

Horrobin, D.F., 1980, Prolactin as a regulator of fluid and electrolyte metabolism in mammals, *Fed. Proc.* 39:2567.

Houdebine, L.-M., and Dijane, J., 1980, Effects of lysomotropic agents and of microfilament- and microtubule-disrupting drugs on the activation of casein gene expression by prolactin in the mammary gland, *Mol. Cell. Endocrinol.* 17:1.

Lavelle, C., 1992, Prolactin - a hormone with immunoregulatory properties that leads to new therapeutic approaches in rheumatic diseases, *J. Rheumatol.* 19:839.

Loretz, C.A., and Bern, H.A., 1982, Prolactin and osmoregulation in vertebrates. An update, *Prog. Neuroendocrinol.* 35:292.

Mircheff, A.K., Warren, D.W., Wood, R.L., Tortoriello, P.J., and Kaswan, R.L., 1992, Prolactin localization, binding, and effects on peroxidase release in rat exorbital lacrimal gland, *Invest. Ophthalmol. Vis. Sci.*, 33:641.

Olivier-Bousquet, M., 1978. Early effects of prolactin on lactating rabbit mammary gland, *Cell Tiss. Res.* 187:25.

Robertson, M.T., Alho, H.R., and Martin, A.A., 1989, Localization of prolactin-like immunoreactivity in grafted human sweat glands, *J. Histochem. Cytochem.* 37:625.

# SUBCELLULAR ORGANIZATION OF ION TRANSPORTERS IN LACRIMAL ACINAR CELLS: SECRETAGOGUE-INDUCED DYNAMICS

Austin K. Mircheff, Ross W. Lambert,
Robert W. Lambert, Carol A. Maves,
J. Peter Gierow, and Richard L. Wood

Departments of Physiology and Biophysics
Ophthalmology, and Anatomy and Cell Biology
University of Southern California
School of Medicine
Los Angeles, CA 90033

## INTRODUCTION

The electrolyte-driven secretion of water is one of the major functions of the lacrimal gland, and impairment of this function is an obvious cause of dry eyes. As reviewed elsewhere (Mircheff, 1986, 1989), Alexander et al. showed in 1972 that the lacrimal gland produces fluid in two stages, secretion of a NaCl-rich solution in the acini, and secretion of a KCl-rich solution in the ducts. Studies of ductal transport have begun only recently (Saito, this volume). In contrast, modern concepts of acinar secretory mechanisms began to emerge more than ten years ago, when Dartt and coworkers (1981) presented the first evidence that the lacrimal glands conform to the principles of epithelial electrolyte secretion formulated by Silva and coworkers (1977).

The Silva model posits that $Cl^-$ ions enter secretory epithelial cells via a $Na^+$-coupled entry mechanism in the basal-lateral membrane and exit via $Cl^-$-selective channels in the apical plasma membrane. The sodium pump enzyme, $Na^+,K^+$-ATPase, drives the entire process by establishing and maintaining a large $Na^+$ electrochemical gradient across the basal-lateral membrane. In addition to ATP, this apparatus requires a mechanism that would allow $K^+$ ions, pumped into the cell by $Na^+,K^+$-ATPase, to flow back into the interstitium. It also requires a $Na^+$-selective pathway that would allow $Na^+$ ions to flow between adjacent cells and join the secreted $Cl^-$ ions in the acinar lumen. Many of this model's predictions were quickly verified, including the presence of basal-lateral membrane $Na^+,K^+$-ATPase (Dartt et al., 1981), $Cl^-$-selective channels (Marty et al., 1984), and $K^+$-selective channels (Peterson, 1986). However, the nature of the $Na^+$-$Cl^-$ coupled transport system remained undefined, largely because the overall transport process it mediates is invisible to electrophysiological measurements.

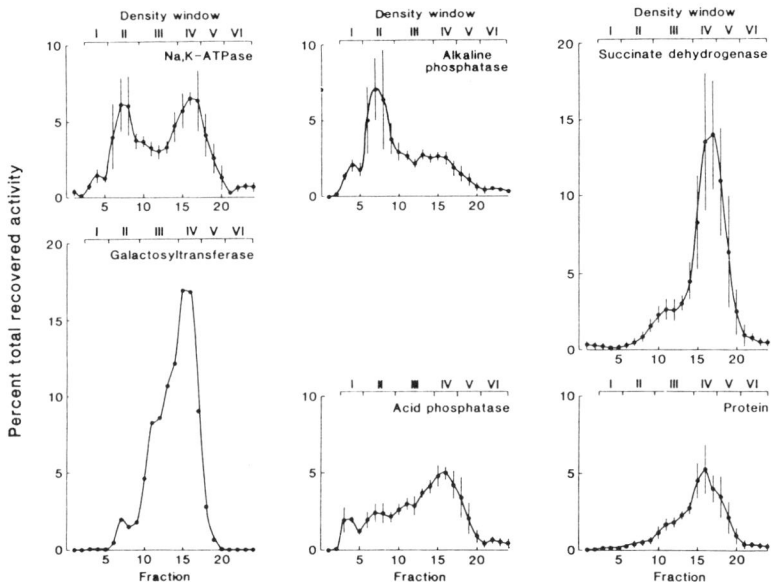

**Figure 1.** Density gradient analysis of rat acinar total membrane fraction (from Yiu et al., 1990).

## FRACTIONATION OF ACINAR CELLS

One can characterize electroneutral transport mechanisms by isolating a sample of sealed membranes vesicles and performing experiments with tracer flux and pH-sensitive dye techniques. The strategy for membrane isolation we have advocated (Mircheff, 1986, Bradley et al., 1993) is to subject a cell lysate to a comprehensive fractionation analysis, delineate the various membrane populations, and identify the basal-lateral membranes on the basis of their content of typical plasma membrane constituents and their low levels of markers for other organelles. When total membrane fractions from rat exorbital gland acini are analyzed on sorbitol density gradients (Figure 1), the distribution of $Na^+,K^+$-ATPase, the prototypical basal-lateral plasma membrane marker, exhibits a series of peaks, troughs, and shoulders which suggest that this enzyme is associated with a number of distinct membrane populations. Discerning which of these represented the basal-lateral membranes proved a challenge, and we have learned that several phenomena account for this complex distribution.

### Golgi-Associated $Na^+,K^+$-ATPase

As illustrated in Figure 1, a significant component of the rat acinar cell's $Na^+,K^+$-ATPase diverges from other basal-lateral membrane constituents, such as alkaline phosphatase and muscarinic cholinergic receptors, and parallels the Golgi enzyme, galactosyltransferase, in the regions of the gradient designated *window III*, *window IV*, and *window V* (Bradley et al., 1990). The $Na^+,K^+$-ATPase continues to parallel galactosyltransferase during partitioning analyses of *density windows III*, *IV*, and *V* in aqueous dextran-polyethyleneglycol two-phase systems (Bradley et al., 1993, Yiu et al., 1990), suggesting that the Golgi complex contains much of the $Na^+,K^+$-ATPase catalytic activity. Immuno-gold cytochemistry corroborates this conclusion (Azuma et al., 1990).

Fractionation analyses of acini from rabbit lacrimal glands yield a picture which is qualitatively similar to the picture of $Na^+,K^+$-ATPase subcellular distribution in the rat lacrimal gland (Figure 2), although it has been convenient to divide these gradients into four, rather than six, *density windows*. There are notable quantitative differences

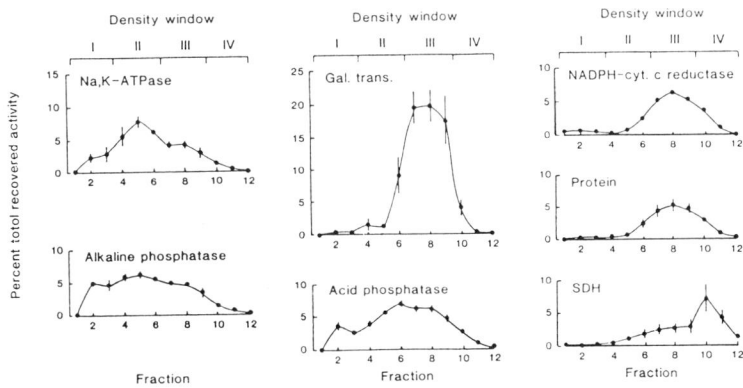

**Figure 2.** Density gradient analysis of rabbit acinar membranes (from Bradley et al., 1990).

between the two species. The distribution of $Na^+,K^+$-ATPase in rabbit lacrimal gland does not diverge from the distributions of other plasma membrane markers; accordingly, the Golgi-associated $Na^+,K^+$-ATPase pool is relatively smaller in rabbits than in rats, and it does not differ substantially from the Golgi-associated pools of other basal-lateral membrane constituents. Some fraction of the Golgi membrane $Na^+,K^+$-ATPase must be attributable to newly-synthesized glycoproteins en route to the plasma membranes. However, preliminary experiments suggest that a differentially targeted population of molecules with truncated $\alpha_1$-subunits and uniquely-processed $\beta_1$-subunits may account for the excess of $Na^+,K^+$-ATPase over other basal-lateral membrane constituents in the rat acinar cell Golgi complex (Azuma et al., 1990).

## Basal-Lateral, Apical, and Endosomal $Na^+,K^+$-ATPase

An additional feature consistently observed in density gradient analyses of rat and rabbit lacrimal acinar cell membrane fractions is the presence of two more-or-less distinct $Na^+,K^+$-ATPase peaks, a smaller peak localized in *density window I* and a much larger peak localized in *density window II*. While these peaks have similar, relatively large $Na^+,K^+$-ATPase specific activities, a close examination indicates that *density window II* has a substantially larger specific activity of galactosyltransferase. Why there should be two such plasma membrane-like vesicle populations remained unanswered for a number of years. On one hand, it seemed possible that they might represent two biochemically distinct microdomains of the basal-lateral membrane. On the other, it did not seem too far-fetched to suggest that one population represented the basal-lateral membrane itself, while the other represented an intracellular organelle involved in assembly or recycling of basal-lateral membrane constituents.

The hypothesis that the basal-lateral membranes are localized in *density window I* has very recently received direct support from experiments in which surface-labeling reagents were used to mark the plasma membranes of isolated acinar cells (C.A. Maves, R.W. Lambert, J.P. Gierow, and A.K. Mircheff, unpublished). When incorporated at 4°, a temperature which does not permit significant intracellular membrane trafficking, these reagents preferentially label the membrane population equilibrating in *density window I*. In addition to identifying the basal-lateral membranes, this result leads to the surprising conclusion that most of the cell's $Na^+,K^+$-ATPase is associated with intracellular structures.

When *density window II* samples from rat and rabbit lacrimal tissues are analyzed by phase partitioning, it becomes apparent that they contain at least three populations of membranes, each of which exhibits measurable $Na^+,K^+$-ATPase catalytic activity. One of these, accounting for minor fractions of the total $Na^+,K^+$-ATPase and protein, is characterized by a relatively large ratio of alkaline phosphatase to $Na^+,K^+$-ATPase and by relatively large specific contents of $Na^+$-amino acid transport systems typically found in intestinal and proximal tubular brush borders. On the basis of these characteristics, this population appears to represent acinar cell apical plasma membranes (Mircheff, Lu, and Conteas, 1983). The identity of this isolated membrane population has not been confirmed independently, but immunocytochemical studies provide independent evidence that some $Na^+,K^+$-ATPase is present at the acinar cell apical surface (Wood and Mircheff, 1986).

Additional experiments indicate that the other two $Na^+,K^+$-ATPase-containing membrane populations equilibrating in *density window II* have been derived from one or more endosomal compartments. Extracellular fluid phase markers rapidly label *density window II* when isolated rabbit lacrimal acinar cells are incubated at temperatures which permit membrane trafficking to proceed. Fluid phase markers also reach *density window III*, but over a significantly longer time-course (Gierow, Wood, and Mircheff, unpublished). Since little, if any, of the fluid phase marker reaches the Golgi complex under the conditions of these experiments, we hypothesize that membranes derived from a late endosomal compartment or from the *trans*-reticular Golgi network equilibrate, along with the Golgi membranes, in *density window III*. In any event, studies with Lucifer Yellow as a fluid phase marker (discussed elsewhere in this volume by Gierow et al.) clearly indicate the presence of distinct early and late endosomal compartments. The same studies also indicate that various secretagogues, most notably carbachol, modulate recycling traffic between the basal-lateral plasma membrane and the endosomes.

## RAPID RECYCLING TRAFFIC

Experiments with the membrane surface labeling reagent, sulfo-N-hydroxysuccinimidyl biotin, illustrate how rapidly membrane constituents move to and from the basal-lateral membrane (Lambert et al., 1993). When acinar cells are chilled to 4°, biotinylated, and incubated with avidin-Lucifer Yellow conjugate, the label is uniformly incorporated onto the surface membranes of intact cells. Upon warming to 37°, the label is internalized, yielding a punctate intracellular pattern. If cells are warmed to 37° for various intervals before exposure to avidin-Lucifer Yellow, the time-course of internalization is apparent from a temperature-dependent decrease in the cells' capacity to bind avidin-Lucifer Yellow during subsequent incubations at 4°. This phenomenon is, to us, astoundingly rapid: More than half the total initial biotinyl groups are internalized well within 30 sec. Since only a negligible fraction of the label remains at the cell surface after 5 min, this process must involve basal-lateral membrane, as well as apical membrane, constituents.

The rapid removal of biotin from the cell surface represents, at least in part, the equilibration between a small pool of surface-expressed constituents and a large pool of intracellular constituents. Cells were biotinylated at 4°, warmed to 37° for 60 min, then, still at 37°, incubated with avidin-Lucifer Yellow conjugate in the presence and absence of excess avidin. The premise of this experiment was that cells would take up avidin-Lucifer Yellow by several processes, including fluid phase endocytosis, non-specific adsorption, and binding to biotinyl groups, either at the cell surface or within endosomal compartments. Fluid phase endocytosis and non-specific adsorption should not be saturable, while binding to biotinyl groups should be competitively inhibited by excess avidin. Biotin-dependent avidin-Lucifer Yellow uptake ultimately represented an amount equivalent to roughly 60% of the total initial binding capacity. This process was complete within 30 min, and it had a half-time of 2 to 3 min. The biotinyl groups accounting for this uptake must have been located in intracellular compartments which were in recycling communication with the extracellular fluid. Carbachol at concentrations of 10 $\mu$M and 1 mM are equally effective at accelerating the communication between internalized biotinyl groups and extracellular avidin-Lucifer Yellow.

## COUPLING SODIUM AND CHLORIDE FLUXES

### $Na^+/H^+$ and $Cl^-/HCO_3^-$ Antiporters

Studies designed to distinguish the activities of various ion transporters were first performed with *density window II* samples from rat exorbital lacrimal glands at a time when these samples were believed to represent the basal-lateral membranes (Mircheff et al., 1987). Several transport phenomena expected of $Na^+/H^+$ antiporters could be observed, including acceleration of $^{22}Na^+$ influx by outwardly-directed $H^+$ gradients and acceleration of proton influx by outwardly-directed $Na^+$ gradients. Density gradient analyses of amiloride-sensitive, pH gradient dependent-$Na^+$ transport activities indicate that the $Na^+/H^+$ antiporters in both rat and rabbit lacrimal gland preparations are distributed between the basal-lateral plasma membranes and endosomal compartments in the same proportions as other typical basal-lateral membrane constituents (Mircheff et al., 1987, Lambert et al., 1991).

It was also possible to demonstrate that lacrimal acinar cells contain anion exchangers capable of catalyzing $Cl^-/HCO_3^-$ antiport. In *density window II* preparations from rat lacrimal glands, outwardly-directed gradients of $HCO_3^-$ stimulate the influx of $^{36}Cl^-$ (Lambert et al., 1988). This phenomenon is sensitive to several typical anion exchange inhibitors, including sulfonic stilbenes and furosemide. $^{36}Cl^-$ influx is also stimulated by outwardly-directed gradients of $^{35}Cl^-$, indicating that the anion exchanger catalyzes $Cl^-/Cl^-$ self-exchange as well as $Cl^-/HCO_3^-$ antiport. Like $Na^+/H^+$ antiporters, anion exchangers are present in the basal-lateral membranes and in the same intracellular compartments where other basal-lateral membrane constituents are found. Similar experiments failed to yield any evidence of transport activity attributable to Na-Cl or Na-K-2Cl symporters, although it must be noted that such symporters may be present in the intact cell but inactivated during membrane isolation procedures.

The influence of intracellular pH on anion exchange activity on has been analyzed in rat lacrimal acini (Lambert, Bradley, and Mircheff, 1991). $Cl^-/Cl^-$ self-exchange in the nominal absence of $HCO_3^-$ is negligible at intracellular pH values less than 7.0 but increases markedly as $pH_i$ increases above this value. Thus, cytoplasmic alkalinization must increase the anion exchange rate independently of changes in intracellular substrate concentration. In the presence of physiological $CO_2$ tensions, intracellular $HCO_3^-$ increases exponentially with intracellular pH, so that the anion exchanger is very well-suited for dissipating alkaline loads. It seems reasonable to predict that the intersection between the $Na^+/H^+$ antiporter and anion exchanger pH dependence curves would define a stable intracellular pH set-point.

### Activation of Ion Fluxes

Several lines of evidence suggest that intracellular mediators regulate the basal-lateral membrane $Na^+/H^+$ antiport activity and, therefore, intracellular pH. Carbachol at a concentration of 10 $\mu$M increases $Na^+$ flux into isolated rat or rabbit lacrimal acini at least 2.5- to 3-fold (Lambert, Bradley, and Mircheff, 1991, Bradley et al., 1992). This response is accompanied by an 0.11 unit increase of cytoplasmic pH (Golchini et al., 1991); alkalinizations as large as 0.2 pH unit have been observed in mouse lacrimal acinar cells by Saito et al. (1988). Preparations of rabbit lacrimal acini incubated for 30 min in the absence or presence of 10 $\mu$M carbachol have been analyzed by subcellular fractionation. $Na^+/H^+$ antiporter-mediated $^{22}Na^+$ influx was 2.5-fold greater in *density window I* samples from stimulated preparations than in corresponding samples from unstimulated preparations, but not significantly altered in other regions of the density gradient. Thus, cholinergic stimulation appears to trigger a selective activation of $Na^+/H^+$ antiporters residing in the basal-lateral membranes (Lambert, Maves, and Mircheff, 1991). Because the amount of material available in *window I* samples from acinar preparations is severely limited, it has not been possible to dissect the kinetic basis of the stimulation-induced activation. However, the intravesicular pH-dependence of *window I* samples from lacrimal gland fragments suggests that the stimulation-induced increase is not likely to be attributable to an increase in the affinity of an intravesicular $H^+$-binding regulatory site.

It proved instructive to compare the $Na^+$ and $Cl^-$ unidirectional fluxes into isolated acini. In resting preparations, $Cl^-$ influx exceeds $Na^+$ influx by a factor of 2.3. While cholinergic stimulation accelerates $Na^+$ influx 3-fold, it increases $Cl^-$ influx only by 25%, a factor expected for a cytoplasmic alkalinization of ~0.2 pH unit (Lambert et al., 1991). We suggest that most of the $Cl^-$ flux into resting cells represents $Cl^-/Cl^-$ self-exchange. Upon stimulation, the intracellular $HCO_3^-$ concentration increases, due to $Na^+/H^+$ antiporter-mediated cytoplasmic alkalinization, and the intracellular $Cl^-$ concentration decreases, due to efflux via activated apical $Cl^-$-selective channels (Saito et al., 1985). Consequently, the fraction of the $Cl^-$ unidirectional influx occurring as exchange for intracellular $HCO_3^-$ increases at the expense of $Cl^-/Cl^-$ self-exchange, increasing net anion exchanger-mediated $Cl^-$ influx.

### Involvement in Lacrimal Insufficiency

Now that many features of the acinar cell's mechanism for secreting electrolytes have been identified, it becomes possible to ask whether they are altered during lacrimal insufficiency. As Warren et al. argue elsewhere in this volume, it appears that the hormonal milieu influences the levels of expression of muscarinic cholinergic receptors and $Na^+,K^+$-ATPase pumps. It remains to be seen how other components of the secretory machinery are regulated by hormonal factors, but one can already begin to envision scenarios that plausibly account for the greater incidence of lacrimal insufficiency in women.

## BASAL-LATERAL MEMBRANE REMODELING

The $Na^+,K^+$-ATPase endosomal pool appears to provide the cell with a reserve which it can recruit to the plasma membranes in response to increased $Na^+$-pumping requirements. Stimulation with carbachol for 30 min increases the $Na^+,K^+$-ATPase activity of *density window I* samples by 40% in both rat (Yiu et al., 1988, 1991) and rabbit (Lambert, Maves, and Mircheff, 1991) lacrimal acini. In rabbit cells, the $Na^+,K^+$-ATPase activity of *density window II* decreases significantly during this period; the magnitude of this decrease is greater than the increase in *density window I*, so that the total activity decreases significantly.

$Na^+,K^+$-ATPase recruitment in rat lacrimal acinar cells is induced by secretagogues which significantly accelerate $Na^+$ transport and not by secretagogues which induce only protein secretion (Yiu et al., 1991), suggesting that this phenomenon is coordinated with accelerated $Na^+$ influx. However, it appears that cholinergic stimulation triggers recruitment even when the cytoplasm is alkalinized to a level that fails to support

Na$^+$/H$^+$ antiport activity, suggesting that the recruitment must be part of the cell's programmed response to the intracellular signalling cascade activated by cholinergic stimulation, rather than an *ad hoc* response to elevated cytoplasmic Na$^+$ concentration.

The net stimulation-induced translocation of Na$^+$,K$^+$-ATPase pump units and decrease of total cell Na$^+$,K$^+$-ATPase content in rabbit lacrimal acinar cells occur during an overall acceleration of the basal-lateral membrane recycling traffic. These phenomena may be related. That is, the accelerated recycling traffic may be the vehicle for functional remodeling of the basal-lateral membrane. One possible explanation for the decrease in total Na$^+$,K$^+$-ATPase is that stimulation also accelerates traffic of membrane constituents to lysosomes, where they are degraded. In a preliminary study, overnight culture of acinar cells in the presence of 10 $\mu$M carbachol significantly decreased the total contents of Na$^+$,K$^+$-ATPase and other basal-lateral membrane constituents. These decreases were localized primarily in the intracellular compartments, suggesting that the cell maintains its plasma membrane pools of various constituents at the expense of its intracellular pools (Maves et al., 1992).

## THE DYNAMIC BASAL-LATERAL MEMBRANE

While we have known for some time that stimulation initiates a cycle of secretory vesicle membrane fusion with and retrieval from the apical plasma membrane, the extent of the traffic to and from the basal-lateral membrane was unexpected. In addition to serving as a vehicle for membrane remodeling, we imagine that this traffic might also participate in other acinar cell functions. It may mediate the internalization steps in transcytotic secretion of IgA and of prolactin (Wood et al., this volume). The fact that this traffic is accelerated during secretory stimulation could account for the observation that total IgA secretion increases with lacrimal gland fluid flow rate (Fullard and Tucker, 1991). The volumes of some secretory cells have been shown to oscillate under certain circumstances, apparently due to the periodic activation and inactivation of basal-lateral ion influx and apical ion efflux mechanisms. The endosomal compartment may provide the depot and reserve which allows the cell to adjust its membrane surface area as it shrinks and swells. Finally, it is possible that when agonist-receptor complexes reach the endosomal compartment, they dissociate, allowing the receptors to bind trimeric G-proteins and return to the plasma membrane. Such receptor recharging could prolong the time that the cell remains in the actively secreting state. However, while the recycling traffic between the basal-lateral plasma membrane and endosomes is emerging as a crucial aspect of the acinar cell's normal functioning, it may also be the aspect of the cell's physiology that makes it particularly apt to initiate the local autoimmune responses that progress to the autoimmune disease of Sjögren's Syndrome (Mircheff et al., this volume).

## ACKNOWLEDGEMENTS

Work in the authors' laboratories has been supported by NIH Grant EY 05801, by grants from the Wright Foundation, Fight-for-Sight Prevent Blindness, Inc., and the Sjögren's Syndrome Foundation, and by a gift from Ronald H. Akashi, M.D.

## REFERENCES

Alexander, J.H., van Lennep, E.W., and Young, J.A., 1972, Water and electrolyte secretion by the exorbital gland of the rat studied by micropuncture and microcatheterization techniques, *Pfluegers Arch.* 337:299.

Azuma, K.K., Bradley, M.E., Wood, R.L., McDonough, A.A., and Mircheff, A.K., 1990, Subcellular distribution of Na,K-ATPase in rat exorbital lacrimal gland, *J. Cell Biol.* 190a.

Bradley, M.E., Lambert, R.W., Lambert, R.W., Lee, L.M., and Mircheff, A.K., 1992, Isolation and subcellular fractionation analysis of acini from rabbit lacrimal glands, *Invest. Ophthalmol. Vis. Sci.* 33:2951.

Bradley, M.E., Lambert, R.W., and Mircheff, A.K., 1993, Isolation and identification of plasma membrane populations. *Meth. Enzymol.*, in press.

Bradley, M.E., Peters, C.L., Lambert, R.W., Yiu, S.C., and Mircheff, A.K., 1990, Subcellular distribution of muscarinic acetylcholine receptors in rat exorbital lacrimal gland, *Invest. Ophthalmol. Vis. Sci.* 31:977.

Dartt, D.A., Møller, M., and Poulsen, J.H., 1981, Lacrimal gland electrolyte and water secretion in the rabbit: Localization and role of $(Na^+ + K^+)$-activated ATPase, *J. Physiol.* 321:557.

Fullard, R.J., and Tucker, D.L., 1991, Changes in human tear protein levels with progressively increasing stimulus, *Invest. Ophthalmol. Vis. Sci.* 32:2290.

Golchini, K., Lambert, R.W., Ghadishah, E., Yasharpour, F., Gierow, J.P., and Mircheff, A.K., 1991, Sodium-proton exchange in rabbit lacrimal acinar cells characterized with a $pH_i$-sensitive dye, *Invest. Ophthalmol. Vis. Sci.* 32s:726.

Lambert, R.W., Bradley, M.E., and Mircheff, A.K., 1988, $Cl^-/HCO_3^-$ antiporters in rat lacrimal gland, *Am. J. Physiol.* 257:G637.

Lambert, R.W., Bradley, M.E., and Mircheff, A.K., 1991, pH-sensitive anion exchanger in rat lacrimal acinar cells, *Am. J. Physiol.* 260:G517.

Lambert, R.W., Maves, C.A., Gierow, J.P, Wood, R.L., and Mircheff, A.K., 1993, Plasma membrane internalization and recycling in rabbit lacrimal acinar cells, *Invest. Ophthalmol. Vis. Sci.* 34: in press.

Lambert, R.W., Maves, C.A., and Mircheff, A.K., 1991, Cholinergic stimulation induces activation of Na/H antiporters and recruitment of Na,K-ATPase in rabbit lacrimal acinar cells, *Invest. Ophthalmol. Vis. Sci.* 32S:728.

Marty, A., Tan, Y.P., and Trautmann, A., 1984, Three types of calcium-dependent channels in rat lacrimal glands. *J. Physiol.* 357:293.

Maves, C.A., Rismondo, V., and Mircheff, A.K., 1992, Chronic simulation of rabbit lacrimal acinar cells decreases intracellular pools of Na,K-ATPase and other surface enzymes, *Invest. Ophthalmol. Vis. Sci.* 33s:1289.

Mircheff, A.K., 1986, Comprehensive subcellular fractionation of rat exorbital gland: An approach to studying lacrimal gland fluid formation, *in:* "The Preocular Tear Film," F.J. Holly, ed., Dry Eye Institute, Lubbock.

Mircheff, A.K., 1989, Lacrimal fluid and electrolyte secretion: A review, *Current Eye Res.* 8:607.

Mircheff, A.K., Ingham, C.E., Lambert, R.W., Hales, K.L., Hensley, C.B., and Yiu, S.C., 1987, Na/H antiporter in lacrimal acinar cell basal-lateral membranes, *Invest. Ophthalmol. Vis. Sci.* 28:1726.

Mircheff, A.K., Lu, C.C., and Conteas, C.N., 1983, Resolution of apical and basal-lateral membrane populations from rat exorbital lacrimal gland, *Am. J. Physiol.* 245:G661.

Petersen, O.H., 1986, Calcium-activated potassium channels and fluid secretion by exocrine glands. *Am. J. Physiol.* 251:G1.

Saito, Y., Ozawa, T., Hayashi, H., and Nishiyama, A., 1985, Acetylcholine-induced change in intracellular $Cl^-$ activity of the mouse lacrimal acinar cells, *Pfluegers Arch.* 405: 108.

Saito, Y., Ozawa, T., and Nishiyama, A., 1988, Intracellular pH regulation in the mouse lacrimal gland acinar cells, *J. Membrane Biol.* 101:73.

Silva, P., Stoff, J., Field, M., Fine, L., Forrest, J.N., and Epstein, F.H., 1977, Mechanism of active chloride secretion by shark rectal gland: role of $Na^+,K^+$-ATPase in chloride transport, *Am. J. Physiol.* 233:F298.

Wood, R.L., and Mircheff, A.K., 1986, Apical and basal-lateral Na,K-ATPase in rat exorbital gland, *Invest. Ophthalmol. Vis. Sci.* 27: 1293.

Yiu, S.C., Lambert, R.W., Bradley, M.E., Ingham, C.E., Hales, K.L., Wood, R.L., and Mircheff, A.K., 1988, Stimulation-associated redistribution of Na,K-ATPase in rat lacrimal gland. *J. Membrane Biol.* 102:185.

Yiu, S.C., Lambert, R.W., Tortoriello, P.J., and Mircheff, A.K., 1991, Secretagogue-induced redistributions of Na,K-ATPase in rat lacrimal acini, *Invest. Ophthalmol. Vis. Sci.* 32:2976.

Yiu, S.C., Wood, R.L., and Mircheff, A.K., 1990, Analytic fractionation of acini from rat lacrimal gland, *Invest. Ophthalmol. Vis. Sci.* 31:2437.

# EFFECT OF ACETYLCHOLINE ON THE MEMBRANE CONDUCTANCE OF THE INTRALOBULAR DUCT CELLS OF THE RAT EXORBITAL LACRIMAL GLAND

Yoshitaka Saito and Soichiro Kuwahara

National Iwate Hospital
48 Dorotayamashita, Yamanome
Ichinoseki, Iwate 021, Japan

## INTRODUCTION

The two step theory of exocrine secretion predicts that the ionic composition of the primary secretion is modified while flowing through the duct system. The transport functions and the regulatory role of the main duct of the salivary gland (Young et al., 1987) and exocrine pancreas (Argent et al.1986) have been clearly demonstrated. On the other hand, very little is known about the transport functions of the duct cells of the lacrimal gland.

In the present study, we explored the electrical properties of the intercalated duct cell membrane of the rat exorbital lacrimal glands, using the patch-clamp techniques. We found that there exists a few distinct type of ion channels and the membrane conductance of the duct cells changes in response to a stimulation with acetylcholine.

## MATERIALS AND METHODS

The exorbital lacrimal glands of the adult male Wister rats killed by inhalation of diethylether in accordance to ARVO resolution on the Use of Animals in Research were excised, minced and incubated in a balanced salt solution containing collagenase and hyarulonidase each at 100 unit/ml concentration for 20 min and further digested in a trypsin (100 unit/ml) containing solution for 5 min. After trituration and filtration, the tissue fragments were washed free of digestive enzymes, placed in a chamber mounted on the stage of an inverted microscope and superfused with warmed, oxygenated physiological saline solution.The composition of the solution was (in mM) 139.2 Na, 4.7 K, 2.6 Ca, 1.1 Mg, 136.1 Cl, 4.9 pyruvate, 5.4 fumarate, 4.9 glutamate, 2.8 glucose and 5.0 Tris/HEPES mixture (pH 7.2 at 37C). The fragments of intercalated ducts were easily identified under the microscope and electric current across the plasma membrane of the duct cells was recorded according to the method by Hamil et al. (1982) in either cell attached, inside-out or outside-out cell-free or whole-cell configuration. The pipette was filled with a solution consisting of 140 mM KCl, 5 mM NaCl, 1 mM $MgCl_2$, 2.5 mM EGTA, 15 mM HEPES/tris mixture and desired amount of $CaCl_2$ to yield free $Ca^{2+}$ concentrations ranging from nominal free to $10^{-5}$ M and the pH was titrated to 7.20.

## RESULTS

The fragments of the intercalated ducts studied were tubules of 10-30 µm O.D. and >100 µm long, and some were branched or associated with secretory endpieces. The presence of the longitudinal tubular lumen and individual small cuboidal cells of several to 10 µm in diameter in the tubular wall were visible under the microscope.

### Single Channel Recordings

Figure 1 shows a trace of single channel current steps recorded in an inside-out patch of the excised basolateral membrane. The pipette potential was held initially at +30 mV then increased to +40 mV. From the difference in the step size, it seemed that the membrane patch contained at least three different types of ion channel. Further studies revealed that 1) the largest steps represented current through a c.a. 300 pS Cl channel, 2) the intermediate one through a K channel and 3) the smallest steps through another Cl channel.

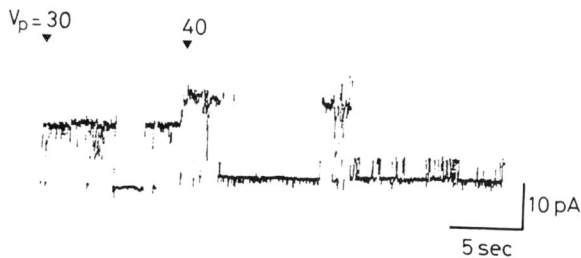

Fig.1. A trace of single channel current steps in an inside-out membrane patch excised from the basolateral membrane of the intercalated duct cell into a physiological saline. Pipette solution was a KCl-rich solution. By the difference in the magnitude of the step size, it is evident that three different types of ion channel were incorporated in this membrane patch. Vp denotes pipette potential in mV. Scales for time and current amplitude are given in the lower right.

### K Channel

The opening of the K channel was voltage-dependent. When membrane was hyperpolarized, the opening was rare, and as membrane was depolarized the open probability was progressively increased. The current-voltage relationship was curvilinear giving a maximal slope conductance of 80 pS and the zero-current potential was about -80 mV indicating a high selectivity for K over Na in an asymmetric ionic condition across the membrane. The conductance of the K channel was inhibited by quinidine (0.5 mM) and tetraethylammonium (5 mM), while $Ba^{2+}$ (5 mM) and 4-aminopyridine (0.5 mM) added to the cytosolic side of the membrane did not change the current steps significantly. The opening of the K channel did not show a $Ca^{2+}$-dependence and was present even in a solution containing free $Ca^{2+}$ of less than $10^{-9}$ M.

## Cl Channel

The open probability of the Cl channel was voltage-dependent and was high around 0 mV and was decreased by either hyper- or depolarization of the membrane. The I-V relationship was linear with a slope of 12 pS and the zero-current potential was 0 mV consistent with $E_{Cl}$. To rule out the possibility of a non-selective cation channel, we replaced all Na in the superfusate with less permeable organic cation N-methyl-D-glucamine leaving the Cl concentration unchanged. We obtained essentially the same I-V relationship as that obtained in NaCl solution. The Cl channel steps were almost instantly abolished when $Ca^{2+}$ concentration of the superfusate was decreased to less than pCa 6.5. The open probability was also decreased by a Cl channel blocker NPPB (>50 um). We did not explore the characteristics of the large conductance ( c.a. 300 pS) Cl channel in detail in this study.

## Whole Cell Current

Upon establishment of a whole-cell configuration, the membrane voltage was held at -80 mV close to the K equilibrium potential of -83 mV, and rectangular voltage pulses of various magnitude were superimposed. Figure 2 shows the traces of whole-cell current and the I-V relationship at the steady state. A remarkable outward-going rectification was observed when the membrane was depolarized. The outward-going current was inhibited by the tetraethylammonium or $Ba^{2+}$. The zero-current potential ranged from -40 to -65 mV suggesting that the membrane was predominantly selective for K. With a free $Ca^{2+}$ concentration of $>10^{-8}$ M in the pipette, the membrane conductance in the control condition was high and it was difficult to see a clear effect of acetylcholine on the conductance. Therefore, we used a pipette solution with no $Ca^{2+}$ and 0.1 mM EGTA. Immediately after the start of whole-cell recording, the membrane potential being clamped at -80 ($E_K$) and 0 ($E_{Cl}$) mV alternately, both inward- and outward-going current decreased significantly and stabilized within a few min. The magnitude of current steps induced by voltage jumps from -80 to 0 mV was decreased to almost 1/10 of the the initial phase of the recording. Figure 3 shows an example demonstrating the effect of acetylcholine under such conditions (in this particular case 5 nM free $Ca^{2+}$ was included in the pipette solution). Both inward and outward-going currents were increased by acetylcholine. This type of response to acetylcholine was not always the case. In some cases, a significant increase in current in response to addition of acetylcholine was seen only on either inward or outward current.

## DISCUSSION

The present results demonstrate that the basolateral membrane of the intercalated duct cells of the rat lacrimal gland possesses three types of ion channels namely 1) a 300 pS Cl channel, 2) a 12 pS Cl channel and 3) a 80 pS K channel. The open probability of these channels was increased when the membrane was depolarized. Effects of acetylcholine on the whole cell current suggested that this portion of the lacrimal duct has a cholinergic receptor and changes its electrical properties in response to cholinergic stimulations.

The effect of acetylcholine on both inward and outward-going currents suggested that the membrane conductance for both K and Cl was increased. Removal of $Ca^{2+}$ from the pipette solution significantly decreased the currents in both directions. These results suggest that both K and Cl conductances observed in the whole-cell configuration are regulated by the cytosolic $Ca^{2+}$ concentration.

The location of acetylcholine-induced KCl conductance, whether in the luminal or basolateral membrane, is not known. The 12 pS Cl channel in the excised basolateral membrane patch exhibited a clear activation by $Ca^{2+}$. On the other hand, the 80 pS K channel excised from the basolateral membrane did not show a significant $Ca^{2+}$-dependence of the opening. Therefore, we speculate that the luminal membrane may contain a K channel activated by cytosolic $Ca^{2+}$.

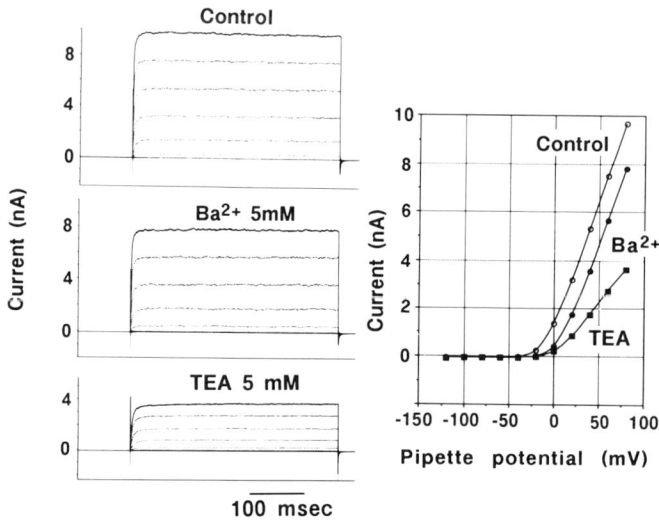

Fig 2. Superimposed traces of whole-cell current induced by various clamp voltage pulses (left panels) and I-V relationship (right panel) in the control condition and in the presence of K channel inhibitors. The pipette solution contained $10^{-8}$ M free $Ca^{2+}$. Note that the magnitude of the current was huge and there was a significant outward rectification when the membrane was depolarized.

Also, whether the conductance change induced by acetylcholine takes a role in transcellular ion transport is not clear at present. Although,indirect evidence suggests that the intralobular duct plays a role in transcellular ion transport. The final lacrimal secretion from the main duct of the carbachol-stimulated gland has been reported to contain a very high K concentration (49 mM). Also, the K concentration in the primary secretion collected from the acinar-intercalated duct region of the rat exorbital lacrimal gland was significantly higher than that in the interstitial fluid and a stimulation with carbachol further increased the K concentration. (Alexander et al. 1972). On the other hand, the current model of acinar secretion suggests that the primary secretion is brought about by Cl transport across the apical membrane of the acinar cells via the $Ca^{2+}$-activated Cl channel and Na transport from the interstitium into the acinar lumen through the paracellular shunt pathway (Marty et al. 1984; also see Petersen 1993 in this volume).

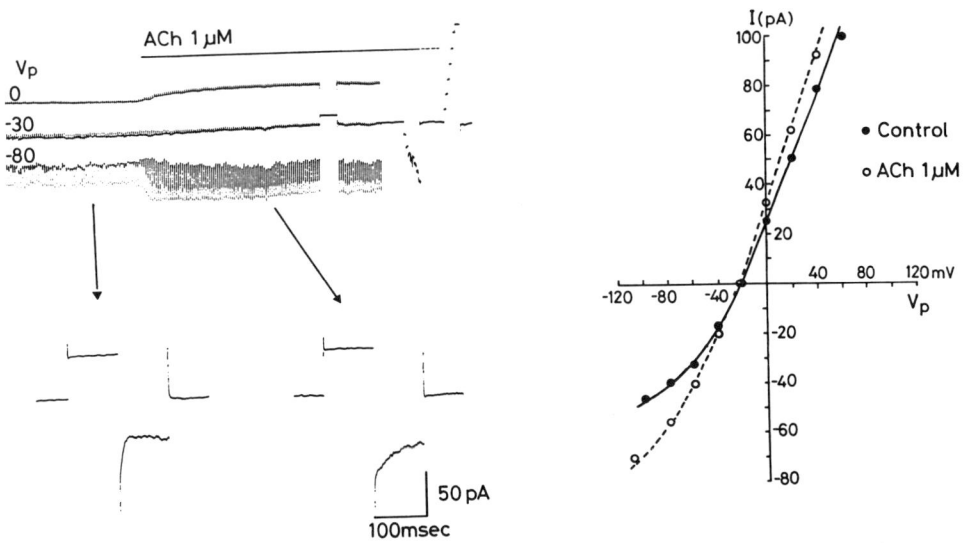

**Fig.3.** Effect of acetylcholine on the whole-cell current. Pipette solution contained 5 nM free $Ca^{2+}$. The membrane potential was clamped at -80 ($E_K$), -30 and 0 ($E_{Cl}$) mV alternately (left panel). Addition of acetylcholine increased both inward and outward current. Lower traces show fast recordings and a rapid inactivation of inward current induced by acetylcholine. I-V relationship is given in the right.

According to this model, high concentrations of Na and Cl but not K in the very primary secretion are presumed. Therefore, it is possible that a substantial amount of K is added to the primary secretion while flowing through the intercalated duct. Another evidence suggestive of transcellular ion transport in the intercalated duct is that this portion of the mouse exorbital lacrimal gland showed a high affinity for both type I and type II carbonic anhydrase antibodies (Henniger et al. 1983). High deposition of carbonic anhydrase is characteristic of epithelia with active ion transport functions.

Our present knowledge on the functions as well as morphology (see Jakobiec and Iwamoto, 1979) of the lacrimal duct system is very limited and superficial. However, employment of recently developed experimental techniques such as immunohistochemical methods, patch-clamp and microscopic fluorimetry methods will provide us with valuable information on the cellular and molecular basis of the ductal functions.

## REFERENCES

Alexander, J.H., Van Lennep, E.W., and Young, J.A., 1972, Water and electrolyte secretion by the
   exorbital lacrimal gland of the rat studied by micropuncture and catheterization techniques.
   *Pflugers Archiv.* 337:299-309.
Argent, B.E., Arkle, S., Cullen, M.J. and Green, R. 1986, Morphological, biochemical and secretory
   studies on rat pancreatic ducts maintained in tissue culture. *Quat. J. Exp. Physiol.*, 71:633-648.
Hamil, O.P., Marty, A., Neher, E., Sakmann, B., and Sigworth, F.J., 1981, Improved patch clamp
   techniques for high resolution current recording from cell and cell-freee membrane patches.
   *Pflugers Archiv.* 391:85-100.
Henniger, R.A., Schulte, B.A. and Spicer, S.S., 1982, Immunolocalization of carbonic anhydrase
   isozymes in rat and mouse salivary and exorbital lacrimal gland. *Anatomical Record.* 207:605- 614.

Jakobiec, F.A., and Iwamoto, T., 1979, The ocular adnexa: lids, conjunctiva, and orbit, *in* "Ocular Histology" B.S. Fine and M. Yanoff eds., Harper and Row, Hagerstown.

Marty, A., Tan, Y.P., and Trautmann, A., 1984, Three types of calcium-dependent channel in rat lacrimal glands. *J. Physiol. (London)*. 357:293-325.

Petersen, O.H., 1992, Cytoplasmic $Ca^{2+}$ signals, $Ca^{2+}$-dependent ion channels and tear formation. Abstract, International Conference on the Lacrimal Gland, Tear Film and Dry Eye Syndromes: Basic Science and Clinical Relevance, p104.

Young, J.A., Cook, D.L., Van Lennep, E.W., and Roberts, M., 1987, Secretion by the major salivary glands. *in* "Physiology of the Gastrointestinal Tract".2nd ed., R.Johnson ed. Raven Press, New York, pp773-815.

# THE CIRCADIAN RHYTHM OF LACRIMAL SECRETION AND ITS PARAMETERS, DETERMINED IN A GROUP OF HEALTHY INDIVIDUALS, AND ITS POTENTIAL DIAGNOSTIC AND THERAPEUTIC SIGNIFICANCE

Amalia Romano, Adrian Peisich and Bogidar Madjarov

Maurice and Gabriela Goldschleger Eye Institute
Sheba Medical Center
Tel-Hashomer
Sackler Faculty of Medicine
Tel-Aviv University
Ramat-Aviv, Israel

## INTRODUCTION

The tear film has an outstanding role in protecting the cornea, especially during the activity period of the day, from potentially noxious factors generally present in the environment. A reduction of the rate of lacrimal secretion was observed, following the application of local anesthetics to the conjunctiva[1,2], or the induction of general anesthesia[3-6].

Do to the paucity of local and photic stimuli at night, a lower rate of tear secretion is also expected during sleep[7,8]. Hence, we evaluated the temporal pattern of lacrimal secretion by using Schirmer test, in a group of healthy individuals, entrained (synchronized)[9] by natural photoperiodic signals. Despite criticism regarding especially, the lack of precision of Schirmer test[10], a repeatedly low level of tear secretion, assessed by it, is the most commonly accepted evidence, obtainable in the office, for the diagnosis of dry eyes[2].

The anticipated reflex lacrimal secretion was controlled, by exposing the subjects to constant illumination (250 lux), half an hour before each examination, and was limited by topical application of an aesthetics agent (benoxinate hydrocloride 0.4%), five minutes prior to each examination. Our intention was to determine the parameters of the expected circadian (circa=about, dias=day) rhythm, if present.

## MATERIALS AND METHODS

Fourteen young subjects, eight men and six women, were examined. Their age ranged from 16 to 22 years. No participant had a history of ocular disease or trauma, and none used any medication, or had any ocular symptoms. Each participant received an initial thorough eye examination, including a slit-lamp evaluation.

The group was entrained by natural photoperiodic dark and light alternations and by a common pattern of daily activity and social contacts. The subjects were exposed to the same environmental conditions for two weeks prior to the start of the study.

Before each examination, the participants were exposed to constant illumination for half an hour (250 lux), and two drops of 50 µl of benoxinate hydrocloride 0.4% were instilled in the

**Table 1A.** The circadian parameters of lacrimal secretion of eleven individuals (mean, amplitude, timing of rhythm trough and amplitude per mean ratio) and their standard errors (SE). The volunteers share a similar rhythm trough timing: at approximately midnight (mean: $5^{35}$ hours).

| Rhythm parameters of tear secretion of the early phase subgroup | | | | |
|---|---|---|---|---|
| Subject index | Mesor mm Schirmer | Amplitude mm Schirmer | Hour of Rhythm trough | Amplitude per mean ratio |
| 1st | 13.23 | 4.21* | 6.94h | 0.32 |
| 3rd | 5.47 | 4.0** | 5.74h | 0.73 |
| 4th | 14.53 | 7.35** | 4.12h | 0.51 |
| 6th | 12.93 | 5.87* | 6.27h | 0.45 |
| 8th | 3.67 | 2.27** | 4.23h | 0.62 |
| 9th | 8.7 | 5.21** | 4.12h | 0.6 |
| 10th | 7.23 | 6.21** | 5.78h | 0.86 |
| 11th | 10.93 | 8,48** | 7.21 | 0.78 |
| 12th | 14.43 | 6.33* | 4.9h | 0.44 |
| 13th | 8.47 | 4.19* | 3.53h | 0.49 |
| 14th | 7.63 | 3.14* | 6.86h | 0.41 |

**Table 1B.** The same circadian parameters and their SEs as in Table 1A, pertaining to the remaining three subjects: a subgroup characterized by a statistically significant 3 hours (h) rhythm phase delay (mean: 9 hours am.). *P<0.01; **P<0.001.

| Rhythm parameters of tear secretion of the delayed phase subgroup | | | | |
|---|---|---|---|---|
| Subject index | Mesor mm Schirmer | Amplitude mm Schirmer | Hour of Rhythm trough | Amplitude per mean ratio |
| 2nd | 10.43 | 4.93** | 8.15h | 0.47 |
| 5th | 9.87 | 7.48** | 10.02h | 0.76 |
| 7th | 8.97 | 4.43* | 9.97h | 0.49 |

inferior cul-de-sac simultaneously, in both eyes, 5 minutes prior to each test. The examination order of the subjects was random. Informed consent was obtained from each subject after the nature of the procedures has been fully explained to him.

The tear flow of both eyes was measured in each volunteer at four hours intervals for eighteen consecutive examinations. Sterile Schirmer test strips of Whatman no. 41 filter paper 5x35 mm in size, folded 5 mm from the end were used. The folded portion of the strip was placed at the outer third of the lower lid. We measured wetting after 5 minutes.

## Statistical Methods

The data was analyzed by the multiple regression method, using the cosine $(\cos[t*2\pi/24])$ and sine $(\sin[t*2\pi/24])$ transformations of the day time (t), as independent variables. The period was assumed to be 24 hours. The defining parameters of the expected circadian rhythm of lacrimal secretion of each subject were: mesor (mean); rhythm trough (stage of minimal lacrimal secretion rate); amplitude; and amplitude per mean ratio. The parameters were computed from the regression coefficients and the corresponding variances of the fitted regression functions Table (1A,1B).

The expected values of the regression functions, and their upper and lower 95% confidence limits were also computed. Multiple comparisons between the parameters of the fitted regression functions were performed using the approximate method, described by Gabriel[11,12].

The data was analyzed using the Statistix II programs on an IBM-PC computer.

## RESULTS

All fitted regressions (Table 1,2) were found to be highly significant (P=0.0--0.0006), pointing to a circadian rhythmic pattern of lacrimal secretion (R2 ranging between 0.426 andTT=0.801, with a mean value of 0.667, where $R^2$ is the coefficient of determination, or the proportion of explained sum of squares due to regression, from the total sum of squares). The average of the individually computed rhythm trough values was approximately at $5^{35}$ hour. The calculated amplitude values ranged between 2.26 and 8.48 mm, with a mean value of 5.29 mm of Schirmer strip.

The stages of the minimal lacrimal secretion rate were similar in a subgroup of eleven subjects (Table 1A), being approximately at 6 am. (range: $3^{30}$ hours-$7^{10}$ hours, mean:$5^{30}$ hours), while the circadian oscillations of the remaining three individuals (Table 1B) showed a significant phase delay of 3 hours (range: $8^{10}$-10 hours, mean: $9^{10}$ hours) (Fig. 1B). The @A=VW= circadian oscillations were characterized by high amplitude per mean ratios (range: 0.32-0.86; mean value: 0.57).

## CONCLUSIONS

The reduced tear volume and the relatively dry eyes during sleep (a time period when the irritative and photic eye stimulation are significantly limited), were interpreted to be consequences of the reduced reflex stimulation of the lacrimal secretion. Our present findings clearly outline the existence of a circadian rhythm of lacrimal secretion, peaking at 6 pm., and describing a trough approximately, at 6 am. A previous report[13] suggested the presence of a circadian rhythm of lacrimal secretion peaking however, in the morning.

As most biologic rhythms are endogenous and do not owe their basic pattern to external influences, we assume that the overt rhythm monitored by us, is the sum of endogenous and exogenous components[14].

The present study was particularly designed to minimize the variability of external influences: the locally instilled anesthetic, while not completely eliminating the effect of irritative agents[2], limited their influences, and the maintenance of constant illumination before each examination controlled the photic response. By using such a standardization, we did not attempt to quantify the questionable basal tear secretion[2], but expected to determine the inherent reactivity of the lacrimal glands and its pattern of change in time.

Thus, as expected, the presence of an overt circadian rhythm of lacrimal secretion, explaining a significant proportion of the actual data, was found in all examined subjects.

Presumably, the circadian oscillations found, regarding the constraint imposed on the variability of external stimuli during the performance of Schirmer test, might result from rhythmic oscillations in the excitability of the afferent reflex neuron pathways to the lacrimal glands, or stem from an internal timing mechanism regulating the reactivity of these glands to neuron stimulation.

Two groups were clearly delineated: one, whose rhythm trough was at midnight (Table 1A), and another showing a phase delay of approximately three hours (Table 1B).

Possibly, the exposure to a common pattern of daily activities and social contacts, recognized as prevalent synchronizers in humans[15,16], lead to the phase synchronization between the circadian rhythms in these two delineated subgroups.

The fitted regression explained a large percent of the variance (approximately 66%). By defining the circadian rhythm of lacrimal secretion and its parameters, the timing and the amplitude of its trough could be predicted, eliminating the hazard of the false positive and the false negative errors in the diagnosis of dry eyes.

Hence, the reported variable accuracy of the Schirmer test[10], performed without knowledge of the coincident tear secretion rhythm phase, might be explained, at least partially, by the large circadian oscillations (large amplitude per mean ratios: mean value=0.57) encountered. However, large variations of the rhythm parameters may be present in the population. Thus, in our group, synchronized for two weeks, a phase difference of 3 hours between two delineated subgroups was found.

Dry eyes represent a chronic condition whose definite cure is unlikely, but by a proper management, effective control can be achieved, vision preserved and the patient can lead a comfortable life. The topical application of artificial tear solutions in diverse dry eye disorders is the mainstay of treatment by now[17, 18]. As a consequence of the previously mentioned reasons, a chronodiagnostic and chronotherapeutical approach, whose role in numerous human pathologies is well established, may be more appropriate than the conventional one[19].

Nonetheless, though the implications stemming from the presence of an overt rhythm of lacrimal secretion are straightforward, caution is recommended in extrapolating these results, defining the physiologic rhythm of lacrimal secretion in healthy individuals, to disease states.


## REFERENCES

1. G. Crabtree and R.A. Dobie, The effect of unilateral corneal anesthesia on the Schirmer test, Otoralingol Head Neck Surg 100:631, (1989).
2. A. Jordan and J. Baum, Basic tear flow. Does it exist? Ophthalmology 87:920, (1980).
3. S.S. Chrai, T.F. Patton, A. Mehta and J.R. Robinson, Lacrimal and instilled fluid dynamics in rabbit eyes, J Pharm Sci 62:1112, (1973).
4. D.A. Cross and T. Krupin, Inplications of the effects of general anesthesia on basal tear production. Anesth Analg 56:35, (1977).
5. T. Krupin, D.A. Cross and B. Becker B, Decreased basal tear production with general anesthesia, Arch Ophthalmol 95:107, (1977).
6. A.H. Brightman II, J.P. Manning, G.J. Benson and E.E. Musselman, Decreased tear production associated with general anesthesia in the horse, J Am Vet Med Assoc 182:243, (1983).
7. J.L. Baum, Clinical implications of basal tear flow. In: Holly F.J., ed. The preocular tear film in health, disease, and contact lens wear. Lubbock, Texas: Dry Eye Institute, (1986):649.
8. J. Baum, A Relatively Dry Eye During Sleep. Cornea 9:1, (1990).
9. D.S. Minors and J.M. Waterhouse, Circadian rhythms and their mechanisms. Experientia 42:1, (1986).

10. A. Shapiro and S. Merrin, Schirmer test and break up time of tear film in normal subjects. Am J Ophtalmol 88:752, (1979).

11. K.R. Gabriel, A procedure for testing the homogeneity of all sets of means in analysis of variance, Biometrics 20:459, (1964).

12. K.R. Gabriel, A simple method of multiple comparisons of means. J Am Stat Assoc 73:724, (1978).

13. W.R.S. Webber, D.P. Jones and P. Wright, Measurement of tear turnover in normal healthy persons by fluorophotometry suggest a circadian rhythm. IRCS Med Sci 12:683, (1984).

14. J. Aschoff, Exogenous and endogenous components in circadian rhythms. Cold Spring Harb Symp Quant Biol 25:11, (1960).

15. J. Aschoff, M. Fatranska, P. Doerr, D. Stamm and H. Wisser, Human circadian rhythms in continuous darkness: entrainment by social cues. Science 171:213, (1971).

16. R.A. Wever, The Circadian System of Man: Result of Experiments under Temporal Isolation. Springer-Verlag, New York (1979).

17. J.P. Gilbard, Topical therapy for dry eyes. Trans Ophthalmol Soc UK 104:485, (1985).

18. M.A. Lemp, Recent developments in dry eye management. Ophthalmology 94:1299, (1987).

19. M.H. Smolensky and G.E. D'Alonzo, Biologic rhythms in medicine. Am J Medicine 85 (suppl 1B):34, (1988).

# EFFECTS OF DIHYDROTESTOSTERONE AND PROLACTIN ON LACRIMAL GLAND FUNCTION

Dwight W. Warren[1], Ana Maria Azzarolo[1],
Laren Becker[1], Kirsten Bjerrum[1],
Renee L. Kaswan[2], and Austin K. Mircheff[1]

[1]Department of Physiology and Biophysics
University of Southern California School of Medicine
Los Angeles, CA  90033
[2]Department of Small Animal Medicine
College of Veterinary Medicine
University of Georgia, Athens, GA  30602

## INTRODUCTION

Because both Sjögren's and non-Sjögren's lacrimal insufficiencies affect women much more frequently than men, it is reasonable to predict that there is a hormonal basis for both conditions, and attention turns to the estrogens as plausible candidates. However, the hormonal states during which women are most likely to experience lacrimal insufficiency are characterized by widely differing estrogen levels.  Both Sjögren's and non-Sjögren's lacrimal insufficiencies are commonly regarded as afflictions of the post-menopausal state, which is characterized by dramatic decreases in ovarian estrogen production.  However, women also tend to experience lacrimal insufficiency particularly frequently during estrogen-based oral contraceptive use, which is characterized by high estrogen levels; pregnancy, which is characterized by high estrogen and increasing prolactin levels; and lactation, which is characterized by low estrogen and high prolactin levels.

The conundrum that both high- and low estrogen states increase the incidence of lacrimal insufficiency might, plausibly, be resolved if androgens, rather than estrogens, play a critical role in maintaining the lacrimal glands' functional capacity.  Although androgens are typically thought of as "male hormones," they are also produced in significant amounts by the ovaries and adrenal glands of females.  Like estrogen production, ovarian androgen production declines markedly at menopause.  The elevated estrogens of pregnancy stimulate the liver to produce sex hormone binding globulin (SHBG), which then decreases the fraction of the total circulating androgens that are unbound, or active.  During estrogen treatment, the exogenous estrogens not only stimulate SHBG production, but, by suppressing production of luteinizing hormone (LH), they also decrease ovarian androgen production.  Therefore, one factor shared by the post-menopausal and the high estrogen states is a decreased level of free (active) androgens. The greater likelihood that androgen levels may fall below some critical value during the female life experience might then account for much of the excessive incidence of lacrimal insufficiency among women. Additional factors are necessary to account for increased lacrimal insufficiency during lactation; as we outline below, prolactin appears as a likely candidate for this role.

## SEXUAL DIMORPHISMS OF THE LACRIMAL GLAND

Some preliminary clues to the possible hormonal basis of the increased female incidence of lacrimal insufficiency have come from studies of lacrimal gland sexual dimorphisms. Sullivan and coworkers have documented numerous morphological, biochemical, and functional differences between the lacrimal glands of males and females of various species. For example, lacrimal acini are significantly smaller in female than in male rats (Cornell-Bell et al., 1985). Lacrimal glands of male rodents express higher specific contents of β-adrenergic receptors (Pangerl et al., 1989), while lacrimal glands of female rodents express larger amounts of leucine aminopeptidase (Lauria and Porcelli, 1979). Lacrimal glands of male rats secrete significantly greater amounts of IgA, and acini isolated from male rats synthesize significantly more secretory component (*i.e.*, SC, the polymeric IgA receptor) than their female counterparts (Sullivan, Bloch, and Allansmith, 1984, Sullivan and Allansmith, 1985).

Most of the parameters which previous investigators have identified as sexually dimorphic are not involved, in any simple or direct way, in the lacrimal glands' ability to secrete fluid. Secretion of water in the lacrimal acini is the osmotic consequence of the secretion of $Na^+$ and $Cl^-$. Work reviewed elsewhere in this volume by Mircheff, Lambert, et al. indicates that acinar NaCl secretion is driven by $Na^+,K^+$-ATPase pumps, operating in concert with an array of additional transport proteins. $Na^+,K^+$-ATPase catalytic activity can be measured relatively easily as the $K^+$-dependent hydrolysis of *p*-nitrophenylphosphate (Mircheff, 1989), and this activity is likely to provide a measure of the lacrimal gland's capacity to secrete fluid. The machinery for secreting NaCl is largely quiescent in the absence of secretomotor stimulation. Thus, the lacrimal gland's content of receptors for the neurotransmitters and neuropeptides which stimulate secretion may determine how well the gland can respond to secretomotor stimulation. Of the receptors which have been characterized in lacrimal acinar cells, muscarinic cholinergic receptors (MAChR) and β-adrenergic receptors (βAR) can be measured relatively easily in lacrimal membrane samples (Bradley et al., 1992).

In preliminary studies we have found that the lacrimal glands of pre-pubertal (1 kg) male and female rabbits are remarkably similar to each other with respect to total mass, total protein, and cell number as estimated from the total DNA content. They are also similar with respect to their total contents of $Na^+,K^+$-ATPase, MAChR, βAR, and alkaline and acid phosphatases, two catalytic activities whose physiological significance is unknown but which have proven useful as membrane markers during subcellular fractionation analyses. As rabbits grow, their lacrimal glands increase in mass, cell number, and contents of all the parameters we have measured. However, the lacrimal glands of males and females diverge from each other, and at sexual maturity (4 kg) the lacrimal glands of males are nearly 30% larger and contain 45% more cells than the lacrimal glands of females. The total number of βAR increases relatively more in males than in females, so that at maturity male lacrimal glands contain nearly 80% more βAR than females. There are no significant differences between the total amounts of $Na^+,K^+$-ATPase, alkaline phosphatase, acid phosphatase, MAChR, and βAR of mature male and female rabbits. However, since the females contain less total protein, the specific content of each of these parameters is significantly greater in females than in males. The difference with respect to $Na^+,K^+$-ATPase is particularly striking: the specific activity of $Na^+,K^+$-ATPase in females is nearly 75% greater than in males. These observations alone would convey little information as to why lacrimal insufficiency occurs more frequently in women. In fact, they suggest that the lacrimal glands of females are, in a sense, more efficiently adapted for responding to parasympathetic stimulation by secreting NaCl. However, in attempting to define the hormonal basis of these dimorphisms, we have obtained evidence which suggests that interactions between androgens and prolactin may not only influence the gender-typic characteristics of the lacrimal glands but also be critical for maintaining its functional capacity.

## ROLE OF ANDROGENS

There is already convincing evidence that androgens contribute to the regulation and/or maintenance of the sexually dimorphic characteristics of the lacrimal glands. Sullivan and coworkers have found that castrating male rats causes their lacrimal gland morphology to become similar to that of females. Castration also significantly decreases SC production in acinar cells and IgA secretion by the whole lacrimal gland. Treatment with testosterone reverses these decreases, and it restores the acini to their larger, male-like appearance (Cavallero, 1967). If female rats are treated with testosterone, the lacrimal acini attain a male-like appearance (Sullivan, Bloch, and Allansmith, 1984).

### Effects of hypophysectomy

In a study designed to indicate whether androgens also influence the parameters related to the lacrimal glands' ability to respond to secretomotor stimulation and secrete NaCl, we analyzed exorbital glands from sexually mature, hypophysectomized female rats. The rationale of these measurements was that hypophysectomy would eliminate the trophic stimulation of ovarian and adrenal steroid production and markedly reduce circulating androgen and estrogen levels. Within 7 days after hypophysectomy, the lacrimal gland protein content decreased to 62% of the control value, and the contents of DNA, $Na^+,K^+$-ATPase acid phosphatase, and $\beta$AR decreased to between 48% and 57% of their control values. In contrast, alkaline phosphatase and MAChR contents only declined to between 70% and 75% of their control values.

### Maintenance by androgens

A low (0.25 mg/rat) dose of the potent androgen dihydrotestosterone (DHT), given for 2 days, *i.e.*, on days 5 and 6 after hypophysectomy, prevented or reversed many of the hypophysectomy-induced changes. The most dramatic effect was on DNA content, which was 2.6-fold greater than in hypophysectomized animals and nearly 50% greater than in control animals. DHT did not significantly change the total gland protein, but it increased the membrane-bound protein by 26%. Thus, giving hypophysectomized females DHT at a dose which is approximately 25% of that necessary to maintain male secondary sex characteristics increased the number of lacrimal gland cells but, in contrast with androgen effects in pituitary-intact, male rats, markedly decreased the average cell size.

DHT increased the total $Na^+,K^+$-ATPase, alkaline phosphatase, and acid phosphatase activities by roughly the same degree that it increased membrane-bound protein. Thus, DHT completely restored alkaline phosphatase activity to the control value, and it increased the acid phosphatase and $Na^+,K^+$-ATPase activities to 70% of the control values. In contrast, DHT had no significant effect on the number of MAChR. Increasing the DHT dose from 0.25 mg to 1.0 mg/rat yielded a $\beta$AR number which was significantly greater than in hypophysectomized, untreated animals. In all other respects the responses to the higher and lower DHT doses did not differ significantly. In contrast, administration of diethylstilbestrol (DES) had no significant effect on any measured parameter.

### Androgen administration after ovariectomy

If the cessation of ovarian androgen production at menopause decreases the lacrimal gland's secretory capacity, the same result should follow ovariectomy, and administration of exogenous androgens should prevent this lacrimal gland regression. Results of a preliminary experiment in which rabbits were ovariectomized, then treated with DHT, appear to support this hypothesis. Compared to sham-operated controls, lacrimal gland protein, DNA and MAChR decreased significantly within 12 days after ovariectomy. $Na^+,K^+$-ATPase and $\beta$AR and appeared to decrease but the changes were not statistically significant. Administration of DHT, beginning at the time of surgery, completely prevented the decreases observed in DNA and protein but not MAChR. Furthermore, DHT increased $\beta$AR and $Na^+,K^+$-ATPase to levels significantly above controls.

### Effects of excess estrogen

An additional experiment provides preliminary support for the hypothesis that exogenously elevated estrogen levels compromise the lacrimal gland's functional capacity, presumably by suppressing ovarian androgen production as well as stimulating SHBG production. Administration of 100 $\mu$g/kg·day of DES for 8 days did not alter lacrimal gland DNA or protein content, but it significantly decreased $Na^+,K^+$-ATPase activity and MAChR number. This same dose of DES significantly decreased both alkaline and acid phosphatase. Interestingly, DES significantly increased $\beta$AR number.

## ROLE OF PROLACTIN

Several observations suggest that hormonal factors other than the androgens also influence lacrimal secretory capacity. In our experiments with hypophysectomized rats, DHT was completely ineffective in restoring MAChR number after hypophysectomy. Sullivan and colleagues have found evidence that some pituitary hormone plays an absolutely essential role in the regulation of SC production by the lacrimal glands of male rats (Sullivan et al., 1988). That is, either hypophysectomy or transplantation of the anterior pituitary to the kidney capsule prevents the ability of DHT to restore SC after castration, and hormones from pituitary-dependent endocrine glands, such as thyroxine and cortisol, fail to compensate for disruption of the hypothalamic-pituitary axis.

There is some basis for predicting that prolactin might be the unidentified pituitary factor and that it must be present within a specific range of concentrations. When the anterior pituitary is separated from its connection to the hypothalamus, it responds to the removal of inhibitory input by greatly increasing its production of prolactin. Moreover, acinar cells from male and female rats contain receptors for prolactin, and, *in vitro*, prolactin appears to exert modest but significant effects on the cells' response to cholinergic stimulation (Mircheff et al., 1992).

### Restoration of Na$^+$,K$^+$-ATPase and MAChR by prolactin

The hypophysectomized female rat model provided an opportunity to examine the effects of circulating prolactin on lacrimal gland functional parameters. When either a low dose (1 mg/kg) or a high dose (5 mg/kg) of prolactin was administered for 2 days, lacrimal gland DNA content remained unchanged. Neither dose significantly changed either the total or membrane-associated protein content or the βAR number. However, the high dose of prolactin increased Na$^+$,K$^+$-ATPase activity to 70% of the control value, and it restored alkaline phosphatase activity to 100% of the control value. Most notably, it increased the MAChR number to 90% of the control value.

## INTERACTIONS BETWEEN PROLACTIN AND DHT

The hypothesis that some optimal level of prolactin is essential for normal lacrimal function, while excess prolactin is detrimental, offers a plausible explanation for the observation that an intact hypothalamic-pituitary axis is essential for restoration of SC production by exogenous androgens. We obtained preliminary data that may be pertinent to this hypothesis by administering combinations of low- and high-doses of DHT and prolactin to hypophysectomized female rats. Given alone, the high dose of prolactin was less effective than the high dose of DHT in restoring Na$^+$,K$^+$-ATPase catalytic activity. The Na$^+$,K$^+$-ATPase response to the high doses in combination was significantly less than the response to DHT alone, suggesting that prolactin impaired the response to DHT. Conversely, given alone, prolactin significantly increased MAChR number, while DHT had no effect. When the two hormones were given in combination, the MAChR number did not change significantly from the value in untreated, hypophysectomized animals. Thus, the high dose of DHT appears to suppress the MAChR response to prolactin. Clearly, a good deal of additional work will be required to establish whether or not optimal prolactin levels contribute to normal lacrimal gland function. On the other hand, the observed interactions between DHT and prolactin suggest the general hypothesis that the balance between prolactin and free androgen levels influences the lacrimal glands' gender-typic characteristics.

## FUTURE DIRECTIONS

If the hypothesis that a minimal level of androgens is critical for normal lacrimal secretory function proves to be correct, then a strategy for decreasing non-Sjögren's lacrimal insufficiency after menopause and during oral contraceptive use may be at hand. This is to administer a low-dose androgen supplement sufficient to replace the lost ovarian androgens or to restore free androgen levels to their normal levels. As Mircheff, Gierow et al. have argued elsewhere in this volume, it is conceivable that such a therapy would also decrease the likelihood of subcellular events which provoke local autoimmune responses that may progress to the autoimmune disease of Sjögren's Syndrome. This would presumably involve mechanisms quite independent of the ability of androgens to suppress ongoing autoimmune activity (Vendramini et al., 1991).

While work remains to be done to test our tentative explanations for the excessive incidence of lacrimal insufficiency among women, it is already appropriate to begin attempting to learn whether androgens and prolactin influence other transport proteins, *e.g.*, the $Na^+/H^+$ and $Cl^-/HCO_3^-$ antiporters and $Cl^-$ channels that mediate lacrimal NaCl secretion, the other neurotransmitter and neuropeptide receptors which modulate lacrimal secretion, and the intracellular messenger cascades by which such signals are transduced. It is also appropriate to begin examining the cellular and molecular mechanisms by which androgens and prolactin exert the actions which we have already documented.

The nature of the role pituitary prolactin plays in maintaining the lacrimal gland's functional status will be a particularly challenging question, complicated by the possibility that prolactin or prolactin-like peptides may also function as local mediators within the lacrimal gland. Prolactin-like immunoreactivity is present in acinar cells of human (Frey et al., 1986), rat (Mircheff et al., 1992), and rabbit (Wood et al., this volume) lacrimal glands. Acinar cells from both male and female rats contain similar numbers of receptors for prolactin, and the vigorous recycling traffic between the basal-lateral plasma membrane and endocytic compartments provides a pathway by which receptor-bound prolactin might be taken up. However, lacrimal glands also possess the ability to transcribe prolactin message and, presumably, to synthesize prolactin. Of particular interest, prolactin mRNA is present at detectible levels in the lacrimal glands of male, but not female, rats. This suggests that lacrimal production of prolactin increases as circulating prolactin levels decrease. The physiological implications of local prolactin production *i.e.*, its possible autocrine and paracrine actions, have yet to be investigated.

## REFERENCES

Bradley, M.E., Lambert, R.W., Lambert, R.W., Lee, L.M., and Mircheff, A.K., 1992, Isolation and subcellular fractionation analysis of acini from rabbit lacrimal glands, *Invest. Ophthalmol. Vis. Sci.* 33:2951.

Cavallero, C.,, 1967, Relative effectiveness of various steroids in an androgen assay using the exorbital lacrimal gland of the castrated rat, *Acta Endocrinol.* 55:119.

Cornell-Bell, A.H., Sullivan, D.A., and Allansmith, M.R., 1985, Gender-related differences in the morphology of the lacrimal gland, *Invest. Ophthalmol. Vis. Sci.* 26:1170.

Frey, W. H., Nelson, J.D., Frick, M.L., and Elde, R.P., 1986, Prolactin immunoreactivity in human tears and lacrimal gland: Possible implications for tear production, *In:* "The Preocular Tear Film in Health, Disease, and Contact Lens Wear," Holly F.J., ed., Lubbock, Dry Eye Institute, p. 798.

Lauria, A., and Porcelli, F., 1979, Leucineaminopeptidase (LAP) activity and sexual dimorphism in rat exorbital lacrimal gland, *Bas. App. Histochem.* 23:171.

Mircheff, A.K., 1989, Isolation of plasma membranes from polar cells and tissues: Apical/basolateral separation, purity, function, *Meth. Enzymol.* 172:18.

Mircheff, A.K., Warren, D.W., Wood, R.L., Tortoriello, P. J., and Kaswan, R.L., 1992, Prolactin localization, binding, and effects on peroxidase release in rat exorbital lacrimal gland, *Invest. Ophthalmol. Vis. Sci.* 33:641.

Pangerl, A., Pangerl, B., Jones, D.J., and Reiter, R.J., 1989, Beta-adrenoreceptors in the extraorbital lacrimal gland of the Syrian hamster. Characterization with [125I]-iodopindolol and evidence of sexual dimorphism, *J. Neural Trans.* 77:153.

Sullivan, D.A., 1988, Influence of the hypothalamic-pituitary axis on the androgen regulation of the ocular secretory immune system, *J. Steroid Biochem.* 30:429.

Sullivan, D.A., and Allansmith, M.R., 1985, Hormonal regulation of the secretory immune system in the eye: Androgen modulation of IgA levels in tears of rats, *J. Immunol.* 134:2978.

Sullivan, D.A., Bloch. K.J., and Allansmith, M.R., 1984, Hormonal influence on the secretory immune system of the eye: Androgen control of secretory component production by the rat exorbital gland, *Immunol.* 52:239.

Vendramini, A. C. L. M., Soo, C., and Sullivan, D.A., 1991, Testosterone-induced suppression of autoimmune disease in lacrimal tissue of a mouse model (NZB/NZW F1) of Sjögren's syndrome, *Invest. Ophthalmol. Vis. Sci.* 32:3002.

# A Na:H EXCHANGER SUBTYPE MEDIATES VOLUME REGULATION IN BOVINE CORNEAL EPITHELIAL CELLS

Peter Reinach[1], Vadivel Ganapathy[2] and Viviana Torres-Zamorano[1]

[1]Physiology and Endocrinology
[2]Biochemistry and Molecular Biology
Medical College of Georgia
Augusta, GA 30912

## ABSTRACT

To identify a role for a Na:H (NHE) exchanger subtype in volume regulation in bovine corneal epithelium, we determined: 1.its sensitivity to inhibition by amiloride analogues 2. the effects of either Cl removal or hypertonicity on the intracellular pH. Our results suggest that volume regulatory responses elicited by stimulation of NHE-2 may help preserve epithelial barrier function in the face of increases in tear film osmolarity.

## INTRODUCTION

For the corneal epithelium to maintain itself as an effective physical and immunological barrier, there must be close cell to cell apposition between neighboring cells.[1]  By maintaining this configuration, an increase in tear film osmolarity is prevented from shrinking the cells which could otherwise compromise these barrier functions. Such increases have been described as part of the dry eye syndrome, which makes it important to determine if the corneal epithelial cells can volume regulate so as to counter any threat to barrier preservation.[2]  One of the transport mechanisms which may be responsible for volume regulation in many cells is the Na:H exchanger (NHE).[3]  NHE activity was identified in the rabbit and bovine, but its role in volume regulation has not been studied.[4,5]

Based on molecular cloning, four different subtypes of this antiport were identified.[6] Two of these four subtypes (NHE-1 and NHE-2) have been characterized more extensively based on pharmacological and physiological studies. The pharmacological approach has involved determining the sensitivity of the exchanger to inhibition by amiloride and its analogues.[7] These subtypes are different because NHE-1 is more sensitive to inhibition by amiloride than NHE-2 (i.e. "amiloride insensitive"). This classification has been helpful in linking type to function because NHE-1 mediates transepithelial $H^+$ flux whereas NHE-2 is the housekeeping type.

However, there have been no reports linking volume regulation to either one of these subtypes.

In this study, we identify a subtype of Na:H exchange activity and consider its role in eliciting volume regulation in response to either loss of KCl solute or to a hypertonic shock. Our results show that it is the NHE-2 or the relatively amiloride insensitive subtype that elicits volume regulation in response to hypertonicity.

## METHODS

### [³H]MIA, 5-(N-Methyl-N-isobutyl)amiloride, Binding

A plasma membrane fraction ($B_2$) which is 14-fold enriched in Na:K ATPase activity with respect to the homogenate was obtained as described.[8] Measurements of [³H]MIA binding to the $B_2$ fraction were performed using a rapid filtration technique (GF/B glass fiber filters). The reaction was started by adding the $B_2$ fraction to a solution containing 25 mM HEPES/ Tris, 300 mM mannitol, 1 mM $MgSO_4$, 0.1 mM PMSF, 25 nM [³H]MIA, pH 8.5 for 90 min at 4°C in the presence or absence of an excess of unlabeled MIA (0.1 mM). The reaction was stopped and the filters were washed with an ice-cold solution containing 160 mM $MgSO_4$, 10 mM HEPES/Tris (pH 8.5). The radioactivity retained by the filter was counted. Specific [³H]MIA binding was determined from the difference between total binding (presence of only [³H]MIA) and non-specific binding (presence of [³H]MIA and excess unlabeled 0.1 mM MIA).[9]

### Monitoring of Intracellular pH

Primary cultures of bovine corneal epithelium were grown to confluence on coverslips and loaded for 6 min with 10 μM BCECF-AM at room temperature. A coverslip was vertically mounted in a thermostatically controlled compartment in a dual wavelength excitation spectrofluorometer. Fluorescence excitation was alternately obtained at 500 and 439 nm and emission was observed at 525 nm. To compensate for any dye loss, $F_{500}/F_{439}$ was continuously recorded.[10]

## RESULTS

### Subtype Identification

To use the extent of specific binding [³H]MIA as an indicator of Na:H exchange, we established the appropriate conditions to maximize its specific equilibrium binding to this amiloride analogue. The largest binding took 60 min to reach equilibrium and occurred at 4° C (pH of 8.5). Therefore, all of our binding experiments were done under these conditions for 90 min. Specific binding under these conditions was about 80% of the total binding. To verify that this binding occurred at a single class of high affinity sites, we performed Scatchard analysis. The relationship between bound/free [³H]MIA and bound MIA (pmoles/mg protein) was linear and highly significant with a $K_d=61$ nM and $B_{max}=271$ pmoles/mg. Therefore, specific [H³]MIA binding is a meaningful indicator for characterizing the Na:H exchanger.

In order to distinguish between the amiloride sensitive (NHE-1) exchanger and its insensitive subtype, we determined the rank order of potency of various amiloride analogues to displace the specific binding of [³H]MIA. To do this, the dose dependent inhibition of [³H]MIA binding by the following amiloride analogues was determined: MIA, dimethylamiloride (DMA) and benzamil. The results shown in Figure 1 indicate that the rank order of potency for these inhibitors was: MIA ($IC_{50}$= 0.79 μM)> DMA($IC_{50}$=3.9 μM)>benzamil ($IC_{50}$ = 79 μM)> amiloride ($IC_{50}$ = 199 μM). This rank order of potency shows that the corneal epithelium contains NHE-2 exchange activity. In most other tissues where such activity was found it is linked to a role in eliciting transepithelial $H^+$ flux across the apical membrane.[11]

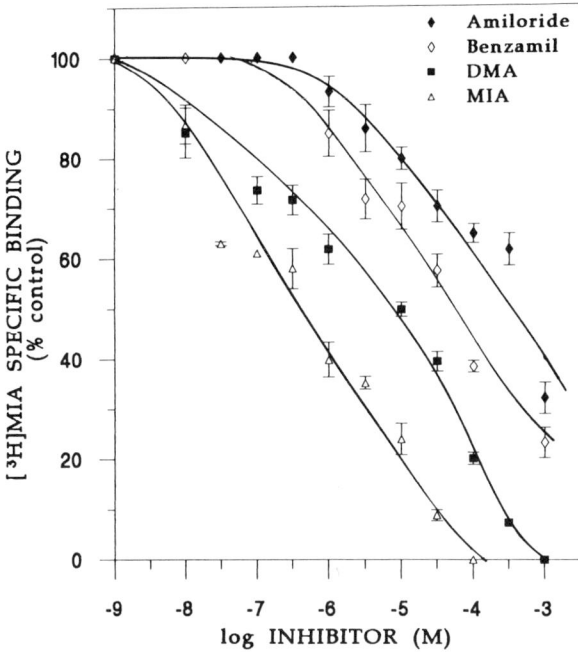

Figure 1.  Concentration-dependent inhibition of specific MIA binding by amiloride and its analogs.  The values represents 3 different membrane preparations.


## Cell Volume Regulation

Primary cultures of epithelial cells generally reached confluence after 7 days and had a typical epithelial shape. They were free of contaminating keratocytes because the culture medium was valine deficient which is an essential amino acid for keratocyte proliferation. To monitor the Na:H exchange activity, the dye loaded cells were bathed in a $HCO_3$-free NaCl Ringers.

Figure 2 compares the typical responses to acid loading with  a prepulse of 20 mM $NH_4Cl$ in isotonic (300 mOsm) with that in hypertonic (360 mOsm or 600 mOsm) Ringers. Following washout of $NH_4Cl$, the  pH initially fell from 7.2 to 6.6, but then recovered after about 5 min to 7.2. If the recovery occurred instead with a hypertonic Ringers, the Na:H exchanger was stimulated because the pH recovery was about 20% more rapid with an initial overshoot.

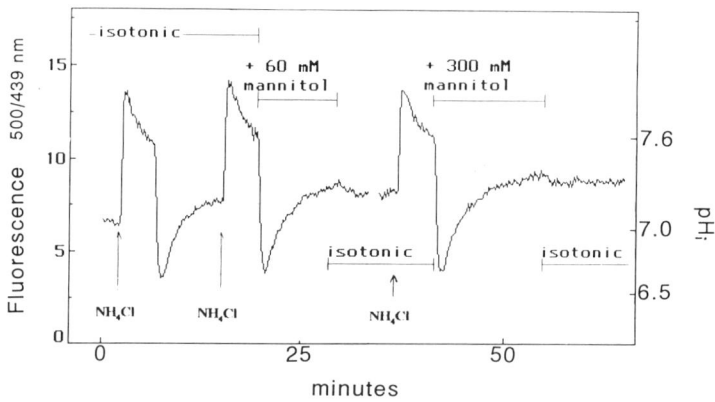

Figure 2. Effect of hyperosmolarity on $pH_i$ recovery following $NH_4Cl$ acid loading: With either 60 or 300 mM mannitol supplementation of NaCl ($HCO_3$-free) Ringers, transient overshoots in $pH_i$ recovery were observed.

Another way of testing for stimulation of the Na:H exchanger was to determine if the isosmotic substitution of NaCl with Nagluconate (Cl-free) Ringers could alkalinize the pH. Such an effect would stem from the presumed loss of intracellular KCl to the medium resulting in shrinkage of apparent cell volume. The response to this substitution shown in Figure 3 indicates that the exchanger was stimulated because the pH increased from 7.4 to 8.0. The initial acidification which preceded this increase was caused by an inhibition of V-type $H^+$ pump activity.[8] Further proof that the alkalinizing response reflected stimulation of the Na:H exchanger is that the pH fell to its control level following exposure to 100 μM MIA.

## DISCUSSION

The corneal epithelium acts as a physical and immunological barrier which is essential for the maintenance of transparency because it prevents the stroma from

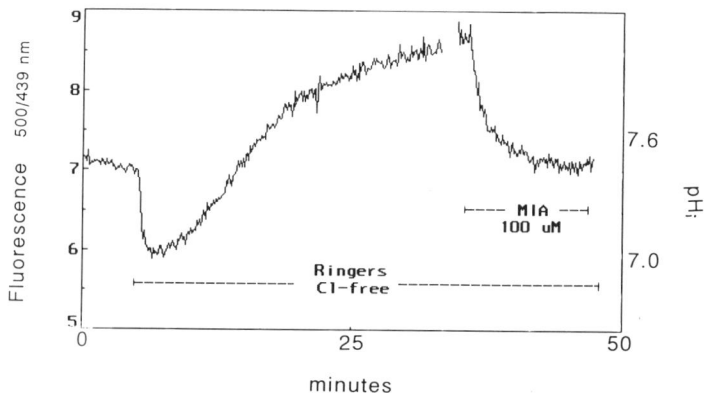

Figure 3. Effect of isosmotic Cl substitution with sodium gluconate ($HCO_3$-free) Ringers on $pH_i$: Initial acidification due to $H^+$ pump inhibition followed by large alkalinization above the resting pH owing to stimulation of $Na^+$-$H^+$ exchange activity.

imbibing excess fluid and becoming translucent. This barrier function depends on close to cell to cell apposition between neighboring cells which can be challenged if cell volume decreases owing to an increase in tear film osmolarity. In dry eye, there is a reported elevation of about 30 mOsm in tear film osmolarity which could compromise the refractive properties of the cornea unless there is a volume regulatory mechanism which restores the normal volume. Since there is Na:H exchange activity in the corneal epithelium and in some other tissues it is linked to volume regulation,we sought to determine if this epithelium also mediates such a response. Another objective was to characterize the Na:H exchange subtype by determining its sensitivity to inhibition by amiloride analogues because this is a criterion for distinguishing between NHE-1 (amiloride sensitive) and NHE-2 (amiloride insensitive) subtypes. Such a characterization enabled us to posit the membrane sidedness of the exchange activity because in other tissues NHE-2 is essentially only in the apical membrane and mediates transmural $H^+$ flux whereas the NHE-1 subtype performs a "housekeeping" function on the basolateral side.

We tested for a volume regulatory response to hyperosmolarity by either elevating the bath osmolality (i.e 60 or 300 mOsm) or isosmotically substituting NaCl ($HCO_3$-free) with $Na_2SO_4$ Ringers. With the first alternative, the cells were acid loaded to stimulate the Na:H exchanger and their rate and extent of pH recovery to the control was measured. Since either of the two increases in osmolality not only shortened the time needed to recover the control pH but also caused a slight overshoot, the Na:H exchanger was stimulated. More convincing evidence for a volume regulatory function for the Na:H exchanger is that it was stimulated enough after $Cl^-$ removal to increase the pH by 0.4 units. These responses are associated with the stimulation of the NHE-2 subtype suggesting its presence in the apical tear-side facing membrane. This orientation makes teleological sense because prompt volume regulatory responses to an increase in tear film osmolality could be most effectively mediated by stimulating an apical membrane Na:H exchanger. Such stimulation would most rapidly reequilibrate the intracellular osmolality with that of the tears.

## REFERENCES

1.  S.D. Klyce and R.W. Beuerman, Structure and function of cornea, in: "The Cornea," H.E. Kaufman, B.A. Barron, and M. Mcdonald, eds. Churchill Livingstone, New York (1988).
2.  J.P. Gilbard, R.L. Farris, and J. Santamaria, Osmolarity of tear microvolumes in keratoconjunctivitis sicca, *Arch. Ophthalmol.* 96:677 (1978).
3.  S. Grinstein. "$Na^+/H^+$ Exchange," CRC Press, Boca Raton (1988).
4.  J.A. Bonanno and T.E. Machen, Intracellular pH regulation in basal corneal epithelial cells measured in corneal explants: characterization of $Na^+/H^+$ exchange, *Exp. Eye Res.* 49:129 (1989).
5.  C. Korbmacher, H. Helbig, C. Foster, and M. Wiederholt, Evidence for $Na^+/H^+$ exchange and pH sensitive membrane voltage in cultured bovine corneal epithelial cells, *Curr. Eye Res.* 7:619 (1988).
6.  J. Orlowski, R.A. Kandasamy, and G.E. Shull, Molecular cloning of putative members of the $Na^+/H^+$ exchanger gene family, *J. Biol. Chem.* 267:9331 (1992).
7.  J.D. Clark and L.E. Limbird, $Na^+/H^+$ exchanger subtypes: a predictive review, *Am. J. Physiol.* 261:C945 (1991).
8.  V. Torres-Zamorano, V. Ganapathy, M. Sharawy, and P. Reinach, Evidence for an ATP-driven $H^+$ pump in the plasma membrane of the bovine corneal epithelium, *Exp. Res.* 55:269 (1992).

9. D. Rosskopt, C. Barth and W. Siffert, Estimation of carrier density and turnover rate of the $Na^+/H^+$ exchanger in human platelets using 5-(N-methyl-N-[$^3$H] isobutyl)-amiloride, *Biochem. Biophys. Res. Comm.* 176:601 (1991).

10. J.A. Bonanno, $K^+$-$H^+$ exchange a fundamental cell acidifier in corneal epithelium, *Am. J. Physiol.* 260:C618 (1991).

11. V. Casovola, C. Helmle-Kolb, and H. Murer, Separate regulatory control of apical and basolateral $Na^+$-$H^+$ exchange in renal epithelial cells, *Biochem. Biophys. Res. Comm.* 165:833 (1989).

# ENDOCYTOSIS AND EXOCYTOSIS IN RABBIT LACRIMAL GLAND ACINAR CELLS

J. Peter Gierow,[1] Richard L. Wood,[2] and Austin K. Mircheff[1]

Departments of [1]Physiology and Biophysics,
[1]Ophthalmology, and [2]Anatomy and Cell Biology
University of Southern California School of Medicine
Los Angeles, CA 90033, USA

## INTRODUCTION

The main function of the lacrimal gland is to maintain the surface of the eye moist, well lubricated, and free of irritants.  It achieves this function by secreting ions, proteins, and water.  Recent studies have shown that stimulation of lacrimal cells by carbachol, an acetylcholine analogue, not only triggers release of secretory products across the apical plasma membrane and activates various ion transporters involved in the trans-epithelial secretion of $Na^+$ and $Cl^-$, but it also causes $Na^+,K^+$-ATPase pump units to be mobilized from a cytoplasmic pool and inserted into the baso-lateral plasma membrane (Yiu et al., 1988).  Recent studies suggest that carbachol also increases the rate of endocytosis at the baso-lateral surface (Lambert et al., 1993).   We have analyzed the internalization and release of a fluorescent fluid phase marker, Lucifer Yellow, in order to learn more about the control of membrane trafficking in acinar cells from rabbit lacrimal glands and to discern the roles of distinct intracellular compartments.  In an attempt to evaluate the extent to which fluid phase marker internalization reflects the retrieval of secretory vesicle membranes, we have also measured protein secretion.

## METHODS

### Preparation of single cells.

Single acinar cells from rabbit lacrimal glands are isolated by a modification of the procedure of Hann et al. (1989).  Briefly, the glands are cut into 1 $mm^3$ fragments, and the fragments are then subjected to two cycles of treatments with collagenase, hyaluronidase, and EDTA.   After centrifugation through a discontinuous Ficoll gradient, the cells thus obtained are washed several times in Hank's medium and finally equilibrated at 37°C for 20 min before use.

### Internalization of fluid phase marker

In a typical experiment, the cells are incubated at 37°C for 20 min in HEPES-buffered Hank's solution, pH 7.6, supplemented with 1 mM $CaCl_2$ and 1.9 mM Lucifer Yellow. The uptake is terminated by a 10-fold dilution with ice-cold buffer, and the cells are pelleted by low-speed centrifugation. The supernatant is collected and analyzed for released protein. Excess dye is removed by 3 washes with 5 ml ice-cold buffer. Lucifer Yellow taken up by the cells is released by treatment with 0.1% Triton X-100. Protein, which otherwise would interfere with the measurement of fluorescence, is precipitated with 6% trichloroacetic acid and pelleted by centrifugation. The supernatant is diluted with buffer and the fluorescence (Ex 428, Em 534) measured directly.

### Efflux of internalized fluid phase marker

Cells are loaded by incubation with 3.8 mM Lucifer Yellow at 37°C for 15 min, then chilled and washed as described above. The rinsed cells are suspended in 37° buffer containing varying concentrations of carbachol, and at appropriate time intervals aliqouts are removed and diluted 10-fold with ice-cold buffer. The cells are pelleted by gentle centrifugation, washed once, and then analyzed for retained Lucifer Yellow fluorescence.

### Calculations

Lucifer Yellow uptake is normalized to total cell-associated protein, which is determined by the method of Lowry et al. (1951) after precipitation with 6% trichloroacetic acid.

## RESULTS

### Stimulation of fluid phase endocytosis and protein secretion by carbachol

Fluid phase endocytosis at 37°, as measured by internalization of Lucifer Yellow, is stimulated by carbachol. The initial rate of uptake is increased by 80% by 10 $\mu$M carbachol, and this difference is maintained during the first 20 min of incubation. A semi-logarithmic analysis of unstimulated uptake indicates a single rate-limiting step with a $t_{0.5}$ of 10 min. However, the carbachol-stimulated uptake can be resolved into two components, one with a $t_{0.5}$ of less than 5 min, and one with a $t_{0.5}$ similar to the unstimulated uptake. Increasing the carbachol concentration to 1 mM causes a 50% decrease in the amount of Lucifer Yellow taken up after 20 min. No significant differences are observed between 10 $\mu$M and 1 mM carbachol during the first 2 min, indicating that increasing the carbachol concentration decreases the total volume of extracellular fluid taken up rather than the rate of the uptake process. If the incubation temperature is lowered to 18°C, a condition that is known to slow endocytosis and to completely block traffic from early to late endosomes (Gruenberg and Howell, 1989), the basal rate of Lucifer Yellow uptake is suppressed. The uptake rate at 18° is accelerated 2.5-fold by 10 $\mu$M and 1 mM carbachol. The time-courses of uptake in the presence of the two carbachol concentrations are identical. Since the supramaximal phenomenon observed at 37° is not seen at 18°, increasing the carbachol concentration from 10 $\mu$M to 1 mM must either decrease the volume of or impair traffic to a late endocytic compartment, i.e., a compartment which is inaccessible at 18°.

Carbachol also stimulates protein secretion, but with a different dose-response relationship than that for Lucifer Yellow uptake. Protein release is increased by 46% at 10 $\mu$M carbachol. Raising the carbachol concentration to 1 mM further increases protein release to 82% over the basal level. In addition, the larger response to the greater carbachol concentration is apparent within 2 min of incubation at 37°.

Both isoproterenol, a $\beta$-adrenergic agonist, and vasoactive intestinal peptide (VIP) stimulate fluid phase endocytosis, resulting in a 32%-35% increases over the basal rate. No supraoptimal phenomenon can be observed with either secretagogue. Only isoproterenol significantly affects protein secretion, resulting in a stimulation of ~35%.

**Release of endocytosed fluid**

Semi-logarithmic analyses of the release of Lucifer Yellow from cells that had been loaded with fluid phase marker in the absence of carbachol suggest the involvement of at least two intracellular compartments. The initial rate of release is stimulated 2.8-fold by carbachol at concentrations of 10 $\mu$M or 1 mM. Approximately 30% of the internalized fluid remains inside the cells after 30 min of release. When cells are loaded in the presence of 10 $\mu$M carbachol, the subsequent time-courses of release in the absence and in the presence of carbachol indicate that a greater fraction of the accumulated Lucifer Yellow reaches a compartment from which it can be released only slowly in the absence of carbachol. VIP at a concentration of 10 nM is as effective as carbachol in permitting release from this compartment. When cells are loaded in the presence of 10 $\mu$M carbachol at 18°, then chilled, washed, and resuspended in 37° medium containing 10 $\mu$M carbachol, all of the internalized Lucifer Yellow is released within 10 min. This observation indicates that endocytosed fluid does not have to be processed through late endocytic compartments before it is released again; thus, there is a considerable recycling traffic between the early endocytic compartment and the plasma membrane. It further indicates that the carbachol-dependent compartment in which Lucifer Yellow can be retained after loading at 37° in the presence of an optimal carbachol concentration lies distal to the 18° block.

**Regulation of endocytosis and secretion**

The mechanism for regulation of uptake into early endosomes can be examined by utilizing the temperature block at 18°C, thus preventing transport from early to late endosomes. Both ionomycin (15 $\mu$M), a $Ca^{2+}$-ionophore, and phorbol 12,13-dibutyrate (0.1 $\mu$M), a protein kinase c activator, stimulate Lucifer Yellow internalization, suggesting that both the diacylglycerol and $Ca^{2+}$/calmodulin pathways triggered by cholinergic receptor activation are involved in regulating fluid phase internalization. Combining ionomycin and the phorbol ester results in an enhancement of the Lucifer Yellow uptake in a manner suggesting additivity. The effects on protein secretion are similar. The endocytic and secretory responses to either ionomycin or the phorbol ester are greater than the responses to 10 $\mu$M carbachol, suggesting that the generation of intracellular messengers may be a limiting factor in the responses to secretagogues.

**CONCLUSIONS**

The results we have obtained indicate that carbachol stimulates both the endocytic internalization of Lucifer Yellow and the release of stored protein. Protein secretion is optimally stimulated by 1 mM carbachol, and the subsequent retrieval of secretory vesicle membranes probably contributes to the measured fluid phase endocytosis. The observation that fluid phase endocytosis is affected differently by increasing the carbachol concentration suggests that the two events are at least partly separate from each other and, therefore, that much of the fluid phase internalization seen in the presence of 10 $\mu$M carbachol must proceed independently of the need to retrieve secretory vesicle membrane constituents. This conclusion is further supported by the observation that carbachol accelerates recycling traffic between the plasma membrane and the early endosome and by our recent observation that labeled surface membrane constituents equilibrate with intracellular pools that are 10- to 20-fold larger than the surface-expressed pool (Lambert et al., 1993). In addition, our results suggest that fluid phase endocytosis involves several different intracellular compartments and that translocations between these compartments are affected differently by carbachol. Uptake into the early endosome is stimulated equally by 10 $\mu$M and 1 mM carbachol, and it is completely and rapidly reversible. In contrast, transfer to the late endosome is optimally stimulated by carbachol at a concentration of 10 $\mu$M and impaired at a concentration of 1 mM. Moreover, uptake into the late endosome is much more slowly reversible, and efflux from the late endosome is stimulated equally well both by 10 $\mu$M and by 1 mM carbachol.

## ACKNOWLEDGEMENTS

This work has been supported by NIH Grant EY 05801, by grants from the Wright Foundation, Fight-for-Sight Prevent Blindness, Inc., and the Sjögren's Syndrome Foundation, Inc.

## REFERENCES

Gruenberg, J., and Howell, K.E., 1989, Membrane traffic in endocytosis: Insights from cell-free assays. *Ann. Rev. Cell Biol.* 5:453.

Hann, L.E., Tatro, J.B., and Sullivan, D.A., 1989, Morphology and function of lacrimal gland acinar cells in primary culture. *Invest. Ophthalmol. Vis. Sci.* 30:145.

Lambert, R.W., Maves, C.A., Gierow, J.P., Wood, R.L., and Mircheff, A.K., 1993, Plasma membrane internalization and recycling in rabbit lacrimal acinar cells. *Invest. Ophthalmol. Vis. Sci.*, in press.

Lowry, O.H., Rosebrough, N.J., Farr, A.L., and Randall, R.J., 1951, Protein measurement with the Folin phenol reagent. *J. Biol. Chem.* 193:265.

Yiu, S.C., Lambert, R.W., Bradley, M.E., Ingham, C.E., Hales, K.L., Wood, R.L., and Mircheff, A.K., 1988, Stimulation-associated redistribution of Na,K-ATPase in rat lacrimal gland. *J. Membrane Biol.* 102:185.

# THE INOSITOL PHOSPHATE-CALCIUM SIGNALLING SYSTEM IN LACRIMAL GLAND CELLS

James W. Putney, Jr. and Gary St. J. Bird

Laboratory of Cellular and Molecular Pharmacology
National Institute of Environmental Health Sciences - NIH
P.O. Box 12233
Research Triangle Park, NC 27709

## INTRODUCTION

In lacrimal gland cells, receptor-activated $Ca^{2+}$ mobilization involves two phases: 1) $Ca^{2+}$ release from an intracellular store, and 2) a more prolonged phase of extracellular $Ca^{2+}$ entry (Parod and Putney, 1978; Putney, 1987). These two phases of the $Ca^{2+}$ response can be dissected by stimulating cells with a phospholipase C-linked agonist in the presence or absence of extracellular $Ca^{2+}$. In the absence of extracellular $Ca^{2+}$, such stimulation results in a transient increase in $[Ca^{2+}]_i$, originating from the release of a finite intracellular $Ca^{2+}$ pool. In the presence of physiological (mM) extracellular $Ca^{2+}$, the elevation in $[Ca^{2+}]_i$ is sustained due to the additional component of extracellular $Ca^{2+}$ influx. Extracellular $Ca^{2+}$ influx is also the source for refilling of the intracellular $Ca^{2+}$ pool upon termination of receptor activation.

## THE SIGNALS FOR CALCIUM RELEASE AND ENTRY

The available evidence now strongly indicates that the initial internal $Ca^{2+}$ release is signalled by $(1,4,5)IP_3$ (Berridge and Irvine, 1984). The mechanism by which $(1,4,5)IP_3$ releases intracellular $Ca^{2+}$ involves its interaction with a specific intracellular membrane receptor and opening of a $Ca^{2+}$ channel on the responsive organelle. This organelle was originally believed to be the endoplasmic reticulum (ER), but more recently it has been suggested to be either a highly specialized ER component or, in fact, a novel organelle termed "calciosome" (Volpe et al.1988). Much is now known about the mechanisms within the cell which mediate this initial, (1,4,5)IP$_3$-induced phase of $Ca^{2+}$ mobilization; in contrast, the regulation of the second, $Ca^{2+}$ entry phase of $Ca^{2+}$ mobilization is poorly understood. There is strong evidence that $Ca^{2+}$ entry is activated as a result of the emptying of a (1,4,5)IP$_3$-sensitive pool (Putney, 1986). This mechanism, which has been termed capacitative $Ca^{2+}$ entry (Putney, 1986), is supported by two lines of experimental evidence (Putney, 1990): 1) an observed transient elevation in plasma membrane $Ca^{2+}$ permeability

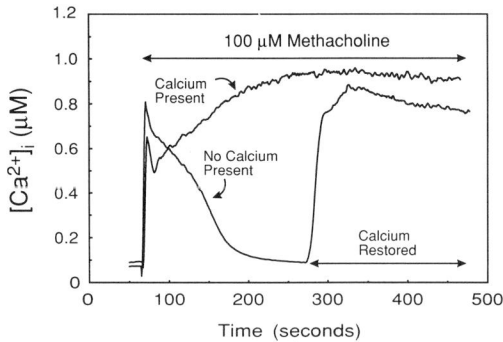

**Figure 1.** Phases of $[Ca^{2+}]_i$ signalling in a single, mouse lacrimal acinar cell. In the trace indicated "Calcium Present", extracellular $Ca^{2+}$ was present throughout. In the trace indicated "No Calcium Present", $Ca^{2+}$ was absent from the extracellular medium, and restored where indicated.

during $Ca^{2+}$ pool refilling under certain conditions, and 2) the activation of $Ca^{2+}$ influx by agents such as thapsigargin, which deplete the intracellular $Ca^{2+}$ pool by a mechanism independent of (1,4,5)IP$_3$. Thapsigargin inhibits certain microsomal $Ca^{2+}$-ATPases including the $Ca^{2+}$ pump responsible for sequestration of $Ca^{2+}$ into the (1,4,5)IP$_3$-sensitive pool. Thapsigargin, presumably solely by virtue of its ability to deplete intracellular pools of $Ca^{2+}$, quantitatively mimics the ability of surface membrane agonists to activate $Ca^{2+}$ entry. The activation of $Ca^{2+}$ entry by thapsigargin has been observed in a wide variety of cell types (Thastrup, 1990), including lacrimal gland cells (Kwan et al.1990; Bird et al.1992b). This action of thapsigargin provides strong support for the capacitative $Ca^{2+}$ entry hypothesis.

## THE ROLE OF (1,3,4,5)IP$_4$

It has been suggested that, together with (1,4,5)IP$_3$, the phosphorylated (1,4,5)IP$_3$ metabolite, (1,3,4,5)IP$_4$, modulates Ca$^{2+}$ mobilization and/or may play a role in the regulation of Ca$^{2+}$ influx (Morris et al.1987). Data in support of this role for (1,3,4,5)IP$_4$ has been obtained from experiments using sea urchin eggs as well as mouse lacrimal cells (Morris et al.1987). In this latter system, it was found that (1,4,5)IP$_3$, when dialyzed into cells through patch pipettes, was capable of inducing a transient mobilization of Ca$^{2+}$, but the response could be sustained only if (1,3,4,5)IP$_4$ was also added. On the other hand, in this same cell type, thapsigargin induces Ca$^{2+}$ influx without elevating inositol phosphates (Kwan et al.1990). Furthermore, Bird *et al.* (Bird et al.1992b) reported that direct injection into these cells of the non-phosphorylatable (1,4,5)IP$_3$ analogue, (2,4,5)IP$_3$, was capable of fully activating both Ca$^{2+}$ entry as well as intracellular Ca$^{2+}$ release. Thus, an obligatory role for (1,3,4,5)IP$_4$ in Ca$^{2+}$ signalling must be questioned until an explanation for these apparent paradoxes is obtained.

## CALCIUM POOLS IN LACRIMAL CELLS

As discussed above, stimulation of lacrimal gland cells with a phospholipase C-linked agonist in the absence of extracellular Ca$^{2+}$ causes release of Ca$^{2+}$ from intracellular stores (Kwan et al.1990; Bird et al.1992a). In studies on single mouse lacrimal acinar cells, following a supramaximal stimulation with methacholine, no additional Ca$^{2+}$ was mobilized by subsequent application of the intracellular Ca$^{2+}$-ATPase inhibitor, thapsigargin, the stable inositol 1,4,5-trisphosphate ((1,4,5)IP$_3$) analog, inositol 2,4,5-trisphosphate ((2,4,5)IP$_3$) (by microinjection), or the Ca$^{2+}$ ionophore, ionomycin (Bird et al.1992a). This indicates that in normal, naive lacrimal cells, a single pool of intracellular Ca$^{2+}$ exists, and this pool is uniformly sensitive to agonists and the Ca$^{2+}$ mobilizing signal, (1,4,5)IP$_3$. However, in this same study (Bird et al.1992a), following prolonged activation of cells by MeCh in the presence of extracellular Ca$^{2+}$, Ca$^{2+}$ was accumulated into a pool which was released by ionomycin but not by thapsigargin. Ca$^{2+}$ accumulation into this latter pool was blocked by prior microinjection of ruthenium red, and therefore is presumed to be the mitochondria. In saponin-permeabilized mouse lacrimal cells, two distinct Ca$^{2+}$-sequestering pools were detected; (i) a ruthenium red-sensitive, thapsigargin-insensitive pool, presumed to be the mitochondria, and (ii) a ruthenium red-insensitive, thapsigargin-sensitive pool. Calcium accumulated into both pools at high (1-10 µM) [Ca$^{2+}$], but only into the thapsigargin-sensitive pool at a [Ca$^{2+}$] similar to that in unstimulated cells (150 nM). The thapsigargin-sensitive Ca$^{2+}$ pool was sensitive to (1,4,5)IP$_3$; however, in contrast to findings in intact cells, only 44 % of this Ca$^{2+}$ pool was released by maximal concentrations of (1,4,5)IP$_3$ or (2,4,5)IP$_3$. Thus, permeabilization of the cells with saponin apparently leads to a fragmentation of the non-mitochondrial pool, resulting in two pools, one sensitive and one insensitive to (1,4,5)IP$_3$ (Bird et al.1992a).

# CALCIUM OSCILLATIONS IN LACRIMAL CELLS

Stimulation of mouse lacrimal acinar cells with submaximal concentrations of methacholine results in a increase in intracellular calcium ($[Ca^{2+}]_i$) which generally takes the form of regular sinusoidal oscillations (Bird et al.1993). These oscillations generally occur on an elevated baseline of $[Ca^{2+}]_i$, they tend to diminish with time, and the frequency of the oscillations is relatively constant (ca. 4-5/min) regardless of the concentration of methacholine. This constancy of frequency suggests that the oscillations may arise from an oscillating negative feed back mechanism somewhere in the signal transduction pathway. This negative feed back mechanism appears to involve oscillations in protein kinase C activity because the oscillations were prevented by either pharmacological activation or pharmacological inhibition of protein C, or by down regulation of protein kinase C. Activation of protein kinase C with phorbol ester drugs inhibited the methacholine-induced $[Ca^{2+}]_i$ signal, as well as the rise in the $Ca^{2+}$ mobilizing messenger, inositol 1,4,5-trisphosphate. $[Ca^{2+}]_i$ signals elicited by intracellular introduction of inositol phosphates did not oscillate, and these signals were not affected by pharmacological activators or inhibitors of protein kinase C. Thus, the constant frequency oscillations in $[Ca^{2+}]_i$ seen in lacrimal cells arise as a result of a negative feed back loop involving inhibition of $[Ca^{2+}]_i$ signalling by the diacylglycerol - protein kinase C limb of the phospholipase C pathway. This inhibition appears to occur at, or proximal to the activation of phospholipase C (Bird et al.1993).

## SUMMARY

From the above discussion, it is clear that the regulation of $Ca^{2+}$ signalling in exocrine cells is a complex process involving activation of both intracellular $Ca^{2+}$ release as well as the entry of $Ca^{2+}$ across the plasma membrane. A poorly understood mechanism links these two phases of $Ca^{2+}$ signalling thereby providing both rapid as well as sustained signals for the initiation and maintenance of appropriate exocrine responses. Further work is needed to better understand the mechanisms controlling this important and ubiquitous signalling system.

## REFERENCES

Berridge M.J. and Irvine R.F., 1984, Inositol trisphosphate, a novel second messenger in cellular signal transduction. *Nature* 312:315.

Bird G.St.J., Obie J.F., and Putney J.W.,Jr., 1992a, Functional homogeneity of the non-mitochondrial $Ca^{2+}$-pool in intact mouse lacrimal acinar cells. *J. Biol. Chem.* 267:18382.

Bird G.St.J., Rossier M.F., Hughes A.R., Shears S.B., Armstrong D.L., and Putney J.W.,Jr., 1992b, Activation of $Ca^{2+}$ entry into acinar cells by a non-phosphorylatable inositol trisphosphate. *Nature* 352:162.

Bird G.St.J., Rossier M.F., Obie J.F., and Putney J.W.,Jr., 1993, Sinusoidal oscillations in intracellular calcium due to negative feedback by protein kinase C. *J. Biol. Chem.* submitted

Kwan C.Y., Takemura H., Obie J.F., Thastrup O., and Putney J.W.,Jr., 1990, Effects of methacholine, thapsigargin and $La^{3+}$ on plasmalemmal and intracellular $Ca^{2+}$ transport in lacrimal acinar cells. *Am. J. Physiol.* 258:C1006.

Morris A.P., Gallacher D.V., Irvine R.F., and Petersen O.H., 1987, Synergism of inositol trisphosphate and tetrakisphosphate in activating $Ca^{2+}$-dependent $K^+$ channels. *Nature* 330:653.

Parod R.J. and Putney J.W.,Jr., 1978, The role of calcium in the receptor mediated control of potassium permeability in the rat lacrimal gland. *J. Physiol. (Lond. )* 281:371.

Putney J.W.,Jr., 1986, A model for receptor-regulated calcium entry. *Cell Calcium* 7:1.

Putney J.W.,Jr., 1987, Calcium-mobilizing receptors. *Trends Pharmacol. Sci.* 8:481.

Putney J.W.,Jr., 1990, Capacitative calcium entry revisited. *Cell Calcium* 11:611.

Thastrup O., 1990, Role of $Ca^{2+}$-ATPases in regulation of cellular $Ca^{2+}$ signalling, as studied with the selective microsomal $Ca^{2+}$-ATPase inhibitor, thapsigargin. *Agents and Actions* 29:8.

Volpe P., Krause K.-H., Hashimoto S., Zorzato F., Pozzan T., Meldolesi J., and Lew D.P., 1988, "Calcioisome," a cytoplasmic organelle: The inositol 1,4,5-trisphosphate-sensitive $Ca^{2+}$ store of nonmuscle cells? *Proc. Nat. Acad. Sci. USA* 85:1091.

# CHARACTERIZATION OF RAT LACRIMAL GLAND PROTEIN KINASE C: EFFECTS OF PHORBOL ESTERS AND PKC INHIBITORS ON HISTONE KINASE ACTIVITY AND LABELLED PROTEIN DISCHARGE

Driss Zoukhri[1], Philippe Mauduit and Bernard Rossignol

Biochimie des Transports Cellulaires, CNRS URA 1116, Bat. 432
Universite Paris XI, 91405 Orsay Cedex, France

## INTRODUCTION

In the rat lacrimal gland, cholinergic agonists induce a large protein secretion through muscarinic receptors. The activation of these receptors stimulates phospholipase C that hydrolyses phosphatidylinositol 4,5-bisphosphate producing inositol 1,4,5-trisphosphate ($IP_3$) and diacylglycerol (DAG). The roles of $IP_3$ and DAG as second messengers are now well documented. $IP_3$ releases calcium from a non mitochondrial intracellular store which increases the cytosolic free calcium concentration. The released calcium may stimulate secretion directly, or may act with calmodulin to activate specific kinases that phosphorylate specific proteins to induce secretion. DAG activates protein kinase C (PKC), a calcium and phospholipid-dependent protein kinase, which also may phosphorylates specific proteins to evoke secretion.

PKC, first described by Takai et al. (1979), is in fact a family of multiple subspecies sharing closely related structures. The PKC family can be grouped into two categories, based on the calcium requirement for activation. The first group contains PKC-$\alpha$, -$\beta$I, -$\beta$II, and -$\gamma$ isozymes which are calcium and phospholipid-dependent kinase. The second group contains PKC-$\epsilon$, -$\delta$, -$\zeta$ and -$\eta$ (L) isoforms which are calcium-independent kinases. It is postulated that different PKC isoforms could be involved in various biological processes (Jaken, 1990). Castagna et al. (1982) showed that phorbol esters can substitute for DAG to activate PKC. Thus, phorbol esters, mainly 4$\beta$-phorbol 12-myristate 13-acetate (PMA), have been widely used to assess the role of PKC in the regulation of many biological processes.

The role of calcium in regulating lacrimal gland protein secretion has been well documented (Herman et al., 1978; Dartt et al., 1982; Mauduit et al., 1984 and Dartt et al., 1988a) but the role of PKC is less clear. Mauduit et al., (1987) showed that PMA evokes the discharge of newly synthesized proteins. Dartt et al. (1988b) showed that another phorbol ester, 4$\beta$-phorbol 12,13-dibutyrate (PdBu) and a DAG analog, OAG (1-oleoyl 2-acetyl glycerol) induced peroxidase secretion in this gland. However, characterization of the lacrimal gland PKC and a comparative study of the effects of phorbol esters and PKC inhibitors on histone kinase activity and labelled protein secretion needed.

In this study we used histone III-S to characterize PKC from the lacrimal gland cytosolic fraction. We tested the effects of several PKC activators and inhibitors on histone kinase activity. We also determined the effects of phorbol esters on labelled protein discharge from lacrimal gland lobules and compared the effects of PKC inhibitors on protein secretion stimultaed by phorbol esters and cholinergic agonists. Our results

---

[1] Present address: Schepens Eye Research Institute, 20 Staniford street., Boston, MA 02114.

demonstrate that the rat lacrimal gland expresses three isoforms of PKC and that PKC might not be the sole effector of the phorbol ester-induced exocytosis. We hypothesize that another mechanism might be involved, i.e., a phospholipase D activity.

## MATERIALS AND METHODS

### Materials

Phorbol esters, histone III-S, phosphatidylserine, 1,2-diolein, H7, arachidonic acid, carbamylcholine chloride (carbachol), staurosporine and sphingosine were purchased from Sigma, St. Louis, USA. Chelerythrine hydrochloride was from Extrasynthese, France. Trifluoperazine dichlorhydrate (TFP) was generously given by Rhone-Poulenc. [$\gamma^{32}$P]-ATP (3000 Ci/mmol) was purchased from Dupont de Nemours, NEN products, France. [$^3$H] L-leucine was from CEA, Saclay, France.

### Methods

Separation of PKC Isoforms. Male Sprague Dawley rats 5-7 weeks old were used in this study. Rat lacrimal gland cytosolic fraction was prepared as previously described (Mauduit et al., 1989). For purification of PKC activity, the cytosolic fraction was loaded onto a column (0.9 x 18 cm) packed with DE52 resin (DEAE-cellulose, Whatman). Column washing, protein elution from the column, and PKC activity assay were performed as previously described (Mauduit et al., 1989). Active fractions from the DE52 column were further analysed in order to separate PKC isoforms by chromatography on Phenyl-Sepharose CL-4B (0.9 x 18 cm, Pharmacia) and hydroxyapatite (0.76 x 10 cm, Mitsui Toatsu Chemicals Inc., Japan) using an HPLC system (Waters 501) as previously described (Zoukhri et al., 1992).

Electrophoresis and Immunoblot Analysis. Partially purified PKC from the DE52 column or purified PKC from the hydroxyapatite column were subjected to SDS-PAGE (10 %) according to the method of Laemmli. Proteins on the gels were electrophoretically transferred onto nitrocellulose filters (Schleicher & Schuell, 0.45 μm). Immunobloting was carried out with specific antibodies to PKC isoforms (Gibco BRL) as previously described (Zoukhri et al., 1992).

Measurement of Protein Secretion. Incubation procedures for pulse labelling of secreted proteins and measurement of protein discharge were performed as previously described (Mauduit et al., 1984 and Zoukhri et al., 1993). Briefly, lacrimal gland lobules from 10 rats were pulse labelled for 10 min in 2 x 5 ml Krebs Ringer bicarbonate (KRB) buffer in the presence of [$^3$H]-leucine (50 μCi, 1.1 MBq). Glands fragments were washed three times with KRB buffer containing 1 mM [$^1$H]-leucine and incubated for 60 min in the same buffer. Fragments were then washed and separated into 50-60 mg portions. For experiments in which a fixed time period was used, fragments were incubated for 40 or 50 min in 3 ml of KRB buffer. For kinetic experiments, fragments were incubated for 100 min in 12 ml KRB buffer. To evaluate discharge of [$^3$H]-proteins, fragments were stimulated or not in the presence or absence of calcium (± 0.5 mM EGTA).

## RESULTS AND DISCUSSION

In the first series of experiments, we partially purified the lacrimal gland cytosolic fraction by anion exchange chromatography on a DE52 column, because no PKC activity could be detected directly in this cytosolic fraction. Using histone as substrate, we identified two peaks of protein kinase activity eluting from this column at salt concentrations of 80 mM and 220 mM NaCl, respectively. The peaks corresponded to those reported in the literature for PKC and PKM, the form of PKC constitutively activated by calpain proteolysis. Moreover, the PKC kinase activity showed cofactor dependency for its activation that matched that described in the literature (calcium+phosphatidylserine+diolein).

To characterize further this calcium and phospholipid-stimulated histone kinase activity, we studied its dependency on histone, ATP and magnesium. The apparent Km for ATP and histone were 8.7 μM and 2.4 μM, respectively. These values were in good agreement with those obtained with PKC preparations from other tissues. No histone kinase activity could be detected below a threshold concentration of 0.3 mM magnesium, and 5-10 mM concentration gave the maximal activity. These results indicate that the calcium and phospholipid-stimulated protein kinase activity purified from the rat lacrimal gland cytosolic fraction might be PKC activity.

In the course of our study, we tested the effects of phorbol esters on PKC partially purified from the lacrimal gland. Only the active phorbol esters PMA and PdBu stimulated this histone kinase activity. In the presence of phospholipids $EC_{50}$ for PMA and PdBu were $2.10^{-9}$ M and $2.10^{-7}$ M, respectively (Table I). We also tested the effects of an unsaturated fatty acid, arachidonic acid (AA), and showed that free AA stimulated this kinase activity with an $EC_{50}$ of 12 μM (Table I). Moreover, AA and PMA, but not PdBu, stimulated, in a dose dependent manner and in the absence of calcium and phospholipids (PKC cofactors) a histone kinase activity that coeluted with PKC from the DE52 column.

Table I. Lacrimal gland cytosolic fraction was partialy purified by chromatography upon DE52 column as described in Materials and Methods. Protein kinase activity was measured using histone III-S and $[\gamma^{32}P]$-ATP as a phosphate acceptor and a phosphate donor, respectively.

|  | PMA | PdBu | AA |
|---|---|---|---|
| $EC_{50}$ | $2\times10^{-9}$ M | $2\times10^{-7}$ M | $1.2\times10^{-7}$ M |

In order to test the specificity of action of PMA and AA, i.e., is it PKC that is activated by these agents in the phospholipid-free system or a protein kinase that copurify with PKC, we conducted the following experiments. First, we tested the effects of two reported PKC inhibitors, staurosporine and sphingosine. Staurosporine, one of the most potent inhibitors of PKC, inhibited with the same potency the PKC activity stimulated by calcium+phosphatidylserine+diolein, PMA, PdBu and AA. The apparent $IC_{50}$ was 2 nM for these activities and is the same as the one reported in the literature for PKC from other tissues. The same results were obtained with sphingosine, another PKC inhibitor with an $IC_{50}$ of 12 μM. Second, we further purified lacrimal gland PKC by chromatography on Phenyl-Sepharose. Even after this second step of purification, the phospholipid-independent and PMA/AA-stimulated protein kinase activity still copurifies with PKC activity. These results indicate that it is PKC, rather than another protein kinase that might copurify with PKC, which is stimulated by PMA and AA in the phospholipid-free system.

To our knowledge, only the γ isoform of PKC has been shown to be activated by low concentrations of free AA in the absence of calcium and phospholipids. Thus, we introduced a third step of purification using a hydroxyapatite column which allows the separation of the different isoforms of PKC. With this column, only one peak of PKC activity could be detected at an elution position corresponding to the α isoform of rat brain PKC. This result was further confirmed by immunoblot analysis using a specific antibody raised against rat brain PKC-α. However, according to a report from Dartt et al. (1990) who identified PKCε in rat lacrimal gland, we conducted immunoblot analysis on partialy purified PKC (DE52 column) using antibodies directed against ε, δ, η and ζ isoforms of PKC. We showed that, indeed, lacrimal gland expresses two other isoforms, PKC-ε and PKC-δ in addition to PKC-α (Zoukhri et al., 1993).

In another series of experiments, we studied the effects of phorbol esters. and PKC inhibitors on labelled protein secretion. Both PMA and PdBu stimulated the discharge of $[^3H]$-proteins in a dose dependent manner. However, PdBu seemed to be more efficient than PMA as the maximal labelled protein discharge, measured after a 40 min incubation period, obtained with PdBu was 13 % compared to 6 % with PMA. In fact, PMA, in contrast to PdBu, showed a 30 min latency before protein secretion reached its maximal rate, similar to that obtained with PdBu. We also studied the dependency of the phorbol ester-induced secretion on calcium and showed that the omission of this cation from the incubation medium resulted in a 50 % decrease of the phorbol ester response.

We then tested the effects of PKC inhibitors on the cholinergic- and the phorbol ester-stimulated [$^3$H]-protein secretion to study the role of PKC in this process. With three PKC inhibitors, sphingosine ($2.10^{-4}$ M), chelerythrine ($3.10^{-5}$ M) and H7 ($10^{-4}$ M), no inhibition of exocytosis stimulated by phorbol esters or carbachol was obtained (Table II). Only staurosporine, up to 1 µM inhibited, but not completely, both the cholinergic- and the phorbol ester-stimulated protein secretion. However, the $IC_{50}$ obtained (Table II) are more than 100 fold greater than the one obtained with purified lacrimal gland PKC (2 nM). It is worth nothing that with these concentrations, staurosporine is no longer specific for PKC.

A report from Dartt et al. (1982) showed that calcium and calmodulin-dependent protein kinase activity is present in the lacrimal gland. Moreover, a previous report from our laboratory (Mauduit et al., 1983) showed that trifluoperazine (TFP) completely inhibited the cholinergic-stimulated exocytosis suggesting a role for the calcium and calmodulin-dependent kinases in this process. We then tested the effects of TFP on the phorbol ester-induced secretion. TFP completely inhibited PMA-induced protein secretion and almost completely that induced by PdBu with observed $IC_{50}$ of 20 µM and 60 µM, respectively (Table II)

**Table II.** Effects of PKC inhibitors on phorbol esters- and cholinergic-stimulated labelled protein secretion. Rat lacrimal gland fragments were prepared as described in Materials and Methods and protein secretion was measured for a period of 40 min. A 10 min incubation period with the inhibitors preceeded the addition of the agonists.

| | PMA (1 µM) | PdBu (1 µM) | Carbachol (1 µM) |
|---|---|---|---|
| chelerythrine | N.I* ($3.10^{-5}$ M) | N.I ($3.10^{-5}$ M) | N.I ($3.10^{-5}$ M) |
| sphingosine | N.I ($2.10^{-4}$ M) | N.I ($2.10^{-4}$ M) | N.T☆ |
| H7 | N.I ($10^{-4}$ M) | N.I ($10^{-4}$ M) | N.I ($10^{-4}$ M) |
| Staurosporine ($IC_{50}$) | 650 nM | 200 nM | 800 nM |
| TFP ($IC_{50}$) | 20 µM | 60 µM | 30 µM |

N.I*. No Inhibition up to the concentration indicated between brackets, N.T☆. Not Tested.

In summary, our results show that three isoforms of PKC are expressed in the rat lacrimal gland. Partially purified PKC can be activated by either calcium + phospholipid + Diolein, active phorbol esters or free arachidonic acid. Reported PKC inhibitors are potent inhibitors of these activities. Lack of effect of any of the PKC inhibitors on the cholinergic- and the phorbol ester-stimulated protein secretion suggests that this kinase might not be the sole effector of the phorbol ester-induced exocytosis and might not be implicated in that stimulated by cholinergic agonists. We hypothesize that another mechanism, perhaps involving a phospholipase D activity, might be implicated in the mechanisms regulating protein secretion in rat lacrimal gland.

# ACKNOWLEDGEMENT

Thanks are due to Jocelyne Dujancourt for her skillful technical assistance. This work was supported by the Centre National de la Recherche Scientifique.

# REFERENCES

Castagna, M., Takai, Y., Kaibuchi, K., Sano, K., Kikkawa, U. and Nishizuka, Y., 1982, Direct activation of calcium-activated, phospholipid-dependent protein kinase by tumour-promoting phorbol esters, *J. Biol. Chem.* 257:7847.

Dartt, D.A., Guerina, V.J., Donowitz, M., Taylor, L. and Sharp, G.W.G., 1982, Calcium and calmodulin-dependent protein phosphorylation in rat lacrimal gland, *Biochem. J.*, 202:799.

Dartt, D.A., Baker, L.V., Rose, P.E., Murphy, S.E., Ronco, L.V. and Unser, M.F., 1988a, Role of cyclic AMP and calcium in potentiation of rat lacrimal gland protein secretion, *Invest. Ophtalmo.* 29:1732.

Dartt, D.A., Ronco, L.V., Murphy, S.A. and Unser, M.F., 1988b, Effect of phorbol esters on rat lacrimal gland protein secretion, *Invest. Ophtalmo.* 29:1726.

Dartt, D.A., Hodges, R.R. and Dicker, D.M., 1990, Translocation of protein kinase C-$\varepsilon$ in the rat lacrimal gland, *J. Cell. Biol.* 111:85a.

Herman, G., Busson, S., Ovtracht, L., Maurs, C. and Rossignol, B., 1978, Regulation of protein discharge in two exocrine glands : rat parotid and exorbital lacrimal glands, *Biol. Cell.* 31:255.

Jaken, S., 1990, Protein kinase C and tumour promoters, *Cur. Op. Cell. Biol.* 2:192.

Mauduit, P., Herman, G. and Rossignol, B., 1983, Effect of trifluoperazine on $^3$H-labeled protein secretion induced by pentoxyfilline, cholinergic or adrenergic agonists in rat lacrimal gland . a possible role of calmodulin ?, *FEBS Lett.* 152:207.

Mauduit, P., Herman, G. and Rossignol, B., 1984, Protein secretion induced by isoproterenol or pentoxyfilline in lacrimal gland : calcium effects, *Am. J. Physiol.* 264:C37.

Mauduit, P., Herman, G. and Rossignol, B., 1987, Newly synthesized protein secretion in rat lacrimal gland : post-second messenger synergism, *Am. J. Physiol.* 253:C514.

Mauduit, P., Zoukhri, D. and Rossignol, B., 1989, Direct activation of a protein kinase activity from rat lacrimal gland by PMA in a phospholipid-free system, *FEBS Lett.* 252:5.

Takai, Y., Kishimoto, U., mori, T. and Nishizuka, Y., 1979, Unsatturated diacylglycerol as a possible messenger for the activation of calcium-activated, phospholipid-dependent protein kinase system, *Biochem. Biophys. Res. Commun.* 91:1218.

Zoukhri, D., Pelosin, J.M., Mauduit, P., Chambaz, E., Sergheraert, C. and Rossignol, B., 1992, The rat lacrimal gland expresses the $\alpha$ isoform of PKC : further evidence for the PMA-activated and phospholipid-independent protein kinase activity, *Cell. Signal.* 4:111.

Zoukhri, D., Sergheraert, C. and Rossignol, B., 1993, Phorbol ester-stimulated exocytosis in lacrimal gland : PKC might not be the sole effector, *Am. J. Physiol.* In press.

# INHIBITION OF LACRIMAL FUNCTION BY SELECTIVE OPIATE AGONISTS

Michele M. Cripps and D. Jean Bennett

Department of Physiology
Louisiana State University Medical Center
New Orleans, LA 70119

## INTRODUCTION

The purpose of the work presented here is two-fold: 1) to define the role of opioid neuropeptides in the control of lacrimal secretory function and 2) to identify the cellular mechanisms involved in this regulatory process. The opioid peptides met-enkephalin and leu-enkephalin are two of several neuropeptides that have been identified in fibers that terminate near the basal surface of the acini in lacrimal glands of humans (Frey et al., 1986), guinea pigs (Lehtosalo et al., 1989) and rats (Walcott, 1990). In some cases these neuropeptides have been demonstrated in parasympathetic fibers innervating the gland, thus they are colocalized with acetylcholine and VIP. This colocalization and the close association of the peptidergic fibers with the secretory structures of the gland suggested that the enkephalins might be important neuromodulators of lacrimal secretion

Indeed, the met-enkephalin analogue, d-ala-$^2$-methionine enkephalinamide (DALA) was found to inhibit cholinergic and VIPergic stimulation of protein secretion by in vitro lacrimal gland preparations. This was the result of a direct postsynaptic effect on the acinar cells (Cripps and Patchen-Moor, 1989). The enkephalin analogue also inhibited basal adenylyl cyclase activity as well as receptor-mediated activation of adenylyl cyclase by VIP and non-receptor mediated activation by forskolin (Cripps and Bennett, 1990). This effect of the opioid indicated homeostatic modulation of secretion that involves inhibitory control of exocytosis via receptor activated alterations in adenylyl cyclase activity and intracellular cAMP.

In previous work, we demonstrated the presence of all four derivatives of pro-enkephalin A in lacrimal gland extracts by specific radioimmunoassay and the ability of these endogenous peptides to inhibit adenlyl cyclase (Cripps and Bennett, 1992a). The potency and efficacy of the endogenous opiates were identical to that of DALA. The specificity of the inhibition of adenylyl cyclase activity was tested by the addition of increasing doses of a specific δ-receptor antagonist. Reversal of inhibition by met- and leu-enkephalin as well as by the heptapeptide met-enk arg-phe suggested that in lacrimal gland the opiates activate the $\delta$-receptor subtype. However, reversal of inhibition of adenylyl cyclase by the heptapeptide was not complete. Because the extended met-enkephalins may not distinguish between opioid receptor sites (Paterson et al., 1983), this indicated a possible activation of receptors other than the $\delta$ subtype.

In the present study, therefore, the receptor subtypes were functionally characterized by the use of endogenous opiates not only for $\delta$, but also for $\mu$ and $\kappa$ opiate receptors. In addition, selective ligands were also tested for their ability to inhibit adenylyl cyclase in lacrimal membrane preparations.

## MATERIALS AND METHODS

### Animals

Male Sprague-Dawley rats (250-300 g) were obtained from Charles Rivers, Wilmington, MA. The animals were housed in a controlled environment, exposed to a 12 hour light-dark cycle and provided with standard laboratory chow and water ad libitum for at least 1 wk before they were used. Animals were killed by intraperitoneal injection of pentobarbital sodium (200 mg/kg body weight) followed by an intracardiac bolus of the same drug (100 mg/kg body weight).

### Membrane Isolation

Lacrimal glands from two animals were removed immediately into ice-cold medium containing 5% sorbitol, histidine-imidazole buffer (pH 7.5), 9 $\mu$g/ml aprotinin, 3 mM dithiothreitol (DTT) 0.5 mM EDTA and 0.2 mM phenylmethylsulfonyl fluoride (PMSF) according to the membrane isolation procedure by Mircheff et al. (1983) for characterization of lacrimal acinar membranes. The capsule and main secretory duct were excised and the glands were sliced into 3-5 mm$^3$ fragments. The fragments were washed 3X and then homogenized in 5 ml medium with a Tekmar Tissumiser (Cincinnati, OH) at low speed for 15 min. The supernatants from two 1000 x g, 10 min centrifugations were combined and centrifuged for 15 min with a final suspension in 2.75 ml medium. The membrane preparation was aliquoted, frozen in liquid nitrogen and stored at -70°C. Membrane protein was determined by the method of Lowry et al. (1951).

### Adenylyl Cyclase Determination

Membranes were quick-thawed at 37°C and diluted with 40 mM Tris, pH 7.5. Enzyme activity was determined in polypropylene tubes in a total volume of 100 $\mu$l of 40 mM Tris (pH 7.5), 4 mM MgCl$_2$, 0.25 mM ATP, 1 mM DTT, 1% BSA, 0.1 mM GTP, 0.1 mM EDTA, 0.1 mM IBMX and 20 $\mu$g membrane protein. Forskolin was prepared as a 20 mM stock in 95% ethanol. Opioid agonists were reconstituted in 1% BSA. D-ala$^2$-methionine enkephalinamide (DALA), [D-Pen$^{2,5}$]-enkephalin (DPDPE), and [D-ala$^2$-N-Me-Phe$^4$ Gly$^5$-

ol]-enkephalin (DAMGO) were obtained from Sigma (St. Louis, MO). β-endorphin and dynorphin A were obtained from Peninsula (Belmont, CA). U50,488 was obtained from Research Biochemicals, Inc. (Natick, MA). Adenylyl cyclase activity was determined in triplicate tubes at 37°C for 10 min and the reaction was terminated in a boiling bath for 2 min. cAMP was measured by the protein binding assay method of Brown et al. (1971). Assay tubes contained 100 μl sample or standard, 10 μl $^3$H-cAMP (20,000 cpm/tube) and 25 μl protein kinase diluted in 50 mM Tris (pH 7.5) and 4 mM EDTA to give approximately 30% binding. The tubes were incubated at 4°C for 60 min. After this, 500 μl of ice cold 20 mM $KH_2PO_4/K_2HPO_4$ (pH 6.0), 0.4% BSA and 0.4% activated charcoal were added to each tube. The tubes were centrifuged at 4°C for 15 min at 3000 rpm and the supernatant containing the bound fraction was counted by liquid scintillation. Adenylyl cyclase specific activities were calculated as pmol/mg membrane protein. Statistical significance was determined by Student's t-tests.

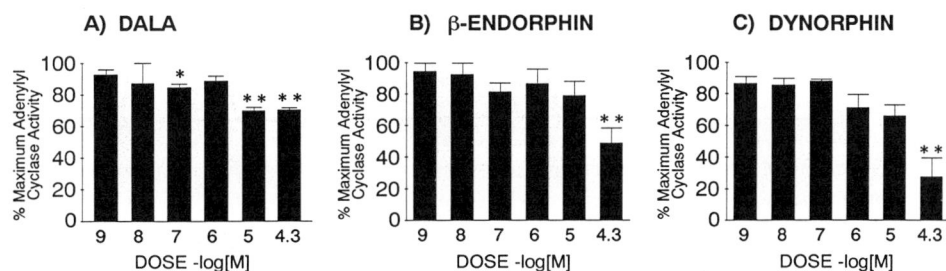

Figure 1. Effect of increasing concentrations of endogenous opioids on forskolin-stimulated adenylyl cyclase activity. Membranes were incubated for 10 min in the presence of 40 μM forskolin with the indicated concentrations of DALA (A), β-endorphin (B), or dynorphin A (C). Values are the mean ± SE of the percent stimulated adenylyl cyclase activity. Statistically significant difference of *p < .05 or **p < .01.

## RESULTS

To determine the opiate receptor subtypes that may be activated in lacrimal gland, endogenous ligands that bind preferentially but not exclusively to μ, δ, or κ receptors were tested for their ability to inhibit forskolin stimulated adenylyl cyclase activity in lacrimal membrane preparations. The endogenous opioids tested were β-endorphin (μ-selective) and dynorphin (κ-selective) in addition to the met-enkephalin (δ-selective) analog DALA. The receptor-selective agonists used were [D-Pen[2,5]]-enkephalin (DPDPE), a delta receptor agonist, [D-Ala[2], N Me-phe[4], Gly[5]-ol]-enkephalin (DAMGO), a ligand for the mu binding site and U50,488, a kappa opiate receptor agonist.

Adenylyl cyclase activity was measured in the presence of forskolin or forskolin and increasing doses of the opioids from $10^{-9}$ to $5 \times 10^{-5}$ M. Forskolin-stimulated activity in the absence of the opiates was expressed as 100% maximum adenylyl cyclase activity with the effect of the opioids expressed as a percent of the forskolin-stimulated values.

The met-enkephalin analogue DALA inhibited forskolin stimulated activity in a dose-dependent manner (Fig. 1). Significant inhibition to approximately 70% of the maximum forskolin- induced cAMP production occurred at doses of $10^{-5}$ M and $5 \times 10^{-5}$ M. $\beta$-endorphin and dynorphin A caused significant inhibition that was in both cases greater than the inhibitory effect of DALA. At a dose of $10^{-5}$ M $\beta$-endorphin or dynorphin A, forskolin-stimulated activity was reduced to 78% and 66%, respectively. At a dose fivefold higher, $\beta$-endorphin inhibited adenylyl cyclase activity to a level of 48% of the maximum and dynorphin reduced cAMP production to approximately 30%.

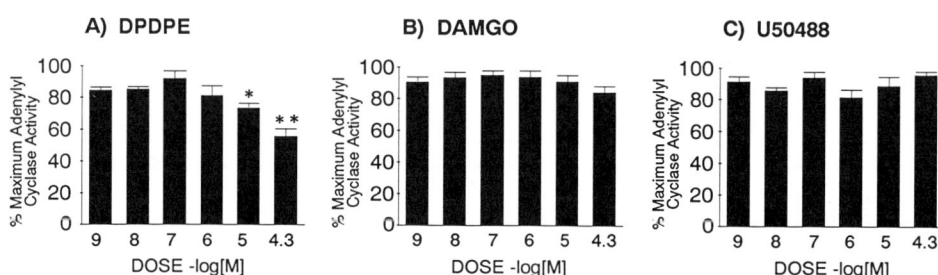

**Figure 2.** Effect of increasing concentrations of selective ligands on forskolin-stimulated adenylyl cyclase activity. Membranes were incubated for 10 min in the presence of 40 µM forskolin with the indicated concentrations of DPDPE (A), DAMGO (B), or U50,488 (C). Values are the mean $\pm$ SE of the percent stimulated adenylyl cyclase activity. Statistically significant difference of *p < .05 or **p < .01.

The addition of the delta selective receptor agonist DPDPE also significantly inhibited forskolin-stimulated activity in a dose dependent manner (Fig. 2). The maximum effect was a reduction to 56% of stimulated activity at a dose of $5 \times 10^{-5}$ M DPDPE. Thus, the specific $\delta$ agonist was more effective at blocking adenylyl cyclase activation than DALA. In contrast to the effect of $\beta$-endorphin and dynorphin A, however, the mu and kappa selective receptor agonists had no effect on forskolin-stimulated adenylyl cyclase activity in lacrimal membranes at any dose tested.

## DISCUSSION

The results we have included here support the hypothesis that cAMP-dependent lacrimal function is dually regulated with both stimulation and inhibition of adenylyl cyclase activity. This dual regulation is achieved by receptor activation coupled to adenylyl cyclase activity. Several receptors linked to adenylyl cyclase and cAMP production are activated by peptides present in the gland. These include VIP, the enkephalins and NPY. VIP causes increases in protein secretion and in intracellular cAMP (Dartt et al., 1984). The enkephalins also inhibit protein secretion and inhibit enzyme activity (Cripps and Bennett 1989, 1992a).

NPY appears to have a stimulatory effect on secretion as shown at this meeting (Williams et al., 1992). Its effect on cAMP production has only been tested at high doses; however, at the higher doses the peptide inhibits adenylyl cyclase activity in lacrimal membrane preparations (Cripps and Bennett, 1992b).

In this study, the opiates β-endorphin and dynorphin A both caused a reduction in stimulated adenylyl cyclase activity. The effect of these opiates was, in fact, greater than that induced by enkephalin. The use of selective ligands, however, indicated the presence only of delta receptor subtypes in lacrimal membranes and supports the conclusion that β-endorphin and dynorphin A act at those sites. Of the receptor subtypes β-endorphin binds, it typically has the highest affinity for mu receptor sites. However, this opiate can activate delta receptors with an $IC_{50}$ equivalent to that of the delta selective ligand DPDPE (Clark et al., 1989). Similarly, dynorphin is characterized by its preferential binding to kappa receptors; however, it also will bind to delta receptors. Thus, the inhibition by β-endorphin and dynorphin could be due to activation of mu, kappa or delta receptor subtypes. The use of the selective ligands indicated that the inhibition of adenylyl cyclase is due to activation of delta opiate receptors. Because DPDPE (δ-selective), but neither DAMGO (μ-selective) nor U50,488 (κ-selective) inhibited adenylyl cyclase activity, we concluded that the effective endogenous peptides are most likely the enkephalins that preferentially activate delta receptors.

The demonstration of negative regulation of lacrimal secretion by endogenous neuropeptides provides several avenues for further investigation. Of particular relevance is the possibility of interactions between neuronal and non-neuronal controls that could result in deficiencies in the protein content of the tears. These interactions may be of major importance in a pathophysiologic condition in which elaboration of inhibitory agents may be enhanced. The possibility that lymphocytic elaboration of inhibitory peptides might account for decreased secretion of tear proteins associated with aging or with autoimmune disorders such as Sjogren's syndrome has been proposed by Mircheff et al. (1992). Because T-lymphocytes are known to synthesize and release a number of peptides including opioids when activated in vivo (Blalock, 1989), our observations that enkephalins inhibit lacrimal secretion suggest a mechanism by which lymphocytic infiltrates could impair lacrimal function. An infiltrate of lymphocytes may result in abnormally high levels of inhibitors including opiates, NPY or prolactin (Mircheff et al., 1992) as well as other as yet unidentified peptide modulators which may act independently or synergistically to override stimulatory neural control. We believe that the characterization of negative control as well as the determination of the ability of non-neuronal sources of synthesis and release of these inhibitory regulators is of possible significance in understanding the pathophysiologic states associated with lacrimal dysfunction.

## REFERENCES

Blalock, J.E., 1989, A molecular basis for bidirectional communication between the immune and neuroendocrine system, *Physiol. Rev.* 69:1.

Brown, B.L., Albano, J.D.M., Ekins, R.P., and Sgherzi, A.M., 1971, A simple and sensitive saturation assay method for the measurement of adenosine 3':5'-cyclic monophosphate, *Biochem. J.* 121:561.

Clark, J.A., Liu, L., Price, M., Hersh, B., Edelson, M., and Pasternak, G., 1989, Kappa opiate receptor multiplicity: evidence for two U50,488-sensitive $\kappa_1$ subtypes and a novel $\kappa_3$ sybtype, *J. Pharmacol. Exp. Ther.* 251:461.

Cripps, M.M., and Patchen-Moor, K.,1989, Inhibition of stimulated lacrimal secretion by [D-ala$^2$] met-enkephalinamide, *Am. J. Physiol.* 257:G151.

Cripps, M.M., and Bennett, D.J., 1990, Peptidergic stimulation and inhibition of lacrimal gland adenylate cyclase, *Invest. Ophthalmol. Vis. Sci.* 31:2145.

Cripps, M.M., and Bennett, D.J., 1992a, Proenkephalin A derivatives in lacrimal gland: occurrence and regulation of lacrimal function, *Exp. Eye Res.* 54:829.

Cripps, M.M., and Bennett, D.J., 1992b, Neuropeptide Y influence on lacrimal gland function, *Invest. Ophthalmol. Vis. Sci.* 33:(Suppl)1290.

Dartt, D.A., Baker, A.K., Vaillant, C., and Rose, P.E., 1984, Vasoactive intestinal polypeptide stimulation of protein secretion from rat lacrimal gland acini. *Am. J. Physiol.* 247:G502.

Frey, W.H., III, Nelson, J.D., Frick, M.L., and Elde, R.P., 1986, Prolactin immunoreactivity in human tears and lacrimal gland: possible implications for tear production, *in*: "The Preocular Tear Film in Health, Disease and Contact Lens Wear", F.J. Holly, ed., Dry Eye Institute, Lubbock, TX.

Lehtosalo, J., Uusitala, H., Mahrberg, T., Pannula, P., and Palkama, A., 1989, Nerve fibers showing immunoreactivities for proenkephalin A-derived peptides in the lacrimal glands of the guinea pig, *Graefe's Arch. Clin. Exp. Ophthalmol.* 227:455.

Lowry, O.H., Rosebrough, N.J., Farr, A.L., and Randall, R.J., 1951, Protein measurement with the Folin phenol reagent, *J. Biol. Chem.* 193:265.

Mircheff, A.K., Conteas, C.N., Lu, C.C., Santiago, G., Gray, M., and Lipson, L.G., 1983, Basal-lateral and intracellular membrane populations of rat exorbital lacrimal gland, *Am. J. Physiol.* 254:G133.

Mircheff, A.K., Warren, D.W., Wood, R.L., Tortoriello, P.J., and Kaswan, R.L., 1992, Prolactin localization, binding, and effects on peroxidase release in rat exorbital lacrimal gland, *Invest. Ophthalmol. Vis. Sci.* 33:641.

Paterson, S.J., Robson, L.E., and Kosterlitz, H.W., 1983, Classification of opioid receptors. *Br. Med. Bull.* 39:31.

Walcott, B., 1990, Leu-enkephalin-like immunoreactivity and the innervation of the rat exorbital lacrimal gland, *Invest. Ophthalmol. Vis. Sci.* 31:(Suppl)44.

Williams, R.M., Lea, R.W., Singh, J., and Sharkey, K.A., 1992, The interactions of VIP and neuropeptide Y with rat lacrimal acini: receptor binding, enzyme secretory studies and immunohistochemistry, *International Conference on the Lacrimal Gland, Tear Film and Dry Eye Syndrome Basic Science and Clinical Relevance*:148.

# SECOND MESSENGER MODULATION OF IgG SECRETION FROM CHICKEN LACRIMAL GLAND

Roger H. Cameron, Benjamin Walcott, S.F. Fan, Monica Pastor, Elizabeth Roemer, Elizabeth Grine and Peter R. Brink

Departments of Physiology and Biophysics, Neurobiology and Behavior, and Pathology, SUNY at Stony Brook, Stony Brook, New York 11794

## INTRODUCTION

Avian tears originate from two sources: a relatively small lacrimal gland located at the lateral canthus of the eye, and a much larger Harderian gland located medial to the eyeball within the orbit. In the chicken, the Harderian gland is the dominant orbital gland[1], producing a mucoid secretion[2,3] that aids in corneal lubrication.

The Harderian gland is also involved in the local immune response of the eye[4-6]. Contained within the interstitium of the gland are vast numbers of lymphoid cells, most of which fit the morphological criteria of plasma cells[7-8]. The majority of these lymphoid cells possess immunoglobulin surface determinants[9] and are responsible for most tear immunoglobulins[6], the remainder being derived from plasma cells of the conjunctiva. In the chicken, these plasma cells secrete several classes of immunoglobulin, with IgG being the most abundant[6]. Thus, unlike the mammalian eye, it is IgG, and not IgA that is the predominant immunoglobulin of tears. This fact is significant from a regulatory point of view in that IgG is not taken up as dimers by the epithelial cells as is the case for IgA. Instead, plasma cells presumably release their IgG into the extracellular space of the gland where it then diffuses between the epithelial cells to gain access to the ducts and eventually tears in a manner similar to IgG secretion in the liver[10].

The avian Harderian gland receives an extensive nerve supply that includes both divisions of the autonomic nervous system as well as nerve fibers containing a variety of neuropeptides (reviewed elsewhere[11]). With respect to the parasympathetic system, cholinesterase-positive fibers have been found within these glands[12,13]. Specifically, few cholinergic fibers supply secretory acini, but instead penetrate the gland interstitium where they end in close spatial proximity to plasma cells[13]. In addition, a muscarinic cholinergic receptor has been demonstrated on the surface of both isolated plasma cells and plasma cells in situ[14]. Thus there is anatomical evidence for a functional link between the autonomic nervous system and the secretory immune system.

In the present investigation, we provide further evidence for a functional link by demonstrating that plasma cells contain a secretagogue mechanism that can be activated by the cholinergic agonist carbachol. Furthermore we have evidence[15] that this secretagogue response requires $Ca^{2+}$ and is mediated by voltage-gated $Ca^{2+}$-channels. Finally, we address the role of second messenger systems in regulating IgG secretion by investigating the effects of forskolin on plasma cell secretion and membrane channel activity.


## METHODS

### IgG Secretion Studies

To measure the rate of IgG secretion from plasma cells, we used an in vitro preparation of chicken Harderian gland. Prior to and during our experiments animals were maintained and treated in accordance with NIH guidelines for animal usage. All experiments were conducted using adult hens (7-10 kg) anesthetized with ketamine (20 mg/kg, i.m.) and sacrificed by decapitation. Once removed from the orbit, the Harderian gland was carefully cleaned of adherent soft tissues and weighed. To expose the gland interstitium, isolated Harderian glands were cut into approximately 10 pieces, and placed into small (10x7x7 mm) baskets which were immediately immersed in 5 ml of an incubation media[16] that was maintained at 37 C and gassed continuously with $95\%O_2/5\%CO_2$. In order to measure the rate of IgG secretion from plasma cells, it was necessary to first wash out the initial protein content of the extracellular space of the gland during the preincubation period. In a typical experiment, the preincubation period consisted of baskets being transferred through 5 ml volumes of incubation media at 10 min intervals. These supernatants were normally discarded. At this point baskets were transferred at 10 min intervals to 14-15 experimental volumes containing 5 ml of incubation media with or without 100 µM carbachol. In the forskolin experiments, 20 µM forskolin was added to all preincubation and experimental volumes. At the end of each experimental incubation, experimental supernatants were collected and assayed for total protein and IgG.

Total protein was assayed using the Bradford method which involves an absorbance shift when Commassie Blue G-250 binds to protein in an acidic solution (Pierce Chemical Company, Rockford, IL). Protein standards were prepared using BSA (Sigma, St. Louis, MO) spanning a concentration range of 1-25 ug/ml.

The concentration of IgG in experimental supernatants was determined by enzyme-linked immunosorbent assay (ELISA) using a double antibody sandwich method. For these assays, the primary antibody was a goat anti-chicken IgG (Fc fragment) and the secondary antibody was a goat anti-chicken IgG (Fc fragment) conjugated to peroxidase (Bethyl Labs, Montgomery, TX). The bound antibody-enzyme conjugate was detected using the substrate o-phenylenediamine dihydrochloride (Sigma, St. Louis, MO) and 0.15% hydrogen peroxide. Plates were read in a microplate reader (Molecular Devices, Palo Alto, CA) using a kinetic analysis (37 C, 450 nm) over a time span of 10 min. IgG standards, prepared using purified IgG (Sigma, St. Louis, MO) were included in each assay.

## Patch Clamp Studies

Plasma cells were isolated as described elsewhere[17]. Patch clamp studies were conducted using methods described in Brink et al.[15]. For the forskolin experiments, the external bathing media was an isotonic KCl saline (135 mM KCl, 2.0 mM $MgCl_2$, 10 mM HEPES, and 1 mM EGTA, pH 7.2-7.3).

## RESULTS AND CONCLUSIONS

Figure 1 shows the time course of protein release from our in vitro preparation of chicken Harderian gland. In addition to IgG, we have also included the rate of total protein release from the gland to assess the exocrine function of the gland. In these preparations, we first include a number of preincubation steps designed to washout the initial soluble components of the extracellular matrix so that we may distinguish between stored immunoglobulin and actively secreted immunoglobulin. Analysis of the preincubation supernatants (data not shown) indicate that initial IgG and total protein concentrations are very high, but within three washes attain a baseline value that continues throughout the experiment in the absence of any added drugs. In addition, assays for soluble intracellular proteins (in this case, lactate dehydrogenase) indicate that there is little cell lysis, even in the initial preincubation washes, nor do any of the treatment effects induce cell lysis.

In the presence of normal media, isolated gland fragments release approximately 20-30 ug/min of IgG, which represents over half the total protein released (Figure 1). Upon stimulation with 100 µM carbachol, there is a rapid increase in the rate of IgG secretion to approximately 70 ug/min which is sustained for the first 20 min, and then gradually declines for the remainder of the 50 min exposure. Once the carbachol is removed, the rate of IgG secretion returns to baseline levels. It is interesting to note that while total protein follows the same pattern of carbachol stimulation, the IgG-independent increase in total protein is only modest. Finally, including 10 µM atropine in the incubation media completely abolishes the carbachol-stimulated increase in IgG (data not shown).

Pretreatment with 20 µM forskolin had two effects (Figure 2). Under these conditions, the baseline secretion rate for both IgG and total protein was reduced by a factor of two over normal conditions. Secondly, 20 µM forskolin completely suppressed the carbachol-stimulated increase in IgG, and total protein, although somewhat more variable, did not increase significantly above baseline upon stimulation with carbachol.

In a similar set of experiments[15], we have demonstrated that the carbachol-stimulated increase in IgG requires $Ca^{2+}$ and probably involves activation of voltage-gated $Ca^{2+}$-channels. Therefore to provide additional information concerning the secretagogue mechanism, we investigated forskolin effects on plasma cell membrane channels.

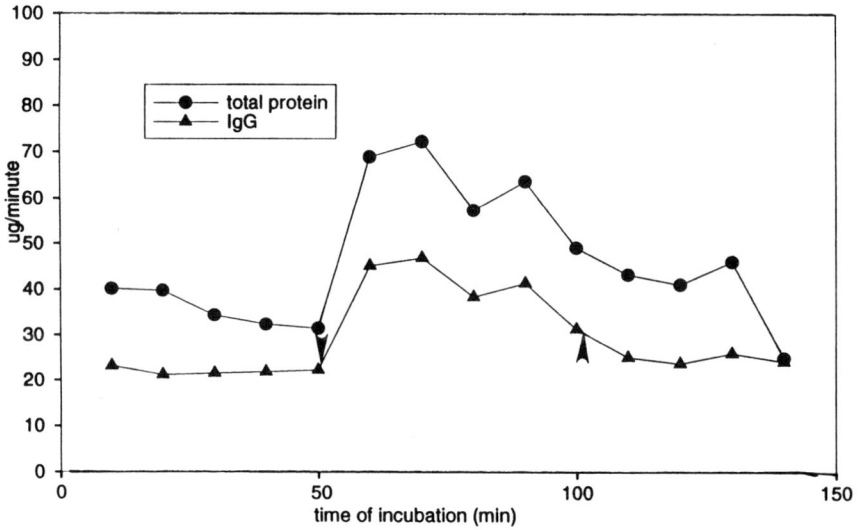

Figure 1. Time course of IgG and total protein release from isolated gland fragments stimulated with 100 μM carbachol. Each curve was obtained from a single isolated gland experiment in which carbachol was introduced into the incubation media at the time indicated by the down arrow and removed at the time indicated by the up arrow along the IgG curve. For IgG, each data point represents the mean rate of IgG secretion determined from repeated measurements of IgG concentrations of experimental supernatants (n =12-20). The standard error of the mean for these data (not shown) had a range of 10-20% of the mean. For both total protein and IgG, the data are expressed in units per gram of gland tissue.

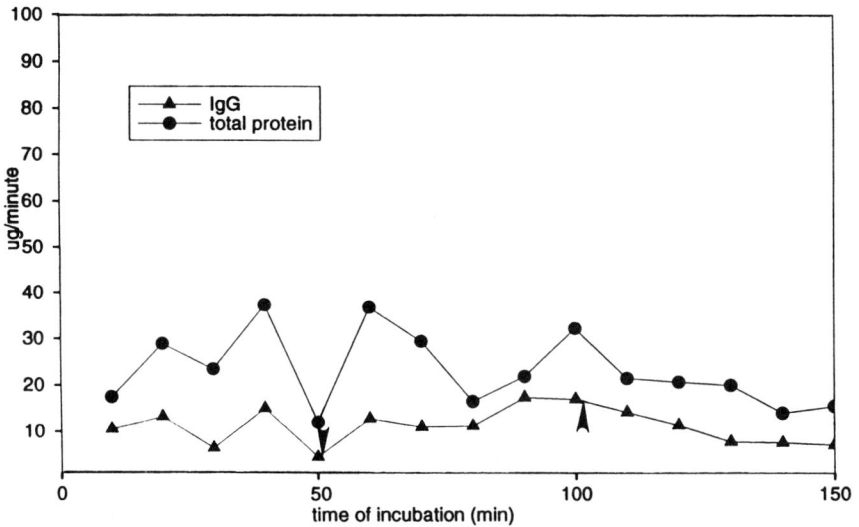

Figure 2. Same as Figure 1 except that isolated gland fragments were exposed to 20 μM forskolin throughout the incubation period.

136

In avian plasma cells, the dominant membrane channel is the maxi-K channel[17]. Figure 3 shows the effects of 20 µM forskolin and 20 µM 1,9-dideoxyforskolin on maxi-K channel activity. From these experiments, it is evident that maxi-K activity in the presence of 1,9-dideoxyforskolin (Figure 3B) cannot be distinguished from control (Figure 3A) or washout (Figure 3C) recordings. In the presence of 20 µM forskolin, however, there is an increase in the number of channels in the patch to 2 or possibly 3, indicating that forskolin has activated maxi-K channels (Figure 3D).

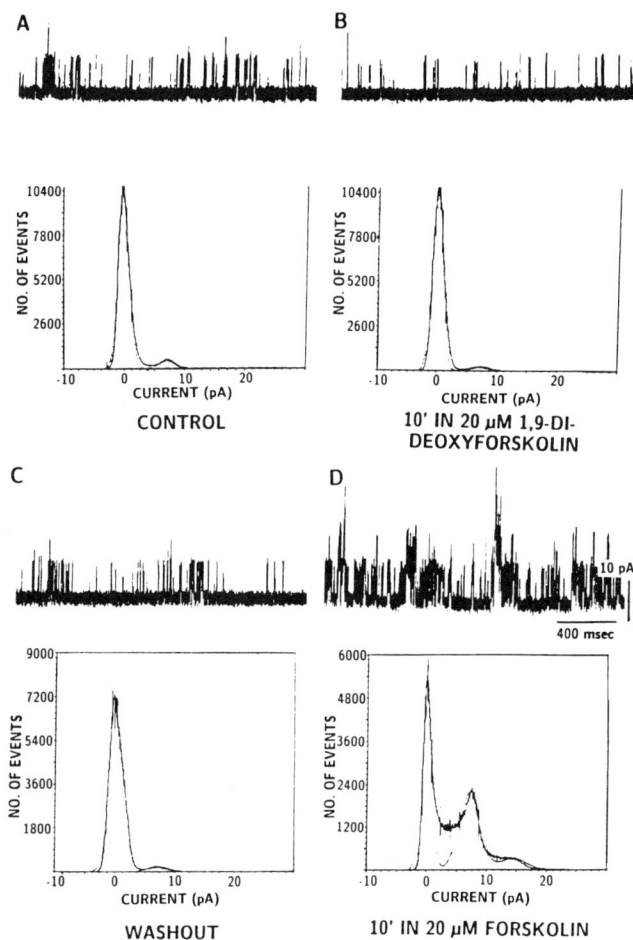

Figure 3. Membrane channel activity of avian plasma cells in the presence of 20 µM forskolin or 20 µM 1,9-dideoxyforskolin. In these experiments, the patch recording mode was cell-attached patch. The holding potential was +30 mV, and was kept constant throughout the experiments. Panels A-D show a short (2 sec) duration of channel records demonstrating unitary channel activity. In addition, each panel includes an amplitude histogram of the current distribution over at least 2 min. See Brink et al.[15] for sampling records.

Maxi-K channels can be activated by at least two different mechanisms:  increasing cytosolic Ca$^{2+}$ [18,19]; or by phosphorylation of the channel[20]. In the present experiments, we have effectively excluded external calcium stores by keeping the [Ca$^{2+}$] of the bathing media in the nanomolar range (see methods section). Furthermore, we do not believe that the cytosolic [$Ca^{2+}$] is increased by liberating intracellular stores since forskolin is not known to activate phospholipase C, thereby increasing levels of inositol 1,4,5-triphosphate ($IP_3$) leading to a release of intracellular $Ca^{2+}$. Forskolin does however, increase intracellular cAMP by direct stimulation of adenylate cyclase[21,22], and this cAMP could inhibit the synthesis of $IP_3$ as demonstrated in other systems[23-25]. Instead, we feel that it is much more likely that phosphorylation of the maxi-K channel occurs in the presence of forskolin via a cAMP-dependent protein kinase.

In summary, we have provided further evidence for a functional link between the autonomic nervous system and the secretory immune system by demonstrating that the cholinergic agonist carbachol stimulates a secretagogue response in avian plasma cells that can be inhibited by atropine. In addition, pretreatment with forskolin suppresses the carbachol-stimulated increase in the rate of IgG secretion, and in the absence of carbachol activates maxi-K channels. Furthermore, the observation that 1,9-dideoxyforskolin was ineffective in activating maxi-K channels suggests that the forskolin effect is specific. Increased activity of maxi-K channels in the presence of forskolin will hyperpolarize the cell, which should lower the probability of opening for $Ca^{2+}$-channels, thus inhibiting secretion. On the basis of these observations, we suggest that maxi-K channels act as intrinsic down-regulators of IgG secretion from plasma cells.

## ACKNOWLEDGMENT

This work was supported by NIH grants EY09406 and HL31299.

## REFERENCES

1. P.A.L. Wight, R.B. Burns, B. Rothwell, and G.M. MacKenzie, The Harderian gland of the domestic fowl. I. Histology, with reference to the genesis of plasma cells and Russell bodies, *J. Anat.* 110:307 (1971).
2. P.A.L. Wight, G.M. MacKenzie, B. Rothwell, and R.B. Burns, The Harderian gland of the domestic fowl. II. Histochemistry, *J. Anat.* 110:323 (1971).
3. I.D. Aitken, and B.D. Survashe, Lymphoid cells in avian paraocular glands and paranasal tissues, *Comp. Biochem. Physiol.* 58A:235 (1977).
4. A.P. Mueller, K. Sato, and B. Glick, The chicken lacrimal gland, gland of Harder, caecal tonsil, and accessory spleens as sources of antibody-producing cells, *Cell. Immunol.* 2:140 (1971).
5. R.B. Burns, Specific antibody production against a soluble antigen in the Harderian gland of the domestic fowl, *Clin. Exp. Immunol.* 26:371 (1976).
6. T. Baba, K. Masumoto, S. Nishida, T. Kajikawa, and M. Mitsui, Harderian gland dependency of immunoglobulin A production in the lacrimal fluid of chicken, *Immunology* 65:67 (1988).
7. B.G. Bang, and F.B. Bang, Localized lymphoid tissues and plasma cells in para-ocular and paranasal organ systems in chickens, *Am. J. Path.* 53:735 (1968).
8. U. Schramm, Lymphoid cells in the Harderian gland of birds. An electron microscopical study. *Cell Tiss. Res.* 205:85 (1980).
9. B. Albini, G. Wick, E. Rose, and E. Orlans, Immunoglobulin production in chicken Harderian glands, *Int. Arch. Allergy* 47:23 (1974).
10. B.M. Mullock, L.J. Shaw, B. Fitzharris, J. Peppard, M.J.R. Hamilton, M.T. Simpson, T.M. Hunt, and R.H. Hinton, Sources of proteins in human bile, *Gut* 26:500 (1985).
11. B. Walcott, R.H. Cameron, and P..R. Brink, The anatomy and innervation of lacrimal glands, this volume.

12. B. Walcott, and J.R McLean, Catecholamine-containing neurons and lymphoid cells in a lacrimal gland of the pigeon, *Brain Res.* 328:129 (1985).

13. B. Walcott, P.A. Sibony, and K.T. Keyser, Neuropeptides and the innervation of the avian lacrimal gland, *Invest. Opthalmol. Vis. Sci.* 30:1666 (1989).

14. B. Walcott, R.H. Cameron, E. Grine, E. Roemer, and P.R. Brink, Anti-muscarinic acetylcholine receptor-like immunoreactivity in lacrimal glands, this volume.

15. P.R. Brink, B. Walcott, E. Roemer, R.H. Cameron and M. Pastor, The role of membrane channels in IgG secretion by plasma cells in the lacrimal gland of chicken, this volume.

16. B.B. Bromberg, Automonic control of lacrimal protein secretion, *Invest. Opthalmol. Vis. Sci.* 20:110 (1981).

17. Brink, P.R., E. Roemer and B. Walcott. Maxi-K channels in plasma cells. Phlugers Archiv. 417:349. 1990.

18. O.H. Petersen, and Y. Maruyama, Calcium-activated potassium channels and their role in secretion, *Nature* 307:693 (1984).

19. O.H. Petersen, Calcium-activated potassium channels and fluid secretion by exocrine glands, *Amer. J. Physiol.* 251:G1 (1986).

20. D. Saviara, C. Lanoue, A. Cadieux, and E. Rousseau, Large conducting potassium channels reconstituted from airway smooth muscle, *Amer. J. Physiol.* 262:L327 (1992).

21. K.B. Seamon, and J.W. Daly, Forskolin: a unique diterpene activator of cyclic AMP-generating system, *J. Cyclic Nucleotide Res.* 7:201 (1981).

22. H. Metzger, and E. Lindner, The positive inotropic-acting forskolin, a potent adenylate cyclase activator, Arzneimittelforschung 31:1248 (1981).

23. S.P. Watson, R.T. McConnel, and E.G. Lapetina, The rapid fromation of inositol phosphates in human platelets by thrombin is inhibited by prostacyclin, *J. Biol. Chem.* 259:13199 (1984).

24. T. Takenawa, J. Ishitoya, and Y. Nagai, Inhibitory effect of prostaglandin $E_2$. forskolin, and dibutyryl cAMP on arachidonic acid release and inositol phospholipid metabolism in guinea pig neutrophils, *J. Biol. Chem.* 261:1092 (1986).

25. J.M. Madison, and J.K. Brown, Differential inhibitory effects of forskolin, isoproterenol, and dibutyryl cyclic adenosine monophosphate on phosphoinositide hydrolysis in canine tracheal smooth muscle, *J. Clin. Invest.* 82:1462 (1988).

# IDENTIFICATION OF SIGMA RECEPTORS IN LACRIMOCYTES AND THEIR THERAPEUTIC IMPLICATION IN DRY EYE SYNDROME

Ronald D. Schoenwald, Charles F. Barfknecht,
Satish Shirolkar, Erning Xia and Christopher C. Ignace

Division of Pharmaceutics
Division of Medical and Natural Products Chemistry
College of Pharmacy
The University of Iowa
Iowa City, IA 52242

## INTRODUCTION

Bromhexine (BH), an oral mucolytic, has shown potential for treating Sjögren's Syndrome and particularly Keratoconjunctivitis Sicca (KCS). Studies of both oral formulations (Linstow et al., 1990; Avisar, 1988) and ophthalmic drops (0.2% w/v: Roßman, 1974; Thumm, 1978) have been reported. The low aqueous solubility ($\approx$ 0.2% w/v) of BH precludes adequate ophthalmic bioavailability of BH. In this work, we have designed soluble BH derivatives to rapidly penetrate the accessory glands in the conjunctival stroma in order to improve to topical effectiveness. Histologically, the accessory lacrimal glands are identical to the main lacrimal glands (Gillette et al., 1980) making the former a logical therapeutic target. The *in vitro* and *in vivo* protein and tear secretion resulting from stimulation of rabbit lacrimal gland is reported here for our BH derivatives.

## STRUCTURES OF BROMHEXINE AND ITS DERIVATIVES

The general procedure by which the derivatives were synthesized, purified and structurally confirmed have been described elsewhere (Schoenwald and Barfknecht, 1989). BH is 2-amino-3,5-dibromo-N-methylbenzenemethamine. Other N,N-disubstituted benzenemethanamines are 1 (N-methyl-N-cyclohexyl), 2 (N-formyl-N-cyclohexyl), 4 (N-methyl-N-isopropyl), 7 (N,N-dimethyl), 10 (N-ethyl-N-cyclohexyl) and 8 (2-methyl-N-cyclohexyl). Ammonium halides are 5 (N,N-dimethyl-N-cyclohexylbenzenemethanammonium chloride) and 9 (N,N-dimethyl-N-cyclohexyl-2-phenylethylammonium chloride. N,N-disubstituted 2-phenylethylamines are 3 (N-methyl-N-cyclohexyl) and E7A (N,N-dimethyl), 13 (N-methyl-N-isopropyl), 14 (N-methyl-N-n-hexyl), and 15 (N-ethyl-N-cyclohexyl). Other compounds include 11 (N-methyl-N-cyclohexyl-2-phenyloxyethylamine) and 12 (N-methyl-N-cyclohexyl)-3-phenylpropylamine.

*Lacrimal Gland, Tear Film, and Dry Eye Syndromes*
Edited by D.A. Sullivan, Plenum Press, New York, 1994

## PROTEIN SECRETION RATE (*IN VITRO*)

New Zealand white rabbits of either sex, without observable eye defects, weighing 1.8-2.2 kg and 2 to 3 months of age were purchased for the study (Iowa Ecology Farms, Wilton, IA). All procedures performed on the rabbits conform to the "ARVO Resolution on the Use of Animals in Research".

Protein secretion rates were measured *in vitro* with the use of a flow-through or perifusion system according to the method of Bromberg (1981) following drug stimulation to either lacrimal gland slices or to isolated, intact lacrimocytes. The experiments that were performed with BH, carbachol and derivatives 1-15 (usually $10^{-12}$ to $10^{-2}$ M) used lacrimal gland slices. The protein secretion of the preferred derivative, **E7A**, was studied using isolated, intact lacrimocytes.

In order to compare each compound, the area under the averaged dose-response curve was calculated (AUC $= \int(-log[C])$) and tabulated (see Table 1). The dose-response curves showed an increase in protein secretion rate with an increase in the concentration of agent used to stimulate protein release. However, each curve also showed a decrease once a maximum response (usually $10^{-4}$ or $10^{-6}$ M) was reached. Table 1 shows that **7**, in particular, yielded a much higher response than any of the other derivatives and therefore was considered a good candidate for molecular modification in order to improve its effect in stimulating the release of protein from lacrimal gland tissue. Derivative **E7A** is an outcome of that effort.

Figure 1 shows a plot of the amount of protein measured over a two minute infusion period for carbachol and **E7A** ($10^{-4}$M) using lacrimocytes. The secretion rate for this experiment is 8.1 μg of protein/million lacrimocytes/min, which cannot be compared to secretion rates obtained for lacrimal gland slices because of differences in the method of expressing the units. The data summarized in Table 1 represents μg of protein/gram of lacrimal gland/min. However, carbachol is common to both sets of experiments and can be used as a reference standard. In table 1 carbachol showed a greater maximum protein secretion rate when compared to **12, 14**, or BH, but its rate is about 10% less (8.1 vs. 7.2 μg of protein/million lacrimocytes/min) when compared to **E7A** using lacrimocytes (figure. 1).

## IDENTIFICATION OF SIGMA RECEPTORS IN THE LACRIMAL GLAND AND $IC_{50}$ DETERMINATIONS OF BROMHEXINE DERIVATIVES

Previous work focused on sigma receptors (Walker et al., 1990) has been primarily associated with the location of the receptors in the brain of the rat and guinea pig. In order to eliminate the possibility that BH and its derivatives were acting *via* a cholinergic mechanism, classical cholinergic bioassay procedures were conducted on these agents (Schoenwald et al. 1990) and indicated that no compounds were acting as $M_1$ or $M_2$ muscarinic agonists or antagonists, nicotinic agonists or antagonists, choline acetylase stimulators or anticholinesterases.

A very efficient and effective method of obtaining insight into the mechanism of a biologically active molecule is to determine to determine the receptor binding profile. Nova Pharmaceutical Corporation (NovaScreen®, Baltimore, MD) offers a screen of approximately 35 bioreceptors. Table 2 summarizes the results for BH and the most active derivatives (*in vivo* secretion rate) for which binding was significant. These results suggested that protein release is occurring as a result of binding to sigma and possibly serotonin receptors. Clearly, the profile for **E7A** is more favorable for clinical application than other derivatives which exhibited a wider profile. Therefore E7A was investigated further.

The presence of sigma receptors was identified in isolated, but intact lacrimocytes obtained from the main lacrimal gland of white rabbits). $^3$H-Haloperidol was used as the ligand ($10^{-4}$ to $10^{-10}$ M). Using the technics of Wolfe et al. (1989) for intact lacrimocytes and McCann and Su

(1991) for membrane suspensions, sigma receptors were isolated and tested for binding to $^3$H-DTG or to $^3$H-haloperidol in the presence of spiperone to prevent binding to serotonin receptors. A Scatchard plot revealed two binding sites with both $^3$H-DTG and $^3$H-haloperidol exhibiting a high

**Table 1.** Ranking of Area Under Dose-Response Curves (AUC) for Carbachol, Bromhexine (BH) and Derivatives following In Vitro Protein Secretion Rates of Drug Added to Incubation Media Containing Rabbit Lacrimal Gland Slice

| Ranking | Series Code | AUC | Percent of Maximum |
|---|---|---|---|
| 1 | 7 | 511.4 | 100. |
| 2 | 6 | 344.5 | 67. |
| 3 | Carbachol | 334.7 | 65. |
| 4 | 2 | 334.5 | 65. |
| 5 | BH | 309.3 | 60. |
| 6 | 1 | 291.4 | 57. |
| 7 | 15 | 223.7 | 44. |
| 8 | 12 | 211.3 | 41. |
| 9 | 4 | 189.7 | 37. |
| 10 | 8 | 182.0 | 36. |
| 11 | 9 | 180.0 | 35. |
| 12 | 3 | 176.6 | 35. |
| 13 | 5 | 140.0 | 27. |
| 14 | 14 | 137.2 | 27. |
| 15 | 11 | 95.8 | 19. |
| 16 | 13 | 53.2 | 10. |
| 17 | 10 | 47.8 | 9. |

affinity site ($K_d$ = 22.7 and 1.04 nM, respectively and $B_{max}$ = 50.7 and 68.0 fmol/mg of protein) and a low affinity site ($K_d$ = 557.0 and 75.3 nM, respectively, and $B_{max}$ = 43.8 and 172.0 fmol/mg of protein). The weaker site was suspected to be intracellular. These results for the rabbit lacrimocytes indicate a similar density of sigma sites and binding capacity to that found in rat exocrine glands and other tissues (Wolfe et al., 1989). $IC_{50}$ values for **12** and **14** ranged from 3.2 to 8.8 nM for both radioligands. The derivative **E7A** yielded an $IC_{50}$ of 7.8 nM for $^3$H-haloperidol but shows less capacity to displace $^3$H-DTG ($IC_{50}$=900 nM).

## PROTEIN SECRETION RATE (IN VIVO)

The concentrations studied in rabbits were 0.0, 0.075, 0.15, 0.2, 0.3 and 0.60 % w/v. Baseline tear secretion was measured by placing a Schirmer tear test strip (Clement Clarke Int., Ltd., Harlow, Essex, U.K.) under the lower lid of each eye for five minutes. The right eye of the animal received 50 μl of the drug formulation and the other eye received the vehicle (borate buffer pH 7.4 with 1.5% w/v of Methocel™ ELV 50 Premium, Dow Corning Corp., Midland, MI). Proparacaine (25 μl of 0.5 % w/v) was instilled 8 minutes before inserting the Schirmer strip. The tear proteins from the Schirmer strip were extracted with 1 ml of 0.9% sodium chloride in pH 6.24 phosphate buffer (0.066M). A volume of 0.1 ml of the extract was reacted with 2.5 ml of modified Coomassie reagent and the absorbance was measured at 595 nm.

Figure 2 shows the averaged percent change from baseline for both treated and control eyes for E7A at various concentrations. The treated eyes show a statistically significant increase (p<.05) at 10 minutes post-dosing for 0.075, 0.15, 0.2, and 0.3%, but not 0.6%. The protein secretion rate in the treated eye at 0.15 % at 60 minutes post-dosing is not statistically significant (p<0.05). The dose-effect curves for all agents tested *in vitro* show an increase in protein release, reaching a

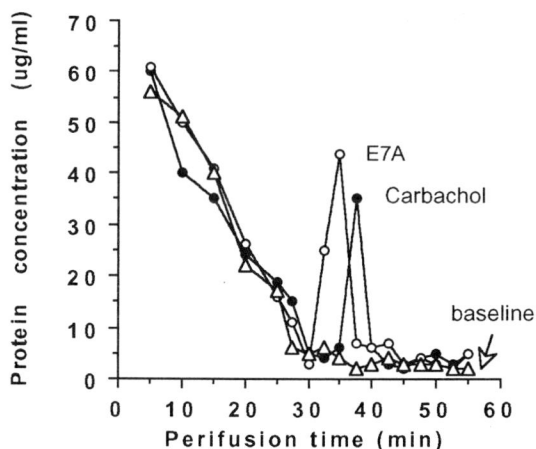

**Figure 1.** Stimulation of Protein Secretion Following the Addition of **E7A** and Carbachol ($10^{-4}$ M) to Isolated, Intact Lacrimocytes.

**Table 2.** Receptor Binding Profile* (NovaScreen®) for Bromhexine (BH) and Sigma Ligands

| Agent | Sigma | $5HT_1$ | $5HT_2$ | $Alpha_2$ | $Dopaminergic_2$ | $Muscarinic_2$ |
|-------|-------|---------|---------|-----------|------------------|----------------|
| BH | ++ | 0 | 0 | 0 | 0 | ++ |
| E7A | ++ | +++ | + | ++ | 0 | + |
| 12 | ++ | 0 | ++ | ++ | 0 | ++ |
| 14 | ++ | ++ | ++ | ++ | ++ | ++ |

0=less than 20% inhibition at 10 μM.
+=marginal (20-49 % inhibition at 10 μM)
++=active (50-100 % inhibition at 10 μM)
+++=very potent (50-100 % inhibition at 1 nM)
(Guinea pig brain was the source of sigma receptors)

maximum at $10^{-4}$ to $10^{-8}$ M followed by a decrease. It is interesting that the results in figure 2 follow a similarly shaped dose-effect curve as the *in vitro* dose-effect curves. None of the control eyes show a statistically significant increase in protein secretion rate suggesting that the effect is local (presumably the accessory lacrimal gland) and not systemic.


TEAR SECRETION RATE (IN VIVO)

In the rabbits for which protein secretion rate was measured, we also measured the tear secretion rate by noting the distance (mm) fluid traveled in five minutes on the Schirmer strip. At

10 minutes post-dosing and for the lowest concentration instilled (0.075% w/v), a statistically significant ($p < 0.05$) increase in the tearing rate (30%) was observed in the treated eye. At other concentrations (or in any of the control eyes), no increase in tearing was significant. At 60 minutes post-dosing, a significant increase (30 %) was measured in the treated eye at 0.2% w/v only.

From the receptor bioassay profile for **E7A** (table 2), it is apparent that sigma may not be the only receptor responsible for tearing and/or protein secretion. In goblet cells of anesthetized rat

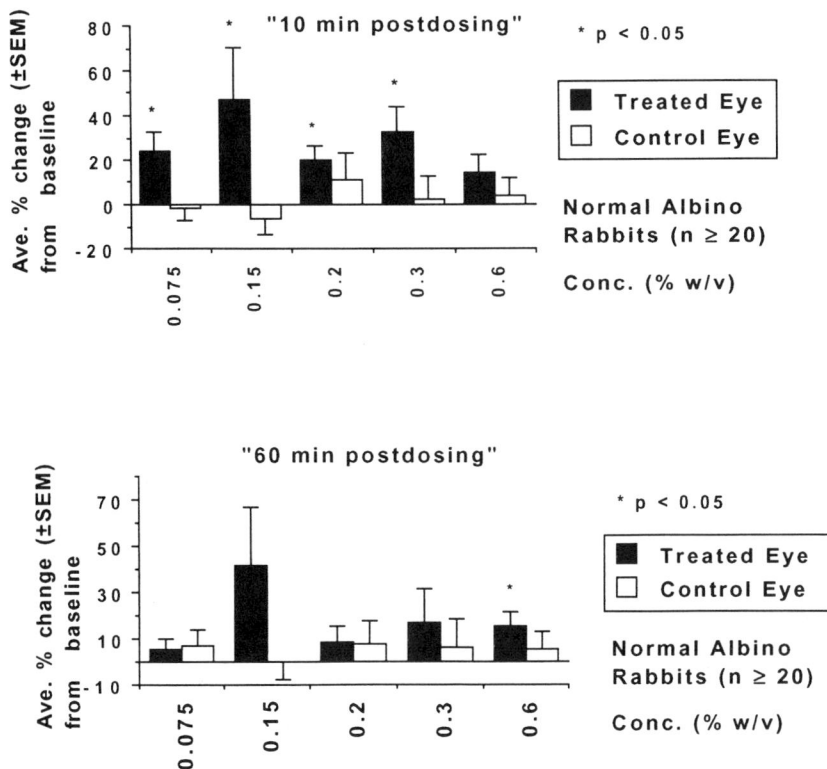

**Figure 2.** % Change in Average Protein Secretion Rate ($\mu$g/5 min) for E7A 10 and 60 minutes after Topical Instillation (50 $\mu$l) to the Rabbit Eye.

eyes, Kessler et al. (1992) observed the release of mucus (i.e., glycoproteins) in response to serotonin stimulation (Kessler et al., 1992). Adenylate cyclase is the second messenger following $5HT_1$ stimulation. After stimulation by vasoactive intestinal peptide, adenylate cyclase is responsible for protein release in acinar cells (Dartt, 1989).

## AGONIST/ANTAGONIST ACTIVITY

Incubation of lacrimocytes with **E7A** alone ($10^{-4}$ M), with haloperidol ($10^{-4}$ M) alone or in combination showed that E7A was acting as an agonist to stimulate protein release (28% increase compared to baseline), whereas, haloperidol elicited a decrease (-19%). The combination yielded a 11% decrease. Stimulating protein release from either lacrimocytes or membrane preparations may provide a simple experimental procedure for identifying agonist/antagonist behavior and for providing a lead regarding the pharmacology and function of sigma receptors in the body, the latter of which is not clear.

## CONCLUSION

Topical instillation of E7A causes a release of secretory proteins from the dosed eye, most likely by stimulation of sigma receptors (and/or possibly by stimulation of serotonin receptors). More clearly, sigma agonist ligands stimulate protein release in lacrimocytes. The increase in protein secretion upon topical dosing is expected to improve tear film stability in dry eye.

## REFERENCES

Avisar, R., and Savir, H. 1988. Our further experience with bromhexine in Keratoconjunctivitis Sicca. Ann. Ophthalmol. 20:382.

Bromberg, B.B. 1981. Autonomic control of lacrimal protein secretion. Invest. Ophthalmol. Vis. Sci., 20:110.

Dartt, D.A. 1989. Signal transduction and control of lacrimal gland protein secretion: a review. Curr. Eye Res., 8:619.

Gillette, T.E, Allansmith, M.R., Greiner, J.V., and Janusz, M. 1980. Histologic and immunohistologic comparison of main and accessory lacrimal tissue. Am. J. Ophthalmol. 89:724.

Kessler, T.L., and Dartt, D.A. 1992. Neural stimulation of conjunctival goblet cell mucous secretion in rats. International Conference on the Lacrimal Gland, Tear Film and Dry Eye Syndromes: Basic Science and Clinical Relevance, Nov. 14-17, Bermuda, p. 73.

Linstow, M., Kriegbaum, N.J., Backer, V., Ulrik, C., and Oxholm, P. 1990. A follow-up study of pulmonary function in patients with primary Sjögren's Syndrome. Rheumatol. Int. 10:47.

McCann, D.J. and Su, T-P. 1991. Solubilization and characterization of haloperidol-sensitive (+)-[3H] SKF-10,047 binding sites (sigma sites) from rat liver membranes. J. Pharmacol. Exp. Ther. 257: 547.

Roßman, H., 1974, Treatment of decreased tear secretion with bromhexine eyedrops, Dtsch. med. Wschr. 99:408.

Schoenwald, R.D., and Barfknecht, C.F. 1989. Lacrimal secretion stimulant (LSS). U.S. Patent 4,820,737.

Schoenwald, R.D., Long, J.P., Barfknecht, C.F., Wei, T., Ignace, C., and Flynn G. 1990. Tear Stimulants: evaluation and mechanism studies of bromhexine analogs. Invest. Ophthalmol. Vis. Sci. 31:402.

Takeda, H., Misawa, M., and Yanaura, S. 1983. A role of lysosomal enzymes in the mechanism of mucolytic action of bromhexine. Japan. J. Pharmacol., 33:455.

Thumm, H.-W. 1978. On the therpy of the "dry eye" with bromhexine HCl drops. Klin. Mbl. Augenheilk. 172:200.

Walker, J.M., Bowen, W.D., Walker, F.O., Matsumoto, R.R., de Costa, B.B., and Rice, K.C. 1990. Sigma receptors: biology and function. Pharmacol. Rev. 42: 355.

Wolfe, S.A., Culp, S.G., and de Souza, E.B. 1989. s-Receptors in endocrine organs: identification, characterization, and autoradiographic localization in rat pituitary, adrenal, testis, and ovary. Endocrinology. 124:1160.

# ROLE OF PROTEIN KINASE C IN $\alpha_1$-ADRENERGIC AND CHOLINERGIC AGONIST STIMULATED PROTEIN SECRETION

Robin R. Hodges, Deanna M. Dicker, and Darlene A. Dartt

Cornea Unit
Schepens Eye Research Institute
&
Department of Ophthalmology
Harvard Medical School
Boston, MA

## INTRODUCTION

In the lacrimal gland, cholinergic agonists produce 1,4,5-inositol trisphosphate ($1,4,5$-$IP_3$) and concomitantly diacyglycerol (DAG) from the hydrolysis of phosphatidylinositol bisphosphate[1]. $1,4,5$-$IP_3$ causes the release of intracellular $Ca^{2+}$ which causes secretion[1]. DAG activates protein kinase C (PKC). It has not been determined if cholinergic agonists produce DAG although cholinergic agonists transiently translocate PKC activity and the PKC isozyme PKC-$\varepsilon$ from the cytosol[2].

In most tissues, $\alpha_1$-adrenergic agonists also transmit their extracellular signal by producing $1,4,5$-$IP_3$ and DAG, by releasing intracellular $Ca^{2+}$, and by activating PKC[3]. Little is known about the cellular mechanisms utilized by $\alpha_1$-adrenergic agonists to stimulate lacrimal gland protein secretion. It is known that $\alpha_1$-adrenergic agonists do not increase cAMP levels, $1,4,5$-$IP_3$, or $[Ca^{2+}]$ [2]. Protein secretion stimulated by the simultaneous addition of a cholinergic agonist and an $\alpha_1$-adrenergic agonist is additive suggesting that the two agonists use distinct and separate pathways[2]. In addition, the simultaneous addition of a phorbol ester, which mimics DAG, and an $\alpha_1$-adrenergic agonist is not additive suggesting that the two share a common pathway and that PKC is involved in this pathway[4].

This study was undertaken to determine the role of PKC in $\alpha_1$-adrenergic and cholinergic agonist-induced stimulation of lacrimal gland protein secretion.

## METHODS

### Preparation of Acini

Both exorbital lacrimal glands were removed from anesthetized and decapitated male Wistar rats. The lacrimal glands were minced and incubated in Krebs-Henseleit buffer (KHB) containing 75 U/ml collagenase. Acini were dispersed by trituration and filtered through a 220-mm pore nylon mesh.

The dispersed acini were preincubated in KHB containing 1% bovine serum albumin (BSA) for 45 min, washed and divided into aliquots.

## Measurement of Peroxidase Secretion

Acini were incubated for 0-20 min in 1% BSA containing agonists. The reaction was terminated by centrifugation and the supernatant removed. The supernatant was analyzed spectrophotometrically for peroxidase, a marker for protein secretion. Total amount of cellular protein was measured using the method of Bradford[5].

## Downregulation of Protein kinase C

Acini were incubated for 2h in 1% BSA containing $10^{-6}$ M phorbol 12,13 dibutyrate (PDBu) at $37^{\circ}$C. The acini were washed and stimulated with agonist. The reaction was stopped by centrifugation and the supernatant analyzed for peroxidase.

## Phosphatidylcholine Hydrolysis

Acini were incubated for 3h with 5 mCi $^{3}$H-choline chloride at $37^{\circ}$C. The acini were washed and stimulated with phenylephrine ($10^{-4}$ M) for 0-20 min. The reaction was terminated by the addition of cold methanol:water (1:0.8 v/v). Lipids were extracted with chloroform and washed twice with methanol:water. The aqueous fraction was dried under $N_2$, resuspended in water and spotted on Silica gel G thin layer chromatography (TLC) plates. The plates were developed in 0.5% NaCl, ethanol, methanol, and ammonium hydroxide (50:30:20:5 v/v) for 2h. The standards were visualized with $I_2$ vapors. The plates were exposed to X-ray film overnight. The bands were analyzed by densitometry.

## RESULTS AND DISCUSSION

## Effect of PKC Downregulation on Peroxidase Secretion

To determine the role of PKC in $\alpha_1$-adrenergic agonist-stimulated peroxidase secretion, PKC was downregulated by PDBu pretreatment. Acini with and without PDBu pretreatment were stimulated 0-20 min with phenylephrine at a concentration of $10^{-4}$M which was maximal for secretion[2]. In cells not pretreated, phenylephrine caused an linear increase in peroxidase secretion (Fig. 1). A 2h PDBu pretreatment increased, although not significantly, peroxidase secretion stimulated with phenylephrine from an untreated value of $1.1 \pm 0.5$ to $3.3 \pm 0.6$ units/mg protein at 1 min, and from $2.1 \pm 0.7$ to $5.1 \pm 1.7$ units at 5 min, before decreasing from an untreated value of $4.8 \pm 0.8$ to $3.3 \pm 1.4$ units at 20 min (n=4).

In similar experiments the cholinergic agonist carbachol ($10^{-5}$ M) caused a rapid increase in peroxidase secretion at 1 min and a slower sustained release at 5 and 20 min[6]. Treatment with PDBu significantly decreased secretion in these cells at 1 min and deceased secretion at 5 and 20 min[6].

These experiments imply that PKC is involved in both cholinergic and $\alpha_1$-adrenergic stimulation of lacrimal gland protein secretion. However, PKC activation appears to inhibit phenylephrine-induced peroxidase secretion while activation of PKC is necessary for stimulation of secretion by carbachol. It is possible that PKC plays a modulatory, inhibitory role in

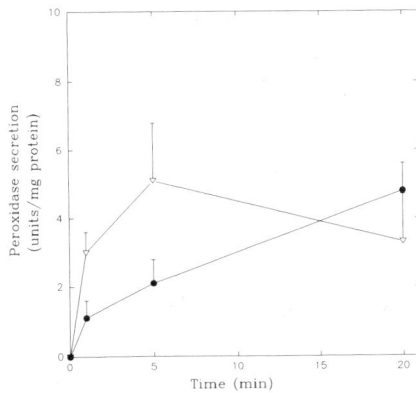

Figure 1. Effect of 2h phorbol dibutyrate treatment on phenylephrine ($10^{-4}$M) induced peroxidase secretion. Circles-control; Triangles-phorbol dibutyrate treated. Values are means $\pm$ SE; n=4

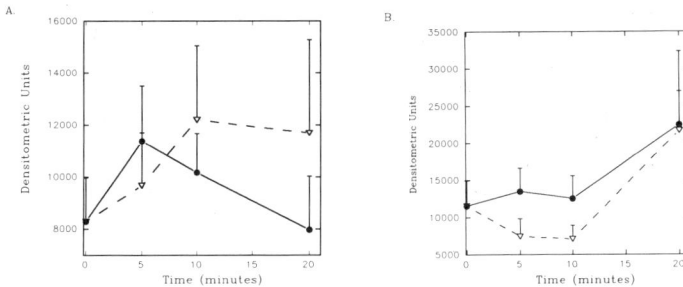

Figure 2. Effect of phenylephrine ($10^{-4}$M) on phosphatidylcholine hydrolysis. A. GPC production. B. Choline production. Triangles-control; Circles-Phenylephrine treated. Values are means $\pm$ SE; n=4-10.

the secretion process stimulated by phenylephrine. This is consistent with the observation that phorbol ester induced peroxidase secretion is not additive with phenylephrine induced peroxidase secretion.

## Effect of $\alpha_1$-Adrenergic Agonists on Phosphatidylcholine Hydrolysis

DAG is necessary for the activation of PKC. Because a possible source for DAG is phosphatidylcholine, the effect of phenylephrine ($10^{-4}$M) on phosphatidylcholine hydrolysis was determined after 0-20 min incubation. After separation of the aqueous fraction by TLC, two components were identified as glycerophosphocholine (GPC) and choline. Phenylephrine slightly increased GPC levels from 9670 $\pm$ 2033 units in untreated cells to 11369 $\pm$ 3591 units at 5 min, before decreasing it from 12197 $\pm$ 2850 units to 10171 $\pm$ 1495 units at 10 min and from 11691 $\pm$ 3591 units to 7969 $\pm$ 2057 at 20 min (Fig. 2). Phenylephrine also increased, although not significantly, choline levels from 7425 $\pm$ 2409 units in untreated cells to 13500 $\pm$ 3158 units at 5 min and from 7067 $\pm$ 1855 units to 12564 $\pm$ 3008 units at 10 min. Phenylephrine did not change choline levels at later times from an untreated value of 21713 $\pm$ 5234 units to 22519 $\pm$ 9856 units at 20 min. The appearance of GPC and choline implies that phospholipase

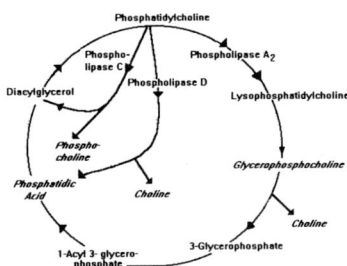

Figure 3. Phosphatidylcholine Hydrolysis Pathways.


$A_2$ activity may be slightly increased upon phenylephrine stimulation (Fig. 3) Another product of phospholipase $A_2$ hydrolysis of phosphatidylcholine is arachidonic acid. It is possible that arachadonic acid or one of its metabolites could play a role in $\alpha_1$-adrenergic agonist stimulated protein secretion.

We conclude that PKC plays a role in cholinergic agonist stimulated protein secretion in lacrimal gland cells[6]. $\alpha_1$-Adrenergic agonist stimulated protein secretion is not dependent on PKC activation. In fact, activation of PKC appears to inhibit $a_1$-adrenergic induced protein secretion. It is possible that PKC could play a role in modulating $\alpha_1$ adrenergic agonist-induced protein secretion although the mechanism is still unknown. $\alpha_1$-Adrenergic agonists appear to increase phosphatidylcholine hydrolysis slightly by activating phospholipase $A_2$ to produce DAG, GPC, choline, and arachidonic acid. Arachidonic acid or its metabolites could also play a role in $\alpha_1$-adrenergic agonist-induced protein secretion although activation of phospholipase $A_2$ cannot completely account for $\alpha_1$-adrenergic stimulation of protein secretion. The remaining mechanism is still unknown.

## References

1. D.A. Dartt, D.M. Dicker. L.V. Ronco, I.M Kjelsen, R.R. Hodges, and S.A. Murphy, Lacrimal gland inositol trisphosphate isomer and tertrakisphosphate production, Am. J. Physiol. 259:G274 (1990).
2. R.R. Hodges, D.M. Dicker, P.E. Rose and D.A. Dartt, $\alpha_1$-Adrenergic and cholinergic agonists use separate signal transduction pathways in lacrimal gland, Am. J. Physiol. 262:G1087 (1992).
3. J.W. Putney, Jr.,Phosphoinositides and alpha-1 adrenergic receptors, in: "The Alpha-1 Adrenergic Receptor," R.R, Ruffolo, Jr. ed., Humana Press, Clifton, NJ (1987).
4. D.A. Dartt, L.V. Ronco, S.A. Murphy, and M.F. Unser, Effect of phorbol esters on rat lacrimal gland protein secretion, Invest. Ophthalmol. Vis. Sci.29:1726(1988).
5. M.M. Bradford, A rapid and sensitive method for the quantitation of microgram quanities of protein utilizing the principle of protein-dye binding, Anal. Biochem. 72:248.
6. D.A. Dartt, R.R. Hodges, and D.M. Dicker, Protein kinase C, PKC-$\varepsilon$, and $Ca^{2+}$ in cholinergic activation of lacrimal gland protein secretion, Submitted.

# THE ROLE OF MEMBRANE CHANNELS IN IgG SECRETION BY PLASMA CELLS IN THE CHICKEN LACRIMAL GLAND

Peter R. Brink, Benjamin Walcott, Elizabeth Roemer, Roger Cameron and Monica Pastor

Departments of Physiology and Biophysics, Neurobiology and Behavior, and Pathology, SUNY at Stony Brook, Stony Brook, New York 11794

## INTRODUCTION

One of the more ubiquitous cell types within the parenchyma of the lacrimal gland is the plasma cell. It is solely responsible for the production of immunoglobulins which are secreted into tears. While it is true that immunoglobulins such as IgA are modified by epithelial cells, it is still the plasma cell which acts as the source. In the case of IgG there is no involvement of the epithelium other than to act as a passive filter via the lateral interstitial space[1]. In terminally differentiated B-lymphocytes or plasma cells the secretion of immunoglobulin has been thought of as a constitutive-like process. The plasma cells simply secrete at a constant rate[2]. The processes which cause differentiation into plasma cells and hence increased output of immunoglobulin have been major focal points. It is thought that alteration in immunoglobulin output is affected more by changes in cell number rather than dynamic changes within individual cells' secretory ability.

Many studies have shown, though, that B-lymphocytes possess membrane receptors and channels which imply a dynamic ability, possibly with regards to secretion[3,4,5,6]. While these results are of great interest and highly suggestive, they do not address the role membrane channels might play as signal transduction sites which might trigger modulation of the secretion of immunoglobulins in plasma cells for example.

The avian lacrimal gland (Harderian gland) is ideal for immunoglobulin secretion studies as it contains an inordinate number of plasma cells within the parenchyma of the gland relative to typical mammalian lacrimal glands[7]. Thus plasma cells can be isolated and studied via patch clamp methods[6] and it is also possible to analyze IgG secretion by whole populations of these same cells.

We have sought to demonstrate whether avian plasma cells utilize channel-regulated influxes, such as $Ca^{2+}$ inward currents, to alter cellular processes like IgG secretion.

*Lacrimal Gland, Tear Film, and Dry Eye Syndromes*
Edited by D.A. Sullivan, Plenum Press, New York, 1994

# METHODS

Patch Clamp methods included whole cell and cell attached recording modes[9]. Isolated cells[6] were perfused in RPMI (15 mM HEPES) or PBS saline. The PBS saline contained 135 mM NaCl, 6 mM KCl, 1 mM $CaCl_2$, 2 mM $MgCl_2$, and 10 mM HEPES at ph=7.2-7.3. All solutions were filtered with 0.22 μm filters. The patch pipette was filled with 135 mM KCl, 0.6 mM EGTA, 0.09 mM $CaCl_2$, 0.5 mM $MgCl_2$, 1 mM ATP and 10 mM HEPES at pH=7.2 (type I pipette solution). In some experiments the pipette solution contained 135 mM CsCl,, 0.6 mM EGTA, 0.09 mM $CaCl_2$, 0.5 mM $MgCl_2$, 1 mM ATP and 10 mM HEPES (type II solution), and bathing media contained 110 mM KCl, 20 mM $BaCl_2$ and 10 mM HEPES at pH =7.2. For cell-attached experiments in addition to PBS the two following solutions were used as bathing medias: Isotonic KCl, 135 mM KCl, 5 mM NaCl, 2 mM $MgCl_2$, and 10 mM HEPES at ph=7.2- 7.3 or Isotonic KCl, 135 mM KCl, 3mM $CaCl_2$, 2 mM $MgCl_2$, and 10 mM HEPES at ph=7.2- 7.3. All data were acquired, analyzed and stored as previously described [10]. The measurement of total protein was as described by Cameron et al. [11]. For $Ca^{2+}$ free experiments on secretion 1 mM EGTA was added to Bromberg Media[12] with no $Ca^{2+}$ added.

# RESULTS AND CONCLUSIONS

Typical whole cell currents for the avian plasma cells are shown in Figure 1A. The whole cell current is outward rectifying and is carried by $K^+$ [6]. The currents in Figure 1A were generated by a cell where the pipette solution was type I and the bathing media was PBS. The holding potential was -70 mV and the applied voltage steps brought the membrane potential to +100, +80, +60, +40, +20, 0, and -20 mV, top to bottom. The record shown in figure 1B was taken in the whole cell configuration where the pipette solution was type II and the bathing media was the Barium saline described in the methods. Under these conditions most of the $K^+$ current has been suppressed. The smaller $Ca^{2+}$ currents can then be seen. The type of $Ca^{2+}$ channel is not presently known but the form of the current is most likely that produced by L-type $Ca^{2+}$ channels [13,14]. The holding potential was -70 mV and the three step potentials used brought the membrane potential to +70, +30 and -20 mV, top to bottom.

The maxi-K channel can be monitored in the cell-attached mode[6] and as such is an excellent assay of changes in intracellular $Ca^{2+}$. Figure 2 (left-hand panel, labelled C) shows control data in the form of an example of the unitary channel activity (inset) and an all points amplitude histogram of the current distribution for 1 minute of data, holding potential equal +10 mV. The larger peak represents the closed state. Clearly the channel activity is low. Application of 1uM BayK 8644 to the media altered the activity dramatically within 30 s (right-hand panel). The amplitude histogram shifted to the right indicating a very significant increase in channel open time. This result is consistent with the known effect of BayK 8644 which is to increase the open time probability of $Ca^{2+}$ channels. The increase in $Ca^{2+}$ influx resulting from the action of BayK 8644 increases the open time probability of the maxi-K channel. This is yet another illustration of a $Ca^{2+}$ influx in the avian plasma cell.

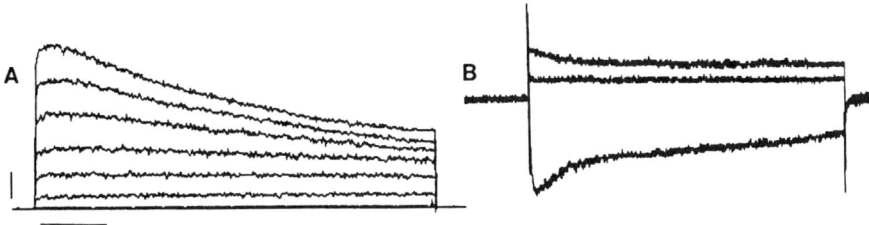

Figure 1. Whole cell currents from two cells both held at -70 mV are shown. Voltage steps were applied as given in the text. (A) Whole cell currents in PBS with pipette solution I. (B) Whole cell currents under ionic conditions used to illustrate $Ca^{2+}$ inward currents, pipette solution II. Scale bars: vertical 500 pA in A, 5 pA for B, horizontial bars 1.5s for A, 2.3s for B.

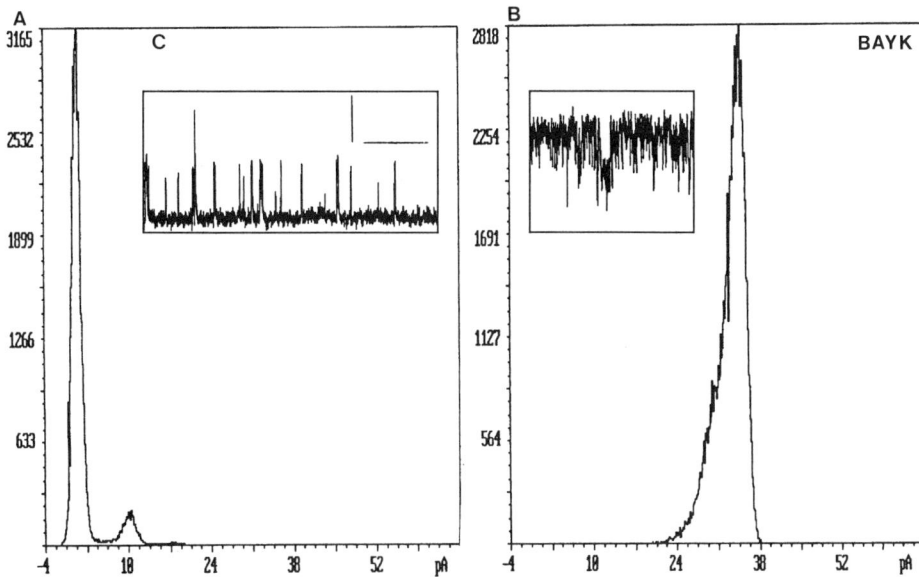

Figure 2. Amplitude histograms of maxi-K channel activity before and after application of 1 µM BayK 8644, cell-attached recording mode. Insets are examples of the channel activity. Vertical bar=8 pA, Horizaontal bar=0.25s.

153

Monitoring the maxi-K channel in the cell attached mode can also illustrate that the $Ca^{2+}$ influx is mediated by a voltage dependent channel which is inactivated by prolonged depolarization. Figure 3 shows the open time probability of a single maxi-K channel monitored in the cell-attached mode under a number of different media conditions. Activity was monitored continuously as the cell was perfused in sequence with the media indicated on the X-axis. Open time was determined by analysis of records of 2 minute duration[10]. In all cases the holding potential was held at +25 mV. In media the open time probability was very low, $P_o = 0.005$. Depolarization by perfusion with isotonic KCl saline with no $CaCl_2$ (free $Ca^{2+}$ less than 1 µM) caused an increase in $P_o$ consistent with voltage dependent changes in $P_o$ typical of maxi-K channel when intracellular $Ca^{2+}$ is not changing. Further perfusion with isotonic KCl but with 3 mM $CaCl_2$ added had no effect on $P_o$. Reperfusion with PBS restored $P_o$ to the initial control level. Once the control $P_o$ was reestablished the cell was again perfused with isotonic KCl which contained 3 mM $CaCl_2$ and the $P_o$ increased to near 10%. Reperfusion with media lowered the $P_o$ to control levels. Each perfusion took 30-60 seconds (30 ml/minute) and the dish volume was 3 ml. After 1 minute of perfusion 2 minutes of channel activity was recorded.

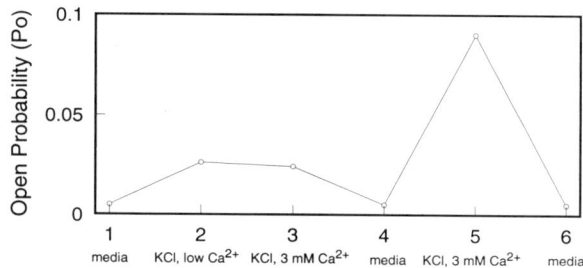

Figure 3. Changes in Maxi-K open time probability with variously perfusates, recording mode: cell attached. The patch was established while the cell was in media and control $P_o$ determined (1 on the X-axis). Subsequently the bath was perfused with isotonic KCl, low $Ca^{2+}$ (2 on t he X-axis). Channel activity was recorded for two minutes after each perfusion. 3 thru 6 represent successive washes from one solution to the next. Note, only when the cell is simultaneously depolarized with mM levels of $Ca^{2+}$ does the activity of the maxi-K channel increase. Holding potential =+25 mV. With each wash from media to KCl the unitary current decreased from 14 pA to 5 pA indicated that the plasma cells normally maintain a resting potenital of some size, ~ -50 mV. Maxi-K channel conductance has been determined on these cells to be in the 250 pS range [6].

Is the $Ca^{2+}$ of the avian plasma cells involved in anyway with the secretion of immunoglobulins such as IgG? To determine this we used an assay method described by Cameron et al.[11]. Under normal conditions the secretion of immunoglobulins appears to occur at a constant rate [2]. The avian plasma cells can respond to stimuli such as the agonist carbachol by transiently increasing their output of IgG as figure 4 shows (triangles). Total protein has been plotted as a function of time where the time interval is 10 minutes. The secretion occurs within a defined volume and samples can be taken and total protein assayed. Note baseline secretion in time points 1-5, this represents an example of the constitutive secretion of IgG. Here it is assumed that total protein is proportional to IgG [11]. The arrow indicates the application of carbachol. The secretion rate increases but is not sustained and washout with media (second arrow) brings the secretion rate down to control levels.

Figure 4 also illustrates what the absence of extracellular $Ca^{2+}$ does to the carbachol mediated secretion (circles). The arrows again indicate the application and withdrawal of the agonist carbachol. There is no transient response to carbachol in the absence of extracellular $Ca^{2+}$, a strong indicator that the $Ca^{2+}$ current already illustrated is an important link in the secretagogue-like secretion elicited by carbachol for IgG.

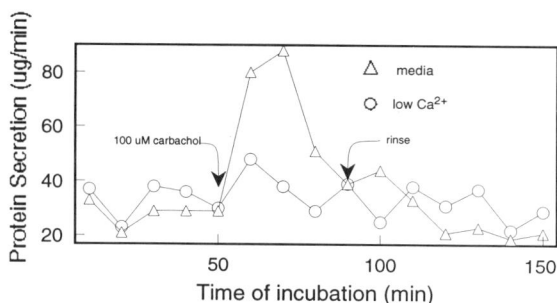

Figure 4. Total protein secretion from gland fragments in Bromberg's media (triangles) and low $Ca^{2+}$Brombergs (circles). The arrows indicate the application of 100 μM carbachol and subsequent rinse with appropriate media.

The data presented illustrate the presence of a $Ca^{2+}$ influx which appears to be voltage dependent and is inactivated by prolonged depolarization. The most likely channel form is the L-type $Ca^{2+}$ channel. Further the $Ca^{2+}$ influx is linked to immunoglobulin secretion. The influx is necessary for the secretagogue triggered IgG secretion of avian lacrimal plasma cells. The $K^+$ outward currents appears to be involved in regulating membrane potential and therefore indirectly affecting the magnitude and duration of any $Ca^{2+}$ influx. The $K^+$ channels act as intrinsic downregulators of the secretagogue activated secretion of the avian plasma cell.

## ACKNOWLEDGMENT

This work was supported by NIH grant EY09406.

## REFERENCES

1. B.M. Mullock, L.J. Shaw, B. Fitzharris, J. Peppard, M.J.R. Hamilton, M.T. Simpson, T.M. Hunt, and R.H. Hinton, Sources of proteins in human bile, *Gut* 26:500 (1985).

2. A.M. Tartakoff, and P. Vassalli, Plasma cell immunoglobulin secretion arrest is accompanied by alteration of the Golgi Apparatus, *J.Exp.Med.* 146:1332 (1977).

3. B.E. Loveland, B. Jarrot, and I.F.C. McKenzie, The detection of ßadrenoreceptors on murine lymphocytes, J. *Immunopharmacology* 3:45 (1981).

4. L.G. Costa, G. Kaylor, and S.D. Murphy, Muscarinic cholinergic binding on rat lymphocytes, *Immunopharmacology* 16:139 (1988).

5. S.L. MacDougall, S. Grinstein, and E.W. Gelfand, Activation of $Ca^{2+}$-dependent $K^+$ channels in human B-lymphocytes by anti-immunoglobulin, *J. Clin. Invest.* 81:449 (1988).

6. P.R. Brink, E.J. Roemer, and B. Walcott, Maxi-K channels in plasma cells, *Pflugers Arch.* 417:349 (1990).

7. B. Walcott, R. Cameron, and P.R. Brink, The anatomy and innervation of lacrimal glands, this volume (1993).

8. S. Grinestein, and S.J. Dixon, Ion transport, membrane potential, and cytoplasmic pH in Lymphocytes: Change during activation, *Physiol. Rev.* 69:417 (1989).

9. O.P. Hamill, A. Marty, E. Neher, B. Sakaman, and F.J. Sigworth, Improved patch clamp techniques for higher-resolution current recording for cells and cell-free membrane patches. *Pflugers Archiv.* 391:85 (1981).

10. K. Manivannan, S.V. Ramanan, R.T. Mathias, and P.R. Brink, Multichannel recordings form membranes which contain gap junctions, *Biophys. J.* 61:216 (1992).

11. R. Cameron, B. Walcott, S.F. Fan, M. Pastor, E. Grine, and P.R. Brink, Second messenger modulation of IgG secretion from chicken lacrimal gland, this volume. (1993).

12. B.B. Bromberg, Autonomic control of lacrimal protein secretion, *Invest. Opthalmol. Vis. Sci.* 20:110 (1981).

13. D.A. Brown, R.J. Docherty, and I. McFadzean, Calcium channels in vertebrate neurons: Experiments on a Neuroblastoma hybrid model, *Ann. N.Y.A.S.* 560:358 (1989).

14. S. Kongsamut, D. Lipsombe, and R. W. Tsein, The N-type Ca channel in Frog Sympathetic Neurons and its Role in alpha-Adrenergic Modulation of Transmitter Release, *Ann.N.Y.A.S.* 560:312 (1989).

# BINDING CHARACTERISTICS, IMMUNOCYTOCHEMICAL LOCATION AND HORMONAL REGULATION OF ANDROGEN RECEPTORS IN LACRIMAL TISSUE

Flavio Jaime Rocha, Robin S. Kelleher, Joan A. Edwards, Janethe D.O. Pena, Masafumi Ono and David A. Sullivan

Department of Ophthalmology, Harvard Medical School and
Immunology Unit, Schepens Eye Research Institute
20 Staniford Street, Boston, MA  02114

## INTRODUCTION

During the past five decades, researchers have found that distinct, gender-related differences exist in the morphology, histochemistry, biochemistry, immunology and molecular biology of the lacrimal gland in a variety of species, including mice, rats, guinea pigs, hamsters, rabbits and humans.[1] These differences include striking variations in acinar cell characteristics (e.g. area, shape, membrane appearance, vesicle and nucleoli densities, nuclear size), lymphocyte populations, messenger RNA levels, enzyme and glycoprotein content, collagen amounts, adrenergic receptor expression, hormone responsiveness, and specific protein secretion.[1] The underlying basis for this sexual dimorphism appears to be due almost entirely to the selective influence of androgens on the lacrimal gland.[1] In contrast, sex steroids such as estrogens or progestins seem to have minimal, or no, direct effect on lacrimal tissue.[1]

The mechanism(s) by which androgens modulate the structure and function of the lacrimal gland remain to be determined, but most likely involves interactions with specific androgen receptors: these binding sites, which are members of the steroid/thyroid hormone/retinoic acid family of ligand-activated transcription factors, appear to mediate all known activities of androgens in other tissues.[2] Therefore, to examine the endocrine basis for the androgen-lacrimal gland interrelationship, the following experiments were designed to: [a] evaluate whether specific, high affinity binding sites for dihydrotestosterone (DHT) occur in rat lacrimal tissue; [b] identify the cellular location of lacrimal androgen receptors; and [c] assess whether changes in the endocrine environment may alter the tissue distribution of androgen receptors. For comparative purposes, we also explored whether specific, high affinity, estrogen binding sites are present in the rat lacrimal gland.

## MATERIALS AND METHODS

Adult male and female Sprague-Dawley rats (6-8 weeks old) were purchased from Zivic-Miller Laboratories and housed in constant temperature rooms with light/dark periods of 12 hours duration. Orchiectomies, ovariectomies, hypophysectomies, and sham-operations were performed by surgeons at Zivic-Miller, and rats were allowed to recover for at least 7 days before further experimentation. To compensate for the electrolyte imbalance in hypophysectomized animals, these rats were given a solution containing sodium chloride (2.03 g/l), potassium chloride (0.083 g/l), magnesium chloride (0.017 g/l) and calcium chloride (0.035 g/l), as previously described.[3] When indicated, operated rats were treated with vehicle or physiological concentrations of testosterone (Innovative Research of America), as reported.[4]

Our method for the determination of the number and affinity of androgen binding sites in rat lacrimal glands has been explained in detail.[5] Briefly, lacrimal tissues were homogenized in ice-cooled TEG buffer (0.05 M Tris-HCl, 1 mM EDTA, 1 mM DTT, 10 mM sodium molybdate, 10% glycerol, pH 7.4) and centrifuged to yield a cytoplasmic supernatant (cytosol). Aliquots of cytosol were incubated with varying amounts of tritiated DHT, testosterone, methyltrienolone (R1881; androgen analogue) or $17\beta$-estradiol (NEN), in the presence or absence of unlabeled competing steroids, for predetermined time intervals at 4°C. Following this incubation period, unbound tritiated steroid was removed by the use of 1% Norit A charcoal/0.5% dextran in TEG. Samples were processed, and data analyzed by Scatchard plots, as previously reported.[5]

To identify the cellular distribution of androgen receptors in lacrimal glands, we utilized an immunoperoxidase protocol, as recently described.[4] In brief, frozen sections (6 μm) of lacrimal tissue were transferred to Vectabond-coated glass slides, fixed in acetone and 4% paraformaldehyde, and blocked with a 2% 'normal' goat serum solution (Vector Laboratories). Sections were then exposed to purified rabbit polyclonal antibody (gift from Dr. Gail Prins, Chicago, IL) to the androgen receptor or appropriate control preparations (e.g. irrelevant rabbit IgG antibodies or rabbit anti-androgen receptor with specific competitors, including purified androgen receptor peptides, which also were gifts from Dr. Prins). Following incubation with first antibody in a humidified chamber, sections were exposed sequentially to separate avidin D and biotin solutions, then to a secondary antibody (biotinylated goat anti-rabbit IgG, Vector). The secondary antibody had been preincubated overnight with rat liver acetone powder at a concentration of 60 mg/ml in a 0.1% gelatin/phosphate buffered saline buffer before use. After antibody treatment, sections were incubated with Vectastain Elite ABC reagent (Vector) and developed with an acetate buffer containing 3-amino-9-ethylcarbazole (Sigma), N, N-dimethylformamide and hydrogen peroxide. Sections were then postfixed in 2% paraformaldehyde, dipped in a lithium carbonate solution and preserved in Crystal Mount (Biomeda). With few exceptions, the application of various reagents to sections was interspersed with either air drying or rinsing of slides with PBS or warm water.

## RESULTS

### Analysis of androgen and estrogen binding sites in rat lacrimal tissue [5]

Our studies[5] demonstrated that specific, high-affinity and saturable binding sites for androgens exist in lacrimal tissues of both castrated male and female rats; these sites were evident after incubation of cytosol with tritiated steroid from 1 to 20 hours. The affinity and number of DHT binding sites, as determined by Scatchard analysis, equaled $1.10 \pm 0.05$ nM and $132 \pm 11$ fmoles/mg protein, respectively, in lacrimal tissues (n = 19) of orchiectomized rats. Similar findings were observed in lacrimal tissues of ovariectomized rats. These binding sites were stereochemically selective and specific for androgens, as shown by competition studies with increasing levels of unlabeled steroids (1.25 nM to 0.3 μM) in castrated male lacrimal glands: DHT, testosterone and R1881 inhibited tritiated DHT binding. In contrast, physiological amounts of estradiol, progesterone and dexamethasone demonstrated little or no competition for androgen receptors. With regards to potential, high-affinity estradiol binding sites, these were undetectable in rat lacrimal gland cytosol.

### Distribution of androgen receptors in the rat lacrimal gland [4]

Immunocytochemical analysis[4] showed that androgen receptors are present, and located almost exclusively, in nuclei of acinar cells in the rat lacrimal gland (n = 8 tissues/group). The distribution of these binding sites was far more extensive in glands of males, as compared to females, and extremely susceptible to alterations in the endocrine environment. Thus, orchiectomy, but not sham castration, led to a precipitous decline in the number of cells containing androgen receptors. Moreover, ablation of the pituitary gland dramatically interfered with androgen receptor expression: in lacrimal tissues of many hypophysectomized rats it was difficult to detect any receptor-positive cells.

The suppressive influence of castration on androgen binding site appearance in the lacrimal gland could be reversed by androgen administration. Administration of testosterone, but not placebo compounds, to orchiectomized rats for 7 days resulted in a marked increase in the frequency and distribution of lacrimal androgen receptors;[4] their location was analagous to that in tissues of intact males. Similarly, exposure of orchiectomized and hypophysectomized rats to androgen therapy (e.g. 7 days) induced a significant up-regulation of androgen receptor expression in lacrimal tissue.[4]

## DISCUSSION AND ACKNOWLEDGMENTS

The present findings demonstrate that specific, high affinity binding sites for androgens, but not estrogens, exist in lacrimal tissues of castrated male and female rats. These results, which are consistent with those of other investigators,[6,7] indicate that androgen action on lacrimal tissue is most likely mediated through the association with these receptors. In support of this hypothesis, the administration of anti-androgens, which competitively bind to androgen receptors, inhibit androgen-induced effects in lacrimal cells both in vivo and in vitro.[8,9]

Our results also show that: [a] the location of androgen receptors in the rat lacrimal tissue is predominantly intranuclear. This compartmentalization, which is similar to that observed in other tissues,[2,10,11] may be due to the presence of a nuclear targeting signal, that occurs in the receptor hinge region immediately following the DNA-binding domain;[11] [b] the density and distribution of lacrimal androgen receptors are considerably greater in glands of males, as compared to females. This reduced content of androgen receptors in female lacrimal tissues, when coupled to a lower circulating androgen level, may well represent the principal basis for the lacrimal gland's sexual dimorphism; [c] alteration in the endocrine environment (e.g. orchiectomy or hypophysectomy) results in a dramatic decrease in the number of androgen receptors in rat lacrimal tissue.[4,5] This receptor diminution correlates well with the reported decline in lacrimal gland function and/or androgen responsiveness in male rats after castration or interruption of the hypothalamic-pituitary axis;[1] and [d] androgens upregulate the amount of androgen receptor protein in lacrimal tissue of orchiectomized rats; this autoregulatory process has also been observed in other androgen target tissues.[10,12] To extend these findings, our ongoing research is examining the mechanisms underlying the endocrine modulation of androgen receptor concentrations, as well as the innate processes involved in the androgen control of transcriptional and posttranscriptional events, in the lacrimal gland.

We express our appreciation to Dr. Gail Prins for her gift of the antibody to, and peptide fragments of, the rat androgen receptor. This research was supported by NIH grant EY05612 and a grant from the Massachusetts Lions' Research Fund.

## REFERENCES

1. D.A. Sullivan, in: "The Neuroendocrine-Immune Network," S. Freier, ed., CRC Press, Boca Raton, FL, pp199-238 (1990).
2. J.H. Clark, W.T. Schrader, and B.W. O'Malley, in: J.D. Wilson, and D.W. Foster, eds., "Williams Textbook of Endocrinology," WB Saunders, Philadelphia, pp 35-90 (1992).
3. D.A. Sullivan, and M.R. Allansmith, Immunology 60:337 (1987).
4. F.J. Rocha, L.A. Wickham, J.D.O. Pena, J. Gao, M. Ono, R.W. Lambert, R.S. Kelleher, and D.A. Sullivan, article submitted (1993).
5. D.A. Sullivan, J.A. Edwards, and R.S. Kelleher, article submitted (1993).
6. M. Ota, S. Kyakumoto, and T. Nemoto, Biochem. Internat. 10:129 (1985).
7. M. Laine, and J. Tenovuo, Arch. Oral Biol. 28:847 (1983).
8. J.D. Hahn, J. Endocr. 45:421 (1969).
9. R.W. Lambert, R.S. Kelleher, L.A. Wickham, J.P. Vaerman, and D.A. Sullivan, article submitted (1993).
10. G.S. Prins, and L. Birch, Endocrinology 132:169 (1993).
11. J.A. Simental, M. Sar, M.V. Lane, F.S. French, E.M. Wilson, J. Biol. Chem. 266:510 (1991).
12. A. Krongrad, C.M. Wilson, J.D. Wilson, D.R. Allman, and M.J. McPhaul, Mol. Cell Endocr. 76:79 (1991).

# REGULATION OF LACRIMAL GLAND IMMUNE RESPONSES

P.C. Montgomery[1], N.L. O'Sullivan[1], L.B. Martin[1], C.A. Skandera[1], J.V. Peppard[2] and A.G. Pockley[3]

[1]Immunology and Microbiology
Wayne State University Medical School
Detroit, MI 48201 USA
[2]Pharmaceuticals Division
CIBA-GEIGY
Summit, NJ 07091 USA
[3]Professorial Surgical Unit
St. Bartholomew's Hospital
London, UK

## INTRODUCTION

The purpose of this review is: (1) to consider the relationship of the lacrimal gland to the mucosal network and general factors regulating lacrimal gland immune responses, (2) to present current data from selected studies on lacrimal gland immunoregulation, and (3) to discuss the future directions of lacrimal gland immunobiology and regulation.

A distinct localized immune system, which is generally characterized by the appearance of antibodies in external secretions bathing mucosal surfaces, is now well documented. Although the detailed parameters of the secretory immune system vary between species, the general predominance of the IgA isotype in external secretions, the distinct structural features of secretory IgA (S-IgA) and the increased numbers of IgA producing plasmacytes in tissues underlying mucosal epithelium have been major hallmarks. The mucosal immune network concept has received support from two main lines of evidence: antibody induction studies and cell migration experiments. It is now well recognized that S-IgA antibodies can be elicited by both local antigen administration as well as stimulation at selected distal mucosal sites. Cellular analysis has documented that IgA committed cells from gastrointestinal (GALT) and bronchial associated lymphoid tissue (BALT) seed remote glandular mucosal tissues (mammary, salivary and lacrimal glands) as well as the lamina propria of the small intestine, bronchial and urogenital tissues, establishing linkage of mucosal sites and the mucosal network. Figure 1 summarizes the major features of the mucosal immune network and the events occurring following antigen stimulation of GALT.

As noted above, lacrimal glands are functional components of the mucosal immune network and contribute to barrier defense at the ocular surface by producing IgA antibodies which are ultimately expressed as S-IgA in tears. Although it is known that lacrimal glands are a repository for cells precommitted to IgA production, the precise

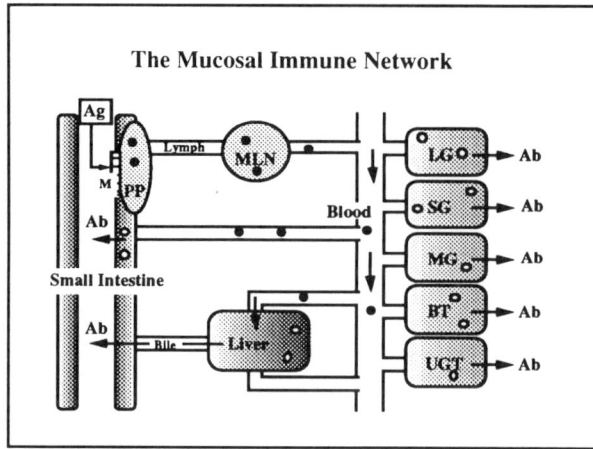

The Mucosal Immune Network

Figure 1.   Intestinal antigens (Ag) are taken up by microfold (M) cells overlying Peyer's patches (PP) and are delivered to lymphoid cells in the PP.   T cells and IgA committed B cells (●) migrate to the mesenteric lymph nodes (MLN), enter the circulation and traffic to the lacrimal (LG), salivary (SG), mammary (MG) glands as well as to the lamina propria of the small intestine, the bronchial (BT), urogenital (UGT) tracts and liver.   B lymphocytes lodging in mucosal tissues differentiate into IgA secreting plasma cells (0), producing polymeric IgA antibodies   (Ab), which are transported into external secretions bathing mucosal surfaces. In some species the liver directly transports significant quantities   of polymeric IgA from  the circulation into bile.

regulatory events leading to the induction and expression of   lacrimal gland IgA antibody responses are not completely defined.   Several reviews deal with the molecular and cellular parameters involved in the regulation of mucosal immunity.[1-3]   Table 1 summarizes factors regulating mucosal immune responses which are generally applicable to lacrimal gland immunobiology.   For non-replicating antigens, both local (ocular topical, OT) and remote-site (oral/gastrointestinal, GI)   stimulation have   been effective at inducing   lacrimal   gland immune responses, with OT being most   effective.[4]   GI followed by OT stimulation down-regulates tear IgA antibody responses, while   OT   followed by   GI stimulation maintains the response. For these immunization   routes, particulate antigens are more effective   than   soluble   antigens. Traditional mucosal adjuvants (cholera

Table 1.   Regulation of mucosal immune responses.

|   |
| --- |
| - Physical state of the antigen |
|     soluble *vs* particulate |
| - Immunization routes |
|     local *vs* remote site |
|     sequence and combination of routes |
| - Immune potentiators |
|     mucosal adjuvants[1] |
|     cytokines |
| - T cells |
| - Lymphocyte traffic |
| - Idiotypes[1] |
| - Neuroendocrine interactions |
|     hormones/androgens |

[1]Not documented for lacrimal glands.

toxin, avridine, MTP) administered by the OT route have not enhanced tear IgA responses. T cells are known to play a role in certain mucosal IgA antibody responses and lacrimal gland T cells appear to promote IgA synthesis in co-culture systems.[5] Studies on the neuroendocrine and hormonal regulation of lacrimal gland responses are reviewed elsewhere.[6,7] With respect to secretory component (SC) production and IgA synthesis, the effects of sexual dimorphism and aging have been documented.[8] Androgens are also known to affect SC synthesis in lacrimal gland acinar cell cultures[9] and IgA production in vivo.[10] The background information for lymphocyte traffic, cytokines and anti-idiotypic antibodies is detailed subsequently.

## LYMPHOCYTE TRAFFIC

An issue central to regulation in the mucosal immune network and, in particular, to lacrimal gland immunobiology, is the mechanism accounting for preferential localization or retention of specific lymphoid populations. In organized lymphoid tissue lymphocyte–high endothelial venule (HEV) interactions control lymphocyte distribution and movement into peripheral lymph nodes and organized mucosal lymphoid tissues.[11-14] In general, these interactions involve lymphocyte homing receptors (HRs) which interact with vascular addressins on HEVs and other specialized endothelial cells. Our laboratory has utilized an in vitro binding assay employing sections from glandular mucosal tissues and shown that circulating lymphocyte populations preferentially adhere to lacrimal gland acinar epithelial cells.[15] Table 2 summarizes the similarities and differences between lymphocyte adherence to lacrimal gland and lymph node tissues. While many characteristics are shared with those reported for organized lymphoid tissue, there are important distinctions. Our recent focus has been to classify lymphocyte–lacrimal interactions within the context of defined families of adhesion molecules. Presently, it appears that lymphocyte HRs which mediate HEV binding are involved in lymphocyte binding to lacrimal tissue, although it is not yet clear whether the lymph node HR (LNHR), the Peyer's patch HR (PPHR) or a distinct receptor is directly involved.[16] Carbohydrate inhibition patterns[17] are similar to those noted for HEV binding and are suggestive of a selectin interaction. Additionally, the lacrimal gland interaction does not appear to involve $\beta_1$ or $\beta_2$ integrins or the CD44 molecule (unpublished data) arguing against the involvement of the integrin or cartilage link adhesion molecule families. Finally, isolated lacrimal gland lymphocytes have been shown to display significantly greater amounts of the rat lymph node HR (compared to Peyer's patch HR), which is known to be 85% homologous to LECAM-1, a member of the selectin family.[18] Although further studies are required to precisely define the nature of the molecular interactions regulating lymphocyte traffic to lacrimal tissues, it is clear that specific lymphocyte–acinar epithelial cell interactions account for their accumulation in lacrimal glands.

## CYTOKINES

Cytokines are known to exert a variety of important regulatory effects on cells participating in immune responses[19] and recombinant molecules are now available in quantities that permit detailed functional and therapeutic evaluations. Systemic administration of interleukin–2 (IL-2) has increased antibody responses in mice[20,21] and has been shown to enhance the protective effects of vaccines in other animal models. These and other observations showing that IL-5 and IL-6 augment in vitro IgA responses in murine B cell cultures[22-25] have

**Table 2.** Lymphocyte adherence: lacrimal gland <u>vs</u> lymph node tissue.

---

Similarities:
- requires viable, metabolically active cells
- requires intact cytoskeletal function
- calcium dependent
- not MHC restricted[1]
- mediated by a trypsin-sensitive surface molecule
- inhibited by antibodies to homing receptors
- involves carbohydrate recognition

Differences:
- epithelial cells target [vs endothelial cell]
- preferential adherence of B lymphocytes[1]
- enhanced adherence with immature tissues

---

[1]O'Sullivan, Chin and Montgomery, in preparation.

suggested that certain cytokines exert important regulatory influences on mucosal IgA responses. Our laboratory has shown that IL-5 and IL-6 can specifically enhance IgA production in a rat lacrimal gland tissue culture system.[26] Further, IL-5 and IL-6 in combination with antigen were shown to enhance tear IgA antibody responses following ocular topical (OT) administration in the rat.[27] Serum IgG responses were also monitored and were unaffected by IL treatment. These studies have now been extended to assess the longevity of the IL induced enhancement of tear IgA antibody responses. Table 3 summarizes the initial response data and extends these observations through 5 immunization cycles over a 10 month time period. Tears (and serum) were obtained at 5 time points during the month following immunization. For this analysis, the mean group tear IgA antibody level was calculated for each immunization cycle using the data obtained from individual time points for the two immunization groups: antigen alone [Ag] or antigen + IL-5/IL-6 [Ag(IL)]. A response ratio was calculated to facilitate comparison of the

**Table 3.** Effect of IL-5 and IL-6 on tear IgA antibody induction following ocular topical immunization.

| Immunization Cycle | Month | Protocol | Tear IgA Antibodies ($\mu g/ml$)[3] | Response Ratio[5] |
|---|---|---|---|---|
| 1 | 1 | Ag[1] | ND[4] | |
| | | Ag+IL[1,2] | ND | ND |
| 2 | 2 | Ag[1] | 2.26 | |
| | | Ag+IL[1,2] | 8.54 | 3.8 |
| 3 | 3 | Ag[1] | 3.86 | |
| | | Ag(IL)[1] | 10.66 | 2.8 |
| 4 | 5 | Ag[1] | 1.89 | |
| | | Ag(IL)[1] | 5.53 | 2.9 |
| 5 | 10 | Ag[1] | 3.04 | |
| | | Ag(IL)[1] | 2.86 | 0.9 |

[1]500$\mu$g DNP-pneumococcus administered over 3 days.
[2]IL-5(50 U) + IL-6(50 U) administered each day over 7 days.
[3]Anti-DNP antibodies determined by RIA; mean values for 5 time points for each immunization group (4-5 rats/group).
[4]Not Determined.
[5]Ag(IL) group mean response ÷ Ag group mean response.

group responses. It should be noted that the Ag(IL) group received IL-5/IL-6 during immunization cycles 1 and 2, with subsequent restimulation (cycles 3,4,5) carried out with antigen alone. Following the second immunization cycle, the Ag(IL) group exhibited a 3.8-fold enhancement in tear IgA antibody levels when compared to the Ag group. When both groups were given antigen only (cycle 3), the Ag(IL) group continued to display an enhanced response (2.8-fold). This enhancement continued following a 1 month rest and restimulation with antigen (cycle 4). After a subsequent 4 month rest and restimulation (cycle 5), tear IgA antibody responses of both groups were equivalent. At all times during this study the serum IgG antibody responses of both groups did not differ. These data show that the OT administration of IL-5/IL-6 in combination with antigen elicits an enhanced tear IgA antibody response. This enhancement continues with subsequent antigen stimulation for at least 3 months without continued IL administration.

## IDIOTYPES

Another class of molecules involved in immune regulation are anti-idiotypic antibodies.[28,29] Operationally and functionally, two major categories are of relevance to our current investigations: 1. anti-idiotypic antibodies that recognize framework epitopes (Ab2α) and 2. anti-idiotypic antibodies that recognize binding site epitopes in the antibody variable region (Ab2ß). Figure 2 diagrammatically depicts essential idiotypic concepts. Anti-idiotypic antibodies (Ab2α) generally mark families of antibody molecules within inbred species and are known to exert regulatory influences on the immune system.[28] Anti-idiotypic antibodies reacting with combining site associated epitopes (Ab2ß) carry the internal image of the antigen reactive site of Ab1 and are able to induce Ab3 which can interact with the original antigenic stimulus.[30,31] Ab2ßs have been of interest in that they can be used for vaccine purposes.[30,31] Although recent reports have now shown that mucosally targeted anti-idiotypic vaccination elicits protective salivary[32] and gastrointestinal[33] responses, the regulatory role and the response to anti-idiotypic antibodies have not been detailed in the ocular compartment. Using monoclonal antibody technology, which provides a stable supply of anti-idiotypic reagents, we have prepared framework specific (Ab2α) and site specific (AB2ß) antibodies. The classification

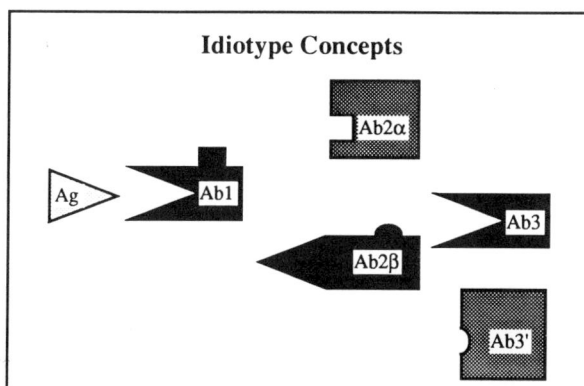

Figure 2. Antigen (Ag) elicits antibody 1 (Ab1). Ab1 elicits Ab2α which interacts with framework epitopes and Ab2ß which interacts with combining site epitopes. Ab2ß elicits Ab3 and Ab3' in a similar fashion. Ab3 has he capacity to interact with the original antigenic stimulus.

of these antibodies as Ab2α or Ab2ß was based on their interaction with a panel of monoclonal IgA anti–DNP antibodies (Ab1) and their inhibitory effects on Ab1–DNP interactions. In a preliminary study we have assessed the ability of these two types of anti–idiotypic antibodies to elicit tear IgA anti–DNP responses. Initial findings indicate that Ab2ß (site specific, internal image) can induce antigen specific (anti–DNP) tear IgA responses comparable to those obtained with antigen, while Ab2α (framework specific) does not. Although these data require confirmation, they represent the first documentation that anti–idiotypic antibodies can elicit antigen specific responses in tears.

## CONCLUSIONS AND FUTURE DIRECTIONS

**Lymphocyte Traffic.** The current data indicate that lymphocyte accumulation in lacrimal tissue is governed by specific interactions with acinar epithelial cells. Although there is suggestive evidence that the lymphocyte receptor may be a member of the selectin adhesion molecule family, the precise relationship of this receptor to existing adhesion molecules has not yet been established. It is also not known if LNHRs, PPHRs or a distinct molecule mediates the adherence to lacrimal tissues. Further, the nature of the acinar cell ligand has not been defined. Experimentation in these areas is required to gain a complete understanding of the molecular interactions responsible for lymphocyte traffic not only to lacrimal glands, but also to the other non–lymphoid glandular tissues of the mucosal immune network. While there is information regarding cell populations that reside in lacrimal tissues in animal models,[16,34] this work needs to be extended to humans. In addition, it is also essential to compare lymphocyte traffic patterns that occur during antigen or disease perturbation.

**Immune Potentiators.** It is now known that OT administration of IL–5 and IL–6 in combination with antigen can enhance tear IgA antibody responses and that this effect continues in the absence of IL stimulation. Presently, little is known regarding the mechanism of this enhancement. The regulatory roles of other cytokines also needs to be investigated not only from the standpoint of exogenous therapeutic administration, but also from the perspective of endogenous production in lacrimal tissue. Additional studies are required to determine the usefulness of other mucosal adjuvants and investigations should be initiated to explore the potential use of antigen delivery systems (eg. liposomes, microparticles) for topical immunization. In dealing with these issues it will be necessary to be selective in choosing immune potentiators that do not promote ocular inflammation or pathology.

**Idiotypes.** It is encouraging that anti–idiotypic antibodies can be used to elicit specific tear IgA antibody responses. In order to fully assess the vaccine potential of anti–idiotypic antibodies, it is necessary to assess the protective capacity of these reagents in an infection model. Further, the regulatory capacity of anti–idiotypic antibodies in the ocular compartment has yet to be explored.

**Other Issues.** Interestingly, the mechanism by which OT antigen administration elicits tear IgA antibody production is not known. The effect of various types of replicating antigens on both the B and T cell compartments also needs to be fully explored. The exciting work regarding neuroendocrine effects on lacrimal gland immunoregulation needs to be extended to determine the precise effects on lymphoid subsets that participate in lacrimal gland immune responses. While a number of investigators are addressing the effects of aberrant responses (eg. autoimmunity) on lacrimal gland immunobiology, it is important to begin to apply mechanistic information to therapeutic modalities.

It is exciting to witness the new data emerging from studies on the biology of the lacrimal gland and encouraging to see the extrapolation of these findings to other mucosal compartments. Although, by design, this review was not intended to be comprehensive, the description of selected areas of lacrimal gland immunoregulation is intended to provide a focus for stimulating additional questions and future investigations.

## ACKNOWLEDGEMENTS

This work was supported by NIH grants EY05133 and EY07093.

## REFERENCES

1. J. Mestecky and J.R. McGhee, Immunoglobulin A (IgA): Molecular and cellular interactions involved in IgA biosynthesis and immune responses, *Adv. Immunol.* 40:153(1987).
2. J.M. Phillips-Quagliata and M.E. Lamm, Migration of lymphocytes in the mucosal immune system, *in* "Migration and Homing of Lymphoid Cells," A.J. Husband, ed., CRC Press, Boca Raton(1988).
3. N.K. Childers, M.G. Bruce and J.R. McGhee, Molecular mechanisms of immunoglobulin A defense, *Ann. Rev. Microbiol.* 43:503(1989).
4. J.V. Peppard, R.V. Mann and P.C. Montgomery, Antibody production in rats following ocular topical or gastrointestinal immunization: kinetics of local and systemic antibody production, *Curr. Eye Res.* 7:471(1988).
5. R.M. Franklin, D.W. McGee and K.F. Shepard, Lacrimal gland-directed B cell responses, *J. Immunol.* 135:95(1985).
6. A.M. Stanisz, J. Bienenstock and A. Agro, Neuromodulation of mucosal immunity, *Reg. Immunol.* 2:414(1989).
7. D.A. Sullivan, Hormonal influence on the secretory immune system of the eye, *in* "The Neuroendocrine - Immune Network," S. Freier, ed., CRC Press, Boca Raton(1990).
8. D.A. Sullivan and M.R. Allansmith, The effect of aging on the secretory immune system of the eye, *Immunology* 63:403(1988).
9. R.S. Kelleher, L.E. Hann, J.A. Edwards and D.A. Sullivan, Endocrine, neural and immune control of secretory component output by lacrimal gland acinar cells, *J. Immunol.* 146:3405(1991).
10. D.A. Sullivan, and M.R. Allansmith, Hormonal influence on the secretory immune system of the eye. Androgen modulation of IgA levels in tears of rats, *J. Immunol.* 134:2978(1985).
11. E.C. Butcher, The regulation of lymphocyte traffic, *Curr. Top. Microbiol. Immunol.* 128:85(1986).
12. J.J. Woodruff, L.M. Clarke, and Y.H. Chin, Specific cell-adhesion mechanisms determining migration pathways of recirculating lymphocytes, *Ann. Rev. Immunol.* 5:201(1987).
13. L.M. Stoolman, Adhesion molecules controlling lymphocyte migration, *Cell* 56:907(1989).
14. S. Jalkanen, G.S. Nash, J. De los Toyes, R.P. MacDermott and E.C. Butcher, Human lamina propria lymphocytes bear homing receptors and bind selectively to mucosal high endothelium, *Eur. J. Immunol.* 19:63(1989).
15. N.L. O'Sullivan and P.C. Montgomery, Selective interactions of lymphocytes with neonatal and adult lacrimal gland tissues, *Invest. Ophthalmol. Vis. Sci.* 31:1615(1990).
16. N.L. O'Sullivan, C.A. Skandera and P.C. Montgomery, Expression of homing receptor molecules on rat lacrimal gland lymphocytes, *"Recent Advances in Mucosal Immunology,"* J. McGhee, J. Mestecky, H. Tlaskalova and J. Sterzl, eds., Plenum, New York (in press).
17. N.L. O'Sullivan, R. Raja and P.C. Montgomery, Lymphocyte adhesive interaction with lacrimal gland acinar epithelium involves carbohydrate recognition, *in* "Lacrimal Gland, Tear Film and Dry

Eye Syndromes," D.A. Sullivan, ed., Plenum, New York (in press).

18. Y.H. Chin, R. Sackstein and J.P. Cai, Lymphocyte homing receptors and preferential migration pathways, *FASEB J.* 2:2462(1988).

19. K. Arai, F. Lee, A. Miyajima, S. Miyatake, N. Arai and T. Yokota, Cytokines: Coordinators of immune and inflammatory responses, *Ann. Rev. Biochem.* 59:783(1990).

20. H. Kawamura, S.A. Rosenberg and J.A. Berzofsky, Immunization with antigen and interleukin 2 *in vivo* overcomes Ir gene low responsiveness, *J. Exp. Med.* 162:381(1985).

21. C.M. Weyand, J. Goronzy, M.J. Dallman and C.G. Faithman, Administration of recombinant interleukin 2 *in vivo* induces a polyclonal IgM response, *J. Exp. Med.* 163:1607(1986).

22. D.Y. Kunimoto, G.R. Harriman and W. Strober, Regulation of IgA differentiation in CH12LX B cells by lymphokines. IL-4 induces membrane IgM-positive CH12LX cells to express membrane IgA and Il-5 induces membrane IgA-positive CH12LX cells to secrete IgA, *J. Immunol.* 141:713(1988).

23. D.A. Lebman and R.L. Coffman, The effects of IL-4 and IL-5 on the IgA response by murine Peyer's patch B cell subpopulations, *J. Immunol.* 141:2050(1988).

24. S. Schoenbeck, T. McKenzie and M.F. Kagnoff, Interleukin 5 is a differentiation factor for IgA B cells, *Eur. J. Immunol.* 19:965(1989).

25. K.W. Beagley, J.H. Eldridge, F. Lee, H. Kiyono, M.P. Everson, W,J, Koopman, T. Hirano, T. Kishimoto, and J.R. McGhee, Interleukins and IgA synthesis. Human and murine interleukin 6 induce high rate IgA secretion in IgA-committed B cells, *J. Exp. Med.* 169:2133(1989).

26. A.G. Pockley and P.C. Montgomery, The effects of interleukins 5 and 6 on immunoglobulin production in rat lacrimal glands, *Reg. Immunol.* 3:242(1991).

27. A.G. Pockley and P.C. Montgomery, In vivo adjuvant effect of interleukin 5 and 6 on rat tear IgA antibody responses, *Immunology* 73:19(1991).

28. K. Rajewsky and T. Takemor, Genetics, expression, and function of idiotypes, *Ann. Rev. Immunol.* 1:569(1983).

29. J.M. Davie, M.V. Seiden, N.S. Greenspan, C.T. Lutz, T.L. Bartholow, and B.L. Clevinger, Structural correlates of idiotopes, *Ann. Rev. Immunol.* 4:147(1986).

30. F.G.C.M. Uytde-Haag, H. Bunschoten, K. Weijer and A.D.M.E. Osterhaus, From Jenner to Jerne: Towards idiotype vaccines, *Immunol. Rev.* 90:93(1986).

31. J.B. Hernaux, Idiotypic vaccines and infectious diseases, *Infect. Immunity,* 56:1407(1988).

32. S. Jackson, J. Mestecky, N.K.. Childers and S.M. Michalek, Liposomes containing anti-idiotypic antibodies: an oral vaccine to induce protective secretory immune responses specific for pathogens of mucosal surfaces, *Infect. Immunity,* 58:1932(1990).

33. G.P. Lucas, C.L. Cambiaso and J.P. Vaerman, Protection of rat intestine against cholera toxin challenge by monoclonal anti-idiotypic antibody immunization via central and parenteral routes, *Infect. Immunity* 59:3651(1991).

34. P.C. Montgomery, J.V. Peppard and C.A. Skandera, A comparison of lymphocyte subset distribution in rat lacrimal glands with cells from tissues of mucosal and non-mucosal origin, *Curr. Eye Res.* 9:85(1990).

# THE LACRIMAL ANTIBODY RESPONSE TO VIRAL
# AND CHLAMYDIAL INFECTIONS

Maureen G. Friedman

Virology Unit, Faculty of Health Sciences
Ben Gurion University of the Negev
Beer Sheva, Israel

## INTRODUCTION

Secretory IgA (sIgA) has been detected in tears of humans in response to both viral and chlamydial infections (reviewed by Friedman, 1990). These antibodies may be synthesized in the lacrimal gland as a result of stimulation of the common mucosal immune system, which leads to homing of sensitized lymphocytes to the various exocrine glands (Russell and Mestecky, 1988), or possibly as a result of local sensitization by as yet undetermined means. Antibodies of other isotypes specific for infectious agents (IgA lacking secretory component, IgG, IgM, and IgE) have also been detected in lacrimal fluid (McClellan et al., 1973). The purpose of this report is to survey some of the literature relevant to the lacrimal antibody response, to formulate some hypotheses, and to point out directions for future investigations.

## SAMPLING AND LABORATORY TECHNIQUES

### Tear sampling.

Several techniques have been used to obtain tear samples including use of polished capillaries (without eye contact, with or without stimulation), saturation of sterile surgical sponges placed into the conjunctival sac, and saturation of Schirmer test strips with subsequent elution into a solution such as phosphate buffered saline. Techniques which do not involve physical contact with the ocular tissue are least likely to cause exudation of serum proteins; however, under conditions of strong stimulation, serum exudation of antibodies and other proteins into the lacrimal fluid may occur (Stuchell et al., 1984).

**Laboratory techniques for isotype-specific detection of antibodies**

Immunoperoxidase (IPA), immunofluorescence (IFA) and microimmuno-fluorescence (MIF), enzyme-linked immunosorbent assay (ELISA) and radioimmuno-assay (RIA) techniques have been used for detection and titration of virus-specific or chlamydia-specific antibodies in tears (Friedman, 1990). The immunoblotting technique (IB) has been used to determine the antigenic specificity of lacrimal antibodies (Caldwell et al., 1985; Shallal et al., 1992; Liotet et al., 1987; Renom et al., 1990). In general, the RIA is the most sensitive technique and will often detect antibodies at sample dilutions which test negative by IFA and IPA. The ELISA can under certain conditions almost approach the sensitivity of the RIA. The relevant techniques are described more fully in the articles cited in this review.


**RESULTS AND DISCUSSION**

Tables 1 and 2 present a summary of data obtained with respect to different antibody isotypes elicited in response to viral and chlamydial infections, respectively. The techniques used for detection may in some cases have determined which antibodies could or could not be detected. For example, IFA was used to detect IgA antibodies to adenovirus in conjunctivitis and epidemic keratoconjunctivitis. At the level of sensitivity of this assay, IgA detection appeared to be specific and indicative of current infection. However, secretory IgA could not be reliably detected (using antibody to secretory component--SC).

The MIF test has been used specifically for serologic detection of chlamydial infections. IgG antibodies have been detected in tears of affected eyes, often at titers higher than in corresponding serum samples, implying at least some local production (Treharne, 1986). IgA antibodies were detected in tears using this technique, and titers declined following treatment. Detection of sIgA by MIF (using anti secretory component antibody) has not been reported. However, detection of specific sIgA by ELISA has been reported by Elsana et al. (1990) as well as by Buisman et al.(1992). Both groups reported secretory antibodies in lacrimal fluid in more cases than were culture positive, and recommended the determination of chlamydia-specific sIgA in tears as an adjunct to chlamydia detection tests in cases of chronic conjunctivitis.

While IgG antibodies found in tears were originally thought to be a serum exudate (McClellan et al., 1973), several more recent investigations indicate that there is probably a local component of IgG synthesis as well. The evidence for this includes MIF detection of chlamydia-specific IgG titers higher in tears than in serum (Treharne et al., 1986), and ELAVIA® detection of IgG antibodies against the AIDS virus (HIV) in tears of seropositive patients with normal lacrimal levels of albumin (Liotet et al., 1987). It may be noted that IgG antibody to the major outer membrane protein of *Chlamydia trachomatis* was found only transiently in tears of isolation positive patients, while these antibodies persisted in sera (Shallal et al., 1992). Also, in persons with herpes simplex virus (HSV) keratitis, the tear:serum ratio of HSV-specific IgG1 antibodies was different than the tear:serum ratio of HSV-specific IgG4 antibodies (McBride and Ward, 1987).

## Table 1. Summary of virus-specific antibodies detected in human tears

| Infectious agent | Method | Results and comments | Reference | Year |
|---|---|---|---|---|
| HSV | RIA | Detection of HSV-specific sIgA in tears of 7/7 HSV keratitis patients and 0/7 controls | Pederson Norrild | 1982 1982 |
| HSV | RIA | HSV specific IgA detected in tears of 12/14 adults with dendritic herpes keratitis and in 2/21 controls. Titers as high as 20,000 in tears and 25,000 in serum. Anti-measles IgA detected in serum, but not in tears. | Shani | 1985 |
| HSV | RIA | Unequal tear:serum ratios of IgG subclasses of virus-specific antibodies | McBride | 1987 |
| Adenovirus | IFA | Detection of adenovirus specific IgA in tears of 5/7 patients with adenovirus keratoconjunctivitis but not HSV- or Chlamydia-specific IgA. During EKC epidemic, 22 patients followed; all lost adeno-specific tear IgA by 9 mo. (most by 4-5 mo.). Detection using anti-SC antibody unreliable (very weak). | Nordbø | 1986 |
| HIV | ELISA, IB | Detected IgG by ELAVIA (Instit. Pasteur) in tears of 33-83% of 16 HIV seropositive patients and none of 25 controls. Lacrimal albumin levels were normal. | Liotet | 1987 |
| HIV | ELISA, IB | Of 68 children born to seropositive mothers, tested at 18 mo of age, 9 were HIV seropositive and all had HIV IgA in tears. Eight had lacrimal HIV IgA already at 9 mo of age. Only children without lacrimal HIV at 9 mo of age became HIV seronegative at 18 mo. | Renom | 1990 |
| Measles | RIA | Measles specific tear IgA detected in all children with laboratory confirmed natural measles infection; similar to serum IgA, tear IgA persisted up to 14 mo or longer after infection. However, SC-bearing antibody in tears was usually no longer detectable after 3 mo, similarly to salivary IgA. | Friedman | 1989 |

Abbreviations used in the table: HSV: Herpes simplex virus; HIV: Human immunodeficiency virus; RIA: Radioimmunoassay; IFA: Immunofluorescence assay; ELISA: Enzyme-linked immunosorbent assay; IB: Immunoblotting; SC: secretory component; sIgA: secretory IgA.

Local synthesis of IgA which does not bear SC as well as sIgA may occur in ocular tissues, although this has not been proven due to the lower sensitivity of SC measurement. Evidence for such synthesis includes the description by Friedman et al. (1989) of tear antibodies in measles virus infections. Measles specific lacrimal sIgA decreased with time and disappeared in about a month after onset of rash in natural measles infections (similarly to salivary IgA), while measles specific lacrimal IgA remained detectable with anti-alpha chain antibodies for long periods of time (up to14 months post onset of rash) by the RIA used, similarly to measles specific serum IgA. In some cases the tear titers were higher than the corresponding serum titers. The apparent long term persistence of the antibodies may have been due, at least in part, to the sensitivity of the RIA used. However, this should not place into doubt their existence. On the other hand, as noted above, ELISA techniques are generally somewhat less sensitive than RIA, while IFA and MIF are still less sensitive and would not be expected to detect very low antibody levels.

### Table 2. Summary of Chlamydia-specific antibodies detected in human tears

| Method | Results and comments | Reference | Year |
| --- | --- | --- | --- |
| MIF | Adult paratrachoma. IgG detected in tears of affected eye. In 43 of the 60 IgG positive patients, tear titers were equal to those detected in serum; 10 had higher IgG tear than serum titers; and 7/60 had serum titers higher than tear titers. Of 62 isolation positive patients, 60 had Chlamydia-specific IgG in tear fluids, and 42 had IgA. | Treharne | 1986 |
| MIF | Of 52 culture positive adults with chlamydial conjunctivitis, 42 had specific IgA in tear samples, and titers declined following treatment. In neonates, 24/67 culture positives had lacrimal IgA (titer $\geq 8$). No culture negative infants had specific tear IgA. | Herrman | 1991 |
| ELISA | Chlamydia-specific sIgA in tears of patients with conjunctivitis correlated well with chlamydia isolation. All initially positive (lacrimal sIgA) patients became negative 1 mo or more after treatment. However, not all became chlamydia IgA seronegative. | Elsana | 1990 |
| ELISA | Chlamydia specific sIgA detected in tears of all patients with positive culture (8 of 30 with chronic conjunctivitis) and an additional 4 culture negative, but clinically suggestive cases. | Buisman | 1992 |
| IB | IgG antibody to major outer membrane protein was shortlived in tears, but persistent in sera (18 mo) of patients with isolation or clinical evidence of infection. | Shallal | 1992 |

Abbreviations used in table: MIF: Microimmunofluorescence; ELISA: Enzyme linked immunosorbent assay; IB: Immunoblotting; sIgA:secretory IgA.

## HYPOTHESIS AND DIRECTIONS FOR FUTURE INVESTIGATIONS

Since it has been shown that long term antibody production is probably the result of periodic antigenic stimulation of memory B cells by antigen sequestered on follicular dendritic cells (Tew et al., 1980; Donaldson et al., 1986; Gray and Skarvall, 1988), it may be that such cells could stimulate long term IgA production in the ocular environment. The presence of follicular dendritic cells in the lacrimal gland but not in the salivary gland has been reported (R. Fox, personal communication). If these cells were located outside of the lacrimal gland, perhaps in conjunctival lymphoid aggregates, the result might be long term local IgA (or IgG) synthesis, with the IgA not necessarily bearing secretory component.

Specific lacrimal IgA should be tested for its molecular form as well as for the presence of SC since if it is polymeric, it is highly unlikely to be a serum exudate. Furthermore, it may be informative to test biopsy material or animal tissues for the presence of IgA subclass mRNA and protein at the single cell level by the combined in situ hybridization/immuno-chemistry technique described recently by Islam et al. (1992).

Further characterization of the lacrimal immune system may lead to better diagnostic methods for ocular infections as well as to clues to the nature of possible hypersensitivity- inducing microbial antigens.

# REFERENCES

Buisman, N.J., Ossewaarde, J.M., Rieffe, M., van Loon, A.M., and Stilma, J.S., 1992, Chlamydia keratoconjunctivitis determination of *Chlamydia trachomatis* specific secretory immunoglobulin A in tears by enzyme immunoassay, *Graefes. Arch. Clin. Exp. Ophthalmol.* 230:411.

Caldwell, H.D., Stewart, S., Johnson, S., and Taylor, H., 1987, Tear and serum antibody response to *Chlamydia trachomatis* antigens during acute chlamydial conjunctivitis in monkeys as determined by immunoblotting, *Infect. Immun.* 55:93.

Donaldson, S.L., Kosco, M.H., Szakal, A.K., and Tew, J.G., 1986, Localization of antibody-forming cells in draining lymphoid organs during long term maintenance of the antibody response. *J. Leukocyte Biol.* 40:147.

Elsana, S., Friedman, M.G., Friling, R., Sarov, B., Shaked, O., Yassur, Y., and Sarov, I., 1990, The local and serum immune response to Chlamydia in paratrachoma infections, *Serodiag. Immunotherapy Infect. Dis.* 4:201.

Friedman, M.G., 1990, Antibodies in human tears during and after infection, *Surv. Ophthalmol.* 35:151.

Friedman, M.G., Phillip, M., and Dagan, R., 1989, Virus specific IgA in serum, saliva, and tears of children with measles, *Clin. Exp. Immunol.* 75:58.

Gray, D. and Skarvall, H., 1988, B cell memory is short-lived in the absence of antigen, *Nature* 336:70.

Herrmann, B., Stenberg, K., and Mårdh, P.-A., 1991, Immune response in chlamydial conjunctivitis among neonates and adults with special reference to tear IgA, *APMIS* 99:69.

Islam, K.B., Christensson, B., Hammarström, L., and Smith, C.I.E., 1992, Analysis of human IgA subclasses by in situ hybridization and combined in situ hybridization/immunocytochemistry, *J. Immunol. Methods* 154:163.

Liotet, S., Hartmann, C., Batellier, L., Chaumeil, C., and Frottier, J., 1987, Anti-HIV-antibodies in tears of patients with AIDS, *Fortschr. Ophthalmol.* 84:340.

McBride, B.W., and Ward, K.A., 1987, Herpes simplex-specific IgG subclass response in herpetic keratitis, *J. Med. Virol.* 21:179.

McClellan, B.H., Whitney, C.R., Newman, L.P., and Allansmith, M.R., 1973, Immunoglobulins in tears, *Am. J. Ophthalmol.* 76:89.

Nordbø, S.A., Nesbakken, T., Skaug, K., and Rosenlund, E.F., 1986, Detection of adenovirus-specific immunoglobulin A in tears from patients with keratoconjunctivitis, *Eur. J. Clin. Microbiol.* 5:678.

Norrild, B., Pedersen, B., Møller Anderson, S., 1982, Herpes simplex virus specific secretory IgA in lacrimal fluid during herpes keratitis, *Scand. J. Clin. Lab. Invest.* 42, *Suppl.* 161:29.

Pedersen, B., Møller Anderson, S., Klauber, A., Ottovay, E., Prause, J.U., Zhong, C., and Norrild, B., 1982, Secretory IgA specific for herpes simplex virus in lacrimal fluid from patients with herpes keratitis--a possible diagnostic parameter, *Brit. J. Ophthalmol.* 66:648.

Renom, G., Bouquety, J.C., Lanckriet, C., Georges, A.J., Siopathis, M.R., and Martin, P.M.V., 1990, Detection of anti-HIV IgA in tears of children born to seropostive mothers is highly specific, *Res. Virol.* 141:557.

Russell, M.W., and Mestecky, J., 1988, Induction of the mucosal immune response, *Rev. Infect. Dis.* 10:S440.

Shallal, A., Treharne, J., and Viswalingham, M., 1992, Immunoblotting analysis of sequentially collected sera and tear fluids in ocular Chlamydial infections of man, *in:* "Proceedings of the European Society for Chlamydia Research," P.-A. Mårdh, M. LaPlaca, and M.Ward, eds., Uppsala University Centre for STD Research, Uppsala, Sweden, p. 91.

Shani, L., Szanton, E., David, R., Yassur, Y. and Sarov, I., 1985, Studies on HSV-specific IgA antibodies in lacrimal fluid from patients with herpes keratitis by solid phase radioimmunoassay, *Curr. Eye Res.* 4:103.

Stuchell, R.N., Feldman, J.J., Farris, R.L., and Mandel, I.D., 1984, The effect of collection technique on tear composition, *Invest. Ophthalmol. Vis. Sci.* 25:374.

Tew, J.G., Phipps, R.P., and Mandel, T.E., 1980, The maintenance and regulation of the humoral immune response: persisting antigen and the role of follicular antigen-binding dendritic cells as accessory cells. *Immunol Rev.* 53:175.

Treharne, J.D., Viswalingham, N.D., and Darougar, S., 1986, Development and persistence of chlamydial antibodies in adult paratrachoma infections, *in:* D. Oriel, G. Ridgway, J. Schachter, et al., eds., "Chlamydial Infections," Cambridge University Press, Cambridge, p. 158.

# NEURAL-ENDOCRINE CONTROL OF SECRETORY COMPONENT SYNTHESIS BY LACRIMAL GLAND ACINAR CELLS: SPECIFICITY, TEMPORAL CHARACTERISTICS AND MOLECULAR BASIS

Ross W. Lambert, Robin S. Kelleher, L. Alexandra Wickham, Jianping Gao and David A. Sullivan

Immunology Unit, Schepens Eye Research Institute
Department of Ophthalmology, Harvard Medical School
20 Staniford Street, Boston, MA 02114

## INTRODUCTION

The ocular surface appears to be protected from bacterial and viral pathogens by polymeric IgA antibodies.[1] These antibodies, which are produced by plasma cells in the lacrimal gland, bind to secretory component (SC) in the basolateral membrane of acinar epithelial cells, and are then transported to the apical membrane and secreted into tears.[2] Of interest, the lacrimal secretion of sIgA, as well as free SC, appears to be significantly influenced by gender and hormones from the hypothalamic-pituitary-gonadal axis.[2] Thus, almost 10 years ago, it was found that the concentrations of free SC and IgA in the tears of male rats were 2 to 5-fold higher than those in tears of female rats. These gender-associated differences in tear SC and IgA were shown to be caused by androgens.[2] For example, castration led to a significant reduction in SC levels in tears of males, while having no impact on SC content in tears of females. In addition, administration of testosterone for 4 days to castrated male or female rats significantly increased tear SC levels. These studies showed that exposure of castrated rats to androgens resulted in alterations of lacrimal SC production.[2] Moreover, later experiments demonstrated that this androgen control of SC, as well as IgA, is modulated by factors from the hypothalamus and pituitary.[2] However, this research did not address whether androgens act directly on lacrimal tissue to change SC or IgA, or indirectly via an androgen-sensitive site, which in turn acts to modulate lacrimal SC and IgA.

Therefore, to more thoroughly examine the processes involved in this hormone regulation, our laboratory's experimental approach focused upon the endocrine control of SC and prompted the development of procedures for the isolation and long-term maintenance of funtional lacrimal acinar cells in vitro.[3] Homogenous suspensions of acinar cells were prepared by the dissociation and digestion of lacrimal gland fragments, followed by filtration and Ficoll step-gradient centrifugation. Isolated cells were then cultured in serum-free conditions on Matrigel extracellular matrix. Initial studies with cultured acinar cells showed that androgens could directly influence lacrimal SC production.[4] Thus, dihydrotestosterone (DHT)-treated cultures produced significantly more SC than control (vehicle)-treated cultures. The hormonal control of lacrimal acinar cell SC production in vitro was shown to be dose-dependent and specific for androgens.[4]

In addition, continuing studies from our laboratory also explored the neural and immune regulation of SC, given that the nervous and immune systems may play a role in the modulation of SC synthesis throughout the body.[2] Cultured acinar cells were treated with a variety of compounds for 4 days to determine their effect on SC production. These studies demonstrated that certain neural compounds (e.g. vasoactive intestinal peptide [VIP], and β-adrenergic agonists) and lymphokines (IL-1α, IL-1β, and TNF-α) significantly increased SC output by lacrimal acinar cells.[5] Of particular interest, the muscarinic agonist carbachol significantly inhibited both basal and DHT-stimulated SC production.[5]

Given this background, the following studies were performed to examine: [a] the possible involvement of cAMP in the neural and endocrine regulation of acinar cell SC; [b] the temporal association between acinar cell signal reception and the eventual SC response; [c] the role of specific receptors and transcription in the control of SC production; and [d] the site-specificity of the acinar cell SC response to neural, endocrine and immune agents.


## RESULTS

### Involvement of cAMP in the Neural and Endocrine Regulation of Acinar Cell SC Production

Many compounds (e.g. DHT, VIP, carbachol) that stimulate lacrimal SC output are known to modify cAMP levels in other tissues.[5] In addition, cAMP analogues (e.g. 8-bcAMP), cAMP inducers (e.g. cholera toxin, PGE$_2$) and phosphodiesterase inhibitors (e.g. IBMX) all elevate lacrimal SC production.[5,6] Therefore, we performed preliminary studies to determine whether several compounds that alter lacrimal SC also change intracellular levels of cAMP. Lacrimal acinar cells (5 x 10$^6$ cells/well, n = 6 wells/group) were treated for 3 hours with cholera toxin (10 μg/ml), DHT (10$^{-6}$ M) or carbachol (10$^{-4}$ M), after which time intracellular levels of cAMP were quantitated by RIA (Advanced Magnetics Inc). In these studies, cholera toxin significantly increased cAMP by a 2.9-fold amount. However, DHT and carbachol had no consistent effect on cAMP levels, suggesting that cAMP may not necessarily be involved in mediating all changes in lacrimal SC production.

### Temporal Association Between Signal Reception and Acinar SC Response [7]

In previous studies, lacrimal acinar cells were exposed to endocrine, neural and immune compounds for the duration of the experiment (e.g. 4 days).[4-6] Thus, for example, 4 day exposure to DHT or cholera toxin significantly increased media SC, while 4 day exposure to carbachol significantly decreased media SC. The earliest time after the initiation of treatment that changes in acinar cell SC production (i.e., DHT-stimulated) could be detected, though, was 2 days.[4] Considering the typically rapid association of these compounds with their cellular receptors, we sought to determine if continuous exposure of acinar cells to these regulatory agents was necessary to elicit the observed changes in SC production. To test this possibility, we compared the effects of acute (i.e. 3 hours) versus chronic (i.e. 4 days) administration of DHT ($10^{-6}$ M), cholera toxin (10 μg/ml) or vehicle on the extent of acinar cell SC output following a 4 day culture period. In addition, we determined the temporal requirements for carbachol's ($10^{-4}$ M) inhibition of SC production. In confirmation of previous studies,[4-6] 4 day treatment of acinar cells with DHT or cholera toxin led to a significant increase in SC, while similar treatment with carbachol significantly decreased SC.[7] The magnitude of these changes in SC production, though, was almost duplicated by cells that were exposed to compounds for the abbreviated 3 hour interval and then cultured for the remaining 4 days in vehicle-containing (control) media.[7] Thus, an acute 3 hour treatment with DHT or cholera toxin resulted in a significant increase in media SC after 4 days, while a 3 hour treatment with carbachol led to a significant decrease in media SC after 4 days. These findings indicate that a pronounced delay exists between the delivery of regulatory signals and the eventual SC response.

An additional observation from these studies was suprising. In contrast to its inhibitory effect on acinar SC production over 4 days, carbachol treatment transiently increased media SC levels within 3 hours of acinar cell exposure to this compound. The cellular basis for this apparent dual effect of carbachol on lacrimal SC is unknown. However the acute (i.e. 3 hour), carbachol-induced stimulation of SC output may be associated with the pronounced changes in lacrimal acinar cell physiology that follow muscarinic stimulation.[8]

**Specific Receptor Involvement in the Control Of Acinar Cell SC [7]**

To determine if the androgen stimulation of lacrimal SC is receptor-mediated, cultured acinar cells were pretreated with cyproterone acetate ($10^{-6}$ M), a competitive androgen receptor antagonist, for 30 minutes before addition of DHT ($10^{-8}$ M) to the incubation media. In these studies, cyproterone acetate did not affect constitutive SC production. However, the antagonist completely abolished DHT-stimulated SC production, demonstrating the involvement of androgen receptors in the stimulation of lacrimal SC by DHT.[7]

To examine whether the cholinergic suppression of androgen-stimulated SC output is mediated through muscarinic receptors, acinar cells were treated with DHT ($10^{-6}$ M) and atropine ($10^{-5}$ M) for 30 minutes prior to the addition of carbachol ($10^{-4}$ M) to the culture media. In these studies, atropine, which competitively binds muscarinic sites, did not alter DHT-stimulated SC production. However, atropine abolished carbachol's inhibition of androgen-induced SC output, demonstrating that the effect of carbachol on lacrimal SC output involves muscarinic receptors.[7]

**Androgen Control Of Acinar Cell SC Involves Transcription [7]**

The classic mechanism of action of androgens involves a hormone-induced alteration in the expression of specific genes, resulting in an increased or decreased production of specific proteins.[9] To determine if the androgen stimulation of SC production might involve such a mechanism, lacrimal acinar cells were cultured for 2 days in the presence of vehicle, DHT ($10^{-6}$ M) and/or the transcriptional inhibitor, actinomycin D (0.1 µg/ml). For these studies, a shorter duration of treatment (i.e. 2 days) was utilized in order to minimize possible toxic effects associated with actinomycin D. Results demonstrated that actinomycin D significantly decreased DHT-stimulated SC production.[7] This finding indicates that DHT-induces changes in SC by a process that requires transcription, possibly of the mRNA for SC. In support of this hypothesis, testosterone administration for 7 days to castrated rats significantly increased SC mRNA content in lacrimal glands, compared to that in placebo-treated controls.[10]

## Endocrine, Neural And Immune Regulation Of SC Is Site-Specific [7]

Depending on the particular mucosal site, a given compound may significantly increase, decrease, antagonize or have no effect on the production of SC. For example, glucocorticoids stimulate SC elaboration by the salivary gland and liver, depress SC output by intestinal epithelial cells and elicit no change in uterine or vaginal SC.[2] We have already seen that a specific array of endocrine, neural and immune compounds alter lacrimal acinar cell SC production.[4-6] The following studies were performed to determine if the endocrine, neural and immune control of SC production in lacrimal acinar cells is site-specific.

Acinar cells from rat submandibular and parotid glands were isolated and cultured by the techniques described for lacrimal acinar cells. Epithelial cell lines included IEC-6 (rat small intestinal epithelium), HT-29 (human colon adenocarcinoma), A-253 (human submaxillary carcinoma) and LNCaP (human prostate carcinoma) and were purchased from ATCC. Cells were exposed to one or more of the following compounds for 4 days: DHT ($10^{-6}$ M), cholera toxin (10 µg/m), carbachol ($10^{-4}$ M), bcAMP ($10^{-3}$ M), bcGMP ($10^{-3}$ M), VIP ($10^{-6}$ or $10^{-7}$ M), γ-IFN (1000 or 3000 U/ml), IL-1α (2 ng/ml) and IL-1β (2 ng/ml). Media were assayed for SC content by using species-specific RIAs. Unless noted, changes in SC were not accompanied by significant variations in either cell number or viability.

In studies with rat salivary acinar cells, cholera toxin, bcAMP, bcGMP and IL-1α, but not DHT, significantly ($P < 0.0001$) increased SC output by cultured submandibular cells (n >10 wells/group), compared to that of control. In addition, SC production by submandibular cells was stimulated in 2 of 4 experiments with carbachol, which did not inhibit basal or cholera toxin-induced SC output.[7] In contrast, cholera toxin had an equivocal effect on SC elaboration by rat parotid acinar cells (n = 14 wells/group). In two studies, cholera toxin slightly elevated media SC levels ($129 \pm 2.0\%$ of control) after 4 days of treatment. However, more cells were recovered from cholera toxin-containing cultures than from control cultures.[7] Based on these data, it cannot be determined if cholera toxin led to a specific increase in SC production per cell. With regard to rat intestinal epithelial cells, treatment of confluent IEC-6 cells with the lymphokines γ-IFN, IL-1α or IL-1β increased media SC levels by 79 to 108%, relative to control. In addition, cholera toxin increased SC production by 48% in 3 out of 5 studies, while DHT, VIP, carbachol, bcAMP and bcGMP did not change IEC-6 SC output during the 4 day experiments.[7] Overall, these findings demonstrate that various endocrine, neural or immune factors modulate SC production by rat salivary or intestinal epithelial cells. However, as shown in Table 1, the pattern of this regulation is different than that observed with rat lacrimal acinar cells.

**Table 1.** Endocrine, neural and immune influence on SC production by rat epithelial cells

| Treatment | Lacrimal | Epithelial Cell SC Production in Vitro Submandibular | Parotid | Intestinal |
|---|---|---|---|---|
| DHT | ↑ | – | ND | – |
| bcAMP | ↑ | ↑ | ND | – |
| bcGMP | - | ↑ | ND | – |
| cholera toxin | ↑ | ↑ | ↑ | –↑ |
| VIP | ↑ | ND | ND | – |
| carbachol | ↓ | –↑ | ND | – |
| γ-IFN | - | ND | ND | ↑ |
| IL-1α | ↑ | ↑ | ND | ↑ |
| IL-1β | ↑ | ND | ND | ↑ |

Epithelial cells, isolated from rat lacrimal, submandibular, parotid or intestinal (IEC 6 cell line) tissues, were cultured, and exposed to various concentrations of one of the above agents, for 4 days. Media SC levels were measured by RIA, and results were compared to those obtained from vehicle-treated cell cultures. Terminology: ↑, increase; ↓, decrease; -, no change; ND, not determined. Data from references (5,7).

As concerns possible endocrine, neural or immune effects on SC production by human epithelial cells, experimental results were inconclusive. In 4 day studies with confluent intestinal HT-29 cells (100% viability, 2 x $10^6$ cells/well, n = 5 wells/treatment group), SC could not be detected in culture media (McCoy's 5A plus 10% FCS or Leibovitz's L-15 plus 10% FCS), irrespective of whether cells were treated with vehicle, cholera toxin, bcAMP bcGMP, DHT, carbachol or γ-IFN. Similarly, SC could not be detected in: [a] control or DHT-containing media (RPMI 1640 plus 10% FCS) from near-confluent prostatic LNCaP cell cultures (100% viability, n = 5 wells/group); or [b] control or cholera toxin-containing media (McCoy's 5A plus 10% FCS) from near-confluent salivary A-253 cultures (100% viability, n = 5 wells/group). Of interest, application of cholera toxin to A-253 cells appeared to reduce cell viability and cause a striking decrease in cell attachment. Our inability to detect SC in culture media of HT-29, LNCaP and A-253 cells, although unexpected, was not surprising. Investigators have documented that intestinal HT-29 cells are relatively undifferentiated[11] and extremely heterogenous in terms of SC expression, with only a very small percentage of cells capable of synthesizing SC.[12,13] In fact, basal SC production by parent HT-29 cells may be undetectable by immunoblotting[14] and the published magnitude of lymphokine-induced SC secretion[13] approaches the lower limit of sensitivity in our human SC RIA. To circumvent these difficulties with SC measurement, researchers have generated cloned derivatives through immunoselection[15] or maintenance in Leibovitz's media[16] to yield differentiated populations of SC-producing cells. However, culture of HT-29 cells in the glucose-free Leibovitz's media does not invariably lead to subclones that secrete SC,[16] which may account for our lack of success in detecting media SC under these conditions. As concerns the human LNCaP cell line, these prostatic cells apparently increase SC synthesis in response to androgens.[17] Yet, this hormone reaction was analyzed by immunoperoxidase staining methods[17] and the amount of SC secreted, if any, is below our assay sensitivity. With regard to the human salivary A-253 cells, this adenocarcinoma line expresses epithelial-like morphology, but may possibly represent a mixture of different cell types and has yet to be shown to produce SC.

## DISCUSSION AND ACKNOWLEDGMENTS

The present studies explored the endocrine, neural and immune regulation of SC production in rat mucosal epithelial cells, with a particular focus upon acinar cells from the lacrimal gland. Our findings demonstrated that: (1) striking, temporal differences exist between regulatory signal reception and the lacrimal SC response: acinar cell processing of androgen or cholera toxin stimulatory signals requires only 3 hours of exposure to these agents, but associated SC secretory responses are delayed at least 48 hours. Moreover, the extent of acinar cell SC output following 4 days of culture is similar if cells are treated acutely or chronically with androgens or cAMP-inducing secretagogues; (2) androgen stimulation and cholinergic suppression of acinar SC are mediated through interactions with specific cellular receptors; (3) androgen control of acinar SC synthesis appears to involve modulation of SC mRNA levels; and (4) the nature of the neuroendocrinimmune regulation of SC output by lacrimal gland acinar cells appears to be unique, when compared to factors controlling submandibular acinar and intestinal epithelial cell SC production. Overall, our results indicate that the hormonal, neural and lymphokine modulation of SC dynamics by lacrimal acinar cells is site-selective and may be mediated through the regulation of genomic processes.

This research was supported by NIH grants EY07074 and EY05612 and a grant from the Massachusetts Lions Research Fund.

## REFERENCES

1. M.G. Friedman, *Surv. Ophthalmol.* 35:151 (1990).
2. D.A. Sullivan, *in*: "The Neuroendocrine-Immune Network," S. Freier, ed., CRC Press, Boca Raton, FL, p199-238 (1990).
3. L.E. Hann, J.B. Tatro and D.A. Sullivan, *Invest. Ophthalmol. Vis. Sci.* 30:145 (1989).
4. D.A. Sullivan, R.S. Kelleher, J.P. Vaerman, and L.E. Hann, *J. Immunol.* 145:4238 (1990).
5. R.S. Kelleher, L.E. Hann, J.A. Edwards, and D.A. Sullivan, J. Immunol. 146:3405 (1991).
6. L.E. Hann, R.S. Kelleher, and D.A. Sullivan, *Invest. Ophthalmol. Vis. Sci.* 32:2610 (1991).
7. R.W. Lambert, R.S. Kelleher, L.A. Wickham, J.P. Vaerman, and D.A. Sullivan, article submitted (1993).
8. A.K. Mircheff, *Curr. Eye Res.* 8:607 (1989).
9. J.H. Clark, W.T. Schrader, and B.W. O'Malley, *in*: J.D. Wilson, and D.W. Foster, eds., "Williams Textbook of Endocrinology," WB Saunders, Philadelphia, pp 35-90 (1992).
10. J. Gao, R.W. Lambert, L.A. Wickham, and D.A. Sullivan, article submitted (1993).
11. C. Huet, C. Sahuquillo-Merino, E. Coudrier, and D. Louvard. 1987, *J. Cell. Biol.* 105:345 (1987).
12. S.S. Crago, R. Kulhavy, S.J. Prince, and J. Mestecky, *J. Exp. Med.* 147:1832 (1978).
13. L.M. Sollid, D. Kvale, P. Brandtzaeg, G. Markussen, and E. Thorsby, *J. Immunology* 138:4303 (1987).
14. J.C. Huff, *J. Invest. Dermatol.* 94:74S (1990).
15. D. Kvale, J. Bartek, L.M. Sollid, and P. Brandtzaeg, *Int. J. Cancer* 42:638 (1988).
16. C.K. Rao, C.S. Kaetzel, and M.E. Lamm, *Adv. Exp. Med. Biol.* 216:1071 (1987).
17. P. Weisz-Carrington, M. Farraj, F. Asadi, R. Sharifi, P.R. Keleman, L. Hwang, and R.J. Buschmann, *FASEB J.* 4:A1709 (1990).

# LYMPHOCYTE ADHESIVE INTERACTION WITH LACRIMAL GLAND ACINAR EPITHELIUM INVOLVES CARBOHYDRATE RECOGNITION

Nancy L. O'Sullivan, Rajiv Raja and Paul C. Montgomery

Immunology and Microbiology
Wayne State University Medical School
Detroit, MI 48201 USA

## INTRODUCTION

The tear film provides a barrier against pathogenic invasion of the ocular surface. One component of this defense is secretory immunoglobulin A (S-IgA), the predominant immunoglobulin in tears in many species, which is known to be inhibitory for bacterial adhesion.[1] IgA is produced by lacrimal gland plasma cells and then transported to the tears by transcytosis following an interaction with secretory component expressed on the surface of epithelial cells. In the mucosal immune network, the precursors of IgA producing plasma cells migrate from sites of antigen encounter to distal mucosal sites such as lacrimal glands. T cells are involved in regulating mucosal immunity and are also retained in the lacrimal gland. The purpose of this study was to further delineate mechanisms responsible for the preferential accumulation of specific lymphoid populations in lacrimal glands.

Our approach was to use the Woodruff-Stamper *in vitro* binding assay[2] to study adhesive interactions between lymphocytes and lacrimal gland tissue. This assay has been valuable for studying the lymphocyte homing receptor and endothelial cell addressin interactions involved in regulating lymphocyte traffic to lymph nodes and mucosal lymphoid tissue such as Peyer's patches.[3] We have adapted the assay to study lymphocyte-lacrimal gland adhesive interactions. Earlier studies showed that lymphocytes preferentially bound to lacrimal gland acinar epithelial cells under conditions which allow specific adherence to HEV of lymph nodes and Peyer's patches. Thoracic duct lymphocytes (TDLs) bound in much greater numbers than did thymocytes, implying that adherence required mature, circulating cells. TDL adherence required metabolic activity on the part of the lymphocyte, was calcium dependent, required cytoskeletal function and was mediated by a trypsin-sensitive lymphocyte surface molecule.[4,5] B cells bound in much larger numbers than did T cells.[6] Isolated lacrimal gland lymphocytes were shown to express homing receptor molecules and TDL adherence to lacrimal gland sections could be inhibited by monoclonal antibodies against lymphocyte homing receptors.[7] Since the lymph node homing receptor is known to be a cellular lectin, recognizing carbohydrate determinants on lymph node endothelium during recirculation,[8] the present studies were designed to determine whether carbohydrate recognition plays a role in lymphocyte-lacrimal gland adhesive interactions.

*Lacrimal Gland, Tear Film, and Dry Eye Syndromes*
Edited by D.A. Sullivan, Plenum Press, New York, 1994

## MATERIALS AND METHODS

TDLs were collected from male F344 rats (Charles River Breeding Laboratories, Kingston, NY), washed with RPMI-1640 medium, then resuspended in RPMI containing the test reagents. All sugars and lectins, except for polyphosphomannan ester (PPME 1842), were obtained from SIGMA Chemical Co. (St. Louis, MO). PPME 1842 was a gift from Dr. Myron Leon (Wayne State University, Detroit, MI). Eight micron cryostat sections of lacrimal gland or lymph node were lightly fixed using glutaraldehyde (3% for 10 minutes), washed and then blocked using lysine. Adherence was assayed by incubating TDLs over the sections for 30 minutes at 7°C and 80 rpm on a clinical rotator. Nonadherent lymphocytes were washed off and the adherent cells were fixed to the sections with glutaraldehyde. The sections, with attached TDLs, were stained with methyl green-thionin and the adherent cells were counted using a microscope fitted with a calibrated ocular reticle. For lacrimal gland, binding was quantitated as number of cells bound per $mm^2$ of tissue. Adherence to lymph node sections was determined as the percent of high endothelial venules positive. HEV with two or more adherent lymphocytes were considered positive. To allow comparison between lacrimal and lymphoid tissue, the data were expressed as percent of control binding (adherence of lymphocytes in the absence of test reagents). The significance of differences from control binding was determined using the Student's $t$-test.

## RESULTS

Data in Figure 1 show that the presence, during the assay, of either fucoidin, a poly-saccharide consisting entirely of sulfated fucose, polyphosphomannan ester (PPME-Y1842) or the monosaccharide mannose-6-phosphate significantly inhibited TDL adherence to lacrimal gland and, as reported,[9] to lymph node sections. Neither

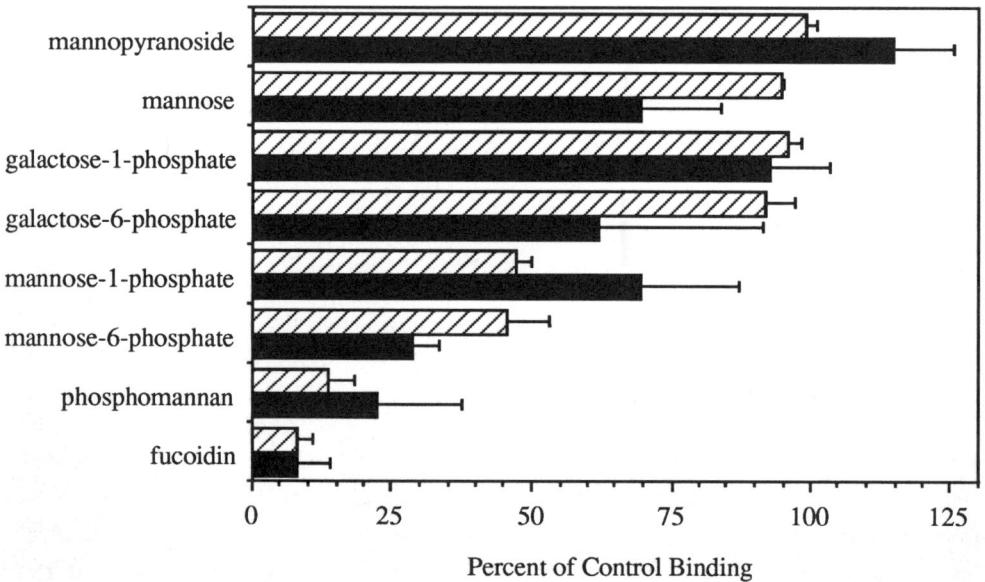

**Figure 1.** Effect of various carbohydrates on TDL adherence to lacrimal and lymph node tissues. Binding to lacrimal gland acinar epithelium (solid bars) or lymph node HEV (hatched bars) was assayed in the presence of carbohydrates. Data represent the mean ± SEM percent of binding in the absence of the sugars (control).

mannose-1-phosphate, galactose-6-phosphate nor galactose-1-phosphate were inhibitory for TDL binding to lacrimal gland acinar epithelium. The phosphate group had a role since neither the non-phosphated monosaccharide, mannose  nor the disaccharide, $\alpha$-methyl -D-mannopyranoside significantly affected binding to either lacrimal or lymph node tissues. In the case of fucoidin, the blockade occurred on the lymphocyte rather than the target tissue since pretreatment of TDL reduced adherence while pretreatment of the sections had no effect.[4]

To further delineate the mechanisms involved in lymphocyte adhesive interactions with lacrimal tissue, we tested the ability of several plant lectins to interfere with binding. Concanavalin A, which has specificity for $\alpha$-D-mannose and $\alpha$-D-glucose, did not significantly inhibit binding to lacrimal or lymph node tissues. Wheat germ agglutinin, which recognizes N-acetyl-D-glucosamine and N-acetyl-neuraminic acid, significantly reduced adherence of TDL to both lacrimal and lymph node sections. Sialic acid has been reported to have a role in both *in vitro* lymphocyte-HEV interaction and in *in vivo* trafficking to lymph nodes[10] and also appeared to be involved in the lymphocyte-lacrimal interaction. Peanut agglutinin greatly increased TDL adherence to lacrimal sections but did not affect lymph node HEV binding. This quadravalent lectin recognizes galactose and N-acetyl-D-galactosamine and may have crosslinked lymphocyte surface galactose to similar determinants on the lacrimal sections. Ulex europaeus-1 lectin binds to $\alpha$-L-fucose and was able to partially inhibit TDL adherence to lacrimal gland sections. Binding to HEV was not significantly affected.

## DISCUSSION

Oligosaccharides are well positioned to function as recognition molecules due to their cell-surface location and structural diversity. Lymphocytes express an array of receptors for sulfated and/or phosphated carbohydrate. The current data suggest that carbohydrate recognition is also involved in lymphocyte-lacrimal gland adhesive inter-

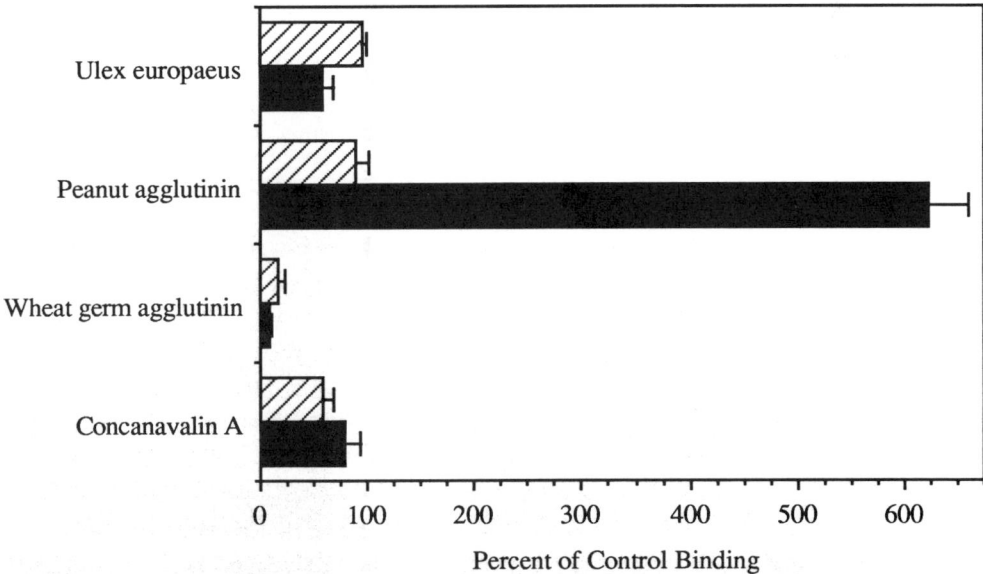

**Figure 2.** Effect of lectins on TDL adherence to lacrimal and lymph node tissues. Binding to lacrimal gland acinar epithelium (solid bars) or lymph node HEV (hatched bars) was assayed in the presence of lectins. Data represent the mean ± SEM percent of binding in the absence of the lectins (control).

actions. The recognition patterns were similar to those reported for the lymphocyte homing receptor - vascular addressin interactions.[9] Trafficking of lymphocytes to normal or inflamed ocular compartments is a complex journey involving exit from one tissue, circulation and exit from the vasculature, movement into extracellular spaces and finally arrest and accumulation in lacrimal glands. Further work is needed to identify and define both lymphocyte and lacrimal gland acinar epithelial cell molecules which regulate mucosal lymphoid cell distribution. Carbohydrate recognition is likely to be a useful property both for characterizing and isolating the receptor/ligand pairs involved and ultimately for therapeutic intervention during inflammatory lacrimal disorders.

## ACKNOWLEDGEMENTS

This work was supported by NIH grants EY05133 and EY07093.

## REFERENCES

1.  J. Mestecky and J.R. McGhee, Immunoglobulin A (IgA): Molecular and cellular interactions involved in IgA biosynthesis and immune response, *Adv. Immunol.* 40:153 (1987).
2.  H.B. Stamper and J.J. Woodruff, Lymphocyte homing into lymph nodes: *in vitro* demonstration of the selective affinity of high-endothelial venules, *J. Exp. Med.* 144:828 (1976).
3.  H.B. Stamper and J.J. Woodruff, An *in vitro* model of lymphocyte homing. I. characterization of the interaction between thoracic duct lymphocytes and specialized high-endothelial venules of lymph nodes, *J. Immunol.* 119:1603 (1977).
4.  N.L. O'Sullivan and P.C. Montgomery, Selective interactions of lymphocytes with neonatal and adult lacrimal gland tissues, *Invest. Ophthalmol. Vis. Sci.* 31:195 (1990).
5.  N.L. O'Sullivan and P.C. Montgomery, Characterization of the lymphocyte-glandular mucosal tissue interaction using an *in vitro* adherence assay. *in:* "Advances in Mucosal Immunology," T.T. MacDonald, J.J. Challacombe, P.W. Bland, C.R. Stokes, R.V. Heatly and A.M. Mowat, ed., Kluwer Academic Publishers, Dordrecht, Boston and London, p.222 (1990).
6.  N.L. O'Sullivan and P.C. Montgomery, T and B cell adhesive interactions with lacrimal gland epithelium, *Invest. Ophthalmol. Vis. Sci.* 32:939 (1991).
7.  N.L. O'Sullivan, C.A. Skandera and P.C. Montgomery, Expression of homing receptor molecules on rat lacrimal gland lymphocytes, *in* "Recent Advances in Mucosal Immunology," J. McGhee, J. Mestecky, H. Tlaskalova and J. Sterzl, eds., Plenum, New York (in press).
8.  L.A. Laskey, M.S. Singer, T.A. Yednock, D. Dowbenko, C. Fennie, H. Rodriguez, T. Nguyen, S. Stachel and S.D. Rosen, Cloning of a lymphocyte homing receptor reveals a lectin domain, *Cell* 56:1045 (1989).
9.  B.A. Braaten, G.J. Sprangrude and R.A. Daynes, Molecular mechanisms of lymphocyte extravasation. II. studies of *in vitro* lymphocyte adherence to high endothelial venules, *J. Immunol.* 113:117 (1984).
10. S.D. Rosen, M.S. Singer and T.A. Yednock, Involvement of sialic acid on endothelial cells in organ-specific lymphocyte recirculation, *Science,* 228:1005 (1985).

# SUBCLASS EXPRESSION OF IgA IN LACRIMAL GLANDS OF PATIENTS WITH SJÖGREN'S SYNDROME

Masafumi Ono,[1] Kenichi Yoshino,[1] Kazuo Tsubota,[1,2] and Ichiro Saito[3]

[1]Department of Ophthalmology, Keio University
School of Medicine, Tokyo, Japan
[2]Department of Ophthalmology
Tokyo Dental College, Chiba, Japan
[3]Medical Research Institute, Tokyo Medical
and Dental University, Tokyo, Japan

## INTRODUCTION

Normally, the human lacrimal gland contains a high proportion of IgA secreting plasma cells. The secreted IgA has two subclasses, IgA1 and IgA2, the expression of which may be controlled by T cells and/or cytokines. A high proportion (approximately 40%) of total lacrimal IgA occurs as IgA2, while IgA2 comprises about 10% of total serum IgA[1].

Sjögren's syndrome (SS) is an immunologic disease characterized by decreased aqueous tear (dry eye) and saliva production, as well as lymphocytic infiltration of exocrine organs such as the lacrimal and salivary glands. The infiltrates may cause changes in the cytokines produced within the glands. In addition, since the number of differentiated B cells is increased in SS, differences may exist between IgA subtypes produced in the lacrimal glands of normal and SS patients.

Given this background, the following experiments were designed to determine whether differences in IgA subclass distribution exist between lacrimal glands of normal and SS patients. In addition, we studied whether the upregulation of interleukin-6 (IL-6), which induces terminal differentiation of B cells, occurs in lacrimal glands from SS patients.

## MATERIALS AND METHODS

### Tissues

Lacrimal gland biopsies from patients with primary SS (n=10; male:female = 1:9) and autopsies from cadavers without autoimmune diseases and ocular diseases (n=4;

male:female = 2:2) were studied. Diagnostic criteria for primary SS are modifications of proposed criteria for classification by Fox et al.[2], and include all of the following:

1) 5-minute Schirmer test with anesthesia of less than or equal to 5mm strip wetting,
2) conjunctival and corneal fluorescein or rose bengal staining,
3) serum auto-antibodies (rheumatoid factor 1:160 or ANA 1:160 or positive SS-A SS-B),
4) extensive lymphocytic infiltrate in minor salivary gland or lacrimal gland biopsy,
5) absence of criteria sufficient for a diagnosis of rheumatoid arthritis, SLE, mixed connective tissue disease, or scleroderma.

The tissue was frozen in liquid nitrogen and embedded in OCT (Tissue Tek, Naperville, IL). For immunochemical procedures, five micron sections were cut from the frozen blocks, fixed in cold acetone, and frozen at -80°C in airtight containers. The remainder was frozen at -80°C until RT-PCR procedures.

## Immunochemical Procedure

The sections were stained by direct paired immunofluorescence. The mouse monoclonal anti-human antibodies to IgA, IgA1 or IgA2 were conjugated with fluorescein isothiocyanate (FITC) or tetramethylrhodamine isothiocyanate (RITC) (kindly provided by Dr. I. Moro, Nihon University School of Dentistry, Tokyo JAPAN), and characterized as described previously[3]. Slides were co-stained with one of the following combinations:

1) FITC-labeled anti-IgA1 and RITC-labeled anti-IgA,
2) RITC-labeled anti-IgA2 and FITC-labeled anti-IgA.

All plasma cells with cytoplasmic fluorescence were counted for each tissue slice.

## RT-PCR

**RNA Extraction.** Tissues were released from OCT with melting at room temperature. OCT-extracted tissues were placed in a sterile foil packet and minced with a razor blade. RNA was extracted from the minced tissue as described by Chomecznski et al.[4,5]. Extracted RNA was qualified by absorption meter (DU-5, Beckman) at 260nm and 10ng of the total samples was used for RT-PCR.

**Oligonucleotides Primers.** The human IL-6 primers were synthesized on a DNA synthesizer (model 391 DNA Synthesizer PCR-MATE, Applied Biosystem, Inc. Foster City, CA) as described by Gendelman et al.[6,7], and their sequences are given below.

IL-6 oligonucleotides were:
5'-primer: GCGCCTTCGGTCCAGTT,
3'-primer: CATGTTACTCTTGTTACATGT and
internal probe: CTCCTTTCTCAGGGCTGAG.

**Reverse Transcription and PCR.** cDNA was synthesized by using Rous associated virus 2 reverse-transcriptase (RTase; Takara-shuzo, Inc. Kyoto, Japan). PCR was performed essentially as previously described[8-10]. cDNA was mixed with PCR buffer, $MgCl_2$, dNTP and Taq polymerase (Takara-shuzo), overlaid with mineral oil to prevent evaporation and then amplified by PCR in a repeated 3-temperature cycle on the thermocycler programmable heating block (Perkin-Elmer Cetus, Norwalk, CT). Heat denaturation started the cycle over again and was repeated 35 times. The amplified products were analyzed by 1.7% agarose gel electrophoresis (Bio-Rad Laboratories, CA) and visualized with ethidium bromide by using size marker (φx174DNA/Hae III cut).

**Analysis of PCR-Amplified Products.** DNA separated on agarose gels was transferred to nylon membranes (Zeta-Probe, Bio-Rad Laboratories) by the method of Southern. The membranes were hybridized with $^{32}$P-labeled internal probe as shown above. Blots were autoradiographed by using X-ray film (Diagnostic X-ray Film, Kodak Inc.) and measured by densitometry. In this study, in excess of 100 fg/ml IL-6 mRNA was detected by the RT-PCR.

## RESULTS AND DISCUSSION

A combination of genetic, endocrine, neural, viral and environmental factors may be involved in SS's pathogenesis. A potential cause of SS may be primary infection by, and reactivation of, Epstein-Barr virus (EBV), cytomegalovirus (CMV), or herpes virus-6. The majority of lymphocytes in the SS lacrimal gland are B cells. Recently, research has demonstrated that EBV may be involved in polyclonal B cell activation and release of inflammatory cytokines such as IL-6.

Previously, IgA subclass distributions have been reported only for normal lacrimal glands. In those studies, the ratio of IgA1:IgA2-containing cells was 56:44[11,12], while the ratio of IgA1:IgA2 in normal tears was 59:41[1]. This study determined if changes in IgA subclass distribution occur in the lacrimal glands of SS patients. We found that the ratio of IgA1:IgA2-containing plasma cells in the SS lacrimal gland was 52:48 (Figure 1). This ratio is identical to that previously reported for normal lacrimal glands.

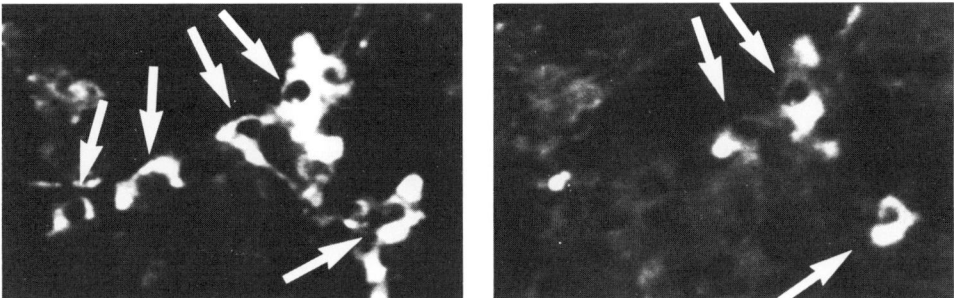

**Figure 1.** Paired immunofluorescence staining

**LEFT:** Staining for total IgA-containing cells in a lacrimal gland biopsy from a patient with Sjögren's Syndrome. Tissue was stained with FITC-labeled IgA monoclonal antibody. Arrows note IgA-containing plasma cells in the lymphocytic infiltrations.

**RIGHT:** The same section was stained for the IgA2 subclass of IgA antibodies. This tissue was stained with RITC-labeled IgA2 monoclonal antibody. Arrows note IgA2-containing plasma cells.

Previously, IgA subclass distributions have been reported only for normal lacrimal glands. In those studies, the ratio of IgA1:IgA2-containing cells was 56:44, while the ratio of IgA1:IgA2 in normal tears was 59:41. This study determined if changes in IgA subclass distribution occur in the lacrimal glands of SS patients. We found that the ratio of IgA1:IgA2-containing plasma cells in the SS lacrimal gland was 52:48 (n=10). This ratio is identical to that previously reported for normal lacrimal glands.

**Table 1.** Detection of IL-6 expression from lacrimal gland by RT-PCR assay

| | Density | | | | | |
|---|---|---|---|---|---|---|
| | 5 | 4 | 3 | 2 | 1 | 0 |
| SS LG | 4 | 4 | 1 | 1 | 0 | 0 |
| Normal LG | 0 | 0 | 0 | 0 | 3 | 1 |

0: no detectable. 1-5: increasing amount of signal detectable.
IL-6 expression was significantly increased compared to the normal lacrimal gland.

In this study, we were also interested to determine if IL-6 plays a role in the infiltration of IgA-containing B cells into lacrimal glands of SS patients. The expression of IL-6 in the all of SS lacrimal glands was detected by RT-PCR and was significantly increased compared to the normal lacrimal glands (Table 1).

Previous research has suggested that IgA-containing cells of the human lacrimal gland are derived from remote lymphoid tissues of the intestinal and bronchial tracts and possibly tonsils and are part of a common mucosal immune system[13]. In addition, a recent study has shown that the human appendix contains B cell subsets that constitutively express IL-6 receptor, that a high proportion of these B cells are committed to the IgA isotype. Furthermore, high numbers of IL-6 responsive IgA2 B cells are present in the human appendix[14]. These studies suggest that IL-6 could be important in the terminal differentiation of B cells within the SS lacrimal gland. In the current study, IL-6 mRNA levels were measured in lacrimal glands of SS patients; IL-6 expression was significantly increased compared to control (normal) lacrimal glands.

Therefore, these results suggest that B cells, which are committed to IgA1 or IgA2 expression, increase in SS and may result in the over-expression of inflammatory cytokines in an autoimmune condition.

# REFERENCES

1. D.L. Delacroix, C. Dive, J.C. Rambaud, and J.P. Vaerman, IgA subclasses in various secretions and in serum, *Immunology* 47:383 (1982).
2. R.I. Fox, C.A. Robinson, J.G. Curd, F. Kozin, and F. Howell, Sjögren's syndrome, proposed criteria for classification, *Arthritis Rheum.* 29:577 (1975).
3. S.S. Cargo and J. Mestecky, Secretory component: interactions with intracellular and surface immunoglobulins of human lymphoid cells, *J. Immunol.* 122:906 (1979).
4. P. Chomczynski and N. Sacchi, Single-step method of RNA isolation by acid guanidium thiocyte-phenol-chloroform extraction, *Analyt. Biochem.* 162:156 (1987).
5. D.A. Rappolee, A.M. Wang, D. Mark, and Z. Weber, Noble method for studying mRNA phenotype in single or small numbers of cells, *J. Cellul. Biochem.* 39:1 (1989).
6. H.E. Gendelman, R.M. Friedman, S. Joe, L.M. Baca, J.A. Turpin, G. Dveksler, M.S. Meltzer, and C. Dieffenbach, A selective defect of interferon α production in human immunodeficiency virus-infected monocytes, *J. Exp. Med.* 172:1433 (1990).
7. E.S. Kawasaki, S.S. Clark, M.Y. Coyne, S.D. Smith, R. Champlin, O.N. Witte, and F.P. McCormick, Diagnosis of chronic myeloid and acute lymphocytic leukemias by detection of leukemia-specific mRNA sequences amplified in vitro, *Proc. Acad. Sci. USA* 85:5698 (1989).
8. H.L. Chang, M.H. Zaroukian, and W.J. Esselman, T200 alternate exon use in murine lymphoid cells determined by reverse transcription-polymerase chain reaction, *J. Immunol.* 143:315 (1989).
9. R.K. Saiki, T.L. Bugawan, G.T. Horn, K.B. Mullis, and H.A. Erich, Analysis of enzymatrically amplified β-globulin and HLA-DQα DNA with allele-specific oligonucleotide probes, *Nature* 324:163 (1986).
10. R.K. Saiki, D.H. Gelfand, S. Stoffel, S.J. Scharf, R. Higuchi, G.T. Horn, K.B. Mullis, and H.A. Erich HA, Primer-detected enzymatic amplification of DNA with a thermostable DNA polymerase, *Science* 239:487 (1988).
11. M.R. Allansmith, J. Radl, J.J. Haaijiman, and J. Mestecky, Molecular forms of tear IgA and distribution of IgA subclasses in human lacrimal glands, *J. Allergy Clin. Immunol.* 76:569 (1985).
12. S.S. Crago, W.H. Kutteh, I. Moro, M.R. Allansmith, J.J. Haaijiman, and J. Mestecky, Distribution of IgA1-, IgA2-, and J chain-containing cells in human tissues, *J. Immunol.* 132:16 (1984).
13. J. Metstecky and M.W. Russei MW, IgA subclasses, *in:* "Basic and Clinical Aspects of IgG, Monographs in Allergy," F. Shakib, ed., American Elsevier Publishing Co., New York (1986).
14. K. Fujihashi, J.R. McGhee, C. Lue, K.W. Beagley, T. Taga, T. Hirano, T. Kishimoto, J. Mestecky, and H. Kiyono, Human appendix B cells naturally express receptors for and respond to interleukin-6 with selective IgA1 and IgA2 synthesis, *J. Clin. Invest.* 88:248 (1991).

# INFLUENCE OF THE ENDOCRINE ENVIRONMENT ON HERPES VIRUS INFECTION IN RAT LACRIMAL GLAND ACINAR CELLS

Zhiyan Huang, Ross W. Lambert, L. Alexandra Wickham and David A. Sullivan

Department of Ophthalmology, Harvard Medical School
Immunology Unit, Schepens Eye Research Institute
20 Staniford Street, Boston, MA, USA 02114

## INTRODUCTION

Throughout the world, one of the greatest single causes of keratoconjuctivitis sicca is Sjögren's syndrome:[1] this disease is an exceedingly complex autoimmune disorder, which occurs almost exclusively in females, and involves a progressive, immune-mediated destruction of lacrimal tissue and consequent dry eye.[2] The etiology of Sjögren's sydrome is unclear, although recent research has suggested that viral infection and hormone action may play significant roles in the pathogenesis and/or expression of this autoimmune disease.[3,4] Thus, primary infection by herpes viruses (e.g. Epstein-Barr virus, cytomegalovirus, herpes virus-6), or retroviruses, may elicit the lymphocytic accumulation, immune cell activation and associated inflammation evident in affected lacrimal tissues.[3,4] Moreover, sex steroids may: [a] influence the onset and/or severity of lacrimal gland immunopathology in Sjögren's syndrome;[4] and [b] potentially alter viral activity in infected lacrimal tissues, given that such hormones are known to modulate herpes virus transcription.[5,6]

However, whether these putative viral and hormonal actions are independent, or possibly synergize, to accelerate the development of lacrimal autoimmunity in Sjögren's syndrome, is unknown. In fact, almost no information is available concerning the infectivity and replication capacity of herpes viruses in the lacrimal gland. Therefore, to advance our understanding of possible virus-endocrine interactions in lacrimal tissue, the purpose of this investigation was two-fold: (1) to explore the ability of herpes viruses to invade and replicate in epithelial (acinar) cells of the rat lacrimal gland; and (2) to examine whether the endocrine environment may influence the extent of this viral infection. To perform these experiments, we utilized rat cytomegalovirus (RCMV), an epitheliotropic, DNA herpes virus, which has previously been shown to invade exocrine (e.g. salivary) tissues, induce a distinct, periductular infiltration of lymphocytes and cause a marked inflammation in glandular parenchyma.[7,8] These sequelae are similar to those observed in lacrimal tissue of Sjögren's patients. Of interest, CMV has also been detected in human lacrimal glands[9] and implicated in this tissue's immune-related dysfunction and associated aqueous tear deficiency in Sjögren's syndrome.[10]

## MATERIAL AND METHODS

To generate RCMV stocks for acinar cell studies, the English (from Dr. G. Sandford, Baltimore, MD) and Dutch (from Dr. C.A. Bruggeman, Maastricht, The Netherlands) strains of RCMV were propagated in cultures of rat embryonic fibroblasts (REF), according to reported procedures.[11] RCMV was then partially purified from the media of these infected monolayer cultures by sequential centrifugations, and the resulting pellet, which contained infectious virus, was resuspended in phosphate-buffered saline and stored at -80°C. For control purposes, an REF cellular extract was also prepared, as previously described.[9] To measure viral content in experimental samples, RCMV titers were accurately quantitated by plaque assay. This assay, which showed a linear relationship between viral dilution and plaque number, employed REF cells, double agarose overlays, a 10 day incubation period, eventual fixation in 10% neutral buffered-formalin, and staining with 1% methylene blue.[11]

Primary cultures of lacrimal gland acinar cells were established, then challenged with RCMV, as previously reported.[11] In brief, acinar cells were isolated from lacrimal tissues of intact, sham-operated or castrated male or female rats (6-8 weeks old; Zivic Miller Laboratories) by serial enzymatic digestions, sequential filtrations and centrifugation through a Ficoll 400 (Pharmacia) step gradient (2-4%). Acinar cell viability was examined by trypan blue exclusion and cell numbers were counted with a hemocytometer. Cells were plated at an average density of $2 \times 10^6$ cells/well on Matrigel (Collaborative Research) in 35 mm Primaria culture dishes (Falcon). After an overnight incubation, attached cells (n = 5-10 wells/group) were inoculated with varying amounts of RCMV or control REF cell antigens. Following a 60 minute viral adsorption, acinar cells were rinsed to remove remaining inoculum, then cultured in defined, serum-free media for 4, 8 or 12 days, with medium replacement every 4 days. All collected media were volumetrically measured, centrifuged and stored at -80°C. When indicated, acinar cells were enzymatically dissociated from Matrigel, analyzed for numerical recovery and viability, resuspended in a bovine serum albumin (Calbiochem-Behring)-containing PBS buffer, sonicated and stored at -80°C. Viral titers in culture media or sonicated cell extracts were determined by the RCMV plaque assay.

Statistical analysis was performed by using the unpaired, two-tailed Student's t test.

## RESULTS

### RCMV Invasion and Replication in Acinar Cells From the Rat Lacrimal Gland [11]

To determine whether RCMV might invade, and replicate, in acinar cells from the rat lacrimal gland, cells were challenged in vitro with RCMV (English strain, $1 \times 10^4$ plaque forming units [PFU]/well), then cultured for an additional 12 days. As shown in figure 1, RCMV exposure led to viral invasion, as well as a significant, time-dependent viral replication. Within 4 days of infection, viral titers were almost undetectable in both incubation media and cells. In contrast, by 8 days after challenge, viral content had increased dramatically, such that titers averaged over 30-fold higher than those present in the original inoculum. Moreover, this RCMV replicative process continued to rise from days 8 to 12 of the experimental time course. For comparison, no RCMV could be recovered from cultures wells treated with REF cell antigen.[11]

These studies, as well as additional research, demonstrated that: [a] RCMV distribution (i.e. cells vs. media) in acinar cultures was primarily cell-associated; [b] RCMV infection induced specific, time-dependent alterations in acinar cell morphology, including changes in cell shape and appearance; [c] the extent of RCMV replication depended significantly upon the viral concentration in the initial inoculum; and [d] the magnitude of RCMV infection was strain-dependent: inoculation with the English, but not the Dutch, strain yielded high titers of viral progeny.[11]

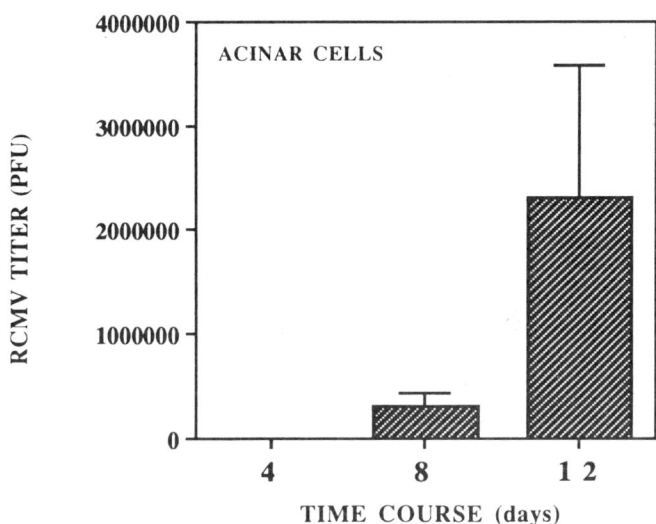

Figure 1. Lacrimal gland acinar cells ($\sim 2 \times 10^6$ cells/ well; n = 5-10 wells/group) were challenged with RCMV ($1 \times 10^4$ PFU/well) or REF cell control antigens on Day 0, then cultured for 4, 8 or 12 days. The RCMV titer reflects the viral recovery from sonicated acinar cell extracts. No RCMV was detected in control cultures. Data from reference (11).

## Influence of the Endocrine Environment on RCMV Replication in Acinar Cells From the Rat Lacrimal Gland [11]

To assess whether alterations in the sex steroid environment might modify viral infectivity and replication capacity in the lacrimal gland, several experiments were performed.[11] To summarize, age-matched male and female rats were subjected to sham-surgery or castration, and following a 10 day recovery period, lacrimal tissues were obtained and processed for the generation of acinar cell cultures (n = 5 to 10 wells/group). Acinar cells were challenged with RCMV (English strain, $1 \times 10^4$ PFU/well) or REF cell antigen and cultured for 8 days. Analysis of viral titers in culture wells demonstrated that: [a] RCMV infects lacrimal acinar cells from sham-operated or castrated male and female rats; and [b] the extent of viral replication in lacrimal acinar cells in vitro may be significantly influenced by prior rat castration in vivo. Moreover, the nature of this castration response appears to be gender-dependent. For example, in 3 separate studies, RCMV production in acinar cells from orchiectomized rats was much higher than that in cells from sham-operated animals. In contrast, in 2 out of 3 experiments, ovariectomy led to an apparent decline in the magnitude RCMV replication.[11] These findings suggest that sex steroids may modulate the magnitude of herpes virus infection in the lacrimal gland.

## DISCUSSION AND ACKNOWLEDGMENTS

The present study shows that: [a] RCMV invades, and undergoes a time-, dose- and strain-dependent replication, in acinar epithelial cells from the rat lacrimal gland; and [b] the extent of RCMV replication in lacrimal acinar cells in vitro may be significantly influenced by earlier changes in the sex steroid environment in vivo. Given these findings, it is possible that viral infection in hormonally-predisposed individuals may promote the onset and development of lacrimal autoimmune disease in Sjögren's syndrome. Future clarification of this viral-endocrine interrelationship may permit the design of new, therapeutic strategies to safely and effectively treat this currently incurable eye disease.

We would like to express our appreciation to Drs. Sandford and Bruggeman for their generous donation of viral reagents. This research was supported by NIH grants EY02882, EY05612 and EY07074, a grant from the Massachusetts Lions' Research Fund and a Postdoctoral Fellowship award from the Sjögren's Syndrome Foundation.

## REFERENCES

1. J.P. Whitcher, *Internat. Ophthalmol. Clin.* 27:7 (1987).
2. N. Talal, H.M. Moutsopoulos, and S.S. Kassan, eds., "Sjögren's Syndrome. Clinical and Immunological Aspects," Springer Verlag, Berlin (1987).
3. R.I. Fox, M. Luppi, H.I. Kang, and P. Pisa, *Springer Semin. Immunopathol.* 13:217 (1991).
4. D.A. Sullivan, and E.H. Sato, *Clin. Immunol. Immunopath.* 64:9 (1992).
5. B.A. Forbes, C.A. Bonville, and N.L. Dock, *J. Infect. Dis.* 162:39 (1990).
6. D. Kleinman, I. Sarov, and V. Insler, *Gynecol. Obstet. Invest.* 21:136 (1986)
7. P.K. Priscott, and D.A.J. Tyrrell, *Arch. Virology* 73:145 (1982).
8. C.A. Bruggeman, H. Meijer, F. Bosman, and C.P.A. van Boven, Intervirology 24:1 (1985).
9. S.C. Pflugfelder, S.C.G. Tseng, J.S. Pepose, M.A. Fletcher, N. Klimas, and W. Feuer, *Ophthalmology* 97:313 (1990).
10. J.C. Burns, *Med. Hypotheses* 10:451 (1983).
11. Z.H. Huang, R.W. Lambert, L.A. Wickham, and D.A. Sullivan, article submitted (1993).

# SIALODACRYOADENITIS VIRUS INFECTION OF RAT LACRIMAL GLAND ACINAR CELLS

L. Alexandra Wickham, Zhiyan Huang, Ross W. Lambert, and David A. Sullivan

Department of Ophthalmology, Harvard Medical School and
Immunology Unit, Schepens Eye Research Institute
20 Staniford Street, Boston, MA 02114

## INTRODUCTION

The secretory immune system of the eye is designed to protect the ocular surface against microbial challenge and infectious disease.[1] This immunological role is mediated primarily through secretory IgA (sIgA) antibodies, which are produced by plasma cells in interstitial areas of the lacrimal gland and are selectively transported to tears by secretory component (SC), the polymeric IgA receptor.[1] After delivery to the eye's anterior surface, sIgA antibodies may act to prevent viral internalization, inhibit bacterial colonization, curtail parasitic infestation and attenuate toxin-related damage.[2]

However, despite the importance of this local immune protection, the mechanisms by which various antigens stimulate the secretory immune system of the eye have not been clarified. For example, active viral infection of the ocular, or even nasopharyngeal or enteric, surfaces, is known to elicit a striking tear IgA antibody response.[1] Yet, whether viruses actually gain access to the lacrimal gland, the principal effector tissue of ocular mucosal immunity,[1] and therein induce an IgA response, is not known. Indeed, it is quite possible that viruses first initiate an immune reaction in distant mucosae, resulting in the migration to lacrimal tissue of IgA-producing lymphocytes,[3] which then serve to provide ocular defense. Identification of the processes and specific sites involved in antigen-induced ocular immune responses is critical, in order to permit the development of optimal vaccination strategies for the effective immunization of the anterior segment.

Therefore, to begin to address this issue, we sought to determine first whether viruses may, in fact, invade, and replicate in, lacrimal tissue. Towards that end, we evaluated the capacity of sialodacryoadenitis virus (SDAV), an epitheliotropic, RNA coronavirus, to infect lacrimal acinar cells in vitro. Our rationale for the selection of this virus was two-fold: [a] SDAV is known to exert a profound impact on the lacrimal gland in vivo: when administered intranasally, SDAV causes a pronounced infiltration of plasma cells, lymphocytes and macrophages into lacrimal tissue, a distinct, non-suppurative periductular inflammation, extensive interstitial edema, degenerative, atrophic and/or necrotic alterations in acinar and ductal epithelia, reduced tear flow and keratoconjunctivitis sicca.[4-8] However, whether this viral action involves invasion and replication in lacrimal tissue is unknown; and [b] SDAV may serve as a model virus with which to examine antigen-immune interactions in the lacrimal gland, to explore the functional role of the ocular secretory immune system against viral challenge, and to develop vaccination strategies for immunization of the anterior surface of the eye. In this regard, it should be noted that sIgA antibodies appear to modulate coronaviral infections in other mucosal sites.[9] To complement these initial studies, we also: [a] assessed whether short term SDAV infection interferes with the viability and function of acinar cells in vitro; and [b] compared the infectivity of SDAV in acinar cells from submandibular and parotid glands, given that salivary tissues are highly susceptible to SDAV infection in vivo.[5-8]

## MATERIALS AND METHODS

For the preparation of viral stocks, SDAV (L2-adapted strain; gift from Dr. Diane Gaertner, New Haven, CT, and originating from Dr. Dean Percy, Guelph, Ontario, Canada) was propagated in mouse L2 cells, according to reported procedures.[10] For the quantitation of SDAV levels in experimental samples, titers were accurately measured by a plaque assay, which involved the use of 4 day old L2 cell monolayers, application of defined overlay media containing Sea Plaque agarose (FMC Inc.), maintenance of inverted culture plates for 3 days at 37°C in a humidified atmosphere containing 5% $CO_2$, fixation of cells with 10% neutral-buffered formalin (Sigma Chemical Company) and staining with methylene blue.[11]

The isolation and culture of acinar cells from lacrimal, submandibular and parotid glands, as well as the exposure of acinar cells to SDAV, have been described in detail.[11-13] Briefly, tissues were obtained from pathogen-free, male Sprague-Dawley rats (6-7 weeks old, Zivic Miller Laboratories), rinsed in a soybean trypsin inhibitor (Worthington Biomedical) solution and disrupted through a series of incubations in EDTA (Gibco), or collagenase (Calbiochem-Behring), hyaluronidase (Calbiochem-Behring), and DNase I (Boehringer Mannheim) in DMEM- or HBSS (Gibco)-based buffers. The resulting digest was filtered consecutively through 500 μm and 25 μm Nitex meshes (Tetko Inc.), gently centrifuged and the cell pellet was resuspended in DMEM containing 20% heat-inactivated fetal calf serum (Hyclone). After centrifugation through a Ficoll 400 (Pharmacia) step gradient (2-4%), the final acinar cell pellet was resuspended in DMEM and plated at an average density of 1.5 to 2 x $10^6$ cells/well on the reconstituted basement membrane, Matrigel (Collaborative Research) in 35 mm Primaria culture dishes (Falcon). Acinar cell viability, which was typically greater than 80%, was determined by trypan blue exclusion and cell numbers were enumerated with a hemocytometer. Following an overnight incubation at 37°C in a humidified incubator containing 95% air/5% $CO_2$, unattached acinar cells were removed and counted and attached cells (n = 5-10 wells/group) were inoculated with SDAV (1 to 3 x $10^4$ plaque forming units [PFU]/well) or control L2 cell antigens. After a one hour adsorption, acinar cells were rinsed to remove residual inoculum, then cultured in serum-free modified Oliver's media containing defined supplements for 4 or 8 days; media was replaced on day 4. When indicated, media was aspirated, volumetrically measured, centrifuged and stored at -80°C. At experimental termination, attached cells were harvested from Matrigel by exposure to trypsin/EDTA (Gibco) and Dispase (Collaborative Research) solutions. Viral titers in culture supernatants or sonicated cell extracts were then measured by plaque assay.

To evaluate the impact of acute SDAV infection on acinar cell function, cells were challenged with SDAV or L2 cell antigen, then cultured for 4 days in the presence or absence of dihydrotestosterone ($10^{-6}$ M DHT; Sigma). This androgen is known to stimulate the synthesis and secretion of SC by lacrimal gland acinar cells.[14] After a 4 day culture interval, media was processed for the measurement of SC levels by RIA.[15] Statistical analysis was performed by utilizing the unpaired, two-tailed Student's t test.

## RESULTS [11]

Exposure of lacrimal gland acinar cells to SDAV, but not L2 cell control antigens, resulted in a definite viral invasion, as well as a dramatic, time-dependent increase in viral replication (Table 1).[11] Within 4 days after viral challenge, SDAV titers had risen significantly, such that total SDAV levels in culture wells averaged almost 5-fold higher than those present in the original inoculum. Moreover, during the following 4 day period, SDAV titers underwent an additional 3-fold increase (day 8). Throughout this 8 day time course, infectious progeny remained almost entirely within acinar cells, as compared to SDAV content in the incubation media. Thus, cell-associated SDAV typically accounted for over 96% of the total recovered virus per culture well (Table 1).

**Table 1.** SDAV infection in acinar epithelial cells from the rat lacrimal gland

| Time Course (days) | (SDAV Titer in Total Cell Culture)/ (SDAV Titer in Initial Inoculum) | (Acinar Cell SDAV Titer)/ (Total Cell Culture SDAV Titer) - % |
|---|---|---|
| 4 | 4.88 ± 1.05 | 97.5 ± 0.6 |
| 8 | 17.16 ± 0.58 | 96.3 ± 0.5 |

Rat lacrimal gland acinar cells (~ 2 x $10^6$ cells/well; n = 5 wells/group) were inoculated with SDAV (3 x $10^4$ PFU/well) or L2 cell control antigens on Day 0. After a one hour incubation period, residual virus was removed and cells were cultured for 4 or 8 days. The total SDAV titer in cell cultures equals the viral levels in both incubation media and sonicated cell extracts. No SDAV was detected in control cultures. Data from reference (11).

Acute SDAV infection had minimal impact on acinar cell viability. Following 4 or 8 days of viral exposure, the percentage of live cells was either similar to, or slightly below, that of control cells, whose viability often exceeded 94%.[11] In addition, short-term SDAV infection (i.e. 4 or 8 days) did not prevent a functional acinar cell SC response to DHT ($10^{-6}$

M). In fact, in certain experiments, the extent of androgen-induced SC production by SDAV-infected cells was almost identical to that observed in control cells.[11]

The infectious capacity of SDAV in acinar cells from the lacrimal gland appeared to differ from that in acinar cells from salivary tissues. Thus, significant variations existed in the kinetics and magnitude of SDAV infection in lacrimal, submandibular and parotid acinar cells.[11] Overall, these comparative studies demonstrated that parotid cells were most susceptible to SDAV replication under the experimental culture conditions.[11]

## DISCUSSION AND ACKNOWLEDGMENTS

These studies show that the coronavirus, SDAV, invades, and replicates in, lacrimal gland acinar cells in vitro. Moreover, our recent preliminary research has indicated that ocular exposure to SDAV may result in lacrimal infection in vivo (Z Huang, LA Wickham, DA Sullivan, unpublished data). Given this information, SDAV may serve as a very useful model virus to: [a] explore the impact of viral infection on the capacity, function and role of the ocular secretory immune system; and [b] evaluate potential vaccination strategies for the effective immunization of the ocular surface.

We wish to thank Drs. Gaertner and Percy for their gift of viral material, as well as their extremely helpful comments. We also express our appreciation to Drs. C. Wira (Hanover, NH), J.P. Vaerman (Brussels, Belgium) and B. Underdown (Hamilton, Ontario) for their provision of SC-related reagents. This research was supported by NIH grants EY02882, EY05612, EY07074 and a grant from the Massachusetts Lions' Research Fund.

## REFERENCES

1. D.A. Sullivan, *in*: "Mucosal Immunology," P.L. Ogra, J. Mestecky, M.E. Lamm, W. Strober, J. McGhee, and J. Bienenstock, eds., Academic Press, Orlando, FL, in press (1993).
2. T.T. MacDonald, S.J. Challacombe, P.W. Bland, C.R. Stokes, R.V. Heatley, A. McI Mowat, eds., "Advances in Mucosal Immunology," Kluwer Academic Publishers, London (1990).
3. P.C. Montgomery, A. Ayyildiz, I.M. Lemaitre-Coelho, J.P. Vaerman, and J.H. Rockey, *Ann. N.Y. Acad. Sci.* 409:428 (1983).
4. Y.L. Lai, R.O. Jacoby, P.N. Bhatt, and A.L. Jonas, *Invest. Ophthalmol. Vis. Sci.* 15:538 (1976).
5. R.O Jacoby, P.N. Bhatt, and A.M. Jonas, *Vet. Pathol.* 12:196 (1975).
6. D.H. Percy, P.E. Hanna, F. Paturzo, and P.N. Bhatt, *Lab. Anim. Sci.* 34:255 (1984).
7. D.L. Eisenbrandt, G.B. Hubbard, and R.E. Schmidt, *Lab. Anim. Sci.* 32:655 (1982).
8. D.H. Percy, Z.W. Wojcinski, and M.K. Schunk, *Vet. Pathol.* 26:238 (1989).
9. K. Callow, *J. Hyg. (London)* 95:173, 1985.
10. D. Percy, S. Bond, and J. MacInnes, *Arch. Virol.* 104:323, 1989.
11. L.A. Wickham, Z. Huang, R.W. Lambert, and D.A. Sullivan, article submitted (1993).
12. R.W. Lambert, R.S. Kelleher, L.A. Wickham, J.P. Vaerman, and D.A. Sullivan, article submitted (1993).
13. L.E. Hann, R.S. Kelleher, and D.A. Sullivan, *Invest. Ophthalmol. Vis. Sci.* 32:2610 (1991).
14. D.A. Sullivan, R.S. Kelleher, J.P. Vaerman, and L.E. Hann, *J. Immunol.* 145:4238 (1990).
15. D.A. Sullivan, and C. R. Wira, *J. Immunol.* 130:1330 (1983).

# GROWTH FACTOR AND RECEPTOR MESSENGER RNA PRODUCTION IN HUMAN LACRIMAL GLAND TISSUE

Steven E. Wilson

Department of Ophthalmology
University of Texas Southwestern Medical Center at Dallas
5323 Harry Hines Blvd.
Dallas, TX 75235

## INTRODUCTION

The lacrimal gland produces the complex aqueous portion of the tear film which contains numerous components, including many proteins that are discussed in other chapters of this proceedings. The components of this tear film are essential for maintaining a healthy ocular surface that is needed to provide the optical quality necessary for optimal visual function. The primary interest in our laboratory has been the production of growth factors, cytokines, and receptors in human lacrimal tissue and how the production of these modulators relates to the maintenance of the ocular surface and to corneal wound healing.[1] The direction of our work has been to initially identify the different modulators that are produced in lacrimal tissue as a prelude to the more difficult task of determining their functions in the lacrimal tissue and at the ocular surface. The method that we have used to begin this task has been to apply the reverse transcriptase-polymerase chain reaction (RT-PCR) method, with a blotting technique to demonstrate the specificity of the amplified products, in order to provide evidence that a particular modulator or receptor may be produced in the lacrimal tissue. Prior to detailing the specific results obtained in our laboratory, as well as the work of others, it is important to point out the limitations of our method and other methods that have been applied to this difficult problem.

In order to demonstrate that a specific protein is produced in a cell or tissue it is necessary to detect both the messenger RNA and the protein. Detection of the protein alone is not conclusive evidence. For example, immunohistologic methods have identified retinol-binding protein (RBP) in the corneal endothelium.[2] In situ hybridization, however, detected RBP messenger RNA only in the retinal pigmented epithelial cells.[2] This suggested that endothelial RBP was synthesized elsewhere and localized to the corneal endothelium through some other process. In addition, many proteins share similar domains. A particular antibody, even if monoclonal, may identify epitopes on several proteins depending on the experimental conditions. For example, many proteins share EGF-like regions and some of these could provide a false-positive signal with a particular antibody. Thus, the use of immunohistologic techniques provides evidence that a particular molecule is present in the tissue, but is not conclusive. Similarly, identification of the messenger RNA coding for a particular protein does not demonstrate that the message is translated to produce the protein. Many examples have been published of messenger RNAs being identified in a tissue in which the protein is not detected, suggesting that there is post transcriptional regulation of translation of the proteins. Of course one could also argue, since methods available for detection of a nucleic acid sequence are far more sensitive than those used to identify proteins, that messages such as these are translated and function at levels below our current ability to detect the protein. For example, recent evidence has suggested that some growth factors such as basic FGF can produce intracrine effects within the cell by binding to nuclear or other intracellular receptors.[3] The levels of the intracellular growth factor required for this type of mechanism may be exceedingly low. Similarly, one could also argue that the messenger RNAs that are within the intracellular pool are translated in response to an as yet unidentified stimulus and, therefore, are of functional significance to the cells in which they are transcribed. This would provide a mechanism for a more rapid response to an environmental cue.

The RT-PCR method has several advantages that make it attractive as an initial method for assessing the possibility that a growth factor, cytokine, or receptor is produced in the lacrimal gland. The RT-PCR method is very sensitive and, therefore, only the limited amounts of tissue that can be obtained from small biopsies are needed to detect the messenger RNA coding for a particular protein. Also, specific messenger RNAs can be detected even though they are present at very low levels. When performed with the proper controls and in conjunction with a blotting technique using a specific internal probe, the RT-PCR technique is exceedingly specific in detecting a particular mRNA. If the genomic organization of the gene is known, the PCR primers should be designed so that they are within different exons. Thus, amplifications from genomic sequences will be differentiated from those from mRNA. The PCR priming sequences should be specific for the modulator of interest. This is facilitated by eliminating potential primer sequences that have high homology to other genes by searching against all of the known sequences in the Genbank and EMBL nucleic acid banks. Similarly, confirmation of the specificity of the amplified products should be obtained using a probe that does not contain primer sequences for Southern or hot blotting[4] of the PCR products. The hot start method should be used to improve the specificity and sensitivity of the PCR reaction. In this technique the reactants of the PCR reaction are at 60-85° C when they are mixed and maintained at that temperature until cycling is begun.[5] To facilitate this method, essential PCR reactants can be separated by a plug of paraffin wax that melts when the temperature of the reaction reaches 65-70° C.[6] Also, because of the possibility of sequence contamination, it is important that PCR reactions be prepared in a laminar airflow hood using only positive displacement pipettes with negative controls for each primer set in every amplification experiment. Another major advantage of the PCR method is that in conjunction with a blotting method it can identify alternatively spliced mRNAs obtained from a single genomic sequence. This has been demonstrated for the FGF receptor-1 messenger RNA.[7] By isolating specific amplification products of a different size than the expected, based on the published sequences, and performing nucleic acid sequencing we are also identifying previously unreported alternate splicing products in lacrimal tissue for modulators such as the HGF receptor and KGF receptor (Weng J., Wilson S.E, unpublished data). Finally, the RT-PCR method can be used to suggest that modulators such as keratinocyte growth factor and its receptor are produced in a tissue, although antibodies for the modulators are not yet available. The major disadvantage of the RT-PCR method is that it cannot determine which cells within a tissue sample produce the mRNA.

Thus, several methods must be used to conclusively establish that a particular protein is produced in a tissue and to establish which cells produce the protein. The scheme that we intend to follow is to initially screen the tissue for the production of a particular mRNA using the RT-PCR method. As we will discuss throughout the remainder of this chapter, we have done this in lacrimal gland for a number of modulators. Protein immunohistologic methods are used to provide evidence that the protein is produced in the tissue and to indicate which cell types participate in the synthesis of the protein. We are currently using these methods to search for proteins corresponding to the messenger RNAs we have identified in lacrimal tissue. Once we have identified the growth factor, cytokine, or receptor immunohistologically in the tissue, we are applying the Western blotting method to confirm that the protein that is identified in the tissue sections is of the appropriate size to definitely correspond to the modulator of interest. This method may also detect precursor proteins that may not be processed post-translationally to produce the functional growth factor. The Western blotting method may also detect nonspecific proteins that may have been identified by immunohistology and alternately-sized specific proteins that may correspond to the alternative mRNAs identified by PCR. Differentiating between these two possibilities can be difficult and may require protein sequencing. Finally, in situ hybridization methods may be used to demonstrate which cells within a tissue contain a specific messenger RNA. In situ hybridization can be fraught with difficulties that include nonspecific binding and inadequate sensitivity to detect genes expressed at low levels.[8] We are attempting to use the reverse transcriptase-in situ PCR method to increase the sensitivity of in situ detection.[9] None of the in situ hybridization methods, however, can differentiate between alternative transcripts of a single gene that contain the same probing sequence. Thus, several methods must be used to obtain strong evidence for the production of specific growth factors, cytokines, and receptors in the lacrimal gland. Once conclusive evidence is obtained of synthesis, we are still left with the even more difficult task of determining function.

I hope that this introduction serves as a framework for illustrating the difficulties that are inherent in demonstrating the presence of growth factors, cytokines, and receptors in the lacrimal gland. Other difficulties will be discussed throughout this chapter. Obviously, this area of investigation could occupy the complete effort of several large laboratories for many years. We hope that our recent observations will stimulate other investigators to work in this area. The information that is obtained from these studies promises to lead to a better understanding of the physiology of the lacrimal gland and to determine what roles, if any, the growth factors and cytokines produced by the lacrimal gland have in the health and disease of the ocular surface.

## LACRIMAL GLAND GROWTH FACTOR PRODUCTION

Over the past few years there has been increasing interest in the hypothesis that the lacrimal gland actively participates in the maintenance of the ocular surface and corneal wound healing by producing growth factors that interact with specific receptors on the surface of corneal and conjunctival epithelial cells.[1] Several groups of investigators have demonstrated with immunologic methods that EGF or an EGF-like molecule is present in the tears of mice[10] and humans.[11,12] These studies suggested that the lacrimal gland may be the source of the EGF that is present in the tears. Subsequent studies demonstrated that mouse,[13] rabbit (Wilson SE, Cook SD, Clarke AF, Thompson H, Beuerman RW, unpublished data), and human[14] primary lacrimal gland tissue contained messenger ribonucleic acid (mRNA) coding for EGF precursor. In addition, EGF-like protein was identified by immunohistologic methods within the lumen of the acini, cells of the tubular ducts, and a low levels in the acinar cells of the rat lacrimal gland.[15] Western blotting methods have been used to demonstrate EGF protein in rabbit lacrimal tissue.[16] Figure 1 demonstrates the detection of mRNA coding for EGF precursor in human lacrimal tissue using PCR with Southern blotting of the PCR products.[14]

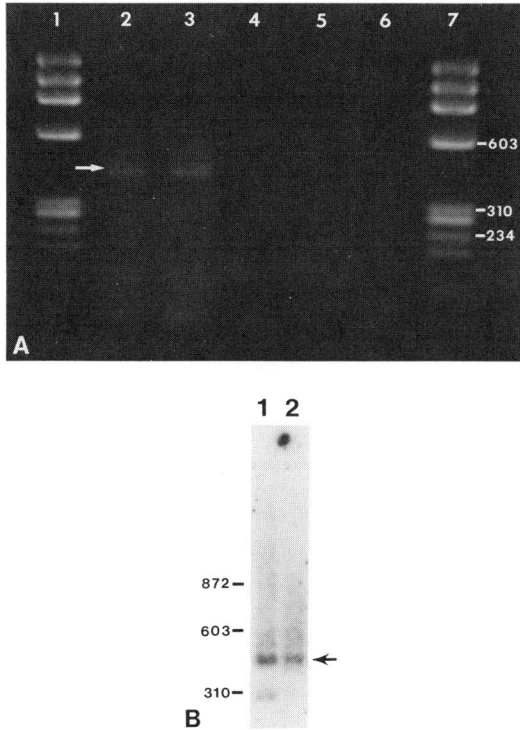

**Figure 1. A.** Polymerase chain reaction (PCR) products for the EGF precursor amplified from complementary DNA prepared from human lacrimal gland. Amplified products were resolved on a 2% agarose gel containing ethidium bromide. Lanes 1 and 7 contain Phi X 174 DNA/*Hae* III standard with the sizes indicated to the right in base pairs. Lanes 2 and 3 show the expected 412 base pair amplification product (arrow) from EGF precursor cDNA, based on the design of the PCR primers. Lanes 4 and 5 were controls with RNA used directly as target in PCR. Lane 6 is a control with water as target. **B.** Southern blot of the PCR products from Fig. 1 A. lanes 2 and 3 using an end-labeled oligonucleotide probe that was internal to the PCR primers demonstrating the specificity of the amplified EGF sequence. Reprinted by permission from Wilson SE, et. al., Cornea 10:519-24, 1991.

EGF in tears and evidence for EGF production in the lacrimal gland has suggested a potential regulatory role for the gland in maintaining the ocular surface, control of corneal wound healing, and diseases of the ocular surface.[1,17] Although this remains a tenable hypothesis, this relationship between lacrimal gland and tear EGF has become complicated by the recent discovery that corneal epithelium, both ex vivo and in culture, produces EGF mRNA[18,19] and protein (Schultz GS, Wilson SE, unpublished data). Corneal epithelial EGF is likely to have autocrine and/or paracrine effects on the epithelial cells since they are known to produce EGF receptors.[18] Therefore, a large portion of the EGF found in tears collected from the ocular surface could be derived from the ocular surface epithelium. If ocular surface epithelial cells do produce functional EGF, as recent studies suggest, then the localized concentrations of the autocrine EGF at the level of the EGF receptors on the surface of the epithelial cells may be much greater than the concentrations produced by the lacrimal gland. In fact, we have an alternate working hypothesis regarding the role of the

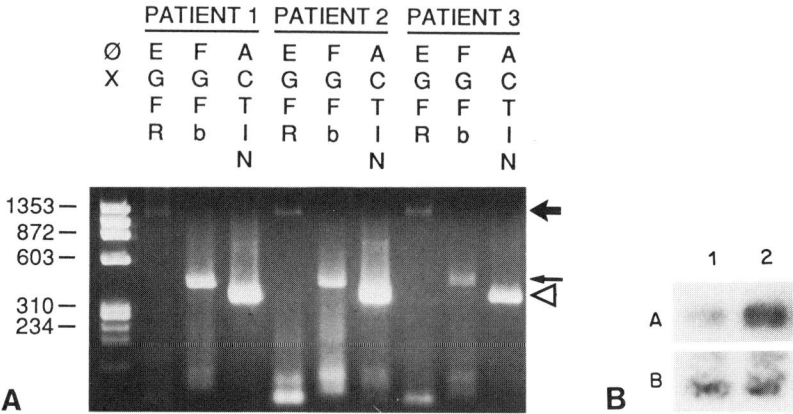

**Figure 2. A.** PCR amplification products for EGF receptor, basic FGF, and beta actin from 3 human lacrimal samples resolved on a 2.0 % agarose gel. EGF receptor amplification products were 1156 base pairs (large arrow). Basic FGF receptor amplification products were 421 base pairs in size (small arrow). Control beta actin amplification products were 350 base pairs in size (arrowhead). Each amplification was the appropriate size based on the published sequence. Concurrent control amplifications were run with each individual sample and were negative (not shown). TGF beta-1 mRNA was not detected in any of the lacrimal specimens. The amplification products were determined to be of the appropriate size by regression analysis using standard PhiX174/HAEIII size markers. **B.** Southern blot of the PCR products for EGF receptor and basic FGF. Lanes 1 and 2 correspond to patients 1 and 2, respectively. Oligonucleotide probes were used to detect the specific amplification product. Row A shows the bands at approximately 1156 base pairs for EGF receptor and row B shows the bands at approximately 421 base pairs for basic FGF. The hybridizing bands were determined to be of the appropriate size by regression analysis using standard PhiX174/HAEIII size markers that were included on the original agarose gel used for resolution of the PCR products for Southern blotting. Reprinted by permission from Wilson SE, et. al. Invest Ophthalmol Vis Sci 32:2816-20, 1991.

lacrimal gland in regulating ocular surface proliferation. We hypothesize that autocrine EGF and other growth factors are constantly produced by the corneal and conjunctival epithelial cells and are normally at high concentration at the level of the growth factor receptors on the epithelial cells. Constant exposure to EGF and other autocrine-produced growth factors at the epithelial cell surface may result in down regulation of the corresponding receptors. This down regulation of the receptors may serve to maintain the proliferation of the epithelial cells at the normal maintenance level. With injury to the ocular surface there could be an up regulation of the epithelial receptors, at least part of which could be due to increased lacrimal tear production that lowers the autocrine growth factor concentrations at the level of the epithelial cell receptors. Up regulation of the growth factor receptors results in an augmented mitogenic response to the autocrine produced growth factors and an increased rate of epithelial cell proliferation. Although there is no direct evidence supporting this hypothesis, this is also true of the hypothesis that the ocular surface epithelial cells, lacrimal gland, and/or other cells increase growth factor production in response to ocular surface injury. Note that the observation that there is a decrease in the concentration of EGF in the tears with reflex increases in production of the aqueous component of the tears from the lacrimal gland[20] could be consistent with the alternative hypothesis.

Recently, we have used the RT-PCR method to demonstrate that human lacrimal tissue contains mRNA coding for basic fibroblast growth factor (bFGF).[19] Although the messenger RNA was detected (Figure 2), further work is needed to determine whether bFGF protein is produced in the lacrimal gland and which cells produce the growth factor. If bFGF protein is produced in lacrimal tissue, then additional work will be needed to determine whether bFGF is present in tears. Since basic FGF does not have a classic secretory sequence,[21] however, it is unclear how bFGF could be released from lacrimal cells in the absence of cell death. Again, this will be complicated by the discovery of bFGF RNA and protein in ex vivo and cultured corneal epithelial cells (Schultz GS, Wilson SE, unpublished data).[17,18]

In the latter study, we were unable to detect transforming growth factor beta-1 (TGF beta-1) mRNA in lacrimal tissue using the PCR-RT method. Almost all cell types have been shown to produce transforming growth factors of one type or another. Thus, I believe that the cells of the lacrimal gland produce another type of TGF beta or our assay was somehow not efficient in amplifying lacrimal TGF beta-1 mRNA despite detection of this message in all the cell types of the cornea using the same methods.[18]

## LACRIMAL GLAND GROWTH FACTOR AND HORMONE RECEPTOR EXPRESSION

As was discussed previously, the discovery of EGF mRNA and protein and bFGF mRNA in the lacrimal gland and the known role of the lacrimal gland in producing other components of the aqueous portion of tears suggested that the lacrimal gland EGF and other growth factors function to modulate the epithelium of the ocular surface.[1,17] It is possible, however, that the growth factors produced in the lacrimal gland have autocrine or paracrine effects on the cells of the lacrimal gland. In fact, even if growth factors are detected in the aqueous component of tears collected by cannulation of the lacrimal ducts, it is difficult to prove that these substances are produced by the lacrimal gland to modulate the cells of the ocular surface instead of being detected in the lacrimal fluid after having localized effects in lacrimal tissue. Supporting the possibility that lacrimal gland-produced EGF and other growth factors have autocrine or paracrine effects in the lacrimal tissue is the recent detection of mRNAs coding for EGF receptor and FGF receptor-1 in lacrimal tissue.[22,23] Further investigation is needed to determine whether the EGF receptor and FGF receptor-1 proteins are produced in the lacrimal tissue, which lacrimal cells produce the receptors, and to determine the functions of the receptors. The PCR primers for FGF receptor-1 used in that study were designed by Hou, et al, to distinguish between mRNAs coding for 3 amino-terminal motifs of the FGF receptor -1 proteins that are derived by alternative splicing of RNA transcribed from a single genomic sequence.[7] The 1100 base pair PCR amplification product corresponds to the alpha amino-terminal motif of FGF receptor-1 that contains three extracellular IgG-like disulfide loops. The 800 base pair product corresponds to the beta amino-terminal motif of FGF receptor-1 that contains two extracellular IgG-like disulfide loops. Finally, the 1000 base pair amplification product corresponds to the gamma amino-terminal motif of FGF receptor-1. This motif does not appear to contain a signal sequence for membrane translocation and is thought to represent an intracellular form of FGF receptor-1. Interestingly, only the mRNA corresponding to the beta amino-terminal motif of FGF receptor-1 was detected in lacrimal tissue despite the detection of all three FGF receptor-1 amino-terminal motif mRNAs in corneal epithelium using the same primers (Figure 3).[23] The importance of this finding to the physiology of the lacrimal gland is unknown.

In addition to EGF receptor and FGF receptor-1 mRNAs, we have also used the RT-PCR method to detect mRNAs coding for the interleukin-1 receptor and glucocorticoid receptor in lacrimal tissue.[23] To date, IL-1 alpha production has not been detected in lacrimal tissue. Further investigation is needed to determine if the IL-1 receptor and glucocorticoid receptor proteins are produced in the lacrimal tissue and to determine the functions regulated by these receptors in the lacrimal gland.

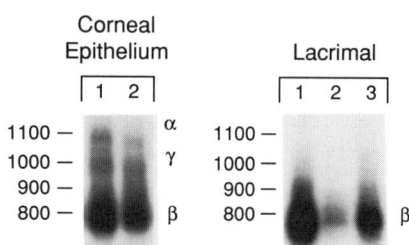

**Figure 3.** Hot blots of PCR amplification products for FGF receptor-1 in human primary corneal epithelial cells and lacrimal tissue. Only the 800 base pair amplification product for the beta amino-terminal motif of FGF receptor-1 (large arrowhead) was detected in each human lacrimal gland sample. Amplification products for the alpha (1100 base pairs) and gamma (1000 base pairs) amino-terminal motifs were not detected in lacrimal tissue, despite exposure of the hot blot for 2 weeks longer than the blot shown and detection of mRNA for all three amino-terminal motifs in the primary corneal epithelial cells.[23] Reprinted by permission Wilson SE, et al, FGF Receptor-1, IL-1 Receptor, and Glucocorticoid Receptor mRNA Production in Human Lacrimal Gland. Invest Ophthalmol Vis Sci, in press, 1993.

## EXPRESSION OF HEPATOCYTE GROWTH FACTOR, KERATINOCYTE GROWTH FACTOR, AND THEIR RECEPTORS IN LACRIMAL TISSUE

Recently, we have become interested in the expression of keratinocyte growth factor (KGF), hepatocyte growth factor (HGF), and their receptors in lacrimal tissue. Hepatocyte growth factor (HGF)[24] and keratinocyte growth factor (KGF)[25] have been shown to stimulate the proliferation of skin keratinocytes in vitro. HGF also modulates motility[24] and KGF regulates differentiation[25] in these epithelial cells. HGF has also been shown to stimulate the proliferation of melanocytes and vascular endothelial cells.[26] Investigators have suggested that HGF[26,27] and KGF[28] may act as a paracrine mediator that is secreted by fibroblast cells to regulate the functions of epithelial cells.[26,27] HGF has been shown to be identical to scatter factor, a fibroblast-derived factor that disperses cohesive colonies of epithelial cells.[29] HGF and KGF produce their effects via the HGF[30] and KGF[31] receptors, respectively. KGF receptor and FGF receptor-2 mRNAs have been found to be alternative transcripts of the same gene.[31]

In recent investigations we have obtained evidence that these growth factor systems may function in the cornea (Wilson SE, Walker J, Chwang EL, and He Y-G, unpublished data). In these studies, we used the RT-PCR method followed by hot blotting to determine that messenger RNAs coding for HGF, HGF receptor, KGF, KGF receptor, and FGF receptor-2 were produced in primary cultures of human corneal epithelial, stromal fibroblast, and endothelial cells, as well as ex vivo corneal epithelium. HGF and KGF mRNAs appeared to be present at barely detectable levels in corneal epithelial cells. We also examined the effects of exogenous HGF and KGF, compared to EGF, on the proliferation of first passage corneal cells. HGF and KGF stimulated proliferation in a dose response manner in first passage corneal epithelial and endothelial cells, but not stromal fibroblast cells. The responses to exogenous HGF and KGF suggest that the HGF and KGF receptor proteins are expressed in corneal epithelial and endothelial cells.

We have recently used similar methods to examine the expression of mRNAs coding for KGF, HGF, HGF receptor, KGF receptor, and FGF receptor-2 in human lacrimal tissue (Wilson SE, Walker J, Kennedy RH, unpublished data). Messenger RNAs coding for HGF, HGF receptor, KGF receptor, and FGF receptor-2 were detected in each cDNA sample derived from human lacrimal tissue. We were unable to detect mRNA coding for KGF in any sample of lacrimal tissue, although KGF mRNAs were detected in corneal endothelial samples that were run simultaneously as positive controls. In addition, we designed another set of PCR primers for KGF that amplified a different region of the KGF mRNA and the results were identical.

Further work is needed to determine whether the corresponding proteins are present in lacrimal tissue once antibodies are available for HGF, HGF receptor, and KGF receptor and to determine which cells produce the modulators. HGF growth factor seems unlikely to be produced in the epidermal derived lacrimal acinar cells. It would appear to be more likely, based on previous work on the cell types that produce and respond to HGF,[26,27] that the HGF is secreted by other fibroblastic cells in the lacrimal tissue as a paracrine modulator of the functions of lacrimal cells of epidermal origin. At this point, however, this is speculative and further investigation is needed. The function of KGF receptor in the lacrimal tissue in the absence of cells secreting the growth factor ligand is unclear. This uncertainty is similar to that for IL-1 receptor. Further work is needed to verify that the KGF receptor and IL-1 alpha proteins are produced in lacrimal tissue.

# CONCLUSIONS

Only recently has evidence been obtained of growth factor and receptor production in lacrimal tissue. Further work is needed to conclusively demonstrate that the majority of these growth factors and receptors are synthesized in lacrimal tissue, determine their functions in the lacrimal gland, and to ascertain their relevance to the ocular surface in health and disease. Table 1 lists the growth factors and receptors for which there is at least limited information available from our work and the investigations of others of production within the lacrimal tissue. This table also indicates areas where further investigation is needed to conclusively demonstrate synthesis within the tissue. We are continuing to work to obtain conclusive evidence and I hope that the interest of other investigators will stimulated by these preliminary studies.

**Table 1.** Growth factor and receptor production in the lacrimal gland

| Modulator | Messenger RNA | | Protein | |
|---|---|---|---|---|
| | RT-PCR | In situ | Immunohistology | Western blot |
| EGF | + | - | + | + |
| EGF receptor | + | - | - | - |
| Basic FGF | + | - | - | - |
| FGF receptor-1 | + | - | - | - |
| FGF receptor-2 | + | - | - | - |
| IL-1 receptor | + | - | - | - |
| HGF | + | - | - | - |
| HGF receptor | + | - | - | - |
| KGF receptor | + | - | - | - |

# REFERENCES

1. S.E. Wilson SE, Lacrimal Gland Epidermal Growth Factor Production and the Ocular Surface, *Am J Ophthalmol.*. 111:763 (1991).
2. J. Herbert, T. Covallaro, R. Martone, The distribution of retinol-binding protein and its mRNA in the rat eye, *Invest Ophthalmol Vis Sci.*. 32:302 (1991).
3. J. Hou, M. Kan, K. McKeehan, G. McBride, P. Adams, W.L. McKeehan, Fibroblast growth factor receptors from liver vary in three structural domains, *Science.* 251:665 (1991).
4. J.D. Parker and G.C. Burmer, The oligomer extension "hot blot": A rapid alternative to Southern blots for analyzing polymerase chain reaction products. *Biotechniques.* 10:94 (1991).
5. R.T. D' Aquila, L.J. Bechtel, J.A. Videler, J.J. Eron, P. Gorczyca, J.C. Kaplan, Maximizing sensitivity and specificity of PCR by preamplification heating, *Nucleic Acid Res.* 19:3749 (1991).
6. D. R. Sparkman, Paraffin wax as a vapor barrier for the PCR, *PCR methods and Applications.* ;2:180 (1992).
7. J. Hou, M. Kan, K. McKeehan, G. McBride, P. Adams, W.L. McKeehan, Fibroblast growth factor receptors from liver vary in three structural domains, *Science.* 251:665 (1991).
8. G.J. Nuovo, P. MacConnell, A. Forde, P. Delvenne, Detection of human papillomavirus DNA in formalin fixed tissue by in situ hybridization after amplification by PCR, *Am J Pathol.* 139:847 (1991).
9. G.J. Nuovo, G.A. Gorgone, P. MacConnell, P. Margiotta, P.D. Gorevic, In situ localization of PCR-amplified human and viral cDNAs, *PCR Methods and Applications.* 2:117 (1992).
10. O. Tsutsumi, A. Tsutsumi, and T. Oka: Epidermal growth factor-like, corneal wound healing substance in mouse tears, *J Clin Invest.* 81:1067 (1988).
11. Y. Ohashi, M. Motokura, Y. Kinoshita, T. Mano, H. Wananabe, S. Kinoshita, R. Manabe, K. Oshiden, and C. Yanaihara, Presence of epidermal growth factor in human tears, *Invest Ophthalmol Vis Sci.* 30:1879 (1989).
12. G.B. van Setten, L. Viinikka, T. Tervo, K. Pesonen, A. Tarkkanen, and J. Perheentupa, Epidermal growth factor is a constant component of normal human tear fluid, *Graefe's Arch Ophthalmol.* 227:184 (1989).
13. S. Kasayama, Y. Ohba, and T. Oka, Expression of the epidermal growth factor gene in mouse lachrymal gland: comparison with that in the submandibular gland and kidney. *J Mol Endocrinol.* 4:31 (1990).
14. S.E. Wilson, S.A. Lloyd, R.H. Kennedy, Epidermal growth factor messenger RNA production in human lacrimal gland. *Cornea.* 10:519 (1991).

15. G.B. van Setten, K. Tervo, I. Virtanen, A. Tarkkanen, and T. Tervo, Immunohistochemical demonstration of epidermal growth factor in the lacrimal and submandibular glands of rats. *Acta Ophthalmol.* 68:477 (1990).

16. T.L Steinemann, H.W. Thompson, K.M. Maroney, C.H. Palmer, L.A. Henderson, J.S. Malter, D. Clarke, B. Bromberg, M. Kunkle, and R.W. Beuerman, Changes in epithelial epidermal growth factor receptor and lacrimal gland EGF concentration after corneal wounding, *Invest Ophthalmol Vis Sci.* 31(Suppl):55 (1990).

17. G.B. Van Setten, T. Tervo, L. Viinikka, K. Pesonen, Ocular disease leads to decreased concentrations of epidermal growth factor in the tear fluid. *Curr Eye Res.* 10:523 (1991).

18. S.E. Wilson, S.A. Lloyd, Epidermal Growth Factor and its Receptor, Basic Fibroblast Growth Factor, Transforming Growth Factor beta-1, and Interleukin-1 Alpha Messenger RNA Production in Human Corneal Endothelial Cells. *Invest Ophthalmol Vis Sci.*. 32:2747 (1991).

19. S.E. Wilson, Y-G. He, and S.A. Lloyd, EGF, basic FGF, and TGF beta-1 messenger RNA production in rabbit corneal epithelial cells. *Invest Ophthalmol Vis Sci.*. 33:1987 (1992).

20. G.B. van Setten, Epidermal growth factor in human tear fluid: increased release but decreased concentrations during reflex tearing, *Curr Eye Res.* 9:79 (1990).

21. W.H. Burgess, T. Maciag, The heparin-binding (fibroblast) growth factor family of proteins, *Annual Rev Biochem.* 58:575 (1989).

22. S.E. Wilson, S.A. Lloyd, and R.H. Kennedy, Basic fibroblast growth factor (FGFb) and epidermal growth factor (EGF) receptor messenger RNA production in human lacrimal gland, *Invest Ophthalmol Vis Sci.* 32:2816 (1991).

23. S.E. Wilson, S.A. Lloyd, and R.H, FGF Receptor-1, IL-1 Receptor, and Glucocorticoid Receptor mRNA Production in Human Lacrimal Gland, *Invest Ophthalmol Vis Sci*, in press.

24. K. Matsumoto, K. Hashimoto, K. Yoshikawa, T. Nakamura, Marked stimulation of growth and motility of human keratinocytes by hepatocyte growth factor, *Exp Cell Res.* 196:114 (1991).

25. C. Marchese, J. Rubin, D. Ron, A. Faggioni, M.R. Torrisi, A. Messina, L. Frati, and S.A. Aaronson, Human keratinocyte growth factor activity on proliferation and differentiation of human keratinocytes: Differentiation response distinguishes KGF from EGF family, *J Cell. Physiol.* 144:326 (1990).

26. J.S. Rubin, A.M-L. Chan, D.P. Bottaro, W.H. Burgess, W.G. Taylor, A.C. Cech, D.W. Hirshfield, J. Wong, T. Miki, P.W. Finch, and S.A. Aaronson, A broad-spectrum human lung fibroblast -derived mitogen is a variant of hepatocyte growth factor, *Proc. Natl. Acad. Sci.* 88:415 (1991).

27. R. Montesano, K. Matsumoto, T. Nakamura, and L. Orci, Identification of a fibroblast-derived epithelial morphogen as hepatocyte growth factor, *Cell.* 67:901 (1991).

28. P.W. Finch, J.S. Rubin, T. Miki, D. Ron, S.A. Aaronson, Human KGF is FGF-related with properties of a paracrine effector or epithelial cell growth, *Science.* 245:752 (1989).

29. K.M. Weidner, N. Arakaki, G. Hartmann, J. Vandekerçkhove, S. Weingart, H. Reider, C. Fonatsch, H. Tsubouchi, T. Hishida, Y. Daikuhara, and W. Birchmeier, Evidence for the identity of human scatter factor and human hepatocyte growth factor, *Proc Natl Acad Sci.* 88:7001 (1991).

30. M. Park, M. Dean, K. Kaul, M.J. Braun, M.A. Gonda, and W.G. Vande, Sequence of MET protooncogene cDNA has features characteristic of the tyrosine kinase family of growth factor receptors, *Proc Natl Acad Sci.* 84:6379 (1987).

31. T. Miki, D.P. Bottaro, T.P. Fleming, C.L. Smith, W.H. Burgess, A.M-L. Chan, and S.A. Aaronson, Determination of ligand-binding specificity by alternative splicing: two distinct growth factor receptors encoded by a single gene, *Proc Natl Acad Sci,* 89:246(1992).

## ACKNOWLEDGMENTS

Supported in part by US Public Health Service grant EY09379 from the National Eye Institute, National Institutes of Health, Bethesda, Maryland, a grant-in-aid from the Fight for Sight Research Division of the National Society to Prevent Blindness, Schaumburg, IL, (Dr. Wilson), and an unrestricted grant from Research to Prevent Blindness, Inc., New York, New York. Dr. Wilson is a Research to Prevent Blindness William and Mary Greve International Research Scholar.

# HUMAN LACRIMAL GLAND SECRETES PROTEINS BELONGING TO THE GROUP OF HYDROPHOBIC MOLECULE TRANSPORTERS

A.M. Françoise Gachon

Laboratoire de Biochimie Médicale
Faculté de Médecine
28, Place Henri Dunant, B.P. 38
63001 Clermont-Ferrand Cedex, France

## INTRODUCTION

Many orbital glands and ocular surface epithelia participate to the formation of the human lacrimal tear film: the main lacrimal gland, the accessory lacrimal glands, the goblet cells and conjunctival epithelial cells, the cornea and the meibomian glands. Consequently, human tears are constituted of a complex mixture of proteins and glycoproteins, with or without enzymatic activities, metabolites, electrolytes and lipids. Furthermore, serum proteins, generally accepted as derived from the ocular surface vessels and indicative of collection trauma, are also present. Three lacrimal gland proteins, lactoferrin, lysozyme and tear specific prealbumin account for the major part of protein content. Our presentation will deal with tear specific prealbumin, which is, as further demonstrated, an inappropriate term. Bonavida, Sapse and Sercarz (1969) described, for the first time, a tear specific prealbumin, present at high levels in tears of humans and several animal species (rabbit, rat, monkey). Its electrophoretic mobility was larger than albumin and it was absent from serum and other biological fluids such as cerebrospinal fluid, saliva, nasal secretion and sweat. Its local synthesis by main lacrimal gland was shown by culturing lacrimal gland slices in the presence of radiolabelled amino-acids (Bonavida et al., 1969). Using higher resolution electrophoretic methods, we demonstrated that Bonavida's team had not described a single protein but a group of at least six proteins whose molecular weight ranged from 15 to 20 kDa with isoelectric points ranging from 4.6 to 5.4 (Gachon et al., 1979; Gachon, Lambin and Dastugue, 1980); we called them 'Proteins Migrating Faster than Albumin' (PMFAs) in order to clearly differentiate them from transthyretin (current denomination of prealbumin).

During many years'work on these tear specific proteins, it has become apparent that it was necessary to elucidate their complete amino-acid sequence to allow determination of their as yet unknown role. Indeed, it is now possible to model a putative protein if it is homologous with a protein of known structure and so find a relation between structure and function.The complete amino-acid sequence would permit better understanding of their mechanism of secretion and confirm whether they are synthetized as pre protein or not. At this congress, we presented the work of our last two years: the first part dealing with amino-acid sequencing and hypothesis of new members of the lipocalin family in human tear fluid, has led to publication just before this congress (Delaire, Lassagne and Gachon, 1992) and the second part, on molecular cloning (Lassagne and Gachon, 1993), is in the press.

---

Since the time of the congress, the same cloning study has been published by Redl, Holzfeind and Lottspeich (1992).

## MATERIALS AND METHODS

### Protein Analysis

Tears were collected from only one male donor. Tear proteins were separated by two dimensional electrophoresis performed according to Walsh et al. (1988) for avoidance of protein N-terminal blockage and tryptophan destruction. The transfer was performed according to Bauw et al. (1989) and PVDF membrane bound proteins were visualized by amidoblack staining. The sequence analysis was carried out with a gas phase sequencer (model 470A, Applied Biosystem) equipped with an on line phenylhydantoin amino-acid derivative analyser.

### Nucleic Acid Analysis

Lacrimal glands were obtained post mortem from 6 different male donors. Unless otherwise stated, all the techniques were performed following the recommendations of Sambrook, Fritsch and Maniatis (1989). Only one lacrimal gland gave good quality RNAs. The lacrimal gland tissue was disrupted at -80°C. Total RNA was extracted by phenol chloroform. Poly $A^+$ RNA was purified on an oligo dT column and a cDNA library was constructed in a λ Zap II vector, as recommended by the manufacturers; both oligo dT and random primers were used. Southern blots were revealed using a radiolabelled fragment from clone 16; Northern blots were hybridized to a $^{32}$P radiolabelled riboprobe synthesized from clone 16 Bluescript plasmid; for size determination, both commercial size markers and internal markers (18S RNA and 28 S RNA from total RNA slot) were used. Sequencing of clone 16 was carried out on both strands by the dideoxy chain-termination method as modified by Hsiao (1991).

## RESULTS

### N Terminal Sequences

Anticipating a role for these proteins and in order to normalize the nomenclature, we proposed to call them "tear lipocalin" , followed by their molecular weight and their isoelectric point: TL MW/pI.The term tear lipocalin aggregate (TLA) followed by the molecular weight in kDa of the aggregate would be used for TL association. The first 32 amino-acid of TL 18/5.3, the first 28 amino-acids of TL 17/4.9 and the first 27 amino-acids of TL I8/5.2 were determined (Delaire, Lassagne and Gachon, 1992). Amino-acid sequence alignments indicate that TL 17/4.9 is 5 amino-acid shorter on the $NH_2$ terminal size. The two first amino-acids in both TL 18/5.3 and TL 18/5.2 were uncertain throughout all the successive analysis. The sequencing of TL 18/5.3 after tryptic digestion allowed us to clearly identify the following peptide: TTLEGGNLEA. The N terminal sequence of TL 18/5.3, matching the best with the sequence further deduced from the cDNA, is given in table 1.

**Table 1.** N terminal amino-acid sequences of TL 18/5.3 (single letter code).

---

H S L L A S D E E I Q D V S G T W Y L K A M T V D / E F P E M N

---

### Nucleic Acid Sequence and Deduced Protein Sequence

We used a pool of 20 mer oligonucleotide (96 possible sequences) coding for the peptide stretch (DEEIQDV) common to the three previously sequenced TL as a hybridization probe on nylon membrane. Thirty positive clones with inserts ranging from 0.5 to 1.2 kb were selected and among them, seven presented similar restriction patterns and cross hybridize. The size of the mRNA, deduced from Northern blots, is 1.1 kb.. Clone 16 sequencing revealed a 770 nucleotide long molecules, starting, on the 5' side, with a 43 nucleotide long untranslated fragment (Table 2).

**Table 2.** Sequence of human tear lipocalin cDNA: nucleotides are numbered from the 5' terminus of cDNA clone 16 to the 770 residue. The complete amino acid sequence of tear lipocalin is shown above the DNA sequence. Amino acid numbers from 1 to 18 refer to the leader peptide, whereas numbers from 19 to 176 refer to the mature protein. Fragments for which direct amino acid sequence data were obtained are underlined. The sequence used to construct oligonucleotide probe which is discussed in the text is enclosed by a solid line.

```
                                                      1
                                                      M   K   P   L   L   L
CAGCAAGCGACCTGTCAGGCGGCCGTGGACTCAGACTCCGGAG ATG AAG CCC CTG CTC CTG
                                                      19
      A   V   S   L   G   L   I   A   A   L   Q   A   H   H   L   L   A
     GCC GTC AGC CTT GGC CTC ATT GCT GCC CTG CAG GCC CAC CAC CTC CTG GCT

      S   D   E   E   I   Q   D   V   S   G   T   W   Y   L   K   A   M
     TCA GAC GAG GAG ATT CAG GAT GTG TCA GGG ACG TGG TAT CTG AAG GCC ATG
                                                                      ↑
      T   V   D   R   E   F   P   E   M   N   L   E   S   V   T   P   M
     ACT GTG GAC AGG GAG TTC CCT GAG ATG AAT CTG GAA TCG GTG ACA CCC ATG

      T   L   T   T   L   E   G   G   N   L   E   A   K   V   T   M   L
     ACC CTC ACG ACC CTG GAA GGG GGC AAC CTG GAA GCC AAG GTC ACC ATG CTG

      I   S   G   R   C   Q   E   V   K   A   V   L   E   K   T   D   E
     ATA AGT GGC CGG TGC CAG GAG GTG AAG GCC GTC CTG GAG AAA ACT GAC GAG

      P   G   K   Y   T   A   D   G   G   K   H   V   A   Y   I   I   R
     CCG GGA AAA TAC ACG GCC GAC GGG GGC AAG CAC GTG GCA TAC ATC ATC AGG

      S   H   V   K   D   H   Y   I   F   Y   C   E   G   E   L   H   G
     TCG CAC GTG AAG GAC CAC TAC ATC TTT TAC TGT GAG GGC GAG CTG CAC GGG

      K   P   V   R   G   V   K   L   V   G   R   D   P   K   N   N   L
     AAG CCG GTC CGA GGG GTG AAG CTC GTG GGC AGA GAC CCC AAG AAC AAC CTG

      E   A   L   E   D   F   E   K   A   A   G   A   R   G   L   S   T
     GAA GCC TTG GAG GAC TTT GAG AAA GCC GCA GGA GCC CGC GGA CTC AGC ACG
                                                                      176
      E   S   I   L   I   P   R   Q   S   E   T   C   S   P   G   S   D
     GAG AGC ATC CTC ATC CCC AGG CAG AGC GAA ACC TGC TCT CCA GGG AGC GAT

      ★
     TAG  GGGCAGGGGACACCTTGGCTCCTCAGCAGCCAAGGACGGCACCATCCAGCACCTCCGTCATTC

     ACAGGGACATGGAAAAAGCTCCCCACCCCTGCAGAACGCGGCTGGCTGCACCCCTTCCTACCACCCC

     CCGCCTTCCCCCTGCCCTGCGCCCCCTCTCCTGGTTCTCCATAAAGAGCTTCAGCAGTTAAAAAAAA
```

Then, a long open reading frame encodes a protein of 176 amino acids; first a 54 nucleotide long fragment codes for a signal peptide; then a 468 long fragment codes for the mature protein; the 32 first amino acids of this protein present 94 % homology with the N terminal sequence of the purified protein TL 18/5.3.; one discrepancy is observed for the second residue (His instead of Ser). No glycosylation site is present. Translation of the nucleotide sequence perfectly matches (100 % homology) the internal TTLEGGNLEA sequenced peptide. The 3' extremity contains the polyadenylation signal and the poly A tail. The computer alignment of the mature form with the sequences deposited in NBRF protein data bank is presented in Table 3.

An unpublished sequence, recently deposited in the EMBL data library by Schmale, indicates that the amino acid sequence of a human Von Ebner's gland protein, in human, is identical to the amino-acid sequence of the TL 17.4/5.5. However, minor differences are observed upstream the translated region.

**Table 3.** Homology of TL with members of the lipocalin family: Human Tear Lipocalin (H-TL), Rat Von Ebner's Gland protein (R-VEG), Goat β Lactoglobulin (GB Lac), Human α1 Microglobulin (H-HC), Rat Androgeno-dependent 18.5 kD protein (R-And), Frog Bowman's gland protein (F-BG), Human Retinol Binding Protein (H-RBP), Mouse-Major Urinary Protein (MO-MUP), Human Placental protein (H-Plac). The consensus sequences are between vertical lines.

```
                    1
1. H-TL      ------------------------HHLLASDEEIQDVS|GTWYI|KAMTVDREFPE--
2. R-VEG     MKALLLTFGLSLLAALQAQAFPTT-EENQDVS|GTWYI|KAAAWDKEIPD--
3. G-B Lac   ----------------IIVTQTMKGLDIQKVA|GTWYS|LAMAASDISL-L-
4. H-HC      ----------GPVPTPPDNIQVQENFNISRIY|GKWYN|LAIGSTCPWL-K-
5. R-And     --------------------AVVKDFDISKFI|GFWYE|IAFASKMGTPGLA
6. F-BG      -------------QCQADLPPVMKGLEENKVT|GVWYG|IAAASNCKQF-L-
7. H-RBP     ----------ERDCRVSSFRVKENFDKARFS|GTWYA|MAKKDPEGLF-L-
8. MO-MUP    --------------EEASSTGRNFNVE--KIN|GEWHT|IILASDKREK-I-
9. H-Plac    ----------------MDIPQTKQDLELPKLA|GTWHS|MAMATNNISL-M-

   51
1. MNLE---SVTPMTLTT-LEGGNLEAK-VTMLI-S-GR|QEV-KAVLEKTDEPGK-Y----
2. KKFGS-VSVTPMKIKT-LEGGNLQVK-FTVLI-A-GR|KEM-STVLEKTDEPAK-Y----
3. DAQSAPLRVYVEELKP-TPEGNLEIL-L--QKWENGE|AQ-KKIIAEKTKIPAV-FKI--
4. KIMDR-MTVSTLVLGE--GATEAEIS-MTSTRWRKGV|EE-TSGAYEKTDTDGK-F-L-Y
5. HKEEK-MGAMVVELKE--NLLALTTT-YYSEDH----|VL-EKVTATEGDGPAK-FQV--
6. QMKSD--NMPAPVNIYSLNNGHMKSS-TSFQT-EKG-|QQM-DVEMT-TVEKGH-Y--KW
7. QDNIV-AEFSVDETGQMSATAKGRVR-LL-NNWDV--|ADM-VGTFTDTEDPAK-FKMKY
8. EDNGN-FRLFLEQIHV-L-ENSLVLK-FHTVR-DE-E|SEL-SMVADKTEKAGE-Y----
9. ATLKAPLRVHITSLLP-TPEDNLEIV-LHRWE-NN-S|VE-KKVLGEKTGNP-KKFKINY

   111
1. -------------------|TAD|GGKHVA-YIIRSHVKDHYIFYCEGELHGKPVRGVK
2. -------------------|TAY|SGKQVL-YIIPSSVEDHYIFYYEGKIHRHHFQIAK
3. DALN-ENKVL--------VL-D|TDY|KKY-LL-F-CMENSAEPEQSL-----ACQ----CL
4. HKSK-WNITMES-----YVV-H|TNY|DEYAI--FLTK---KFS--RH-----HGPTITA-K
5. TRLSG-KKEV--------VVEA|TDY|LTYAIIDI-TSL--VAG----------AVHRTM-K
6. KMQQGDSETI--------IV-A|TDY|DAF-LMEF-TKI--QMG--AE-----VCV-T-V-K
7. WGV---ASFLQKGNDDHWIV-D|TDY|DTYAVQ-YSCRLLNLDG--------TCADS-YSF
8. -------SVTYDGFNTFTIPK|TDY|DNF-LMAHLINE--KDG--------ETFQL-MG-
9. T--------V--ANEAT-LL-D|TDY|DNF-L--FLCL---QDT-------TTPIQSM-MCQ

   171
1. LVGRDPKNNLEALEDFEKAAGAR-GLSTESILI-PRQS--ET|C|SPGSD---------
2. LVGRDPEINQEALEDFQSVVRAG-GLNPDNIFI-PKQS--ET|C|PLGSN---------
3. V--RTPEVDNEALEKFDKALKAL-PMH-IRLAFNPTQ-LEGQ|C|HV-----------
4. LYGRAPQLRETLLQDFRVVAQGV-GIPEDSIFTMADR---GE|C|VPGEQEPEPILIPR
5. LYSRSLDDNGEALYNFRKITSDH-GFSETDLYILKHD---LT|C|VKVLQSAAESRP--
6. LFGRKDTLPEDKIKHFEDHIEEKVGLKKEQYIRFHTK---AT|C|-VPK---------
7. VFSRDPNGLPPEAQKIVRQRQEE-LCLARQYRLIVHN---GY|C|DGRSERNLL-----
8. LYGREPDLSSDIKERFAQLCEEH-GILRENIIDLSNA---NR|C|LQARE--------
9. YLARVLVEDDEIMQGFIRAFRPL-PRHLWYLLDLKQME--EP|C|RF-----------
```

## CONCLUSION

The assignment of this 176 amino acid protein to the superfamily of lipophilic molecule carriers was performed by searching for similarity between the complete sequence and any group of sequence deposited in NBRF-protein Data Bank. Lipocalins are a group of 18-25 kDa proteins (about 20 are currently known), characterized as extracellular proteins sharing a common framework for binding and transport of small hydrophobic ligands. Based on three dimensional structure of three proteins (retinol binding protein, β lactoglobulin and bilin binding protein) and on stretches of sequences which are strongly conserved, Godovac-Zimmermann (1988) suggested they were highly appropriate for binding a variety of hydrophobic ligands such as retinoids, steroids, bilins and lipids. In such a family, ligands are bound within the central cavity consisting of a barrel formed by many β sheets and an α helix. The putative role of the α helix is to induce protein aggregates, previously observed with α1 microglobulin (protein HC) and several of the other lipocalins (Akerström and Lögdberg, 1990). The ligand specificity could be obtained by changing the residue types inside the central cavity. The tear lipocalin presents similarities to many members of the family; the best score was obtained for VEG protein (rat (Schmale, Holtgreve-Grez and Christiansen, 1990)) with 58.4% identity in a 178 amino acid overlap; then the score decreased: for β lactoglobulin (sheep (Halliday, Bell and Shaw, 1991)), 28.4 % identity in a 176 amino acid overlap; for α1 microglobulin (human (Lopez et al, 1981) 26.7 % identity in a 180 amino acid overlap; for androgen-dependent epididymal 18.5K protein (rat (Brooks et al, 1986)), 23.5 % identity in a 170 amino acid overlap; for olfactory protein precursor (frog (Lee, Wells and Reed, 1987)), 18.2 % identity in a 176 amino acid overlap. Lower scores were observed for plasma retinol binding protein (human (Papiz et al, 1986)), Mouse Major Urinary Protein (MUP) (Shaw, Held and Hastie, 1983)) and placental protein (human (Julkunnen, Seppala and Janne, 1988)). Most of the members have only 20-25% sequence identities with at least one of the members, and overall, there are very few residues that are identical in all the sequences when they are positioned for maximum alignment. On the basis of possession of these motifs and its resemblance to VEG protein, the 176 amino acid protein may tentatively be assigned to this lipocalin family and named "Tear Lipocalin" (TL).

N terminal amino-acid sequences of the mature forms of tear lipocalins, previously reported, allowed us to identify the processed form of this TL; Fullard's group (Fullard and Kissner, 1991) had isolated five isoelectric forms; among them, the arbitrarily named $P_1$ (90% identity in a ten amino acid overlap) is probably identical to the TL described in this paper. The reported 32 amino acid long sequence of the N terminal extremity of purified TL 18/5.3 (Delaire, Lassagne and Gachon, 1992), matched also with part of the 176 amino acid sequence (94% identity in a 32 amino acid overlap), suggesting that the unprocessed 176 amino-acid protein contains a 18 amino acid signal peptide presenting the general features of signal peptides: a long hydrophobic core, a basic amino acid in position 2, and small apolar amino-acids, alanine and leucine, at positions 16 and 18. The molecular weight and isoelectric point of the mature protein, calculated from the amino acid composition, being respectively 17.445 (corrected to 17.4) and 5.52 (corrected to 5.5), tear lipocalin is named TL 17.4/5.5. The TL 17.4/ 5.5 described in this paper might be the same as P1 isoform described by Fullard's group and the same as TL 18/5.3 described by Gachon's group.

In most cases, the three-dimensional structure of a lipocalin is unknown and the biologically relevant ligands uncertain. So, elucidation of TL function may be difficult. However, two interesting observations may aid future investigation of TL biological function: (i) among the lipocalin family members, the Mouse Major Urinary Proteins (MUPs) are an antigenically related group of proteins encoded by a multigene family. The tissues of expression include, besides the liver, salivary, lacrimal and mammary glands. These glands present commun features: they are anatomically comparable glands, whose secretions are destined to external "cavities"; all three synthesize and export antibacterial proteins such as lysozyme and lactoferrin. It has been suggested that these proteins could have a possible function in chemical communication (Shaw, Held and Hastie, 1983). (ii) In minor salivary glands, the VEG protein is supposed to be involved in the transport of hydrophobic odorants across the hydrophilic mucus covering the sensory neurones (Schmale, Holtgreve-Grez and Christiansen, 1990).

As only one mRNA species is present, the various post translationnal modifications leading to the various isoforms have to be investigated, as well as the study of the expression of TL in different human tissue and the identification of specific ligands.

## REFERENCES

Bonavida, B., Sapse, A. T. and Sercarz, E. E., 1969, Specific tear prealbumin: a lacrimal protein absent from serum and other secretions. *Nature* 221: 375.

Gachon, A.M., Lambin, P. and Dastugue, B., 1980, Human tears: Electrophoretic characteristics of specific proteins. *Ophthalmic Res.* 12: 277.

Gachon, A.M., Verrelle, P., Betail, G. and Dastugue, B., 1979, Immunological and electrophoretic studies of human tear proteins. *Exp. Eye Res.* 29: 539.

Delaire A., Lassagne H. and Gachon A.M.F.,1992, New members of the lipocalin family in human tear fluid. *Exp. Eye Res.* 55: 645.

Lassagne H. and Gachon A.M.F., 1993, Cloning of a human lacrimal lipocalin secreted in tears. *Exp. Eye Res.* in the press.

Redl B., Holzfeind P. and Lottspeich F., 1992, cDNA cloning and sequencing reveals human tear prealbumin to be a member of the lipophilic-ligand carrier protein superfamily, *J. Biol. Chem.*, 267: 20282.

Walsh, M.J., McDougall, J. and Wittmann-Liebold, B., 1988, Extended N terminal sequencing of proteins of archaebacterial ribosomes blotted from two-dimensional gels onto glass fiber and poly(vinylidene difluoride) membrane. *Biochemistry* 27: 6867.

Bauw, G., Van Damme, J., Puyre, M., Vandekerkhove, J., Gesser, B., Ratz, G.P., Lauridsen, J.B. and Celis, J.E., 1989, Protein electroblotting and micro sequencing strategies in generating protein data base from two dimensional gels. *Proc. Natl. Acad. Sci. USA* 86: 7701.

Sambroock, J., Fritsch, E.F. and Maniatis, T., 1989, Molecular cloning: a laboratory manual. Sd Ed. Cold Spring Harbor Laboratory Press.

Hsiao K.,1991, A fast and simple procedure for sequencing double stranded DNA with sequenase. *Nucleic Acids Research* 19: 2787.

Pervaiz, S. and Brew, K., 1987, Homology and structure function correlations between α1-acid glycoprotein and serum retinol-binding protein and its relatives. *FASEB J.* 1: 209.

Godovac-Zimmermann, J.,1988, The structural motif of β Lactoglobulin and retinol binding protein: a basic framework for binding and transport of small hydrophobic molecules. *TIBS* 13: 64.

Akerström, B. and Lögdberg, L., 1990, An intriguing member of the lipocalin protein family: α1 microglobulin. *TIBS* 15: 240.

Schmale, H., Holtgreve-Grez, H. and Christiansen, H.,1990, Possible role for salivary gland protein in taste reception indicated by homology to lipophilic-ligand carrier protein. *Nature* 343: 366.

Halliday, J.A., Bell, K. and Shaw, D.C., 1991, The complete amino acid sequence of feline β lactoglobulin II and a partial revision of the equine β lactoglobulin II sequence. *Biochim. Biophys. Acta* 1077: 25.

Lopez, C., Grubb, A., Soriano, F. and Mendez, E., 1981, The complete amino acid sequence of human complex forming glycoprotein heterogeneous in charge (Protein HC). *Biochem. Biophys. Res. Commun.* 103: 919.

Brooks, D.E., Means, A.R., Wright, E.J., Singh, S.P. and Tiver, K.K., 1986, Molecular cloning of the cDNA for two major androgen-dependent secretory proteins of 18.5 kilodaltons synthetized by the rat epididymis. *J. Biol. Chem.* 26: 4956.

Lee, K.H., Wells, R.G. and Reed, R.R., 1987, Isolation of an olfactory cDNA. Similarity to retinol binding protein suggests a role in olfaction. *Science* 235: 1053.

Papiz, M.Z., Sawyer, L., Eliopoulos, E.E., North, A.C.T., Findlay, J.B.C., Sivaprasadarao, R., Jones, T. A., Newcomer, M.E. and Kraulis, P.J., 1986, The structure of β lactoglobulin and its similarity to plasma retinol binding protein. *Nature* 324: 383.

Shaw, P., Held W.A. and Hastie N.D.,1983, The gene family for the Major Urinary Proteins: Expression in several secretory tissues in the mouse. *Cell* 32: 755.

Julkunen, M., Seppala, M. and Janne, O.A., 1988, Complete amino acid sequence of human placental protein 14: a progesterone-regulated uterine protein homologous to β lactoglobulins. *Proc. Natl. Acad. Sci. USA* 85: 8845.

Fullard, R.J. and Kissner, D.M.,1991, Purification of the isoforms of tear specific prealbumin. *Curr. Eye Res.* 10: 613.

# TRANSCRIPTION OF MESSAGE FOR TUMOR NECROSIS FACTOR-alpha BY LACRIMAL GLAND IS REGULATED BY CORNEAL WOUNDING

Hilary W. Thompson,[1] Roger W. Beuerman,[1] Julie Cook,[2] and Lauren W. Underwood,[1] and Doan H. Nguyen[1]

[1]Laboratory of the Molecular Biology of the Ocular Surface
LSU Eye Center, New Orleans, LA 70112
[2]Ochsner Medical Foundation, Division of Research
New Orleans, LA 70121

## INTRODUCTION

Tumor necrosis factor-alpha (TNF), or cachectin, has been implicated in the pathophysiology of wasting diseases and inflammation.[1] Although most cells have receptors for this hormone, the receptor-mediated events may range from a promotion of fibrosis to cell death.[1,2] The properties of TNF suggest that it could play a role in the progress of lacrimal gland disease, such as Sjögren's syndrome, in which fibrosis and lymphocytic infiltration are both present.[3]

The lacrimal gland is part of the secretory immune system that interacts with the front of the eye through the secretion of products, such as enzymes and growth factors, into the tears.[4,5] In turn, the cornea is connected to the lacrimal gland by multisynaptic, neuronal pathways of trigeminal origin that can induce secretion after corneal stimulation.[6,7]

Within the lacrimal gland, the complex interactions of cytokines and growth factors are being unraveled "one molecule at a time" using molecular biology techniques.[8,9] Primary response genes, as well as the genes for epidermal growth factor and its receptor, have been identified in the lacrimal gland.[10,11] Recent work has shown that the expression of these genes can be modified by wounds to the cornea. In the acinar cells, the secretory component can be modulated by hormones, cytokines such as TNF, and growth factors.[12,13]

The technique used in this study, the nuclear run-on assay, has more physiological relevance and allows the simultaneous sampling of the relative transcriptional levels of a number of molecules of interest. Northern blots showed that TNF message increased following corneal wounds and nuclear run-on assays showed that TNF is present in the normal gland as well. In situ hybridization indicated that macrophages, as well as the acinar cells, produce TNF.

## MATERIALS AND METHODS

### Corneal Stimulation: Wounding

A standard 6-mm diameter wound has been used extensively in healing studies in this laboratory and the rate of closure has proven to be highly reproducible.[14] Enhanced neural activity from corneal nerves after the standard wound has also been described in our laboratory.[7] All animal procedures were carried out in accordance with the ARVO Resolution on the Use of Animals in Research.

Lacrimal glands (LG) were harvested over time after corneal wounding from animals that were sacrificed by lethal injection. Tissues for RNA analysis were immediately minced and snap frozen in liquid nitrogen. Tissues for in situ hybridization were processed as described below. Run-on assays used normal glands from rabbits without corneal wounds.

### RNA Procedures

Total cellular RNA from rabbit tissues was obtained as previously described.[14] Total cellular RNA was quantified by spectrophotometry and inspected for freedom from degradation on an ethidium bromide agarose gel.

RNA from rabbit tissues was used in four different procedures: 1) reverse transcriptase (RT) production of cDNA for polymerase chain reaction (PCR) amplification to determine the presence of TNF mRNA in lacrimal gland, 2) slot blot analysis to determine the relative amounts of TNF mRNA in different rabbit tissues, 3) Northern blot analysis to determine the change in the amount of TNF mRNA in the lacrimal gland following wounding, and 4) nuclear mRNA labeled in run-on assays to determine whether the TNF mRNA in the lacrimal gland was newly transcribed or part of a stable message pool. In situ hybridization was used to determine the cellular localization of TNF mRNA in the lacrimal gland, both at rest and after wounding.

### RT-PCR

RT cDNA production was conducted using methyl mercury (100 Mm) in 1 $\mu$g of RNA to denature mRNA secondary structure and allow association of the random heximer primer (1 $\mu$g/$\mu$l). The PCR reaction, using primers derived from the rabbit TNF sequence, was used to amplify the cDNAs produced in the RT reaction. The thermal cycler conducted the sequence of a one-minute 94°C denaturation step, a one-minute 50°C annealing step, and a one-minute 72°C elongation step 30 times.

### In Situ Hybridization

Lacrimal gland tissues from wounded or control rabbits were divided into quarters and fixed in 4% paraformaldehyde at 4°C overnight. Tissues were then dehydrated in an ethanol series, moved into chloroform, and paraffin infiltrated and embedded. Five to seven micron sections were cut and the slides were stored at 4°C in a desiccator. After counter-staining in hematoxylin and eosin, slides were examined and photographed at 100 x. For quantitation, the positive cells in ten randomly chosen fields were counted by a masked, trained observer.

## PCR

The polymerase chain reaction (PCR) was used to produce oligomeric probes for TNF. Primers used for TNF were selected from exon 1 of the TNF gene. The PCR primers were synthesized by Integrated DNA Technology (Coraville, IA). The TNF cDNA target was purchased from American Type Culture Collection (ATCC, Rockville, MD). Solutions and methods not specifically described were conducted as in Sambrook et al.[15]

## Northern Blots

For Northern blots, an equal quantity of each RNA sample to be compared was run on formaldehyde-denaturing gels at constant current with buffer recirculation. Nylon membranes (Zeta-Probe, Biorad) were blotted from the gels, prehybridized at 42°C for 5 hrs, and hybridized overnight at 42°C.[14]

## Nuclear Run-on Assays

Run-on assays were conducted using previously described methods.[16] In order to adapt these methods to the rabbit lacrimal gland, we dissected lacrimal glands from rabbits and 1 gram of lacrimal tissue was rinsed in ice-cold 0.14 M NaCl, 10 mM Tris-HCl, pH 8.0. The level of activity, counted on a Molecular Dynamics Phosphoimager, was compared to the binding of labeled RNA to plasmid DNA without insert. DNA probes were obtained from the American Type Culture Collection or were in use in one of our laboratories (JC). Probes used included: transforming growth factor-beta1 (TGF-ß-1), transforming growth factor-beta2, colony stimulating factor-granulocyte macrophage, platelet derived growth factor-alpha (PDGF-AA), platelet derived growth factor-beta (PDGF-BB), tumor necrosis factor-alpha, receptor for PDGF-AA, receptor for PDGF-BB, epidermal growth factor, receptor for epidermal growth factor, angiotensin, and glial fibrillary acidic protein.[17]

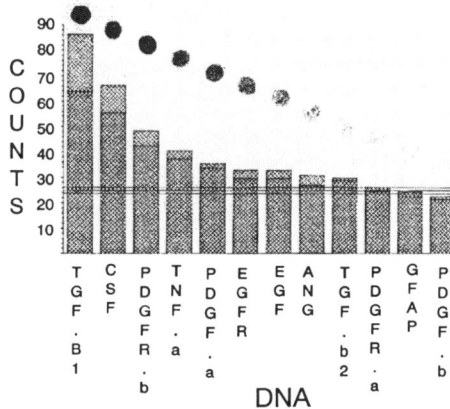

**Figure 1.** Run-on analysis dot intensity. Dots from one trial shown above bars of means and standard errors (stacked bars) of the transcription of each gene in three repeated trials.

# RESULTS

In Figure 1, the densities of the dots above the bars of the histogram show the variable amounts of messenger RNA from lacrimal gland acinar cells transcribed in response to a number of DNA probes immobilized on a nitrocellulose membrane. The three horizontal lines on the histogram represent the mean ± standard error of the counts binding to the plasmid DNA without insert in which all the other probes were produced. This level of binding was considered to indicate the level of non-specific binding. Binding below this level, or not distinguishable from it, was not identified as newly transcribed messenger RNA from the lacrimal gland.

The highest amount of transcription of any message tested in the lacrimal gland was for transforming growth factor beta1 (Fig. 1). Northern blot analysis confirmed that messenger RNA for TNF is present in constitutive amounts in the lacrimal gland (time 0), and increases to a maximum at 12 hours following corneal wounding (Fig. 2A). Beginning at 60 hours, there is a steady increase in the amount of TNF messenger RNA up to the final measurement at 84 hours. RT-PCR results (Fig. 2B, left side) definitively identified the rabbit LG TNF mRNA, and a Southern blot (Fig. 2B, right side) demonstrated the homology of the rabbit gene product with the human TNF gene.

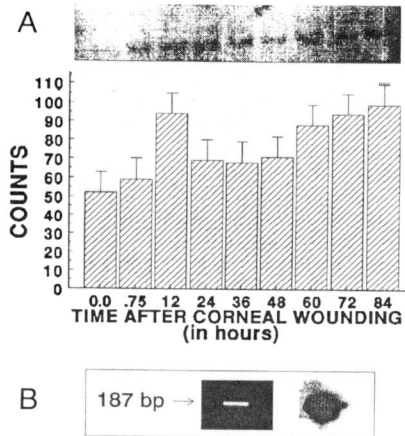

**Figure 2.** (A) Northern blot of LG TNF. Histogram shows results of three repeated experiments. (B) RT-PCR product of TNF primers and LG-RNA. Southern blot (right) of product with human TNF probe.

## In Situ Hybridization

Tissue sections of glands removed from rabbits immediately after wounding were compared with those obtained from animals at 45 minutes, 12 hours, and 96 hours after wounding. Control corneas from sham-operated animals that were processed in parallel with corneas from the wounded animals using the DNA probe for the rabbit sequence for the latency associated transcript (LAT) did not show specific label (Fig. 3A). In contrast, sections probed for the rabbit sequence for TNF mRNA showed label in association with two cell types. Large cells found within the connective tissue space of the lacrimal gland and identified as macrophages had accumulations of label (Fig. 3B). In addition, label was identified over the cytoplasm of the acinar cells (Fig. 3C). Quantitation of 10 random fields at 100x showed that the total number of labeled cells increased over time, from 10 positive cells at 12 hours to 19 positive cells at 96 hours.

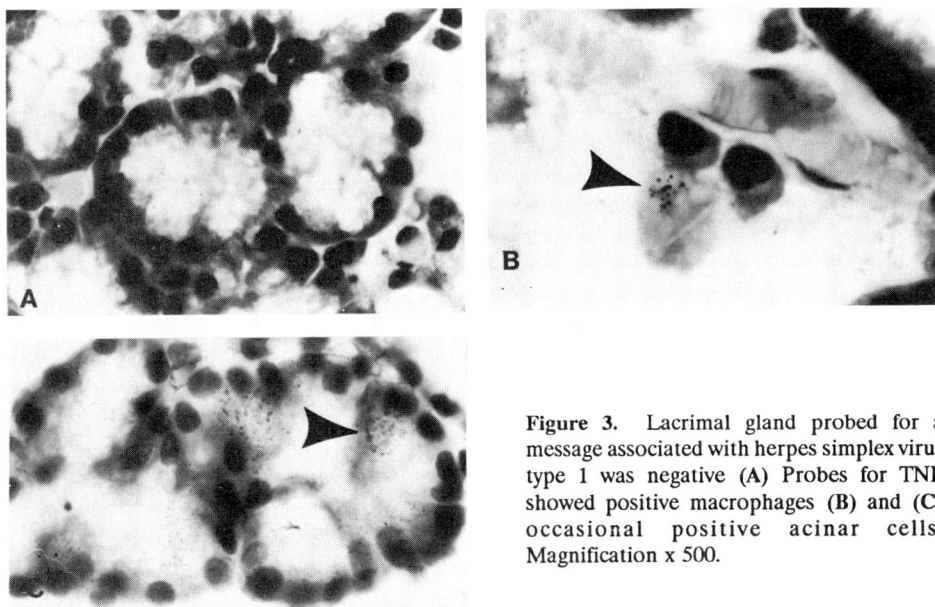

**Figure 3.** Lacrimal gland probed for a message associated with herpes simplex virus type 1 was negative **(A)** Probes for TNF showed positive macrophages **(B)** and **(C)** occasional positive acinar cells. Magnification x 500.

The control slides had only one positive cell. The data indicate that the wounding procedure may lead to a change in vascular permeability in the lacrimal gland, thereby allowing more macrophages to enter the gland.

## DISCUSSION

The run-on assay allows the simultaneous study of the transcription of a number of messages in small amounts of tissue. It is particularly useful with physiological stimulation and over time. Repeated probings of Northern blots or multiple parallel quantitative RT-PCRs would yield much less information with more effort. In the present study, the run-on assay provides information about messages that are transcribed on a constitutive basis, suggesting that expression of these genes may be required for autocrine or paracrine functions in the normal gland, as well as secretion. We found no transcription in response to primers for angiotensin and glial fibrillar protein; however, these substances have not been associated with the lacrimal gland.

The focus on TNF-alpha in this study was based on our interest in the involvement of this cytokine in inflammatory diseases.[1] The association with fibrotic tissue destruction and lymphocyte infiltration suggests a role in lacrimal gland diseases such as Sjögren's.[3] Increased transcription of TNF following a wound to the cornea may underlie a mechanism that has cumulative effects over time and may be part of the pathogenesis of lacrimal gland disorders. TNF, as well as interleukin-1, which can be induced by TNF, has been shown to increase secretory component in cultured acinar cells.[1,12] The monocyte origins of TNF were confirmed in this study by the observation of increased numbers of macrophages positive for TNF after wounding. The association of macrophage invasion of lacrimal gland with wounds to the anterior segment may result from a change in vascular permeability caused by activation of nerve endings in the gland and release of transmitter substances. The production of TNF mRNA in acinar cells is a new observation and parallels the observation of TNF mRNA in renal tubule cells, where both T cells transcribe message.[18] In addition to

215

macrophage TNF production within lacrimal gland tissue, production by acinar cells may represent an important part of the action on target tissues and may be evidence of accessory immune tissue function in lacrimal gland.

Northern blot results could differ from run-on assay transcription results. A message which is positive in a run-on assay is being actively transcribed but may have a short half-life and not appear in Northern blots. We found TNF in Northerns from unstimulated glands both active transcription and a stable pool of TNF mRNA. Cytokines, such as CSF and TNF, like many primary response oncogenes, have AUUUA motifs repeated in their 3'-untranslated regions. This region binds proteins that tag mRNAs for rapid degradation.[19] Such messages may have no detectable cytoplasmic pools but can still be transcribed at a high rate with a very short message half-life. This may be the case in particular for TGF beta which, in the present results, had the greatest level of transcription in the resting lacrimal gland. This is interesting, especially compared to the EGF mRNA level, since the two growth factors have opposite effects on cell proliferation in some systems.[1,2] Using PCR techniques, Wilson et al[20] failed to find this message in biopsy material from human lacrimal gland, but did find TGF-ß-1 in the rabbit corneal epithelium; this could be due to species differences. The discrepancies between our results and those of Wilson et al[20] suggest that, for studies in which the physiology of gene expression is important, both Northern blots and run-on assays should be performed.

In comparison to the cytokines, the EGF message is relatively stable and is detectable by Northern blots in normal and stimulated lacrimal gland,[11] but is transcribed at a level only slightly above background. This message lacks a long A+U exclusive region and apparently has a longer half-life and accordingly a lower level of transcription. Run-on assays comparing the unstimulated and postwounding transcriptional activity may show that the observed increase in mRNA on Northern blots is accompanied by an increase in transcription.[11]

Both the A and B forms of PDGF, as well as the A and B subunits of the PDGF receptor (PDGF-R), were probed. The PDGF-BB subunit had a high level of transcription, while the PDGF-A monomer was also transcribed above background levels. The A subunit of the receptor and the B subtype of PDGF were not transcribed above background levels. These findings represent the first experimental results indicating that PDGF mRNA is found in the lacrimal gland. Its production in the corneal epithelium has recently been suggested, and it was shown that both receptors are present on corneal epithelial cells, stromal fibroblasts, and endothelial cells, with the beta receptor being most abundant.[21] Our results suggest that the production of PDGF receptor subunits and the production and secretion of PDGF monomers in lacrimal gland are also important in lacrimal gland function.

## REFERENCES

1. B. Beutler and A. Cerami, The biology of cachectin/TNF: a primary mediator of the host response, *Ann Rev Immunol.* 7:625 (1989).
2. B.J. Sugarman, B.B. Aggarwal, P.E. Hass, I.S. Figari, M.A. Palladino, Jr., and H. M. Shepard, Recombinant human tumor necrosis factor-alpha: effects on proliferation of normal and transformed cells in vitro, *Science.* 230:943 (1985).
3. J. Williamson, A.A.M. Gibson, T. Wilson, J.V. Forrester, K. Whaley, and W.C. Dick, Histology of the lacrimal gland in keratoconjunctivitis sicca, *Br J Ophthalmol.* 57:852 (1973).
4. Y. Ohasi, M. Motokura, Y. Kinoshita, T. Mano, H. Watanabe, S. Kinoshita, R. Manabe, K. Oshidin, and C. Yamaihara, Presence of epidermal growth factor in human tears, *Invest Ophthalmol Vis Sci.* 30:1879 (1989).
5. R. L. Farris, R.N. Stuchell, and I.D. Mandel, Basal and reflex human tear analysis.

I. Physical measurements: osmolarity, basal volumes, and reflex flow rate. *Ophthalmology.* 88:852 (1981).

6. R.N. Stuchell, R.L. Farris, and I.D. Mandel: Basal and reflex human tear analysis. II. Chemical analysis: Lactoferrin and lysozyme, *Ophthalmology.* 88:852 (1981).

7. R.W. Beuerman, A.J. Rozsa, and D.L. Tanelian, Neurophysiological correlates of posttraumatic acute pain, *Adv Pain Res Therapy.* 9:73 (1985).

8. S.E. Wilson, S.A. Lloyd, and R.H. Kennedy, Epidermal growth factor messenger RNA production in human lacrimal gland, *Cornea.* 10:519 (1991).

9. S.E. Wilson, S.A. Lloyd, and R.H. Kennedy, Basic fibroblast growth factor (FGFb) and epidermal growth factor receptor messenger RNA production in human lacrimal gland, *Invest Ophthalmol Vis Sci.* 32:2816 (1991).

10. H.W. Thompson and R.W. Beuerman, Tumor necrosis factor alpha mRNA is present in lacrimal gland and increases following wounds to the cornea, *Invest Ophthalmol Vis Sci* 33(Suppl):951 (1992).

11. H.W. Thompson and R.W. Beuerman, Regulation of growth factor production correlates in lacrimal gland wound responses from the eye, *J Cell Biochem.* 16B:187 (1992).

12. R.S. Kelleher, L.E. Hann, J.A. Edwards, and D.A. Sullivan, Endocrine, neural and immune control of secretory component output by lacrimal gland acinar cells, *J Immunol.* 146:3405 (1991).

13. L.E. Hann, R.S. Kelleher, and D.A. Sullivan, Influence of culture conditions on the androgen control of secretory component production by acinar cells from the rat lacrimal gland, *Invest Ophthalmol Vis Sci.* 32:2610 (1991).

14. R.W. Beuerman and H.W. Thompson, Molecular and cellular responses of the corneal epithelium to wound healing. *Acta Ophthalmol.* 70(Suppl. 202):7 (1992).

15. J. Sambrook, E.F. Fritach, and T. Manatu, "Molecular Cloning - A Laboratory Manual," Cold Spring Harbor Press, Cold Spring Harbor, New York (1987).

16. J.L. Cook, S. Irias-Donaghey, and P.L. Deininger, Regulation of myelin proteolipid protein gene expression. *Neuroscience Letters* 137:56 (1992).

17. J. Denereux, P. Haeberti, and O. Smithies, A comprehensive set of sequence analysis programs for the VAX. *Nucleic Acid Res.* 12:387 (1984).

18. A.M. Jevnikar, D.C. Brennan, G.G. Singer, J.E. Heng, W. Maslinski, R.P. Wuthrich, L.H. Glimcher, V.E. Kelly Rubin, Stimulated kidney tubular epithelial cells express membrane associated and secreted TNF alpha, *Kidney International.* 7:203 (1991).

19. G. Shaw and R. Kamen, A conserved AU-sequence from the 3'-untranslated region of GM-CSF mRNA mediates selective mRNA degradation. *Cell.* 46:659 (1986).

20. S. Wilson, EGF, basic FGF, and TGF beta-1 messenger RNA production in rabbit corneal epithelial cells. *Invest Ophthalmol Vis Sci.* 33:1987 (1992).

21. V.P.T. Hoppenreijs, E. Pels, G.F.J.M. Vrensen, P.C. Felten, W.F. Treffers, Platelet-derived growth factor: receptor expression in corneas an effects on corneal cells. *Invest Ophthalmol Vis Sci.* 35:637 (1993).

# ANDROGEN REGULATION OF SECRETORY COMPONENT mRNA LEVELS IN THE RAT LACRIMAL GLAND

Jianping Gao, Ross W. Lambert, L. Alexandra Wickham and
David A. Sullivan

Department of Ophthalmology, Harvard Medical School and
Immunology Unit, Schepens Eye Research Institute
20 Staniford Street, Boston, MA 02114

## INTRODUCTION

The functional expression of the secretory immune system of the eye is critically dependent upon secretory component (SC),[1] the polymeric IgA receptor.[2] This glycoprotein, which is produced by lacrimal gland epithelial cells, controls the transfer of secretory IgA (sIgA) antibodies to the ocular surface, whereupon sIgA defends against microbial agents and toxic compounds.[1] Given this pivotal role of SC in ocular mucosal immunity, our research has sought to elucidate the processes involved in the synthesis and secretion of this lacrimal protein. Such studies have demonstrated that: [a] SC production by rat lacrimal tissue displays distinct, gender-related differences (i.e. glands of males produce significantly more SC than those of females);[3,4] and [b] SC synthesis by rat lacrimal acinar cells is uniquely regulated by the endocrine, nervous and immune systems.[5,6] Thus, SC production is stimulated by androgens, vasoactive intestinal peptide (VIP), $\beta$-adrenergic agonists (e.g. isoproterenol), interleukin-1$\alpha$ (IL-1$\alpha$), interleukin-1$\beta$ (IL-1$\beta$), tumor necrosis factor-$\alpha$ (TNF-$\alpha$) and prostaglandin $E_2$ (PGE$_2$), and suppressed by cholinergic agonists (e.g. carbachol choline). Moreover, these gender-associated and neuroendocrinimmune effects on lacrimal SC synthesis appear to be unique to the eye, given that SC production by other mucosal sites: [a] may show no sexual dimorphism (e.g. salivary, respiratory, intestinal, D.A. Sullivan, unpublished data); and [b] may be enhanced or inhibited, depending upon the tissue, by estrogens, progestins, glucocorticoids, prolactin, thyroxine, substance P and $\gamma$-interferon ($\gamma$-IFN),[7-13] which agents have no demonstrable impact on the constitutive SC

*Lacrimal Gland, Tear Film, and Dry Eye Syndromes*
Edited by D.A. Sullivan, Plenum Press, New York, 1994

output by lacrimal acinar cells.[6] In fact, these tissue-specific variations in SC modulation represent but one example of the numerous, site-selective influences of hormones, neurotransmitters and lymphokines on the secretory immune system.[14]

At present, the cell biological mechanisms underlying the neuroendocrinimmune control of SC synthesis by the lacrimal gland are unknown. Theoretically, androgen action, which appears to account for the gender-related differences in lacrimal SC production,[5] may well involve hormone association with specific, nuclear receptors in acinar cells, binding of these androgen/receptor complexes to genomic acceptor sites and the promotion of SC mRNA transcription and eventual translation. In support of this hypothesis: [a] saturable, high affinity and steroid specific binding sites for androgens exist in lacrimal acinar cells;[15-17] [b] androgen-receptor protein complexes in lacrimal tissue adhere to DNA;[15] [c] androgens modulate the levels of various mRNAs in the lacrimal gland;[18-20] and [d] androgen-induced SC synthesis by acinar cells may be curtailed by androgen receptor (cyproterone acetate), transcription (actinomycin D) or translation (cycloheximide) antagonists.[3,21] In contrast, the neural and immune regulation of acinar cell SC dynamics may involve, in part, alterations in intracellular adenylate cyclase and cAMP activity. Thus, VIP, adrenergic agents, IL-1 and TNF-$\alpha$ are known to augment cAMP accumulation in various tissues.[22,23] Furthermore, treatment of lacrimal gland acinar cells with cyclic AMP analogues (e.g. 8-bromoadenosine 3':5'-cyclic monophosphate), cyclic AMP inducers (e.g. cholera toxin) or phosphodiesterase inhibitors (e.g. 3-isobutyl-1-methylxanthine), but not cGMP modifiers, may increase SC elaboration.[6] Of interest, the kinetics of these neuroimmune and secretagogue effects are also consistent with an influence on SC gene transcription.

The purpose of the current investigation was to begin to identify the molecular biological mechanisms involved in the neuroendocrinimmune regulation of lacrimal gland SC. Accordingly, we endeavored in this study to determine whether the gender-associated variation in, and androgen modulation of, acinar SC synthesis are mediated through alterations in SC gene expression.

## MATERIALS AND METHODS

Young adult male and female Spraque-Dawley rats were purchased from Zivic-Miller Laboratories and maintained in temperature-controlled rooms with light and dark intervals of 12 hours length. Orchiectomies, ovariectomies or sham castration procedures were performed by surgeons at Zivic-Miller Laboratories on 6 week old animals. Rats were allowed to recover for a minimum of one week after surgery before experimental treatment. When indicated, castrated animals were administered subcutaneous implants of placebo- or testosterone (50 mg)-containing pellets (Innovative Research of America) in the subscapular region. This method of hormone exposure ensures a slow, but continuous, release of androgens, resulting in the generation of physiological serum testosterone concentrations (i.e. for an adult male rat). Lacrimal glands were removed from sacrificed rats and processed for various molecular biological protocols.

For the determination of mRNA levels in experimental samples, Northern blot procedures were utilized, as previously described.[24] Briefly, total cellular RNA was isolated from rat lacrimal tissue by using an acid guanidinium-thiocyanate-phenol-chloroform extraction method,[25] and poly(A) RNA was purified from total RNA by using the Micro-Fast Track mRNA isolation kit (Invitrogen). The RNA preparations were quantitated by spectrophotometry at 260 nm, resolved (5 to 25 µg) on 1.2% agarose gels containing 6.6% formaldehyde, transferred to Immobilon-N (Millipore) or GeneScreen (Dupont/NEN) membranes by positive pressure and fixed by UV cross-linking; all gels contained an RNA molecular weight ladder (Gibco). Blots were then incubated with a 5' @950 base pair *Bam*H 1 fragment of the rat SC cDNA, which was radiolabeled (specific activity > 5 x $10^8$ cpm/µg DNA) with [alpha-$^{32}$P] dCTP by random priming. This fragment contains most of the SC coding region and was obtained from the full length SC cDNA (3083 base pairs, subcloned in the pGEM4 plasmid, gift from Dr. George Banting, United Kingdom). Following an overnight hybridization with the SC cDNA probe, blots were washed (high stringency), processed for autoradiography (XAR X-ray film, Kodak) and analyzed by densitometry. For control purposes, extensive studies were conducted with sense (no reactivity) and anti-sense probes and with total RNA or mRNA from SC-positive and SC-negative tissues, as previously reported.[24] In addition, all experimental blots were rehybridized with labeled mouse β-actin cDNA probes (cDNA provided by Dr. Lan Hu, Boston, MA) to verify that similar amounts of total RNA were analyzed in the various groups.

To evaluate the distribution of SC mRNA in lacrimal tissue, lacrimal glands were processed for *in situ* hybridization, as described in detail.[24] In brief, tissues were fixed in 4% paraformaldehyde, exposed to varying ethanol and xylene solutions, embedded in paraffin and cut into 6 µm sections, which were placed on slides pretreated with 2% 3-amino propyltriethoxysilane in acetone. Sections were deparaffinized, hydrated, refixed in 4% paraformaldehyde, subjected to proteinase K digestion, acetylated, rinsed, and incubated with anti-sense or sense $^{35}$S-labelled SC mRNA probes. Labeled riboprobes were prepared by utilizing the linearized (with *Hind* III or *Ava* I) pGEM4 plasmid containing the SC cDNA, [$^{35}$S]-UTP (NEN), a riboprobe transcription kit from Promega, and either SP6 (anti-sense) or T7 (sense) RNA polymerases, followed by treatment with RNase-free DNase I (to remove the SC cDNA template) and purification by phenol-chloroform extraction and ethanol precipitation. Radiolabeled probes (@ 5 x $10^7$ cpm/ml) were denatured by heating, chilled, mixed in hybridization buffer and applied to sections, which were then coverslipped and incubated for 16 to 18 hours at 37°C. Following hybridization, sections were washed in a series of buffers, exposed to RNase A, washed, dehydrated, dried, dipped in NTB-2 Kodak emulsion and exposed for 10 to 14 days in a tightly sealed, black box at -20°C. After this period, sections were developed with D-19 Kodak developer at 19°C, stained with hematoxylin and mounted with permount. The distribution and labeling intensity of grains were then compared between anti-sense and sense exposed sections.

Statistical analysis of the data was performed with Student's two-tailed t test.


## RESULTS

### Influence of Gender on the Amount and Distribution of SC mRNA in Rat Lacrimal Tissue [24]

Analysis of SC mRNA levels in lacrimal glands of young adult male and female rats (n = 5/group) demonstrated that tissues of males contained significantly (p < 0.0001) greater amounts of SC mRNA than those of females.[24] Thus, when standardized to β-actin levels, SC mRNA content in male lacrimal glands was approximately 4-fold higher than that in female tissues. These results could not be accounted for by possible gender-related fluctuations in total RNA or β-actin concentrations, given that no significant differences in these variables existed between male and female rats.

The impact of gender on SC mRNA levels also appeared to extend to the distribution of this message in lacrimal tissues. Preliminary findings with *in situ* hybridization procedures indicated that the frequency and extent of SC mRNA-positive cells was considerably greater in lacrimal glands of male rats, as compared to those of females.[24]

### Effect of Castration on the Level of SC mRNA in the Rat Lacrimal Gland [24]

To determine whether these gender-related differences in lacrimal SC mRNA content were possibly due to the influence of sex steroids, age-matched male and female rats (n = 5/treatment group) were subjected to either castration or sham-operations, and lacrimal glands were obtained 7 days after animal surgery to permit quantitation of SC mRNA levels by Northern blots. Results[24] showed that orchiectomy led to a significant (p < 0.0001), 3-fold decline in SC mRNA amounts, relative to that in lacrimal tissues of sham-operated males. In contrast, ovariectomy exerted no significant effect on lacrimal SC mRNA content. In these studies, neither orchiectomy nor ovariectomy induced any alteration in total RNA or β-actin levels in lacrimal tissue, as compared to levels in glands of sham-operated controls.

### Impact of Androgen Administration on SC mRNA Content in Lacrimal Tissues of Castrated Male and Female Rats [24]

To assess whether the orchiectomy-induced decrease in lacrimal SC mRNA levels might be attributable to the loss of androgens, castrated male rats (n = 5/group) were administered placebo compounds or physiological amounts of testosterone for 7 days, and then lacrimal tissues were processed for SC mRNA measurements. For comparative purposes, we also examined the influence of androgen administration on SC mRNA content in lacrimal glands of ovariectomized rats. Findings[24] demonstrated that testosterone treatment stimulated a significant (p < 0.01), 2-fold rise in SC mRNA amounts in lacrimal tissues of both castrated male and female rats, as compared to levels in tissues of placebo-treated controls. These results were apparent irrespective of whether data was expressed in terms of total SC mRNA, or normalized to β-actin content.

### DISCUSSION AND ACKNOWLEDGMENTS

Our previous research has shown that androgens directly stimulate the synthesis and secretion of SC by lacrimal gland acinar cells,[5] significantly augment the transfer of SC from the lacrimal gland to tears[26] and induce a striking increase in the tear SC concentration of rats.[27] In addition, this hormone action appears to mediate, in part, the pronounced gender-associated differences in SC production by rat lacrimal tissue.[5,14] To extend these findings, the present study was designed to explore the molecular biological mechanisms underlying these androgen effects. Our results provide evidence that the gender-related variations in, and the androgen control of, lacrimal gland SC production involve significant alterations in SC gene expression. Thus, SC mRNA levels in lacrimal tissues of male rats are considerably higher than those in glands of females, and this sexual dimorphism appears due to the influence of androgens.

However, despite these findings, many additional questions remain concerning the processes involved in the androgen, and neuroendocrinimmune, regulation of SC in the rat lacrimal gland. For example: [a] does the androgen modulation of SC gene expression include influences on the transcriptional rate and/or stablity of SC mRNA?; [b] do the known modifications of androgen-induced SC production by insulin, glucocorticoids, high-density lipoprotein, extracellular calcium[28] or factors from the thyroid, adrenal and hypothalamic-pituitary axis[29] involve SC gene interactions?; [c] do the mechanisms by which VIP, β-adrenergic agonists, IL-1α, IL-1β, TNF-α, PGE$_2$, and cAMP analogues or inducers enhance, and carbachol abrogates, SC synthesis by lacrimal gland acinar cells[6] involve transcriptional and/or posttranscriptional events?; [d] are possible endocrine, neural and/or immune effects on SC mRNA paralleled by changes in IgA mRNA in lacrimal tissue?; and [e] do analagous regulatory processes for SC and IgA occur in lacrimal glands of other species?

Overall, our observation that neuroendocrinimmune agents may modulate SC or other components[1,14] of the ocular secretory immune system is not unusual. During the past several years, it has become increasingly recognized that the endocrine, nervous and immune systems exert a profound, regulatory influence on mucosal immunity.[1,14,30] The extent and diversity of this regulation is quite comprehensive and includes control of the: [1] immigration, retention, proliferation and/or activity of IgA, IgG- and IgM-containing cells, T cells, eosinophils, basophils, mast cells, natural killer cells, polymorphonuclear leukocytes and/or macrophages; [2] production and/or secretion of IgA, IgG and IgM antibodies and cytokines, the expression of MHC Class II antigens, the synthesis and output of SC and the SC-mediated transport of polymeric IgA into external secretions; and [3] attachment and presentation of microorganisms to luminal surfaces, the extent of neurogenic inflammation and the magnitude of local immune defense against infectious agents.[1,14] The exact nature (i.e. stimulation, inhibition, or no effect) of this neuroendocrinimmune modulation is highly site-specific and, in certain tissues, may be significantly modified by antigenic challenge.[1,14] Clearly, further research is required to elucidate the basic mechanisms, and potential clinical relevance, of these endocrine, neural and mucosal immune interrelationships in the eye, as well as throughout the body.

We would like to express our appreciation to Drs. Banting and Hu for their generous provision of cDNA probes. This research was supported by NIH grants EY05612 and EY07074, and a grant from the Massachusetts Lions' Research Fund.

## REFERENCES

1. D.A. Sullivan, in: "Mucosal Immunology," P.L. Ogra, J. Mestecky, M.E. Lamm, W. Strober, J. McGhee, and J. Bienenstock, eds., Academic Press, Orlando, FL, in press (1993).
2. J.E. Casanova, Ann. N.Y. Acad. Sci. 664:27 (1992).
3. D.A. Sullivan, K.J. Bloch, and M.R. Allansmith, Immunology 52:239 (1984).
4. D.A. Sullivan, and M.R. Allansmith, Immunology 63:403 (1988).
5. D.A. Sullivan, R.S. Kelleher, J.P. Vaerman, and L.E. Hann, J. Immunol. 145:4238 (1990).
6. R.S. Kelleher, L.E. Hann, J.A. Edwards, and D.A. Sullivan, J. Immunol. 146:3405 (1991).

7. D.A. Sullivan, B.J. Underdown, and C.R. Wira, *Immunology* 49:379 (1983).

8. C.R. Wira, and D. A. Sullivan, *Biol. Reprod.* 32:90 (1985).

9. C.R. Wira, and R. M. Rossoll, *Endocrinology* 128:835 (1991).

10. P. Weisz-Carrington, S. Emancipator, and M.E. Lamm, *J. Reprod. Immunol.* 6:63 (1984).

11. J.P. Buts, J.P. Vaerman, and D.L. Delacroix, *Immunology* 54:181 (1985).

12. D. Kvale, P. Brandtzaeg, and D. Lovhaug, *Scand. J. Immunology* 28:351 (1988).

13. S. Freier, D. McGee, M. Eran, and J. McGhee, *in*: "Abstracts of the 7th International Congress of Mucosal Immunology," Prague, Czechoslovakia, p. 76 (1992).

14. D.A. Sullivan, *in*: "The Neuroendocrine-Immune Network," S. Freier, ed., CRC Press, Boca Raton, FL, p199-238 (1990).

15. M. Ota, S. Kyakumoto, and T. Nemoto, *Biochem. Internat.* 10:129 (1985).

16. D.A. Sullivan, J.A. Edwards, and R.S. Kelleher, article submitted (1993).

17. F.J. Rocha, L.A. Wickham, J.D.O. Pena, J. Gao, M. Ono, R.W. Lambert, R.S. Kelleher, and D.A. Sullivan, article submitted (1993).

18. P.H. Shaw, W.A. Held, and N.D. Hastie, Cell 32:755 (1983).

19. R.M. Gubits, K.R. Lynch, A.B. Kulkarni, K.P. Dolan, E.W. Gresik, P. Hollander, and P. Feigelson, *J. Biol. Chem.* 259:12803 (1984).

20. J. Winderickx, K. Hemschoote, N. De Clercq, P. Van Dijck, W. Rombauts, G. Verhoeven, and W. Heyns, *Mol. Endocr.* 4:657 (1990).

21. R.W. Lambert, R.S. Kelleher, L.A. Wickham, J.P. Vaerman, and D.A. Sullivan, article submitted (1993).

22. D. Dartt, *Curr. Eye Res.* 8: 619 (1989).

23. Y. Zhang, J.X. Lin, Y.K. Kip, and J. Vilcek, Proc. Nat. Acad. Sci. USA 85:6802 (1988).

24. J. Gao, R.W. Lambert, L.A. Wickham, and D.A. Sullivan, article submitted (1993).

25. P. Chomczynski, and N. Sacchi, Anal. Biochem. 162:156 (1987).

26. D.A. Sullivan, and L.E. Hann, *J. Steroid Biochem.* 34:253 (1989).

27. D.A. Sullivan, K.J. Bloch, and M.R. Allansmith, *J. Immunol.* 132:1130 (1984).

28. L.E. Hann, R.S. Kelleher, and D.A. Sullivan, *Invest. Ophthalmol. Vis. Sci.* 32:2610 (1991).

29. D.A. Sullivan, and M.R. Allansmith, *Immunology* 60:337 (1987).

30. R.H. Stead, M.H. Perdue, H. Cooke, D.W. Powell, and K.E. Barrett, eds. "Neuro-Immuno-Physiology of the Gastrointestinal Mucosa. Implications for inflammatory diseases," Ann. N.Y. Acad. Sci. vol. 664 (1992).

# VASOACTIVE INTESTINAL PEPTIDE INDUCES PRIMARY RESPONSE GENE EXPRESSION IN LACRIMAL GLAND

Michele M. Cripps[1], Hilary W. Thompson[2] and Roger W. Beuerman[2]

[1]Department of Physiology
[2]Department of Ophthalmology
Louisiana State University Medical Center
New Orleans, LA 70119

## INTRODUCTION

The transcription of the proto-oncogene c-fos represents an initial event in the response of diverse tissues to physiologic stimulation. The product of translation of c-fos mRNA is a nuclear phosphoprotein that functions as an activator of transcription by forming a c-fos, c-jun protein heterodimer that possesses DNA binding activity. Activation of the c-fos proto-oncogene and the subsequent cascade of sequential gene expression is initiated by stimulus-transcription coupling events that may involve interactions of ligands with cell-surface receptors and signal transduction by known second messenger pathways (Morgan and Curran, 1991). Many stimuli that are associated with differentiation, proliferation or specific cellular processes that require regulated biosynthesis are capable of eliciting induction of c-fos mRNA and protein.

In a recent study, lacrimal gland c-fos mRNA increased rapidly following corneal wounding in New Zealand albino rabbits (Thompson et al, 1991). These results suggest that primary response gene expression is among the first events of the lacrimal gland response to corneal wounding. Presumably this response was initiated by autonomic input to the gland mediated via neuropeptide or neurotransmitter release. The purpose of this study was to determine whether neurotransmitter or neuropeptides present in fibers that innervate the lacrimal gland influence c-fos expression. In these initial experiments, the effect of a muscarinic agonist and the parasympathetically co-localized peptide (VIP) on primary response gene expression was assessed in lacrimal gland preparations. In addition, forskolin, a non-receptor mediated activator of adenylate cyclase was used to determine if the effects on c-fos expression are cAMP-mediated.

*Lacrimal Gland, Tear Film, and Dry Eye Syndromes*
Edited by D.A. Sullivan, Plenum Press, New York, 1994

## MATERIALS AND METHODS

### Secretion

Lacrimal glands from male New Zealand white rabbits (2.0 to 3.0 kg) were prepared as fragments 3 to 5 mm$^3$ in size. Approximately 200 mg wet weight of the fragments were incubated in oxygenated, Hepes buffered medium of the following composition (mM): 116 NaCl, 5.4 KCl, 0.81 MgSO$_4$, 1.01 Na$_2$HPO$_4$, 1.3 CaCl$_2$, 5.6 dextrose, 1 hydroxybutyrate and 10 Hepes, pH 7.4, as previously described for rat lacrimal preparations (Cripps and Patchen-Moor, 1989). The fragments were incubated in Erlenmeyer flasks at 37°C in a shaking water bath with continuous O$_2$ gassing and at the appropriate times the medium was removed for the determination of secreted protein by the method of Lowry (1951). Total tissue wet weight was also measured for each sample. A basal rate of secretion was determined for each flask during an initial 15 min period. Protein secreted above the basal level during the subsequent periods under control conditions and in the presence of 100 nM VIP (Peninsula, Belmont, CA), 100 μM carbachol (Sigma, St. Louis, MO) or 40 uM forskolin (Calbiochem, LaJolla, CA) was measured and expressed as μg protein secreted per mg tissue wet weight.

### Total Cellular RNA Extraction and Blots

After removal of medium for assay of secreted proteins, the fragments were weighed, snap frozen and stored in liquid nitrogen. Freshly prepared unincubated fragments were also collected and used to determine baseline levels of c-fos mRNA expression. Total cellular RNA was determined as modified from Chirgwin et al. (1979) by grinding the fragments in a 4 M guanidinium solution and layering over 6.5 M CsCl. Separation of total cellular RNA was obtained by centrifugation at 50,000 x g at 21°C for 5 hours. The RNA pellet was washed 2X with ice-cold 80% ethanol to remove residual CsCl. The pellet was solubilized in DEPC H$_2$O with the addition of .04 volumes 5 M NaCl and 2.5 volumes 100% ethanol and stored at -70°C. RNA was precipitated by centrifugation at 14,000 x g for 30 min and solubilized in DEPC H$_2$O for determination of RNA concentration by spectrophotometry.

RNA slot blots were conducted on 0.45 μm nitrocellulose filters (Schleicher and Schuell, Keene, NH) with a 7.5% formaldehyde, 10X SSC blotting buffer (15 or 20 μg RNA/lane). In some experiments, additional external standard blots (2 μg/lane) were prepared. The blots were air-dried, followed by baking at 80°C in a vacuum oven for 2 hr.

Membranes were incubated overnight at 42°C in prehybridization buffer with a final concentration of 50% d-formamide, 5X SSPE, 5X Denhardt's solution, 0.1% SDS and 0.15 mg/ml salmon sperm DNA (5'-3' Inc., Boulder, CO). This was followed by hybridization overnight at 42°C by adding heat-denatured $^{32}$P labelled c-fos or ß-actin probe to the prehybridization buffer to give 10$^7$ cpm/ml. C-fos (1 Kb PSTI fragment from pcfos-1, human, ATCC, Rockville, MD) or ß-actin (plasmid DNA) were labelled with the PROBE-EZE (5'-3', Inc., Boulder, CO) random primer DNA labelling kit and dCTP[α-$^{32}$P] (New

England Nuclear, Boston, MA) followed by the removal of unincorporated nucleotide triphosphates with Sephadex Select D G-50 columns (5'-3', Inc., Boulder, CO). At the end of hybridization the blots were washed 3X for 15 min each with 2X SSC and 0.1% SDS at room temperature and 2X for 30 min each with 0.1X SSC and 0.1% SDS at 45°C. The blots were sealed and activity was determined by scanning densitometry of exposed film or by storage phosphor imaging (PhosphoImager, Molecular Dynamics, Model 400E).

## RESULTS

Protein secretion was measured under control conditions and in the presence of either 100 nM VIP (Table 1) in the first series of experiments or with 100 µM carbachol, or 40 µM forskolin (Table 2) in a second series of experiments. In both instances, secretion of protein under control conditions increased with increasing incubation times. Stimulation of the rabbit fragments by the peptide VIP (Table 1), resulted in increased protein secretion when compared with controls at all time points. Significant increases ($p < .05$ or $p < .01$) also occurred in the presence of carbachol at 15, 60 and 120 min. Thus, stimulation of rabbit lacrimal gland fragments by VIP, carbachol or forskolin resulted in a significant time-dependent increase in protein secretion when compared with controls.

Table 1.     Effect of VIP on protein secretion. Lacrimal gland fragments were incubated in the presence or absence of 100 nM VIP. Values are the mean ± SE of 3 experiments Significance levels of *P < .05 were determined by t-tests.

|  | Secreted Protein (µg/mg tissue) | |
|---|---|---|
|  | 15 min | 60 min |
| Control | 1.8 ± 0.6 | 3.7 ± 0.1 |
| 100 nM VIP | 4.2 ± 0.4* | 16.3 ± 4.2* |

To determine whether VIP influences primary response gene expression in addition to protein secretion, fragments were collected, weighed, and frozen in liquid nitrogen for total cellular RNA extraction. RNA slot blots were probed with human c-fos probes and activity was measured by scanning densitometry. The expression of c-fos mRNA remained relatively constant under control conditions up to 1 hr inculation. In the VIP stimulated fragments (Fig. 1) there was an initial 2-fold increase in c-fos expression compared with non-stimulated tissue at 15 min. The expression of c-fos in the VIP stimulated fragments declined at 1 hr but still remained above the non-stimulated controls.

**Table 2.** Effect of carbachol or forskolin on protein secretion. Lacrimal gland fragments were incubated in the presence or absence of 100 μM carbachol or 40 μM forskolin. Values are the mean ± SE of 3 experiments. Significance levels of *P < .05 or **P < .01 were determined by t-tests

|  | | | Secreted Protein (μg/mg tissue) | |
| --- | --- | --- | --- | --- |
|  | 15 min | 30 min | 60 min | 120 min |
| Control | 1.8 ± 0.5 | 3.2 ± 1.5 | 2.5 ± 1.5 | 5.4 ± 1.6 |
| 100 μM carbachol | 7.0 ± 0.9** | 6.9 ± 2.1 | 13.6 ± 1.6** | 64.6 ± 11.1** |
| 40 μM forskolin | 5.6 ± 3.4 | 6.4 ± 1.3 | 12.9 ± 4.6* | 18.6 ± 8.0* |

Levels of c-fos mRNA were also measured in fragments incubated in additional experiments in which 100 μM carbachol and 40 μM forskolin resulted in significant time-dependent increases in protein secretion when compared with controls. In these experiments, expression of c-fos under control conditions remained constant up to 1 hr with an increase at 2 hr. Expression of c-fos mRNA also remained constant up to 1 hr in the presence of carbachol with a decrease in the level of c-fos mRNA at 2 hr. With the addition of forskolin to the medium, at 1 hr and 2 hrs c-fos mRNA was higher than levels measured at earlier time points in the presence of the adenylate cyclase activator. These levels of mRNA were also elevated when compared with c-fos mRNA expression under control conditions or in the presence of carbachol at either 1 hr or 2 hr.

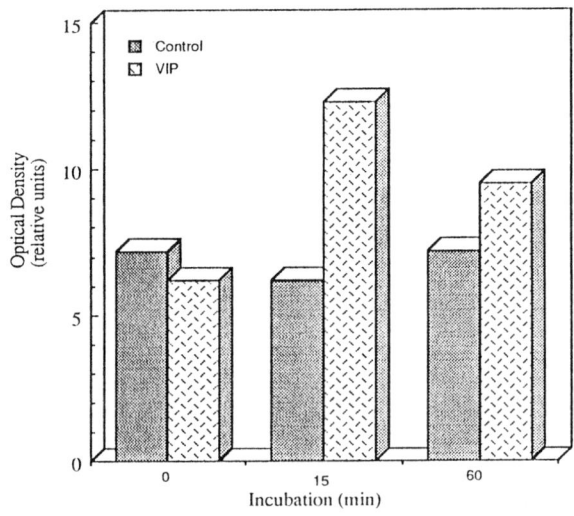

**Fig. 1.** Effect of VIP on c-fos expression. Lacrimal gland fragments were incubated in the presence or absence of 100 nM VIP for the times indicated. Relative optical densities were determined by scanning densitometry. Zero times are unincubated controls.

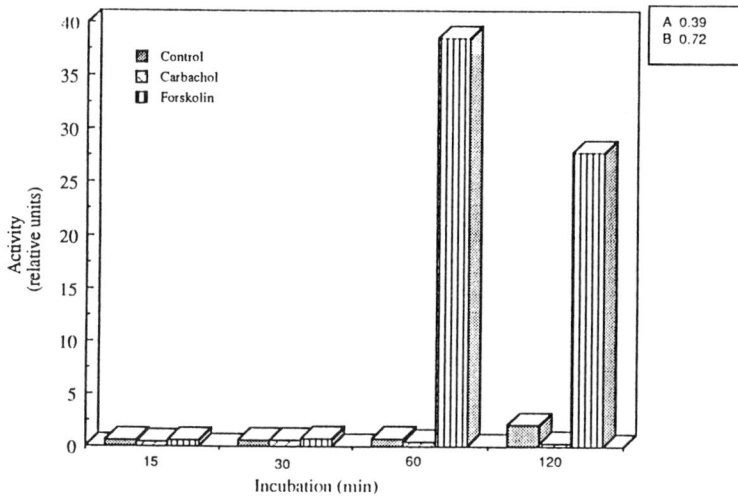

Fig. 2. Effect of carbachol or forskolin on c-fos expression. Lacrimal gland fragments were incubated in the presence or absence of 100 μM carbachol or 40 μM forskolin for the times indicated. Incorporation of $^{32}P$ was measured by phosphor imaging and expressed as activity of c-fos expression as a function of β-actin determined on separate external standard blots. A and B are unincubated controls.

## DISCUSSION

The results presented in this study indicate that activation of VIPergic receptors in rabbit lacrimal gland results in increased c-fos mRNA expression. The increase in the level of c-fos mRNA was rapid with a 2-fold increase at 15 min. Specificity of the response is suggested by the lack of increased c-fos expression in response to cholinergic receptor activation. These results are consistent with those reported for parotid acinar cells after in vitro stimulation of β-adrenergic receptors (Kousvelari et al, 1988). In that study, isoproterenol resulted in an elevation of c-fos mRNA and c-fos protein that was a specific response to β-adrenergic stimulation. Activation of the primary response gene by forskolin suggests that induction of c-fos expression in lacrimal gland is mediated via cAMP, most likely through a cAMP response element (CRE), a regulatory sequence of the gene that results in enhanced induction of c-fos mRNA in the presence of elevated cAMP (Morgan and Curran, 1991).

While this study demonstrates a specific cAMP dependent activation of c-fos expression by VIP, the physiologic significance remains to be determined. The induction of c-fos expression as a response to stimuli that initiate cellular proliferation is well-established (Kaczmarik and Kaminska, 1989). However, the accumulation of c-fos mRNA is not limited to this process. In fact, in the study in parotid acini (Kousvelari et al, 1988) there was no correlation of isoproterenol induced c-fos with mitogenic responses of the acinar cells as measured by DNA synthesis. Whether the c-fos induction by VIP in lacrimal acinar cells is associated with proliferation remains to be determined.

Alternative consequences of c-fos induction by VIP may be related to regulation of transcription of other genes whose expression alters the degree or character of secretion of the biosynthetic products of the lacrimal gland. Alteration in biosynthesis driven by changes in neural input to the gland is supported by elevation of c-fos mRNA followed by an increase in epidermal growth factor (EGF) in response to corneal wounding (Thompson et al., 1991). Those results suggest a possible relationship between the primary response gene c-fos and the later expression of EGF in response to neural input initiated by corneal wounding. Finally, as suggested by Morgan and Curran (1989) transcriptional activation of and by c-fos could have a number of consequences related to stimulus secretion coupling. The primary response gene products may influence transcription of components of the signal transduction cascades such as receptors, specific G proteins, kinases or phosphatases which would permit the tissue to modify its response to repeated or multiple stimuli, a condition that is common in the regulation of secretion by the gland in both health and disease.

## REFERENCES

Chirgwin, J.J., Przbyla, A.E., MacDonald, R.J., and Rutter, W.J., 1979, Isolation of biologically active ribonucleic acid from sources enriched in ribonuclease, *Biochemistry* 18:5294.

Cripps, M. M., and Patchen-Moor, K., 1989, Inhibition of stimulated lacrimal secretion by [D-ala$^2$] met-enkephalinamide, *Am. J. Physiol.*, 257:G151.

Kaczmarck, L., and Kaminska, B., 1989, Molecular biology of cell activation, *Exp. Cell Research* 183:24.

Kousvelari, E. Louis, J. M., Huang, L-H., and Curran, T., 1988, Regulation of proto-oncogenes in rat parotid acinar cells in vitro after stimulation of β-adrenergic receptors, *Exp. Cell Research* 179:194.

Lowry, O.H., Rosebrough, N.J., Farr, A. L., and Randall, R.J., 1951, Protein measurement with the Folin phenol reagent, *J. Biol. Chem.* 193:265.

Morgan, J.I., and Curran, T., 1989, Fos and the immediate-early response in the central nervous system. In "Genes and Signal Transduction in Multistage Carcinogenesis", M.H. Colburn, ed., New York :Dekker.

Morgan, J.I., and Curran, T., 1991, Stimulus-transcription coupling in the nervous system:involvement of the inducible proto-oncogenes fos and jun, A*nnu. Rev. Neurosci.* 14:421.

Thompson, H.W., Griffith, S., and Beurerman, R.W., 1991, Changes in the expression of the c-fos oncogene and the EGF precursor gene in lacrimal gland following keratectomy wounds of the cornea, *Investigative Ophthalmol. Vis. Sci.* 32:(Suppl)1112.

# COMPOSITION AND BIOPHYSICAL PROPERTIES OF THE TEAR FILM: KNOWLEDGE AND UNCERTAINTY

John M. Tiffany

Nuffield Laboratory of Ophthalmology
University of Oxford
Oxford OX2 6AW, U.K.

## COMPOSITION AND STRUCTURE OF THE TEAR FILM

Although other authors in this Symposium volume will also cover the structure of the tear film, and the nature and origins of its components, it is worth mentioning the current picture briefly here, since our view of this will influence the nature and interpretation of biophysical studies on the tear film. Several different structural models have been presented over the years, but all agree on the principal components. Not all of these are produced by the lacrimal glands, so there may be considerable differences between the stable, unirritated precorneal tear film and that in non-equilibrium states such as lacrimation or progressive disease. This must always be borne in mind when studying the interactions of components or collecting fluid tears for other study or analysis. The actual collection technique is very important since results may vary according to whether the tears are reflexly stimulated or collected unstimulated from a quiescent eye.

Despite differences of opinion on the details, it is generally agreed that the tear film is a three-layered structure, consisting of an outermost layer of meibomian oil, a fluid aqueous layer containing soluble proteins and mucins, and an underlying layer of mucus in the gel form. Much information can be revealed by direct optical means: light reflection from the tear film surface (with narrowing of the palpebral aperture where necessary) can give interference colours from the lipid layer; slit lamp microscopy can give an estimate of overall thickness, while fluorometry following instillation of fluorescein can (with some reservations) give the thickness of the aqueous layer; both the overall thickness and that of the mucoid layer can be given by laser interference or confocal microscopy. The general principle of establishment of the film is that (a) the lids secrete and sweep out a wettable mucus layer, (b) aqueous tears spread on it, (c) meibomian oil spreads on the aqueous surface. The nature of the underlying epithelial surface has influenced various models. In particular the demonstration by Holly and Lemp (1971) of low epithelial wettability gave rise to a model requiring the presence of a mucus layer to give a wettable surface; this model influenced thought on the physiological functions of the tear film for many years, and as a result we should consider the epithelial surface (at least from the point of view of its wettability and the nature of its surface groups for anchoring of the tear film) as an essential part of the structure.

## BIOPHYSICAL PROPERTIES

A study of the biophysical properties of the tear film covers a very large number of different aspects; a survey such as this can deal with only a partial list, and then very briefly. Some of these are properties of the individual components (lipids, proteins, mucins, etc.), while others depend upon some combination of components. Starting with the corneal epithelial surface and moving outwards, we shall consider some (though by no means all) of the properties encountered; in doing this, one notices immediately how many large gaps there are in our present knowledge.

### Epithelial Surface

The principal biophysical property associated with this structure (considered as an integral part of the tear film) is wettability. Earlier measurements by Holly and Lemp (1971), using the critical surface tension method (Zisman, 1964) with a series of pure "indicator liquids", suggested that the corneal epithelial surface of the rabbit, after wiping off adherent mucus, had a wettability comparable with polyethylene. Models of tear film function based on this interpretation required the presence of a mucus layer to mask the hydrophobic surface; if either this layer was disrupted so as to expose the underlying non-wettable epithelium, or the mucus itself became contaminated (e.g. by lipid), a local instability could trigger break-up. It was also proposed that lipid might diffuse across a thinning tear film to cause this contamination.

This view of wettability was questioned by Cope et al. (1986) on the grounds that scanning electron microscopy showed severe surface damage caused by the "indicator liquids" and by wiping, so no conclusion could be drawn about in vivo wettability of the epithelium. Other electron-microscopic studies (Nichols et al., 1985) where precautions were taken to preserve the mucus layer showed a much thicker mucus blanket overlying a highly-glycosylated glycocalyx, which would be expected to be intrinsically wettable because of its strongly polar nature.

Tiffany (1990 a,b) used methods for wettability measurement tested and recommended by Neumann and co-workers (Neumann et al., 1983; Absolom et al., 1986) as non-denaturing and suitable for a variety of cell types, as well as removal of adherent mucus by mucolytics rather than by wiping; these showed high wettability of the rabbit cornea (surface tension against air 67.5-72 mN/m, interfacial tension against saline 0.02-0.33 mN/m). There is hence no strict requirement for a mucus covering for tear film stability, although this is present in the healthy eye and will help to overcome temporary non-wetting due to desquamation or surface damage.

This demonstrates the need for great care in the planning and interpretation of in vitro experiments. Methods which impose non-physiological conditions such as drying, abrasion and cell damage, or incompatible reagents on the tissue should not be used as the basis for far-reaching models of physiological function.

### Mucus Layer

As mentioned above, the electron-microscopic studies of Nichols et al. (1985) showed a mucus layer at least 1 μm thick on the cornea and conjunctiva in the guinea pig, adherent to a clearly-distinguishable glycocalyx. In some quarters this was taken as proof that the mucus layer was no more than 1 μm thick, but this may be only an artefact of preparation, leaving a residual layer of this thickness from a much thicker *in vivo* layer.

Studies by Prydal (1992 a,b) and R. Jeacocke (personal communication), using laser interferometry and confocal microscopy, indicate that there is a much thicker mucus layer even than this: in the human tear film, approximately 30 µm of thickness is mucus, and 10 µm is aqueous layer. Removal of mucus by N-acetylcysteine treatment leaves only the aqueous layer, but the mucus layer gradually reforms over about 40 minutes.

These findings must have profound consequences for our model of tear film formation, stability and break-up, as well as for many aspects of contact lens practice, but as yet no new coherent theory has appeared. The tear film breaks up to form dry spots, but we do not know whether these are at the aqueous/mucus interface (perhaps as suggested by Proust et al., 1983), or at the mucus/epithelium interface. If the latter, their appearance may be connected with the turnover and sloughing rate of surface epithelial cells, and possibly with the rate of development of glycocalyx on newly-revealed cells, since these cell-surface membrane structural glycoproteins are only expressed in the outermost layer of cells (Gipson et al., 1992). Sharma and Coles (1990) have suggested that such small areas could act as nucleation points for tear film break-up, depending on interfacial tension, cell area and tear contact angle. Feenstra and Tseng (1992) have determined that Rose Bengal staining depends in part upon the presence or absence of a mucus coating; possibly resistance to staining may also depend on the presence of a fully-developed or intact glycocalyx, and one or both of these cases could help to explain rapid and random tear film break-up in many dry-eye conditions. Gipson et al. (1992) also showed that glycocalyx-like material forms part of the goblet-cell mucus granule membranes, and that the amount associated with corneal epithelial cell surfaces declines with age. Thus goblet cell function, as well as the quantity or nature of the mucus produced, and ageing of epithelial cells, may influence the wettability of the corneal surface.

We still know little about the way in which the mucus layer is anchored to the epithelial surface. Liotet et al. (1986) suggest specific binding between IgA and mucus, and between IgA and epithelium. Scott et al. (1991) have shown that hydrophobic binding between polysaccharide chains can contribute to shear-sensitive solution viscosity or (at suitable concentrations) to gel formation. It seems possible that similar interactions between oligosaccharides of mucins and glycocalyx may help to stabilise the mucus layer in the tear film.

It is important to distinguish between mucus in different physical forms and degrees of mucus degradation. Holly (1986) has classified ocular mucus into "visible" and "invisible". "Visible" mucus can be collected as the mucus thread or by conjunctival scrapings, and contains mucin molecules in various stages of degradation, complexed or associated with virtually all tear proteins, meibomian lipids and other components. At this stage it has lost its solution properties and takes little further part in fluid or lubricating functions in the eye. "Invisible" mucus is either that which is contained in solution in collected tears, responsible in large part for their viscoelastic and surface tension properties, or the gel-form forming the inner layer of the tear film. and in equilibrium with mucus in solution (although the dynamics of this sol-gel transformation are at present unknown). There are many possibilities still to be explored covering the development and maintenance of the aqueous/mucus interface, possibly by coacervation, and the importance of disturbance or possibly thixotropic behaviour as part of the blinking process.

During blinking the rate of shear (relative velocity of lid and globe, divided by the thickness of the fluid layer) can be very high (Dudinski et al., 1983; Bron and Tiffany, 1991). If shearing forces are transmitted to the epithelial surfaces, cell damage and painful dragging sensations may occur (Hammer and Burch, 1984). Non-Newtonian shear-thinning behaviour can help by reducing viscosity as the shear rate increases, and the elastic component of tears can help to absorb energy during small rapid eye movements. These viscoelastic properties of mucus solutions (in the aqueous layer) or mucus gels (close to the epithelial surfaces) will cushion and lubricate during all types of eye movement. In the presence of a mucus blanket, a gradient of stress will be established across its thickness, effectively reducing shear forces at the cell surfaces to near zero. The exact form of the stress gradient may depend on whether under these conditions there is a sharp boundary between mucus and aqueous fluid, or whether the apposed interfaces are to some extent mingled together. (See also papers by Tiffany and by Holly, this Symposium).

## Aqueous Layer

The generally-accepted estimate of $7 \pm 2$ µl for the total retained tear volume in the open eye, consisting of about 1 µl in the actual tear film and the remainder in the marginal meniscuses (depending in part on the surface tension of the tears), was determined by a fluorescein-dilution method (Mishima et al., 1966). At present little is known about the volume of tear fluid under the lids, in either the open or closed eye, although Mishima et al. (1966) estimated this volume in the rabbit to be about 4.5 µl. However, such methods have been criticised by Macdonald and Maurice (1991) on the grounds that the mixing and distribution of fluorescein may be far from even or rapid. We have no figures for the volume of tear fluid under the lids in the human, but no reason to suppose that any considerable volume is so retained. Hence the entire tear volume is probably swept towards the lid margins during a blink, and the aqueous part of the tear film respread from the marginal meniscus as the lids reopen. These figures for tear volume and the capacity of the conjunctival sac must be taken into consideration in relation to the drop size and retention time of ocular drugs.

A number of methods have been used in the past to estimate the surface tension of tears; because of the volume of tears required, most of these were impractical for clinical or repeated research use, and especially on dry eye patients, where less than 1 µl may be available. Tiffany et al. (1989) adapted the method of Ferguson and Kennedy (1932), using as little as 0.3 µl, and making the measurement in the same capillary tube in which the tears were collected. For normal eyes the surface tension was $43.6 \pm 2.7$ mN/m, and for dry eyes $49.6 \pm 2.2$ mN/m. As previously shown by Holly (1973), both mucus and proteins substantially influence surface tension, but mucus has the greater effect. It is possible that surface tension could be used as a measure of the concentration of soluble mucus in dry eye patients.

There is much interest in the overall concentration of tears, usually reported in units of osmolarity, both as a diagnostic sign for dry eye disorders and as an indication of malfunction of some components of the tear film. The "nanolitre osmometer", as used by Gilbard et al. (1978) and by Benjamin and Hill (1983), requires only about 0.03 µl for measurement of freezing-point depression and hence tear samples can be collected with minimal stimulation. Other instruments may use the lowering of vapour pressure. The major part of the response is due to the electrolytes present, so the measurement is of tonicity rather than a true osmotic pressure. Direct measurement of osmotic pressure is possible, but the currently-available instruments require at least 100-200 µl of sample, so are impracticable for the individual patient. The significance of the osmolarity reported is as an indicator of over-concentration of the tears. Continued exposure to hyperosmotic solutions is reported to cause loss of conjunctival goblet cells (Gilbard et al., 1985), and should presumably be followed by reduction of mucus output, although there is some difference of opinion on this point (Huang et al., 1989; Kinoshita et al., 1983). Hypotonic solutions may in some cases be helpful in treatment of dry eyes, but if carefully controlled other factors such as viscosity and pH appear to be of equal importance (Wright et al., 1987).

The rheological properties of human and rabbit tears have begun to be studied (Tiffany 1991; Hamano & Mitsunaga 1973; Tiffany, this Symposium). They are very largely determined by mucus or possibly mucus/protein/lipid complexes (Kaura & Tiffany, 1986). At present there are difficulties in collecting enough unstimulated tears from individual patients; the smallest usable volume on any of the commercially-available rheometers of sufficient sensitivity for the task is 50-70 µl. The tears show appreciable shear-thinning, with a 5- or 6-fold drop in the coefficient of viscosity between the extreme low-shear and high-shear states. Elasticity is also appreciable, and capable of cushioning the epithelial surfaces against the strongly-transmitted shearing effect of rapid eye movements (Tiffany, this Symposium).

It has been generally assumed that successful artificial tears should incorporate as many as possible of the desirable rheological properties of natural tears; non-Newtonian solutions of polymers such as hyaluronic acid or polyacrylic acid have been found to be effective and well-tolerated, although many refinements of formulation are still possible. However, many of the older type of Newtonian solutions are likely to continue in use for some years. Bothner et al. (1990) compared the viscosities of many artificial tears over a wide range of shear rates, but did not include natural tears, so it is not yet clear which formulation most closely mimics the natural product, or whether this will indeed prove to be the best pharmaceutic strategy.

## Lipid Layer

Much of our knowledge of the surface chemistry and spreading ability of meibomian oil on the outer surface of the tear film is still that provided by Holly (1973). Jaeger et al. (1986) and Kaercher et al. (1992) have extended this by use of an automated Langmuir trough. Their results show to some extent that the behaviour of meibomian lipid mixtures can be most closely simulated by a phospholipid. This agrees with the idea that the phospholipids spread first, followed by the less polar components, but no extensive study has yet been made of the individual and cooperative spreading properties of the lipid classes of the secretion, as monomolecular or multimolecular layers.

The lipid layer may vary considerably in thickness between individual subjects, and to some extent this may depend on the amount of oil secreted onto the lid margin, as well as its composition. A device has recently been developed to measure the lid-margin casual level in normals and detect changes in dry-eye or blepharitis patients before and after treatment (Chew et al., 1993 a,b). Oil layer thickness is still estimated by the semi-quantitative method of Norn (1979), by narrowing the palpebral aperture until interference colours are seen. Tiffany (1986) has measured refractive index as a function of wavelength, and this could be used to refine estimates of thickness. Olsen (1985) used calculated reflectance values to derive a mean thickness of 36-48 nm, but it is clear from palpebral-narrowing studies that considerable local variations can be present simultaneously. Thickness may also depend on toxic, possibly surface-active, substances in the air, and there is much interest in the "sick building syndrome" resulting from such substances in the work-place. Franck (1991) has shown some reduction of thickness from morning to afternoon in sufferers; exposure to solvent vapours (Norn, 1986) or cigarette smoke (Basu et al., 1978) are also thought to diminish or disrupt the lipid layer.

Measurements of rate of evaporation from the open eye have mostly concentrated on the difference between normal and dry eyes (Tsubota and Yamada, 1992; Rolando and Refojo, 1983, Hamano et al., 1980) or the effect of instillation of artificial tears or preservatives (Tomlinson and Trees, 1991). Results of different workers show great variation, and even disagree on whether the rate rises or falls in the dry eye relative to normal. At present little is known about the variation of evaporation with meibomian layer thickness, either in vitro or in vivo, or the correlation of this with onset of wetting disorders.

Other limited studies have been made of biophysical properties of meibomian oil. Thus the dependence of both melting point and viscosity on composition have been reported (Tiffany and Marsden, 1986). Preliminary studies (Tiffany, unpublished) suggest that such complex lipid mixtures close to their melting points have viscoelastic properties which may be important in extrusion of the material from the glands. No investigation of surface rheology of the lipid layer in relation to damping effects on the tear film surface has yet been reported. In view of the findings of McCulley and co-workers (Shine and McCulley, 1991) on differences in composition of meibomian oil in both normals and chronic blepharitis patients, the interrelationships of composition and properties are of considerable interest.

## Combined Biophysical Properties

Many of the properties so far referred to have involved more than one component of the tear film. In some cases all the components are involved. The most important of these is the overall stability of the film. Any fully satisfactory theory must be able to explain the formation and structure of the film; how its stability is compromised; how, when and why it breaks up; the consequences of break-up; and how these consequences may be alleviated by supplementation of one or more components. Several stability theories have been produced (Holly, 1973; Sharma and Ruckenstein, 1986; Liotet et al., 1987b), although of these only that of Holly could claim to cover all of these aspects. However recent findings, particularly on wettability of the corneal surface, on the thickness of the mucus layer lipid migration, cell sloughing, and the development of the glycocalyx, call for the development of a new theory which will incorporate all of these and other properties.

Unfortunately, the only measure of stability we have at the moment is the break-up time. As currently practiced this has many drawbacks; if used with fluorescein there is the problem of introducing an alien substance which may itself influence stability, while the non-invasive method suffers from problems of interpretation (what is a dry spot?) and is to some extent subjective. We still need some objective measure of the effectiveness of the tears and the tear film in carrying out its basic functions of protecting and nourishing the surface of the eye while maintaining a high-quality optical surface.

Tear ferning has been considered as a useful means of assessing the "quality" of tears, if not of their performance in the tear film. It has been variously claimed to be an indicator of either mucus concentration (Tabbara & Okumoto, 1982; Rolando, 1984), protein (Golding and Brennan, 1989, Liotet et al., 1987a) or electrolyte balance (Kogbe et al., 1991). Ferning is in fact a very complex phenomenon which does not depend on any one factor; it is certainly not, as widely imagined, a semi-quantitative measure of mucus availability, although quality of ferning shows a strong correlation with severity of keratoconjunctivitis sicca (Rolando, 1984). Possibly this test could be further developed as a clinically-useful indicator of tear component functions and interactions.

## Conclusion

A considerable number of different physical properties of the tear film and its components have been referred to above, such that only brief mention can be made of any single aspect. On closer inspection, what is most striking is how little we still know in detail about any of these. There are still a very large number of possible avenues open for study, and we can still add enormously to our understanding of the complex interactions and functions of the tear film.

## REFERENCES

Absolom, D.R., Zingg, W., and Neumann, A.W., 1986, Measurement of contact angles on biological and other highly hydrated surfaces, *J. Colloid Interface Sci.* 112: 599-601.

Basu, P.K., Pimm, P.E., Shephard, R.J., and Silverman, F., 1978, The effect of cigarette smoke on the human tear film, *Canad. J. Ophthalmol.* 13: 22-25.

Benjamin, W.J., and Hill, R.M., 1983, Human tears: osmotic characteristics, *Invest. Ophthalmol. Vis. Sci.* 24: 1624-1626.

Bothner, H., Waaler, T., and Wik, O., 1990, Rheological characterization of tear substitutes, *Drug. Dev. Ind. Pharm.* 16: 755-768.

Bron, A.J., and Tiffany, J.M., 1991, Pseudoplastic materials as tear substitutes. An exercise in design, *in*: "The Lacrimal System", O.P. van Bijsterveld, M.A. Lemp and D. Spinelli, eds., Kugler and Ghedini, Amsterdam, pp. 27-33.

Chew, C.K.S., Jansweijer, C., Tiffany, J.M., Dikstein, S., and Bron, A.J., 1993a, An instrument for quantifying meibomian lipid on the lid margin: the Meibometer, *Curr. Eye Res.* 12: 247-254.

Chew, C.K.S., Hykin, P.G., Jansweijer, C., Dikstein, S., Tiffany, J.M., and Bron, A.J.. 1993b, The casual level of meibomian lipids in humans, *Curr. Eye Res.* 12: 255-259.

Cope, C., Dilly, P.N., Kaura, R., and Tiffany, J.M., 1986,. Wettability of the corneal surface: a reappraisal, *Curr. Eye Res.* 5: 777-785.

Dudinski, O., Finnin, B.C., and Reed, B.L., 1983, Acceptability of thickened eye drops to human subjects, *Curr. Therapeut. Res.* 33: 322-337.

Feenstra, R.P.G., and Tseng, S.C.G., 1992, What is actually stained by Rose Bengal? *Arch. Ophthalmol.* 110: 984-993.

Ferguson, A., and Kennedy, S.J., 1932, Notes on surface-tension measurement, *Proc. Phys. Soc.* 44: 511-520.

Franck, C., 1991, Fatty layer of the precorneal film in the "office eye syndrome", *Acta Ophthalmol.* 69: 737-743.

Gilbard, J.P., Farris, R.L., and Santamaria, J., II., 1978, Osmolarity of tear microvolumes in keratoconjunctivitis sicca, *Arch. Ophthalmol.* 96: 677-681.

Gilbard, J.P., Carter, J.B., Sang, D.N., Refojo, M.F., et al., 1985, Morphologic effect of hyperosmolarity on rabbit corneal epithelium, *Ophthalmology* 91: 1205-1212.

Gipson, I.K., Yankauckas, M., Spurr-Michaud, S.J., Tisdale, A.S., and Rinehart, W., 1992, Characteristics of a glycoprotein in the ocular surface glycocalyx. *Invest. Ophthalmol. Vis. Sci.* 33: 218-227.

Golding, T.R., and Brennan, N.A., 1989, The basis of tear ferning, *Clin. Exp. Optom.* 72: 102-112.

Hamano, H., and Mitsunaga, S., 1973, Viscosity of rabbit tears, *Jap. J. Opthalmol.* 17: 290-299.

Hamano, H., Hori, M., and Mitsunaga, S., 1980, Application of an evaporimeter to the field of ophthalmology, *J. Japan. Contact Lens Soc.* 22: 101-107.

Hammer, M.E., and Burch, T.G., 1984, Viscous corneal protection by sodium hyaluronate, chondroitin sulfate, and methylcellulose, *Invest. Ophthamol. Vis. Sci.* 25: 1329-1332.

Holly, F.J., 1973, Formation and rupture of the tear film, *Exp. Eye Res.* 15: 515-525.

Holly, F.J., 1986, Dry eye and the Sjögren's syndrome, *Scand. J. Rheumatol.* Supp 61: 201-205.

Holly, F.J., and Lemp, M.A., 1971, Wettability and wetting of corneal epithelium, *Exp. Eye Res.* 11: 239-250.

Huang, A.J.W., Belldegrün, R., Hanninen, L., Kenyon, K.R., Tseng, S.C.G., and Refojo, M.F., 1989, Effects of hypertonic solutions on conjunctival epithelium and mucinlike glycoprotein discharge, *Cornea* 8: 15-20.

Jaeger, W., Möbius, D., and Kaercher, T., 1986, Biophysical experiments with lipid-layers formed with meibomian gland secretion, *in*: "The Preocular Tear Film", F.J. Holly, ed., Dry Eye Institute, Lubbock, Texas, pp. 609-621.

Kaercher, T., Möbius, D., and Welt, R., 1992, Biophysical characteristics of the Meibomian lipid layer under *in vitro* conditions, *Intl. Ophthalmol.* 16: 167-176.

Kaura, R., and Tiffany, J.M., 1986, The role of mucous glycoproteins in the tear film, *in*: "The Preocular Tear Film", F.J. Holly, ed., Dry Eye Institute, Lubbock, Texas, pp. 728-732.

Kinoshita, S., Kiorpes, T.C., Friend, J., and Thoft, R.A., 1983, Goblet cell density in ocular surface disease. A better indicator than tear mucin, *Arch. Ophthalmol.* 101: 1284-1287.

Kogbe, O., Liotet, S., and Tiffany, J.M., 1991, Factors responsible for tear ferning, *Cornea* 10: 433-444.

Liotet, S., Leloc, M., and Glomaud, J., 1986, Lacrimal secretory IgA fixation on conjunctival mucus, *in*: "The Preocular Tear Film", F.J. Holly, ed., Dry Eye Institute, Lubbock, Texas, pp. 770-775.

Liotet, S., Kogbe, O., and Schemann, J.F., 1987a, Cristallisation des larmes: un test de qualité du film lacrymal? *Bull. Soc. Opht. France* 87: 321-324.

Liotet, S., van Bijsterveld, O.P., Kogbe, O., and Laroche, L., 1987b, A new hypothesis on tear film stability, *Ophthalmologica* 195: 119-124.

MacDonald, E.A., and Maurice, D.M., 1991, The kinetics of tear fluid under the lower lid, *Exp. Eye Res.* 53: 421-425.

Mishima, S., Gasset, A., Klyce, S.D., Jr., and Baum, J.L., 1966, Determination of tear volume and tear flow, *Invest. Ophthalmol.* 5: 264-276.

Neumann, A.W., Absolom, D.R., Francis, D.W., Omenyi, S.N., Spelt, J.K., Policova, Z., Thomson, C., Zingg, W., and van Oss, C.J., 1983, Measurement of surface tensions of blood cells and proteins, *Ann. N.Y. Acad. Sci.* 416: 276-298.

Nichols, B., Chiappino, M.L., and Dawson, C.R., 1985, Demonstration of the mucous layer of the tear film by electron microscopy, *Invest. Ophthalmol. Vis. Sci.* 26: 464-473.

Norn, M.S., 1979, Semiquantitative interference study of fatty layer of precorneal film, *Acta Ophthalmol.* 57: 766-774.

Norn, M.S., 1986, The effect of drugs on tear secretion, *in*: "The Preocular Tear Film", F.J. Holly, ed., Dry Eye Institute, Lubbock, Texas, pp. 221-229.

Olsen, T., 1985, Reflectometry of the precorneal film, *Acta Ophthalmol.* 63: 432-438.

Prydal, J.I., and Campbell, F.W., 1992, Study of precorneal tear film thickness and structure by interferometry and confocal microscopy, *Invest. Ophthalmol. Vis. Sci.* 33: 1996-2005.

Prydal, J.I., Artal, P., Woon, H., and Campbell, F.W., 1992, Study of human precorneal tear film thickness and structure by interferometry, *Invest. Ophthalmol. Vis. Sci.* 33: 2006-2011.

Proust, J.E., Arenas, E., Petroutsos, G., and Pouliquen, Y., 1983, Le film lacrymal, structure et stabilité, *J. Fr. Ophtalmol.* 6: 963-969.

Rolando, M., 1984, Tear mucus ferning test in normal and keratoconjunctivitis sicca eyes, *Chibret Intl. J. Ophthalmol.* 2: 32-41.

Rolando, M., and Refojo, M.F., 1983, Tear evaporimeter for measuring water evaporation rate from the tear film under controlled conditions in humans, *Exp. Eye Res.* 36: 25-33.

Scott, J.E., Cummings, C., Brass, A., and Chen, Y., 1991, Secondary and tertiary structures of hyaluronan in aqueous solution, investigated by rotary shadowing-electron microscopy and computer simulation, *Biochem. J.* 274: 699-705.

Sharma, A., and Coles, W.H, 1990, Physico-chemical factors in tear film breakup, *Invest. Ophthalmol. Vis. Sci.* 31 (Supp.): 552.

Sharma, A., and Ruckenstein, E., 1986, The role of lipid abnormalities, aqueous and mucus deficiencies in the tear film breakup, and implications for tear substitutes and contact lens tolerance, *J. Colloid Interface Sci.* 111: 8-34.

Shine, W.E., and McCulley, J.P., 1991, The role of cholesterol in chronic blepharitis, *Invest. Ophthalmol. Vis. Sci.* 32: 2272-2280.

Tabbara, K.F., and Okumoto, M., 1982, Ocular ferning test. A qualitative test for mucus deficiency, *Ophthalmology* 89: 712-714.

Tiffany, J.M., 1986, Refractive index of meibomian and other lipids, *Curr. Eye Res.* 5: 887-889.

Tiffany, J.M., 1990a, Measurement of wettability of the corneal epithelium. I. Particle-attachment method, *Acta Ophthalmol.* 68: 175-181.

Tiffany, J.M., 1990b, Measurement of wettability of the corneal epithelium. II. Contact-angle method, *Acta Ophthalmol.* 68: 182-187.

Tiffany, J.M., 1991, The viscosity of human tears, *Intl. Ophthalmol.* 16: 371-376.

Tiffany, J.M., and Marsden, R.G., 1986, The influence of composition on physical properties of meibomian secretion, *in*: "The Preocular Tear Film", F.J. Holly, ed., Dry Eye Institute, Lubbock, Texas, pp. 597-608.

Tiffany, J.M., Winter, N., and Bliss, G., 1989, Tear film stability and tear surface tension, *Curr. Eye Res.* 8: 507-515.

Tomlinson, A., and Trees, G.R., 1991, Effect of preservatives in artificial tear solutions on tear film evaporation, *Ophthal. Physiol. Optics* 11: 48-52.

Tseng, S.C.G., Huang, A.J.W., and Sutter, D., 1987, Purification and characterization of rabbit ocular mucin, *Invest. Ophthalmol. Vis. Sci.* 28: 1473-1482.

Tsubota, K., and Yamada, M., 1992, Tear evaporation from the ocular surface, *Invest. Ophthalmol. Vis Sci.* 33: 2942-2950.

Wright, P., Cooper, M., and Gilvarry, A.M., 1987, Effect of osmolarity of artificial tear drops on relief of dry eye symptoms: BJ6 and beyond, *Brit. J. Ophthalmol.* 71: 161-164.

Zisman, W.A., 1964, Relation of the equilibrium contact angle to liquid and solid constitution, *in*: "Contact Angle, Wettability and Adhesion", F.M. Fowkes, ed., *ACS Advan. Chem. Ser.* 43: 1-51.

# STRUCTURE AND FUNCTION OF THE TEAR FILM

P.N. Dilly

Department of Anatomy
St. George's Hospital Medical School
Cranmer Terrace, Tooting
London SW17 ORE

## INTRODUCTION

The existence of the tear film is well known but its structure is less well understood. It is unwise to consider the tear film in isolation from blinking. Blinking has a profound influence upon the structure, stability, and function of the tear film. Many glands from different sites contribute to the clear fluid that bathes the surface of the eye. Chemically tears are very similar to dilute blood, with a reduced protein content. (The pH of tears approximates to that of blood plasma but it has a slightly greater osmotic pressure.) The film covers the exposed surface of the eye and provides an optically smooth interface with the atmosphere. Lacrimation is well known throughout the animal kingdom, but crying with sorrow and laughter are probably confined to man. This is probably an adjunct to the vast range of facial expressions available to man. The parasympathetic nerve fibers that are secretomotor to the lacrimal gland are distributed for much of their course with the facial nerve, the motor nerve of facial expression.

It has been known for some time that the tear film has three layers (Wolff, 1954). There is a mucus layer close to the epithelium, an aqueous layer outside that and the external surface of the aqueous layer is covered with an oily layer. The oily and aqueous layers are easy to demonstrate. The aqueous being obvious, and the oily layer can be made to produce a whole host of diffraction colors depending upon its thickness. The mucus layer is more difficult to visualize simply because it has an identical refractive index with the aqueous phase of the tear film. It can be deduced from the vast numbers of goblet cells that discharge onto the surface of the eye from the conjunctival surface and also from the mucus thread that is found in the lower canthus. Postmortem it can be demonstrated by a whole host of mucus revealing stains. Electron microscopic investigations have confirmed

*Lacrimal Gland, Tear Film, and Dry Eye Syndromes*
Edited by D.A. Sullivan, Plenum Press, New York, 1994

its presence and subsequent workers have reported an increasing thickness for this layer as the methods of preservation have advanced.

Many protective substances are present in tears including lactoferrin, lysozyme, non-lysozyme antibacterial factor, complement and anti-complement factor and interferon, as well as the immunoglobulins. There are also lymphocytes in tears. Lactoferrin works by chelating iron and thus deprives microorganisms of iron. It modulates complement activity, at least in vitro. It is thought that it is synergistic with a specific antibody, and may also enhance the action of lyzozyme.

Lysozyme is lytic to glycosaminoglycans and in the presence of complement it facilitates IgA bacteriolysis. The immunoglobulin A neutralizes viruses and inhibits bacterial adherence to the epithelial surface. IgG promotes phagocytosis and complement-mediated bacteriolysis. IgE is increased in allergic responses.

The closed eye presents a special set of problems in the understanding of the tear film. When the eye is closed, the precorneal tear film is in osmotic equilibrium with the aqueous humor, and no osmotic flow occurs; as a consequence the corneal stroma thickens. During extended periods of lid closure such as sleep the concentrations of immunoglobulins increase, only to decrease again in the tear film of the open eye. An increase in the concentration of plasmin and a considerable increase in the numbers of polymorphonuclear leucocytes is also associated with the closed eye of sleep.

**BLINKING**

Blinking functions above all as a protective mechanism, at its crudest interposing the tissue of the eyelid between a potential insult to the eye and the eye. There are other known functions of blinking. It has been shown that a blink is associated with a discharge of secretion from the meibomian glands. The normal human blinks between two and ten times a minute in a normal environment. These blinks have been shown to change the distribution of the lipid layer covering the aqueous layer of the tears. The mucus for the eye comes mainly from the conjunctival goblet cells. Although there is some autonomic control of mucus secretion, it is difficult to see how this control is effected since the goblet cells are not associated with myoepithelial cells. It is known that a blink exerts a force of at least 5 grammes on the globe, and that firmly closing the eye can displace the eye backwards in the orbit several millimeters. It is possible that this squeezing could cause mucus to be released from the goblet cells. Any painful object is probably painful because it has penetrated the mucus layer and is stimulating the superficial nerve endings of the cornea and conjunctiva directly, evoking the response of tightly screwing the eyelids together. Perhaps this is a mechanism to replace rapidly the contaminated mucus layer with a fresh layer from beneath and thus removing the foreign body from contact with the globe. The increased load applied to the tear film during this powerful lid activity probably also serves to distribute the fresh mucus evenly. Prydal (1992) has shown that it takes about 30 minutes for the tear film mucus to be replaced after it has been destroyed by acetyl cystein. This probably represents the physiological maximum rate at which the tear film mucus can be replaced from the goblet cell source by normal unconscious blinking mechanisms. The similar well known vast increase in aqueous flow is effective in washing foreign material from the surface of the mucus phase of the tear film.

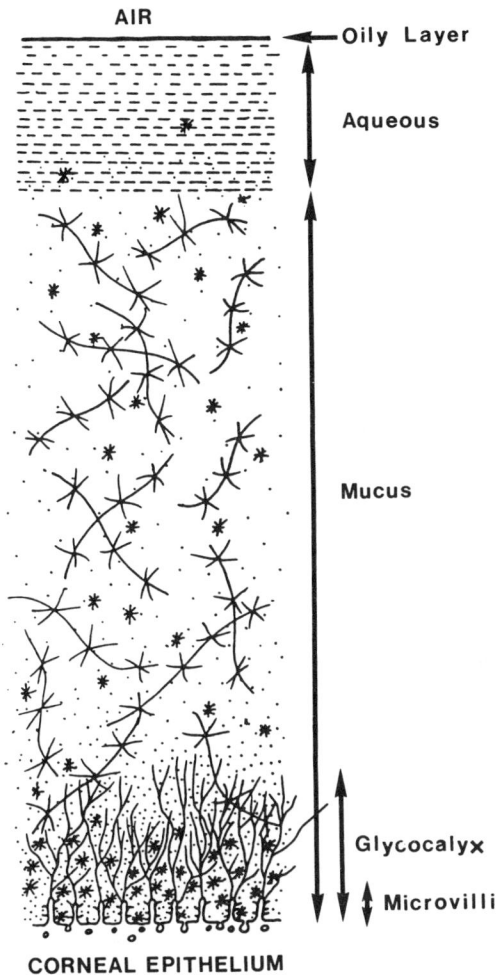

**Figure 1.** Diagram of the structure of the tear film. The film is 30-40 μm thick. The immunoglobulins are represented by stars, and the glycoproteins by the long lines with crosses. It is probably a much more stable structure than has previously been proposed, and measurable flow probably only occurs after stimulation. The components are known to be secreted from many different sources. How their relative proportions and distribution is controlled is unknown. Most components probably exist in varying concentrations in most layers.

241

Another function of the blink is to smooth the anterior surface of the mucus layer and thus improve its optical properties. Indeed the physical forces applied to the tear film during blinking are probably essential not only for spreading the oily layers, but also for maintaining the structural integrity of the deeper layers.

## OILY LAYER

The thickness of the oily layer varies with the width of the palpebral fissure and in varying places across the exposed part of the globe. It is made up of many oily substances derived from the meibomian glands. It is said that meibomian secretions help stabilize the tear film, but when meibomian secretion is added to a thin water film it seems to have little effect. Meibomian secretions collected directly from the glands are solid or semisolid, having a melting point above body temperature, tear temperature is about 3 or 4°C below this, so it must need some additional factors to allow melting at this lower temperature and spreading across the aqueous tears.

The oily layer behaves as if it is a sheet suspended from the upper lid, thickening as the eyelid closes and thinning again on opening. It is an inhomogenous layer containing complexes of lipids and other substances. The layer is not of uniform thickness, but these variations seem to have little effect on visual activity. Part of the lipid layer's function is to reduce the evaporation of the aqueous tears. The normal rate of evaporation from the tear film is 0.085 µl/minute, the superficial oily layer is probably responsible for this low value. Mishima has estimated that without the oily layer the rate would be between 0.85 to 1.7 µl/minute. It is known that this layer dispels and breaks up if a drop of sebum is added and the rate of evaporation increases. The other probable effect is to reduce the surface tension of the aqueous phase and thus to reduce wave formation as the aqueous is propelled over the cornea by movements of the eye or eyelids.

The oily part of the tear film is extremely efficient in protecting the eye from small dust particles. This is easily demonstrated by examining the tear film in one eye and the contact lens surface of the other eye in volunteers, who have been exposed to a dusty environment. After as little as ten minutes examination with the slit lamp reveals few particles in the contact lens free eye, whereas the contact lens that destroys the lipid aqueous integrity is heavily contaminated on its exterior surface. It is surprising that the considerable inhomogeneity of the lipid layer, unless gross, seems to have little effect on visual acuity.

## AQUEOUS

The aqueous layer is about 7 µm deep in man. Normally most of it is produced by the accessory lacrimal glands. Its more important functions are to provide a lubricating layer between the moving surfaces of the eye and its adnexa, to remove foreign material and to nurture the corneal and conjunctival epithelia. There is probably an aqueous layer separating the mucus covering of the bulbar and palpebral conjunctivae under the eyelids.

It would be desirable that such a layer exists since otherwise the sheer stress of lid movements would be transmitted via the mucus to the epithelium. Since the lacrimal gland discharges its watery secretions into the upper canthus, it would seem reasonable to assume that the secretions would be released into the aqueous part of the tear film. The osmolarity of the tear film seems to be critical. Gilbard (1978) has shown that concentrations of over 310 milliOsmols are associated with dry eye syndromes. Part of the functions of the aqueous phase may be to supply water to keep osmolarity below this critical level. It is known from Gilbard's work with keratoconjunctivitis sicca that a rise in osmolarity is associated with aqueous tear deficiency. There are many soluble mucins contained in the aqueous layer.

It is a function of aqueous to provide the water for mucus hydration by inward flow from the aqueous into the mucus layer.

The lubrication of the lid movements over the globe and vice versa is provided by the aqueous part of the tear film. The gliding surfaces themselves are covered with mucus. This gliding is enhanced by the exclusion of the lipid layer between lid and globe. The relative velocity of the sliding surfaces is about 15-25 cm/sec and the sheer rate 20,000 sec -1. The viscosity of the aqueous is low about 1.1 cps. The sheer stress at the mucus/aqueous interface is 150 dynes/cm$^2$. Because the mucus is so much more viscous, the sheer will decrease rapidly in mucus and will be negligible at the ocular cellular surface. Thus, this is an excellent protective mechanism stopping the blinking eyelids from damaging the corneal and conjunctival epithelium.

A stable aqueous phase must exit under the lid since the structure of the tear film and the nature of the lid/globe movements are such that a hydrodynamic lubrication mechanism is required. If it were a mucus/mucus interface, the boundary lubrication would be inadequate to prevent ocular surface tissue damage.

Wearing contact lenses destroys the integrity of the oily layer, and Maurice (1961) has shown that the tear evaporation rate then becomes significant. Mishima's (1965) work on evaporation rates from the normal eye show a miniscule loss of water. It would seem wasteful to have a basic tear flow rate greater than this, as the eye would rapidly overflow, or be drained via the nasolacrimal system. It would seem unreasonable to have a basic flow rate ten times that required to maintain the aqueous volume against evaporation, when there exists a massive reflex flow rate for emergencies. There has been much debate about the amount and significance of any basic tear flow in the unstimulated eye (Jordan & Baum 1980). Such flow, if it exists, is very difficult to measure as almost any investigation produces some reflex tearing. Patients are known who have a normal tear film and a blocked nasolacrimal system who do not suffer epiphora. As methods of investigation have become less and less invasive, estimates of a basal secretion rate for tears have reduced in volume. It is now probably worthwhile to reappraise if there is a basal secretion rate. Many investigators using fluorophotometry have assumed that the fluorescein mixing in the tears was homogeneous. Recently Maurice (1993) has shown that after the installation of a micro drop of fluorescein, it can take up to several minutes before it is evenly distributed. As yet the feed back mechanisms that must signal the state and composition of the tear film are virtually unknown. There must be some clues in the distribution of the paccinian-like corpuscles within the eye. They are concentrated along the superior limbus, a site passed over by the upper eyelid during each blink. The fact that the eye is so sensitive to changes and abnormalities in the tear film suggests that there are other receptors sensitive to very subtle changes in the tears.


## MUCUS LAYER

The mucus layer in the eye is unlike the mucus layer lining the lung, in that it is probably less mobile, it is not being moved along by the action of cilia, rather it is static and anchored to the microvilli. The cleansing role of lung mucus is undertaken in the eye by the aqueous layer of the tear film flowing over the mucus layer. There must however be some movement of mucus from the conjunctiva where it is secreted, to the precorneal tear film. It is probably spread across the surface of the eye by the action of the lids. Although the mucus layer is anchored to the cells beneath it, it is not in accurate register with their boundaries. It is possible to observe mucus spreading out from goblet cell apertures, the spread between adjacent goblet cells is confluent, and often is not in register with the underlying cell outlines. The goblet cell mucus has variable staining properties, suggesting some chemical non homogeneity (Adams & Dilly 1989).

The tear film is rich in nutrients, but it is not known how much if any are used for metabolism by the corneal epithelium. It does have the vital function of allowing diffusion of oxygen and other gasses to and from the epithelium. Obstruction of this pathway by an impermeable contact lens rapidly leads to corneal damage.

Whereas the aqueous provides the cleavage plain for movements of the eyelids, it is the mucus layer that protects the underlying epithelium from sheer damage.

The mucus layer itself heals rapidly and effectively. Holes and other defects made in the mucus layer of the tear film repair almost immediately, provided that the underlying cells are intact. The edges of defects are probably brought together by attractive forces between the long chain molecules. It is probable that these attractive forces are enhanced by the compression forces of the eyelids during blinking. The rapid restoration of this layer is vital for the protection of the eye against drying and bacterial invasion. It is the external surface of the mucus layer that is the first solid layer encountered by incoming material. It is known that mucus has an inhibiting effect on bacterial adhesion.

Mucus is a spongelike material with fluid in a meshwork of glycoprotein molecules. Mucus is known to be viscoelastic (Kaura & Tiffany 1986), which probably explains the dents left in the precorneal tear film mucus when tonometer cones or glass fibres are pressed against them (Prydal et al, 1993).

The mucus layer, because of its micellar structure, probably acts as a reservoir for immunoglobulins that allows only the slow release of immunoglobulins during the day when the eye is open and its vulnerability to air borne pathogens and antigens is increased. The overnight increase in concentration probably represents the normal physiological rate of secretion of immunoglobulins by the untroubled eye. When the eye is closed it is unlikely that a layer of lipid exists, its function being taken over by the physical barrier of the opposed eyelids. The aqueous layer would need to be maintained in order to protect the surface epithelium from sheer damage during the rapid eye movements associated with sleep.

Further evidence for a micellar structure of the mucus layer of the tear film is that microdrops of fluorescein introduced into the aqueous rapidly disperse and disappear, whereas drops introduced more deeply take several minutes to become evenly spread throughout the tear film in spite of normal blinking (Maurice 1993).

The mucus layer is sufficiently rigid and elastic for a contact lens to float on it. The movements induced in a contact lens by the upper eyelid would soon destroy the corneal epithelium if it was not for this protective buffer. The potential effects of a contact lens moving in the eye upon the corneal epithelium must be similar to that of an eyelid during a blink and both of them are prevented from damaging the eye by the barrier of the mucus layer.

Doane (1980) has shown that there is a retraction of the eyeball of between 0.7 -1.6 mm during a blink. Riggs (1987) showed that the amount of retraction increased the more powerful the blink. It is known that the eyelids can exert a force of several grammes on the globe and it is likely that it is this push from the eyelids that displaces the globe backwards. Such a force if applied directly to the epithelial cells once every 2-10 seconds while awake would surely disrupt the cells. It is the viscoelastic nature of the mucus layer that prevents this damage from occurring. Besides the eyelids as a potential source of disruption of the tear film, movements of the globe will also cause stresses in the film and the elastic behaviour of the tears will also resist the distortion of the tear film during eye movements.

The mucus of the tears contains very long molecules of over 3 microns in length and with molecular weights of 50 million daltons or more. These long molecules are probably glycoproteins and hyaluronan, and are responsible for trapping the water in the mucus layer, the self repair of this layer and also for the viscoelastic properties of the mucus layer of the tear film. They have the classical 'lampbrush' structure, but there are spaces and naked areas along the backbone that permit aggregation through the disulphide bonds. It is the combination of this molecular structure and the enclosed and trapped water that gives the viscoelastic properties to the ocular mucus layer.

The mucus is probably linked together by hydrophobic bonding. This bonding is weak and probably forms and breaks with minor local changes, the molecular lattice zipping open and shut. This bonding is probably part of the mechanism that produces the self repair properties of the mucus. The stability of this mucus bonding is probably enhanced by hyaluronan. Hyaluronan has the ability to form gels of randomly entangled molecules. It is known to be responsible for the main opposition to water flow in the matrix of connective tissue. It can achieve this at extremely low concentrations of less than 1 µg/ml. The aggregation is probably driven by the hydrophobic bonding between the glycosaminoglycan polymer backbone, which with additional hydrogen bonds balances the electrostatic repulsion between the polyionic aggregants (Scott 1992).

In connective tissues these bondings can take on a semipermanent nature. Such a feature in tear glycosaminoglycans would support the idea of a more permanent nature for tear film mucus than has previously been suggested. If this is so, then hyaluronan should be an effective treatment for some types of dry eye syndromes.

Mucus much reduces the ability of some bacteria to adhere to the surface of the eye. The bacteria seem inhibited from penetrating the mucus layer, and thus prevented from reaching the vulnerable cell layers beneath it. The size of bacteria would suggest that a submicron thick layer of mucus would not have the dimensions to prevent bacterial penetration and that the thicker layer found by Prydal (1992 a & b) and his colleagues would be better suited for this role.

The tears are known to contain many biologically active molecules. Lysozyme, first discovered by Fleming in 1922 and renowned for its bacteriocidal activity, may have another much more important function in tears. It is known that lysozyme destroys bacteria by digesting their mucopolysaccharide coats. A similar activity involving the glycosaminoglycans in the mucus layer might reduce the rate of drying and solidification of the mucus layer of the tear film. It is known that in some cases dry eyes in which there is a mucus abnormality the concentration of lysozyme is reduced. It is a common experience that mucus removed from the eye rapidly solidifies. Just adding water does not cause it to liquify for many hours. This process can be speeded up by adding lysozyme. Lysozyme may have a role in the management of some forms of dry eyes. Perhaps the lysozyme has a role in maintaining the physical properties of the complex chemical mixture that is tear film mucus. The other property of lysozyme that has a role in tear film stability is its molecular shape and electric charge. Besides helping to keep the mucus fluid, lysozyme is known also to increase the viscosity of tear mucus.

## GLYCOCALYX

The attachment of the mucus layer to the surface of the epithelial cells is crucial for the stability of the whole tear film. The epithelial cells upon which the tear mucus layer abuts are covered with a dense layer of microvilli. These microvilli are decorated with long chains of material that appear anchored to the microvilli and extend out into the mucus layer. They are revealed by tannic acid and ruthenium red staining. I have proposed that these filaments are the same material as is found in the sub-surface vesicles, and that their function is to anchor the mucus layer to the cell surface (Dilly 1985). In diseases of the epithelial cells, it is this anchorage system that is destroyed with the consequent destabilisation of the tear film.

The presence of a glycocalyx at the epithelial surface would suggest that the surface is strongly polar, and hence readily wettable by aqueous solutions. This is required for all cellular surfaces according to modern theories of membrane structure with dense arrays of heterogeneous polar and highly hydrated carbohydrate groups attached to membrane glycoproteins giving a very low interfacial tension against aqueous solutions.

## CONCLUSION

I am proposing therefore that the tear film is much thicker than the 7 µm usually quoted. The figure is nearer 30-40 µm, with the 'extra' thickness being contributed by the thick mucus layer. This thick layer had not been detected previously because it is invisible to standard optical methods. I propose that the mucus layer is stable and anchored to the epithelium. It contains many vital biologically active substances that can probably be 'slow released' from its micelles. The aqueous phase contains dissolved mucins and provides the cleavage plane for lid movements and the more viscous mucus layer stops the forces generated by the movements from disrupting the ocular epithelium. It is probably more accurate to describe the mucus and aqueous layers of the tear film as phases with more or less mucus, respectively.


## REFERENCES

Adams, G.G.W., and Dilly, P.N., 1989, Differential staining of ocular goblet cells, *Eye* 3:840.

Cope, C., Dilly, P.N., Kaura, R., and Tiffany, J.M., 1986, Wettability of the corneal surface: a reappraisal, *Curr. Eye Res.* 5:777.

Dilly, P.N., 1985, On the nature and role of the subsurface vesicles in the outer epithelial cells of the conjunctiva, *Brit. J. Ophthalmol.* 69:447.

Dilly, P.N., 1985, Contribution of the epithelium to the stability of the tear film, *Trans Ophthalmol Soc. U.K.* 104:381.

Doane, M.G., 1980, Interaction of eyelids and tears in corneal wetting and the dynamics of the normal human eyeblink, *Amer. J. Ophthal.* 89:507.

Fleming, A., 1922, On a remarkable bacteriolytic element found in tissues and secretions, *Proc. Roy. Soc.* B 93:306.

Gilbard, J.P., Farris, R.L., and Santamaria, J., 1978, Osmolarity of tear microvolumes in keratoconjunctivitis sicca, *Arch. Ophthalmol.* 96:677.

Holly, F.J., 1973, Formation and stability of the tear film. In *The preocular Tear Film and Dry Eye Syndromes*, F.J. Holly and M.A. Lemp (Eds.) International Ophthalmology Clinics, Boston, Little, Brown.

Jordan, A., Baum, J., 1980, Basic Tear Flow Does It Exist? *Ophthalmology* 87:920.

Kaura, R., Tiffany, J.M., 1986, The role of mucus glycoproteins in the tear film, In *The Precorneal Tear Film*, Dry Eye Institute, Lubbock Texas, Holly F.J., Editor, 728-732.

Kijlstra, A., and Veerhuis, R., 1981, The effect of an anticomplementary factor on normal human tears, *Amer J. Ophthalmol.* 92:24.

Mishima, S., 1965, Some physiological aspects of the precorneal tear film, *Arch. Ophthalmol.* 73:233.

Mishima, S., and Maurice, D.M., 1961, The oily layer of the tear film and evaporation from the corneal surface, *Exp. Eye Research* 1:39.

Prydal, J.I., Campbell, F.W., 1992, Study of tear film thickness and structure by interferometry and confocal microscopy, *Invest. Ophthalmol Vis Sci* 33:1996.

Prydal, J.I., Artal, P., Woon, H., Campbell, F.W., 1992, Study of human tear film thickness using laser inteferometry, *Invest. Ophthalmol Vis Sci* 33:2006.

Prydal, J.I., Kerr Muir, M.G., and Dilly, P.N., 1993, Comparison of the tear film thickness in three species determined by the glass fibre method and confocal microscopy, *Exp.. Eye Research.* In Press.

Riggs, L.A., Kelly, J.P., Manning K.A., and Moore, R.K., 1987, Blink related eye movements, *Invest. Ophthalmol.* 28:334.

Scott, J.E., 1992, Supramolecular organisation of extra cellular matrix glycosaminoglycans in vitro and in the tissues, *FASEB J.* 62:639.

Wolff, E., 1954, Anatomy of Eye and Orbit, *New York Blakiston Co.* 4th Edn. 207.

# OCULAR SURFACE CHANGES INDUCED BY REPEATED IMPRESSION CYTOLOGY

Maurizio Rolando, Valeria Brezzo, and Giovanni Calabria

Department of Ophthalmology, University of Genoa, Genoa, Italy

## INTRODUCTION

The correct diagnosis of ocular surface diseases can be difficult because of the overlapping of different specific and nonspecific reactions to the starting event. Direct evaluation of ocular surface conditions, such as the health of epithelial or goblet cells, can be of help in speeding-up the diagnosis of these diseases[1,2]. Moreover adequate knowledge of epithelial conditions is useful for monitoring the evolution of the disease and to verify the efficacy or the possible toxicity of the therapeutic regimen[3-5].

Among the bioptic methods, impression cytology is the most frequently used. It consists of collecting a sheet of superficial cells from the ocular surface by pressing a properly cut cellulose paper on the conjunctiva. Although the trauma that the peeling of a superficial epithelial sheet can cause to the eye is very low, it could induce changes on the ocular surface which, if these changes last for a certain period of time, can alter the appearance of a subsequent sampling in the same area, leading to wrong diagnostic conclusions or to wrong therapeutic approaches.

The aim of this study was to evaluate the reproducibility of impression cytology sampling on the same conjunctival area of healthy eyes, in order to identify the highest frequency of collection able to provide consistent results, not disturbed by the temporary changes induced by the trauma of collection.

## MATERIALS AND METHODS

Twenty healthy volunteers, 13 males and 7 females, aged 34 to 45 years (average 39 ± 3.2 years), were enrolled for this study. Inclusion criteria were:

- no subjective or objective signs of anterior surface pathology;
- no history of ocular diseases;
- no history of long term eye drop treatment;
- no eye drop use for at least 30 days before the day of the inclusion in the study;
- absence of known systemic pathology able to affect the health of the ocular surface;
- Schirmer I test results over 15 mm / 5', break up time (BUT) over 20";
- understanding the aim and the schedule of the study and a willingness to participate.

For all the participants in the study an informed consent was obtained.

A thorough anterior segment examination, including refraction, biomicroscopy, tonometry, flourescein and Rose Bengal staining, Schirmer I test and BUT, was performed on each eye at the time of patient selection, at least two days before the beginning of the study.

Impression cytology was performed, always by the same researcher (V.B.), according to the technique described by Tseng[3] which in summary involves the placing of a 5x8 mm properly cut paper (Millipore, pore size: 0.45 µm) on the bulbar conjunctiva so to include an area located 3 to 6 mm from the limbus. After 5" of gentle and uniform compression, the paper was peeled off and dipped into a fixative solution for less than 24 hours and then stained according to a Gill's modification of Papanicolau staining. The samples were then observed by light microscopy and when necessary a photomicrograph of the cellular morphology was taken.

Both eyes of each subject were studied, and in each eye the sampling was made from the superior (12 o'clock) and temporal (on the axis of 3-9 o'clock) bulbar conjunctiva. Only the samples containing a uniform sheet of cells, filling at least one third of the paper, have been considered valid for evaluation.

Impression cytology was performed on the right eye on days 0 (baseline), 1, 2, 3, 4, 8, and 12; on the left eye on days 0, 3, 8, and 12. In order to evaluate whether the changes revealed by multiple samplings were localized in the sampling area only or were accompanied by other modifications involving the entire ocular surface, an impression cytology sample was obtained from the nasal bulbar conjunctiva of the right eye on day 3.

Goblet cell (g.c.) concentration per square mm (obtained by means of direct or imprint visualization) and nucleus/cytoplasm (N/C) ratio were calculated in at least 6 noncontiguous microscopy fields of each sample and the mean value was considered for statistical evaluation.

The differences from the baseline were observed and their statistical significance was evaluated by means of Student's t test. Furthermore, by means of a computerized stochastic mathematical model, the best fitting curve able to fit the experimental data obtained on the N/C ratio was defined.

## RESULTS

### Right Eye

A progressive loss of goblet cells was apparent when daily sampling was performed in the same location of bulbar conjunctiva (figure 1). In particular, while there was nearly a

30% loss of goblet cells the day after the first sampling (superior conjunctiva = 32% g.c. loss; temporal conjunctiva = 28% g.c. loss), the loss was greater 70% after the third day of sampling (superior conjunctiva = 73%; temporal conjunctiva = 81%) and reached 75% for superior conjunctiva and 90% for the temporal conjunctiva on the fourth day.

After four days without sampling, the impression cytology showed a good repopulation of goblet cells with no significant difference from the baseline (day 0). Similarly, no significant difference of goblet cell concentration from the baseline was present on day 12 when a new impression cytology was performed in the same areas.

Although there was some difference in the degree of goblet cell loss between superior and temporal bulbar locations, with temporal locations showing a more pronounced goblet cell loss with time, this difference was only statistically significant on days 2 (p<0.05) and 4 (p<0.01).

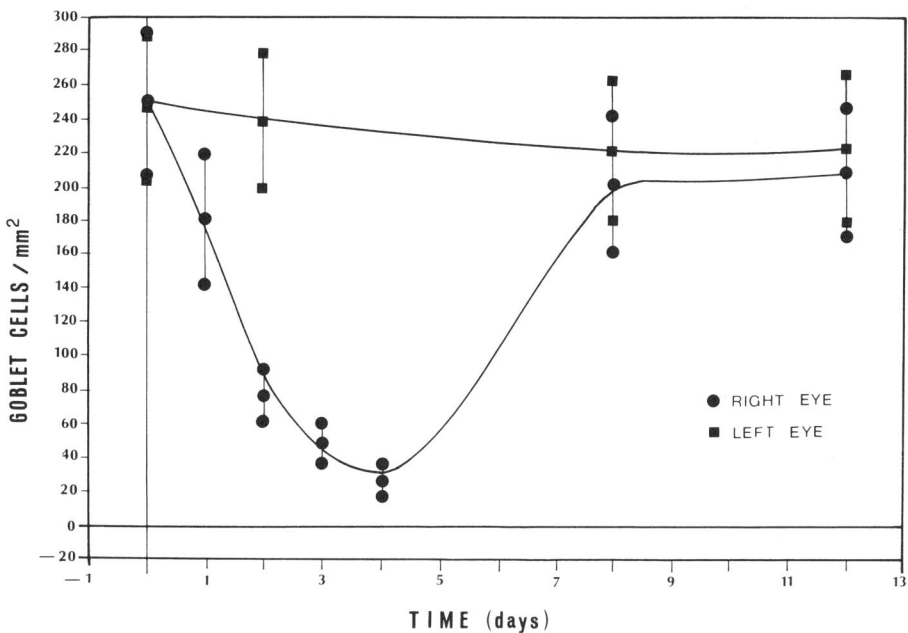

Figure 1. Impression cytology. Trend of goblet cell concentration, measured by direct or imprint visualization, in the bulbar conjunctiva of normal eyes as a function of the sampling interval time. High frequency initial sampling (1 day) in right eyes, as compared with low frequency (3 days) in left eyes. Data are expressed as the mean goblet cell concentration ($\pm$ S.E.) per $mm^2$ of cellulose paper. Computer drawn curves were calculated by a best fit mathematical model.

A behavior similar to goblet cell concentration was also observed for the N/C ratio, which tended to increase with the repetition of sampling (figure 2). Starting from the third day, there was a statistically significant (p<0.01), 52% reduction in the N/C ratio from the baseline, which reached 65% on the fourth day (statistical significance of the difference from the baseline: p<0.001).

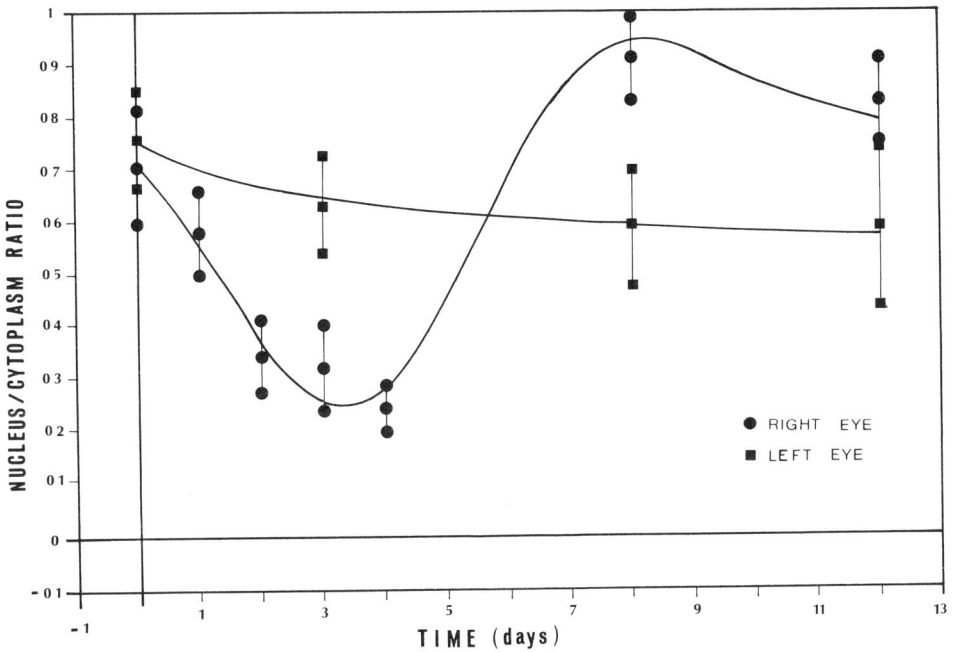

Figure 2. Impression cytology. Trend of nucleus/cytoplasm ratio measured in epithelial cells of the bulbar conjunctiva of normal eyes as a function of the sampling interval time. High frequency initial sampling (1 day) in right eyes as compared with low frequency (3 days) in left eyes. Mean values ± S.E. are shown. Computer drawn curves were calculated by a best fit mathematical model.

The results of the impression cytology collection obtained on day 3 from the nasal bulbar conjunctiva, being the goblet cell density of $276 \pm 41$ (SD) cells per square mm, and the N/C ratio of $1.3 \pm 0.21$ (SD), did not show a statistically significant variation from the samples obtained from other areas of collection at the beginning of the study.

After a four day interval, the new sampling failed to show a significant difference of N/C ratio from the baseline. In addition the new impression cytology repeated after four more days again showed no significant difference of N/C ratio from the baseline. No significant difference in N/C ratio was present between the samples collected from the superior and temporal conjunctiva, even if a tendency to higher signs of deterioration was present in the temporal samples.

**Left Eye**

A very low loss of goblet cells was observed when the sampling was performed on day 3, and the difference from the baseline, which was nearly 20%, was not statistically significant. No difference was observed when a new sampling was repeated after 5 days (day 8) and after four more days (day 12) (figure 1). No differences were observed between the two sampling areas.

No statistically significant difference in the N/C ratio, as compared to the baseline value, was present in the impression cytologies made, both in the superior and temporal bulbar conjunctiva, on day 3, 8 and 12 (figure 2).

## DISCUSSION

The ocular surface is a functional unit constituted by the tear film, the cornea, the conjunctiva and the muco-epidermal junction. There is some evidence that a great number of the diseases occurring on the ocular surface are "pan ocular surface diseases" and not just separate problems of its constituents[6,7]. It has also been demonstrated that conjunctival health can play a major role in maintaining the corneal integrity and in allowing its rapid recovery in case of ulcerations[8].

Characteristic modifications in the conjunctiva can occur in systemic diseases (e.g. early vitamin A deficiency, diabetes, anorexia nervosa, etc.), in the absence of clinically relevant changes[9-12]. The usefulness of a non-traumatic repeatable diagnostic technique, which is able to give direct information about the condition of the conjunctiva, is therefore apparent.

Impression cytology was introduced by Egbert et al. in 1977[1] and since then it has been used to diagnose and stage a number of ocular surface diseases[3]. Given its low cost, good acceptance by the patients and simplicity of execution, it is widely used for monitoring the progression or the regression of ocular surface disease with time and for evaluating the efficacy or the possible toxicity of a topical therapy[4,5,13].

Our results show that impression cytology on the same areas as the bulbar conjunctiva, when performed at interval times shorter than 4 days, will produce localized changes in goblet cell concentration and in the N/C ratio of epithelial cells. Loss of goblet cells and N/C ratio are the parameters most frequently used to indicate the degree of ocular surface disease[4]. Variation of such parameters is a sign of a change in the ocular surface conditions and it is usually attributed to a progression or to a regression of the disease.

The reason why these changes occur is not clear. We could hypothesize that repeated sampling will mechanically remove goblet cells from the area, leading to their reduced concentration. This is possible, but it is not consistent with the diffuse experience of the rapid regeneration and sliding of these cells in healthy conjunctiva[7]. It is known that goblet cell loss is a nonspecific sign of ocular surface suffering and it occurs in dryness as well as in many other diseases of the ocular surface or of the entire eye. Tseng, Hirst, Maumenee et al. have suggested that inflammation and a decrease of vascularization can be the cause of goblet cell loss[14]. Both conditions were difficult to find in our patients, who did not show any clinically detectable sign of inflammation or of vascular problems as a consequence of repeated impression cytology.

More intriguing are the changes occurring to the N/C ratio. If just the effect of mechanical removal was involved, one should expect the appearance of small cells with a relatively big nucleus and a N/C ratio close to 1. It is possible that the removal of the superficial sheet of conjunctival epithelium will expose immature areas of ocular surface, unable to be wetted or to keep a proper tear film on their top. The repetition of sampling does not allow the correct development of cellular characteristics or the sliding of mature cells from the surrounding conjunctiva. These areas will then undergo the typical changes caused by tear film instability and dryness[11], which can explain both the goblet cell loss and the decreased N/C ratio.

Our results show that in order to avoid the risk of introducing artificial variations in the results of impression cytology, when used to monitor the evolution of the ocular surface during a disease or a therapy, an interval of time of at least 4 days should be observed between two subsequent samplings from the same area.

# REFERENCES

1. P.R. Egbert. S. Lauber, and D.M. Maurice, A simple conjunctival biopsy, *Am. J. Ophthalmol.* 84:798 (1977).

2. J.R. Wittpenn, S.G. Tseng, and A. Sommer, Detection of early xeoroftalmia by impression cytology, *Arch. Ophthalmol.* 104:237 (1986).

3. S.G. Tseng, Staging of conjunctival squamous metaplasia by impression cytology, *Ophthalmology* 92:728 (1985).

4. G.G.W. Adams, P.N. Dilly, and C.M. Kirkness, Monitoring ocular disease by impression cytology, *Eye* 2:506 (1988).

5. M. Rolando, M.V. Brezzo, P. Campagna, S. Burlando, and G. Calabria, The effect of antibiotic agents on the ocular surface of normal humans, *Chibret Int. J. Ophthalmol.* 8:46 (1991).

6. C. Cope, P.N. Dilly, R. Kaura, and J.M. Tiffany, Wettability of the ocular surface: a reappraisal, *Curr. Eye Res.* 5:777 (1986).

7. R.A. Thoft RA, and J. Friend, The x,y,z hypothesis of corneal epithelial maintenance, *Invest. Ophthalmol. Vis. Sci.* 23:73 (1982).

8. A.J.W. Huang, and S.G. Tseng, Corneal epithelial wound healing in the absence of limbal epithelium, *Invest. Ophthalmol. Vis. Sci.* 32:96 (1991).

9. D.G. Keenum, R.D. Semba, S. Wirasasmita, G. Natadisastra, K.P. West, and A. Sommer, Assessment of vitamin A status by a disk applicator for conjunctival impression cytology. *Arch. Ophthalmol.* 108:1436 (1990).

10. J.M. Gilbert , J.S. Weiss, A.L. Sattler, and J.M. Koch, Ocular manifestation and impression cytology of anorexia nervosa, *Ophthalmology* 97:1001 (1990).

11. M. Rolando, F. Terranga, G. Giordano, and G. Calabria, Conjunctival surface damage distribution in keratoconjunctivitis sicca. An impression cytology study. *Ophthalmologica* 200:170 (1990).

12. M. Rolando, R. De Marco, D. Paoli, and S. Burlando, Conjunctival impression cytology in diabetic patients with and without vascular damage, *in:* "Ophthalmic Cytology," J. Orsoni, ed., Centro Grafico Editorale dell'Universit, Parma p.101 (1988).

13. S.G. Tseng, Topical tretinoin treatment for dry eye disorders, *Int. Ophthalmol. Clin.* 27: 47 (1987).

14. S.G. Tseng, L.W. Hirst, A.E. Maumenee, et al., Possible mechanisms for the loss of goblet cells in mucin deficient disorders, *Ophthalmology* 91: 547 (1984).

# DEVELOPMENTAL APPEARANCE OF A COMPONENT OF THE GLYCOCALYX OF THE OCULAR SURFACE EPITHELIUM IN THE RAT

Hitoshi Watanabe,[1,2] Ann S. Tisdale,[1] Sandra J. Spurr-Michaud,[1] and Ilene K. Gipson[1,2]

[1]Cornea Unit, Schepens Eye Research Institute and
[2]Department of Ophthalmology
Harvard University Medical School, MA 02114

## INTRODUCTION

The glycocalyx is a carbohydrate-rich cell coat that is present all along the apical membrane of the ocular surface epithelium and that is prominent at the tips of the microplicae.[1-3] A previous study reports that the glycocalyx is adjacent to the mucous layer, the inner most layer of the tear fluid, and that the spread of the mucous layer over the microplicae of the apical cells is believed to be facilitated by the filamentous glycocalyx on the microplicae of the apical cell's apical-most membrane.[1,2] Therefore, the biochemical nature of the glycocalyx may be important to maintaining the tear fluid over the ocular surface epithelium. The biochemical nature of the glycocalyx and its interaction with or in the spread of mucus over the apical cells is, however, still unknown.

Gipson et al. have recently developed a monoclonal antibody that binds to the apical squamous cells of the ocular surface epithelium of the rat.[3] Immunofluorescence studies of adult rat show localization to all the flattened cell layers or squames along the entire ocular surface epithelium and to some goblet cells of the goblet cell clusters.

Immunoelectron microscopic findings showed that the binding was at the apical membrane of apical cells and it was prominent at the tips of the microplicae. The binding, thus, corresponds to the region of the glycocalyx layer, adjacent to the tear film (Fig 1). In subapical squames, the antigen is found in the cytoplasmic vesicles, where it appears to be stored until cells differentiate to take up the apical position adjacent to the tear layer. Thus, the antibody was designated a rat ocular surface glycocalyx or ROSG antibody.

By immunoblot analysis, the rat ocular surface glycocalyx antibody reacts with a prominent band that has a molecular weight more than 205 kD. Alcian blue followed by silver staining of SDS-PAGE gels demonstrates a corresponding band. Periodic acid schiff reagent (PAS) also stains a band in the same region. These previously reported data suggest that the antigen is a highly glycosylated glycoprotein. Subsequent periodate oxidation treatment diminished the binding of the ROSG antibody to the ocular surface epithelium in cryostat sections and to the appropriate weight band in immunoblots. These data suggested

*Lacrimal Gland, Tear Film, and Dry Eye Syndromes*
Edited by D.A. Sullivan, Plenum Press, New York, 1994

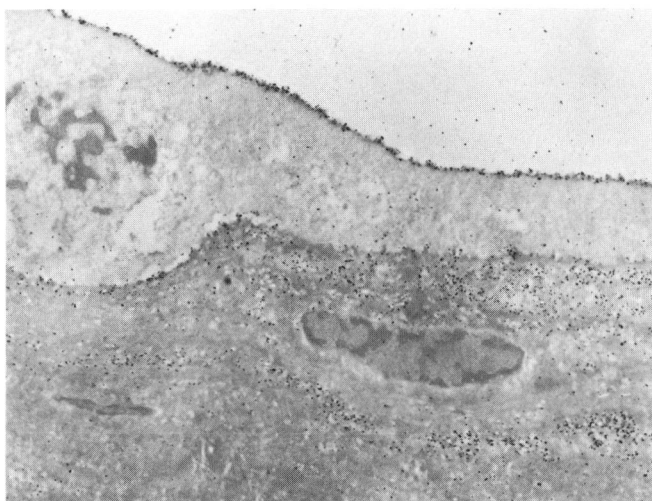

**Figure 1**. Immunoelectron microscopic localization of the ROSG antibody in corneal epithelium. The silver enhanced 1-nm immunogold is present along the apical surface membrane of apical cells. In subapical cells, the label can be seen within the cytoplasm of the cell. (x 31200) (From Gipson I.K. et al.;[3] reprinted with permission of Investigative Ophthalmology and Visual Science)

that the epitope recognized by the ROSG antibody is to a carbohydrate portion of the antigen.

Taken together, these data suggest that the ROSG antibody recognizes a carbohydrate epitope on a highly glycosylated glycoprotein in the glycocalyx of the rat ocular surface. In an additional, previously reported study using this ROSG antibody, we explored developmental appearance of the ROSG antigen in newborn rats.[4] A summary of this data follows.

## MATERIALS AND METHODS

All investigations described in this report conformed to the ARVO Resolution on the Use of Animals in Research. Both female and male Sprague-Dawley rats (Charles River Laboratories, Wilmington, MA) of various ages ( day 1 [newborn] to 5, 7, 11 to 15) were used. The animals were sacrificed by intraperitoneal injection of an overdose of sodium pentobarbital. The eyes and the eyelids were excised from the rat pups (n=6). For immunofluorescence study, cryostat sections (6 μm) of rat eyes with upper and lower lids, were processed following a method previously described.[3] For immunocytochemical study, excised rats eyes with the eyelids were fixed in 4% paraformaldehyde and 0.2% glutaraldehyde in 0.1M phosphate buffer, pH 7.4, for 1 hour at 4°C. Post embedding, immunoelectron microscopic localization of the antigen was done on L.R. white embedded tissues using silver enhanced 1-nm gold conjugated anti-mouse IgG following a method previously described.[3]

### Artificial Eyelid Opening

We performed artificial eyelid opening of one eye of developing rats at days 8-11 (n=3). The contralateral eye was left untouched as control. The eyes with lids were excised within 24 hr of the treatment and were processed for the immunohistochemical study.

**Table 1.** Binding of the ROSG antibody to the ocular surface epithelium of the developing rats.

| Stage | Immunolocalization to Ocular Surface Epithelium | | | | No. of Cell Layers in Corneal Epithelium | |
| | Conjunctiva | | | Cornea | Range | Mean |
| | Lid Margin | Palpebral | Fornix/Bulbar | | | |
|---|---|---|---|---|---|---|
| (Closed) | | | | | | |
| Days 1-4 | + / – | + / – | + / – | – | 1-2 | 1.3 |
| Days 5-7 | +++ | + / – | – ~ + / – | – | 1-2 | 1.6 |
| Days 12-14 | +++ | + / – | – ~ + / – | – | 2-4 | 3.3 |
| (Within 24 hours) | | | | | | |
| Days 12-14 | +++ | +++ | +++ | +++ | 4-6 | 4.7 |

n=6 for antibody binding, n=3 for determination of cell layers. –, no binding. + / –, scattered binding to apical cells only. +++, strong binding to apical and subapical cells (From Watanabe H., et al.;[4] reprinted with permission of Investigative Ophthalmology and Visual Science)

## RESULTS

### Developmental and Immunolocalization Studies

Table 1 summarizes the stratification of corneal epithelium and the binding of ROSG antibody to the ocular surface epithelium of postnatal rats. Rats are born with their eyelids closed, and they remained closed until day 12 to 14.

On day 1 (newborn) to 4, the epithelium of the eyelids was fused. The stratification of conjunctival epithelium was greatest at the lid edge with 6 to 9 cell layers. The number of cell layers decreased to 1 to 2 layers toward central palpebral conjunctiva and cornea (Fig 2). The binding of the ROSG antibody was observed in a very few scattered apical cells in the palpebral conjunctiva, particularly at the lid fusion area. Outer lid epithelium, bulbar conjunctiva, or cornea did not bind the ROSG antibody.

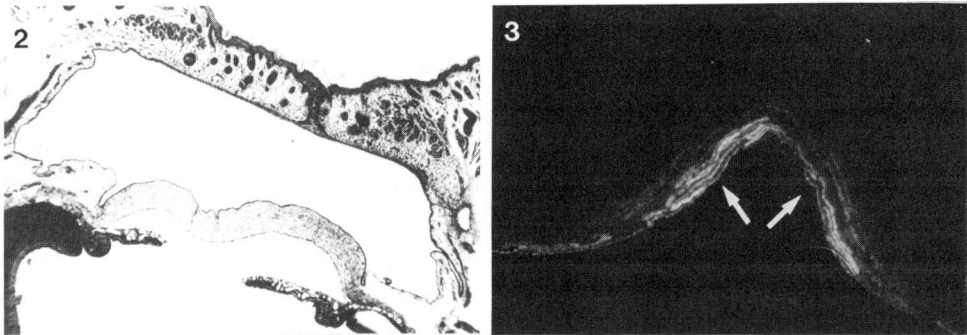

**Figure 2.** (left) Histological photograph of anterior segment of one-day-old rat. Light micrograph demonstrates the fused lid, the conjunctiva, and the cornea (X 50).(From Watanabe H, et al.;[4] reprinted with permission of Investigative Ophthalmology and Visual Science)

**Figure 3.** (right) Palpebral conjunctival epithelium at lid margin of 7-day-old rat. Light micrograph demonstrates the binding of the ROSG antibody is in the several layers of apical flattened squamous conjunctival cells near the lid fusion area (arrows) (x 750) (From Watanabe H., et al.;[4] reprinted with permission of Investigative Ophthalmology and Visual Science)

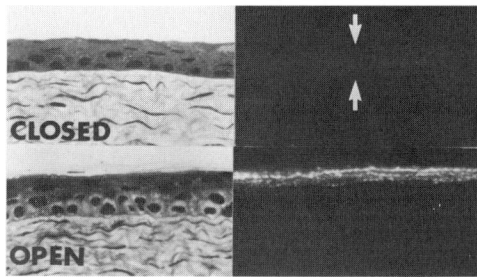

**Figure 4.** The corneal epithelium of a 12 day-old rat with the eyelid closed (top) and open (bottom). (Top, left) Light micrograph demonstrating that the corneal basal cells are still ovoid in the rat with the eyelid closed. (Top, right) Immunofluorescence micrograph demonstrating localization of the ROSG antigen in the corneal epithelium. No binding is present in the cornea. (Bottom, left) Light micrograph demonstrating the central corneal epithelium in the rat with the eyelid open. The corneal basal cells are cuboidal or columnar. The corneal epithelium with the eyelids open is more stratified than that with the eyelid closed. (Bottom, right) Immunofluorescence micrograph demonstrating localization of the ROSG antigen in the corneal epithelium. Binding is present in several apical cell layers (all photographs, x 450). (From Watanabe H., et al.;[4] reprinted with permission of Investigative Ophthalmology and Visual Science)

Days 5 to 7, the eyelids are still fused. At the lid margin, the stratification of conjunctival epithelium increased to 9-12 layers, and the binding of the ROSG antibody was observed in apical most layer and layers 2-4 below that layer at the lid margin of the palpebral conjunctiva (Fig. 3), however, scattered cell binding of the antibody was found in the rest of palpebral conjunctiva and bulbar conjunctiva. No binding was found in the cornea.

At days 12-14, in the rats with the eyelids closed, the stratification of corneal epithelium was still 2-4 layers in the central cornea (Fig. 4). The binding of the ROSG antibody to apical and subapical cells was localized at the lid edge in the palpebral conjunctiva (Fig. 5). In the remaining conjunctiva, the binding was still scattered in the apical cells and was not observed in subapical cells. No binding was present in the corneal epithelium (Fig. 4).

At days 12-15, in the rats with the eyelids open, within 24 hours of eyelid opening, the stratification of corneal epithelium was observed to increase to 4-6 layers in the central

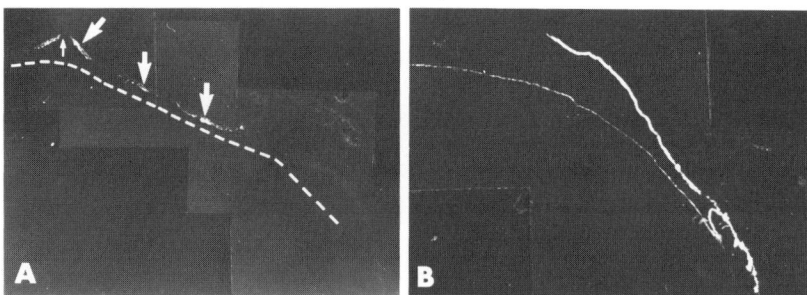

**Figure 5.** Immunofluorescence localization of the ROSG antigen in the ocular surface epithelium of a 12-day-old rat with closed eyelids (A) and with open eyelids (B). (A); Binding is present in the palpebral conjunctiva (shown by large arrows). No binding is present in the cornea (position shown by dotted line). Small arrow indicates eyelid fusion area. (B); Binding is contiguously present all along the ocular surface epithelium. (A,B, x 50) (From Watanabe H., et al.;[4] reprinted with permission of Investigative Ophthalmology and Visual Science)

**Table 2.** Binding of the ROSG antibody to the ocular surface epithelium in artificial, premature eyelid opening.

| Stage | Imunolocalization to Ocular Surface Epithelium | | | | No. of Cell Layers in Corneal Epithelium | |
|---|---|---|---|---|---|---|
| | Conjunctiva | | | Cornea | Range | Mean |
| Day 11 | Lid Margin | Palpebral | Fornix/Bulbar | | | |
| Treated | +++ | +++ | +++ | +++ | 3-5 | 4.0 |
| Control | +++ | + / − | − ~ + / − | − | 2-3 | 2.8 |
| Day 10 | | | | | | |
| Treated | +++ | +++ | +++ | +++ | 3-4 | 3.5 |
| Control | +++ | + / − | − ~ + / − | − | 2-3 | 2.3 |
| Day 9 | | | | | | |
| Treated | +++ | ++ ~ +++ | ++ ~ +++ | ++ ~ +++ | 2-4 | 3.2 |
| Control | +++ | + / − | − ~ + / − | − | 1-2 | 1.9 |
| Day 8 | | | | | | |
| Treated | +++ | ++ ~ +++ | ++ ~ +++ | ++ ~ +++ | 2-4 | 2.9 |
| Control | +++ | + / − | − ~ + / − | − | 1-2 | 1.8 |

n=3, −, no binding. + / −, scattered binding to apical cells only. ++, binding to apical and subapical cells +++, strong binding to apical and subapical cells. (From Watanabe H., et al.;[4] reprinted with permission of Investigative Ophthalmology and Visual Science)

cornea. The basal cells simultaneously changed their shape from flattened or ovoid to cuboidal or columnar (Fig. 4). The binding of the ROSG antibody extended all along the ocular surface epithelium from the lid margin to the cornea (Fig. 5). Moreover, the binding was observed both in apical cells and subapical cells 2 to 4 layers below the apical cell layer (Fig. 4). Binding to some goblet cells of the cluster was also present (data not shown).

Immunoelectron microscopic localization showed binding in the apical glycocalyx region of apical cells, and was particularly prominent at the tips of the microplicae (data not shown). In subapical cells, the binding was also found within small vesicles in the cytoplasm. This binding pattern of the ROSG antibody to the ocular surface epithelium appeared similar to that in the adult rat.[3]

## Artificial Eyelid Opening Experiments

Immunolocalization of the ROSG antigen is shown in Table 2. The binding of the ROSG antibody was found all along the apical and subapical 1-2 cell layers of ocular surface epithelium in the rats treated at days 8-11. In control eyes, eyelids were still fused at the time the animals were sacrificed. The binding pattern of the ROSG antibody to the artificially opened eyes corresponded to the developing rats with eyelids closed. The number of cell layers in corneal epithelium was increased in the treated eyes compared to that in the contralateral eyes in the same rats.

## DISCUSSION

A major observation of this study was that, within 24 hours of eyelid opening, the ROSG antigen appeared all along the ocular surface epithelium. The subsequent artificial eyelid opening studies clearly showed a similar rapid appearance. These two experiments demonstrate that eyelid opening induces the expression of the ROSG antigen in apical cells of the ocular surface epithelium.

The ROSG antigen is observed at the tip of microplicae where the mucous layer is present, and thus appears to be a component of the glycocalyx layer. The glycocalyx layer presumably bind loosely to the mucous layer, which is believed to be very important in maintenance of the tear fluid over the ocular surface epithelium. In regard to the relationship of the mucous layer and the ROSG antigen, the present study is interesting because the expression of the ROSG antigen is coincident with the time when mucous appears on the ocular surface. Hazlett et al. previously showed that, in the developing mouse, mucin is first present at the ocular surface after eyelid opening.[5] Mucin is not present when the eyelids are closed.[5] They also demonstrated that corneal epithelium is resistant to the inoculation of *Pseudomonas aeruginosa* after eyelid opening whereas corneal epithelium was very susceptible to *Pseudomonas aeruginosa* prior to eyelid opening.[9,10] These data may indicate that ROSG antigen is associated with mucous spread and perhaps may have a role in prevention of the entrance or adherence of pathogens.

The study also demonstrated that, at eyelid opening, the corneal epithelium increased in cell stratification and basal cell shape rapidly changed from flattened or ovoid to cuboidal. These observations corroborate previous studies of shape changes of basal cells and stratification of the ocular surface epithelium at the time of eyelid opening.[8,9] Eyelid opening appears to facilitate the proliferation and differentiation of the ocular surface epithelium.

The epitope of the ROSG antibody is to a sugar portion of glycocalyx glycoprotein present on the ocular surface and the present study demonstrates the binding of the ROSG antibody rapidly extends over the ocular surface at eyelid opening. These observations raise two questions. Upon eyelid opening, is the glycoprotein synthesized de novo or does terminal glycosylation of an already existing glycoprotein present in the corneal epithelium occur? Previous reports have shown that both possibilities are possible within 24 hours.[10] Regarding the glycosylation possibility, there have been several reports showing that terminal glycosylation of glycoproteins occurs with development and differentiation.[11-15] To determine which phenomenon occurs at the eyelid opening, the preparation of probes to detect the protein core of the glycoprotein is required.

Another question is what induces the differentiation of the corneal epithelium upon eyelid opening. Previous reports showed that several other changes occur in ocular surface epithelium at this time when the eyelids open. The partial pressure of oxygen increases from 55 mmHg to 155 mmHg,[16] and the activity of enzymes such as lactate dehydrogenase has been shown to increase in accordance with increasing of the partial pressure of oxygen.[16,17] The pH in the cornea changes due to the decreasing of the partial pressure of carbon dioxide.[18] Since pH optima are often critical for the enzyme activity, the pH change upon eyelid opening may activate enzymes involved in glycosylation of glycocalyx glycoproteins in ocular surface epithelium. Blinking stimulation may affect expression because terminal endings of the sensory nerves are in the corneal epithelium. This stimulation may facilitate neural transmitter release which induces the expression of the ROSG antigen. Light stimulation has been shown to affect the mitosis of corneal epithelium.[19] There is thus a possibility that light may affect the differentiation of corneal epithelium. Preliminary experiments showed that the developing rat reared in the dark also expressed the ROSG antigen within 24 hours of eyelid opening. Thus light may not be involved in induction of ROSG antigen expression. To clarify the exact mechanism of induction of expression of the ROSG antigen upon eyelid opening, further investigation is needed.

# REFERENCES

1.  B.A. Nichols, C.R. Dawson, and B. Togni, Surface features of the conjunctiva and cornea. *Invest. Ophthalmol. Vis Sci.* 24:570 (1983).
2.  B.A. Nichols, M.L. Chiappino, and C.R. Dawson. Demonstration of the mucous layer of the tear film by electron microscopy. *Invest Ophthalmol Vis Sci.* 26:464 (1985).
3.  I.K. Gipson, M. Yankauckas, S.J. Spurr-Michaud, A.S. Tisdale, and W. Rinehart. Characteristics of a glycoprotein in the ocular surface glycocalyx. *Invest. Ophthalmol. Vis. Sci.* 33:218 (1992).
4.  H. Watanabe, A.S. Tisdale, and I.K. Gipson. Eyelid opening induces expression of a glycocalyx glycoprotein of rat ocular surface epithelium. *Invest. Ophthalmol. Vis. Sci.* 34:327 (1993).
5.  L.D. Hazlett, B. Spann, P. Wells, and R.S. Berk. Desquamation of the corneal epithelium in the immature mouse: A scanning and transmission microscopy study. *Exp. Eye. Res.* 31:21 (1980).
6.  L.D. Hazlett, D.D. Rosen, and R.S. Berk. Age-related susceptibility to *Pseudomonas aeruginosa* ocular infections in mice. *Infect. Immun.* 20:25 (1978).
7.  L.D. Hazlett, P. Wells, B. Spann, and R.S. Berk. Penetration of the unwounded immature mouse cornea and conjunctiva by *Pseudomonas*: SEM-TEM analysis. *Invest. Ophthalmol. Vis. Sci.* 19:694 (1980).
8.  Y.F. Pei, and J.A.G. Rhodin. Electron microscopic study of the development of the mouse corneal epithelium. *Invest Ophthalmol Vis Sci.* 10:811 (1971).
9.  E-H. Chung, G. Bukusoglu, and J.D. Zieske. Localization of corneal epithelial stem cells in the developing rat. *Invest Ophthalmol Vis Sci.* 33:2199 (1992).
10. R.P. MacDermott, R.M., Jr. Donaldson, and J.S. Trier. Glycoprotein synthesis and secretion by mucosal biopsies of rabbit colon and human rectum. *J. Clin. Invest.* 54:545 (1974).
11. F. Dall'olio, N. Malagolini, G. Di Stefano, M., and F. Seafaring-Cessi. Postnatal development of rat colon epithelial cells is associated with changes in the expression of the $\beta$1,4-N-acetylgalactosaminyl-transferase involved in the synthesis of Sd$^a$ antigen and of $\alpha$2,6-sialytransferase activity towards N-acetyl-lactosamine, *Biochem. J.* 270:519 (1990).
12. J.D. Zieske, and I.A. Bernstein. Modification of cell surface glycoprotein: addition of fucosyl residues during epidermal differentiation. *J. Cell Biol.* 95:626 (1982).
13. E. Dabelsteen, P. Vedtofte, S. Hakomori, and W.W. Young. Carbohydrate chains specific for blood group antigens in differentiation of human oral epithelium. *J. Invest. Dermatol.* 79:3 (1982).
14. E. Dabelsteen, K. Buschard, S. Hakomori, and W.W. Young. Pattern of distribution of blood group antigens on human epidermal cells during maturation. *J. Invest. Dermatol.* 82:13 (1984).
15. J.R. Wilson, D.A. Dworaczyk, and M.M. Weiser. Intestinal epithelial cell differentiation-related changes in glycosyltransferase activities in rats. *Biochem. Biophys. Acta.* 797:369 (1984).
16. J. Friend. Physiology of the cornea: metabolism and biochemistry, In: "The Cornea.," Little Brown and Co., Boston, (1983).
17. G.E. Lowther, and R.M. Hill. Corneal epithelium: recovery from anoxia. *Arch. Opthalmol.* 92:231 (1974).
18. J.A. Bonanno, and K.A. Polse. Measurement of in vivo human corneal stromal pH: open and closed eyes. *Invest. Ophthalmol. Vis. Sci.* 28:522 (1987).
19. S.S. Cardoso, and J.G. Sowell. Control of cell division in the cornea of rats. I. Interaction between isoproterenol and dexamethasone. *Proc. Soc. Exp. Biol. Med.* 147:309 (1974).

# MIXING OF THE TEAR FILM UNDER THE EYELIDS

David M. Maurice

Department of Ophthalmology
School of Medicine
Stanford University
Stanford CA 94306

## INTRODUCTION

Apart from its role in flushing out debris or noxious substances from the conjunctival sac there are two practical reasons why we can be interested in the circulation of the tear fluid over the surface of the conjunctiva beneath the eyelids. The first is that after topical instillation of a drug it could be trapped in a reservoir formed by the fornices of the conjunctiva and this would control its contact time with the eye. The second is that poor mixing of the tear fluid might introduce sampling errors into the estimation of tear flow by the fluorescein dilution method[1].

## EXPERIMENTAL TECHNIQUE

Tear movement around the ocular surface has been identified by observing the redistribution of tracers introduced into the conjunctival sac. Three types of tracer have been used; radioactive, fluorescent or colored, and particulate.

### Scintigraphy

In this technique the two dimensional distribution of a gamma-ray emitting compound is determined by a type of pin-hole camera. It has the advantage, in principle, of being able to view the distribution of material under the lids where it is hidden from visual observation. Such studies were made by Fraunfelder[2], who followed the distribution of a compound of Technetium 99m after applying a 0.5 μl volume of the tracer solution to various points on the conjunctiva. If it was placed under the tarsal plates it generally moved directly to the marginal tear strips. When it was placed deep in a fornix it not infrequently underwent "rivus-like flow" to the lid margins. But occasionally, after application to the fundus of the fornix, it appeared to spread over most of the surface of the globe on the same side. The tracer could not be seen to cross the lid margins into the conjunctival sac, if it was instilled into the interpalpebral space, and Fraunfelder believed that no such movement occurred. He also stated that a depot of the radioactive tracer could remain under the lid for hours suggesting that the tear fluid was virtually stagnant in some regions of the conjunctiva.

Some of these conclusions are questionable because the sensitivity of the scintigraphic method is poor even though a large pin-hole collimator, 3mm diameter, is used. The tear layer is so thin that it is difficult to detect the Technetium that is distributed throughout it, unless doses are instilled so high as to be deemed insanitary by the authorities. Furthermore, the large pinhole results in poor spatial resolution that can lead to misinterpreting the distribution of the tracer material. If it dries on the lid margins, for example, its image may be so smeared as to give the appearance of persisting in the tear fluid[3].

*Lacrimal Gland, Tear Film, and Dry Eye Syndromes*
Edited by D.A. Sullivan, Plenum Press, New York, 1994

Fraunfelder[2] emphasized the effect of gravity on the movement of excess tear fluid, which favors flow to the most dependent point. In this connection made the interesting suggestion that when the lids were closed, freshly secreted fluid could accumulate beneath the upper lid and run from the upper to the lower fornix, presumably around the canthi.

In order to track the flow pathways, which might correspond to furrows in the conjunctiva, for example, it would be desirable to obtain the better resolution provided by a pin-hole 1 mm in diameter, or less. To increase the collection efficiency of the system, a camera on a smaller scale than that available commercially would have to be designed. In addition, a tracer with a half-life shorter than that of Technetium would reduce the radiation falling on the lens and allow a correspondingly greater dosage. These requirements may not be easy to satisfy, and perhaps a different technique, such as MRI, would be more appropriate.

## Fluorometry

Tracers that are fluorescent are more readily quantitated in the tear film than those which are just colored. In my laboratory, tear fluid mixing in human subjects has been studied with fluorescein. Movement in the vertical direction was investigated by placing very small quantities of the solution, less than 0.2 μl, into the fundus of the lower cul-de-sac and noting the time taken for it to appear in the precorneal tear film[4]. This time averaged 3 minutes, but was as long as 9 minutes (Fig. 1). The maximum fluorescence in the film was reached in 8 minutes on the average but was as long as 17 minutes. These times were shortened by rapid blinking.

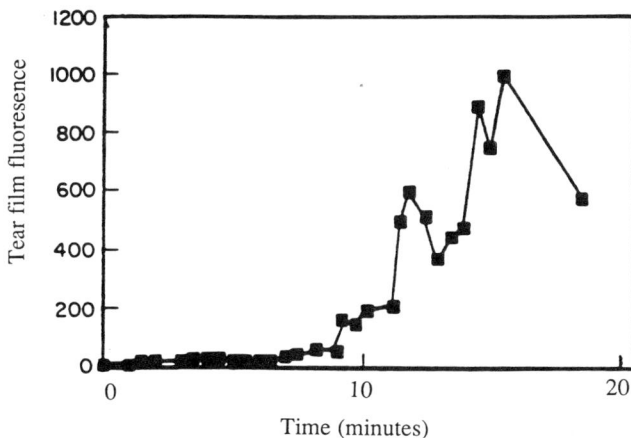

Fig. 1 The appearance of fluorescein in the tear film over the cornea after placement of a small drop in the fundus of the lower cul-de-sac of a volunteer [4]

In more recent experiments the lateral movement of fluorescein was studied[5]. A 1 μl drop of the dye was instilled on either the lateral or medial surface of the bulbar conjunctiva and subsequently the fluorescence of the tear film over the lateral and medial edge of the cornea was followed. Large differences in the fluorescence of the two edges of the cornea were noted to persist for several minutes, showing that mixing was also slow in this direction (Fig. 2). Determination of the tear flow rate by the fluorescein method requires an estimate of the total volume of tears in the conjunctival sac. This is achieved by extrapolating back the curve of tear film fluorescence to zero time. The accuracy of such determinations is severely compromised by the lack of uniformity of the tear film as a result of poor mixing.

264

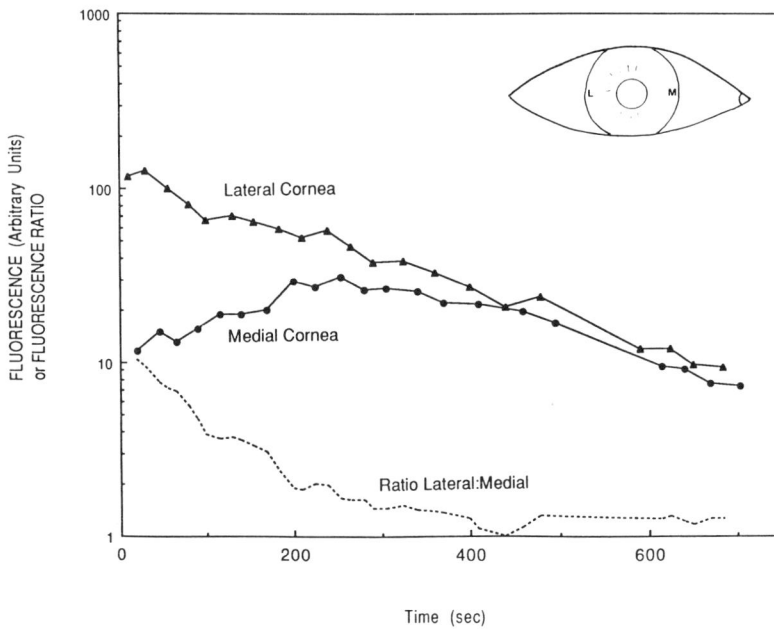

Fig. 2  Fluorescent readings over lateral (L) and medial (M) edges of cornea after instillation of a
1 µl drop of fluorescein laterally at zero time[5]

## Particulate Flow

Particles such as carbon dust are of limited value as tracers but allow the actual flow of
the tears in the lacrimal strips to be discerned under the biomicroscope[6,7]. They show that
the freshly secreted tears that enter the upper lacrimal strip divide into two streams, one
flows round the outer angle to the lower strip and drains out through the lower punctum,
whereas the other moves in the nasal section of the upper strip and drains through the upper
punctum.

## THEORETICAL CONSIDERATIONS

Redistribution of solutes in the tear film can be accomplished by three mechanisms:
diffusion, tear flow, and stirring by movement of the conjunctiva.

## Diffusion

This is a very slow mechanism for the transfer of solutes over the eye. It would take
more than an hour for a substance deep in the conjunctival fornix to spread out over the
corneal surface by diffusional movement alone[4]. On the contrary, distribution is found to
take place in minutes under normal circumstances. Diffusion is not important in general,
therefore, though it could play a role if the eyes were very still, as in sleep. In these
circumstances a drug might be trapped beneath the lids and spread slowly over the surface
of the eyeball. Even then, absorption across the conjunctiva is likely to be so fast that most
drugs would be lost to the bloodstream before they would spread a significant distance.

## Tear Flow

There is no direct evidence of how the basic tear secretion flows beneath the lids.
With the head erect, the freshly formed fluid that passes through the tear ducts could be
expected to drop down to the interpalpebral space, to circulate around the lacrimal strips,
and flow out through the puncta, as made evident by scintigraphy and particulate tracers. In
this case the flow of fluid would not flush out the lower fornix nor the fundus of the upper
fornix.

When the eyes are closed, Fraunfelder's observations suggest that the fresh tear fluid can accumulate under the upper lid and then flow down beneath the lower lid, to dilate the conjunctival sac. The fluid would be expelled by the next blink. This is probably the mechanism that flushes a noxious stimulus away during actual crying. Whether it happens to any extent during normal blinking is conjectural.

## Conjunctival Stirring

Sliding of the conjunctival surfaces over one another could be a potent cause of stirring of the tear fluid under the lids and is probably the major cause of the mixing of solutes dissolved in it. Such sliding can result both from blinks and rotation of the eye.

I am not aware of any discussion in the literature as to how the conjunctiva responds to rotation of the globe. For example, as the eyeball turns up, the conjunctiva could roll around the fold in the fundus of the lower fornix without changing its length (Fig. 3B). Alternatively, the lower bulbar conjunctiva could stretch with no shift of the fundal point (Fig. 3C). In the latter case the flattening of the furrows in the tissue might constitute a pump mechanism for circulating fluid.

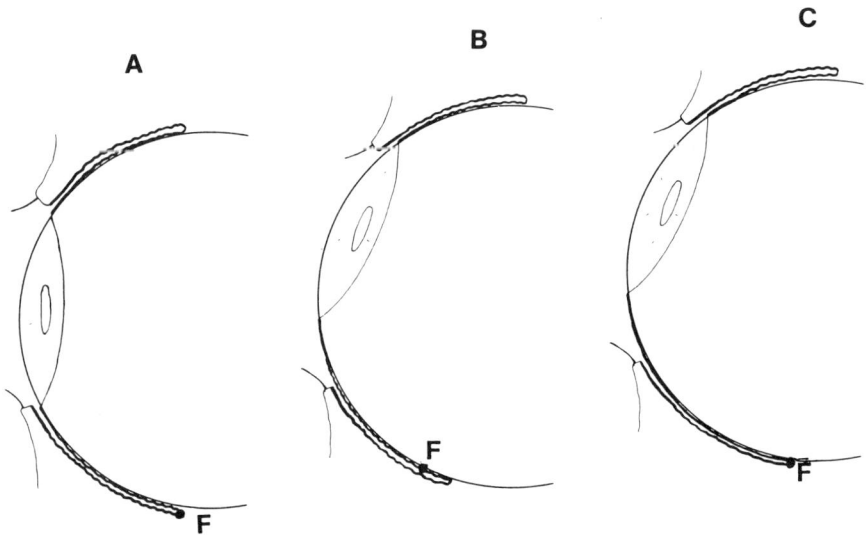

Fig. 3   Possible changes in conformation of the conjunctiva when the eye rolls up. A, Normal configuration. B, Eye turns up, conjunctiva rolls around so that the original fundus F moves onto globe. C, Eye turns up, and bulbar conjunctiva stretches

## References

1. S. Mishima, A. Gasset, S. Klyce, and J. Baum, Determination of tear volume and tear flow, *Invest. Ophthalmol.* 5:264 (1966).
2. F.T. Fraunfelder, Extraocular fluid dynamics: hour best to apply topical ocular medication, *Tr. Amer. Ophthalmol. Soc.* 74:457 (1976).
3. D.M. Maurice and S.P. Srinivas, The use of fluorometry in assessing the efficacy of cation sensitive gel as an ophthalmic vehicle: comparison with scintigraphy. J. Pharm. Sci., 81:615 (1992).
4. E. A. Macdonald, and D.M. Maurice, The kinetics of tear fluid under the lower lid, *Exp. Eye Res.* 53:421
5. N.M. Sang, and D.M. Maurice, Poor mixing of microdrops with the tear fluid reduces the accuracy of tear flow estimates by fluorometry, Submitted, *Exp. Eye Res.* (1993).
6. D.M. Maurice, The dynamics and drainage of tears. *Int. Ophthalmol. Clin.* 13:103 (1973).
7. M.G. Doane, Tear spreading, turnover and drainage, in: "The Preocular Tear Film in Health, Disease and Contact Lens Wear," F.J. Holly, ed., Dry Eye Institute, Lubock (1972).

# VISCOELASTIC PROPERTIES OF HUMAN TEARS AND POLYMER SOLUTIONS

John M. Tiffany

Nuffield Laboratory of Ophthalmology
University of Oxford
Oxford OX2 6AW, U.K.

## INTRODUCTION

There is a need for artificial tears to replace a reduced or inadequate supply of natural tears. These will be more effective if they mimic as closely as possible the desirable characteristics of normal tears. The flow or rheological properties are among the most important of these characteristics, and depend on shearing forces generated by movements of globe or lids, and the rates and amplitudes of their application.

Two distinct rheological regimes can be recognised in the eye. One is a low-shear static condition in the open eye between blinks. The problem is mainly one of resisting drainage of the precorneal film which would lead to film thinning and break-up. A high viscosity will aid this. The other regime is a high-shear condition during blinking. The combination of high lid speed and small film thickness gives a high velocity gradient and a shear rate up to 20,000 sec$^{-1}$ (Dudinski et al, 1983; Bron & Tiffany 1991). Combination of high shear rates with high viscosity will cause unacceptable dragging forces on the epithelial surfaces (Hammer and Burch, 1984), so low viscosity is desirable in this region.

Dilute polymer solutions such as natural or artificial tears can be divided into two types: (1) **Newtonian**, where the coefficient of viscosity is independent of shear rate, or (2) **non-Newtonian**, (usually shear-thinning), where viscosity falls as the rate of shear increases. Natural tears are of non-Newtonian type (Tiffany, 1991); so also are solutions of sodium hyaluronate of high molecular weigh, although many polymers currently used in commercial artificial tears are Newtonian (Bothner et al., 1990). The formulation of Newtonian eyedrops is an uneasy compromise between having a viscosity high enough to resist draining in the open eye, but low enough to avoid sensations of drag and surface damage during blinking.

At present very little is known of the rheology of tears; only two studies report their non-Newtonian shear-thinning nature (Hamano and Mitsunaga, 1973 on rabbit, and Tiffany, 1990, 1991 on human). Artificial tears have been compared with each other (Dudinski et al, 1983; Bothner et al., 1990) but not with natural tears. Until now, no attempt has been made to investigate the possible elastic as well as the viscous nature of tears. Many fluids are found to have elastic as well as viscous properties.

This paper describes preliminary attempts to measure the viscoelasticity of human tears and of a number of other solutions of macromolecules chosen to indicate the effect of different degrees of solution complexity on the viscoelastic behaviour.

*Lacrimal Gland, Tear Film, and Dry Eye Syndromes*
Edited by D.A. Sullivan, Plenum Press, New York, 1994

## MATERIALS AND METHODS

Samples investigated were: (1) normal human pooled tears collected by glass capillary tubes with minimal stimulation; (2) human mucoid saliva diluted 1:1 with distilled water, as an example of a complex biological mixture of mucins and proteins; (3) purified pig gastric mucin dissolved in distilled water, 30 mg/ml and 40 mg/ml, as as an example of mucin without additional complexing materials; (4) sodium hyaluronate (MW = $2.5 \times 10^6$), 2 mg/ml in distilled water, as a defined polyanionic polymer; (5) polyethylene glycol, PEG, (MW = $5 \times 10^6$) 1 mg/ml in distilled water, as a defined nonionic polymer.

### Rheology

Variable-shear viscosity results reported for tears (Hamano and Mitsunaga, 1973; Tiffany, 1990, 1991), were obtained from rotational rheometry, which measures total torque as a function of rotational speed. Using instrument constants, this is converted into an apparent viscosity. The results can be interpreted in terms of a tangled meshwork of linear polymer molecules which tend to be pulled apart by shearing, but which can re-entangle at low speeds (resulting in high viscosity); at high speeds the speed of separation of molecules prevents re-entanglement (resulting in low viscosity).

During viscous flow, energy is applied to the system to cause motion of the molecules, but is entirely taken up in overcoming frictional forces, and is dissipated as heat. This dissipation rapidly reduces flow as the shearing force is removed, so viscosity is determined by the **rate of shear**. A purely elastic body is deformed by shearing, storing energy in the process proportional to the **shear**, and this energy can be recovered on removing the force. In fact, the tangled meshwork of polymer molecules also possesses elasticity as well as viscosity, since energy input can be absorbed by distorting the meshwork in a recoverable manner as well as causing some relative movement of the molecules or viscous flow.

Separation of the viscous and elastic components is most easily achieved by forced sinusoidal oscillation of the sample at a frequency $\omega$ (= $2\pi/f$). Then for a shear rate of amplitude $A_t = A_o \sin \omega t$, the amplitude of the shear stress is given by $B_t = B_o \cos (\omega t + \delta)$. $\delta$ is the phase angle and lies between $0°$ (purely elastic) and $90°$ (purely viscous). Hence $\delta$ can be found from the phase difference between total shear stress and rate of shear. The dynamic viscosity modulus $\eta'$ and the dynamic rigidity or elasticity modulus $G'$ are then calculated from $\omega$, $\delta$ and instrument constants. The ratio $\eta'/G'$ is defined as the relaxation time $\lambda$; at oscillation times longer than this (i.e. low frequencies) the viscous element has time to flow during the cycle, and applied energy will tend to be dissipated, and the material will show largely viscous behaviour. At higher frequencies ($\omega > 1/\lambda$), little energy can be dissipated before the cycle is reversed, so elastic behaviour predominates.

### Rheometer

The Contraves Low-Shear 30 rheometer used is of the concentric-cylinder Couette type in which a fluid sample (in this instrument as little as 70µl) fills the annular space between the moving cup and the bob which detects torque. In rotational mode, the cup revolves at a programmed speed and a servo system records the counter-torque needed to maintain the position of the bob. In oscillation mode, a system of cams converts rotational motion of the cup to to sinusoidal oscillatory motion of $\pm 30°$. The output of the servo system and the rate of shear are plotted on the X- and Y-axes respectively of an X-Y recorder, giving an inclined elliptical Lissajou figure whose shape and inclination depend on $\delta$. Not all speeds were in fact possible with all samples, as turbulence and distortion of the ellipse may occur at frequencies close to the natural frequency of the bob and torsion wire. All readings were at room temperature.

## RESULTS

For each value of frequency $\omega$ in the range 0.14-5.55 rad.sec$^{-1}$, families of ellipses were obtained for each sample, from which the quantities $\delta$, $\eta'$ and $G'$ were determined, and relaxation times $\lambda = \eta'/G'$ calculated. As expected from previous rotational studies on human tears where marked shear-thinning was found (Tiffany 1991), the viscosity modulus $\eta'$ is high at low frequencies (at which only low rates of shear are experienced) but falls until, at high frequencies and higher values of maximum shear rate, a significantly lower viscosity is found. The elastic coefficient $G'$ was expected to rise with frequency, but in fact fell slightly for tears, indicating that energy storage is significant in the low-shear region even where energy-dispersive viscous flow takes place. The relaxation time $\lambda$ falls as frequency increases, but appears substantially constant at the upper end of the frequency range. Table 1 gives values for the three variables at low ($\omega = 0.14$ rad.sec$^{-1}$) and high ($\omega = 1.63$ rad.sec$^{-1}$) frequencies, for all samples.

**Table1.** Rheological data for human tears and model polymer solutions

| SAMPLE | LOW FREQUENCY | | | HIGH FREQUENCY | | |
|---|---|---|---|---|---|---|
| | $\eta'$ mPa.s | $G'$ mPa | $\lambda$ s | $\eta'$ mPa.s | $G'$ mPa | $\lambda$ s |
| Tears | 65.5 | 47.7 | 1.37 | 10.1 | 34.9 | 0.29 |
| Saliva | 7.9 | 18.4 | 0.43 | 7.7 | 22.4 | 0.35 |
| Hyaluronate* | 332 | 239 | 1.39 | 310 | 239 | 1.30 |
| Mucin 30mg/ml | 48.6 | 15.9 | 3.05 | 13.9 | 31.1 | 0.45 |
| Mucin 40mg/ml | 20.3 | 72.4 | 0.28 | 16.0 | 47.7 | 0.34 |
| PEG | 0.3 | --- | --- | 0.3 | --- | --- |

* $\omega$ range 0.14-0.88 rad.sec$^{-1}$ only

## CONCLUSIONS

Not all results so far have proved repeatable for some samples, so these figures must be treated with caution. However, certain general conclusions can be drawn. The PEG viscosity was very low, with little evidence of shear-thinning, and its elastic constants unmeasurable, despite a higher MW and comparable concentration to hyaluronate; charged groups seem to be essential to swelling and interaction of polymers in solution. Hyaluronate seems relatively insensitive to frequency over the range available, possibly due to the "make and break" type of adhesion thought to occur between these molecules in solution (Scott et al., 1991). Saliva, although similar in many respects to tears (especially in mucin content), shows a much lower frequency dependence of viscosity, but its elastic modulus rises steadily. Results of concentration differences for gastric mucin are equivocal: low-frequency viscosity is found to be higher for 30 mg/ml than 40 mg/ml, but this may be artifactual.

Natural tears contain mucus glycoproteins (mucins). Model mucin solutions are non-Newtonian, and comparison of their rheological properties with those of proteins suggests that, by themselves, tear proteins would contribute comparatively little to the overall apparent viscosity of tears (Kaura & Tiffany, 1986). Hence the component of tears which is principally responsible for the observed shear-thinning behaviour is the very large linear mucus polymer molecules (either alone or perhaps complexed with tear proteins). However, some differences have been found between different batches of pooled tears, not to be explained by variations of "normal" and "dry" tear input.

There are clearly practical difficulties in collecting pooled samples of even 70-100 μl, if the contribution from the individual donor is to be genuinely unstimulated, as it must be if the mucin content largely determines rheology, and especially on samples from a single patient. Thus comparison of normal and dry-eye tears remains a relatively distant prospect, in view of the large number of dry-eye donors necessary and the finding that rheological properties are altered by refrigerated storage during pooling.

However, it can certainly be concluded that there is a substantial contribution from the elastic component in normal human tears. This implies a cushioning effect of the globe by the bathing film of tears during relatively rapid eye movements such as micro-saccades and fine tremor, in which energy can be stored elastically and recovered or dispersed as the eye slows down; thus the eye can make rapid movements of small amplitude without the need to induce extensive redistribution of tear fluid by movement. Blinking, as a more prolonged movement, will tend to allow continuous energy dissipation through viscous flow. Measurements such as those reported here on natural or synthetic polymers are potentially of considerable value in helping to design and formulate artificial tear drops to replace or supplement an inadequate supply. Much more work still needs to be done, both in achieving reliability in the measurements, and in studying a wide variety of effects due to combinations of polymers, influence of molecular weights and concentrations, etc.

## Acknowledgements

Thanks are due to The Wellcome Trust and The Royal National Institute for the Blind, London, for equipment grants, and to Elaine Lovelady for technical assistance.

## References

Bothner, H., Waaler, T. and Wik, O. (1990). Rheological characterization of tear substitutes. Drug. Dev. Ind. Pharm. 16: 755-768.

Bron, A.J. and Tiffany, J.M. (1991). Pseudoplastic materials as tear substitutes. An exercise in design. In: The Lacrimal System (van Bijsterveld, O.P., Lemp, M.A. and Spinelli, D., eds.). Kugler and Ghedini, Amsterdam, pp. 27-33.

Dudinski, O., Finnin, B.C. and Reed, B.L. (1983). Acceptability of thickened eye drops to human subjects. Curr. Therap. Res. 33: 322-337.

Hamano, H. and Mitsunaga, S. (1973). Viscosity of rabbit tears. Jap. J. Opthalmol. 17: 290-299.

Hammer, M.E. and Burch, T.G. (1984). Viscous corneal protection by sodium hyaluronate, chondroitin sulfate, and methylcellulose. Invest. Ophthamol. Vis. Sci. 25: 1329-1332.

Kaura, R. and Tiffany, J.M. (1986). The role of mucous glycoproteins in the tear film. In The Preocular Tear Film (Holly, F.J., ed.). Dry Eye Institute, Lubbock, Texas, pp. 728-732.

Scott, J.E., Cummings, C., Brass, A. and Chen, Y. (1991). Secondary and tertiary structures of hyaluronan in aqueous solution, investigated by rotary shadowing-electron microscopy and computer simulation. Biochem. J. 274: 699-705.

Tiffany, J.M. (1991). The viscosity of human tears. Intl. Ophthalmol. 16: 371-376.

Tiffany, J.M. (1990). Rheology of tears and tear substitutes. In 4th Workshop on External Eye Diseases and Inflammation. Fisons Pharmaceuticals, Loughborough, UK, pp. 33-38.

# EFFECT OF AGE ON HUMAN TEAR FILM EVAPORATION IN NORMALS

Alan Tomlinson* and Careen Giesbrecht**

*Department of Vision Sciences, Glasgow Caledonian University, Cowcaddens Road, Glasgow G4 0BA, United Kingdom. **Southern California College of Optometry, 2575 Yorba Linda Boulevard, Fullerton, California 92631, USA

## INTRODUCTION

The evaporation of fluid from the tear film accounts for between 10 and 40% of the elimination of tears from the human eye[1,2]. Attempts have been made to measure the rate of evaporation using different techniques. More than 30 years ago Mishima and Maurice[3] used assessments of tear film thickness as an indicator of the rate of fluid loss. A vapor pressure gradient technique was originally used by Hamano et al[4] and later modified to be less invasive by Tomlinson and Trees[5]. Different methods for measuring the relative humidity within a goggle-cup over the eye have been employed by Tomlinson and Cedarstaff[6], Rolando and Refojo[7] and Tsubota and Yamada[8]. The differences in technique account for some of the variability observed in the determination of human tear film evaporation, values between 4.07 and 67.1 having been recorded for normals.[4,5,7,8]

Human tear evaporation rate has been shown to be affected by the integrity of the lipid layer[3,9], contact lens wear[4,6,10], and the instillation of various fluids into the eye[7,11,12]. The absence of an intact lipid layer increases evaporation by up to 20 times[3] and a thicker and more stable lipid layer in the early morning dramatically reduces tear evaporation[13]. A high evaporation rate is an important factor in the aetiology of keratoconjunctivitis sicca and has been described by Rolando and Refojo[14].

Age related change in tear film evaporation rate was thought worthy of study as it could be a factor in the condition of dry eye which occurs with increasing frequency in older patients. Increased tear film evaporation in an eye which is suffering from a decrease in reflex tear secretion capacity[15,16,17] is likely to be a significant problem. In an attempt to define the effect of the ageing process on tear film evaporation, a study was undertaken to determine if a relationship exists between evaporation rate and age in **normal** eyes.

## SUBJECTS

Forty-seven normal subjects were selected for this study; of these 21 were male and 26 female (age range 7-92 years). All subjects were free of dry eye symptoms, anterior segment pathology and had no contact lens wear for at least 3 months and were not using ocular or systemic medication likely to affect the tear film. The absence of dry eye was defined by the use of a symptomatology directed questionnaire[18].

## MEASUREMENT OF TEAR FILM EVAPORATION RATE

Tear film evaporation rate measurements were made in this study with the Servomed Evaporimeter by the procedue previously described[5]. Some variability exists in this as in all techniques for measurement of tear film evaporation. These variations can include differences due to the depth of the eyeball within the orbital structure of the face. This can lead to variations in the distance from the anterior ocular surface to the sensor in instruments such as the Evaporimeter, because differences in distance from the sensor can affect readings[19].

This technique may also be affected by drafts if they intrude into the goggle area. Avoidance of drafts is achieved for most subjects by producing a contact seal between the orbital region and the cup of the goggle. It is possible with some individuals that a complete seal was not possible, although every attempt was made to avoid this by eliminating such subjects from the study.

In this technique, as in techniques used by others, it is necessary in order to obtain the evaporation from the ocular surface to either eliminate the contribution from the skin of the face and lids within the goggle area by the use of barrier creams, or to factor out this contribution from a knowledge of skin area and skin evaporation rate. Attempts were made to eliminate skin evaporation by the use of various barrier creams, however an effective prevention of more than 70% of skin evaporation was not achieved by any of the creams tried in preliminary experiments. Therefore skin evaporation was factored out by calculation of skin area and skin evaporation rate as described before[5]. However this required an assumption that the skin area was represented by the area obtained from a flat photograph which may not be accurate in all cases, particularly those in which the globe was steeply curved and in which the orbital contours showed a well rounded form. Others have described[20] a method for calculating the "true" area of the ocular surface from the use of standardised measurements for the cornea and sclera but this technique still leaves assumptions about orbital contour form. It is possible that the assumptions made in the technique described in this experiment may well have contributed random variation between individuals which was independent of age. Certainly, over the number of subjects as used in this study, variations due to orbital configuration are unlikely to have systematically affected the results.

The ideal evaporation measurement technique would be totally non-invasive. The techniques used over the past twelve years have varied enormously in this regard. The technique of Hamano et al.[4] required disruption of the tear film by placing a probe on the anterior ocular surface. This technique prohibited normal blinking to restore the tear film as do others.[7] The techniques of Rolando and Refojo[7] and Tomlinson and Cederstaff[6] required eventual or continuous air flow across the ocular surface with the potential for increased evaporation. The least invasive techniques are those of Tomlinson and Trees[5] and Tsubota and Yamada[8] but even these can suffer from the problem of condensation within the goggle (due in part to the absence of 'forced' air flow) and cannot be used for a long series of measurements.

As a result of these important differences in techniques of measurement, comparison of absolute values for human tear evaporation found by different investigators is difficult. Therefore, comparisons should only be made between values obtained on the same instrument and ideally should be restricted to measurements of **change** in evaporation rate in the same individual as a consequence of variations in experimental conditions (e.g. drop instillation, contact lens wear, etc.).

## RESULTS AND CONCLUSION

Tear evaporation rates for all subjects are shown in Figure 1. No correlation was found between the evaporation rate of the tear film and age ($R^2 = 0.002$, for all subjects; $R^2 = 0.06$ for females; $R^2 = 0.14$ for males). The overall mean tear evaporation rates were $44.0 \pm 24.6 g/m^2/hr$ for females and $29.5 \pm 21.2 g/m^2/hr$ for males; a significant difference ($t = 2.2$, $p < 0.05$). However it can be seen from Figure 2 that this apparent gender difference is due to significantly lower evaporation rates in the male subjects over 40 years ($t = 2.95$ $p < 0.01$ vs younger males; $t = 3.52$ $p < 0.002$ vs older females); females did not show significant differences in the two age categories in Figure 2 ($t = 0.92$ $p > 0.05$).

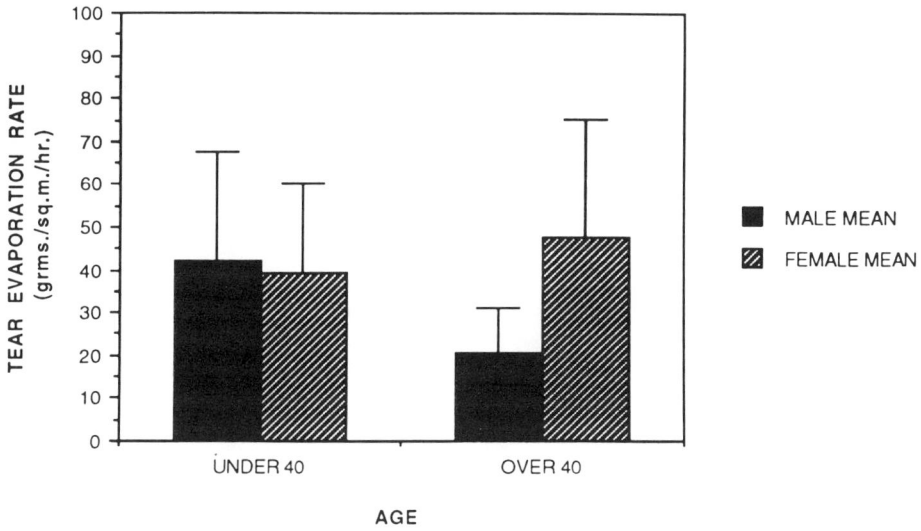

**Figure 1.** The tear evaporation rates for male and female subjects in the age range 7-92 years are shown. The equation for the regression line of evaporation rate against age is given and the line included on the scatter plot. No significant correlation between age and tear evaporation rate was found. (Reproduced with permission from British Contact Lens Association Journal,[23]).

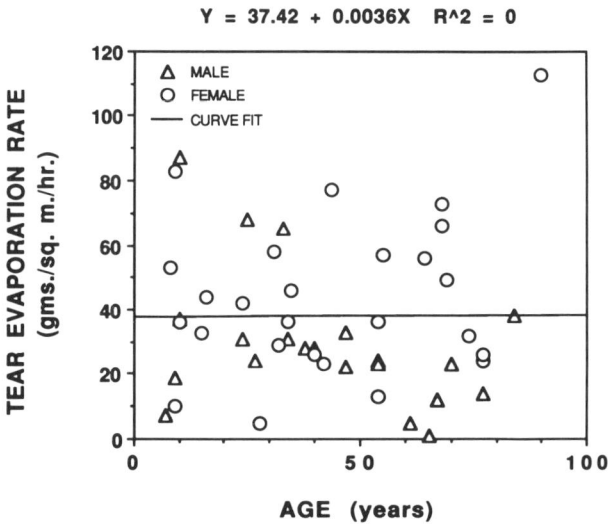

**Figure 2.** The frequency histogram of tear evaporation rates for male and female subjects divided into groups of those under and over 40 years of age. Standard error bars are shown on this plot. A significantly reduced evaporation rate is seen in males over 40 years of age.

The overall results indicate an apparent gender difference in favour of a higher evaporation rates in females. Initially it was thought that this may be a factor in the greater predisposition to dry eye problems in females as they grow older[21]. However the gender difference is a result of a reduced tear evaporation in males over 40 years. This may be a result of the small sample size in this subset of subjects as no explanation on the basis of tear physiology is offered. Our main finding of an overall lack of correlation between age and tear evaporation rate is in agreement with that of Rolando and Refojo[7].

It may have been expected that tear evaporation rate would have increased with age as it is primarily controlled by the lipid layer of the tear film[3,9], and lipid production reduces with age. However, lipid layer thickness appears to be constant for different age groups[22]. Therefore, it appears that a fortuitous compensatory mechanism may exist which combats the potential for dry eyes in the ageing normal eye. Such a mechanism would require reduced tear elimination (drainage from the eye) to compensate for the reduced reflex tear secretion and lipid layer production found in the older **normal** eye[23].

## REFERENCES

1. W. Herold, Die verdunstungsrate der tranenfussigkeit beim menschen verglichen mit einem physiokalischen nmodell, Klin Monatsbl Augenheilkd 190: 176 (1987)
2. B. Milder, The lacrimal apparatus, in: Moses R. A. and Hart Wm., (Eds) "Adlers Physiology of the Eye", 8th. Edition, CV Mosby Co, St Louis (1987)
3. S. Mishima and D. M. Maurice, The oily layer of the tear film and evaporation from the corneal surface, Exp Eye Res 1: 39 (1961)
4. H. Hamano, M. Hori and S. Mitsunaga, Measurement of evaporation rate of water from the pre-corneal tear film and contact lens, Contacto 25(?): 7 (1981)
5. A. Tomlinson and G. Trees, The effect of artificial tear solutions and saline on tear film evaporation, Optom. Vis. Sci. 67: 886 (1990)
6. A. Tomlinson and T.H. Cederstaff, Tear evaporation from the human eye: the effects of contact lens wear, J. Bri. Contact Lens Assoc. 5: 141 (1982)
7. M. Rolando and M.F. Refojo, Tear evaporimeter for measuring water evaporation rate from the tear film under controlled conditions in humans, Exp. Eye Res. 36: 25 (1983)
8. K. Tsubota and M. Yamada, Tear evaporation and the ocular surface, Invest. Ophthalmol. Vis. Sci .33: 2942 (1992)
9. S. Iwata, M.A. Lemp, F.J. Holly and C.H. Dohlman , Evaporation rate of water from the pre-corneal tear film and cornea in the rabbit, Invest. Ophthalmol. Vis. Sci. 8: 613 (1969)
10. T.H. Cedarstaff andA. Tomlinson, A comparative study of tear evaporation rates and water content of soft contact lenses, Am. J. Optom. Physiol. Opt. 60: 167 (1983)
11. T.H. Cedarstaff andA. Tomlinson , Human tear volume, quality and evaporation: a comparison of Schirmer, tear breakup time and resistance hygrometry techniques, Ophthal. Physiol. Opt. 3: 239 (1983)
12. A. Tomlinson and G. Trees, The effect of preservatives in artificial tear solutions on tear film evaporation, Ophthal. Physiol. Opt. 11: 48 (1991)
13. A. Tomlinson and T.H. Cedarstaff, Diurnal variation in human tear evaporation, J. Bri. Contact Lens Assoc. 15: 77 (1992)
14. M. Rolando and M.F. Refojo, Increased tear evaporation in eyes with kerotoconjunctivitis sicca, Arch. Ophthalmol. 101: 557 (1983)
15. M. Norn, Tear secretion in normal eyes, Acta. Ophthalmol. 43: 567 (1965)
16. T. Hamano, S. Mitsunaga, S. Kotani, T. Hamano, K. Hamano, H. Hamano, R. Sakamoto and H. Tamura, Tear volume in relation to contact lens wear and age, CLAO J. 16:57 (1990)
17. R. Furukawa and K. Polse, Changes in tear flow accompanying age, Am. J. Optom. and Phys. Opt. 55: 69 (1978)
18. C. McMonnies and A. Ho, Responses to the dry eye questionnaire from the normal population, J. Am. Optom .Assoc. 58:296 (1987)
19. G.E. Nilsson, Measurement of water exchange through skin, Med. Biol. Eng. Comput. 15: 209 (1977)
20. J. Tiffany, Personal Communication (1992)
21. M.A. Lemp, Diagnosis and treatment of tear deficiencies, in: T.D. Duane (Ed) "Clinical Ophthalmology". (1980)
22. M.S. Norn, Semi-quantitative interference study of fatty layer of pre-corneal film, Acta Ophthalmol 57: 766 (1979)
23. A. Tomlinson and C. Giesbrecht, The ageing tear film, J. Bri. Contact Lens Assoc. 16: 67 (1993)

# ADVANCES IN OCULAR TRIBOLOGY

Frank J. Holly and Thomas F. Holly

Dry Eye Institute
P.O. Box 98069
Lubbock, TX 79499

## INTRODUCTION

Tribology is the science and technology concerned with interacting surfaces in relative motion to each other.[1]  The principles of tribology are also interconnected with the principles of rheology.[2]  Rheology is the science of shape changes in solids and flow in liquids resulting from external forces.  Lubrication and abhesion (antonym of adhesion), as well as friction and adhesion also have surface chemical implications even though those have remained mostly unexplored.[3,4]

We measured the viscosity of whole tears in humans and rabbits and aqueous mucin solutions under uniform shear rate conditions and investigated the effect of shear rate, temperature, and ionic strength on dynamic viscosity.  Using the principles of tribology,[5] we then evaluated the role tears play in lid lubrication in normal eyes and diseased eyes and speculate on the lubrication mechanism operative in blinking.

## BASIC ASPECTS OF TRIBOLOGY

When two solids in contact are moved relative to one another, additional energy has to be expended due to friction (drag) and damage to the surfaces (wear) may occur.[6]  Friction is due to the attractive forces between the molecules of two solid surfaces when they slide over one another.[7]  The fundamental law of friction, the Amontons' law, states that frictional drag is directly proportional to the load.  Hence, the frictional coefficient does not depend on the apparent area of contact or the sliding speed.[8]  This is quite contrary to intuitive thinking,[4] but the law usually holds in the absence of lubrication.

When a fluid layer having low internal friction (viscosity) is placed between such solid surfaces, the situation changes quite drastically.  The frictional coefficient becomes much smaller and becomes a function of the viscosity of the lubricant, the sliding speed, and the load. The fluid comprising the layer is called a lubricant and the process is known as lubrication.

There are two basic types of lubrication operative in sliding friction, namely boundary lubrication and hydrodynamic lubrication.[9] Boundary lubrication occurs when the thickness of the lubricant layer is decreased to the scale of the surface roughness, usually at low gliding speed, high load, or when the lubricant film becomes surface-chemically unstable. The friction in boundary lubrication would depend on the roughness and chemistry of the solid surfaces in contact[10] and would not depend on the viscosity of the lubricant. The magnitude of the frictional coefficient would be between 0.040 and 0.300.[9]

If the thickness of the lubricating layer is greater than the scale of surface roughness, then the lubrication mechanism is hydrodynamic.[9] Complete wetting of the surfaces by the lubricant is a must in achieving hydrodynamic lubrication. Hydrodynamic lubrication diminishes the frictional resistance considerably and also protects the solid surfaces. Friction is then directly proportional to viscosity and also to the velocity of sliding provided that viscosity of the lubricant remains constant. A magnitude of a typical coefficient of friction in hydrodynamic lubrication would be about 0.005.[11]

**Rheology of Lubricants**

In 1668 Sir Isaac Newton stated that "the internal friction of fluid (viscosity) is constant with respect to the rate of shear," thereby establishing the fundamental relation of fluid rheology governing viscous flow.

The force acting on a unit area of a liquid is called the shear force. The adjacent fluid layers move at different velocities, and the difference in these velocities will be greater the lower the internal friction of the fluid. The velocity change per unit distance perpendicular to the direction of motion is called the rate of shear. Newton's law defines viscosity as the ratio of the shear force and the shear rate. The dimension of viscosity can be expressed as the product of pressure and time; $(dyne.sec)/cm^2 = g/(cm.sec) = poise$. The viscosity of water at room temperature is approximately 1 centipoise (cps). Table I contains the viscosities of some common lubricants and biological fluids at room temperature.

**Table 1.** Viscosity of Various Materials at Room Temperature

| Various Materials | Viscosity in cps at 20°C | Various Materials | Viscosity in cps at °C |
|---|---|---|---|
| Synovial fluid* | 1,000-10,000 | Synovial fluid** | 2 - 5 |
| Honey | 2,000 | Human Tears | 1.3 |
| SAE 50 oil | 800 | Saliva | 1.0 - 1.2 |
| Glycerol | 500 | Aqueous Humor | 1.02 - 1.10 |
| SAE 30 oil | 300 | Water | 1.00 |
| Ejaculum | 14.1± 8.1 | Diethyl ether | 0.2 |

\* determined at low shear rate      \*\* determined at high shear rate

The majority of fluids exhibits a non-Newtonian behavior. These latter fluids can be classified according to the type of shear-rate dependence of their viscosity. Polymeric systems, especially at high polymer concentrations, often exhibit viscoelastic behavior which combines liquid-like and solid-like characteristics. When such bodies are subjected to sinusoidally fluctuating stress (shear force) the strain (shear rate) is neither in phase with the stress as it would be with a perfectly elastic body nor 90° out of phase as it would be for a perfectly viscous liquid but somewhere between these two extremes. We found no evidence of pronounced viscoelastic behavior for tears and tear components. Others, however, have attempted the use of such techniques.[12]

## RHEOLOGY OF THE PREOCULAR TEAR FILM

The tear film in the open eye consists of two fluid layers,[13] the superficial lipid layer of about 0.1 micrometer thickness and high viscosity and the aqueous tear layer of about 10 micrometer thickness and low viscosity.[13-15] The preocular tear film rests on a thick, viscous mucus layer covering the epithelial surface. The epithelial cell surface in the normal eye has a surface roughness of no more than 0.5 micrometers consisting of microridges.[16] In the normal eye the microridges are mostly covered with a mucus coacervate[17] or a mucus gel[18] further decreasing surface roughness.[19]

The apparent rigidity of the tear film[20] and its resistance to gravity-induced flow led some lacrimologists to believe that this behavior is indicative of high viscosity or even gel formation. Actually increased viscosity would only slow down fluid flow and gel formation would interfere with blinking. The real reason behind the apparent rigidity of the tear film is its thinness. Langmuir has estimated[21] that the behavior of fluid films is governed by surface forces, which overwhelm gravitational effects, once the film thickness becomes less than 100 micrometers.[22]

During the closure of the eye, the upper eye lid moves downward with an approximate speed of 15 cm/sec.[23] At the same time, the superficial lipid layer is compressed by the moving eye lid so that the lubricating layer between the lid and the eye consists of aqueous tear layer bounded on both sides by a mucus layer of high viscosity.[22]

### Rheological Properties of Aqueous Tears

The viscosity of human aqueous tears apparently has not been well established. In the Geigy Scientific Tables of Body Fluids,[24] the range of the *relative viscosity* for the tears is given as 1.26 - 1.32 from the reference of Liotet and Cochet.[25] If we assume that the determination was conducted at 20 °C (neither the method nor the temperature is stated) where the viscosity of water is 1.002 cps, the range numerically would be practically the same in *absolute viscosity* units. Other French authors determined the value to be in the same range (1.3 cps), about 30% above that of pure water, despite its significant content of mucous glycoprotein.[26,27] Kaura and Tiffany[28] published the average value of 14 determinations in terms of *specific viscosity* to be 1.82 ±0.23, which, assuming 25 °C as the temperature of determination (not stated), converts into 2.51 ±0.26 cps. It is important to note that their value was obtained at very low shear rates (magnitude not given) and while non-Newtonian behavior was claimed, supporting data were not given. In the few articles published on the topic of ocular tribology, the tears are thought to be pseudoplastic due to their mucin content;[11,28-30] but this claim is not supported by data obtained under uniform shear rate conditions.

The viscosity of rabbit tears was more thoroughly investigated mostly by Japanese researchers.[31,32] Their viscosity measurements were conducted under uniform shear conditions and at several temperatures. Considerable shear-thinning was found for rabbit tears obtained by cannulating the rabbit lacrimal gland. The viscosity of the rabbit tears at the shear rate of 3.76 sec[-1] was 16.8 cps. Due to the pronounced shear-thinning, however, at shear rates above 100 sec[-1] the rabbit tear viscosity approached that of water. Our results obtained with rabbit tears collected by glass capillaries subsequent to pilocarpine injection (Figure 1) were quite comparable with the results of Hamano and Mitsunaga.[31]

The viscosity of native conjunctival mucus obtained from rabbits was also found to be low and exhibited minimal shear-thinning at physiologic concentrations.[32] Purified rabbit ocular mucus glycoprotein was found to have higher viscosity than native mucin. The contribution of serum albumin to viscosity was found to be minor. The authors concluded that the tear viscosity is determined by its native glycoprotein (mucin) content.[28]

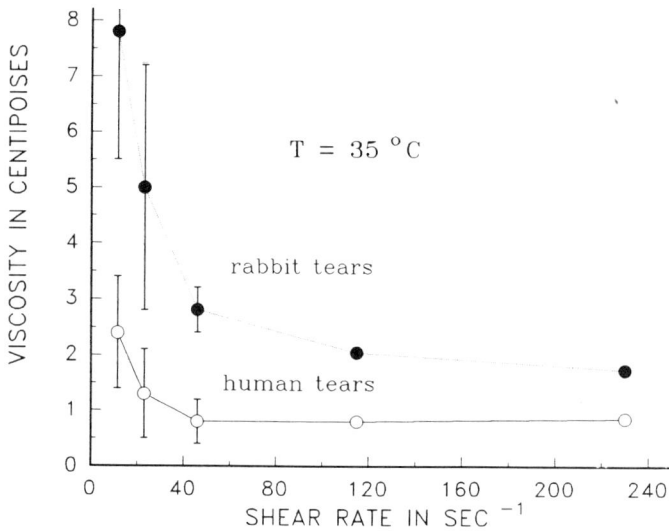

**Figure 1.** Viscosity of human and rabbit tears as a function of shear rate.

Our results, obtained under uniform shear rate conditions, indicate that the viscosity of human tears (Table 2) has a viscosity comparable to that of water when determined at high shear rates ( > 115 sec[-1]) after exhibiting minor or possibly insignificant shear dependence (Figure 1). We found that bovine submaxillary mucin dissolved in 0.9% saline at 1.0 mg/ml concentration, that may be representative of physiological levels of dissolved mucin, does not significantly enhance the viscosity of water at 35°C. When mucin was dissolved in saline at 10 mg/ml concentration, the viscosity was only somewhat higher. Higher viscosity and considerable shear-thinning

**Table 2.** Viscosity and its shear dependence of aqueous tears and mucin solutions

| AQUEOUS TEARS AND MUCIN SOLUTIONS | VISCOSITY in cps [shear rate, sec$^{-1}$] | SHEAR THINNING | TEMPERATURE in °C |
|---|---|---|---|
| Normal human tears | 1.07 ± 0.03  [115-230] | insignificant | 25 |
| Normal human tears | 0.83 ± 0.03  [115-230] | insignificant | 35 |
| Rabbit tears (fresh) | 2.05 ± 0.03  [230] | definite | 25 |
| Rabbit tears (fresh) | 1.74 ± 0.04  [230] | definite | 35 |
| Mucin in water (10 mg/ml) | 14.0 ± 0.38  [115] | definite | 25 |
| Mucin in saline (10mg/ml) | 3.13 ± 0.14  [115-230] | insignificant | 25 |
| Mucin in water (1.0 mg/ml) | 2.94 ± 0.04  [115] | Definite | 25 |
| Mucin in saline (1.0 mg/ml) | 1.28 ± 0.08  [230] | insignificant | 25 |

was observed at room temperature only if the mucin was dissolved in distilled water (Table 2). Apparently, sodium and chloride ions even at 0.9% (physiologic) concentration effectively shield the electric charges of the dissociated sialic acid end-groups in the secondary incomplete helical structure of the mucin molecule which leads to an increase in the flexibility of the otherwise rigid, rod-like molecule diminishing viscosity.

## BOUNDARY LUBRICATION HYPOTHESIS FOR BLINKING

Apparently Ehlers was the first to offer a hypothetical lubrication mechanism operative during blinking,[33] which has not been challenged until now. In his classical thesis Ehlers suggested that the lubrication of the lid-globe system is of the boundary-type and hydrodynamic lubrication is possibly operative only in the mucous folds of the fornices. Ehlers based his suggestion on the mistaken assumption, then widely accepted, that the closing eye lid compresses not only the superficial lipid layer of the tear film but also removes the aqueous tear layer so that intimate contact between the lid and the cornea is achieved. Ehlers in forming his hypothesis was also influenced by his visual observation of the movement of the lacrimal meniscus during lid closure and also his assumption, later found to be erroneous,[22] that the lacrimal surfactant consists of phospholipids similar in composition to lung surfactant.

Ehlers, in formulating the boundary lubrication hypothesis of blinking, may also have been influenced by the industrial tribology data predicting boundary lubrication in the low velocity ( < 3 cm/sec), low viscosity, and high load region,[3-5] and was led to believe that at the low gliding speed of the lid (which at that time was thought to be not more than 10 cm/sec), boundary lubrication should prevail. Kalachandra and Shah[34] accepted Ehlers' conjecture, since upon super-ficial examination, their own data, obtained with artificial tears in an in vitro system where the stability of the lubricating fluid layer was jeopardized, appeared to support that view.

# HYDRODYNAMIC LUBRICATION MECHANISM OF BLINKING

## Tribology of Blinking in Normal Humans

When one considers the hydrophilic nature of mucus-coated epithelial surfaces of tarsal conjunctiva and the globe,[35] and the surface-chemical view of the role of blinking in tear film formation,[36] it is quite certain that a stable, continuous tear layer exists under the lid which would serve as a hydrodynamic lubricant. The force pressing the eye lid against the globe (load) is quite small, not exceeding a few grams. The positive disjoining pressure,[37] that must exist in the tear layer sandwiched between two mucus-coated solid surfaces,[38] will ensure the existence of a stable fluid layer between the globe and the lid.

Since, the continuous tear layer between the lid and the globe could be as thick as 10 micrometers, while the magnitude of the surface asperities consisting of micro-

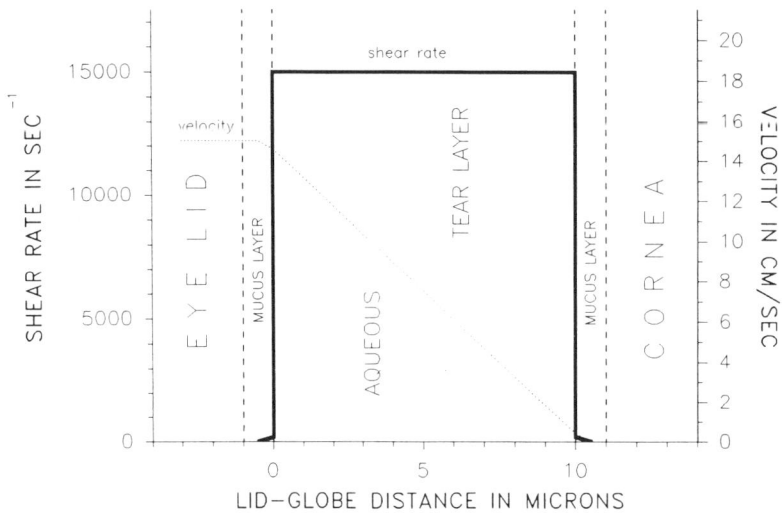

**Figure 2.** Shear rate and velocity distribution under the eye lid in hydrodynamic lubrication

ridges is less than 5% of this value, in the normal eye hydrodynamic lubrication mechanism should, and in all likelihood, does prevail. Then the lid-globe friction would be mainly determined by the viscosity of the tear layer at shear rates estimated to reach 15,000 sec[-1] during blinking.

The viscosity of the lubricant tear especially at the high shear rate created by blinking would be similar to that of water. Hence, the frictional coefficient between the upper lid and the globe would be less than 0.005. The high shear rate is confined mostly to the aqueous tear layer, so most of the slip will occur there. The velocity gradient would become quite small in the highly viscous boundary mucus layers. Hence, the highest shear stress value operative during blinking and confined to the mucus layer - tear layer interfaces would only be approximately 150 dyne cm[-2]. The shear stress at the ocular surface, and, therefore, the "wear" on the superficial epithelium would become negligible (Figure 2).

The drag due to shear forces, however, would be sufficient to remove dying or dead cells and mucous threads contaminated with lipid from the ocular surface especially since these would protrude into the aqueous tear layer undergoing a high shear rate. This is indeed the mechanism of the surface cleaning process in the eye suggested by Holly and Lemp[36] after it was observed and documented by Norn[39] monitoring the movement of mucous globs and fibrils in the palpebral fissure and the inferior fornix.

### Tribology of Blinking in Dry Eye Patients

The aforesaid, however, is only valid for the normal eye. In dry eye states, the ocular surface disease can readily increase the scale of surface roughness ten-fold. In addition, the lacrimating ability of the eye could also be impaired resulting in a thinner tear film and thus a thinner lubricating tear layer. These two factors may change the mechanism of lubrication from the hydrodynamic type to the boundary type thereby increasing the frictional coefficient considerably. This increased force transfer would be capable of further damaging the ocular surface especially since the cohesiveness of cells due to intercellular edema has also been diminished. This lack of proper lubrication no doubt contributes to ocular discomfort and continuous damage to the ocular surface, so that the rate of cell death and cell loss could overwhelm mitotic and other tissue repair processes.

In filamentary keratitis, a particularly painful condition, lipid-laden mucus forms filaments that attach to cells by hydrophobic bonding (contact adhesion). These processes extend into the bulk of the tear film where the velocity gradient is the greatest. The shear forces pulling on the attached cells can cause considerable pain. When the ocular surface is protected against the shear forces by a bandage contact lens,[40,41] or when the filaments are dispersed by a mucolytic agent,[42] the patient becomes much more comfortable and the ocular surface has a chance to heal regaining its previous smoothness and cohesiveness.

### CONCLUSIONS

In industrial tribology water or aqueous solutions are considered to be poor lubricants, because they cannot form stable films between (lipid-contaminated) metal surfaces and also, because water in the presence of oxygen is quite corrosive to metals.

In the human body, however, the lubricants are water-based. The stability of the aqueous film is assured by hydrophilic macromolecules that adsorb at the interfaces. These hydrophilic biopolymers at sufficiently high concentration, are also capable of raising the viscosity of the solution to proper levels. The lubrication of the joint is assured by the aqueous solution of hyaluronic acid, a high molecular weight glycosaminoglycan that exhibits pronounced shear thinning and also a considerable viscosity increase under pressure.

In the eye, a continuous tear film is maintained and the cleansing of the ocular surface is assured by the rapid blinking movement where the eye lid slides over the cornea and the conjunctiva. It is argued here that a fairly substantial aqueous tear layer exists over the ocular surface at all times providing the lubricant for hydrodynamic lubrication. This results in a low frictional coefficient lessening the work by the eye lid muscles (drag) and protects the sensitive ocular surface tissue from abrasion (wear). In a closed eye the lubricating tear layer is sandwiched between two mucus layers. The mucus layer ensures the continuity of the tear layer and diminishes surface roughness. Since high viscosity is detrimental to lubrication and is not needed for achieving tear film stability, mucous glycoproteins are present in the tears at levels sufficiently low so that the viscosity of the tears remains close to that of water. The main role of lacrimal glycoproteins appears to be chiefly surface chemical in ensuring film continuity and stability.

Any pathological process that increases surface roughness and decreases the wettability of the ocular tissues while diminishing its thickness would interfere with the lubrication mechanism. The instillation of a high viscosity solution in the eye could also result in poor lubrication leading to surface damage. Fortunately, eye drops of extremely high viscosity would not mix readily with the tear film and thus would not interfere greatly with lubrication, limiting tissue damage.

The maintenance of sufficient lubrication is very important in contact lens wear in order to avoid poor contact lens tolerance. The assurance of a continuous liquid layer over the lens would not only ensure lubrication but would also protect the lens surface from deposit formation. The tribology of ophthalmic demulcents and contact lens cushioning solutions is important to consider when formulating collyria for instillation into the eye for any purpose.

## REFERENCES

1. A. Cameron, Basic Lubrication Theory, Ellis Horwood Ltd., Surrey, (1981).
2. R. Gohar, Elastohydrodynamics, J. Wiley and Sons, New York, (1988).
3. A. Bondi, Physical Chemistry of Lubricating Oils, Rheinhold, New York, (1951), p. 146.
4. A.W. Adamson, Physical Chemistry of Surfaces, 4th ed., John Wiley and Sons, New York, (1982) p. 402.
5. D.F. Moore, Principles and Applications of Tribology, Pergamon, New York, (1975).
6. P. Freeman, Lubrication and Friction. Pitman, New York, (1962).
7. B.V. Derjaguin, Molecular theory of friction and sliding (in Russian) Zh. Fiz. Khim. 5:1165 (1934).
8. G. Amontons, De la résistance causée dans les machines, Mém. Acad. Roy. Sci., (1699), p. 206
9. D.D. Fuller, Theory and Practice of Lubrication for Engineers, 2nd ed. J. Wiley and Sons, New York, (1984).
10. W.B. Hardy and I. Bircumshaw, *Proc. Roy. Soc. (London)* A108:1 (1925).
11. W.B. Hardy, Collected Scientific Papers, Cambridge University Press, London, (1936).
12. J.M. Tiffany, The viscosity of human tears: Rheological analysis. Abstracts, 2nd Triennial Mtg. Int. Soc. Dakryol., Amsterdam, July 23-25, 1990.
13. F.J. Holly, Tear film formation and rupture: An Update. In "The Preocular Tear Film: In Health, Disease, and Contact Lens Wear", Holly, F.J., Lamberts, D.W., and MacKeen, D.L. eds, Dry Eye Institute, Lubbock, TX 1986, p. 634.
14. F.J. Holly and M.A. Lemp, Tear physiology and dry eyes. *Surv. Ophthalmol.* 22(2):69 (1977)
15. J. Murube del Castillo, J.: Dacriologia Basica, publ. by the Spanish Society of Ophthalmology, Las Palmas, 1981, pp. 438.
16. P.N. Dilly, Contribution of the epithelium to the stability of the tear film. *Trans. ophthalmol. Soc. U.K.* 104:381 (1985).
17. F.J. Holly and M.A. Lemp: Wettability and wetting of corneal epithelium. *Exp. Eye Res.* 11:239 (1971).

18. R. Kaura, R.: Ocular Mucus: A Hypothesis for its Role in the External Eye Based on Physiologic Considerations and on the Properties and Functions Known to be Common to Other Mucous Secretions, In "The Preocular Tear Film: In Health, Disease, and Contact Lens Wear," Holly, F.J., Lamberts, D.W., and MacKeen, D.L. eds, Dry Eye Institute, Lubbock, TX 1986, p. 743.

19. B. Nichols, Dawson, and B. Togui, Surface features of the conjunctiva and the cornea. *Inv. Ophthalmol. Vis. Sci.* 24:570 (1983).

20. J.E. McDonald, Surface phenomena of tear films. *Trans Am. Ophthalmol. Soc.* 66:905 (1968).

21. I. Langmuir, Oil lenses on water and the nature of monomolecular expanded films. *J. Chem. Phys.* 1: 756 (1933).

22. F.J. Holly, Physical chemistry of the normal and disordered tear film, *Trans. ophthalmol. Soc. U.K.* 104:374 (1985).

23. M.G. Doane, Interaction of eye lids and tears and the dynamics of the normal human eye blink. *Am. J. Ophthalmol.* 89:507 (1980).

24. Geigy Scientific Tables of Bodily Fluids, 8th ed. Vol. 1, Ciba-Geigy, West Caldwell, NJ (1981), p. 182

25. Liotet and Cochet, Notions concerning human tears, *Arch. Ophthalmol. (Paris)* 27:251 (1967).

26. J.P. Vignat and G. Gougaud, Lacrimal syndromes, *Arch. Ophthalmol. (Paris)* 36:773 (1976).

27. J.P. Metaireau, G. Baikoff and P. Brun, Lacrimal physiology, *Arch. Ophthalmol. (Paris)* 37:401 (1977).

28. R. Kaura and J.M. Tiffany, The role of mucous glycoproteins in the tear film. In "The Preocular Tear Film: In Health, Disease, and Contact Lens Wear", Holly, F.J., Lamberts, D.W., and MacKeen, D.L. eds, Dry Eye Institute, Lubbock, TX (1986), p. 728.

29. J.E. Tiffany and A.J. Bron, Role of tears in maintaining cornea integrity. *Trans. Ophthalmol. Soc. U.K.* 98(3):335 (1978).

30. A.J. Bron and J. Tiffany, Pseudoplastic materials as tear substitutes, Lecture given at the 6th International Symposium on the Lacrimal System, Singapore, March 17, 1990. Rep. 6th Int. Symp. Lacrim. Sys., Kugler-ghedini Publ. Amsterdam-Berkeley-Milano, 1990, p. 4.

31. H. Hamano, H. and S. Mitsunaga, Viscosity of rabbit tears, *Jap. J. Ophthalmol. (Japan)* 17:290 (1973).

32. H. Hamano and H.E. Kaufman, The Physiology of the Cornea and Contact Lens Applications, Churchill Livingstone, New York (1987), p. 35.

33. N. Ehlers, The Precorneal Film, *Acta Ophthalmol. (Kbh) Suppl.* 81: p. 111 (1965).

34. S. Kalachandra and D.O. Shah, Lubrication and surface chemical properties of ophthalmic solutions. *Ann. Ophthalmol.* 17:708 (1985).

35. M.A. Lemp, C.H. Dohlman and F.J. Holly, Corneal desiccation despite normal tear volume. *Ann. Ophthalmol.* 2:258 (1970).

36. F.J. Holly and M.A. Lemp: Formation and rupture of the tear film. *Exp. Eye Res.* 15:515 (1973).

37. B.V. Derjaguin, Repulsive forces between charged colloid particles and the theory of slow coagulation and stability of lyophobe sols. *Trans. Farad. Soc.* 36:203 (1940).

38. F.J. Holly, Biophysical aspects of the adhesion of corneal epithelium to stroma. *Invest. Ophthalmol.* 17:552 (1978).

39. M.S. Norn, M.S.: Mucus flow in the conjunctiva. Rate of migration of the mucous thread in the inferior conjunctival fornix towards the inner canthus. *Acta Ophthalmol. (Kbh)* 43:557 (1965).

40. A. Gasset and L. Lobo, Simplified contact lens treatment in corneal diseases, *Ann. Ophthalmol.* 9: 843 (1977).

41. D.W. Lamberts, Bandage (therapeutic) Lenses, in "Contact Lens Practice," by R.B. Mandell, C.C. Thomas, Springfield, IL, 4th ed. (1988) p. 644.

42. M.J. Absalon and C.A. Brown, Acetylcystein in keratoconjunctivitis sicca. *Br. J. Ophthalmol.* 52: 310 (1968).

# EFFECT OF SYSTEMIC INGESTION OF VITAMIN AND TRACE ELEMENT DIETRY SUPPLEMENTS ON THE STABILITY OF THE PRE-CORNEAL TEAR FILM IN NORMAL SUBJECTS

Sudi Patel, Colin Ferrier, and Jeff Plaskow

Department of Vision Sciences
Glasgow Caledonian University
Cowcaddens Rd, Glasgow
Scotland, U.K, G4 OBA

## INTRODUCTION

Inadequate ingestion of specific vitamins and trace elements are commonly cited as pre-requisites in the development of some dry eyes. It is well documented that the health of the anterior eye is highly dependant upon diet. In particular, third world studies have identified vitamin A deficiency leads to non-wettability of the ocular surface in turn leading to severe desiccation (xerophthalmia), corneal scarring and a high risk of ocular morbidity. There are several indications that vitamin A is vital for the number, and hence secretory activity, of the conjunctival goblet cells [1-3]. Restoration of systemic vitamin A results in a rapid resolution of the dry eye state in the presence of sufficient protein[2]. Certain systemic conditions with associated dry eye symptoms also allow other dietary factors to be identified as being important for the health or homeostasis of the tear film (e.g. zinc, manganese, niacin, vitamins B6 and C etc.) These are important in relation to the composition of the glandular secretions which constitute the tear film because their absorption either:

i)      directly or indirectly affects the composition  or,

ii)     affects the transport of essential metabolites  for the tear constituents [4-7].

In a diet-healthy normal population, xerophthalmia is not prevalent but the marginal dry eye (e.g. KCS) certainly is. It can be proposed that the marginal dry eye condition is due to sub-clinical marginal dietary imbalances. A far as we know, this has not been proven, this rationale has yet to be established on a scientific basis. A recent review[7] tended to centre around mainly anecdotal evidence that KCS may have a dietary link. Direct application of selected concentrations of vitamin A have been advocated as useful remedies for the dry eye and conjunctival inflammations[8-13]. Perhaps these remedies improve tear properties in 'normals as well as in 'subnormals'. Tear constituents are derived from the blood supply, it is  logical to propose that the tear enhancing modalities which are normally encountered in the diet should be administered orally. Proper clinical trials under strict experimental conditions have not been performed. Such an investigation would result in data amenable to satisfactory statistical scrutiny. Only using this kind of approach can we decide whether additions to the usual diet have any genuine effect on tear film properties and/or status. The present study aims to test the effect of dietary supplements in a group of supposedly healthy normals with normal tear stability.

*Lacrimal Gland, Tear Film, and Dry Eye Syndromes*
Edited by D.A. Sullivan, Plenum Press, New York, 1994

## METHODS

i) 60 normal healthy subjects (30 male and 30 female) chosen from a student population entered the study. The age range was 19 to 23 years, within this range tear film stability does not vary significantly [14]. Subjects had no history of ocular disease, allergies, general health disorders, or contact lens wear.

ii) Tear stability was measured initially and 10 days later. Only the right eye was investigated. Tear stability was estimated by the standard non-invasive technique of Tear Thinning Time measurement (TTT) using a standard Bausch and Lomb keratometer[14-16]. All measurements were taken at mid-afternoon to avoid the effects of any diurnal influences[15].

iii) 2 types of commercially available diet supplements were used, a multi-vitamin and trace element tablet (Supradyn[T.M.], by La Roche) and a purely vitamin C tablet (Redoxon[T.M.], by La Roche). Both tablets are packaged identically (silver foil wrap, 10 tablets per tube), are of similar size, and need to be dissolved in water before swallowing. The exact constituent of the tablets are listed in Table 1. Each Supradyn tablet contains: Vitamins, A(1500µg), B1(15µg), B2(5µg), B6(11.6µg), B12(5µg), C(150µg), D(10µg), E(50µg). Minerals, Calcium (1.25µg), Iron(1.25µg), Magnesium(5µg), Copper(0.5µg), Zinc(0.5µg), Molybdenum(0.1µg),Manganese(0.5µg),Niacin(50µg).Pantothenicacid(11.6µg).Biotin (250µm). Folic acid (300µg). Phosphate(45µg). Redoxon contains 1000mg of vitamin C.

iv) 20 tubes of each were assigned random number labels by one of us, issued by another, and TTT measured by the third member of the team.

v) None of the subjects changed their normal diets or took any additional supplements.

vi) After collecting all data, the identities of the numbered tubes were revealed. Group A (control), group B (Redoxon), group C (Supradyn).

## RESULTS

The average TTT values and results of statistical tests are listed in Table 1.

**Table 1.** Mean tear stabilities and the results of statistical analyses.

| Group | Mean tear stabilities (secs) Before | After | 't' statistic (paired) | Significance |
|---|---|---|---|---|
| A | 16.19 (11.54) | 16.54 (12.16) | -0.81 | N.S. @ 5% |
| B | 14.22 (7.51) | 20.37 (8.72) | -6.52 | Significant @ 5% |
| C | 13.96 (8.46) | 19.94 (10.96) | -2.78 | "      "      " |

Standard deviation,(+/- Ó$_{n-1}$), shown in parenthesis after mean value.

## DISCUSSION

Viewing Table 1, both treatments enhance TTT but, Supradyn (group C) has the greater effect. Perhaps, the actual health of the subjects was below the ideal in terms of diet but there again how do we define 'ideal' ? The question is why and how is the stability of the tear film being improved? vitamin A is essential for the repopulation of conjunctival goblet cells and hence mucus production, but other vitamins and minerals (e.g. zinc) also have important roles in the cycle of events[16,18,19]. Nevertheless, the relative concentrations of the remaining constituents of the tablet and their contributions cannot be ignored. Is it possible that a placebo effect is in operation (2 of the 3 groups were masked)? This can be ruled out because it is difficult to postulate how a group of naive individuals can improve their tear stabilities by cogitation. The synergistic/ antagonistic re-actions coupled with the concentration of the tablet constituents, the individual's own dietary habits and metabolism render the task of attempting to produce a comprehensive hypothesis account for the findings, impossible. Certainly, vitamins A, B6 and C together with manganese and zinc are required for general protein synthesis[20], lacrimal gland and goblet cell activity. It is unclear if the improved TTT is due to increased quality and/or quantity of lacrimal fluid, mucus or both.

The improvement in TTT within the treated groups of allegedly normal healthy subjects suggests that, perhaps more rigorous selective procedures should have been adopted. If these normal groups were used as controls in a comparative trial with dry eyes investigating the true value of a diet treatment, then it is possible for the true value of the treatment to be rejected as purely a placebo effect. The clinical implication of this study is clear, in marginal or occasional dry eyes, perhaps the patient would be better managed after a full assessment of dietary status by a nutritionist before embarking on a course of diet additives.

## CONCLUSIONS

In a group of 'normal, healthy' individuals, the mean stability of the pre-corneal tear film improved by 5.98 secs after a 10 day daily intake of 1000mgm of vitamin C. Similarly, the mean stability improved by 6.15 secs in a second group taking a tablet of Supradyn once a day. The increases were statistically significant. Over the same period, the average stability of a control group did not significantly change.

## REFERENCES

1. A.J.W. Huang, S.C.G. Tseng, K.R. Kenyon, Change of paracellular permeability of ocular surface epithelium by vitamin A deficiency, *Invest Ophthalmol Vis Sci.* 32: 633-639 (1991).
2. A.Sommer and W.R. Green, Goblet cell response to vitamin A treatment for xerophthalmia, *Am J Ophthalmol.* 94: 213-215 (1982).
3. A.Sommer, Effects of vitamin a deficiency on the ocular surface,*Ophthalmology.* 90: 592-600 (1983).
4. R.S. Smith, T. Farrell and T.Bailey, Keratomalacia, *Surv.Ophthalmol.* 20: 213-219 (1975).
5. N.J. van Haeringen, Clinical biochemistry of tears, *Surv. Ophthalmol.* 26: 84-96 (1981).
6. C.A.Patterson and M.C. O'Rourke, Vitamin C levels in human tears, *Arch. Ophthalmol.* 105: 376-377 (1987).
7. B.E. Caffery, Influence of diet on tear film function, *Optom &Vis Sci.* 68:58-72 (1991).
8. J.E. Schultz, Treating giant papillary conjunctivitis while wearing contact lenses, *Int. Cont Lens Clin.* 17: 139-143 (1990).
9. J.L. Ubels, H.F. Edelhauser, K.M. Foley, J.C. Liao and P. Gressel, The efficacy of retinoic acid ointment for treatment of xerophthalmia and corneal epithelial wounds, *Curr Eye Res.* 4: 1049. (1985).
10. D.B. Chandra, S.D. Verma, R. Verma, M.K. Khare and P. Chandra, Topical vitamin A palmitate in dry eyes, *Afro-Asian J Ophthalmol.* 7: 74-80 (1988).
11. J.F. Molinari and R.H. Rengstorff, Management of soft lens-induced GPC with vitamin A aqueous drops, *Contact Lens J.* 16: 169-170 (1988).
12. R.H. Rengstorff, C.C. Krall and D.I. Westerhout, Topical antioxidant treatment for dry-eye disorders and contact lens related complications, *Afro-Asian J Ophthalmol.* 7: 81-83 (1988)
13. B.L.Butts and R.H.Rengstorff, Antioxidant and vitamin A eyedrops for GPC, *Contact Lens J.* 18: 40-43 (1989).
14. S. Patel, J.C.Farrell, Age related changes in precorneal tear film stability. *Am J Optom & Physiol Opt.* 66: 175-178 (1989).
15. S.Patel, R.Bevan and J.C. Farrell, Diurnal variations in precorneal tear film stability. *Am J Optom & Physiol Opt.* 65: 151-154 (1988).
16. S.Patel, S.Laidlaw, L. Mathewson, L.McCallum and C.Nicholson, Iris colour and the influence of local anaesthetics on precorneal tear film stability. *Acta Ophthalmol.* 69: 387-392 (1991).
17. W.Bayer. "Handbook of Tables for Probability and Statistics," CRC Press, Cleveland (1974).
18. I. Rask, Vitamin A supply to the cornea. *Exp Eye Res.* 31: 201-211 (1980).
19. J.L. Ubels and S.M. Macrae, Vitamin A is present in the tears of humans and rabbits, *Curr Eye Res.* 3: 815-822 (1984).
20. R.E. Hodges. "Nutrition and Medical Practice," W.B.Saunders, London (1980).

# PATTERNS OF CYTOKERATIN EXPRESSION IN IMPRESSION CYTOLOGY SPECIMENS FROM NORMAL HUMAN CONJUNCTIVAL EPITHELIUM

Kathleen L. Krenzer and
Thomas F. Freddo

Departments of Ophthalmology and Pathology
Boston University Medical Center
Boston, Massachusetts   02118

## INTRODUCTION

The cytokeratins are a category of intermediate filaments characteristically found in epithelial cells.  Approximately 20 cytokeratins have thus far been described and their normal pattern of expression appears to be dependent on cell-type and differentiation state.[1]  Moreover, in each epithelial cell type, a distinct subset of cytokeratin polypeptides will be present.  Patterns of cytokeratin expression have been shown to change secondary to altered microenvironmental conditions (e.g. disease states) and these changes have become useful in diagnostic pathology.  To this end, cytokeratin expression patterns in the various human epithelia, both normal and diseased, have been extensively examined.  However, similar studies in the conjunctival epithelium are sparse and generally have examined non-human, cultured or limbal conjunctival epithelium and/or have used a limited panel of cytokeratin markers.

The goal of the present study was to broadly examine the cytokeratin expression pattern in human conjunctival epithelium obtained from normal volunteers by impression cytology.  Impression cytology is a sampling procedure by which conjunctival epithelial tissue can be obtained in a non-invasive fashion.[2]  Clinically it is a quick and simple method which allows repeatable and representative samples to be obtained with minimum discomfort for the patient.

By combining impression cytology with a new method that permits immunocytochemical staining of these specimens, we have examined cytokeratin expression patterns, using commercially available antibodies, in normal human conjunctival epithelium obtained directly from patients. In addition we have examined two other structural proteins, filaggrin and involucrin, which are markers for keratinization.[3,4]

*Lacrimal Gland, Tear Film, and Dry Eye Syndromes*
Edited by D.A. Sullivan, Plenum Press, New York, 1994

## MATERIALS AND METHODS

Specimens for this study were obtained from 26 normal volunteers (16 females, 10 males) ranging in age from 23 to 78 years. Impression cytology specimens (ICS) were collected on pure nitrocellulose filter membranes (FMC Bioporducts, Inc.) using the method of Nelson.[5] Briefly, 0.5% proparacaine HCl was instilled in each eye. A circular membrane (6 mm in diameter) was then placed on the bulbar conjunctiva using blunt forceps. Uniform pressure was applied using an ophthalmodynamometer at 40 mmHg for 15 s. The filter membrane was then peeled off the conjunctiva, fixed with Spray-cyte (Clay-Adams, Inc.), and placed specimen-side down on a poly-L-lysine coated slide. Each ICS was cut in half and one half was transferred to another slide. Four ICSs, one from each quadrant, were obtained from each eye. By halving each ICS, a total of eight ICSs from each eye were available for analysis. After drying completely, a drop of acetone was placed on each membrane to assure adherence during transport. Each half was then assigned a marker according to a matrix which ensured representative data from each of the four quadrants for each marker. ICSs were placed in acetone for 30 min. with constant motion in order to dissolve the bulk of the filter membrane material leaving only the cells on the slide. A small amount of membrane material persisted which could have impeded penetration of the antibodies and reagents. Therefore, the specimens were treated with cellulase (Sigma, Inc., 10 units/ml) for two hours at 37°C. This treatment optimized the immunocytochemical staining procedure.

ICSs were stained using a standard immunocytochemical procedure. After a brief wash with water, endogenous peroxidase activity was quenched with 3% hydrogen peroxide in 70% methanol. Non-specific binding sites were blocked in 4% low fat milk[a] for 30 min., then drained. In humidity chambers, ICSs were incubated in appropriately titrated monoclonal antibodies (MAbs) overnight at 4°C. (See Table 1 for MAbs and specificity). The following day, the ICSs were allowed to warm to room temperature for one hour. The specimens were then sequentially incubated with a biotin-labelled secondary antibody (1:200 for 30 min.), horse-radish peroxidase-labelled streptavidin biotin complex (1:500 for 30 min.) and finally in diaminobenzidene yielding a brown precipitate. All steps were followed by thorough washing in PBS. Specimens were counterstained in hematoxylin, dehydrated, mounted and examined with a Zeiss light microscope. A positive tissue control for each marker was included in each run and one ICS from each eye was used as a negative control (MAb exclusion).

## RESULTS AND DISCUSSION

The results for each marker are summarized in Table 1. All positive and negative controls reacted appropriately. For each marker, no age, gender or quadrant differences were noted.

---

[a] Blocking solution, primary antibodies and secondary antibodies were diluted in phosphate buffered saline (PBS) with 0.05% Tween 20.

**Table 1.** Monoclonal Antibodies and Results

| MAb/Source | CK(Moll #) | #Pos./Total |
|---|---|---|
| α-CK AE 1,3 (ICN, Inc.) | 1,2,5,14,16,19 | 20/20 |
| α-CK AE 2 (ICN, Inc.) | 1,2,10,11 | 5/20 |
| α-CK 8.6 (Sigma, Inc.) | 1,10,11 | 5/51 |
| α-CK 8.12 (Sigma, Inc.) | 13,16 | 51/51 |
| α-CK 1C7 (Medscand, Inc.) | 13 | 31/31 |
| α-CK 4.62 (Sigma, Inc.) | 19 | 20/20 |
| α-CK NCL5D3 (Medscand, Inc.) | 8,18,19 | 27/31 |
| α-CK RCK105 (Medscand, Inc.) | 7 | 22/31 |
| α-filaggrin (BTI, Inc.) | N/A | 8/51 |
| α-involucrin (BTI, Inc.) | N/A | 6/51 |

α-CK AE 1,3 is a mix which detects a broad array of cytokeratins and is therefore considered a pan-epithelial marker. Positivity with this marker was demonstrated in all specimens. A uniform intracytoplasmic distribution was found.

α-CK 8.12 and α-CK 1C7 detect an overlapping cytokeratin spectrum (13, 16, and 16 respectively). Both were positive in all specimens, demonstrating a cytoplasmic distribution. No staining was seen in goblet cells. Because of the duality of α-CK 8.12, it is unclear whether K13 or both K13 and K16 are responsible for the positivity noted. The presence of K13, as established by positive staining with both markers is consistent with the K4/K13 pair found in other non-keratinizing , stratified squamous epithelia.

α-CK AE 2 and αCK 8.60, on the other hand, are considered markers for keratinized epithelium. Most of the specimens were negative for these two markers. In specimens where positivity was noted, only a few scattered cells were positive similar to the pattern seen by Gigi-Leitner et al. in cervical epithelium.[6]

α-CK 4.62 and α-CK NCL 5D3, which also detect an overlapping set of cytokeratins (19 and 8, 18, 19, respectively) characteristic for simple epithelia, were positive in most of the specimens examined, demonstrating a diffuse cytoplasmic staining pattern of variable intensity.

α-CK RCK 105, when positive, demonstrated a patchy distribution being concentrated in the immediate vicinity of the goblet cells. Although no definite conclusion can presently be made about this unusual pattern, it is interesting to note that Moll et al. found K7 to be present predominantly in human glandular tissues.[1]

α-filaggrin and α-involucrin are two proteins involved in the keratinization process. These two markers were generally negative. In positive specimens, as with α-CK AE 2 and α-CK 8.60, only a few cells were positive with the remainder of the specimen being negative.

## CONCLUSION

These findings provide a basic understanding of the expression of the cytokeratins, filaggrin and involucrin in normal human conjunctival epithelium. The MAbs which were selected for this study represent a panel which detects a broad array of cytokeratin pairs, the expression of which may change in various disease states. Our findings are similar to those in other mucosae including oral,[7] cervical,[8] urinary,[9] and respiratory[10] epithelia.

In other mucosal epithelia, changes in the expression of

these proteins have been noted in various disease states (e.g. squamous metaplasia and neoplasia). However, this avenue of study has not been pursued in the human conjunctiva. The use of impression cytology combined with immunocytochemistry now allows us to examine changes in the cytokeratins, involucrin and filaggrin in the various ocular surface disorders in a simple, non-invasive manner. This method offers promise to be clinically useful for diagnosing and monitoring disease progression or response to treatment.

## REFERENCES

1. R. Moll, W.W. Franke, D.L. Schiller, B. Geiger and R. Krepler, The catalog of human cytokeratins: patterns of expression in normal epithelium, tumors and cultured cells, *Cell* 31:11(1982).

2. P.R. Egbert, S. Lauber and D. M. Maurice, A simple conjunctival biopsy, *Am. J. Ophthalmol.* 34:798(1977).

3. A.M. Lynley and B.A. Dale, The characterization of human epidermal filaggrin; a histidine-rich, keratin filament-aggregating protein, *Biochem. Biophys. Acta* 744:28(1983).

4. A.E. Walts, J.W. Said, M.B. Siegal and S. Banks-Schlegel, Involucrin, a marker of squamous and urothelial differentiation. An immunohistochemical study on its distribution in normal and neoplastic tissues, *J. Pathol.* 145:329(1985).

5. J.D. Nelson, Impression cytology, *Cornea* 7:71(1988).

6. O. Gigi-Leitner, B. Geiger, R. Levy and B. Czernobilsky, Cytokeratin expression in squamous metaplasia of the human uterine cervix, *Differentiation* 31:205(1986).

7. H. Clausen, P. Vedtofte, D. Moe, E. Dabelsteen and T.T. Sun, Differentiation-dependent expression of keratins in human oral epithelia. *J. Invest. Dermatol.* 86:249(1986).

8. J.R. Whittaker, A.M. Samy, J.P. Sunter, D.P. Sinha and J.M. Monaghan, Cytokeratin expression in cervical epithelium: an immunohistological study of normal, wart virus-infected and neoplastic tissue. *Histopathol,* 14:151(1989).

9. M. Cintorino, M.T. DelVecchio, M. Bugnoli, R. Petracca and P. Leoncini. Cytokeratin pattern in normal and pathological bladder urothelium: Immunohistochemical investigation using monoclonal antibodies, *J. Urol.* 139:428(1988).

10. F. Bejai-Thivolet, J. Viac, J. Thivolet and M. Faure, Intracellular keratin in normal and pathological bronchial mucosa: Immunocytological studies on biopsies and cell suspensions, *Virchows Arch. Pathol. Anat.* 395:87(1982).

# INCREASE IN TEAR FILM LIPID LAYER THICKNESS FOLLOWING TREATMENT OF MEIBOMIAN GLAND DYSFUNCTION

Donald R. Korb[1] and Jack V. Greiner[2,3]

[1]80 Boylston Street, Boston, Mass.
[2]Department of Ophthalmology, Harvard Medical School, Boston, Mass.
[3]Schepens Eye Research Institute, 20 Staniford Street, Boston, Mass.

## INTRODUCTION

Meibomian gland lipid secretions contribute to the formation of a stable tear film. Meibomian gland dysfunction may result in dry eye symptoms, keratoconjunctivitis (Keith, 1967; McCulley and Sciallis, 1977) and contact lens intolerance (Henriquez and Korb, 1981), presumably due to an inadequate tear film lipid layer secondary to the meibomian gland dysfunction (MGD) itself. The presence of dysfunctional meibomian glands can often be indicated by serrated and inflamed eyelid margins, although in some instances, signs of inflammation may be lacking. In contrast to normal meibomian glands whose orifices open easily and secrete transparent sebum upon gentle expression, the dysfunctional meibomian gland requires forceful digital pressure to elicit expression of sebum, which is generally cloudy in appearance. Frequently, sebum may not be released from dysfunctional meibomian glands even with forceful digital pressure. The present prospective study evaluated whether a program of manual expression of the meibomian glands of patients with a diagnosis of MGD could increase tear film lipid layer thickness by relieving meibomian gland obstruction.

## MATERIALS AND METHODS

### Subject Selection

Subjects ranging in age from 25-35 years (n=10) were selected on the basis of fulfilling criteria of (1) a lipid layer thickness (LLT) of 60 nm or less as judged by specular reflection techniques described below, (2) obstructed meibomian glands as evidenced by the appearance of the lid margin (Fig. 1) and the lack of visible secretion upon application of moderate digital pressure, (3) dry eye symptoms including discomfort and/or sandy, gritty, foreign body sensations, (4) no evidence of other acute internal or external hordeola or other ophthalmic conditions, and (5) no current contact lens use.

*Lacrimal Gland, Tear Film, and Dry Eye Syndromes*
Edited by D.A. Sullivan, Plenum Press, New York, 1994

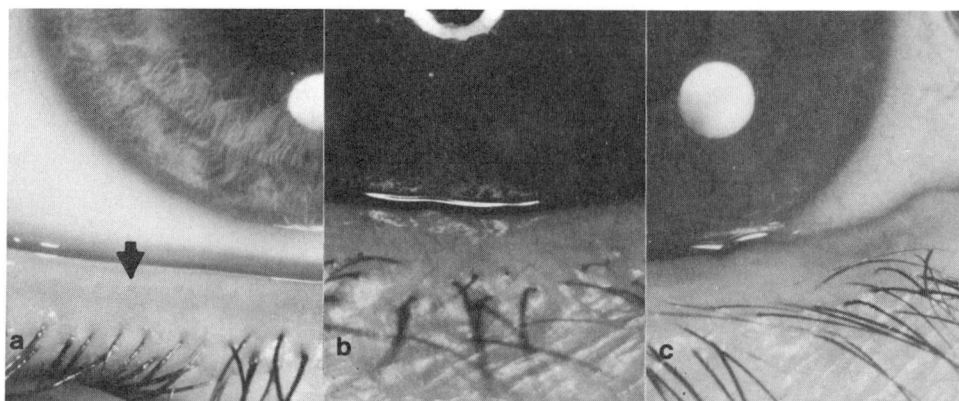

Figure 1. Comparison of the lower eyelid margin of (a) normal and (b,c) dysfunctional meibomian glands. (a) Obvious meibomian gland orifices (arrow). (b) Note dilated blood vessels indicative of acute inflammation, and serrated eyelid margin. (c) Note serrated eyelid margin and absence of dilated blood vessels, indicating chronic inflammation. Meibomian gland orifices are not readily apparent (b,c).

## Meibomian Gland Treatment

A six-month treatment program was established that included four meibomian gland expressions performed as office procedures at six week intervals and daily self-administered meibomian gland treatment. Meibomian gland expressions in the office setting were performed under 16x magnification in order to allow observation of the discharge from the meibomian gland orifices. Two drops of topical anesthetic (0.5% proparacaine hydrochloride ophthalmic solution) were instilled into the conjunctival sac. The eyelid margins were scrubbed with a sterile cotton-tipped applicator soaked in the same anesthetic solution in order to remove surface debris. With the eye directed in upgaze, the lower eyelid was gently drawn away from the globe in order to accommodate a sterile cotton-tipped applicator soaked with the proparacaine solution. The applicator was positioned in the lower conjunctival sac of the right eye, against the lower tarsal conjunctiva adjacent to a meibomian gland and its orifice. Meibomian gland expression was achieved by simultaneous digital pressure on the skin and applicator pressure on the palpebral conjunctiva so as to compress the gland without trauma to the globe. The expressed sebum was removed with the cotton-tipped applicator during the procedure. Each meibomian gland along the lower eyelid was expressed and drained in a similar manner. The characteristics of the expressed sebum, or the lack of it, were noted. The meibomian glands of the upper eyelid were similarly expressed after directing the eye in downgaze and positioning the cotton-tipped applicator against the upper tarsal conjunctiva, adjacent to a meibomian gland and its orifice on the eyelid margin. The entire procedure was then repeated for the left eye. At the conclusion of the procedure the eyelid margins were scrubbed with cotton-tipped applicators soaked with fresh anesthetic in order to remove loose surface cells and any remaining inspissated sebum. Subjects were then instructed to perform a daily treatment regimen that included (1) application of warm compresses to the lower eyelid for two minutes and (2) the use of an eyelid scrubbing solution consisting of baby shampoo.

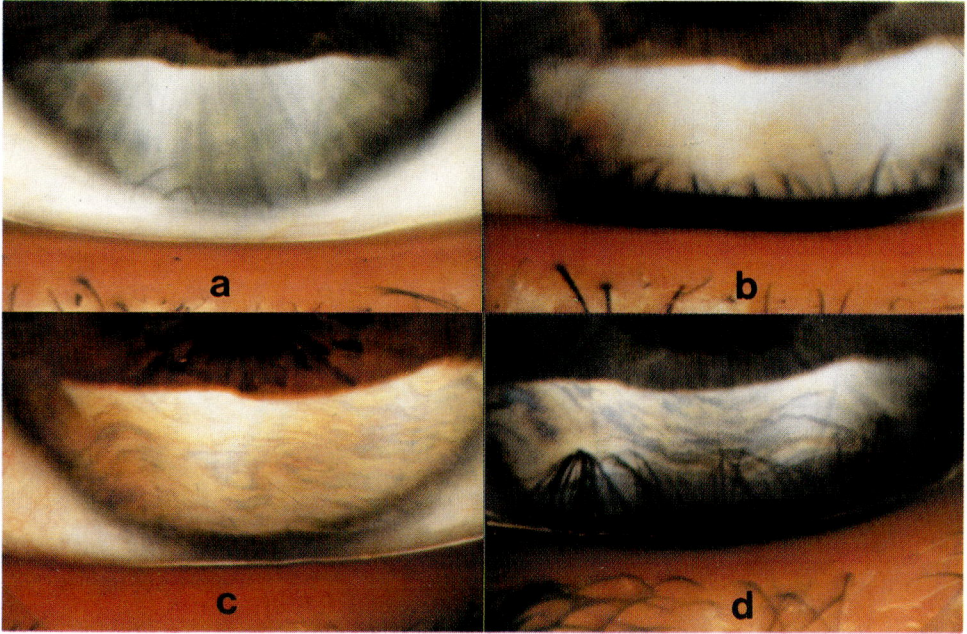

Figure 2. Color patterns of various tear films and their corresponding lipid layer thickness (LLT) values as listed in Table 1. (a) No colors or waves visible in zone of specular reflection. The gray or gray (white) appearance indicates a lipid layer thickness of 60 nm or less. (b) A dominant color of yellow is present on the lower portion of the cornea approaching the meniscus (centrally), representing a lipid layer thickness of 90 nm. (c) An irregular wave pattern with a dominant brown color (135 nm), and secondary colors of brown (yellow) (120 nm) and blue (180 nm). (d) The dominant color of blue (165-180 nm) is seen in the form of a more regular wave pattern, with secondary colors of brown and yellow indicating thinner regions.

## Measurement of Tear Film Lipid Layer

A custom-designed specular reflection microscope system allowed quantification of the tear film LLT based on the interference colors of the lipid layer (Korb et al., 1993). This system included a hemicylindrical broad spectrum illumination source with heat absorbing filters, a binocular microscope with a Zeiss beam-splitter providing 70% light to a high resolution video camera, a VHS recorder, and a high resolution 20-inch color monitor. Following calibration with Eastman Kodak color reference standards (Wratten filters), the static and dynamic appearance of the lipid layer was observed before and after blinking. During this observation period, the subject was instructed to blink naturally while gazing at a fixation target. For purposes of quantitation and standardization, a specific region of the tear film was designated for analysis. This area encompassed a zone approximately one mm above the lower meniscus to slightly below the inferior pupillary margin, averaging 5 mm wide by 2.5 mm in height. The dominant color within this designated area was used as the basis for assigning LLT values. Thickness values were assigned to specific colors on the basis of prior work on tear film lipid layer interference colors (McDonald, 1969; Norn, 1979; Guillon, 1982; Hamano et al., 1982) and are summarized in Table 1. To confirm the LLT values assigned to each subject's tear film, video tape recordings were independently graded by two observers masked as to subject identity. Examples of tear films and their corresponding LLT values are shown in Figure 2 (colorplate). Statistical analysis was performed using Student's t-test.

## Experimental Sequence

Baseline lipid layer thickness values were established for each subject at the time of initial examination and again two weeks later. Following these two baseline measurements, subjects began the six-month treatment protocol which included four meibomian gland expressions performed as office procedures at six week intervals and daily self-administered meibomian gland treatment. Six months after the initial meibomian gland expression, subjects were evaluated for lipid layer thickness (Measurement 1). The subject then continued daily self-treatment for an additional two weeks at which time a

Table 1. Quantification of lipid layer thickness (LLT) according to dominant color of interference pattern. Parentheticals indicate prominent, but less dominant color.

| COLOR | | LLT(nm) |
|---|---|---|
| White | = | 30 |
| Gray (White) | = | 45 |
| Gray | = | 60 |
| Gray (Yellow) | = | 75 |
| Yellow | = | 90 |
| Yellow (Brown) | = | 105 |
| Brown (Yellow) | = | 120 |
| Brown | = | 135 |
| Brown (Blue) | = | 150 |
| Blue (Brown) | = | 165 |
| Blue | = | 180 |

second post-treatment measurement of lipid layer thickness was determined (Measurement 2).

## RESULTS

In contrast to the clear sebum expressed from normal meibomian glands (Fig. 3a), manual expression of dysfunctional meibomian glands often yielded a thickened, cloudy discharge (Fig. 3b). Occasionally, this cloudy discharge was copious in nature (Fig. 3c). Frequently, meibomian material appeared to have undergone solidification and was expressed from the glands in the form of a small, firm mass (Fig. 3d). A paste-like, filamentous material was expressed in some instances (Fig. 3e). Frequently, there was a minimal yield, or no yield, of sebum despite maximum digital pressure. The consistency and morphology of the expressed material varied among meibomian glands within a given individual. Over the course of the six-month treatment period, there was some indication of improved meibomian gland function in all subjects, as evidenced by the increased number of glands from which sebum could be expressed. Concurrently, there was a decrease in the number of glands exhibiting solidified meibomian material.

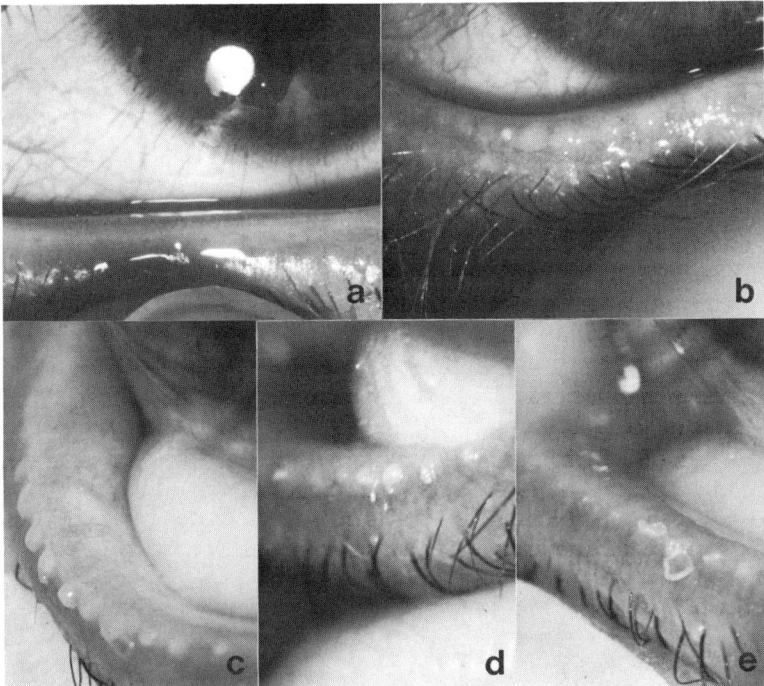

Figure 3. Meibomian gland expression of (a) normal and (b,c,d,e) dysfunctional meibomian glands. (a) Clear sebum. (b) Cloudy discharge. (c) Copious cloudy discharge. (d) Minimal discharge of firm mass. (e) filamentous, paste-like extrusion.

Baseline LLT measurements did not significantly vary at the two time points studied prior to beginning the treatment program, remaining 60 nm or less in all cases. Following the six-month program, tear film LLT values increased in all ten subjects (Table 2). The LLT increased from a value of 60 nm or less to a mean of 111 nm. This difference between pre- and post-treatment values was highly significant ($p < 0.001$). Additionally, all subjects reported symptomatic relief following the treatment program, as characterized by increased comfort and/or decreased dry eye symptoms.

Table 2. Tear film lipid layer thickness following six-month treatment program.

| Subject # | Measurement 1 | Measurement 2 | Average |
|-----------|---------------|---------------|---------|
| 1 | 90 | 90 | 90 |
| 2 | 105 | 135 | 120 |
| 3 | 90 | 120 | 105 |
| 4 | 90 | 90 | 90 |
| 5 | 75 | 75 | 75 |
| 6 | 180 | 120 | 150 |
| 7 | 135 | 75 | 105 |
| 8 | 135 | 135 | 135 |
| 9 | 75 | 75 | 75 |
| 10 | 180 | 150 | 165 |

Pre-treatment values for all subjects were 60 nm or less.
Measurements 1 and 2 determined at a two week interval after the six-month program.
All data expressed as nm.

## DISCUSSION

The present study examined the changes in tear film lipid layer thickness following a six-month treatment program designed to improve meibomian gland function. At the onset of the study, subjects displayed tear film LLT values of 60 nm or less, and the overt appearance of expressed sebum was consistent with that reported by Korb and Henriquez (1980) for patients demonstrating contact lens intolerance. Following treatment, subjects revealed a trend toward less solidified forms of sebum and reported symptomatic relief. These results were accompanied by a significant increase in tear film lipid layer thickness, which may result in a more stable tear film. This treatment regimen may thus be beneficial in cases of keratoconjunctivitis, dry eye, contact lens intolerance, and other syndromes where underlying meibomian gland obstruction may be a significant component in the etiology of the disorder.

## REFERENCES

Guillon, J.-P., 1982, Tear film photography and contact lens wear, J. Brit. Contact Lens Assoc. 5:84.

Hamano, H., and Mitsunaga, S., 1982, Clinical examinations and research on tear, in:"Menicon Toyo's 30th Anniversary Special Compilation of Research Reports," K. Tanaka, N. Anan, M. Mikami, H. Kaneko, S. Matsuoka, E. Kato, and K. Tanaka, eds., Toyo Contact Lens Co., Tokyo.

Henriquez, A.S., and Korb, D.R., 1981,Meibomian glands and contact lens wear, Brit. J. Ophthalmol. 65:108.

Keith, C.G., 1967, Seborrheic blepharo-keratoconjunctivitis, Trans. Ophthalmol. Soc. UK 87:85.

Korb, D.R., and Henriquez, A.S., 1980, Meibomian gland dysfunction and contact lens intolerance, J. Am. Opt. Assoc. 51:243.

Korb, D.R., Baron, D.F., Herman, J.P., Finnemore, V.M., Exford, J.M., Hermosa, J.L., Leahy, C.D., Glonek, T., and Greiner, J.V., 1993, Tear film lipid layer thickness as a function of blinking, Cornea, in press.

McCulley, J.P., and Sciallis, G.P., 1977, Meibomian keratoconjunctivitis, Am. J. Ophthalmol. 84:788.

McDonald, J.E., 1969, Surface phenomena of the tear film, Am. J. Ophthalmol. 67:56

Norn, M.S.,1979, Semiquantitative interference study of fatty layer of precorneal film, Acta Ophthalmol. 57:766.

# ANALYSIS AND FUNCTION OF THE HUMAN TEAR PROTEINS

Aize Kijlstra[1,2] and Abel Kuizenga[1]

[1]Graduate School Neurosciences Amsterdam
The Netherlands Ophthalmic Research Institute
PO Box 12141
1100 AC Amsterdam, The Netherlands
[2]Department of Ophthalmology
University of Amsterdam
Amsterdam, The Netherlands

## INTRODUCTION

The ocular mucosa contains various proteins exerting numerous functions including amongst others: anti-microbial defense, lubrication, wound healing and regulation of the inflammatory response. During evolution the tear film has been adapted, depending upon the environment where certain species live. This explains why the composition of tear proteins is largely different amongst various species. In this minireview we will discuss recent analytical techniques to analyse tear proteins and their function.

## TECHNIQUES FOR ANALYSIS OF PROTEINS IN HUMAN TEAR FLUID

Several methods can be employed to sample tear fluid from the human eye, each accompanied by it's own shortcoming[1]. The effect of collection techniques[2-5] as well as tear flow[6,7] on protein composition in tears is well established. Numerous techniques are used for analysis and characterization of proteins in tears. They include gel filtration and anion exchange chromatography, high performance liquid chromatography (HPLC)[2,6-12], various electrophoretic techniques e.g. sodium dodecyl sulfate polyacrylamide gel electrophoresis (SDS-PAGE), native electrophoresis, immuno-electrophoresis, isoelectric focusing (IEF), two-dimensional (2D) electrophoresis and a recently introduced capillary electrophoretic technique[13]. With these techniques tear proteins can be separated, offering an overall view of the protein composition in tears.

In addition, analysis of the individual proteins in tears can be performed using various immunological techniques e.g. the enzyme linked immunosorbent assay (ELISA), the radioimmunoassay (RIA) as well as immunodiffusion. Finally, various assays are used for the determination of enzymes, known to be present in tear fluid[1].

## High Performance Liquid Chromatography

HPLC analysis of human tears was first reported by Boonstra et al. in 1984 and later elaborated further by the group of Fullard[2,6-12]. This technique offers a rapid view on tear protein composition, using microamounts of tears (2 $\mu$l). Fullard used HPLC separation of tear proteins in combination with specific ELISA's for individual collected tear protein fractions[11]. In these latter studies, HPLC fractionation was necessary to eliminate factors in tears[14], which interefere with the ELISA. The major tear proteins secretory IgA (sIgA), lactoferrin, tear specific prealbumin (TSPA) and lysozyme were identified. These techniques were also used more recently by Fullard for the purification of TSPA[12]. HPLC can therefore be used for preparative purposes as well.

## Electrophoresis

The best available resolution of human tear proteins is currently achieved using polyacrylamide gel electrophoresis. Until recently, the use of this method has been performed using macro-electrophoretic systems. In general, these involved time-consuming procedures and hampered large-scale analysis of tear samples. These drawbacks were overcome by the introduction of an automated minigel electrophoresis system. With this system, various electrophoretic analytical methods can now be performed on a microscale level, offering rapid analysis of biological samples on a large scale, using minute amounts of sample[15]. Staining of SDS-PAGE minigels with silver can detect as little as 0.3 to 0.5 ng per protein band (fig. 1).

Other powerful methods for tear protein analysis include native electrophoresis, isoelectric focusing (IEF) and two dimensional electrophoresis. One example is separation of tear proteins by IEF (denoted as the first dimension), followed by SDS-PAGE of the

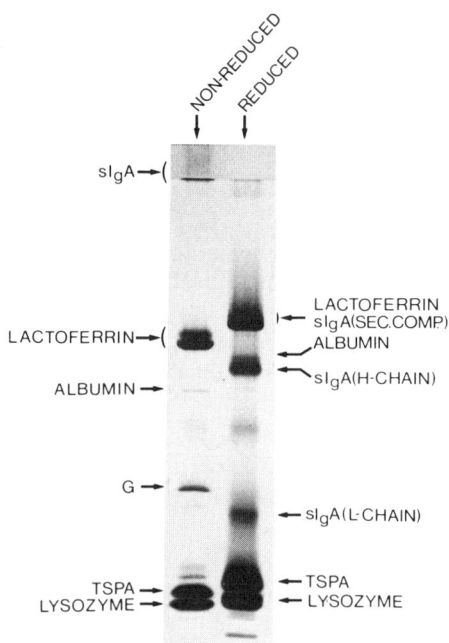

**Figure 1.** Minigel SDS-PAGE of normal human tears under disulfide bridge reducing and nonreducing conditions.

separated proteins in the second dimension. Another example is separation of proteins by native PAGE, followed by their separation in the second dimension with SDS-PAGE or IEF. With these techniques, Gachon et al detected over 60 proteins in tears and also found TSPA to consist of several isoforms[16,17].

After SDS-PAGE of tears, separated proteins can be transferred (blotted) by diffusion and by electric means to suitable carrier materials like nitrocellulose paper. With this combined technique, unknown protein bands on SDS-PAGE profiles can be localized and further identified using e.g a variety of immunological or lectin probes. Lectins are commercially available proteins, which specifically bind to certain carbohydrate residues, thus allowing detection of and differentiation between various glycoproteins. A further characterization and identification of tear proteins using this technique was reported by Kuizenga et al.[18].

## PROTEINS IN HUMAN TEAR FLUID

The principal proteins in human tears known today are secretory immunoglobulin A (sIgA), lactoferrin, tear specific prealbumin (TSPA) and lysozyme. They comprise the major part of total protein content in tear fluid. In particular lactoferrin, lysozyme and sIgA are known to play a protective role against microbial invasion at the surface of the eye.

### Secretory Immunoglobulin A

The ocular secretory immune system is of importance in the defense of the ocular mucosal surfaces against microbial challenge. This protective function is mediated primarily through secretory immunoglobulin A (sIgA), the predominant antibody in tears[19] as well as in other external secretions. IgA in external secretions is usually present in a dimeric form containing the additional polypeptides J (joining) chain and a glycoprotein called secretory component. IgA in secretions is therefore denoted as secretory IgA and has a molecular weight of 385 kilodaltons. This assembled sIgA molecule is a product of two different cell types: IgA with J chain is produced by plasma cells and secretory component by epithelial cells. External secretions contain about equal amounts of the IgA1 and IgA2 subtypes[20].

The primary site of the secretory immune system of the eye seems to be the lacrimal gland. Immunoglobulin A in tears is thought to be locally produced by interstitial plasma cells after homing to main and accessory lacrimal gland tissue[3,21-23]. Intracellular transport of IgA by acinar epithelial cells and secretion to the lacrimal lumen requires dimerisation and combination of IgA by J-chain and subsequent coupling to secretory component. In the lacrimal glands, secretory component is synthesized by mucosal epithelial cells[22-24] and acts as a membrane receptor for IgA[25], picking up IgA produced by plasma cells after which the complex is endocytosed and subsequently secreted into the lacrimal ducts.

Apart from main and accessory lacrimal IgA secretion, there may be additional sources for this protein in tears. This item has been investigated in both animals and humans. Reports on this subject are however controversial. IgA containing plasma cells have been detected by Franklin et al.[26] in rabbit conjunctival stroma along with secretory component in conjunctival epithelial cells. In the human conjunctiva, numerous IgA-containing plasma cells have been detected as well[21] but no conjunctival epithelial cells containing secretory component were observed[24]. In rat conjunctival tissue, both IgA-containing plasma cells and secretory component containing epithelial cells were absent[23]. Despite these controversial reports, the human conjunctiva has been observed to be overlayed with a continuous film of sIgA, which seems to be attached to conjunctival mucus[27].

Levels of sIgA in normal human tears have been measured by ELISA and radial immunodiffusion. Coyle et al used a modified sensitive ELISA for measuring tear immunoglobulins[28]. In general, IgA-levels in tears are found to vary between 0.1 and 0.6 g/L and depend on the used assay method as well as the conditions during tear collection. Fullard has recently emphasized the influence of tear flow rate on protein levels: their sIgA levels were high in nonstimulated tears and levels dropped after stimulation of tear flow[6]. sIgA is the main protein on the eye during closed eye conditions[29].

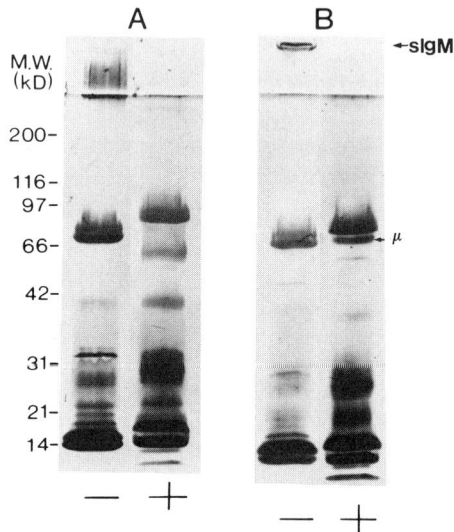

**Figure 2.** SDS-PAGE analysis of a tear sample of a normal individual (A) and an IgA deficient individual with sIgM compensation (B). Samples were treated under reducing (+) or nonreducing (-) conditions. For further details see[30].

With respect to other immunoglobulins, IgG levels in tears are normally very low[19,28]. Transudation of IgG into the tears can however occur in cases of trauma to conjunctival bloodvessels induced during tear collection[5] or in inflammatory conditions[31,32], whereby permeability of conjunctival bloodvessels is increased.

Secretory immunoglobulin M (sIgM) in tears has been found to be virtually absent by many authors, although Coyle found detectable low amounts[28].This immunoglobulin is capable of secretory dynamics similar to sIgA[33]. In IgA-deficiency there is evidence of enhanced synthesis of sIgM at mucosal surfaces and this immunoglobulin may compensate for absence of sIgA on these sites[34-36]. With regard to the eye, absence of sIgA in the tears may suggest absence of IgA-producing plasma cells in the lacrimal gland or the inability of these cells to produce sIgA. Deficiency of sIgA, which is not compensated by sIgM can be observed with HPLC analysis of tears[9]. We recently investigated tear samples of three IgA-deficient individuals by SDS-PAGE and immunoblotting and showed that two individuals contained sIgM in their tears[30] (fig. 2).

## Lactoferrin

Lactoferrin is an iron-binding protein, belonging to the transferrin family and serves to control iron levels in body fluids by sequestering and solubilizing ferric iron[37]. It has been detected in various secretions such as milk, saliva, pancreatic juice and semen[38] and also in specific granules of polymorphonuclear granulocytes[39]. In serum, lactoferrin is practically absent; in this bodyfluid the iron-binding protein is transferrin.

In human tears, lactoferrin was detected by Masson et al.[38] and its presence was further confirmed by Broekhuyse[40]. On SDS-PAGE it migrated similar as human milk lactoferrin with a molecular weight of about 82 kilodaltons[40] and could electrophoretically be distinguished from its transferrin counterpart in serum[17].

With ELISA, lactoferrin levels in normal human tears have been measured[41]. The concentration ranges between 1 and 2 g/L, representing about 25% of total tear protein content. Its abundance in tears and its production in the acinar epithelial cells of the main and accessory lacrimal glands[22,42] makes it one of the major proteins in human tear fluid. Levels in normal human tears were found to vary largely[41] and tend to decrease with older age[43]. Markedly decreased levels of tear lactoferrin have been reported in diseases affecting the lacrimal gland like keratoconjunctivitis sicca and Sjögren syndrome[44-46].

The role of lactoferrin on the ocular surface has recently been reviewed[47]. This protein is thought to act as a defense protein against bacterial invasion, owing to its bacteriostatic and bactericidal properties[40,48]. A recent publication suggests that lactoferrin may alter the membrane of gram negative bacteria allowing subsequent lysis by lysozyme[49]. Due to its iron-binding capacity it can prevent bacterial colonization of the external ocular structures by depriving the ocular mucosa of free iron, which is an essential factor for bacterial growth. Apart from this antimicrobial effect, this protein may also be involved in the regulation of inflammatory disorders. In human tears an anti-complementary effect was found to be associated with tear lactoferrin[50]. On the ocular surface, tear lactoferrin may therefore have an anti-inflammatory function by dampening complement activation during inflammatory processes.

The iron binding capacity of lactoferrin may implicate another function of this protein in the tear film with respect to reactive oxygen species. In biological systems, the production of the highly reactive hydroxyl radical is known to occur via the so-called Fenton type Haber-Weiss reaction, whereby iron is known to function as an effective catalyst[51]. Owing to the iron-binding capacity of lactoferrin, this protein has been shown to influence $OH^{.}$ formation[52] suggesting that this protein may help to protect the ocular mucosal surface against reactive $OH^{.}$ species, which can be formed during inflammatory conditions[52].

## Tear specific prealbumin

The source and nature of tear specific prealbumin was first well described by Bonavida et al[53], finding this protein to be unique for tear fluid and its absence in serum and other secretions. With histochemical and electrophoretic methods it was found that TSPA is synthesized in the main and accessory lacrimal glands[3,54]. Studies on tear samples from normals and keratoconjunctivitis sicca patients showed that TSPA may serve as a marker protein for lacrimal gland function[55]. Levels of this protein in human tears range between 0.5 and 1.5 g/L, making up 10-20% of the total protein content in tears[56]. TSPA has been shown to be a heterogenous protein (15-20 kilodaltons; isoelectric points from 4.6-5.4) which may reflect a genetic polymorphism of this protein[12,16,57]. Under certain conditions TSPA can aggregate resulting in a 31 kD protein band seen on SDS-PAGE denoted as protein G[11,12,16,58]. The origin and concentration of TSPA in tears may reflect an important function of this protein on the ocular surface but till now its role remains speculative.

Terminal amino acid analysis indicates that TSPA shows sequence homologies to a group of proteins called the lipocalin family[59].

## Lysozyme

This protein was first described by Fleming, who discovered its lytic action on bacterial cell-wall material, especially of Micrococcus lysodeicticus and denoted this protein as lysozyme[60]. Lysozyme is an antibacterial enzyme; it acts as a muramidase by hydrolyzing N-acetylmuramic($\beta$1-4)N-acetylglucosamine linkages of the peptidoglycan constituting bacterial cell walls, causing ultimate lysis of these cells. Human tear fluid contains the highest concentration of lysozyme as compared to other body fluids[61]. Reported levels in normal tears vary from 0.5 to 4.5 g/L[56] and are not influenced by tear flow[6]. Apart from tears of some monkeys[62], this enzyme is not present in tears of other species.

Human tear lysozyme (14-16 kilodaltons; isoelectric point: 10.4) is synthesized in the main and accessory lacrimal glands[3] and has been localized in the acinar and ductular epithelial cells[22,63]. The presence of lysozyme in tears is a reliable parameter for lacrimal gland function[55]. In dry eye states like keratoconjunctivitis sicca and Sjögren syndrome, lysozyme levels in tears are markedly reduced[45,46].

## Other proteins in tear fluid

Apart from the tear proteins described above, tears contain numerous enzymes of conjunctival and/or lacrimal origin[1,64]. Amylase in tears is an enzyme, that may be involved in glycogen metabolism of corneal cells during the process of corneal wound healing[65].

Human tears have been shown to contain protease activities[66]. One example is plasminogen activator (PA), a serine protease. It is thought to be derived from the lacrimal gland[67] as well as from the corneal epithelium and vascular endothelial cells of the conjunctiva[68]. Two types of PA have been found in tear fluid, urokinase PA (u-PA) and tissue PA (t-PA)[69]. PA plays an important role in fibrinolysis, catalyzing the conversion of plasminogen to the active proteolytic enzyme plasmin. Levels of plasmin in tears are normally low but become elevated in subjects with corneal disorders[70] and during contact lens wear[71]. Plasmin can cleave an adhesive protein called fibronectin. Fibronectin is a high molecular weight glycoprotein, present in soluble form in biological fluids and in insoluble form on cell surfaces and connective tissue. It has also been detected in human tear fluid[72]. Fibronectin, in combination with the plasminogen activator-plasmin system is thought to regulate healing of the cornea in both normal and pathological conditions[73-76]. Recently it was shown that plasmin activity in tears obtained at eye opening after sleep, was inversely proportional to the level of C3, C3 breakdown products and leucocyte number in tears. This indicates that plasmin-induced C3 activation may have caused recruitment of leucocytes on the ocular surface during sleep[77].

Another protease which can appear in the tears is tryptase. It is released from conjunctival mast cells in patients with ocular allergic inflammation[78].

Protease inhibitors, controlling protease activity have also been detected in tears and include $\alpha$1-antitrypsin, $\alpha$1-antichymotrypsin, inter-$\alpha$-antitrypsin and cystatins[1,79,80]. With regard to the cornea, there is increasing evidence that tear fluid may supply the cornea with growth factors. Recently, epidermal growth factor (EGF), a protein present in a variety of bodyfluids, has been detected in tear fluid and is of lacrimal origin[81,82]. This factor is known to stimulate corneal epithelial proliferation and promotes epithelial wound healing of the cornea[83]. The concentration of EGF in tears is very low (below 10 ng/ml) and seems to depend on dynamics of tear flow[84]. Pathological states of the eye seem to be associated with a decreased presence of EGF in tear fluid[85].

# FINAL CONCLUSIONS AND PERSPECTIVES FOR FURTHER INVESTIGATION

One of the main functions of the proteins present in the human tear film is to combat various microorganisms which are continuously bombarding the eye during the day and which may stay resident during the night in the closed eye situation. The main proteins involved hereby are sIgA, lactoferrin and lysozyme.

Recently developed techniques now offer a rapid qualitative and quantitative analysis of protein composition in human tear fluid. In particular the minigel electrophoresis system, is very suitable for studies on proteins in human tears, requiring only minute amounts of sample. This technique has been combined with immunoblotting, using a set of immunological as well as nonimmunological probes and has lead to a more precise interpretation of the protein pattern after electrophoresis. Despite the fact that there is a large variability of total protein levels in tears between healthy individuals, the minigel electrophoresis technique can be applied as a diagnostic tool for protein analysis of tears from subjects, in whom tear production is disturbed. Sjögren's syndrome, a lacrimal disease associated with decreased levels of the tear specific proteins lactoferrin, TSPA and lysozyme, may possibly be recognized with this method.

Renewed interest in diurnal effects upon tear protein composition may provide more insight in the events occuring during the night especially in relation to granulocyte influx and high sIgA levels. In subjects with IgA-deficiency, sIgA status in tear fluid and possible compensation by sIgM can easily be verified with electrophoretic techniques. Little is known however concerning the relation between ocular disease and sIgA deficiency.

More attention is being paid to proteins in human tears present in low quantities but with powerful biologic activities such as growth factors and proteolytic enzymes. In the near future it is to be expected that this list will be expanded with others such as the various members of the cytokine network.

## REFERENCES

1.  N.J. van Haeringen, Clinical biochemistry of tears, *Surv. Ophthalmol.* 26:84 (1981).
2.  A. Boonstra, A.C. Breebaart, C.J.J. Brinkman, L. Luyendijk, A. Kuizenga, and A. Kijlstra, Factors influencing the quantitative determination of tear proteins by high-performance liquid chromatography, *Curr. Eye Res.* 7:893 (1988).
3.  P.T. Janssen, and O.P. van Bijsterveld, Origin and biosynthesis of human tear proteins, *Invest. Ophthalmol Vis. Sci.*, 24:623 (1987).
4.  P.T. Janssen, and O.P. van Bijsterveld, Blood-tear barrier and tear fluid composition, in: "The Preocular Tear Film: In Health, Disease, and Contact Lens Wear," F.J. Holly, ed., Dry Eye Institute, Lubbock, TX, USA, 471 (1986).
5.  R.N. Stuchell, J.J. Feldman, R.J Farris, and D. Mandel, The effect of collection technique on tear composition, *Invest. Ophthalmol. Vis Sci.* 25:374 (1984).
6.  R.J. Fullard, and A.C. Snyder, Protein levels in non-stimulated and stimulated tears of normal human subjects, *Invest. Ophthalmol. Vis. Sci.* 31:1119 (1990).
7.  R.J. Fullard, and D.L. Tucker, Changes in human tear protein levels with progressively increasing stimulus, *Invest. Ophthalmol. Vis. Sci.* 32:2290 (1991).
8.  A. Boonstra, and A. Kijlstra, Separation of human tear proteins by high performance liquid chromatography, *Curr. Eye Res.* 12:1461 (1984).
9.  R.J. Boukes, A. Boonstra, A.C. Breebaart, D. Reits, E. Glasius, L.Luyendijk, and A. Kijlstra, Analysis of human tear protein profiles using high performance liquid chromatography, *Doc. Ophthalmol.* 67:105 (1987).
10. R.J. Fullard, L.J. DeLucas, and T.S. Crawford, HPLC analysis of proteins in human basal and reflex tears, in: " The Preocular Tear Film: In Health, Disease, and Contact Lens Wear," F.J. Holly, ed., Dry Eye Institute, Lubbock, TX, USA, 482 (1986).
11. R.J. Fullard, Identification of proteins in small tear volumes with and without size exclusion HPLC fractionation, *Curr. Eye Res.* 7:163 (1988).

12. R.J. Fullard, and D.M. Kissner, Purification of the isoforms of tear specific prealbumin, *Curr. Eye Res.* 10:613 (1991).
13. S.T. Lin, R.B. Mandell, R. Dadoo, and R.N. Zare, Nanoliter tear sample analysis by capillary electrophoresis,. *Invest. Opthalmol. Vis. Sci.* 32 (Suppl.):733 (1991).
14. A. Boonstra, N. van Haeringen, and A. Kijlstra, Human tears inhibit the coating of proteins to solid phase surfaces, *Curr. Eye Res.* 11:1137 (1985).
15. A. Kuizenga, N.J. van Haeringen, and A. Kijlstra, SDS-Minigel electrophoresis of human tears, *Invest. Ophthalmol. Vis. Sci.* 32:381 (1991).
16. A.M. Gachon, P. Lambin, and B. Dastugue, Human tears: Electrophoretic characteristics of specific proteins, *Ophthalmic Res.* 12:277 (1980).
17. A.M. Gachon, P. Verrelle, G. Betail, and B. Dastugue, Immunological and electrophoretic studies of human tear proteins, *Exp. Eye Res.* 29:539 (1979).
18. A. Kuizenga, N.J. van Haeringen, and A. Kijlstra, Identification of lectin binding proteins in human tears, *Invest. Ophthalmol. Vis. Sci.* 32:3277 (1991).
19. B.H. McClellan, C.R. Whitney, L.P. Newman, and M.R. Allansmith, Immunoglobulins in tears, *Am. J. Ophthalmol.* 76:89 (1973).
20. J. Mestecky, and J.R. McGhee, Immunoglobulin A (IgA): Molecular and cellular interactions involved in IgA biosynthesis and immune response, *Adv. Immunol.* 40:153 (1987).
21. M.R. Allansmith, G.A Kajiyama, M.B. Abelson, and M.A. Simon, Plasma cell content of main and accessory lacrimal glands and conjunctiva, *Am. J. Ophthalmol.* 82:819 (1976).
22. T.E. Gillette, M.D. Allansmith, J.V. Greiner, and M.A. Janusz, Histologic and immunohistologic comparision of main and accessory lacrimal gland tissue, *Am. J. Ophthalmol.* 89:724 (1980).
23. O.G. Gudmundsson, D.A. Sullivan, K.J. Bloch, and M.R. Allansmith, The ocular secretory immune system of the rat, *Exp. Eye Res.* 40:231 (1985).
24. M.R. Allansmith, and T.E. Gillette, Secretory component in human ocular tissues, *Am. J. Ophthalmol.* 89:353 (1980).
25. D.J. Ahnen, W.R. Brown, and T.M. Kloppel, Secretory component: The polymeric immunoglobulin receptor, *Gastroenterol.* 89:667 (1985).
26. R.M. Franklin, R.A. Prendergast, and A.M. Silverstein, Secretory immune system of rabbit ocular adnexa, *Invest. Ophthalmol. Vis. Sci.* 18:1093 (1978).
27. S. Liotet, M. Leloc, and J. Glomaud, Lacrimal secretory IgA fixation on conjunctival mucus, in: "The Preocular Tear Film: In Health, Disease and Contact Lens Wear," ed., F.J. Holly, Dry Eye Institute, Lubbock, TX, USA, 770 (1986).
28. P.K. Coyle, and P.A. Sibony, Tear immunoglobulins measured by ELISA, *Invest. Ophthalmol. Vis. Sci.* 27:622 (1986).
29. R.A. Sack, K.O. Tan, and A. Tan, Diurnal tear cycle: Evidence for a nocturnal inflammatory constitutive tear fluid, *Invest. Ophthalmol. Vis. Sci.* 33:626 (1992).
30. A. Kuizenga, T.R. Stolwijk, E.J. van Agtmaal, N.J. van Haeringen, and A. Kijlstra, Detection of secretory IgM in tears of IgA deficient individuals, *Curr. Eye Res.* 9:997 (1990).
31. M. Ballow, L. Mendelson, P. Donshik, A. Rooklin, and Z. Rapacz, Pollen-specific IgG antibodies in the tears of patients with allergic-like conjunctivitis, *J. Allergy Clin. Immunol.* 73:376 (1984).
32. Y. Barishak, A. Zavoro, Z. Samra, and D. Sompolinsky, An immunological study of papillary conjunctivitis due to contact lenses, *Curr. Eye Res.* 3:1161 (1984).
33. P. Brandtzaeg, Human secretory immunoglobulin M. An immunochemical and immunohistochemical study, *Immunol.* 29:559 (1975).
34. R.R. Arnold, S.J. Prince, J. Mestecky, D. Linch, M. Linch, and J.R. McGhee, Secretory immunity and immunodeficiency, *Adv. Exp. Med. Biol.* 107:401 (1978).
35. P. Brandtzaeg, I. Fjellanger, and S.T. Gjeraldsen, Immunoglobulin M. Local synthesis and selective secretion in patients with immunoglobulin A deficiency, *Science* 160:789 (1968).
36. I.M. Coelho, M.T. Pereira, G. Virella, and R.A. Thompson, Salivary immunoglobulins in a patient with IgA deficiency, *Clin. Exp. Immunol.* 18:685 (1974).
37. P. Aisen, and I. Listowsky, Iron transport and storage proteins, *Ann. Rev. Biochem.* 357 (1980).
38. P.L. Masson, J.F. Heremans, and C. Dive, An iron binding protein common to many external secretions, *Clin. Chim. Acta.* 14:735 (1966).
39. M. Baggiolini, C. de Duve, P.L. Masson, and J.F. Heremans, Association of lactoferrin with specific granules in rabbit heterophil leucocytes, *J. Exp. Med.* 131:559 (1970).
40. R.M. Broekhuyse, Tear lactoferrin: a bacteriostatic and complexing protein, *Invest. Ophthalmol. Vis. Sci.* 13:550 (1974).

41. A. Kijlstra, S.H.M. Jeurissen, and K.M. Koning, Lactoferrin levels in normal human tears, *Br. J. Ophthalmol.* 67:199 (1983).

42. R.M. Franklin, K.R. Keyon, and T.B. Tomasi, Immunohistologic studies of human lacrimal gland: Localization of immunoglobulins, secretory component and lactoferrin, *J. Immunol.* 110:984 (1973).

43. J.I. Mc Gill, G.M. Liakos, N. Goulding, and D.V. Seal, Normal tear protein profiles and age-related changes, *Br. J. Ophthalmol.* 68:316 (1984).

44. P.T. Janssen, and O.P. van Bijsterveld, Comparison of electrophoretic techniques for the analysis of human tear fluid proteins, *Clin. Chim. Acta* 114:251 (1979).

45. P.T. Janssen, and O.P. van Bijsterveld, Lactoferrin versus lysozyme in the sicca syndrome, in: "The Preocular Tear Film: In Health, Disease and Contact Lens Wear," F.J. Holly, ed., Dry Eye Institute, Lubbock, TX, USA, 167 (1986).

46. P.T. Janssen, and O.P. van Bijsterveld, Tear fluid proteins in Sjögren's syndrome, *Scand. J. Rheumatol.* (Suppl.) 61:224 (1986).

47. A. Kijlstra, The role of lactoferrin in the nonspecific immune response on the ocular surface, *Regional Immunol.* 3(4):193 (1990/1991).

48. R.R. Arnold, M.F. Cole, and J.R McGhee, A bactericidal effect for human lactoferrin, *Science* 197:263 (1977).

49. R.T. Ellison, and T.J. Giehl, Killing of gram-negative bacteria by lactoferrin and lysozyme, *J. Clin. Invest.* 88:1080 (1991).

50. R. Veerhuis, and A. Kijlstra, Inhibition of hemolytic complement activity by lactoferrin in tears, *Exp. Eye Res.* 34:257 (1982).

51. F. Haber, and J. Weiss, The catalytic decomposition of hydrogen peroxide by iron salts, *Proc. Royal Lond. Series A*, 147:332 (1934).

52. A. Kuizenga, N.J. van Haeringen, and A. Kijlstra, Inhibition of hydroxyl radicals by human tears, *Invest. Ophthalmol. Vis. Sci.* 28:305 (1987).

53. B. Bonavida, A.T. Sapse, and E.E. Sercarz, Specific tear albumin: A unique lachrymal protein absent from serum and other secretions, *Nature* 221:375 (1969).

54. K. Inada K, Studies of human tear proteins. 3. Distribution of specific tear albumin in lacrimal glands and other ocular adnexae, *Jpn. J. Ophthalmol.* 28:315 (1984).

55. P. Janssen, and O.P. van Bijsterveld, The relations between tear fluid concentrations of lysozyme, tear-specific prealbumin and lactoferrin, *Exp. Eye Res.* 36:773 (1983).

56. E.R. Berman ER, in: "Biochemistry of the Eye", C. Blakemore, ed., Plenum Press, New York & London (1991).

57. E.A. Azen, Genetic polymorphism of human anodal tear protein, *Biochem. Genet.* 14:225 (1976).

58. J. Baguet, V. Claudon-Eyl, and A.M.F. Gachon, Tear protein G originates from denatured reat specific prealbumin as revealed by two-dimensional electrophoresis, *Curr. Eye Res.* 11:1057 (1992).

59. A. Delaire, H. Lassagne, and A.M.F. Gachon, New members of the lipocalin family in human tear fluid, *Exp. Eye Res.* 55: 645 (1992).

60. A. Fleming, and V.D. Allison, Observations on a bacteriolytic substance ("lysozyme") found in secretions and tissues, *Br. J. Exp. Pathol.* 3:252 (1922).

61. J. Hankiewicz, and E. Swierczek, Lysozyme in human body fluids, *Clin. Chim. Acta* 57:205 (1974).

62. N.J. van Haeringen, and L. Thörig, Enzymology of tear fluid, in: "Protides of the Biological Fluids", H. Peeters, ed., Pergamon Press, Oxford, 39:399 (1984).

63. T.E. Gillette, J.V. Greiner, and M.R. Allansmith MR, Immunohistochemical localization of human tear lysozyme, *Arch. Ophthalmol.* 99:298 (1981).

64. N.J. van Haeringen, and L. Thörig, Enzymatic composition of tears, in: "The Preocular Tear Film: In Health, Disease and Contact Lens Wear", F.J. Holly, ed., Dry Eye Institute, Lubbock, TX, USA, 522 (1986).

65. T. Kuwabara, D.G. Perkins, and D.G. Cogan, Sliding of epithelium in experimental corneal wounds, *Invest. Ophthalmol. Vis. Sci.* 15:4 (1976).

66. P.K. Tsung, and F.J. Holly, Protease activities in human tears, *Curr. Eye Res.* 6:351 (1981).

67. L. Thörig, G. Wijngaards, and N.J. van Haeringen, Immunological characterization and possible origin of plasminogen activator in human tear fluid, *Ophthalmic Res,* 15:268 (1983).

68. E. Lantz, and A. Andersson, Release of fibrinolytic activators from the cornea and conjunctiva, *Graefe's Arch. Clin. Exp. Ophthalmol.* 219:263 (1982).

69. K. Hayashi, and K. Sueishi, Fibrinolytic activity and species of plasminogen activator in human tears, *Exp. Eye Res.* 46:131 (1988).

70. E.M. Salonen, T. Tervo, E. Torma, A. Tarkkanen, and A. Vaheri, Plasmin in tear fluid of patients with corneal ulcers: basis for a new therapy, *Acta Ophthalmol.* (Copenh.), 65:3 (1987).

307

71. T. Tervo, and G.B. van Setten, Plasmin-like activity in tear fluid and contact lens wear, *Contact Lens Journal* 19:142 (1991).

72. O.L. Jensen, B.S. Gluud, and H.O. Eriksen, Fibronectin in tears following surgical trauma to the eye, *Acta Ophthalmol.* (Copenh.), 63:346 (1985).

73. T.M. Phan, S.C. Foster, A.S. Boruchof, L.M. Zagachin, and M.B. Colvin, Topical fibronectin in the treatment of persistent corneal epithelial defects and trophic ulcers, *Am. J. Ophthalmol.* 104:494 (1987).

74. E. Ruoslahti, E. Evgvall, and E. Hayman, Fibronectin: current concepts of its structure and functions, *Coll. Res.* 1:95 (1981).

75. K. Watanabe, S. Nakagawa, and T. Nishida, Stimulatory effects of fibronectin and EGF on migration of corneal epithelial cells, *Invest. Ophthalmol. Vis. Sci.* 28:205 (1987).

76. S. Barlati, E. Marchina, C.A. Quaranta, F. Vigasio, and F. Semeraro, Analysis of fibronectin, plasminogen activators and plasminogen in tear fluid as markers of corneal damage and repair, *Exp. Eye Res.* 51:1 (1990).

77. B.A. Holden, A. Vannas, R.A. Sack, and A. Tan A, Plasmin and complement C3 activation of inflammatory cells in the closed eye, *Invest. Ophthalmol. Vis. Sci.* 32 (Suppl.):732 (1991).

78. S.I. Butrus, K.I. Ochsner, M.B. Abelson, and L.B. Schwartz LB, The level of tryptase in human tears. An indicator of activation of conjunctival mast cells, *Ophthalmol.* 97:1678 (1989).

79. M. Zirm, O. Schmut, and H. Hofman, Quantitative bestimmung der antiproteinasen in der menschlichen tränenflussigkeit, *Alb. v Graefes Arch. Klin. Exp. Ophthal.* 198:89 (1976).

80. T. Barka, P.A. Asbell, H. van der Noen, and A. Prasad, Cystatins in human tear fluid, *Curr. Eye Res.* 10:25 (1991).

81. Y. Ohashi, M. Motokura, Y. Kinoshita, T. Mano, H. Watanabe, S. Kinoshita, R. Manabe, K. Oshiden, and C. Yanaihara, Presence of epidermal growth factor in human tears, *Invest. Ophthalmol. Vis. Sci.* 30:1879 (1989).

82. G.B. van Setten, L. Viinikka, T. Tervo, K. Personen, A. Tarkkanen, and J. Perheentupa, Epidermal growth factor is a constant component of human tear fluid, *Graefe's Arch. Clin. Exp. Ophthalmol.* 227:82 (1989).

83. B.J. Tripathi, P.S. Kwait, and R.C Tripathi, Corneal growth factors: A new generation of ophthalmic pharmaceuticals, *Cornea* 9:2 (1989).

84. G.B. van Setten, Epidermal growth factor in human tear fluid: increased release but decreased concentrations during reflex tearing, *Curr. Eye Res.* 9:79 (1990).

85. G.B. van Setten, T. Tervo, L. Viinikka, K. Personen, J. Perheentupa, and A. Tarkkanen, Ocular disease leads to decreased concentrations of epidermal growth factor in the tear fluid, *Curr. Eye Res.* 10:523 (1991).

# TEAR PROTEIN COMPOSITION AND THE EFFECTS OF STIMULUS

Roderick J. Fullard and Denise Tucker

Department of Physiological Optics
University of Alabama at Birmingham
Birmingham, AL  35294

## INTRODUCTION

Tear protein profiles are significantly affected by stimulus and by the use of invasive tear collection techniques.  The importance of stimulus control has been largely overlooked until recent times.  This is most likely due to the fact that the main three tear proteins, lactoferrin, tear specific prealbumin (TSP) and lysozyme, show very little change over a wide range of stimulus conditions (Fullard and Snyder, 1990; Fullard and Tucker, 1991). Comparing non-stimulated tears collected under carefully controlled conditions with high flow-rate stimulated tears collected after discarding the initial 20 µl reveals very large differences for many tear proteins.  Secretory IgA (both subclasses), IgM and IgG all show a more than 5-fold decrease in stimulated tears (Fullard and Snyder, 1990).  Transferrin and albumin show a smaller, but still significant, decrease in stimulated tears.  The three main lacrimal gland proteins and peroxidase undergo minimal change.  Levels of the enzyme, lactate dehydrogenase (LDH) have also been shown to be significantly lower in stimulated tears (Van Haeringen and Glasius, 1974; Fullard and Carney, 1984).

If stimulated tears are collected at gradually increasing flow-rates, patterns of change according to tear protein type begin to emerge.  The constitutive lacrimal gland protein, secretory IgA (Dartt, 1989), decreases more gradually than the serum proteins, albumin and transferrin (Fullard and Tucker, 1991).  IgM follows a similar pattern of change to secretory IgA.  Therefore, at least in terms of its secretion pattern, IgM appears to be behaving as a constitutive lacrimal gland protein.  With the initial onset of stimulus, levels of all regulated lacrimal gland proteins (Dartt, 1989) decrease slightly relative to non-stimulated tears. However, with increasing intensity of stimulus, their levels remain constant (Fullard and Tucker, 1991).  This suggests that non-stimulated levels of regulated proteins are artificially higher due to evaporation effects.  Therefore, their levels may be independent of tear flow-rate under all stimulus conditions.

Changes in tear secretory IgA with stimulus intensity were studied by approximately doubling tear flow-rate for each of five successive tear samples (Fullard and Tucker, 1991). With this rapidly increasing tear flow-rate, most changes occurred between non-stimulated tears and the first two stimulated samples. In addition, a more rapid decrease in serum protein levels was not clearly evident. Stimulus intensity was increasing with each successive sample, as was cumulative reflex tear volume. This precluded the possibility of determining the relative contribution of each factor. Therefore, the current study was designed to differentiate between flow-rate effects and the influence of cumulative reflex tear volume on changes in stimulated tear protein levels.

## METHODS

Experiments were designed to investigate: (a) the effects of gradual increases in stimulus intensity on tear protein levels, and (b) the effects of increasing, decreasing and variable stimulated tear flow-rates.

Informed consent was obtained from all subjects after the nature of the experimental procedures had been fully explained to them. Reflex tear flow was induced by indirect nasal stimulus using the sneeze reflex or ammonia vapor. Direct ocular irritation and ocular surface contact were avoided throughout the tear collection sequences. All study participants (n = 10) were experienced in tear collection techniques and had been specifically trained in tear flow-rate control. Following tear collection, samples were stored at -20°C. Within 24 hours, samples were fractionated by size exclusion HPLC as described previously (Fullard and Tucker, 1991). ELISA and kinetic assays were then applied to relevant HPLC fractions as described previously (Fullard and Tucker, 1991).

### Study 1 - Effect of Rapidly Increasing Stimulus Intensity

Details of the tear collection methods used in this study are included here as they form the basis for the design of studies 2 and 3. A 10 µl non-stimulated sample was initially obtained (collection rate < 0.5 µl/minute). Five stimulated tear samples were then collected in uninterrupted sequence at progressively increasing tear flow-rates: 5, 10, 20, 40 and > 50 µl/minute. Results of study 1 have been reported elsewhere (Fullard and Tucker, 1991).

### Study 2 - Effect of Gradually Increasing Stimulus Intensity

A total of 20 stimulated tear samples were collected with much more gradual increments in flow-rate between successive samples than in study 1. Starting flow-rate was also slower (mean 1.3 µl/minute). After 8 samples, flow-rate had increased to an average of 5 µl/minute. At the end of the 20 sample sequence tear flow-rate was 20 µl/minute.

### Study 3 - Effect of Variable Stimulus Intensity

Forty tear samples were obtained using a variable flow-rate pattern. Initial tear flow-rate was high ($\approx$ 40 µl/minute) and was maintained for the first eleven tear samples. Flow-rate was then progressively decreased over the next 18 samples (12 - 29) to $\approx$ 1 µl/minute, after which it again increased to 5 µl/minute for the final eleven samples (30 - 40).

## RESULTS

### Study 1

As mentioned above, results of this study have been presented elsewhere (Fullard and Tucker, 1991). Both constitutive lacrimal gland and serum-derived proteins underwent a decrease in concentration over the initial stimulated tear samples, serum protein levels dropping more rapidly. Regulated lacrimal gland proteins (lactoferrin, TSP and lysozyme) showed relatively little change.

### Study 2

Both secretory IgA (constitutive) and albumin (serum-derived) showed a gradual decrease in concentration over the first nine to ten tear samples. Secretory IgA levels remained at a stable minimum thereafter. However, albumin levels subsequently increased, forming a definite peak. Variation in the levels of these two proteins is shown as a function of cumulative reflex tear volume (figure 1) and tear flow-rate (figures 2 and 3). Transferrin followed a very similar pattern to albumin. Throughout the entire collection sequence, levels of the regulated lacrimal gland proteins, lactoferrin, TSP and lysozyme remained constant.

### Study 3

With the high initial tear flow-rate, levels of secretory IgA decreased very rapidly. They then remained constant as tear flow-rate decreased down to 2 µl/minute. At this stage (sample 20), secretory IgA began to increase, forming a second peak. With the final increase in tear flow-rate, secretory IgA levels again dropped. Serum albumin also decreased rapidly with the high initial tear flow-rate. After ten samples, the albumin level began to increase dramatically, forming a broad peak that spanned more than 20 samples, with a maximum concentration exceeding the starting albumin level. Variation in the secretory IgA and albumin levels are again shown as a function of cumulative reflex tear volume (figure 4) and tear flow-rate (figures 5 and 6). Transferrin again varied in much the same fashion as albumin. As in study 2, regulated tear proteins varied little throughout the tear collection sequence and showed no pattern of change with tear flow-rate.

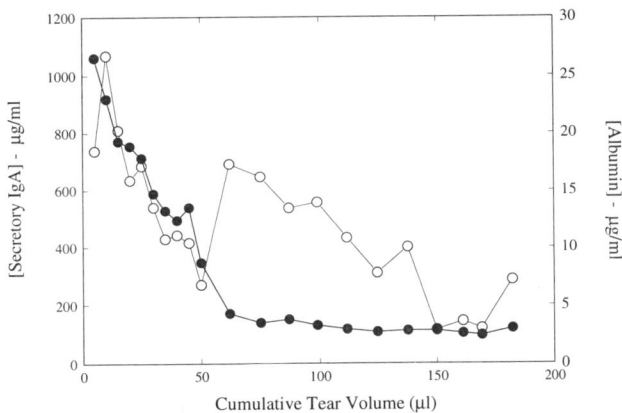

**Figure 1.** Tear secretory IgA (filled circles) and albumin (open circles) concentration vs. cumulative reflex tear volume collected since the onset of stimulus. Flow-rate increasing with each successive sample (study 2).

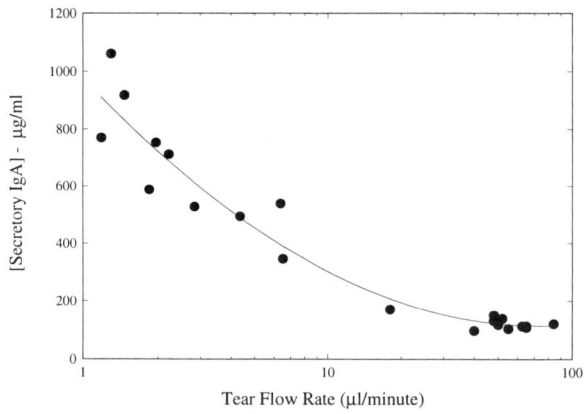

**Figure 2.** Tear secretory IgA concentration vs. stimulated tear flow-rate (study 2).

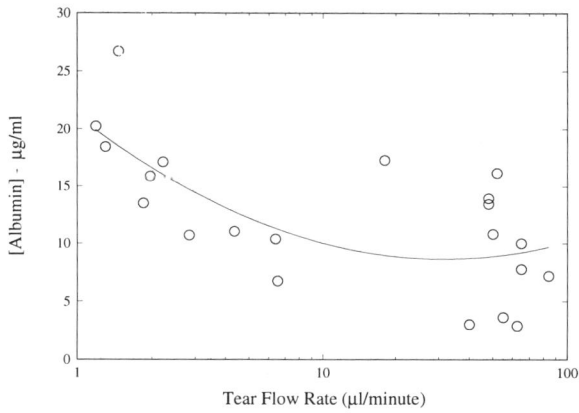

**Figure 3.** Tear albumin concentration vs. stimulated tear flow-rate (study 2).

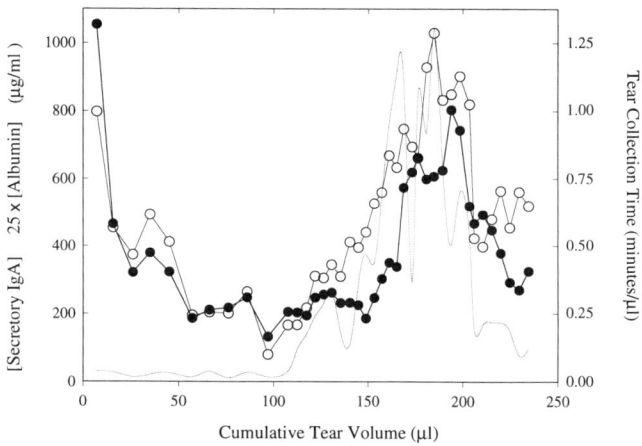

**Figure 4.** Tear secretory IgA (filled circles) and albumin (open circles) concentration vs. cumulative reflex tear volume collected since the onset of stimulus. Dotted line shows reciprocal tear flow-rate (study 3).

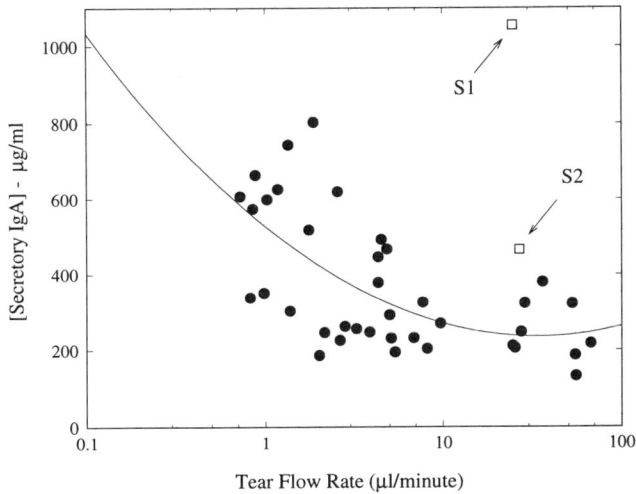

**Figure 5.** Tear IgA-SC concentration vs. tear flow-rate (filled circles) (study 3). First two tear samples (open squares) excluded from regression.

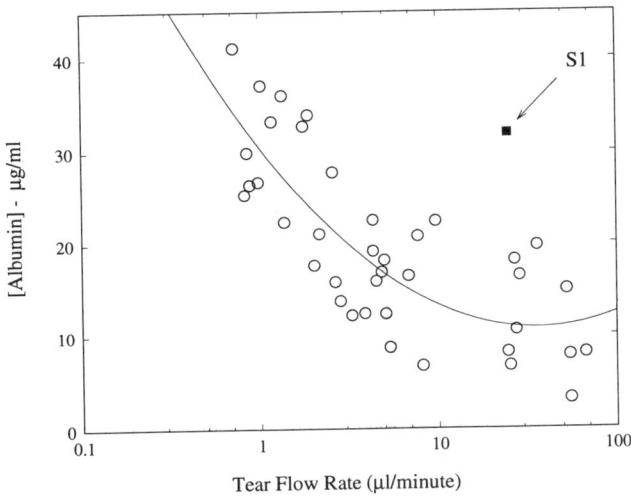

**Figure 6.** Tear albumin concentration vs. tear flow-rate (open circles) (study 3). First sample (filled square) excluded from regression.

## DISCUSSION

Results of the above studies demonstrate the relative importance of tear flow-rate and cumulative tear volume collected after the onset of stimulus on the various types of tear protein. Regulated lacrimal gland proteins, lactoferrin, tear specific prealbumin (TSP) and lysozyme, remained at a more or less constant level throughout all experiments. Tear concentrations of these proteins appear to be totally independent of tear flow-rate and cumulative stimulated tear volume - in the short-term. However, as reported previously (Fullard and Tucker, 1991), with regular collection of large tear volumes (in the order of milliliters per day) levels of all regulated tear proteins drop to as low as 30 - 50% of normal values. This presumably reflects depletion of lacrimal gland secretory granules.

Levels of the constitutive lacrimal gland protein, secretory IgA, were shown in study 3 to be primarily dependent on tear flow-rate, not cumulative tear volume collected after the onset of stimulus. Study 2 demonstrated that a very smooth decline in secretory IgA, protracted over 10 samples (> 50 µl total volume), can be elicited with steadily increasing stimulus intensity. However, as in study 1, this collection pattern did not differentiate between flow-rate and cumulative reflex tear volume effects. The variable flow-rate pattern used in study 3 showed that tear secretory IgA concentration is flow-rate dependent, not a function of cumulative reflex tear volume, as secretory IgA increased to form a second peak late in the stimulated tear sequence. This second peak correlated well (inversely) with decreasing tear flow-rate, while showing no dependence on cumulative reflex tear volume.

Serum albumin results from study 2 showed the greatest departure from secretory IgA patterns. Despite progressively increasing tear flow-rate, albumin levels began to rise sharply after collection of the first 10 samples (50 µl) and remained above the sample 10 level for the subsequent seven samples. In study 3, albumin levels were already rising prior to any significant decrease in tear flow-rate (sample 11) and ≈ 10 - 12 samples before secretory IgA showed any consistent increase. Since transferrin varied in a similar manner to albumin, this pattern probably occurs for all serum-derived tear proteins. Stimulated tear collection was very carefully controlled in this study to ensure that sampling micropipettes did not contact the ocular surface at any stage. This indicates that the pattern of change in tear levels of serum proteins was due to a non-mechanical increase in vascular permeability triggered by sustained reflex lacrimation. We should therefore reconsider the commonly accepted notion that increased tear levels of serum proteins must be due to mechanical trauma (Stuchell et al, 1984).

**ACKNOWLEDGEMENTS**

Supported by National Institutes of Health grants EY-07783, EY-03039 and RR05807.

**REFERENCES**

Dartt, D.A. Signal transduction and control of lacrimal gland protein secretion: A review. Curr. Eye Res. 8:619, 1989.

Fullard, R.J. and Tucker, D.L. Changes in human tear protein levels with progressively increasing stimulus. Invest. Ophthalmol. Vis. Sci. 32:2290, 1991.

Fullard, R.J. and Snyder, C. Protein levels in nonstimulated and stimulated tears of normal human subjects. Invest. Ophthalmol. Vis. Sci. 31:1119, 1990.

Fullard, R.J. and Carney, L.G. Diurnal variations in human tear enzymes. Exp. Eye Res., 38:15, 1984.

Stuchell, R.N., Feldman, J.J., Farris, R.L. and Mandel, I.D. The effect of collection technique on tear composition. Invest. Ophthalmol. Vis. Sci. 25:374, 1984.

Van Haeringen, N.J. and Glasius, E. Lactate dehydrogenase in tear fluid. Exp. Eye Res. 18:345, 1974.

# GROWTH FACTORS IN HUMAN TEAR FLUID AND IN LACRIMAL GLANDS

Gysbert-B. van Setten,[1,2,3] Gregory S. Schultz,[2]
and Shawn Macauley[2]

[1]Karolinska Institute, St. Eriks Eye Clinic, Laboratory for Dacryology
Fleminggatan 22, S-11282 Stockholm, Sweden
[2]University of Florida, Department of Obstetrics and Gynecology, Institute of
Wound Research, P.O. Box 100294, Gainesville, FL 32610, USA
[3]Eye Clinic, Helsinki Central University Hospital
Haartmaninkatu 4c, 00290 Helsinki, Finland

## INTRODUCTION

Epidermal growth factor (EGF) was the first member identified of a family of growth factors which exert their effects via a single 170,000 Mr plasma membrane receptor (for review see Todaro et al. 1990). Other members include transforming growth factor alpha (TGF-$\alpha$), amphiregulin (AR) and several viral growth factors. TGF-$\alpha$ is a peptide with a MW of about 5600 which shares 30% structural homology with EGF and has the three disulfide bonds that are characteristic for this family. However, EGF and TGF-$\alpha$ differ somewhat in their mechanism of binding to the EGF receptor and also have slightly different effects on their target cells (for review see Derynck 1988, Winkler et al. 1989). Another, new member of the EGF family is amphiregulin (AR). AR is a glycosylated, 84-amino acid polypeptide growth regulator which has sequence homology to the EGF family of proteins (Johnson et al. 1992). AR inhibits growth of some tumor cells but also stimulates growth of the normal keratocytes and fibroblasts.

The effects of EGF on corneal tissue has been subject of many investigations during recent years. The clinical interest in EGF derives mainly from the observations that the external application of EGF to the corneal epithelium leads to cellular hypertrophy and stimulates cell division, thus enhancing corneal reepithelialization ( for review, see Burstein

*Lacrimal Gland, Tear Film, and Dry Eye Syndromes*
Edited by D.A. Sullivan, Plenum Press, New York, 1994

1987, Tripathi et al. 1990). The observation, however, that EGF also increases the production of fibronectin (FN) by cultured fibroblasts (Chen et al. 1977) and by corneal tissue (Nishida et al. 1984), implies that EGF has effects on corneal tissue which are not purely mitogenic. The intention to use the effects of EGF on corneal tissue via its topical application as therapeutical agent has to consider the natural presence of growth factors in tear fluid and at the ocular surface. Hence, attention has to be focussed on the question whether there are other growth factors present in tear fluid that could bind to the EGF receptors or interfere with their binding mechanisms at the ocular surface. Only recently it has been found that another memeber of the EGF family, TGF-α, is produced endogenously by corneal cells and may play key an important role in the natural wound healing process by paracrine and autocrine mechanisms (for review see Schultz et al. 1992). Nothing, however, is currently known on the presence of TGF-α in human tear fluid. The results presented in this paper now give initial evidence, that not only EGF, but also TGF-α derive from the lacrimal gland and is naturally present in tear fluid.

## MATERIALS AND METHODS

### Radioimmunoassay

Tear fluid samples were collected as described previously by the use of blunted capillaries (van Setten et al. 1989) from healthy individuals. Half of each sample was analyzed for the determination of TGF-α concentration. The remaining part from all samples was pooled for triplicate determination of TGF-α concentration in four dilutions for the logit transformation. All samples were reflex tears which were collected after stimulation with onion vapour.

The concentration of TGF-α was determined by radioimmunoassay (RIA) which uses specific polyclonal rabbit antibody. Human recombinant TGF-α was used as standard and for $I^{125}$labelling.

In the first experiment the pooled sample was analysed in triplicate at four dilutions. The resulting displacement data were linearized by logit transformation (Chard 1990) and the best-fit lines were determined by linear regression analysis. Slopes of lines generated by TGF-α standard and tears were compared for difference by t-test for slopes. Concentrations of TGF-α were calculated by averaging the amounts of TGF-α interpolated from the linear regression of the standard curve and were expressed as pg of TGF-α immunoreactivity per ml of tear fluid.

### Immunohistochemistry

For immunohistochemical demonstration of TGF-α-like immunofluorescence (TGF-α-LI) in bovine lacrimal glands, sheep anti-TGF-α (Chemicon, Temecula, CA, USA) served as the primary antibody. The term TGF-α- LI was chosen for the immunoreaction detected because immunohistochemical cross-reactivity with substances resembling TGF-α could not be excluded. Fluorescein-isothiuocyanate (FITC) conjugated anti-sheep-IgG was used as the secondary antibody. For immunohistochemical control, either the primary or the secondary antibody was omitted. All specimens were fixed immediately in a descending ethanol series. Subsequently they were rinsed overnight in phosphate buffered saline containing 25% sucrose. Cryostat sections (7 μm) were processed for indirect immunohistochemistry.

## RESULTS

Logit transformation of the competitive binding data produced by diltions of the TGF-α standard and the dilutions of the pool of tears generated lines with slopes that were not statistically different. This indicates the presence of immunoreactive TGF-α protein in the tear fluid. Analysis of individual tear samples indicated that TGF-α was present in all samples analyzed. The concentration was approximatel 80-100 pg/ml for reflex tears.

Immunohistochemistry revealed TGF-α-LI immunofluorescence in the acini of the lacrimal glands, close to the nuclei of the dacryocytes. An intense staining of the interacinar or intraacinar duct cells like shown previously for EGF (van Setten et al. 1990) was not observed.

## DISCUSSION

The present results indicate that TGF-α is present in human tear fluid. Although the number of specimens investigated is so far low, it seems most likely that TGF-α is a constant component of human tear fluid similar to EGF (van Setten et al. 1989). Another similarity between EGF and TGF-α in tears concerns its origin. TGF-α seems to be produced by the lacrimal gland like EGF. Recent immunohistochemical studies using monoclonal antibodies and horseraddish-peroxidase techniques support the initial results.

As reported earlier, the concentrations of EGF in tear fluid collected with minimal stimulation is about 7-8 ng/ml (van Setten 1990). During and after stimulation of reflex tearing this concentration decreases to 2-3 ng EGF/ml (van Setten 1990). Corneal disease induced long-term stimulation of reflex tearing was associated with even lower concentrations of EGF in tear fluid, i.e. 1-2 ng/ml (van Setten et al. 1991). Although the effects of this temporary decrease of EGF concentration in tear fluid on corneal (patho-) physiology is still speculative, the resulting models focus special attention on the possible balance between concentrations of free EGF and the EGF receptors (van Setten et al. 1991). In our current model, the constant presence of EGF in tear fluid together with the presence of specific receptors at the epithelial surface of the cornea led to the assumption that a certain minimum concentration of EGF may be required for the maintenance of corneal surface integrity and cell turnover (van Setten et al. 1990). This model has to consider also the simultaneous presence of TGF-α in tear fluid, which is known to bind to the same receptors, although in a different way. If both growth factors are released by the lacrimal gland and follow the same secretion pattern, it is most probably that this difference becomes more evident in samples of human tears collected with a minimum of stimulation.

The physiological significance of EGF and TGF-α in tear fluid remains to be established. Present experimental results indicate that EGF or TGF-α may penetrate through injured epithelium and stimulate the migration and metabolism of cells in the epithelium and stroma whereas the uninjured epithelium appears to constitute a notable barrier for these growth factors. Recently the coexistance and possible effects of autocrine and paracrine pathways on corneal cells has been reviewed (Schultz et al. 1993).

Although the homology in the primary structure of EGF and TGF-α (Marquardt et al. 1984, Derynck et al. 1984) allows both factors to bind to the same receptors (Marquardt et al. 1984), it remains to be clarified whether and to what extent this structural similarity also leads to similar physiological response in the receptor expressed by corneal cells. These effects may be rather similar as suggested by the observation that in the induction of precocious eyelid opening, EGF (Cohen 1962) can be totally substituted by TGF-α (Smith et al. 1985).

The development of physiological models is, on the other hand furthermore complicated by the recent observation that corneal epithelial cells may express two kinds of EGF receptors, i.e. high affinity, and, more frequent, low affinity receptors (Hongo et al. 1992).


## CONCLUSION

Based on the results of the present study it is conlcuded that both EGF and TGF-α coexist in human tear fluid. Furthermore TGF-α probably is, like EGF, a constant component of human tear fluid. Considering the ability of both growth factors to bind to the same receptors, the physiological role TGF-α in tear fluid remains to be clarified.


## ACKNOWLEDGEMENTS

The authors are very thankful for the financial support of the M.Ehrnrooth Foundation, Helsinki, Finland and the University of Florida, Gainesville, Florida, NEI 05587 as well as for the grant of the honourable Councellor of the University of Helsinki, Finland which made the presented study possible.

## SUMMARY

EGF has been shown to be a constant component of human tear fluid. Its concentration depends on the actual tear fluid flow, as shown for other proteins secreted by the lacrimal gland. This organ has also been considered to be the origin of tear fluid EGF and immunohistochemical evidence for this hypothesis was found. During corneal disease the concentration of EGF in tear fluid considerably decreases to levels even lower than those found during short time stimulation of reflex tearing. Other members of the EGF family, such as TGF-α, have considerable similarity with the EGF molecule and even bind to the same receptor. Currently it is thought that TGF-α may be, in certain phases of cell life, even more important in the regulation of cell metabolism than EGF. In the present study we have investigated the presence of TGF-α in tear fluid and the lacrimal gland. The initial results presented here, show for the first time that TGF-α like EGF, seems to be constant component of human tear fluid and to originate, at least partially, from the lacrimal gland.

# REFERENCES

Burstein, N.L., 1987, Review:growth factor effects on corneal wound healing, *J. Pharmacol.* 1:263.

Chard, T., 1990, An introduction to radioimmunoassay and related techniques. *in:* "Laboratory Techniques in Biochemistry and Molecular Biology," R.H. Burdon, and P.H. Knippenberg, eds., Elsevier, Amsterdam, p.21.

Chen, L.B., Gudor, R.C., Sun, T.-T., Chen, A.B., and Mosesson, M.W., 1977, Control of a cell surface major glycoprotein by epidermal growth factor, *Science* 197:776.

Cohen, S., 1962, Isolation of a mouse submaxillary gland protein accelerating incisor eruption and eyelid opening in the new-born animal, *J. Biol. Chem.* 237:1555.

Derynck, R., Roberts, A.B., Winkler, M.E., Chen, E.Y., and Goeddel, D.V., 1984, Human transforming growth factor-alpha: precursor structure and expression in E.coli, *Cell* 38:287.

Derynck, R., 1988, Transforming growth factor a, *Cell* 54:593.

Hongo, M., Itoi, M., Yamaguchi, N., and Imanishi, J., 1992, Distribution of epidermal growth factor (EGF) receptors in rabbit corneal epithelial cells, keratocytes and endothelial cells, and the changes induced by transforming growth factor-beta 1, *Exp. Eye Res.* 54:9.

Johnson, G.R., Saeki, T., Gordon, A.W., Shoyab, M., Salomon, D.S., and Stromberg, K., 1992, Autocrine action of amphiregulin in a colon carcinoma cell line and immunocytochemical localization of amphiregulin in human colon, *J. Cell. Biol.* 118:741.

Marquardt, H., Hunkapiller, M.W., Hood, L.E., and Todaro, G.J., 1984. Rat transforming growth factor type I: Structure and relation to epidermal growth factor, *Science* 223:1079.

Nishida, T., Tanaka, H., Nakagawa, S., Sasabe, T., Awata, T., and Manabe, R., 1984, Fibronectin synthesis by the rabbit cornea: effects of mouse epidermal growth factor and cyclic AMP analogs, *Jpn. J. Ophthalmol.* 28:196.

Schultz, G., Chegini, N., Grant, M., Khaw, P., and MacKay, S., 1992, Effects of growth factors on corneal wound healing, *Acta Ophthalmol. Copenh. (Suppl)* 70:60.

Schultz, G., Khaw, P., Grant, M., MacKay, S., Chegini, N., and van Setten, G.B., 1993, Corneal wound healing: Role of EGF and TGF-α, Submitted

Smith, J.M., Sporn, M.B., Roberts, A.B. Derynck, R., Winkler, M.E., and Gregory, H., 1985, Human transforming growth factor-alpha causes precocoious eyelid opening in the newborn mice, *Nature* 315:515.

Todaro, G.J., Rose, T.M., Spooner, C.E., Shoyab, M., and Plowman, G.D., 1990, Cellular and viral ligands that interact with the EGF receptor, *Semin. Cancer Biol.* 1:257.

Tripathi, B.J., Kwait, P.S., and Tripathi, R.C., 1990, Corneal growth factors: A new generation of ophthalmic pharmaceuticals, *Cornea* 9:2.

van Setten, G.B., 1990, Epidermal growth factor in human tear fluid: Increased release but decreased concentrations during reflex tearing, *Curr. Eye Res.* 9:79.

van Setten, G.B., Tervo, T., Tarkkanen, A., Pesonen, K., Viinikka, L., and Perheentupa, J., 1989, Epidermal growth factor is a constant component of human tear fluid, *Graefes Arch. Clin. Exp.* Ophthalmol. 227:84.

van Setten, G.B., Tervo, T., Viinikka, L., Pesonen ,K.,Perheentupa, J., and Tarkkanen, A., 1991, Ocular disease leads to decreased concentrations of epidermal growth factor in the tear fluid, *Curr. Eye Res.* 10:523.

van Setten, G.B., Tervo, K., Virtanen, I., Tarkkanen, A., and Tervo, T., 1990, Immunohistochemical demonstration of epidermal growth factor in the lacrimal and submandibular glands of rats, *Acta Ophthalmol.* (Copenh.) 68:477.

van Setten, G.B., Tervo, T., Viinikka, L., Perheentupa, J., and Tarkkanen, A., 1991, Epidermal growth factor in human tear fluid: a minireview, *Int. Ophthalmol.* 15:359.

Winkler, M.E., O'Connor, L., Winget ,M., and Fendly, B., 1989, Epidermal growth factor and transforming growth factor α bind differently to the epidermal growth factor receptor. *Biochem.* 28:6373.

# RE-EXAMINATION OF THE ORIGIN OF HUMAN TEAR LDH

Toshio Tsubai and Masato Murai

Institute of Contact Lens Science
Osaka, 530 Japan

## INTRODUCTION

It is generally accepted that our basic tear fluid is secreted in part by the accessory glands of the conjunctiva.[1] However, human tear lactate dehydrogenase (LDH) is believed to originate from the corneal epithelial cells.[2] This report investigates the possibility that tear LDH is derived principally from the conjunctiva.

LDH released by the cornea would accumulate in the inferior marginal meniscus. It would not accumulate in the superior fornix because of the flow of tears from the lacrimal gland, and it would not accumulate in the inferior fornix because of the seal of the inferior lid margin against the corneal epithelium. Closing of the lids squeezes aqueous tears out and drags mucous down.[3,4] Thus any LDH in the fornix would be wrapped into the mucous thread. We have compared the LDH activities in the superior and inferior fornix, and compared them with the tear LDH in the inferior marginal meniscus. Moreover, LDH isoenzyme patterns of tears taken from each location were examined and compared with activities and isoenzyme patterns of tarsal and bulbar conjunctival epithelial LDH.

## METHODS

### Subjects

Fifty-one normal male (n = 18, ages 19-45 years old, mean ± STD = 25.6 ± 7.0) and female (n = 33, ages 19-40 years old, mean ± STD = 22.5 ± 4.2) volunteers participated in several experiments. A total of 272 measurements were made. Ten participants wore contact lenses (RGP = 8, HEMA = 2).

*Lacrimal Gland, Tear Film, and Dry Eye Syndromes*
Edited by D.A. Sullivan, Plenum Press, New York, 1994

## Collection of Tears

Tears were collected from 6 points on each subject, including right and left inferior marginal meniscus, right and left superior fornix, right and left inferior fornix, and each sample was analyzed individually. A soft polyethylene tube (Handaya) fitted over a micro syringe (Hamilton) was used to collect tears from the inferior meniscus. Under careful observation through a biomicroscope, 3 microliters of tears were collected from the inferior tear meniscus by immersing the 1 mm tip of the polyethylene tube (0.5 mm diam.) without touching the lower lid. The collected tears were dispersed into 5 ml saline.

For forniceal tear collections, a 1 mg piece of cotton, removed from an otological swab and shaped into a spindle, was placed in each fornix for 3 minutes while the subject blinked freely. The tear absorbent cotton, after weighing on an electroic balance, was rinsed in 5 ml saline. Earlier studies showed that this collection method does not alter the proportions of LDH isoenzymes present.

## Collection of Conjunctival Epithelia

After inspection of the cornea and conjunctiva, the ocular surface was anesthetized with 0.4% oxybuprocaine hydrochloride. The lids were everted and gently blotted with a cotton sheet to remove moisture. The tarsal conjunctiva (right and left and superior and inferior) and the bulbar conjunctiva (right and left and superior and inferior) were then lightly rubbed 10 times in 10 seconds with a cotton stick used in urology.

## Measurement of LDH Activities and Isoenzyme Pattern

LDH activity was determined with Hitachi 7350 Auto Analyzer using reagent kit LDH-HR (Wako Pure Chemical Ind, Ltd.). Electrophoretically separated LDH isoenzymes were stained with LDH isozyme Blue (CIBA Corning) and quantitated by densitometry (600 nm, ADC-20EX densitometer, KAYAGAKI Co., Ltd).

## RESULTS

### Tear LDH Activities

The LDH activities in the superior ($39.68 \pm 29.77$ IU/L; mean $\pm$ STD, n = 38) and inferior ($44.95 \pm 32.27$ IU/L; n = 68) fornices were similar, but significantly ($p < 0.01$) greater than the value in the inferior marginal meniscus (about 1.0, n = 30).

### Conjunctival LDH Activity

LDH activity was highest in the superior tarsal conjunctiva, $282 \pm 20$. In the inferior tarsal conjunctiva the value was $119 \pm 11$. The bulbar conjunctival activity was $0.83 \pm 1.67$.

## LDH Activity in Normals and Contact Lens Wearers

LDH activity in normals and contact lens wearers was 44 ± 40 (n = 68) and 54 ± 32 (n = 17) respectively.

## Normal Tear LDH Isoenzyme Patterns

Isoenzyme distributions in tears from superior and inferior fornices of normal individuals are shown in Figure 1 A and B respectively.

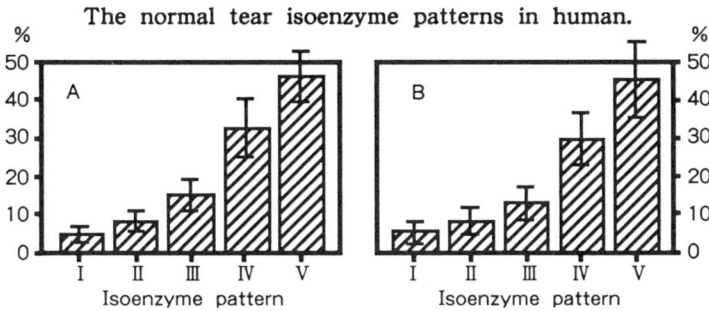

**Figure 1.** The normal tear isoenzyme patterns. A : the tear fluid in the superior conjunctival fornix (M ± STD, n = 29). B : tear fluid in the inferior conjunctival fornix (M ± STD, n = 30). Though the tear mix is negligible between these two spots, the both patterns were very similar. And these patterns resemble to the pattern which was indicated by Kahan & Ottovay.

## LDH Isoenzyme Patterns in High and Low Activity Tears

The isoenzyme distribution was the same in tears with high levels of LDH activity compared to tears with low levels of activity (Figure 2).

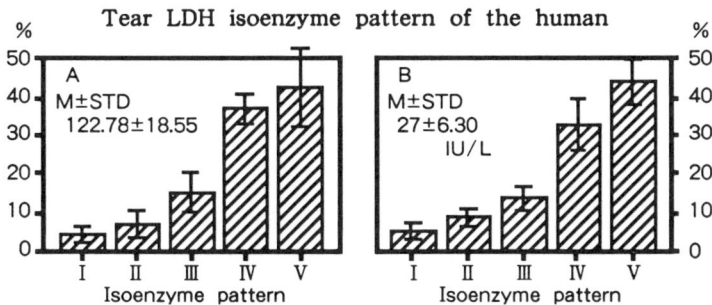

**Figure 2.** A : eyes with high LDH activities (n = 9). B : eyes with low LDH activities (n = 12). The tear LDH isoenzyme patterns do not change, even it there exists about 4 times differences

323

## LDH Isoenzyme Patterns in the Conjunctival Epithelium

The same pattern of isoenzymes seen in tears was present in the superior and inferior conjunctival epithelial cells. The LDH V activity was relatively high (Figure 3).

LDH isoenzyme patterns of the tarsal conjunctival epitheliums.

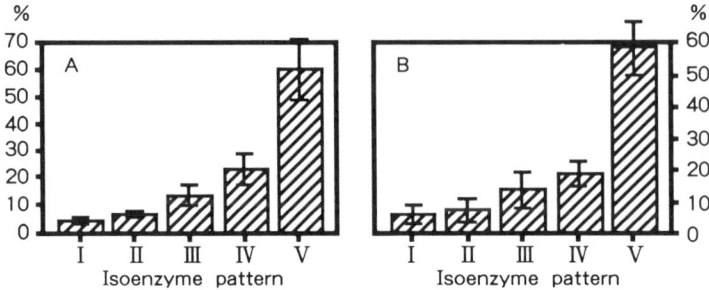

**Figure 3.** LDH isoenzyme patterns of A : the superior conjunctival epitheliums (n = 18) and B : the inferior conjunctival epitheliums (n = 16). These two patterns resemble to the tear's but LDH V is relatively higher. Since the half-life of LDH V is the shortest (1/5 of LDH I and 1/2 of LDH IV), it will dissipate faster in tear fluid.

## LDH Isoenzyme Patterns in Contact Lens Wearers

LDH isoenzyme V was relatively higher (p < 0.05) in contact lens wearers compared to age matched normals (Figure 4).

The isoenzyme patterns of CL wearers.

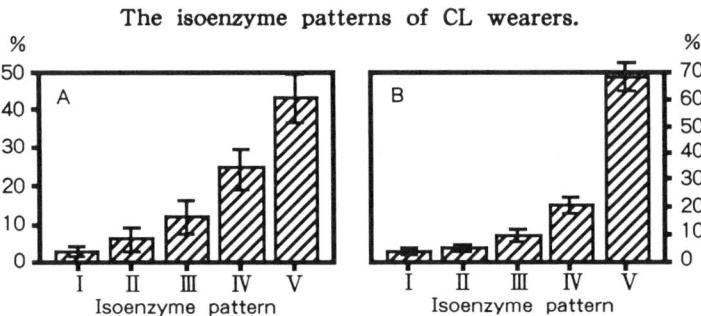

**Figure 4.** A : The tear isoenzyme pattern of CL wearers (RGP = 17, HEMA = 4). LDH V of CL wearers became relatively higher (p < 0.05) than the normalS in age matched. B : The isoenzyme pattern of the tarsal conjunctiva. The LDH V of CL wearers became relatively higher (p < 0.01).

## DISCUSSION

Lactate, the end-product of anaerobic glycolysis is preferentially removed posteriorly from the corneal endothelium to the aqueous humor[5]. Very little lactate reaches the lachrymal fluid under normal conditions due to the relative impermeability of the epithelia to this metabolite[6,7]. The epithelium consists of a basal cell layer of high mitotic and metabolic activity and four additional layers with diminishing metabolic activity in the direction of the tear film. Desquamating epithelial cells probably have minimal metabolic activity[8-10]. If LDH diffuses from the superficial layer, then the concentration of the surface does not become zero. It will suggestively support these facts that the carbonic anhydrase of the cornea exists only in the endothelium[11]. It is difficult to believe that LDH, with a molecular weight of 134,000, diffuses through the epithelium into the tear film in the normal state.[12,13] Thus, when it does leak through the epithelium, it must be due to irreversible damage or necrosis of the epithelial cells.[14]

Lesions which affect the cornea may also affect the conjunctiva.[15] The conjunctiva is involved in secretion and absorption and is also sensitive to environmental stress. Therefore, if LDH is present in the conjunctival epithelium, it is a likely site of the origin of the tear LDH.

Thoft and Friend compared the LDH activities of human cornea and conjunctiva, and established that the LDH activity of conjunctiva was 1/54 that of the cornea.[16] Others have claimed that the conjunctiva contributes very little of the LDH to the preocular tear film.[7] However, Thoft and Friend[15] must have examined the bulbar conjunctiva. Our results show that human tarsal conjunctival epithelium contains more LDH than does the bulbar conjunctiva. So, the comparison of LDH activities between the cornea and the bulbar conjunctiva does not represent the true status in humans.

The LDH isoenzyme pattern may indicate from which source it is derived. The average isoenzyme pattern of the tear is maintained even over a 4 fold variation in LDH activity. This indicates that there is only one main source for the origin of LDH in tears.

The LDH isoenzyme patterns detected here are very similar to those detected by Kahan and Ottovay.[17] However this pattern is different from the isoenzymes derived by them from the cornea where the LDH IV shows the largest peak and LDH I is absent.

The quantity of each of the isoenzyme peaks in the tears is determined in part by the half-life of each isoenzyme. Each enzyme may be diluted and dissipated according to its half-life. The half-lives in blood of the various isoenzymes is as follows: LDH I, 79 hrs; LDH II, 75 hrs; LDH III, 31 hrs; LDH IV, 15 hrs, LDH V, 9 hrs.[14] Because LDH V dissipates 5 times faster than LDH I and twice as fast as LDH IV, the LDH V of the source tissue must be lower than in the tear fluid.

We determined whether the specific change of isoenzyme pattern for the tarsal conjunctiva may affect the tear isoenzyme patterns or not. In contact lens wearers, LDH V increased compared to normals. This may be the result of mechanical stress on the tarsal conjunctiva rather than a metabolic shift because LDH V is the main characteristic of the conjunctival tissue itself.

## CONCLUSION

The human tarsal conjunctival epithelium contains more abundant LDH in comparison with the bulbar conjunctiva. There exists a large gradient of LDH activity from the tarsal conjunctiva to the tear film. The tear LDH isoenzyme pattern is similar to the half-life deformed tarsal conjunctival pattern rather than the cornea. The specific change of LDH isoenzyme pattern for tarsal conjunctiva by the environmental stress does effect the tear LDH pattern. So we cannot ignore the tarsal conjunctiva as a source of normal human tear LDH.

# REFERENCES

1. N.J. van Haeringen and L. Thoring, Enzymatic composition of tears, " The Preocular Tear Film," F.J. Holly ed., Dry Eye Inst., Lubbock, Texas. (1986).
2. R.J. Fullard, L.G. Carney, and T.Hum, Enzymes of carbohydrate metabolism in human tear fluid, "The Preocular Tear Film," F.J. Holly ed., Dry Eye Inst., Lubbock, Texas. (1986).
3. T. Tsubai, H. Matsukawa, et al. Reformation of tear film and role of blinking, 1st Report, Mechanism of Mucin-coating over the corneal surface, *J. Jap. C. L. Soc.* 30:237 (1988).
4. T. Tsubai, N. Sakatani, J.M. Tiffany, et al., Relationship between tear film and blinking, 2nd Report, Protective function and elimination of superficial oily layer, *J. Jap. C. L. Soc.* 32:199 (1990).
5. R.J. Fullard and L.G. Carney, Human tear enzyme changes as indicators of the corneal response to anterior hypoxia, *Acta Ophthalmol.* 63:687 (1985).
6. S.D. Klyce, Stromal lactate accumulation can account for corneal oedema osmotically following epithelial hypoxia in the rabbit, *J. Physiol.* 321:49 (1981).
7. R.J. Fullard and L.G. Carney, Diurnal variation in human tear enzyme, *Exp. Eye Res.* 38:14 (1984).
8. B.R. Masters, A.K.Ghosh, J. Wilson, and M. Matschinsky, Pyridine nucleotides and phosphorylation potential of rabbit corneal epithelium, *Invest. Ophthalmol. Vis.Sci.* 30:861 (1989)
9. J.E. King, A. Augsburger, and R.M. Hill, Quantifying the distribution of lactate acid dehydrogenase in the corneal epithelium with oxygen deprivation, *Am. J. Optom.* 48:1016 (1971).
10. G. E Lowther and R.M Hill, Corneal epithelium: Recovery from anoxi, *Arch. Ophthalmol.* 92:231 (1974).
11. G. Lonnerholm. Carbonic anhydrase in the cornea. *Acta Physiol. Scand.* 90:143 (1974).
12. C.L. Markert, Physicochemical nature of isozymes. *Ann. N.Y. Acad. Sci.* 94:768 (1961).
13. M. Maekawa, Lactate isoenzymes, *J. Chromatogr. Biomed. Appl.* 429:373 (1988).
14. E. Schmidt and F.W. Schmidt, Clinical enzymology, *FEBS Lett.* 62 (suppl):E62 (1976).
15. S. Hirakai, T. Ishida, and M. Kano, Estimation of corneoconjunctival damage due to topical ophthalmic agents by biochemical analysis of tear fluid. *Jap. J. Ophthal.* 92:1553 (1988).
16. R.A. Thoft and J. Friend., Biochemical transformation of regenerating ocular surface epithelium, *Invest. Ophthalmol. Vis. Sci.* 16:14 (1977).
17. I.L. Kahan and E. Ottovay, The significance of tears' lactate dehydrogenase in health, and external eye diseases, *Graefes Arch. Klin. Exp. Ophthal.* 194:267 (1975).

# TEAR PROTEINS AND ENZYMES IN THE CHIMPANZEE

N.J. van Haeringen, V.M.W. Bodelier, and [1] P.S.J. Klaver

Biochemical Laboratory, The Netherlands Ophthalmic Research Institute, Amsterdam, The Netherlands. [1] Zoo, Natura Artis Magistra, Amsterdam, The Netherlands

## INTRODUCTION

Information published during the last decades has identified pronounced species differences, not only in morphological organization of ocular structures, but also in biochemical composition of eye tissues and fluids.

The rabbit, rat and guinea pig have been used primarily, if not exclusively, in most areas of ophthalmic research and the acceptance of their eyes as suitable models for the mammelian eye has apparently been based on the supposition that their expression of mechanisms and constitution are identical to those of other more costly eyes of primates. This is particularly surprising in light of the different composition of their tear fluid as compared to human tear fluid [1-5]. Our knowledge of the composition of tear fluid in animals is derived from studies in a small number of species, seldom including primates.

The objective of the present study is a further investigation into qualitative and quantitative differences in the composition of tears of chimpanzees and man.

On behalf of this study tear fluid could be collected from eight chimps which were anaesthetized for an investigation into a possible pericarditis. Protein concentration, protein composition and activities of the enzymes lysozyme, peroxidase and amylase are analyzed in the tear fluid and compared with those of human tear fluid and earlier findings in tear fluid of rabbit, rat and guinea pig.

## MATERIALS AND METHODS

### Tear Fluid Samples

The chimps were anaesthetized with Zoletil 50 (Virbac, France, containing equal amounts of tiletamin and zolazepam), 15-25 mg/kg body weight, depending on the condition of the animal. During anaesthesia tear fluid was collected from the ocular surface of both eyes by suction in glass capillaries (volume 5-20 $\mu$l). The glass capillaries were sealed with hematocrit wax and stored at -20 °C. Human tear fluid was collected from healthy volunteers without anaesthesia.

## Assays

Total protein was determined by a colorimetric assay according to Bradford[6], using bovine serum albumin as standard.
Lysozyme was determined by a turbidimetric assay[7]. As standard Hen Egg Lysozyme (HEL) was used and the activity is expressed in mg HEL aequivalents per ml tear fluid.
Peroxidase was measured with 2,2-azino-bis-3-ethylbenzthiazoline-6-sulfonic acid (ABTS) as substrate[8]. The enzyme unit (U) is defined as 0.001 increase in absorbance per min at a wave length of 412 nm in a 1-cm lightpath cuvette.
Amylase was measured with a colorimetric test (Boehringer, Mannheim, Germany) with 4-nitro-phenyl-D-maltoheptaoside as substrate[9]. The activity is expressed in International Units (IU) per ml tear fluid. One IU is the amount of enzyme which converts 1 $\mu$mol of substrate per min.
The protein composition was analysed by sodium dodecylsulfate polyacrylamide gel electrophoresis (SDS-PAGE), using an automated minigel system (Phast System, Pharmacia, Sweden) and staining the minigels with silver in the development unit accessory to the Phast System[10].

## RESULTS

The concentration of total protein and the activities of the tear specific enzymes lysozyme, peroxidase and amylase were measured in tear fluid of eight chimps. Associations between total protein and the enzymes have been screened but in neither combination the correlation was found to be significant. In table 1. the mean values of the parameters in tear fluid are presented for the investigated chimps and human individuals; for comparison values of rabbit, rat and guinea pig derived from earlier studies are also given.
In eight chimps SDS-PAGE separation of tear fluid proteins followed by silverstain showed no qualitative differences in the protein patterns. In fig. 1 a protein profile is presented of chimpanzee and human tears. The protein pattern of the chimp is almost identical with that of the human in respect of the presence of secretory immunoglobulin A. (sIgA), lactoferrin (LF), tear specific pre-albumin (TSPA) and lysozyme (LZM), with exeption of an additional band at the site of TSPA.

## DISCUSSION

This study presents the first comprehensive investigation of tear fluid in the chimpanzee and shows qualitative and quantitative species differences as compared to human tear fluid.

The mean values of total protein in chimp tear fluid do not significantly differ from those of human tear fluid. LZM in the chimps is half the value of human tear fluid, peroxidase is about double and amylase is about threefold. In the non-primates enzyme activities sometimes differ more than one unit in magnitude from those of the primates. In the rat the very high total protein and peroxidase values are striking. In the rabbit hardly any activity of the investigated enzymes is detectable. A significant correlation between mean total protein and peroxidase of all species can be demonstrated ($r = 0.90$, $P < 0.05$).

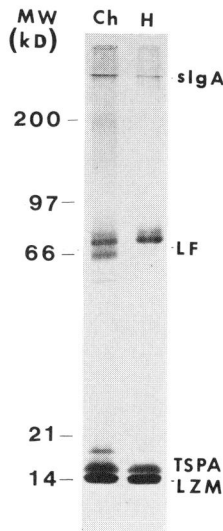

Fig.1. Silver stain of SDS-PAGE (10-15%) protein profile of chimpanzee (Ch) and human (H) tears. sIgA = secretory immunoglobulin A, LF = lactoferrin, TSPA = tear specific prealbumin, LZM = lysozyme.

The protein pattern of tear fluid in chimps resembles strongly to that of human, with only an additional band at the site of TSPA. The double band of LF in the chimp tears probably is caused by an iron-saturated form of tear LF, which also has been demonstrated to occur in human tears.[11] The occurrence of a band at the site of LZM is in correspondence with the measurement of notable LZM-activity in the tears of chimps. In rat, rabbit and guinea pig the protein profile is far more different, showing at 75 kD a band corresponding to transferrin[12] and many bands in the lower molecular range (rat: 11 bands, rabbit 7 bands and guinea pig 23 bands[2,4,5])

The causes for all these differences between species remain unclear, but an explanation might be that they are attributable to adaptation to the changing environment and behavior of each species which occurred during the evolution of the various animals. To verify the existence of a relationship between evolutionary adaptations to the environment and the tear fluid composition of different species, more investigations in a greater number of species are necessary.

Table 1. Mean values ± SEM (standard error of the mean) of total protein and the enzymes lysozyme, peroxidase and amylase in tear fluid of several species.

| | | Total protein (g/l) | | Lysozyme (mg HEL/ml) | | Peroxidase (U/ml) | | Amylase (U/ml) | |
|---|---|---|---|---|---|---|---|---|---|
| Chimp | (n) | 8.8 ± 0.3 | (8) | 6.2 ± 1.5** | (8) | 115 ± 18*(8) | | 3.5 ± 0.4*** | (6) |
| Man | (n) | 10.0 ± 0.6 | (10) | 11.8 ± 1.6 | (22) | 70 ± 5 | (24) | 1.0 ± 0.2 | (14) |
| Rat | (n) | 36 | (3#) | 0.01 | (3#) | 5000 | (3#) | 0.1 | (3#) |
| Guinea pig | (n) | 8.5 | (2) | 0.03 | (2) | 400 | (2) | 0.5 | (2) |
| Rabbit | (n) | 6.8 ± 4 | (15) | 0.015 ± 0.002 | (16) | < 10 | (15) | <0.1 | (15) |

(n)= number, #= pooled tear fluid samples.
Significancy of the difference of the mean values in chimp compared with the mean value of man:
* $p < 0.05$, ** $p < 0.01$, *** $p < 0.001$
Rat values derived from Thörig et al [2], rabbit and guinea pig values derived from Thörig et al [5].

# REFERENCES

1. N.J. van Haeringen, F.T.E. Ensink, and Glasius E, The peroxidase-thiocyanate-hydrogenperoxide system in tear fluid and saliva of different species, *Exp. Eye Res.* 28:343 (1979).
2. L. Thörig, N.J. van Haeringen, and G. Wijngaards, Comparison of enzymes of tears, lacrimal gland fluid and lacrimal gland tissue in the rat, *Exp. Eye Res.* 38:605 (1984).
3. J.L. Ubels and S.M. MacRae, Vitamin A is present as retinol in the tears of human and rabbits, *Curr. Eye Res.* 3:815 (1984).
4. E.J. van Agtmaal, L.Thörig, and N.J. van Haeringen, Comparative protein patterns in tears of several species, *in*: "Protides of the Biological Fluids", H. Peeters, ed., Pergamon Press, Oxford (1985).
5. L. Thörig, E.J. van Agtmaal, E. Glasius, K.L. Tan, and N.J. van Haeringen, Comparison of enzymes of tears and lacrimal gland fluid in the rabbit and guinea pig, *Curr. Eye Res.* 4:913 (1985).
6. M. Bradford, A rapid and sensitive method for the quantitation of microgram quantities of protein utilizing the principle of dye binding, *Anal. Bioch.* 72:248 (1976).
7. F.T.E. Ensink and N.J. van Haeringen, Pitfalls in the assay of lysozyme in human tear fluid, *Ophthalm. Res.* 9:366 (1977).
8. J. S. Shindler and W. G. Bardsley, Steady state kinetics of lactoperoxidase with ABTS as Chromogen, *Bioch. Bioph. Res. Comm.* 67:1307 (1975).
9. E. Rauscher, S. von Bülow, and U. Neumann, Determination of the activity of $\alpha$-amylase by contineous monitoring with 4-nitrophenyl-$\alpha$-D-maltoheptaoside as substrate, *Ber. Österr. Ges. Klin. Chem.* 4:150 (1981).
10. A. Kuizenga, N.J. van Haeringen, and A. Kijlstra, SDS-minigel electrophoresis of human tears, *Inv. Ophthalmol. Vis. Sci.* 32:381 (1991).
11. A. Kijlstra, A. Kuizenga, M. van der Velde and N.J. van Haeringen, Gel electrophoresis of human tears reveals various forms of tear lactoferrin, *Curr. Eye Res.* 8:581 (1989).
12. A. Boonstra and A. Kijlstra, The identification of transferrin, an iron-binding protein in rabbit tears, *Exp. Eye Res.* 38:561 (1984).

# COMPARISONS OF TEAR PROTEINS IN THE COW, HORSE, DOG AND RABBIT

Harriet J. Davidson,[1] Gary L. Blanchard,[2] and Paul C. Montgomery[3]

[1]College of Veterinary Medicine, Kansas State University, Manhattan, KS, USA
[2]College of Veterinary Medicine, Michigan State University, East Lansing, MI, USA
[3]Department of Immunology and Microbiology, Wayne State University Medical School, Detroit, MI, USA

## INTRODUCTION

Tear proteins play a major role in normal corneal health.[1] Several changes in tear proteins have been shown to be related to direct corneal irritation.[2] In veterinary medicine, differences between species often occurs in the clinical response to corneal irritation and external ocular disease. These differences may be a reflection of various factors including tear proteins. There is little information available on tear film properties in veterinary species. Some studies have evaluated the protein concentration for selected species; with little information on the molecular weight distribution of the proteins.[3-7] This project was undertaken to determine normal protein concentration and molecular weight distribution of tear proteins in the cow, the horse, the rabbit and the dog.

## MATERIALS AND METHODS

### Animals and sample collection

Animals used in this study were examined to insure they were in normal physical and ocular health, no animal used was being treated with any form of ocular medication. The cows sampled were 10 adult female Holsteins. The horses used were 10 adults of several breeds: Arabian, Tennessee Walker, Quarter Horse and Thoroughbred; they were females, males or geldings. Ten adult New Zealand White female or male rabbits were sampled. The dogs were female intact or spayed, male intact or castrated Beagles, Akitas or Sheba Inu.

Tear samples were collected without using anesthesia or lacrimal stimulation. A 5μl or 25μl microcapillary pipette was placed at the lateral canthus. Pooled tears from the inferior conjunctival sac were allowed to enter the tube via capillary

*Lacrimal Gland, Tear Film, and Dry Eye Syndromes*
Edited by D.A. Sullivan, Plenum Press, New York, 1994

action.  Tear volumes were measured and individual tear samples were stored at -20°C until evaluated.

## Protein concentration and molecular weight determination

Protein concentration was determined on $2\mu l$ aliquots of individual tear samples using the Bradford technique.[8]  Protein electrophoresis was accomplished using either a 12% sodium dodecyl sulfate-polyacrylamide with a 4% stacking gel or a 4-20% discontinuous gradient gel(Bio-Rad, Richmond, CA). A mini-gel system was used for both types of gels.[9]  Samples were assayed in the reduced and nonreduced form.  Sample reduction was completed with 4% ß-mercaptoethanol.  Lanes were loaded with 20$\mu g$ of protein taken from pooled samples. Electrophoresis was completed at 200V over 45 minutes.  Gels were stained with a 0.1% Commassie blue, then rinsed with 35% methanol and 10% acetic acid to remove excess stain. Electrophoresis was completed in duplicate.  A set of known molecular weight standards was assayed with each set of samples.  Least squares analysis was used to determine the molecular weight distribution of the unknown proteins.[10]  The values reported are the average molecular weight for each protein in the duplicate gels.

## RESULTS

The protein concentration was different for each of the four species evaluated.  Horse had the highest protein concentration, rabbits were second, followed by cows and finally dogs (Table 1).

**Table 1.**  Protein concentration(mg/ml) for normal tears.

|  | Cows | Horses | Dogs | Rabbits |
|---|---|---|---|---|
| Mean | 5.8 | 13.7 | 2.6 | 10.2 |
| Standard deviation | 2.2 | 4.0 | 1.0 | 3.5 |

The molecular weight of the protein bands ranged from 228.1 to 8.9 Kd for nonreduced samples(Tables 2 & 3) and 264.8 to 6.2 Kd(Tables 4 & 5) for reduced samples.  Nonreduced protein samples in both the 12% and 4-20% gels did not completely enter the separating gel.  The protein bands at approximately 76 and 60 kD for 12% and 70 and 61 kD for 4-20% gels were dark staining for each of the four species.  The band at 18 kD for 12% and 22 kD for 4-20% gels stained heavily for horses, dogs and rabbits, with only a faint stain uptake for cows.  In rabbits there was a very heavily stained band in the region between 18-13 kD for 12% and 22-16 kD for 4-20%, that was not easily resolved into separate bands.  There were a greater number of protein bands in the reduced samples, with most of the protein entering the separating gels.  Protein bands in the highest molecular weight ranges stained heavily for cows, horses and dogs, with only pale staining for rabbits. Bands at approximately 76 and 56 kD for 12% and 78 and 53 kD stained heavily for each of the species.  In rabbits the protein band between 20-8 kD was heavily stained and could not easily be resolved into separate bands.

**Table 2.** Molecular weight(Kd) distribution for nonreduced tear samples. 12% gel.

| Cows | Horses | Dogs | Rabbits |
|------|--------|------|---------|
| 228.1 | 228.1 | 228.1 | 228.1 |
| 114.1 | 114.1 | 114.1 | 114.1 |
|  |  | 105.3 | 105.3 |
|  |  |  | 96.3 |
| 75.9 | 75.9 | 75.9 | 75.9 |
| 59.7 | 61.4 | 59.7 | 61.4 |
|  | 42.7 |  |  |
|  | 24.8 |  |  |
| 18.5 | 18.5 | 18.6 | 18.3 |
| 17.0 | 17.0 | 16.9 | 17.0 |
| 12.6 | 12.6 | 12.6 | 12.6 |
| 8.9 |  |  | 8.9 |

**Table 3.** Molecular weight(Kd) distribution for nonreduced tear samples. 4-20% gel.

| Cows | Horses | Dogs | Rabbits |
|------|--------|------|---------|
| 229.6 | 229.6 | 229.6 |  |
| 150.0 | 150.0 | 150.0 | 150.0 |
|  |  |  | 95.2 |
| 70.3 | 70.3 | 70.3 | 70.3 |
| 61.0 | 61.0 | 61.0 | 61.0 |
|  |  | 42.5 | 42.5 |
| 32.9 |  | 32.9 |  |
| 22.4 | 22.4 | 22.3 | 22.4 |
|  |  | 20.0 |  |
| 16.0 | 16.0 |  | 16.0 |
| 13.3 | 13.3 | 13.3 |  |
| 9.0 |  |  |  |

**Table 4.** Molecular weight(Kd) distribution for reduced tear samples. 12% gel.

| Cows | Horses | Dogs | Rabbits |
|------|--------|------|---------|
| 264.8 | 264.8 | 264.8 |  |
| 166.1 | 179.3 | 179.3 | 179.3 |
| 127.2 |  | 127.3 |  |
| 98.4 |  |  |  |
| 74.5 | 77.5 | 77.5 | 71.8 |
| 56.0 | 56.0 | 56.0 | 52.3 |
| 45.7 |  |  | 44.3 |
|  |  |  | 34.5 |
| 26.3 |  |  |  |
| 23.5 | 23.5 | 23.5 |  |
| 20.2 | 20.2 | 20.2 | 19.7 |
| 17.7 | 16.3 |  | 16.1 |
|  |  |  | 15.3 |
| 12.9 | 12.9 | 12.9 | 11.6 |
| 10.5 | 10.9 | 10.9 | 10.9 |

**Table 5.** Molecular weight(Kd) distribution for reduced tear samples. 4-20% gel.

| Cows | Horses | Dogs | Rabbits |
|------|--------|------|---------|
|  |  | 276.0 |  |
| 144.4 |  | 144.4 |  |
|  |  |  | 125.7 |
| 101.0 | 101.0 | 106.0 |  |
| 90.8 |  | 90.8 |  |
| 78.5 | 78.5 | 78.5 | 78.5 |
| 60.1 | 60.5 | 67.8 | 67.8 |
| 53.1 | 53.1 | 54.6 | 54.6 |
| 27.5 | 26.0 |  |  |
| 23.1 | 23.8 | 23.1 |  |
| 16.7 | 16.3 | 16.7 | 17.3 |
| 13.7 | 13.0 | 13.5 |  |
|  |  | 12.5 | 12.2 |
| 10.0 | 8.7 |  | 8.8 |
|  | 6.5 |  | 6.2 |

## DISCUSSION

There is a marked difference in the protein concentration for each of the four species evaluated. There is no apparent difference in the protein concentration between different ages of different sexes although there is insufficient data for statistical analysis. Values for cows and rabbits compare favorably with previously reported values.[3,7] Values for dogs agree with one report[6] and disagree with a second[5], this may be a reflection of tear collection method. In our study values for horses are higher than previously reported values[4], again this may represent a difference in collection method. The overall variation in protein concentration between species may reflect differences in species lacrimal gland compositions.[11]

The molecular weight distribution pattern for both nonreduced and reduced samples compares favorably with values previously reported for humans and rabbits.[7,9,12] There is a noted difference between the species in the density of several of the protein bands. The most pronounced difference was in rabbit tears where the high molecular weight bands stained lightly and the low molecular weight bands stained heavily.

Cow tears also had a variation in the faint stain uptake of the lower molecular weight proteins. These apparent difference in protein composition may also reflect differences in lacrimal gland structure.[11] Specific identification of protein bands was not completed in this study. Based on molecular weight alone the data would suggest the presence of sIgA, lactoferrin, tear specific prealbumin and lysozyme.[9,12]

These data suggest that there are differences between the species in the protein composition of tears. To further document if these protein differences play a role in the response to clinical disease, the specific proteins and concentrations of those proteins need to be identified.

**REFERENCES**

1.  R.A. Moses, W.M. Hart. "Adler's Physiology of the Eye". CV Mosby C., St Louis (1987).
2.  T. Vinding, J.S. Eriksen, N.V. Vielsen. The concentration of lysozyme and secretory IgA in tears from healthy persons with and without contact lens use. *Acta Ophth.* 65:23 (1987).
3.  D.C. Maidment, D.E. Kidder, M.N. Taylor. Electrolyte and protein levels in bovine tears. *Br. Vet. J.* 141:169 (1985).
4.  R.E. Halliwell, N.T. Gorman. "Veterinary Clinical Immunology". WB Saunders, Philadelphia (1989).
5.  R. Barrera, A. Jimenez, R. Lopez, M.C. Mane, J.F. Rodriguez, J.M. Molleda. Evaluation of total protein content in tears of dogs by polyacrylamide gel disk electrophoresis. *Am. J. Vet. Res.* 53:454 (1992).
6.  S.R. Robert, O.F. Erickson. Dog tear secretion and tear proteins. *J. Sm. Animal Prac.* 3:1 (1965).
7.  L. Thorig, E.J. van Agtmaal, E. Glasius, J.L. Tan, N.J. van Haeringen. Comparison of tears and lacrimal gland fluid in the rabbit and guinea pig. *Curr. Eye Res.* 4:913 (1985).
8.  A. Johnstone, R. Thorpe. "Immunochemistry in Practice," Blackwell Scientific Publications, Boston (1988).
9.  A. Kuizenga, N.J. van Haeringen, A. Kijlstra. SDS-minigel electrophoresis of human tears. *Invest. Ophthalmol. Vis. Sci.* 32:281 (1991).
10. H.E. Schaffer, R.R. Sederoff. Improved estimation of DNA fragment length from agarose gels. *Anal. Chem.* 115:113 (1981).
11. J.H. Prince. "Comparative Anatomy of the Eye". Charles C Thomas, Springfield (1956).
12. P.T. Janssen, O.P. Van Biejsterveld. Comparison of electrophoretic techniques for the analysis of human tear fluid proteins. *Clin. Chim. Acta* 114:207 (1981).

# TEAR PROTEIN G ORIGINATES FROM DENATURED TEAR SPECIFIC PREALBUMIN AS REVEALED BY ANTI-TSP ANTIBODY

Joël Baguet, Véronique Claudon - Eyl and Anne-Marie Françoise Gachon[*]

Laboratoire Meuse Optique Contact (MOC)
Centre Hospitalier, Bar le Duc, France
[*]Laboratoire de Biochimie Médicale, Faculté de Médecine
Clermont-Ferrand, France

## INTRODUCTION

Human tears are a complex mixture containing lipids, electrolytes, proteins and glycoproteins, enzymes, in which the identity of some tear components, like Tear Specific Prealbumin (TSP) and protein G, is still undefined. First described by Bonavida[1] TSP is a major tear protein which electrophoretic mobility was larger than albumin. In 1979, Gachon[2] called TSP "Proteins Migrating Faster than Albumin" (PMFAs) and demonstrated that TSP was a group of at least six proteins. TSP was found[3] to present different isoforms whose molecular weight (MW) ranged from 15 to 20-kD and isoelectric points (pHi) ranged from 4.6 to 5.4. Absent from serum and other biological fluids, its function still remains poorly characterized. TSP was suggested to have a bacteriostatic activity[4] and to transport vitamin A to the corneal surface[5]. It was recently demonstrated that TSP belongs to the group of hydrophobic molecule transporters called lipocalins [6]. As suggested by Delaire[6], to normalize the nomenclature, in this article, we call TSP "Tear Lipocalin" (TL: followed by their MW).

Another major tear protein, the role of which is yet unknown is protein G. First described by Gachon[3], its MW was 31-kD and was irregularly found on sodium dodecyl sulfate polyacrylamide gel electrophoresis (SDS - PAGE) patterns of tear proteins. Kissner and Fullard[7,8] have recently suggested that protein G could be an experimental artifact due to change in conformation of TSP under denaturing conditions with SDS treatment. At this congress, we presented the work which clearly demonstrated that protein G originates from TL. The part dealing with the formation of protein G by denaturation of a protein of 17-kD MW and 5.0 pHi under SDS treatment has led to publication just before the congress[9]. In this paper, we present the part of the work which demonstrates that the denatured protein was TL. Production of protein G forms was performed under various SDS dissociation/denaturation procedures and analyzed by one-dimensional (1D) SDS PAGE. The origin of the altered proteins was studied by two-dimensional (2D) method where an entire gel lane was cut from the SDS PAGE on minigels, treated under denaturing conditions and finally SDS PAGE electrophoresed in the second dimension. After separation, the proteins were electroblotted on a PVDF membrane. Then, TL and TL aggregates (TLA) were detected by anti-TL antibody.

## MATERIALS AND METHODS

### 1D-SDS PAGE of Tear Proteins

Reflex tear fluids were collected from 2 members of the MOC laboratory who had never worn contact lenses. Tear samples were diluted in appropriated buffer (phosphate buffered saline pH 7.4 (PBS) or SDS buffer) for 5 min. just before electrophoresis and mildly or hardly treated. Mild treatment of samples was incubation of diluted sample in SDS buffer for 5 min. just before electrophoresis. Under hard treatment, the diluted samples were either boiled for 5 min. or submitted to six freeze-thaw cycles at -20°C (FTC). Then, proteins were separated by 1D-SDS PAGE on 8% - 25 % precast minigels on Pharmacia PhastSystem according to the manufacturer's programmed instructions.

### 2D-SDS PAGE of Tear Proteins

To investigate the origin of protein G due to denaturing treatment, samples were analyzed by using an off diagonal 2D-SDS PAGE on minigels as already described[9]. Before the second dimensional step, the gel strip was mild or hard denaturated as described for sample treatment. Thus, the unmodified proteins have the same mobility in both dimensions and lie on a diagonal. On the contrary, the modified proteins generally have a different mobility in the second dimension and lie off the diagonal.

### Influence of Reducing Agent Treatment

Electrophoreses under reducing treatment were performed as already described[9].

### Electrotransfer and Immunoblotting

After electrophoresis, minigels were transferred electrophoretically onto a 0.45 µm polyvinyldifluoride (PVDF) membrane in the Pharmacia PhastTransfer unit system according to the manufacturer's programmed instructions. Immunodetection was performed using anti-TL antibody and anti-rabbit IgG/horseradish peroxidase (HRP) complexes. To avoid non-specifically immunostaining, anti-TL antibody was preliminary incubated in human milk which is known to contain the same main proteins as tear fluid except TL.

## RESULTS

### Proteins G Group Formation Analysis by 1D-SDS PAGE

When the tear sample was diluted in PBS without SDS before electrophoresis, anti-TL antibody revealed well-stained TL 17 and TL 23 bands, a weak and diffuse 32-kD band and a fine 50-kD band (Fig. 1A, lane 1). Dilution of sample in SDS buffer 5 min. before electrophoresis (mild SDS denaturing conditions) gave a spontaneous decrease of TL 23 band (Fig. 1A, lane 2). By denaturation of the diluted tear sample with six FTC, positive bands were noticed at 32, 36, 42, 50, 55, 60 and 105-kD (Fig. 1A, lane 3). The patterns of denatured tear proteins with boiling treatment (100°C, 5 min.) was similar as with cold denaturation except that TL 18 band and five high MW bands were revealed. Moreover, PBS diluted sample (lane 1) and denatured samples (lanes 3 and 4) showed positive bands at the interface stacking gel-running gel. These positive bands decreased under SDS dissociation.

After reducing treatment of samples, 1D-SDS PAGE showed:
- TL 17 and intense TL 18 immunostained bands
- protein G group mainly disappeared. Protein G formed by denatured treatment were not completely reduced. Residual faint bands were noted at 28-kD (related to Ig light chain fragments) and 32-kD (Fig.1B).
- band related to lactoferrin was due to non-specific binding of anti-rabbit IgG complexes.

## Proteins G Group Formation Analysis by 2D-SDS PAGE

PBS diluted sample was electrophoresed (SDS PAGE) in the first dimension. Then, the gel strip was treated as indicated and electrophoresed (SDS PAGE) in the second dimension.

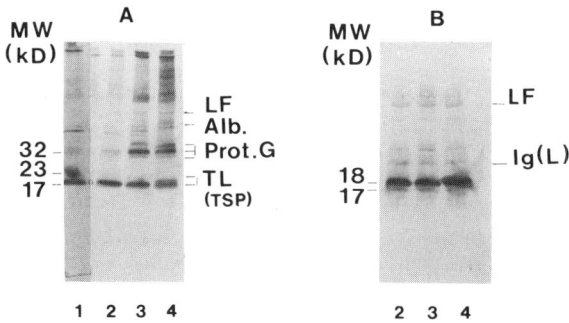

**Figure 1.** Protein profiles of reflex tear fluid obtained after SDS PAGE (8%-25% minigel), electroblotting and immunostaining with anti-TL antibody. **A** : non-reducing conditions. **B** : reducing conditions. *Lanes 1:* sample dilution in phosphate buffered saline, pH 7.4, *Lanes 2:* sample dissociation in SDS buffer, *Lanes 3:* sample denaturation by six freeze-thaw cycles at -20°C, *lanes 4*: sample denaturation by boiling (100°C, 5 min.) treatment. Alb. : albumine, Prot.G : protein G, Ig(L) : Ig ligh chain fragments, LF. : lactoferrine, TL (TSP) : tear lipocalin (previously called TSP). Estimated molecular weights markers are indicated on the left of the patterns.

When the gel strip was submitted to mild denaturing conditions (incubation in SDS buffer for 5 min. at 23°C), TL 23 changed to TL 17 (Fig. 2A). When the gel strip was submitted to hard denaturing conditions (incubation in SDS buffer for 5 min. at 100°C), TL 17 and bands (related to 18, 29 and 32-kD) coming from TL 17 were revealed (Fig. 2B). In the two experiments, positive material was noted on the diagonal.

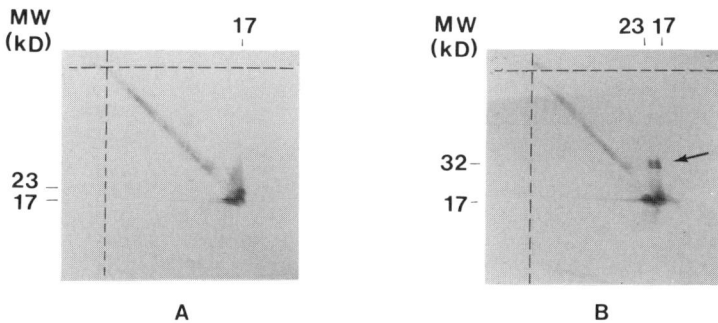

**Figure 2.** Determination of the origin of protein G by off-diagonal 2D electrophoresis. Patterns obtained after 2D SDS PAGE of a tear sample diluted in PBS, electroblotting and immunostaining with anti-TL antibody. **A**: mild denaturing treatment of the gel strip. **B**: boiling (100°C, 5 min.) of the gel strip. Note protein G coming from TL 17 (arrow). Estimated molecular weights markers are indicated on the left and at the top of the patterns.

# CONCLUSION

The experiments described in our study were originally conducted to investigate the formation of protein G in tear fluid. We studied such a phenomenon by using different dissociating/denaturating procedures of tear fluid to produce protein G and revealing TL aggregates with anti-TL antibody.

This study clearly demonstrates:

1 - no TL aggregates appear in reflex tear samples under non-denaturing conditions
2 - multiple TL aggregates related to protein G appear in presence of strong denaturants like SDS. The aggregated forms depend on the denaturing conditions of the sample preparation. Then, as suggested by Delaire[6,] protein G must be called TLA (TL aggregates). It is not possible in our denaturing conditions to aggregate all the TL, which certainly means that only a certain molecular structure of TL is susceptible to aggregate.
3 - disulfide bond formation plays a major role in the aggregation process of TL.

SDS (ionic detergent) treatment induces a change in the conformation of the protein structure and denatures the proteins by destroying their secondary, tertiary and quaternary structure. Such a phenomenon is increased by cold and heat conditions. The most convincing mechanism leading to the formation of protein G group is the following: the denaturing conditions with SDS treatment change TL 23 and part of TL 17 into denatured TL 17 and TL 18. These conformational changes probably expose lysine, cysteine and hydrophobic residues on the surface of TL 17. The presence of these residues close to the inter-subunit contact areas forms lysine, disulfide and hydrophobic bonds. Recent works[6] on TL showed that it belongs to the group of hydrophobic molecule transporters called lipocalins. Interestingly, human milk adsorbed anti-TL antibody also revealed denatured high MW proteins. This suggests that tear lipocalins bind to denatured tear proteins presenting hydrophobic domains on their surface. This hypothesis is under further investigations in our laboratory.

## ACKNOWLEDGMENTS

Supported by grants from Conseil Général et Comité d'Aménagement, de Promotion et d'Expansion du département de la Meuse (CAPEM), Mairie de Bar le Duc, Hôpital de Bar le Duc, Conseil Régional de Lorraine and Essilor International.

## REFERENCES

1. B. Bonavida, A. T. Sapse and E.E. Sercraz, Specific tear prealbumin: a lacrimal protein absent from serum and other secretions, *Nature* 221:375 (1969).
2. A.M. Gachon, P. Verrelle, G. Betail and B. Dastugue, Immunological and electrophoretic studies of human tear proteins, *Exp. Eye Res.* 29:539 (1979).
3. A.M. Gachon, P. Lambin and B. Dastugue, Human tears: electrophoretic characteristics of specific proteins, *Ophthalmic Res.* 12:277 (1980).
4. M.E. Selsted and R.J. Martinez, Isolation and purification of bactericides from human tears, *Exp. Eye Res.* 34:305 (1982).
5. C.W. Chao and S.M. Butala, Isolation and preliminary characterization of tear prealbumin from human ocular mucus, *Curr. Eye Res.* 5:895 (1986).
6. A.Delaire, H. Lassagne and A.M.F. Gachon, New members of the lipocalin family in human tear fluid, *Exp. Eye Res.* 55:645 (1992).
7. D.M. Kissner R.J. and Fullard, Characterisation of the isoforms of tear specific prealbumin and their relationship to protein G, *Invest. Ophthalmol. Vis. Sci.* 32 (Suppl.):733.(1991).
8. D.M. Kissner and R.J. Fullard, Is tear protein G a disulfide-bonded dimer of TSP in vivo? *Invest. Ophthalmol. Vis. Sci.* 29 (Suppl.):1288 (1992).
9. J. Baguet, V. Claudon-Eyl and A.M.F. Gachon, Tear protein-G originates from denatured tear specific prealbumin as revealed by 2-dimensional electrophoresis, *Curr. Eye Res.* 11:1057 (1992).

# TEAR LACTOFERRIN LEVELS AND OCULAR BACTERIAL FLORA IN HIV POSITIVE PATIENTS

Sandra E. Comerie-Smith, Jose Nunez, Marion Hosmer and R. Linsy Farris

Department of Ophthalmology, Harlem Hospital Center, and
Edward S. Harkness Eye Institute, Columbia University, New York, NY

## INTRODUCTION

Lactoferrin is an antibacterial protein secreted by the acinar cells of the lacrimal glands.[1] This protein: (a) has both bacterocidal and bacteriostatic properties; (b) combines and acts with specific antibodies in tears; and (c) inhibits $C_3$ convertase formation in the classical complement system.[2] Lactoferrin therefore plays an important role in the protective mechanisms of the external eye and may be an anti-inflammatory agent by inhibiting the complement pathway.

We looked at external ocular resistant factors using lactoferrin levels and bacterial count on the external eye of asymptomatic, human immunodeficiency virus (HIV)-positive patients. There was a significant difference in these parameters compared to normal. Tear osmolarity and Schirmer tests were also compared and there was no significant difference between HIV positive group and normals. We therefore ruled out keratoconjunctivitis sicca as the cause of decreased lactoferrin levels in these asymptomatic patients.

## MATERIALS AND METHODS

Eighteen eyes of HIV positive patients and eighteen eyes of HIV negative controls were evaluated. Nine HIV positive patients were recruited from an ambulatory infectious disease clinic. Ages ranged from 20-46 years. Seven had Center for Disease Control (CDC) criteria for Acquired Immunodeficiency Syndrome (AIDS) and two were HIV positive without opportunistic infection. None of the patients had ocular symptoms, wore contact lenses or used ocular medications. There was no history of immunosuppressive medications or other immunosuppressed states other than HIV positivity. Controls were age- and sex-matched ocular asymptomatics and all were HIV negative.

Tear osmolarity was performed on the undisturbed eyes in both groups using a fine-tipped pipette to collect tears by capillary action from the inferior marginal tear-strip. The tear was blown into a column of oil, extracted and loaded into 6 wells on a Cryostat Nanoliter Osmometer, after calibration and a standard test sample run. The results were read when the last well was thawed. Values above 312 milliosmole/litre were considered abnormal.[3]

A sterile plastic loop was passed along the lower lid margin of the right eye and plated immediately on a chocolate agar medium. Another sterile loop was passed along the lower fornix and plated separately. Tears were stimulated using a bright light and again plated on separate chocolate agar using a sterile loop. The entire procedure was repeated for the left eye. The culture plates were immediately taken to the bacteriology laboratory and incubated. The number of colonies of each plate was counted and the bacterial species identified.

Tear lactoferrin was measured by a radial immunodiffusion (RID) method using a lactoplate test (Eagle Vision Inc; Memphis, TN). A 4 mm filter paper disc was placed in the inferior fornix of each eye using a sterile blunt forceps. The disc was removed after full saturation, blotted and placed on the reagent gel chamber. The plate was tightly covered and stored at room temperature for 72 hours. The diameter of the antigen-antibody ring was read and converted to mg/dcl using a standard conversion chart. Less than 90 mg/dcl was considered abnormal.[3]

## RESULTS

The mean tear osmolarity of the HIV positive group was 312 mosmol/litre compared to 306 mosmol/litre in the control group. There was no significant difference between the two groups ($P > 0.01$) (Figure 1). Results of the Schirmer tests showed a mean of 11 mm/wetting for HIV positive group and 12.7 mm wetting for the control (Figure 2).

In contrast, decreased concentrations of lactoferrin were found in tears of the HIV-positive group with a mean of 85.8 mg dcl, compared to control with a mean of 156 mg dcl ($P < 0.01$) (Figure 3). In addition, bacterial growth was found on the lids and fornices of 77% of HIV positive patients while only 33% of HIV negative patients had growth. The average colony count on the lids of HIV positive patients was 4.1 colonies/patient compared to 1.5 colonies/patient in the control group (Figure 4). There were 3 patients in each group that had growth in the fornices, and no patient had growth in tears. The bacteria on the lids were mostly coagulase-negative staphyolccocus (normal pathogens), but 3 patients in the study group grew streptoccal viridans and one patient had bacilli. All growths in HIV negative patients were coagulase-negative staphyloccoci.

## DISCUSSION

Lucca et al. [4] reported an increased incidence of dry eyes in symptomatic HIV positive patients but no evidence of dry eyes in asymptomatic HIV positive patients. In our study we found that there was no increased incidence of dry eyes in asymptomatic HIV positive

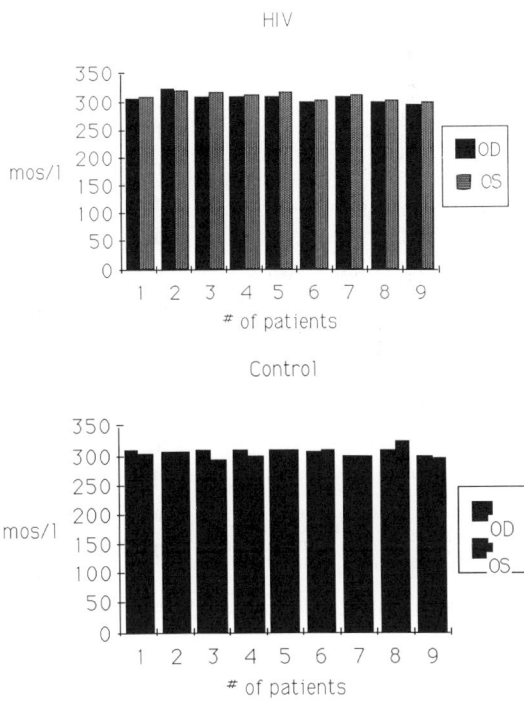

Figure 1. Tear osmolarity concentrations in HIV-positive patients and controls.

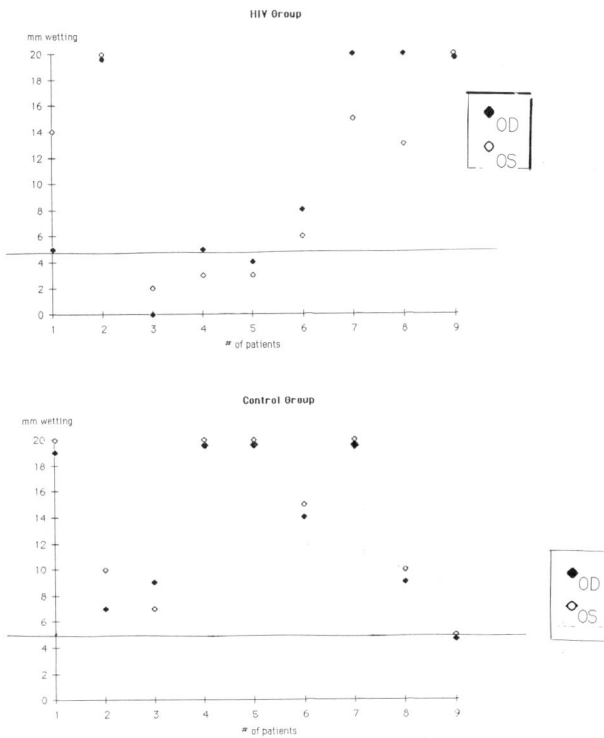

Figure 2. Results of Schirmer tests in mm/wetting in HIV positive patients and control.

LACTOFERRIN CONCENTRATIONS
Asymptomatic

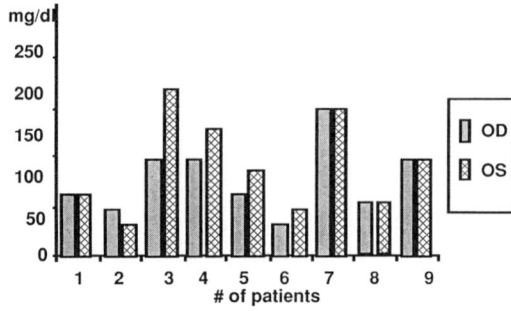

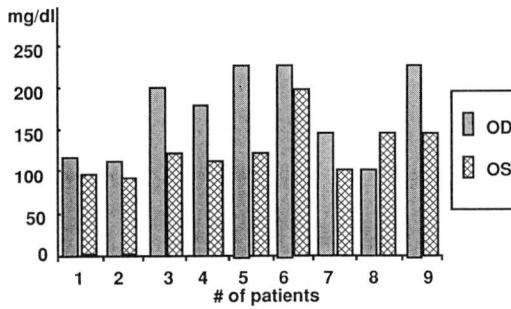

HIV Group

LACTOFERRIN CONCENTRATIONS
Asymptomatic

Control Group

Figure 3. Lactoferrin concentration in tears of HIV positive patients compared to those of controls.

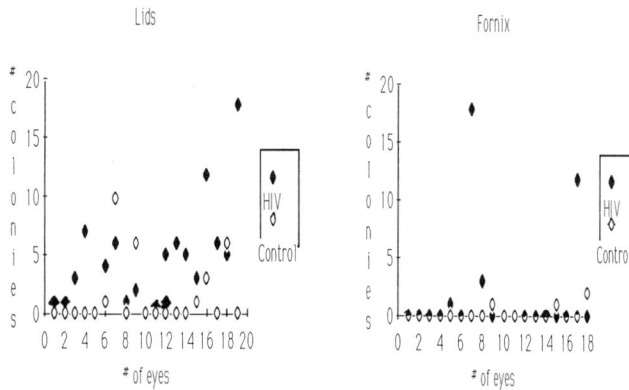

Figure 4. Bacterial flora (number of colonies) in the lids and fornices of HIV-positive patients, as compared to controls.

342

patients using tear osmolarity (95% specificity and 90% sensitivity)[5] and a Schirmer test, in addition to the absence of clinical symptoms and signs.

However, in our study there was a marked decrease in tear lactoferrin levels in HIV-positive patients, compared to those in HIV-negative patients. The decrease in lactoferrin levels could not be attributed to keratoconjunctivitis sicca, as explained in the previous paragraph. Some acute ocular conditions like giant papillary conjunctivitis[6] cause decreased lactoferrin concentration, and some chronic conditions like muscular dystrophy and keratoconjunctivitis sicca[3] also result in decreased concentrations. None of our patients had any of these conditions mentioned above. Our study seems to suggest that HIV infection can be added to the list of conditions that cause decreased tear lactoferrin levels despite no evidence of dry eyes.

There are increased numbers of normal bacterial flora on the lids of HIV positive patients compared to our study group. We do not know if this is a result of lower levels of antibacterial lactoferrin in tears or if the two factors are independent. It is probable that the combination of low lactoferrin levels and increased numbers of bacteria could potentially cause an increased rate of infection in the external eye of HIV positive patients.

Even though our results are conclusive, the number of patients is small. We may also need to look at other antibacterial proteins in tears and the incidence of lid and corneal infections in this group of patients.

## SUMMARY

Keratoconjunctivitis Sicca[4] has recently been reported to occur at a greater rate in HIV-positive symptomatic patients. We looked at HIV positive asymptomatic patients, compared to age matched HIV negative patients to study external ocular resistant factors, namely lactoferrin levels in tears, bacterial flora in lid margins, conjunctiva and tears, and evidence of dry eyes using a Schirmer test and tear osmolarity. Eighteen eyes of nine HIV positive patients and eighteen eyes of HIV negative controls were studied. Results showed markedly decreased lactoferrin levels in HIV positive asymptomatic patients with a mean of 85.8 mgs/dcl compared to HIV negative patients with a mean 156 mgs/dcl (P< 0.01). There were increased numbers of colonies of bacterial flora on the lids of HIV positive asymptomatic patients with an average colony count 4.1 colonies/patient compared to 1.5 colonies/patients in the control group (P< 0.025). Seventy eight percent of the study group had bacterial growth compared to 33% in the control group. The tear osmolarity in both groups had no significant difference; mean in HIV positive being 312 mosml/litre; mean in control 306 mosml/litre. The Schirmer test also showed no significant difference, with the mean in HIV positive patients being 11 mm wetting, and in control patients being 12.7 mm wetting. Therefore, despite no symptomatic or clinical evidence of dry eyes, asymptomatic HIV-positive patients had markedly decreased levels of lactoferrin in tears and increased colony counts of bacterial flora in the lids.

## REFERENCES

1. T. Gillette, and M. Allansmith, Lactoferrin in ocular tissue, *Amer. J. Ophthalmol.* 90:30 (1980).
2. A. Kijlstra, and A. Jerrison, Modulation of classical $C_3$ convertase of complement by tear lactoferrin, *Immunology* 47:263 (1982).

3. J. Lucca, J. Nunez, and R.L. Farris, A comparison of diagnostic tests for KCS -Lactoplate, Schirmer and tear osmolarity. *CLAO J.* 16:109 (1990).
4. J.A. Lucca, R.L. Farris, J. Bielory, and A. Capinto A, Keratoconjunctivitis sicca in male patient infected with HIV Virus Type, *J. Ophthalmol.* 97:10008 (1990).
5. R.L. Farris, R.N. Stuchell, and I.D. Mandel, Physical measurement osmolarity, basal volumes and reflex flow rate, *Ophthalmology* 88:8527 (1981).
6. M. Ballow, P.C. Donshik, P. Rapacz, and L. Samartino, Tear lactoferrin levels in patients with external inflammatory ocular diseases. *Inves. Ophthalmol. Vis. Sci.* 28:543 (1987).

# VITRONECTIN IN HUMAN TEARS - PROTECTION AGAINST CLOSED EYE INDUCED INFLAMMATORY DAMAGE

Robert A. Sack[1,3], Ann Underwood[2,3], Kah Ooi Tan[3], and Carol Morris[4]

[1]SUNY Optometry, USA
[2]CSIRO Div. of Biomolecular Engineering, N Ryde, NSW, Australia
[3]The Cooperative Research Center for eye Research and Technology, NSW, Australia
[4]Renal Laboratory, Prince Henry Hospital, La Perouse, Australia

## INTRODUCTION

Recent studies have shown that eye closure is associated with a shift in the pre-ocular tear film from a reflex fluid-rich layer in dynamic equilibrium to a secretory IgA-rich layer which is stagnant in nature.[1,2,3,4,5] This is accompanied by plasminogen and complement activation and build-up of serum albumin, followed by the recruitment of PMN cells into the tear film.[1-5] All of these changes are indicative of a state of sub-clinical inflammation.

It has been speculated that this shift to a secretory IgA and PMN-cell rich layer protects the highly vulnerable closed eye ocular surfaces from pathogenic invasion.[1] The mechanism by which autologous cytolytic damage is avoided is uncertain. In other tissue, vitronectin (VN), also known as complement protein-S or serum spreading factor, has proved a potent inhibitor of plasmin and complement mediated autolytic damage.[6,7] VN is also a known potentiator of phagocytic processing of pathogenic organisms.[8,9,10] This study was initiated to investigate the possible presence of VN in the external ocular environment.

## MATERIALS AND METHODS

Preparation of monoclonal antibody (MAb) - Human VN was prepared from out-dated serum.[11] After antigenic challenge, hybridomas were made by standard methods with culture fluids screened by ELISA against purified VN. Positive colonies were cloned by single cell picking, and purified MAbs were prepared by A-Sepharose column chromatography (see Underwood, et al. for details[12]).

## Tear Collection

Three types of samples were collected from trained asymptomatic non-contact lens wearing donors by glass microcapillary tubes.

1.  Reflex tear sample - Reflex tearing was stimulated by restricting closure of the upper lid. The initial in situ contaminated fluid was discarded and the eye washed with saline. A relatively pure reflex fluid was then collected at a high rate of flow (>50 ul/minute).
2.  Non-reflex open eye tear samples - 1-2 ul size samples were self-collected from trained donors habituated to minimize reflex tear contamination.
3.  Closed eye tear samples - Non-reflex 1-2 ul size samples were collected immediately upon eye opening after overnight sleep.

## Western Blot Analysis

Tears were separated by SDS-PAGE under reducing conditions, transferred onto nitrocellulose and the immunoblot probed using MAbs against VN and an ABC-HRP amplification system.

## Sandwich ELISA Assay

Tear samples were quantitatively screened for VN using a Mab capture-probe sandwich ELISA assay developed for serological studies employing two MAb clones which reacted with different antigenic sites common to both the 75 and 65kDa forms of VN. This allowed a low probability of reaction with VN fragments.

## Immunohistochemical Staining of Bovine Cornea

Fresh bovine tissue was snap-frozen, sectioned on a cryostat and immunochemically stained for VN and Fibronectin using MAbs and an ABC-HRP amplification system. After counter-staining with 0.2% light green and dehydrating, samples were scanned by confocal microscopy.

## RESULTS

Immunoblot analysis revealed in going from reflex (R) to open (O) to closed (C) eye tear samples, a marked and progressive increase in concentration of VN, accompanied by a shift from the intact 75KD species to the 65 KD biologically active breakdown product with further degradation occasionally encountered. ELISA assay of tears from these subjects in which extensive degradation of VN was noted confirmed these findings.

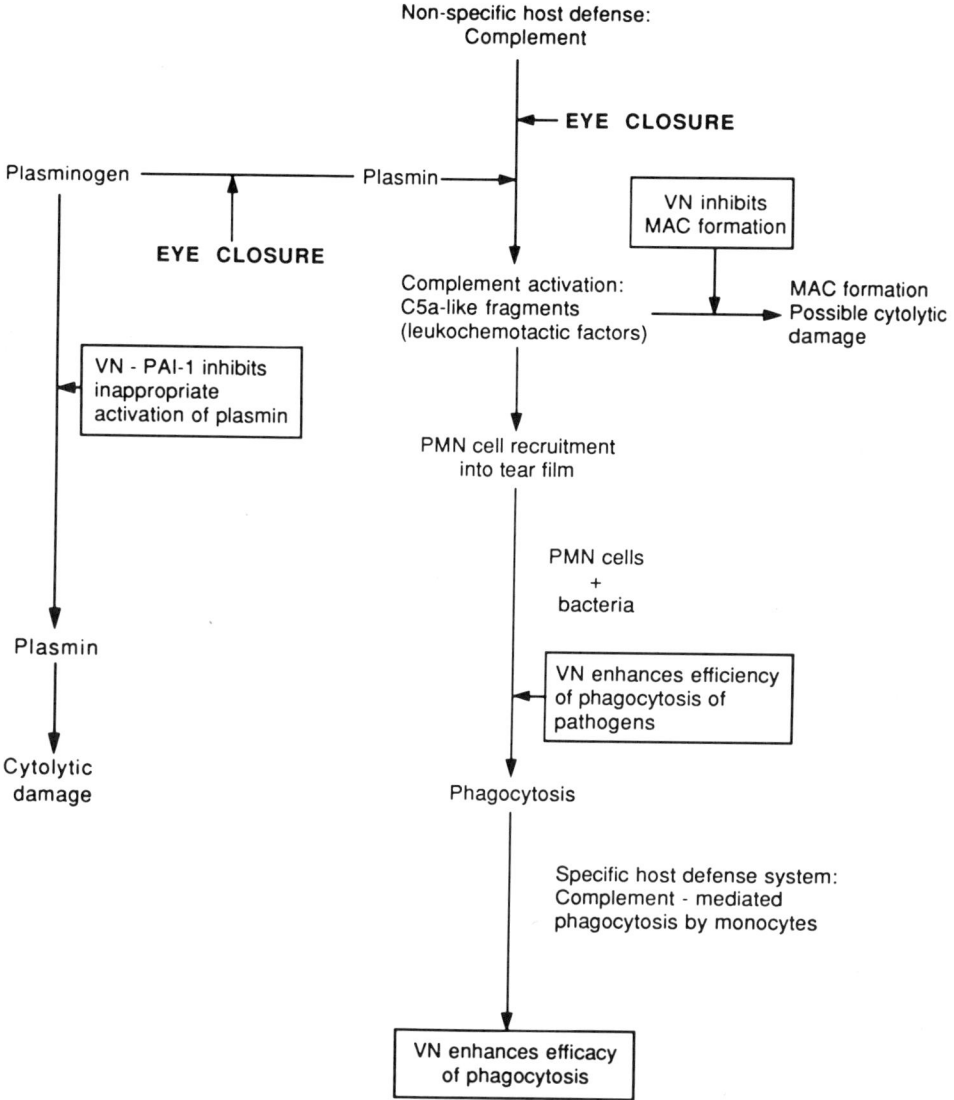

Non-specific host defense:
Complement

◀── **EYE CLOSURE**

Plasminogen ──────── Plasmin───▶

```
                                    ┌──────────────┐
                                    │ VN inhibits  │
                                    │ MAC formation│
                                    └──────────────┘
```

**EYE  CLOSURE**

Complement activation:                    MAC formation
C5a-like fragments  ──────────────▶       Possible cytolytic
(leukochemotactic factors)                damage

```
┌──────────────────┐
│ VN - PAI-1 inhibits │
│ inappropriate      │
│ activation of plasmin│
└──────────────────┘
```

PMN cell recruitment
into tear film

Plasmin                           PMN cells
                                      +
                                   bacteria

```
                              ┌────────────────────┐
                              │ VN enhances efficiency│
                              │ of phagocytosis of   │
                              │ pathogens            │
                              └────────────────────┘
```

Cytolytic
damage                            Phagocytosis

Specific host defense system:
Complement - mediated
phagocytosis by monocytes

```
┌────────────────────┐
│ VN enhances efficacy │
│ of phagocytosis      │
└────────────────────┘
```

Figure 1

| VN (ug/ml) | | (n = 6) |
|---|---|---|
| Reflex | Open | Closed |
| 0.073±.036 | 0.58±0.30 | 6.62±2.95 |

Immunohistochemistry revealed that VN was localized and restricted to corneal epithelium and stromal keratocytes. The details of this work will be published elsewhere.

## CONCLUSIONS

1. Tear fluid contains VN, the concentration of which increases in going from R to O to C eye samples. This pattern of distribution excludes a reflex-type lacrimal gland origin and supports the contention that VN is either of local or serological origin. In the later case, the increase in VN can be attributed to two factors: The closed eye induced an increase in vascular permeability and the build-up of background leakage resulting from decreased fluid turnover. A similar build-up in tear fluid has been demonstrated for the serum albumin, a molecule of similar molecular weight.[1]

2. The shift from the intact molecule to the physiologically active breakdown product is consistent with the known sensitivity of VN to trypsin-like proteolytic cleavage and the presence in the closed eye tear fluid of plasmin, t-PA and urokinase activities.[4]

3. VN is localized in the corneal epithelium and stromal keratocytes. This could be due to either local synthesis or to the presence of hypothetical VN specific receptors.

4. Given the highly stagnant nature of the closed eye tear fluid and the relatively high concentrations of VN in this fluid and on the epithelium, one might suspect a physiological function. Based upon these findings, we propose the following hypothesis (Figure 1): that VN is a component of the external ocular host defense system acting in conjunction with secretory IgA and PMN cells as a first line of defense, minimizing the risk of proliferation of entrapped microorganisms, while protecting against autologous cell damage. Here, VN could function to enhance the efficiency by which PMN cells and monocytes can process microorganisms.[8,9] We further suggest that VN could minimize the risk of autolytic damage resultant from plasmin and complement activation. VN could stabilize PAI-1[8] thereby minimizing inappropriate plasmin activation, and prevent autolytic damage by inhibiting formation of the intact complement membrane attack complex.[6,7]

## REFERENCES

1. R.A. Sack, K.O. Tan KO, and A. Tan, Diurnal tear cycle: evidence for a nocturnal inflammatory constitutive tear fluid, *Invest. Ophthalmol. Vis. Sci.* **33**: 625 (1992).
2. R.A. Sack, K.O. Tan KO, and A. Tan, The closed eye tear film. Evidence for a complement activated nocturnal basal inflammatory tear layer, *Invest. Ophthalmol. Vis. Sci.* 33 (suppl):733 (1991).
3. B.A. Holden, A. Vannas, R.A. Sack, and A. Tan A, Plasmin and complement C3 activation of inflammatory cells in the closed eye, *Invest. Ophthalmol. Vis. Sci.* 34 (suppl):732 (1991).
4. R.A. Sack, K.O. Tan, J. Chuck, B.A. Holden, and A. Vannas, Eye closure induces activation of tear plasminogen and complement systems, *Invest. Ophthalmol. Vis. Sci.* 33 (suppl):849 (1992).

5. K.O. Tan, B.A. Holden, and R.A. Sack, Changes in tear protein composition: sleep versus closed eye, *Invest. Ophthalmol. Vis. Sci.* 33 (suppl):1288 (1992).

6. K.T. Preissner, and Jenne D, Structure of vitronectin and its biological role in haemostasis, *Thrombosis Haemotasis* 66:123 (1991).

7. T.S. Halstensen, T.E. Mollnes, and P. Brandazeg, Terminal complement (TCC) and S-protein (vitronectin) on follicular dendritic cells in human lymphoid tissues, *Immunology* 65:193 (1988).

8. M. Hermann, M.E. Jaconi, C. Dahlgren, F.A. Waldvogel, O. Tendahi O, and D.P. Lew, Neutrophil bactericidal activity against Staphylococcus aureus adherent to biological surfaces. Surface-bound extracellular matrix proteins activate intracellular killing by oxygen-dependent and independent mechanisms, *J. Clin. Invest.* 86:942 (1990).

9. C.J. Parket, R.N. Frame RN, and M.R. Elstad MR, Vitronectin (S-protein) auugments the functional activity of monocyte receptors for IgG and complement C3b, *Blood* 71:68 (1988).

10. H. Kawahira, Clinical and experimental studies on vitronectin in bacterial penumonia, *J. Jap. Assoc. Infectious Diseases* 64:741 (1990).

11. T. Yatohgo, M. Izumi, H. Kashiwagi, and M. Hayashi, Novel purification of vitronectin from human plasma by heparin affinity chromatography, *Cell Struct. Funct.* 33:281 (1988).

12. P.A. Underwood, J.G. Steele, B.A. Dalton, and F.A. Bennett, Solid-phase monoclonal antibodies: a novel method of directing the function of biologically active molecules by presenting specific orientation, *J. Immunol. Meth.* 127:91 (1990).

# MUCOPOLYSACCHARIDE DEGRADING ENZYMES (MPDE) IN THE TEAR FLUID: A NEW DIAGNOSTIC TEST FOR RAPID DETECTION OF ACUTE OCULAR INFECTIONS

Amalia Romano

Goldschleger Eye Research Institute
Sheba Medical Center, Tel Hashomer
Sackler School of Medicine, Tel-Aviv University
Israel

## INTRODUCTION

The majority of the known laboratory methods for in vitro diagnosis of viral and bacterial infections, such as microorganism isolation and serological tests, require the use of well equipped laboratories, skilled personnel and are time consuming.[1] As clinical diagnosis of external ocular inflammatory conditions is not always reliable,[2] new laboratory techniques for early detection of viral infections are required.

The general term "hyaluronidases" comprises a group of enzymes, mucopolysaccharide degrading enzymes (MPDE),[3] which are involved in the processes of dermal permeability, embryonic and tumor development, cellular immune response, and in the control of the angiogenesis factor.[4] Although hyaluronidases have been identified in cornea and iris, they are not normally present in tears. In our previous work, we found that MPDE can be detected in tear fluid in the early stages of acute external ocular infections .[5]

Based on the above findings, we developed a method for early diagnosis of viral and bacterial eye infections.

## METHODS

### Experimental Ocular Infections

Experimental herpetic keratitis was induced in 8 rabbit eyes by inoculation of HSV-1 strains isolated from human corneal herpetic infections. Four eyes were examined during the acute stage of the infection. In another 4 eyes, iontophoresis was performed upon the establishment of latent infection (reactivation procedure). Clinical follow-up, viral isolation and the MPDE tests were performed daily in both infected and control animals (4 eyes). Eleven rabbit eyes were infected with pseudomonas aureginosa.

## Patients

Tear samples were collected from 2500 healthy individuals and from 2565 patients with herpetic keratitis, adenovirus keratoconjunctivitis, varicella zoster ophthalmicus, and acute ocular bacterial infections. Patients were examined with biomicroscopy and anamnestic data was obtained. Clinical diagnosis was confirmed by viral isolation on cell culture, bacterial identification on growth media, and serological evaluation.

## Specimen Collection

The tear fluid was collected directly from the lower fornix of both eyes by cotton swabs or Schirmer test strips. The samples were stored in 4°C test plates until used.

## MPDE Test

Test plates were prepared in sterile conditions and contained matrix (agarose) with various formulations of MPDE degradable substrate (mucopolysaccharide-glucosamine glycane): chondroitin sulfate A-C, hyaluronic acid, heparin sulfate, heparin, or keratin sulfate.[6] The tear fluid absorbent carriers were incubated on the test plates for 3-24 hours at 37°C. After the incubation, the carrier was removed and the plates were incubated for 10 minutes with aqueous toluidine blue solution,[6] rinsed twice with 5% acetic acid solution and dried at room temperature. For evaluation of bacterial infection, basic matrix also contained suitable growth media.

Because the dye is able to bind to both intact substrate and to the degradation products, this test provides indication of the acute infection as well as its severity.[6] The existence of an infection was indicated by the appearance of a discolored zone on those parts of the matrix were the substrate was degraded. The intensity of an infection was judged by the extent of the discoloration.

## RESULTS

The visual detection of the infection indicated by the MPDE test was based on the formation of color patterns, which in many instances were characteristic of the specific type of the microorganism causing the infection:

1. Adenovirus infection was characterized by a relatively large egg shaped discolored zone with well defined edges.
2. Herpes virus infections was characterized by a mosaic like pattern of discoloration when the tear fluid is applied onto a Schirmer paper strip.
3. The type of a bacterial infection could be determined on matrices which include suitable selective growth media.

### Experimental Animal Infections

1. Herpetic infection in rabbit eyes: The tear fluid of the animals was used in accordance with the basic mode of the method. In all cases of the infected rabbits (both acute and reactivated diseases), a characteristic positive reaction was obtained by the first day of the infection. Noninfected eyes provided a negative reaction.
2. Pseudomonas aureginosa infections: In this model, a correlation between the degree of infection and the depth of discoloration was present, and when the infection was reduced by treatment, the depth of discoloration was reduced accordingly.

### Patients

1. Healthy individuals: 98.8% of 5000 samples showed a negative reaction (30 patients with particularly dry eyes yielded false-positive results).
2. Adenovirus infection: 1000 patients with bilateral infection and 500 patient unilaterally infected were examined. All infected eyes produced definite discolored zone typical of adenovirus infection, as described above, while healthy eyes showed a negative reaction. There was a close correlation between the intensity and the size of the reaction and the severity of the infection.
3. Herpes infection: 55 patients with herpes simplex ocular infections and 10 patients with herpes zoster ophthalmicus were evaluated. In all cases, a positive reaction typical of herpes simplex virus (as described above) was detected in the infected eyes only.

## DISCUSSION

Our method is based on the observation that upon infection, mucopolysaccharide degrading enzyme (MPDE) appears in the tear fluid, where it is normally not present. Moreover, MPDE from different sources (cellular or bacterial) have different specificities for various substrates (hyaluronic acid, chondroitin, chondroitin sulfate A-C, heparin sulfate, etc.). This feature may be used for diagnosing the exact type of microorganism by applying tear fluid samples to different matrices with the relevant MPDE substrate.

As distinct from the conventional diagnosis methods in which microorganisms are cultured and identified, our test detect pathogens on the basis of the MPDE they secrete (or induce). This method is suitable for early diagnosis of infection, since the change in MPDE level occurs shortly after the onset of the disease. The technique is simple and does not require the use of elaborate laboratory equipment.

## REFERENCES

1. R.M. Woodland, Laboratory diagnosis of chlamydial and viral ocular infections, *Eye* 2 (suppl):S70 (1988).
2. S. Darougar, R.M. Woodland, P. Walpita, Value and cost effectiveness of double culture tests for diagnosis of ocular viral and chlamydial infections, *Br. J. Ophthalmol.* 71:673 (1987).
3. C. Abramson, Staphylococcal hyaluronatelyase: multiple electrophoretic and chromatographic forms, *Arch. Biochem. Biophys.* 121:103 (1967).

4. B. Fiszer-Szafarz, Hyaluronidase polymorphism detected by polyacrylamide gel electrophoresis, *Ann. Biochem.* 143:76 (1984).
5. A. Romano, J. Moissiev, Bacterial enzyme in viral infections: a new concept, *Metab. Peb. Syst.Ophthalmol.* 3-4:169 (1982).
6. A. Romano. A method of infection by testing of body fluid mucopolysaccharide degrading enzymes. PATENT APPLIED.

# TEAR IMMUNOGLOBULINS AND LYSOZYME LEVELS IN CORNEAL ULCERS

Harbans Lal[1] and A.K. Khurana[2]

Department of Biochemistry[1] and Department of Ophthalmology[2]
Medical College and Hospital, Rohtak 124001 India

## INTRODUCTION

Tear secretions are of utmost importance for the normal functioning of the eyes. The secretions clean the cornea mechanically, mantain an optically uniform surface, provide nourishment and, above all, possess anti-bacterial activity. This antibacterial property of tears was discovered by Fleming, who also discovered lysozyme, called muramidase - a bacteriolytic enzyme (Fleming, 1922). Later, McEven et al. (1958), while analyzing tear proteins electrophoretically, recognized three bands corresponding to serum albumin, globulins and lysozyme. They attributed the antibacterial action of secretions to globulins and lysozyme. Since then many studies have been done on tear proteins and at least 60 components of tear protein fractions, including the specific and nonspecific immunoglobulins, have been demonstrated. These form the first line of defense against external infection and seem to be more effective than the systemically produced antibodies (Knoph et al, 1971). However, there are conflicting results about tear immunoglobulins in patients with corneal ulcers. Katargine et al (1975) have reported increased levels of IgA, IgG and IgM in patients with keratitis, while others have found high levels of IgA only (Bluestone et al, 1975; Sen and Sarin, 1979). Furthermore, there exists a paucity of comparative studies on tear lysozyme and immunoglobulin levels in cases of bacterial, viral and fungal ulcers. Therefore, the present study was performed to determine these levels in patients with corneal ulcers.

## MATERIALS AND METHODS

Forty-five patients with corneal ulcers (15 each of bacterial, viral and fungal type), as well as 15 healthy controls, were included in this study. Patients with malnutrition, pregnancy, malignant diseases (e.g. lukemia) and lymphomas, renal diseases, tuberculosis, leprosy, amoebiasis, malaria, rheumatoid arthritis, splenomegaly and diabetes mellitus were excluded, because these factors are known to influence humoral immunity. Furthermore, none of the patients studied was receiving immunosuppressive therapy, or other clinical treatment, prior to the initiation of these experiments.

A clinical diagnosis of bacterial corneal ulcer was made in patients with a greyish-white central or marginal ulcer associated with marked pain, photophobia, blepharospasm, lacrimation, circumcorneal congestion, purulent/mucopurulent discharge, and presence or absence of hypopyon with or without corneal vascularization. In each case scraped material from the ulcer margins and conjunctival sac was subjected to gram staining and cultured on blood agar medium. If no growth was obtained on culture, the test was repeated 3 times before declaring it negative.

Out of the 15 bacterial ulcer patients studied, 12 showed positive laboratory data. Cultures were positive in 9 and Gram's smears positive in 5 patients. There were 4 patients that were culture-positive, but Gram's smear negative, while 3 patients were positive by Gram's smear, but negative on culture. Out of the 9 patients who were culture-positive, *Staphylococcus aureus* was isolated from scraped material of the 5, *Staphylococcus epidermoidis* from 2 and *Streptococcus pyogenes* and *Pseudomonas pyocyanea* from 1 patient each. No bacteria were isolated on cultures of the scraped material from 6 patients.

Fungal ulceration was suspected in patients with greyish-white dry-looking slough, surrounded by a yellow line of demarcation, satellite infiltrates around the ulcer, early hypopyon and the absence of corneal vascularisation. A history of trauma, especially vegetative, and clinical signs out of proportion to symptoms, i.e. less marked photophobia and lacrimation with intense ciliary and conjuntival congestion, supported a fungal origin. Scrapings from fungal ulcers examined by 10% KOH preparation and by culture on Sabouraud's dextrose agar medium. If no growth was obtained in up to 3 weeks, the culture was declared negative.

Out of the 15 patients with mycotic ulcers, 9 were confirmed as culture-positive by analysis of laboratory data. Five patients were positive both on KOH smear, as well as on culture. Four patients were negative on KOH smear, but were culture-positive. Out of the 9 patients who were culture positive, *Candida albicans* was isolated from 4, *Aspergillus flavus* from 3 and *A. niger* from 2 patients.

Only typical cases of herpetic viral ulcerations (9 dendritic and 6 amoeboid) with high indices of clinical suspicion were included in the study. The ulcers were conspicuous by a superficial and central location, absence of corneal vascularization, decreased or absent corneal sensation, associated with a history of predisposing factors such as fever, emotional stress, exposure to excessive heat, cold, etc. Cytologic evaluation of the scrapings from ulcers was the only procedure used to support the clinical diagnosis. Presence of multinucleated giant cells, monocytes and lymphocytes in Giemsa-stained specimens were considered evidence of herpetic corneal ulcerations and were noted in 12 (80%) cases.

Quantitation of tear immunoglobulins was carried out by radial immunodiffusion as described by Mancini et al. (1965). Tear lysozyme concentrations were determined turbidometrically by spectrophotometry, using *Micrococcus lysodeikticus* as the substrate (Ronen et al., 1975). The concentration of lysozyme in tear samples was calculated from a standard curve which was prepared by using hen egg lysozyme (HEL). Data obtained for the control and different study groups were analyzed by Student's t test.

# RESULTS

The mean IgA, IgG, and IgM concentrations in tears of the control group were 35 ± 0.9, 11 ± 0.7, and 12 ± 0.5 mg/dl. In patients with viral ulcerations, levels of all three immunoglobulins (IgA = 70 ± 2.6 mg/dl; IgG = 35 ± 1.2 mg/dl; and IgM = 29 ± 0.7 mg/dl) were elevated in tears, when compared to values in controls ($p < 0.05$). The rise in IgG and IgM concentrations was nearly 3-fold, whereas the increase in IgA content was 2-fold. In contrast, in patients with bacterial and mycotic ulcerations, only tear IgA values (62 ± 1.7 and 61 ± 1.7 mg/dl, respectively) were raised without any significant difference in the levels of tear IgG (12 ± 0.6 mg/dl in bacterial and 10 ± 0.5 mg/dl in mycotic ulcers) and IgM (11 ± 0.5 mg/dl in bacterial, and 14 ± 1.2 mg/dl in mycotic ulcers).

The mean value of the tear lysozyme observed in controls was 5.14 ± 2.42 mg/ml of HEL. In patients with bacterial and viral corneal ulcers, the tear lysozyme levels (1.70 ± 1.37 and 2.85 ± 1.88 mg/ml of HEL respectively) were significantly lower than the control group ($p<0.05$). The mean value for the group of patients with mycotic ulcers (3.02 ± 2.46 mg/ml of HEL) was not significantly different from the controls.

# DISCUSSION

Corneal ulcers are known to be caused by various factors and present different clinical pictures depending on the causative agent and host defense mechanisms. Whereas Little et al. (1969) have reported that immunoglobulins play an important role in the host defense against bacterial and fungal infections of the eye, Gachon et al. (1979) have stressed the role of tear lysozyme. Therefore, in the present study we have assayed both tear immunoglobulins as well as tear lysozyme levels in patients with corneal ulcers.

When compared to control levels, IgA, IgG and IgM concentrations were found to be significantly raised in patients with viral corneal ulcers, while only IgA levels were found to be increased in bacterial and mycotic ulcerations. The corneal ulcer-associated rise in tear IgA has been reported to be due to transudation (Newman et al., 1972). However, Sen and Sarin (1979) attributed this increase in tear IgA to its local production only.

Low levels of tear lysozyme were observed in patients with infective corneal ulcers, as compared to the controls. The lowest level, however, was seen in patients with bacterial corneal ulcers. The decrease in tear lysozyme levels in these patients has been attributed to the enhanced flow rate of tears.

An increased level of tear immunoglobulins in patients with viral corneal ulcers suggests a breakdown of the blood-tear barrier, while an increase in tear IgA in bacterial and mycotic ulcerations suggests that the immunoglobulin is locally produced. The quantitative differences in the levels of 3 immunoglobulins suggest different humoral immune responses in patients with various types of ulcers, while reduction in tear lysozyme in relation to the increased flow rate of tears suggests that lysozyme production remains constant during different corneal ulcerative diseases. The changes in these tear constituents may aid in the diagnosis of disease; however, more studies are required to confirm the usefulness of their estimations in bacterial, viral and fungal corneal ulcers.

## ACKNOWLEDGEMENT

We thank Scriptor Publisher, Denmark, for permitting us to use material from their publications.

## REFERENCES

Bluestone, R., David, L.E., Leonard, S.G., Barrie, R.J., and Donas, H.P., 1975, Lacrimal immunoglobulins and complements quantified by counter immune electrophoresis, Br. J. Ophthalmol. 59: 279.

Fleming, A., 1922, On a remarkable bacteriolytic element found in tissues and secretions, Proc. R. Soc. Lond. (Biol.) 93: 306.

Gachon, A.M., Verrelle, P., Betail, G., and Dastugue, B., 1979, Immunological and electrophoretic studies of human tear proteins, Exp. Eye Res. 29: 539.

Katargine, L.A., Muravieva, T.V., and Zaitseva, N.S., 1975, The role of immunoglobulins in the clinical picture and pathogenesis of herpetic keratitis in children, Vestn. Ophthalmol. 2:60.

Knoph, H., Blacklow, N., Glassman, M., Cline, W., and Wong, V., 1971, Antibody in tears following intranasal vaccination with inactivated virus. III. Role of tear and serum antibody in experimental vaccinia conjunctivitis, Invest. Ophthalmol. Vis. Sci. 10: 760.

Little, J.M., Centifanto, Y.M., and Kuafman, H.R., 1969, Immunoglobulins in human tears, Am. J. Ophthalmol. 68: 898.

Mancini, G., Carbonara, A.O., and Heremans, J.F., 1965, Immunochemical quantitation of antigens by radial immunodiffusion, Int. J. Immunochem. 2:235.

Mc Even, W.K., Kimura, S.J., and Feenay, M.I., 1958, Filter paper electrophoresis of tears; Human tears and their high molecular weight components, Am. J. Ophthalmol. 45:67.

Newman, M.D., Allansmith, M.R., and Charles R., 1972, Immunoglobulins in tears of normal and pathological cases, Am. J. Ophthalmol. 45: 67.

Ronen, D., Eylan, E., Romano, A., Stein, R., and Modan, M., 1975, A spectrophotometric method for quantitative determination of lysozyme in human tears; description and evaluation of the method and screening of 60 healthy subjects, Invest. Ophthalmol. 14:479.

Sen, D.K., and Sarin, G.S., 1979, Immunoglobulin concentration in human tears in ocular diseases, Br. J. Ophthalmol., 63: 297.

# MUCUS AND *PSEUDOMONAS AERUGINOSA* ADHERENCE TO THE CORNEA

Suzanne M. J. Fleiszig, Tanweer S. Zaidi, and Gerald B. Pier

Channing Laboratory
Department of Medicine
Brigham and Women's Hospital and Harvard Medical School
Boston, MA 02115

## INTRODUCTION

Most of what we currently know about the initial interaction between *Pseudomonas aeruginosa* and the adult cornea has come from studies of bacterial adherence to overtly injured cornea. These investigations have focused on receptor-adhesin interactions between injured corneal cell associated molecules and the microbe.[1] However, not all infections with *P. aeruginosa* are preceded by overt injury. For example, contact lens related keratitis is likely to follow more subtle disruptions to the ocular surface. In vivo, bacteria must pass through and interact with the tear film and the epithelial cell surface glycocalyx before they could adhere to corneal cells. This aspect of keratitis pathogenesis has received little attention.

We have found that mild acid treatment is able to promote bacterial adherence to the cornea without disrupting epithelial cell morphology.[2] This suggests that there may be antiadherence factors at the corneal surface which are disrupted by mild acid. Others have found that in the bladder, the surface mucus layer is able to inhibit bacterial adherence to the underlying healthy uroepithelial layer.[3] The human tear film is composed substantially of mucus.[4] For these reasons we hypothesized that ocular mucus may play a role in preventing bacterial adherence to the healthy cornea. Since the mucus layer of the tear film may be disrupted during contact lens wear, or by certain dry eye conditions, support of the hypothesis could at least partially explain why these populations are predisposed to infectious keratitis.

## MATERIALS AND METHODS

*P. aeruginosa* strain 6294 was maintained in trypticase soy broth (TSB) at -70ºC with 10% glycerol. Bacteria were inoculated onto a trypticase soy agar (TSA) plate covered with a 12,000-14,000 molecular weight pore size dialysis membrane,[5] and grown overnight at 37°C. The inoculum was prepared by resuspension of bacteria from the membrane into Hank's balanced salt solution (HBSS) until the appropriate optical density was achieved.

Male and female adult New Zealand white rabbits and Wistar rats were sacrificed with an overdose of phenobarbital and eyes were immediately removed.

Endogenous ocular mucus was removed from rabbit eyes with a mucolytic agent to determine the effect on *P. aeruginosa* adherence. One eye from each of 11 rabbits was treated with 5% N-acetylcysteine (Sigma Chemical Co., St. Louis MO) dissolved in phosphate buffered saline (PBS) for 20 minutes prior assessing bacterial adherence to the cornea.[6] The other eye was used as a matched control and received only PBS. N-acetylcysteine treated and control rabbit eyes were washed twice in 40 ml PBS and incubated separately in 4 ml of $1.5 \times 10^8$ colony forming units (cfu)/ml *P. aeruginosa*. Following a 20 minute incubation period at 37°C, each eye was washed six times in 40 ml PBS to remove non-adherent microorganisms. The cornea was excised from the globe, washed twice in 40 ml PBS and homogenized in 1 ml TSB. A viable count was performed in triplicate from the homogenate to calculate the number of colony forming units that had been adherent to the cornea.

Ocular mucus was prepared by two different means that were tested separately for their effect on bacterial adherence. To collect whole eye ocular mucus, 178 fresh whole rat globes were suspended in 9 beakers each containing 20 ml HBSS. After incubation for 4 hours at 4°C, the eyes were removed and the supernatant pooled. In order to collect only pre-corneal mucus, another 198 fresh rat eyes were embedded in agar on 10 petri dishes such that only the corneal surface of each eye protruded from the agar. In addition, 10 agar plates were prepared without embedded eyes to control for the effect of eluted agar on bacterial adherence. The surface of each agar plate was soaked with 10ml HBSS for 4 hr at 4°C. Each of the 3 pooled supernatants was filtered through 3μm pore size filters (Nucleopore Corp.), centrifuged 3 times at 5000 rpm for 20 min (to remove cell debris and other solid material), dialyzed, lyophilized and stored at -80°C until required.

For adherence assays, mucus was prepared by suspending 37 mg/ml of porcine stomach mucin, bovine submaxillary gland mucin (both purchased from Sigma Chemical Co., St Louis MO) or one of the two ocular eluate preparations in HBSS in order to produce solutions that might mimic the viscous mucus layer of the tear film.

Following treatment with N-acetylcysteine in HBSS for 20 minutes to remove endogenous ocular mucus, each rat eye was placed cornea upwards onto sterile cotton soaked in HBSS in a petri dish. Eyes were coated with a total of 16 μl of either a mucus preparation or HBSS alone and incubated at 37°C for 30 minutes. The preparations were added as 2 aliquots of 8 μl each at 15 minute intervals to prevent drying of the corneal surface. Corneas were then inoculated with 16 μl of either the bacterial suspension alone or a bacteria/mucus combination and incubated at 37°C for 30 minutes. Again this was delivered as 2 aliquots of 8 μl at 15 minute intervals. Bacteria/mucus combination suspensions were prepared 30 minutes prior to addition to the eye. Each eye was then washed six times in 5 ml PBS to remove non-adherent microorganisms. Corneas were excised from each eye, washed twice in 5 ml and once in 50 ml PBS, and homogenized in 1 ml TSB. A viable count was performed in duplicate to calculate the number of colony forming units that were adherent to each cornea. Two series of experiments were performed with the rat precorneal supernatant material. One series was controlled by treating matched eyes with HBSS alone and the other series involved control eyes coated with agar supernatant without ocular material (prepared as described above).

The Mann-Whitney-U test, the Wilcoxon signed-rank test and ANOVA were used for statistical analysis of adherence data.

## RESULTS

N-acetylcysteine treatment was found to enhance adherence of *P. aeruginosa* 3-fold $(3.1\pm1.0\times10^4/\text{cornea})$ for rabbit cornea as compared to control PBS treated eyes $(9.2\pm3.6\times10^3/\text{cornea, p=0.02})$.

All ocular and non-ocular mucus preparations inhibited *P. aeruginosa* adherence to N-acetylcysteine treated rat cornea (Table 1).[7]

**Table 1.** The effect of mucus on *P. aeruginosa* adherence to rat cornea.

| Type of mucus | No. of eyes | Percentage reduction in adherence compared to control eyes | | |
| --- | --- | --- | --- | --- |
| | | Mucus added before inoculum | Mucus added with inoculum | Mucus added before & with inoculum |
| PSM [1] (paired) | 14 | | | 93% |
| Whole ocular[2] (paired) | 24 | | | 63% |
| Precorneal[3] (unpaired) | 25 | 48% | 52% | 89% |
| PSM[1] (unpaired) | 40 | 55% | 64% | 77% |
| BSGM[4] (unpaired) | 24 | 83% | 82% | 75% |
| Precorneal[3] (paired, agar control)[5] | 24 | | | 78% |

[1]Porcine stomach mucin
[2]Eluate from whole rat eye
[3]Eluate from rat corneal surface only
[4]Bovine submaxillary gland mucin
[5]Control eyes treated with agar eluate

In preliminary experiments using paired eyes, mucus prepared from either porcine stomach mucin or from whole rat eye supernatant was found to inhibit bacterial adherence when added before and with the bacterial inoculum (p=0.002 and 0.02 respectively). The results of subsequent experiments revealed that rat precorneal supernatant was also able to inhibit adherence to cornea as compared to eyes treated with HBSS alone (p=0.001). In these experiments eyes were not paired and mucus was added in three different ways. Some eyes were precoated with mucus prior to the addition of bacteria, some received mucus only with the bacterial inoculum and a third group were treated with mucus both before and with the inoculum. The data revealed that bacterial adherence to the cornea was significantly reduced only if the bacterial inoculum was preincubated with mucus. Corneal pretreatment with mucus alone did not statistically alter bacterial adherence (p=0.1). Experiments using both porcine stomach mucin and bovine submaxillary gland mucin revealed that in contrast to ocular derived material, the manner in which these nonocular mucins were added was not important, since adding mucus before the inoculum, with the inoculum, or at both intervals all significantly reduced adherence (p=0.0001 and 0.0006 respectively). For both types of nonocular mucin, the inhibition of bacterial adherence was similar in magnitude with all three methods of introduction (p=0.16 and 0.58). Experiments using agar eluate rather that HBSS as a buffer to coat control eyes demonstrated that inhibition of bacterial adherence by precorneal supernatant was due to ocular factors and not to factors derived from agar (p=0.0076).

## CONCLUSIONS AND SIGNIFICANCE

The results of this study suggest that mucus is able to modulate *P. aeruginosa* adherence to the cornea. We have demonstrated increased *P. aeruginosa* adherence to cornea following the removal of endogenous surface mucus, and were able to inhibit adherence by the addition of exogenous mucus from various sources, including ocular material. There are changes to mucus production, composition and clearance from the eye during contact lens wear and in certain dry eye conditions — some of which are treated with mucolytic agents.[8] These disruptions to ocular mucus are likely to alter bacterial adherence to the cornea and could contribute to the pathogenesis of infectious keratitis.

The mechanism by which mucus modulates adherence was not investigated. In the bladder repulsion by negatively charged groups on glycosaminoglycans are thought to be responsible for the antiadherence effect of bladder mucus.[3] In contrast, respiratory mucins bind and entrap *P. aeruginosa*.[9] Further studies are indicated to determine whether mucus inhibits adherence to the cornea by either of these pathways. Alternatively, mucus may have inhibited bacterial adherence through a simple viscosity barrier effect, since the experiments described in this report involved the use of highly concentrated mucus. Future studies will need to address whether low concentrations of mucus also inhibit bacterial adherence. In order to further characterize the role of mucus as an antiadherence factor, it would be of interest to determine whether mucus inhibits adherence to injured or diseased cornea, whether mucus modulates adherence of bacteria other than *P. aeruginosa*, and if the mucin component of ocular mucus is involved.

## REFERENCES

1. A. Singh, L.D. Hazlett, and R.S. Berk, Characterization of *Pseudomonas aeruginosa* adherence to mouse corneas in organ culture, *Infect Immun*. 58:1301 (1990).
2. S.M.J. Fleiszig, E.L. Fletcher, and G.B. Pier, Evidence that anti-adherence factors are involved in the non-specific inhibition of bacterial adherence to the cornea, *Invest Ophthalmol Vis Sci (suppl)*. 33:844 (1992).
3. C.L. Parsons, and S.G. Mulholland, Bladder surface mucin. It's antibacterial effect against various bacterial species, *Am J Pathol*. 93:423 (1978).
4. J.I. Prydal, P. Artal, H. Woon, and F.W. Campbell, Study of human precorneal tear film thickness and structure using laser interferometry, *Invest Ophthalmol Vis Sci*. 33:2006 (1992).
5. S.M.J. Fleiszig, N. Efron, and G.B. Pier, Extended contact lens wear enhances *Pseudomonas aeruginosa* adherence to human corneal epithelium, *Invest Ophthalmol Vis Sci*. 33:2908 (1992).
6. F. Thermes, S. Molon-Noblot, and J. Grove, Effects of acetylcysteine on rabbit conjunctival and corneal surfaces, *Invest Ophthalmol Vis Sci*. 32:2958 (1991).
7. S.M.J. Fleiszig, T.S. Zaidi, and G.B. Pier. Ocular mucus and bacterial adherence to the intact and injured cornea (submitted for publication).
8. P. Versura, M.C. Maltarello, M. Cellini, F. Marinelli, Caramazza, R. Laschi. Detection of mucus glycoconjugates in human conjunctiva by using the lectin - colloidal gold technique in TEM. lll. A qualitative study in asymptomatic lens wearers, *Acta Ophthalmol*. 65:661 (1987).
9. S. Vishwanath, R. Ramphal, Adherence of *Pseudomonas aeruginosa* to human tracheobronchial mucin, *Infect Immun*. 45:197 (1984).

# MEASUREMENT OF HUMAN TEAR LYSOZYME USING A NOVEL SYNTHETIC SUBSTRATE

Mark D. Sherman, Vicky Cevallos, Rafat Gabriel, Chandler Dawson and Richard S. Stephens

The Francis I. Proctor Foundation, University of California at San Francisco, San Francisco, California USA

## PURPOSE

Lysozyme is a bacteriolytic enzyme representing approximately 30% of the protein content of human tears.[1] Human tear lysozyme production generally parallels aqueous tear secretion and is a good indicator of lacrimal gland function. However, in some instances of ocular pathology, tear lysozyme concentration (HTL) may be decreased before aqueous tear production is reduced.[2] Various spectrophotometric, turbidimetric, and immunologic methods for quantitating HTL have been proposed. The most common methods for measuring HTL are based on the ability of lysozyme to cleave the cell wall of <u>Micrococcus lysodeikticus</u>. These techniques suffer from poor standardization, variable results, and the need for specialized personnel to maintain reagents or operate equipment. The purpose of this study was to utilize a known biochemical reaction based on the selective action of lysozyme on N-acetyl-oligosaccharides to develop a rapid, reliable, and sensitive colorimetric methods for the measurement of HTL. The chemical substrate selected for this assay was p-nitrophenyl penta-N-acetyl B-chitopentaoside (PNP).[3] The reaction of lysozyme with this substrate releases p-nitrophenyl N-acetyl B-D-glucosaminide, which can be coupled to a reaction with B-N-acetylhexosaminidase (NAHase) to liberate p-nitrophenol. The free p-nitrophenol can then be accurately quantitated with a simple colorimeter.

## METHODS

Standardization of the assay parameters was accomplished in three stages. (1) Serial dilutions of PNP (Seikagaku America, Inc. Rockville, MD) were tested against serial dilutions of hen egg-white lysozyme (HEW). Reagents were mixed in a 96 well ELISA plate

*Lacrimal Gland, Tear Film, and Dry Eye Syndromes*
Edited by D.A. Sullivan, Plenum Press, New York, 1994

and optical density was measured on a Titertek Multiscan ELISA Reader at 405 nm. (2) Serial volumes and dilutions of HEW (100 to 1 uL, 16 to 0.125 mg/ml) were tested based on the results from (1) to determine the smallest "tear" sample size necessary for accurate results. (3) The optimal reaction time for the assay (0 min to 24 hr) was determined based on the parameters from (1) and (2). We then compared the precision of our chemical assay to the standard "lysoplate" method of measuring tear lysozyme.[4] Finally, we collected twenty tear samples from ten volunteers using sterile 3 microliter micropipettes and measured HTL using the PNP chemical assay.

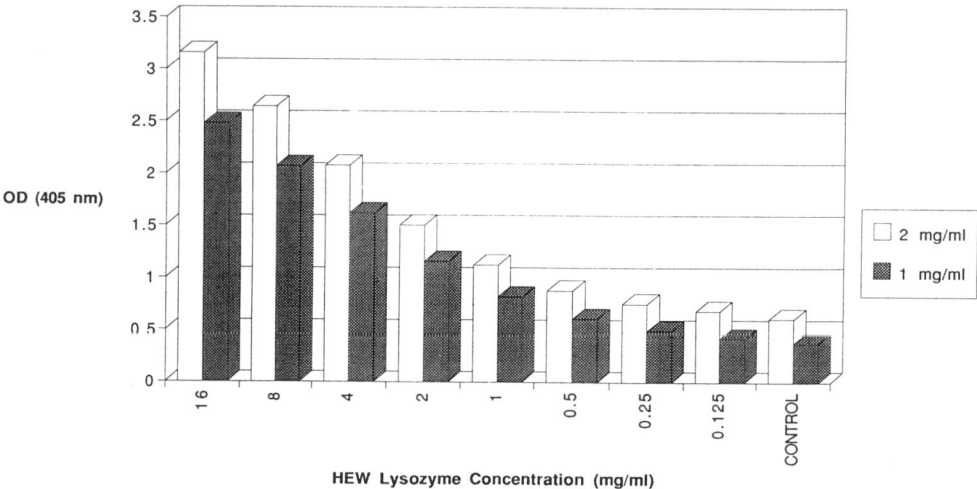

Figure 1. Lysozyme Assay (Chemical Substrate).

## RESULTS

The standard conditions for the chemical assay were established as follows. PNP substrate (2 mg/ml) is diluted in 0.1 M sodium citrate (pH=5.0). A 5 uL lysozyme sample (i.e tear specimen) is added to 100 uL of substrate. Ten microliters of NAHase (diluted 1:100 in buffered saline) is added to the reaction mixture (SIGMA Chemical Co., St. Louis, MO). The mixture is incubated at 37 C for one hour and the reaction is inactivated by the addition of 100 uL of sodium carbonate (1.0 M). The color change is read on a spectrophotometer at 405 nm. Figure 1 demonstrates the standard curve for the PNP assay using both 2 mg/ml and 1 mg/ml of substrate and 5 uL samples of HEW. Figure 2 demonstrates the corresponding curve using the standard "lysoplate" technique. Table 1 demonstrates the accuracy of this chemical substrate assay. The accuracy and precision of the PNP assay was significantly greater than for the "lysoplate" technique. Figure 3 demonstrates a series of twenty random human tear specimens which were measured using the PNP assay.

Table 1. Lysozyme Assay (Chemical Substrate).

|  | 2 mg/ml | | | 1 mg/ml | | |
|---|---|---|---|---|---|---|
| Lysozyme | OD[+] | SDw[*] | SDb[**] | OD | SDw | SDb |
| 16 mg/ml | 3.159 | 0.215 | 0.139 | 2.480 | 0.092 | 0.042 |
| 8 | 2.641 | 0.083 | 0.024 | 2.071 | 0.021 | 0.058 |
| 4 | 2.078 | 0.095 | 0.007 | 1.619 | 0.047 | 0.052 |
| 2 | 1.504 | 0.070 | 0.020 | 1.161 | 0.029 | 0.035 |
| 1 | 1.123 | 0.049 | 0.016 | 0.821 | 0.016 | 0.018 |
| 0.5 | 0.880 | 0.024 | 0.012 | 0.610 | 0.010 | 0.010 |
| 0.25 | 0.753 | 0.018 | 0.010 | 0.495 | 0.007 | 0.011 |
| 0.125 | 0.686 | 0.014 | 0.002 | 0.422 | 0.008 | 0.011 |
| Control | 0.612 | 0.020 | 0.018 | 0.373 | 0.019 | 0.004 |

[+] Optical Density (405 nm)

[*] Standard Deviation within groups

[**] Standard Deviation between groups

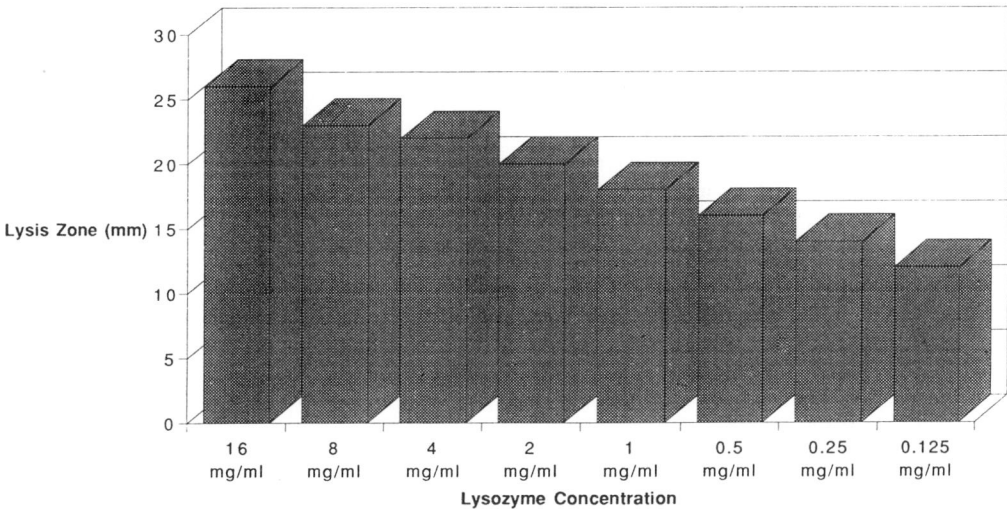

Figure 2. Lysozyme Assay (M. lysodeikticus Zone Lysis).

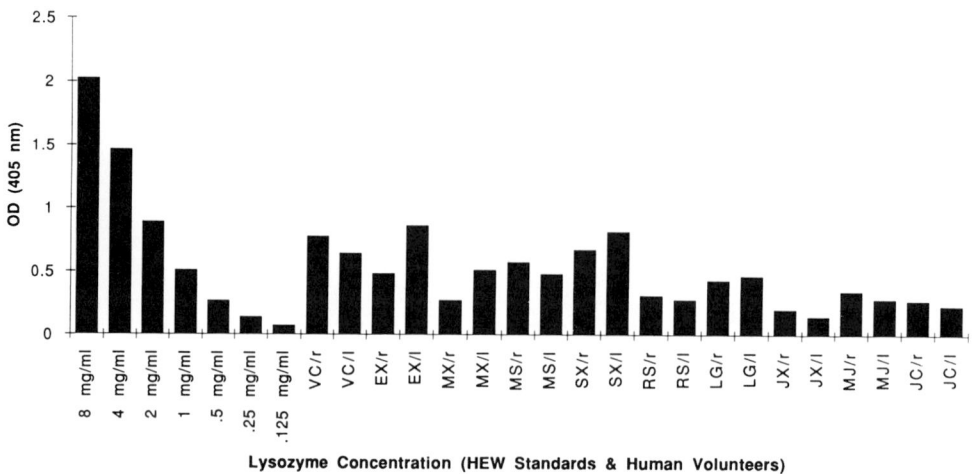

Figure 3. Human Tear Lysozyme Colorimetric Assay Using a Synthetic Chemical Substrate.

## DISCUSSION

We have demonstrated the use of a novel synthetic substrate for the measurement of human tear lysozyme. Further studies are needed to refine the parameters of this assay before it is availabe for routine clinical and diagnostic applications. This assay offers several advantages over current techniques for HTL measurement. It is a homogenous technique that is extremely sensitive, accurate, rapid and easy to perform. We expect this assay to facilitate further studies of HTL and dry eye syndromes. This study was supported in part by a grant from the AIDS Clinical Research Center (UC San Francisco).

## REFERENCES

1. N.J. Van Haeringen, Clinical biochemistry of tears, *Surv. Ophthalmol.* 26:84 (1981).
2. J.W. Whitcher, Clinical diagnosis of the dry eye, *Int. Ophthalmol. Clin.* 27:7 (1987).
3. J. Nanjo, K. Sakai, and T. Usui, p-Nitrophenyl penta-N-acetyl B-chitopentaoside as a novel synthetic substrate for the colorimetric assay of lysozyme, *J. Biochem.* 104:255 (1988).
4. O.P. van Bijsterveld, Standardization of the lysozyme test for a commercially available medium, *Arch. Ophthalmol.* 91:432 (1974).

# RETROSPECTIVE USE OF FROZEN SCHIRMER STRIPS FOR THE MEASUREMENT OF TEAR LYSOZYME

Mark D. Sherman, John P. Whitcher, and Troy E. Daniels

The Francis I. Proctor Foundation, School of Dentistry, and Sjögren's Syndrome Clinic, University of California at San Francisco, San Francisco, California USA

## PURPOSE

The Schirmer test is commonly used to evaluate aqueous tear production in suspected dry eyes.[1] Tear lysozyme concentration can be measured directly from tear samples collected on standard Schirmer test strips using an agar disk diffusion technique.[2,3] In conjunction with results from other diagnostic measurements, such as tear breakup time (TBUT), rose bengal staining pattern, and labial salivary gland cytology, a classification system has been proposed to group patients into categories of severe, moderate, and mild disease.[4] The standardized method for measuring tear lysozyme requires that specimens be incubated within 24 hours of collection on an agar plate flooded with M. lysodeikticus.[3] This is not always feasible, nor practical, in a large referral center. The Schirmer measurement, rose bengal staining pattern, and TBUT can be determined immediately during the clinical encounter. The analysis of tear lysozyme is more cumbersome and lysozyme measurement is not always necessary to make a diagnosis of Sjögrens syndrome in our patients. We routinely freeze all Schirmer strips collected from patients referred to the UCSF Sjögren's Clinic if the specimens are not immediately analyzed for lysozyme content. Over the past 15 years, this has allowed us to accumulate more than 600 Schirmer strip specimens with associated clinical and histopathological data. To our knowledge, there are no studies investigating the activity of lysozyme frozen on Schirmer strips for more than 24 hours. The purpose of this study was to determine whether human tear lysozyme that was frozen on Schirmer strips for greater than one year maintained activity and could be used to to contribute clinically useful information for future retrospective studies.

## METHODS

Twenty-eight frozen Schirmer strips from 14 patients (Table 1) evaluated at the UCSF Sjögren's Syndrome clinic were randomly selected. The patients had been examined between 1989 and 1990 and the Schirmer strips were frozen at O degrees Centigrade immediately following collection. Clinical data was retrieved for each patient, including age, sex, diagnosis, Schirmer measurement and rose bengal staining pattern.[4] (Table 1). For comparison, we prepared a fresh set of lysozyme standards (Table 2). The frozen Schirmer strips were allowed to return to room temperature for ten minutes and were then placed on the "lysoplate" which was prepared as follows. Fresh diagnostic sensitivity test agar was

Table 1. Patient Data - Results.

| Age/Sex | Diagnosis | Schirmer test (mm) | Rose Bengal Stain Pattern | Lysozyme Lysis Zone (mm) |
|---------|-----------|--------------------|---------------------------|--------------------------|
| 53 yr/F | Sjogren's | 2 OD/2 OS | "C" | 11 OD/19 OS |
| 56/F | Sjogren's | 1/3 | "B" | 10/12 |
| 43/F | Sjogren's | 3/2 | "C" | 17/17 |
| 68/F | Sjogren's | 1/2 | "C" | 18/17 |
| 67/F | Sjogren's | 2/1 | "B" | 11/13 |
| 9/F | Sjogren's | 1/2 | "C" | 20/19 |
| 41/F | Sjogren's | 2/2 | "A" | 9/9 |
| 48/F | Sjogren's | 2/1 | "C" | 17/12 |
| 45/F | Normal | 23/27 | No stain | 21/20 |
| 57/F | Normal | 21/23 | No stain | 20/18 |
| 66/M | Normal | 30/30 | No stain | 18/15 |
| 35/M | Normal | 30/30 | No stain | 19/19 |
| 22/F | Normal | 30/30 | No stain | 13/19 |
| 79/F | Normal | 30/30 | No stain | 20/20 |

prepared in petri dishes measuring 9 cm in diameter. The dried medium was then flooded with a 24 hr heart infusion broth culture of Micrococcus lysodeikticus. Excess fluid was removed and the surface was allowed to dry for 20 minutes. The proximal 5 mm tip of the frozen Schirmer strip was stored at room temperature for 10 minutes, and then placed on the Micrococcus plate. The standards were prepared by placing 5 uL aliquots of serial diluted hen egg-white lysozyme (HEW 8 mg/ml to 0.25 mg/ml) on the tip of a Schirmer strip and placing the proximal 5 mm tip of the Schirmer strip on a lysoplate prepared in an identical manner. The plates were incubated for 24 hours at 37 degrees Centigrade. The diameter of the zone of lysis was measured with calipers and recorded in millimeters.

# RESULTS

Sixteen of the specimens selected came from patients with a diagnosis of primary Sjögren's Syndrome (Table 1). All of these patients were female and the average age was 42 years (range 9 to 68 yrs). All of these patients had a Schirmer test 3 mm or less in five minutes, and fourteen measured 2 mm or less. Five of the patients had a "C" pattern of rose bengal staining.[4] Two patients had a "B" pattern, and one patient had an "A" pattern. This distribution is consistent with the frequency of these patterns in a previous study of patients with Sjögren's Syndrome.[4] The average diameter of lysozyme lysis was 14 mm. Six of the patients were diagnosed as non-Sjögren's Syndrome ("normal"). The average age for this group was 51 years (range 22 to 79 yrs). There were four females and two males in this group. The average Schirmer test was 28 mm. None of these patients had rose bengal staining. The average diameter of lysozyme lysis was 18 mm. The lysis zones for the lysozyme standards are demonstrated in Table 2.

Table 2. Fresh Lysozyme Standards.

| Lysozyme Concentration (HEW mg/ml) | Diameter of Lysis Zone (mm) |
|---|---|
| 8 | 23 |
| 4 | 22 |
| 2 | 20 |
| 1 | 18 |
| 0.5 | 17 |
| 0.25 | 16 |

# DISCUSSION

Our data suggest that lysozyme measurements obtained from frozen Schirmer strips can contribute useful information to retrospective clinical studies of patients with Sjögren's Syndrome. Certain conditions must be met, however, before the data from frozen specimens can be used to support generalized clinical observations. First, it is imperative that a group of "normal" specimens be available which were collected in the same manner at the same time period for comparison. Second, our study indicates that there is a downward "drift" of lysozyme activity in all of the frozen specimens when compared to fresh standards. When we compare the lysis diameters for the frozen specimens (normal and Sjögren's patients) to fresh HEW lysozyme standards, it is evident that the frozen "normal" specimens fall below established "normal" values for fresh specimens.[5] This is not necessarily a problem as long as the pool of frozen specimens include both "normal" and Sjögren's specimens for comparison. Retrospective studies that utilize these frozen specimens can be enhanced by including as much clinical data as possible, such as rose bengal staining pattern, Schirmer measurement, and results of labial salivary gland cytology. Further research is necessary to determine whether the lysoplate technique is best suited for lysozyme measurement in previously frozen specimens.

## REFERENCES

1. O. Schirmer, Studien zur Pshychologie und Pathologie der Tranenabsonderung und Tranenabjuhr, *Albrecht von Graefes Arch Ophthalmol.* 56:197 (1903).
2. B. Bonavida, and A.T. Sapse, Human tear lysozyme. II. Quantitative determination with standard Schirmer strips, *Am. J. Ophthalmol.* 66:70 (1968).
3. O.P. van Bijsterveld, Standardization of the lysozyme test for a commercially available medium, *Arch. Ophthalmol.* 91:432 (1974).
4. J.P. Whitcher, Clinical diagnosis of the dry eye, *Int. Ophthalmol. Clin.* 27:7 (1987).
5. N.J. Van Haeringen, Clinical biochemistry of tears. *Surv. Ophthalmol.* 26:84 (1981).

# RAPID MEASUREMENT OF SELECTED TEAR PROTEINS IN HEALTH AND DISEASE USING THE TOUCH TEAR MICROASSAY SYSTEM

Gary N. Foulks, Keith Baratz, and Phil Ferrone

Duke University Eye Center
PO Box 38802
Durham, NC  27710

## INTRODUCTION

Measurement of tear fluid immunoglobulins has been restricted to investigational studies due to the complexity of testing methods, yet many studies have demonstrated association of elevated tear IgE with allergic disease of the external eye[1-7]. The methods of testing have also been hampered by need for a relatively large volume of tear fluid and the attendant induced reflex tearing at the time of collection resulting in artefactually low values of IgE. The Touch Tear Microassay System was designed to adapt solid-phase ELISA testing with sensitive reflectometric determination of color density reaction during the measurement of protein levels in very small volume tear samples (2 µl). The present clinical study attempted to quantitate levels of IgE, IgG and C-reactive protein in normal patients and patients with external ocular signs and symptoms due to allergic and infectious disease using the Touch Tear Microassay System.

## MATERIALS AND METHODS

Patients were selected from the Cornea Service of the Duke University Eye Center to include adults of both sexes who provided informed consent prior to tear testing. Clinical diagnosis of external ocular disease evaluated general medical history, ocular history, and symptoms and signs of external inflammation including itching, mucous discharge, tearfilm stability and debris, conjunctival vascular engorgement and ocular surface changes of vital dye (fluorescein, rose bengal) staining. Evaluation of conjunctival changes of chemosis, papillary reaction, and follicular reaction was done for each patient. Patients with seasonal allergic conjunctivitis, prosthesis or contact lens associated giant papillary conjunctivitis and vernal keratoconjunctivitis were identified and agreed to tear testing. Patients with viral and bacterial conjunctivitis were also tested.

Tear samples were obtained atraumatically without anesthesia at the slitlamp biomicroscope by application of a 2 ul polished capillary tube to the inferior and lateral tear meniscus. This method of sampling did not stimulate reflex tearing and no complications were encountered. The obtained fluid was transferred to one well of the test card and allowed to absorb into the solid-phase membrane that contained mouse monoclonal anti-human IgE antibody while the paired well was inoculated with 2 ul of a standard control solution of IgE. One drop of mouse anti-human IgE:enzyme conjugate was applied to each well and allowed to fix for 60 seconds. Three drops of buffered saline were applied to remove excess conjugate and a final drop of tetramethyl benzidene substrate was added to each well. The card was immediately inserted into the reading chamber and the rate of colorimetric change monitored for five minutes by a highly sensitive reflectometer. Similar procedures were used for analysis of IgG and C-reactive protein by substituting the appropriate binding reagents, and anti-IgG:enzyme or anti-C-reactive protein:enzyme as the ELISA conjugate reagent.

Statistical analysis of the difference in mean values between respective diagnostic groups was performed by unpaired, two-tailed Student-t test analysis assuming normal distribution within groups. Significant difference was assigned at the $p<0.01$ level. A second statisitical analysis used the two-tailed Wilcoxon rank sum test.

## RESULTS

### Tear Immunoglobulin E

Ten asymptomatic normal patients were evaluated by the Touch Tear Microassay System. The mean tear IgE level was 13 ng/ml with a range of 0 to 35 ng/ml (SD 14 ng/ml). Seventeen patients with seasonal allergic conjunctivitis were measured with a mean value of 376 ng/ml and a range of 30 - 1020 ng/ml (SD 394 ng/ml). Nine patients with giant papillary conjunctivitis measured a mean value of 336 ng/ml with a range of 30 - 783 ng/ml (SD 299 ng/ml) while the eight patients with vernal keratoconjunctivitis measured a mean value of 986 ng/ml with a range of 600-1000 ng/ml (SD 309 ng/ml) [Figure 1]. Statistical analysis of the differences of the mean of each group by unpaired, two-tailed Student-t test determined that the difference between the means of normal patients and any of the other groups was greater than chance: allergic conjunctivitis $p< 0.008$, GPC $p< 0.003$, vernal $p<0.00006$. Wilcoxon rank sum test analysis confirmed a statistical difference between normals and the allergic groups: allergic conjunctivitis $p< 0.0009$, GPC $p<0.0003$, vernal $p<0.00005$. When comparing allergic conjunctivitis, GPC and vernal keratoconjuncitivitis to each other, there was no statistically significant difference found between groups, except borderline significance between GPC and vernal (t-test $p=0.04$, Wilcoxon rank sum $p=0.06$). Patients with viral conjunctivitis had a mean value of 23 ng/ml (range 0 - 85 ng/ml, SD 30 ng/ml) that was statistically significantly lower than the allergic disorders but not different from normal patients.

### Tear Immunoglobulin G

In a series of 30 normal patients tear IgG measured 21 μg/ml with a range of 0 - 81 μg/ml (SD 35 μg/ml). Only two patients with culture proven bacterial conjunctivitis were tested and measured a mean of 605 μg/ml with a range of 121 - 1080 (SD 685 μg/ml) but this was significantly different when ANOVA testing was done ($p < 0.0001$).

Tear C-reactive protein

Thirty normal patients were examined for the presence of C-reactive protein in the tear fluid. No C-reactive protein was identfied with a threshold sensitivity of the test being 16.5 μg/ml. One patient with culture proven staphylococcal conjunctivitis was found to have 154 μg/ml of C-reactive protein in the tear fluid.

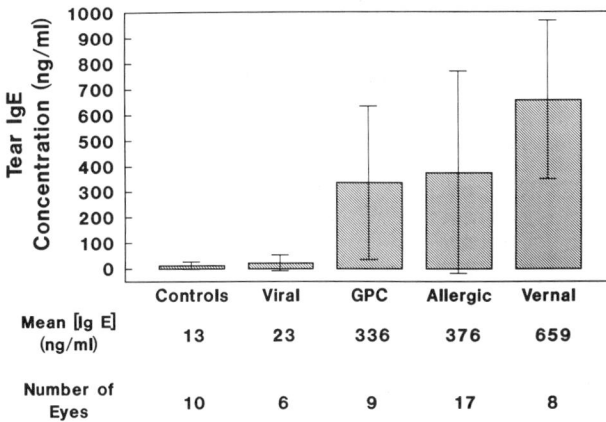

| | Controls | Viral | GPC | Allergic | Vernal |
|---|---|---|---|---|---|
| Mean [Ig E] (ng/ml) | 13 | 23 | 336 | 376 | 659 |
| Number of Eyes | 10 | 6 | 9 | 17 | 8 |

Figure 1. Tear IgE concentrations for each diagnostic category of conjunctivitis.

## DISCUSSION

The remarkable value of the Touch Tear Microassay System in the clinical measurement of tear fluid proteins resides in the rapidity, accuracy,  and small volume requirement of tear fluid to be tested.  Comparison of our results in patients with allergic ocular disease with previous reports of tear IgE levels reveals an inverse linear relationship of measured IgE to volume of fluid required for testing (Figure 2).  This association suggests that collection of larger volumes of tear fluid stimulates reflex tearing and results in a consequent dilution of the measured tear IgE level.  Occasionally collection of tear samples with the microcapillary tube is hampered by mucus clumps in the allergic tear, most patients are readily sampled by this method.  The patients routinely tolerate the procedure well and we have seen no adverse side effect from the tear sampling. In those patients with chronic disease that we have had opportunity to measure at several points in time, there has been good reproducibility of measurement within the ranges described for the disease groups.

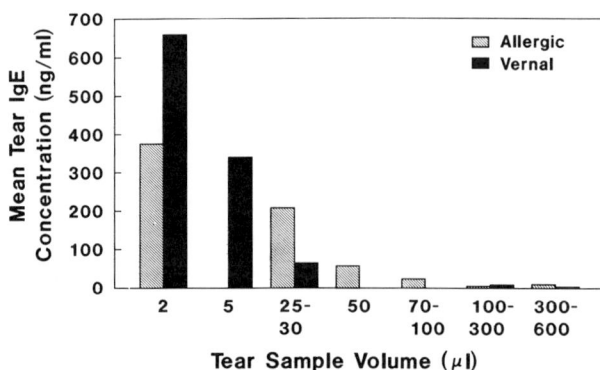

Figure 2. Tear IgE concentrations as a function of tear sample volume. Data is a compilation of the current data and data from previously published studies.[1-7]

While tear IgG levels show some variability, there does appear to be an elevation of IgG level in tears in chronic bacterial surface infections. Elevated tear IgG levels were also noted in one patient tested who had chronic staphylococcal blepharoconjunctivitis (77 µg/ml).

C-reactive protein has been measured in so few diseased patients that statistical analysis is impossible, but the identification of C-reactive protein in our patient with chronic staphylococcal blepharoconjunctivitis while all normal patients have had no measureable C-reactive protein suggests that further study is indicated.

Should reliable and reproducible patterns of multiple analytes in the tear fluid be established and correlated with clinical disease, it is technically possible to configure the Touch Tear Microassay System to measure different analytes at the same time and on the same analysis card providing an objective method of characterizing tear protein profiles. This availability could then be used to verify clinical disease diagnosis and to monitor therapeutic response.

## ACKNOWLEDGMENTS

The authors have no financial interest in the instruments described in this manuscript. This work was, however, supported by a grant from the North Carolina Biotechnology Center to Touch Scientific, Inc., Raleigh, NC and contracted to the Duke University Eye Center. The authors wish to thank Keith Garrett, Pearce Youngbar, and Tim Hamilton, Ph.D., for technical advice in this project.

## REFERENCES

1. B.H. McClellan, C.R. Whitney, L.P. Newman, and M.R. Allansmith, Immunoglobulins in tears, *Amer. J. Ophthalmol.* 76:89 (1973).
2. M. Ballow, and L. Mendelson, Specific immunoglobulin E antibodies in tear secretions of patients with vernal keratoconjuncitivitis, *J. Allergy Clin. Immunol.* 66:112 (1980).

3. P.C. Donshik, and M. Ballow, Tear immunoglobulins in giant papillary conjunctivitis induced by contact lenses, *Amer. J. Ophthalmol.* 96:460 (1986).
4. D. Sompolinsky, Z. Samra, A. Zavaro, and Y. Barishak, Allergen-specific immunoglobulin E antibodies in tears and serum of vernal conjunctivitis patients, *Int. Arch. Allergy appl. Immunol.* 75:317 (1984).
5. J.K.G. Dart, R.J. Buckley, M. Monnickendan, and J. Prasad, Perennial allergic conjunctivitis: definition, clinical characteristics and prevalence, *Trans. Ophthalmol. Soc. U.K.* 105:513 (1986).
6. M.R. Allansmith, and R.N. Ross, Immunology of the tear film. *in:* "The Preocular Tear Film in Health, Disease and Contact Lens Wear," F.J. Holly, ed., Dry Eye Institute, Lubbock, TX, p.750 (1986).
7. M.S. Insler, J.M. Lim, J.T. Queng, C. Wanissorn, and J.P. McGovern, Tear and serum IgE by Tandem-R IgE immunoradiometric assay in allergic patients, *Ophthalmology* 94:945 (1987).

# THE TEAR FILM: PHARMACOLOGICAL APPROACHES AND EFFECTS

Jeffrey P. Gilbard[1,2]

[1]Cornea Research Unit, Schepens Eye Research Institute
[2]Department of Ophthalmology, Harvard Medical School
Boston, MA 02114

## INTRODUCTION

The ability to intervene medically in the treatment of tear film and dry-eye diseases has been evolving rapidly over the past twenty years. Central to the improvement in therapy has been the improvement in our understanding of the pathogenesis and the natural history of dry-eye surface disease. Pharmaceuticals have been designed to act at earlier and earlier points in the disease process.

Of major importance in the improvement of our understanding of the pathogenesis is the link that has been established between increased tear film osmolarity and the ocular surface disease of keratoconjunctivitis sicca (KCS). The data comes from in vitro and in vivo work in rabbits as well as clinical studies. First, in cell culture, increased osmolarity is toxic to corneal epithelium.[1] Second, in studies of rabbit models for lacrimal gland disease and meibomian gland dysfunction the ocular surface changes of dry-eye disease are dependent upon and proportional to increases in tear film osmolarity.[2-7] And, finally, in clinical studies tear film osmolarity shows a positive correlation with rose Bengal staining.[8]

The natural history of the ocular surface changes in human dry eye disease has been confirmed and further clarified by the study of rabbit models of KCS.[2,3,5] For the purposes of this review we will define four major events or "milestones" that occur with progression of the disease process. In these rabbit models and man the first event or milestone is a decrease in tear production, and in the rabbit it has been shown to be associated with an increase in tear osmolarity that occurs within 24 hours of the onset of disease. The second milestone is a loss of mucous-containing conjunctival goblet cells. Conjunctival goblet cell density is an extraordinarily sensitive indicator of ocular surface health, and decreased conjunctival goblet cell density is evident several weeks after onset of disease in the rabbit model for KCS.[2] Corneal glycogen levels also decrease early in the disease and parallel the decreases in conjunctival goblet-cell density. In contrast to biochemical changes in the cornea, corneal morphology remains normal for about one year after onset of disease in the rabbit model for KCS.[5] This is analogous to human disease where morphological changes in the conjunctiva precede and exceed those observed in the cornea.[9,10] The third milestone in KCS is the development of morphological changes in the cornea, first characterized by increased corneal desquamation. At one year after onset of disease in the rabbit, no changes were observed in tear film stability. We can postulate that a decrease in tear film stability requires severe changes in the corneal epithelial cell surface probably including loss of cell membrane microplicae and microvilli and changes in cell membrane glycoproteins. The fourth milestone then is a destabilization of the cornea-tear interface. As we review the progress made in the treatment of dry-eye disease we shall see that treatments have been designed to act more and more effectively on earlier and earlier milestones of dry-eye surface disease. As this has occurred we have moved from treatments that have the potential to provide only transient symptomatic relief, to treatments that minimize toxicity, to treatments that can reverse the basic events and significant milestones of dry-eye surface disease.

## TREATMENT OF THE FOURTH MILESTONE: DESTABILIZATION OF THE CORNEA-TEAR INTERFACE

Demulcents are polymers traditionally added to solutions to improve their lubricant properties. Carboxymethylcellulose, dextran 70, hydroxypropyl methylcellulose, polyethylene glycol, polyvinyl alcohol and povidone are all demulcents commonly used in what are popularly called artificial tear solutions but by FDA regulations are termed "lubricating eye drops." Demulcents are considered the active ingredients in these drops. In 1975 Lemp and co-workers[11] published a classic report that demonstrated that demulcent solutions (all containing a preservative at the time) increase tear film breakup time about twofold in normal controls. The effect lasted between 60 and 90 seconds after drop instillation.

In 1977 erodible insert technology was developed to increase tear film stability. Erodible collagen inserts were described first and reported to increase tear film breakup time by five seconds, about a 33% increase, in normal controls. The effect averaged seven hours.[12] Erodible cellulose polymers appeared more effective in rabbit studies. These cellulose inserts increased breakup time ten times control and three times demulcent solutions in rabbits. In this study the effect of demulcent solutions lasted less than 30 minutes while the effect of erodible cellulose polymers lasted for more than six hours.[13] When these cellulose inserts were tested in clinical studies involving dry-eye patients they were found to increase breakup time about 3 seconds.[14] A recently introduced solution containing polycarbophil, an insoluble polymer, seems to have a long-lasting presence in the tear film and may represent an advance in the tradition of these other Fourth Milestone treatments.

## TREATMENT OF THE THIRD MILESTONE: CORNEAL DESQUAMATION

Morphological changes in the cornea, specifically desquamation and loss of tight junctions between superficial corneal epithelial cells (manifested by a loss of epithelial barrier function), precede destabilization of the cornea-tear interface. The development of treatments targeted at the Third Milestone was heralded by the recognition that preservatives were toxic to corneal epithelium and altered corneal epithelial morphology.[15] Ultimately preservative-free demulcent solutions were introduced into clinical practice. In 1992 Gobbels and Spitznas[16] measured corneal barrier function in vivo by computerized fluorophotometry in dry-eye patients. They found that treatment with a polyvinyl pyrrolidone solution without preservatives improved corneal barrier function, and therefore corneal cell junctions, in dry-eye patients. A control solution with benzalkonium chloride diminished corneal epithelial barrier function. These experiments highlight the limitations of treatments targeted only on the Fourth Milestone. While solutions may improve tear film stability they may simultaneously have the undesired effect of altering corneal morphology and decreasing corneal barrier function.

An important advance in Third Milestone treatment was developed by Wilson and co-workers.[17,18] They found that normal rabbit and human corneas bathed in a preservative-free balanced electrolyte solution show less desquamation than corneas bathed in a preservative-free saline solution. They called their balanced electrolyte solution "BTS" and it contains the major tear ions at the concentrations shown in Table 1. In a subsequent series of experiments Bernal and Ubels[19] studied the effect of a similar ion solution they called "physiologic tear" (PT) (Table 1) on recovery of corneal epithelial barrier function after disruption by benzalkonium chloride. In these experiments barrier function was a surrogate measurement for one aspect of corneal morphology--the tight junctions between superficial epithelial cells. PT was better than the other nonpreserved lubricating solutions they tested in decreasing epithelial permeability. Difficult to interpret is that a commercially-available lubricating solution without electrolytes and preserved with polyquaternium-1 decreased permeability to $5.7 \pm 0.5$ nmol/g after 1.5 hours of exposure, a level comparable to or better than the permeability of $6.5 \pm 0.55$ nmol/g observed after exposure to PT, and better than the permeability observed after exposure to all of the commercially-availiable preservative-free solutions tested.

**TABLE 1.** Electrolyte Balance of Three Electrolyte Solutions

|  | NaCl | KCl | NaHCO$_3$ | NaH$_2$PO$_4$ | CaCl$_2$ | MgCl$_2$ |
|---|---|---|---|---|---|---|
| BTS[17,18] | 116.4 | 18.7 | 25.9 | 0.7 | 0.4 | 0.6 |
| PT[19] | 128.7 | 17.0 | 12.4 | 0 | 0.3 | 0.35 |
| Solution 15[20] | 99.0 | 24.0 | 32.0 | 1.0 | 0.8 | 0.6 |

Based collectively on the data collected in these studies, it appears that the addition of certain electrolytes to a preservative-free lubricating solution reduces corneal desquamation compared with such solutions missing those electrolytes. Third Milestone treatments are directed at changes in corneal morphology. The cornea, compared to the conjunctiva, is relatively resistant to dry-eye disease and changes in corneal morphology occur late into the course of the disease. These preservative-free electrolyte solutions were the best that could be designed while focusing only on issues related to corneal morphology.

## TREATMENT OF THE SECOND MILESTONE: DECREASED CONJUNCTIVAL GOBLET CELL DENSITY AND DECREASED CORNEAL GLYCOGEN

Decreases in conjunctival goblet-cell density and corneal glycogen may be the most sensitive indicators of ocular surface health available today and manifest themselves very quickly with the onset of dry eye disease. Goblet cells discharge mucus within hours of exposure to tear osmolarities seen in dry-eye disease,[7] and in a rabbit model for KCS decreases in conjunctival goblet cell-density and corneal glycogen are evident 4 weeks after the onset of disease.[2]

While Wilson and co-workers were studying the effect of electrolyte solutions on corneal morphology, Gilbard and co-workers were studying the effect of electrolyte solutions on conjunctival goblet-cell density.[20,21] They hypothesized that in order for a solution to restore conjunctival goblet-cell density it would have to be able to maintain normal goblet-cell density with exposure to normal eyes. Based on extended 12-hour experiments in which rabbit eyes were bathed in a variety of electrolyte solutions, Gilbard and co-workers developed a balanced electrolyte solution they called "Solution 15" (Table 1). After 12-hours of continuous exposure to this solution, rabbit eyes maintained normal conjunctival goblet-cell density. For comparison, BTS, developed by Wilson and co-workers based on corneal morphology, maintained only about 60% of the original density of goblet cells. From Solution 15 Gilbard and co-workers developed ATF, a hypotonic version with a demulcent and buffer added. This solution was shown to maintain normal goblet-cell density as well as Solution 15.[21]

In a subsequent series of experiments,[22] rabbits with surgically-induced KCS were treated in a masked fashion with ATF and three commercially-available lubricating solutions. Two of the commercially-available solutions were preservative-free and the third solution was the commercially-available solution preserved with polyquaternium-1 used by Bernal and Ubels[19] in their study of corneal barrier function. Rabbits were treated four times a day, five days a week, for twelve weeks. At the completion of treatment rabbits treated with ATF showed a statistically significant restoration of conjunctival goblet-cell densities and corneal glycogen levels. Rabbits treated with the commercially-available preservative-free solutions showed no changes compared to untreated controls, and rabbits treated with the commercially-available solution preserved with polyquaternium-1 showed a statistically significant decrease in conjunctival goblet-cell densities and corneal glycogen levels (Figures 1 and 2).

These experiments highlight the limitations of treatments targeted only on the Third Milestone. While preservative-free solutions may reduce corneal desquamation and improve corneal barrier function, they may simultaneously have the undesired effect of decreasing conjunctival goblet-cell density and corneal glycogen. Treatment directed at the Second Milestone represents a turning point in the treatment of the surface disease of dry eye. The effect of treatment at this level moves beyond the aim of reducing toxicity to producing an entirely therapeutic effect, and actually reversing the specific changes of dry-eye surface disease.

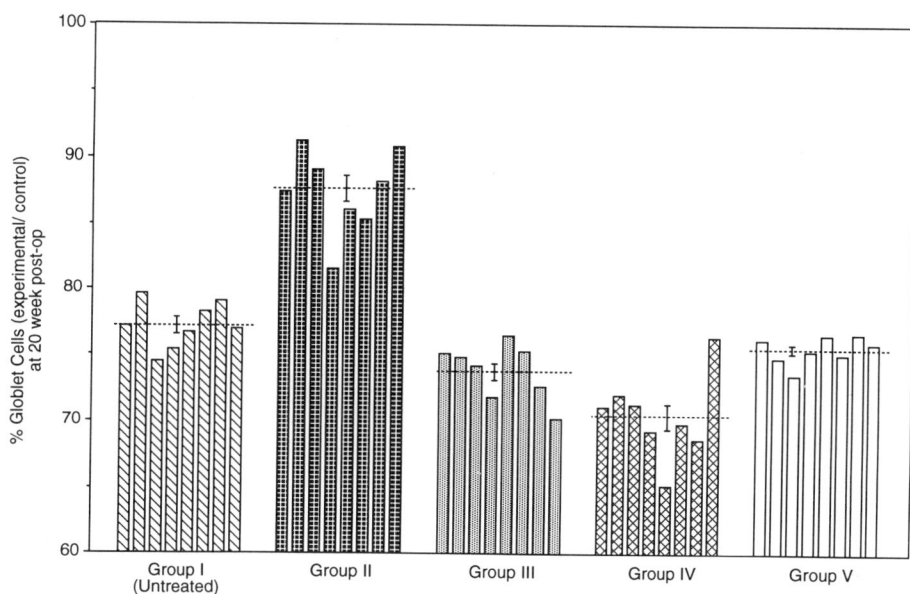

**Figure 1.** The effect of 12 weeks of treatment with four different artificial tear solutions on conjunctival goblet-cell density. Group II rabbits were treated with ATF. Published courtesy of *Ophthalmology* (1992;99:600-4).

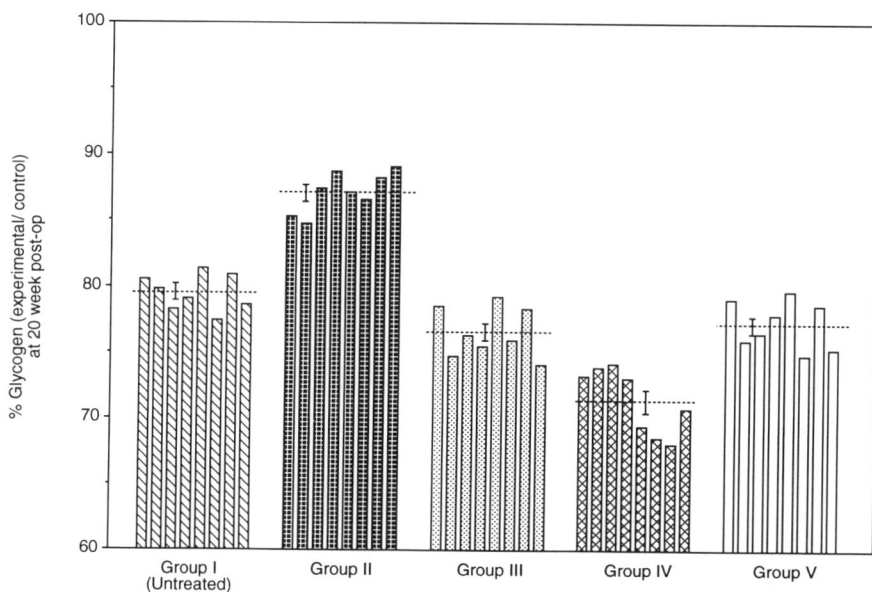

**Figure 2.** The effect of 12 weeks of treatment with 4 different artificial tear solutions on corneal glycogen levels. Group II rabbits were treated with ATF. Published courtesy of *Ophthalmology* (1992;99:600-4).

## TREATMENT OF THE FIRST MILESTONE: DECREASED TEAR PRODUCTION

Decreased tear production causes the ocular surface changes characteristic of dry-eye disease and so, at least theoretically, should precede decreases in conjunctival goblet-cell density and corneal glycogen. While ATF provides the electrolyte balance known to be needed for ocular surface function and health, stimulating tear production itself would also provide those electrolytes along with other agents that might be needed from the lacrimal gland, by the ocular surface, for normal function, maintenance and repair.

In 1990 Gilbard and Dartt and co-workers demonstrated in a rabbit model for KCS that accessory lacrimal gland secretion could be pharmacologically stimulated by the topical application of agents that increase intracellular levels of cAMP or cyclic guanosine monophosphate.[23] 3-Isobutyl-1-metylxanthine (IBMX), a potent phosphodiesterase inhibitor, stimulated tear production producing a significant dose-dependent decrease in tear osmolarity and a significant increase in tear volume.

Topical treatment of a rabbit model for KCS with IBMX produced a statistically significant decrease in tear film osmolarity, and statistically significant increase in conjunctival goblet-cell density and corneal glycogen (unpublished data). In a clinical study of patients with KCS,[24] topical application of IBMX produced a dose-dependent decrease in tear film osmolarity similar to that observed in the earlier rabbit studies. Treatment of KCS patients for four weeks produced a rapid and statistically significant decrease in tear film osmolarity (Figure 3) and was superior to vehicle alone in reducing rose Bengal staining.

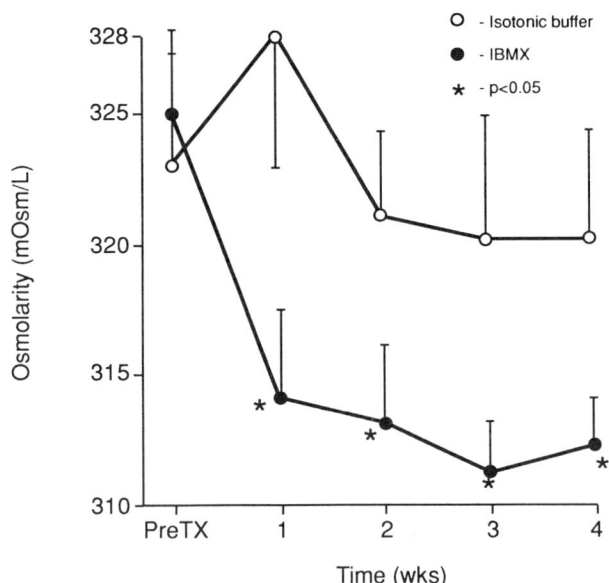

**Figure 3.** Six-times-a-day treatment with topically applied 3.0 mM IBMX (closed circles) produces a rapid and statistically significant reduction in tear film osmolarity in patients with dry-eye disease. Open circles represent treatment with vehicle alone. Published courtesy of *Archives of Ophthalmology* (1991;109:672-6).

Treatment targeted at the First Milestone acts focally on the primary local event in dry-eye disease--decreased tear production. Treatment at this level holds the potential to reverse not only the ocular surface changes of dry-eye disease, but also the decrease in tear production so important in dry-eye disease.

The progress that has been made over the past twenty years in the treatment of dry-eye disease is very exciting, and the technologies developed should provide great relief for patients with dry eyes and great satisfaction for their physicians.

# REFERENCES

1. J.P. Gilbard, J.B. Carter, D.N. Sang, M.F. Refojo, L.A. Hanninen, K.R. Kenyon, Morphologic effect of hyperosmolarity on rabbit corneal epithelium, *Ophthalmology*. 91:1205 (1984).
2. J.P. Gilbard, S.R. Rossi, K.L. Gray, A new rabbit model for keratoconjunctivitis sicca, *Invest Ophthalmol Vis Sci*. 28:225 (1987).
3. J.P. Gilbard, S.R. Rossi, K.L. Gray, L.A. Hanninen, K.R. Kenyon, Tear film osmolarity and ocular surface disease in two rabbit models for keratoconjunctivitis sicca, *Invest Ophthalmol Vis Sci*. 29:374 (1988).
4. J.P. Gilbard, S.R. Rossi, K. Gray Heyda, Tear film and ocular surface changes after closure of the meibomian gland orifices in the rabbit, *Ophthalmology*. 96:1180 (1989).
5. J.P. Gilbard, S.R. Rossi, K.L. Gray, L.A. Hanninen, Natural history of disease in a rabbit model for keratoconjunctivitis sicca, *ACTA Ophthalmol*. (Suppl 192)67:95 (1989).
6. J.P. Gilbard, S.R. Rossi, Tear film and ocular surface changes in a rabbit model of neurotrophic keratitis, *Ophthalmology*. 97:308 (1990).
7. A.J.W. Huang, R. Belldegrun, L. Hanninen, K.R. Kenyon, S.C.G. Tseng, M.F. Refojo, Effects of hypertonic solutions on conjunctival epithelium and mucinlike glycoprotein discharge, *Cornea*. 8:15 (1989).
8. J.P. Gilbard, S.R. Rossi, D.T. Azar, K.L. Gray, Effect of punctal occlusion by Freeman silicone plug insertion on tear osmolarity in dry eye disorders, *CLAO J*. 15:216 (1989).
9. H. Sjogren, *A New Conception of Keratoconjunctivitis Sicca*, (Translation by J.B. Hamilton). Australian Medical Publishing Company, Sydney (1943).
10. H. Sjogren, Some problems concerning keratoconjunctivitis sicca and the sicca-syndrome, *Acta Ophthalmol*. 29:33 (1951).
11. M.A. Lemp, M. Goldberg, M.R. Roddy, The effect of tear substitutes on tear film break-up time, *Invest Ophthalmol*. 14:255 (1975).
12. S.E. Bloomfield, M.W. Dunn, T. Miyata, K.H. Stenzal, S.S. Randle, A.L. Rubin, Soluble artificial tear inserts, *Arch Ophthalmol*. 95:247 (1977).
13. P.D. Gautheron, V.J. Lotti, J.C Le Douarec, Tear film breakup time prolonged with unmedicated cellulose polymer inserts, *Arch Ophthalmol*. 97:1944 (1979).
14. J.I. Katz, H.E. Kaufman, C. Breslin, I.M. Katz, Slow-release artificial tears and the treatment of keratitis sicca, *Ophthalmol*. 85:787 (1978).
15. R.R. Pfister, N. Burstein, The effects of ophthalmic drugs, vehicles, and preservatives on corneal epithelium: a scanning electron microscope study, *Invest Ophtalmol*. 15:246 (1976).
16. M. Gobbels, M. Spitznas, Corneal epithelial permeability of dry eyes before and after treatment with artificial tears, *Ophthalmol*. 99:873 (1992).
17. W.G. Bachman, G. Wilson, Essential ions for maintenance of the corneal epithelial surface, *Invest Ophthalmol Vis Sci*. 26:1484 (1985).
18. R.J. Fullard, G.S. Wilson, Investigation of sloughed corneal epithelial cells collected by non-invasive irrigation of the corneal surface, *Curr Eye Res*. 5:847 (1986).
19. D.L. Bernal, J.L. Ubels, Artificial tear composition and promotion of recovery of the damaged corneal epithelium, *Cornea*. 12:115 (1993).
20. J.P. Gilbard, S.R. Rossi, K. Gray Heyda, Ophthalmic solutions, the ocular surface, and a unique therapeutic artificial tear formulation, *Am J Ophthalmol*. 107:348 (1989).

21. J.P. Gilbard, Non-toxic ophthalmic preparatons, U.S. Patent 4,775,531.

22. J.P. Gilbard, S.R. Rossi, An electrolyte-based solution that increases corneal glycogen and conjunctival goblet-cell density in a rabbit model for keratoconjunctivitis sicca, *Ophthalmol.* 99:600 (1992).

23. J.P. Gilbard, S.R. Rossi, K. Gray Heyda, D.A. Dartt, Stimulation of tear secretion by topical agents that increase cyclic nucleotide levels, *Invest Ophthalmol Vis Sci.* 31:1381 (1990).

24. J.P. Gilbard, S.R. Rossi, K. Gray Heyda, D.A. Dartt, Stimulation of tear secretion and treatment of dry-eye disease with 3-Isobutyl-1-methylxanthine, *Arch Ophthalmol.* 109:672 (1991).

# INFLUENCE OF VARIOUS PHARMACEUTICAL AGENTS ON TEAR FLOW AS ASSESSED BY FLUOROPHOTOMETRY

Martin J. Göbbels and Manfred Spitznas

University Eye Hospital
Sigmund-Freud-Str. 25
D-W-5300 Bonn 1
Germany

## INTRODUCTION

Many pharmaceutical agents have been reported to influence tear production. Most of them, e.g. betablockers, sympathicomimetic agents, contraceptives or tranquilizers, drugs that are all extensively used, are thought to decrease the amount of tear secretion, whereas only few drugs are referred to as tear stimulators, e.g. eledoisin, bromhexine or, still in experimental use, isobutyl-methyl-xanthine.

However, up to now the effect of such pharmaceutical agents on tear flow has been evaluated mostly by means of filter paper strip tests, often leading to controversial results. For, even with standarized procedure, the Schirmer test show considerable intraindividual variation and has a poor reproducibility. In addition, the sensitivitiy of the Schirmer test is very low, ranging between 10% and 30% (1,2).

Thus, even though the Schirmer test may be of some value as a rough screening test, it is undoubtedly not appropriate for the exact determination of tear secretion. That is why, until now only poor and partially even contradictory data on the actual effect of pharmaceutical agents on the amount of tear secretion is available.

More than 25 years ago, Maurice and Mishima and coworkers (3,4) developed a fluorophotometric technique that could successfully be used to determine tear secretion. On the basis of these pioneering studies tear film fluorophotometry has been subtantially improved. Using modern computerized objective fluorophotometers we are now able to determine tear flow objectively and very precisely (5-10).

After instillation of only very small amounts of sodium fluorescein into the tear film , we use only 2 $\mu$g, the fluorophotometer measures the decrease of the tear film fluorescein concentration over a period of 15 min. Thereafter, the computer connected online with the fluorophotometer calculates a regression line for the clearance of tear film fluorescein, the tear film turnover rate as well as the amount of tear production.

This technique is evidently less invasive and less irritative than conventional filter paper strip tests. Whereas most patients find measurements of tear secretion by filter paper strip tests definitely unpleasant, almost all individuals examined by fluorophotometry report no unpleasant sensations whatsoever during fluorophotometry.

Thus, it is not surprising that in control investigations the mean intraindividual variation of tear flow as measured by objective fluorophotomtery could be shown to be as low as 7.4%, whereas Schirmer test results (after topical anesthesia) differed as much as 31.2% on the average (5). Thus, when compared with conventional filter paper strip tests, tear film fluorophotometry can be assumed to obtain more accurate measurements of unprovoked, physiologic aqueous tear flow.

In our studies we investigated the effect on tear flow of four pharmaceutical agents, two of them being very extensively used topical ophthalmic drugs, namely oxymetazoline 0.026% and timolol 0.5%.The other two,namely topical eledoisin 0.1% and systemic bromhexine, are believed to stimulate tear secretion.

## OXYMETAZOLINE

Vasoconstrictive, alpha-sympathicomimetic agents are perhaps the ophthalmic drugs most frequently applied in western countries. However, patients using such eye drops often complain of typical dry-eye symptoms, such as burning and foreign body sensation. Thus, clinical experience suggests that alpha-sympathicomimetic eye drops may induce tear film problems, possibly as a result of a reduction in tear production.

In order to verify this clinical impression, we performed tear film fluorophotomtery on a total of 43 healthy adults after topical application of 20 $\mu$l 0.026% oxymetazoline. Twenty-one of the subjects were checked 30 and 90 min after instillation (Group A), the rest 3, 6, and 24 hours after application (Group B).

The results showed that before instillation of oxymetazoline tear flow was found to be in a normal range (Group A 1.4+/-0.9 $\mu$l/min; Group B 1.6+/-0.6 $\mu$l/min;). Thirty minutes after instillation tear production was significantly diminished (0.8+/-0.5 $\mu$l/min; $P < 0.001$), reaching a minimum of only 29 % of its initial value 90 minutes after instillation (0.4+/-0.2 $\mu$l/min; $P < 0.001$). Three and 6 hours after instillation tear production had slightly recovered, but was still significantly decreased (3h 1.1+/-0.3 $\mu$l/min, $P < 0.001$; 6h 1.1+/-0.2 $\mu$l/min, $P < 0.001$). Twenty-four hours after instillation tear secretion had returned to the previous normal range (1.5+/-0.5 $\mu$l/min, $P = 0.5$; Figure 1).

The speed and the amount of the decrease in tear flow after instillation of a single drop of a commercially available sympathicomimetic drug are striking. Considering that a single drop of 0.026% oxymetazoline is sufficient to reduce the tear secretion of healthy young eyes to less than one third of its previous value and that tear flow remains significantly decreased over a period of at least 6 hours, it is conceivable that long-lasting or frequent application of such eye drops may cause dry-eye symptoms, and even more, may induce secondary degenerative changes of the conjunctival and corneal surface. This process would be similar to changes known to occur in the nasal mucosa where abuse of vasoconstrictive nasal sprays or drops may lead to chronic rhinitis sicca. The present fluorophotometric data emphasize the potential risk of vasoconstrictive, alpha-sympathicomimetic eye drops, the use of which therefore should be strictly limited and - if possible - subjected to medical monitoring.

## TIMOLOL

Topical betablockers have been successfully used in the treatment of glaucoma for more than 25 years. Although topical betablockers are relatively devoid of ocular side effects, many glaucomatous patients so treated complain of typical dry-eye symptoms. Previous studies investigating the influence of topical betablockers on tear secretion by means of filter paper strip tests have yielded contradictory results (11-14).

In the present study we determined the effect of unpreserved 0.5 % timolol, topically applied twice daily, in 24 consecutive patients with bilateral primary open-angle glaucoma. Tear flow fluorophotometry was perfomed before treatment, as well as 7 and 14 days, and 4 to 7 months after the onset of antiglaucomatous treatment.

The results show that before treatment the tear flow of the glaucomatous patients was in a normal range (1.1+/-0.5 $\mu$l/min). However, seven days after the onset of treatment tear flow had decreased significantly to about two thirds of its previous value (0.7+/-0.5 $\mu$l/min; P<0.01). Fourteen days (1.2+/-0.4 $\mu$l/min) as well as 4 to 7 months (1.1+/-0.5 $\mu$l/min) after the beginning of treatment, values of tear flow did not significantly differ from those measured before initiation of treatment (Figure 2).

These results suggest that topical medication with 0.5 % timolol leads to a significant, but transient decrease in tear production. In the long term, however, antiglaucomatuos treatment with 0.5 % timolol eye drops has no effect on tear secretion. Thus, dry-eye complaints induced by such treatment are apparently not the result of a decrease in tear production, but rather a consequence of changes of tear film stability and/or preservatives usually contained in such eye drops.

## ELEDOISIN

Almost 20 years ago Bietti and coworkers (15) first reported the successful treatment of dry eyes with eledoisin, an endekapeptide that can be isolated from the salivary glands of a mediterranian octopus, Eledona maschata. According to most studies investigating the effect of eledoisin on tear production there is a distinct increase in tear flow of dry eyes about 30 to 60 min after the instillation of the oligopeptide; however, some investigators found tear production to be still significantly increased even 2 to 8 hours after application ofeledoisin (16,17).

In the present study we investigated the effect of 0.1 % eledoisin, applied topically as a single dosage of 20 $\mu$l, on both 24 healthy adults and 26 dry-eye patients. Tear film fluorophotometry was performed before instillation, as well as 1, 2, and 4 hours afterward. As for the healthy eyes, we found a significant increase in tear flow over at least 2 hours (1.8+/-0.5 $\mu$l/min; P<0.05), reaching a maximum 1 hour after application when tear secretion was increased to 127 % of its previous value (1.9+/-0.5 $\mu$l/min; P<0.01). Four hours after instillation, tear flow had returned to the range of the initial values (1.4+/-0.2; P=0.3; Figure 3).

Dry eyes also showed a significant increase in production that lasted at least 2 hours after the instillation of eledoisin. However, the relative increase of tear flow is much more drastic in dry-eye patients than that found in healthy controls, showing a threefold and twofold increase 1 and 2 hours (1h 0.7+/-0.4 $\mu$l/min, P<0.001; 2h 0.4+/-0.2 $\mu$l/min, P<0.001) after application of the tear stimulator (Figure 4).

It is, however, still questionable ,whether this effect can actually be used for the management of dry eyes. Even if 1 hour after instillation tear secretion in severely dry eyes is increased to about 60 % of healthy eyes, as soon as 1 hour later it again drops to the range of dry eyes.

Presuming that eledoisin is capable of stimulating tear secretion even after long-term treatment, the drug will have to be applied frequently, i.e. hourly. Moreover, the few subjects of this study, who suffered from extremely dry eyes with no more than residual tear flow showed but a minor increase in tear flow following instillation of eledoisin.

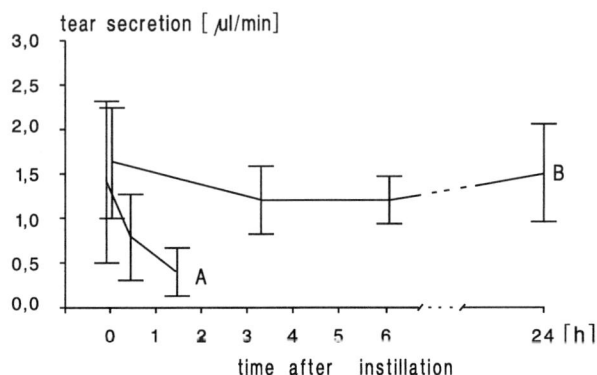

**Figure 1.** Tear flow after topical application of 20 μl 0.026% oxymetazoline. Group A was checked 30 and 90 min after instillation, group B 3, 6, and 24 hours after application.

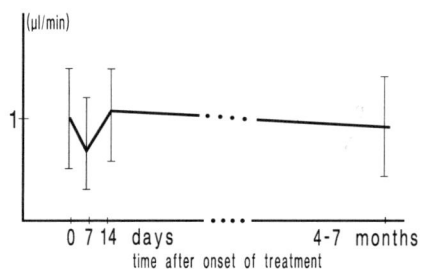

**Figure 2.** Tear flow of glaucomatous patients treated with 0.5% timolol eye drops (b.i.d.).

Tear secretion of healthy eyes
after topical application of
20 µl 0.1% eledoisin

[ µl / min]

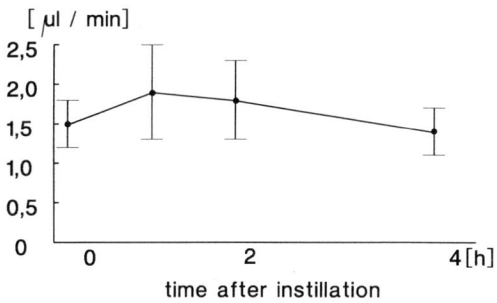

time after instillation

**Figure 3.** Tear flow of healthy eyes treated with 0.1% eledoisin.


Tear secretion of dry eyes
after topical application of
20 µl 0.1% eledoisin

[ µl/min]

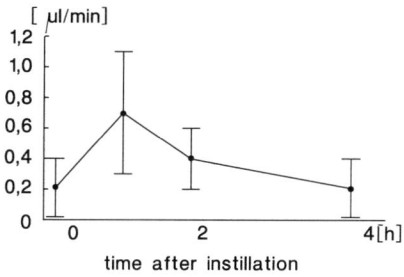

time after instillation

**Figure 4.** Tear flow of dry eyes treated with 0.1% eledoisin.

Apparently eledoisin fails to stimulate the tear flow of lacrimal glands with minimal function. Thus, in contrast to the experience reported by some studies, extremely severe cases of keratoconjunctivitis sicca should not be expected to benefit significantly when treated with eledoisin.

## BROMHEXINE

Oral administration of bromhexine has been reported to have beneficial effects on patients suffering from severe dry-eye disease, especially on those with Sjoegren's syndrome. However, the question of whether or not such treatment leads to an increase in tear secretion is still unanswered, as studies on its efficacy showed controversial results (18,19).

In the present study tear film fluorophotometry was performed on 23 consecutive dry-eye patients with Sjoegren's syndrome, who received an oral medication of 48 mg bromhexine daily. The fluorophotometric measurement were taken before, as well as 7 and 21 days after the onset of treatment.

The results showed that at least within 3 weeks of medication, oral treatment with bromhexine has no effect on the amount of tear production. However, further investigation is needed to evaluate the influence of bromhexine on tear film qualitiy, tear film stability and ocular surface disease in dry-eye patients (Figure 5).

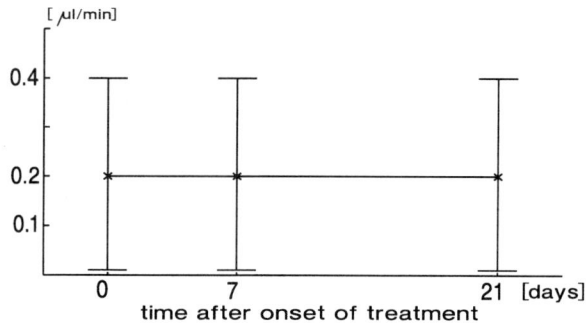

**Figure 5**. Tear flow of dry-eye patientas with Sjögren's syndrome treated with bromhexine 48 mg/d.

## REFERENCES

1. N.W. Pinschmidt. Evaluation of the Schirmer tear test. South Med J 63: 1256-1259 (1970).

2. A. Shapiro, S. Merin. Schirmer test and breakup time of tear film in normal subjects. Am J Ophthalmol 88: 58-66 (1979).

3. D.M. Maurice. A new objective fluorophotometer. Exp Eye Res 2: 33-38 (1963).

4. S. Mishima, A. Gasset, S.D. Klyce. Determination of tear volume and tear flow. Invest Ophthalmol 5: 264-275 (1966).

5.  M. Göbbels, G. Goebels, R. Breitbach, M. Spitznas. Tear secretion in dry eyes as assessed by objective fluorophotometry. German J Ophthalmol 1: 350-353 (1992).

6.  M. Göbbels, C. Achten, M. Spitznas. Einfluß vasokonstriktiver Augentropfen auf Tränenvolumen und Tränensekretion. Fortschr Ophthalmol 88: 173-175 (1991).

7.  M. Göbbels, C. Achten, M. Spitznas. Effect of topically applied oxymetazoline on tear volume and tear flow in humans. Graefe's Arch Clin Exp Ophthalmol 229: 147-149 (1991).

8.  M. Göbbels, J. Selbach, M. Spitznas. Effect of eledoisin on tear volume and tear flow in humans. Graefe's Arch Clin Exp Ophthalmol 229: 549-552 (1991).

9.  M.J. Puffer, R.W. Neault, R.F. Brubaker. Basal precorneal tear turnover in human eyes. Am J Ophthalmol 89:369.376 (1980).

10. W.R.S. Webber, D.W. Jones, P. Wright. Fluorophotometric measurements of tear turnover rate in normal healthy persons: evidence for a circadian rhythm. Eye 1: 615-620 (1987).

11. A. D'Andrea, R. De Natale, A. Mancini, E. Catanese. The influence of different beta blockers in lacrimal secretion of glaucomatous patients. Boll Ocul 66 (2): 235-238 (1987).

12. G. Zavarise, S. Michieletto, E. Noya. Effects of timolol maleate 0.5% on tear flow in normal human eyes. Boll Ocul 64: 923-927 (1985).

13. J. Fromow, V. Manzanilla, M.E. Tapia Diaz, T.O. Mc Donald. Pindolol: Short-term evaluation in glaucoma patients. Glaucoma 6: 17-22 (1984).

14. L. Bonomi, G. Zavarise, E. Noya, S. Michieletto. Effects of timolol maleate on tear flow in human eyes. Graefe's Arch Clin Exp Ophthalmol 213: 19-22 (1980).

15. G.B. Bietti, P. Capra, G. De Caro. Zur Anwendung eines neuen Medikamentes, des Eledoisins, zur Behandlung der Keratokonjunktivitis sicca. Ber Dtsch Ophthalmol Ges 73: 399-407 (1975).

16. M.G. Bucci, P. Capra, M.P. Cichetti, A. Gualano, G.L. Manni. Treatment of the tear hyposecretion after topical beta-blockers. Ann Oftalmol Clin Ocul 114: 1217-1224 (1988).

17. W. Jaeger. Treatment of severe courses of keratoconjunctivitis sicca with eledoisin. Klin Monatsbl Augenheilkd 192: 163-166 (1988).

18. R. Avisar, H. Savir. Our further experience with bromhexine in keratoconiunctivitis sicca. Ann Ophthalmol 20: 382 (1988).

19. N.J. Kriegbaum, M. von Lingstow, P. Oxholm, P. Prause. Keratoconjunctivitis sicca in patients with primary Sjoegren's syndrome. A longitudinal study of ocular parameters. Acta Ophthalmol 66: 481-484 (1988).

# NEURAL STIMULATION OF CONJUNCTIVAL GOBLET CELL MUCOUS SECRETION IN RATS

Timothy L. Kessler and Darlene A. Dartt

Cornea Unit
Schepens Eye Research Institute
20 Staniford Street
Boston, MA 02114
and
Department of Ophthalmology
Harvard Medical School

## INTRODUCTION

The mucous layer of the tear film provides constant protection to the ocular surface. A rapid release of mucus in response to surface irritants, trauma, or toxins (bacterial and environmental) is necessary to replenish the mucous layer and protect the ocular surface. The rapid release of mucus may be controlled by neural regulation.

In other tissues that contain goblet cells, the digestive and respiratory tracts, there is evidence of neural regulation of mucous secretion. Adult rat intestinal goblet cells from the crypt, but not the villus can be stimulated with cholinergic agonists.[1] In rabbit mucosal explants from small and large intestine,[2] Neutra et al. have shown that adrenergic agent, epinephrine, phenylephrine, and dopamine do not stimulate short-term goblet cell secretion. Goblet cell secretion is shown in guinea pig airways to be under neural control by cholinergic, a non-adrenergic-non-cholinergic, and possibly by adrenergic pathways.[3]

Although the conjunctiva is innervated by parasympathetic and sympathetic nerves, conjunctival goblet cells, similarly to the intestinal, colonic, and tracheal goblet cells, are not directly innervated. The simple diffusion of neurotransmitters could stimulate goblet cell secretion. Little is known about the regulation of mucous secretion from the conjunctival goblet cells and nothing about neural regulation. One study showed that $PGE_2$ can stimulate mucous secretion in rabbits.[4] One other report indicated that long-term exposure (1-30 days) to phenylephrine did not effect rabbit conjunctival goblet cell density.[5]

The present study describes the effect of sensory reflex neural stimulation and of topical application of potential secretagogues on conjunctival goblet cell mucous secretion in an *in vivo* rat model. Central corneal debridement, a sensory stimulus, and topical application of VIP, serotonin, epinephrine, phenylephrine, and dopamine stimulated goblet cell mucous secretion in 1 hr-incubation.

*Lacrimal Gland, Tear Film, and Dry Eye Syndromes*
Edited by D.A. Sullivan, Plenum Press, New York, 1994

## METHODS

### Materials

L-phenylephine hydrochloride, (-)-epinephrine, 5-hydroxytryptamine (serotonin), and 3-hydroxytyramine (dopamine) were obtained from Sigma (St. Louis, MO). Vasoactive Intestinal Peptide (VIP) was purchased from Peninsula Laboratories, Inc. All other compounds were purchased from Sigma or Fisher (Pittsburgh, PA) unless indicated otherwise.

### Animals

Male Sprague-Dawley rats at 12 weeks of age (young adults) (Charles River Laboratories, Wilmington, MA and Taconic Laboratory Animals, Germantown, NY) were anesthetized with intraperitoneal injection of 65 mg/kg of sodium pentobarbital or anesthetized with intraperitoneal injection of 100 mg/kg ketamine and 6.7 mg/kg acepromazine. All experiments comformed to the USDA Animal Welfare Act (1985) and SERI Animal Care and Use Committee.

### Corneal Debridement

To determine the effect of central corneal debridement, rats were anesthetized with pentobarbital. The central cornea of one eye was debrided (2-3 mm) and 1 hr later the animal was euthanized with intraperitoneal sodium pentobarbital (1300 mg/kg).

### Neurotransmitters

To determine the effect of topically applied compounds, an experimental buffer containing the test compound (20 µl drops) was placed on one eye at the temporal region every 20 min for 1 hr unless otherwise indicated. The buffer, designed to mimic tear ionic composition, contained 106.5 mM NaCl, 26.1 mM $NaHCO_3$, 18.7 mM KCl, 1.0 mM $MgCl_2$, 1.1 mM $CaCl_2$, 0.5 mM $NaH_2PO_4$, and 10 mM HEPES; it had a pH of 7.45 $\pm$ 0.02 and a calculated osmolarity of 330 mOsm/L. The tear buffer for the neurotransmitter, dopamine contained 1 mM ascorbic acid. The solutions were removed from the nasal region of the eye with a cotton-tipped applicator. After the 1 hr protocol, animals were euthanized with sodium pentobarbital.

### Measurement of Goblet Cell Density

After euthanization the eyes were fixed with half-strength Karnosky's solution (2.5% glutaraldehyde and 2% paraformaldehyde in cacodylate buffer). Because conjunctival goblet cell density is not uniform throughout the conjunctiva, care must be taken to use a specific area for sampling. To mark the area for sampling, a 7-mm cross-hair trephine coated with Vismark surgical skin marker (Viscot Industries, Inc.) was applied to the corneal surface, marking the superior, inferior, nasal and temporal regions of the ocular surface. Then a 2-mm trephine was placed adjacent to the limbus on the central inferior bulbar conjunctiva location to obtain conjunctival buttons. Conjunctival tissue was dissected and tissue placed epithelial side up on gelatin-coated, glass microscope slides. Flat-mount preparations were then fixed with 65% ethanol, 5% acetic acid and 2% formaldehyde overnight. Mucin contained in the secretory granules of the goblet cells was stained with alcian blue and periodic acid-Schiff (PAS) stain.[6] The number of goblet cells was counted in a masked fashion using a standard Zeiss light microscope with an ocular grid at 160x magnification.

## Scoring parameters

The mucin-containing goblet cells stained purple and fuchsia with alcian blue and PAS. Cells with moderate to intense staining and sharp, defined cell borders were considered non-secreted and were counted. Cells with very light stain, absence of color, indistiguishable colors from background staining, or fuzzy borders were considered cells that have secreted mucus and were not counted. Following these criteria, the non-secreted mucin-containing goblet cells in three-0.16 mm$^2$ areas of each button were counted, averaged, and expressed as mucin-containing goblet cell density. A decrease in mucin-containing goblet cell density indicated an increase in mucous secretion.

## Statistical analysis

All data are expressed as means $\pm$ SE. Statistical significance was determined by Student's $t$ test for paired and unpaired data.


RESULTS AND DISCUSSION


Since the density of mucin-containing goblet cells does not differ between right and left eyes,[7] only the right eyes of anesthetized rats are treated and the untreated left eyes served as contralateral controls. Because of variability in conjunctival goblet cell density between individuals, the results are expressed as percent of contralateral control. After corneal debridement, the mucin-containing goblet cell density was significantly decreased by 55 % ($P < 0.005$) (n = 4 - 5) from contralateral controls. Corneal debridement was also significantly different from animals with un-wounded eyes and from animals with sham-wounded eyes, ($P < 0.006$) (Fig. 1) which were decreased 17% from control values. The mucin-containing goblet cell density in sham-wounded eyes was unchanged from either its contralateral controls or from un-wounded eyes.

The stimulation of goblet cell secretion by corneal debridement suggests that ocular damage stimulates the reflex sensory nerves (afferent neurons) of the cornea to activate a local reflex arc. In turn the efferent neurons in the conjunctiva would be activated and at their termini release neurotransmitters, which then diffuse through the stroma to stimulate the conjuctival goblet cells. To determine which neurotransmitters may be responsible for goblet cell secretion, we surveyed potential secretagogues.

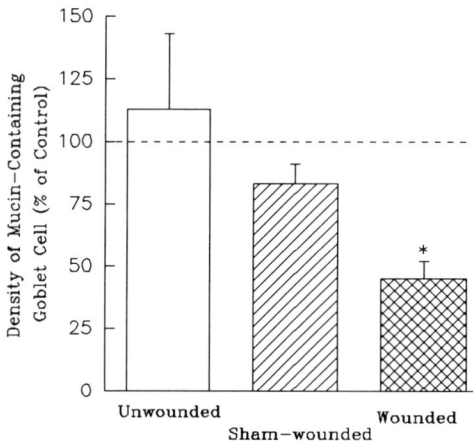

**Figure 1.** Mucin-containing goblet cell density of inferior conjunctiva in untreated (open bar), sham-wounded (right-hatched bar), corneal debridement (cross-hatched bar) treated eyes, and contralateral control eyes (dashed line). Results are expressed as mean $\pm$ SE (n = 4 - 5) ($P < 0.006$).

We first tested VIP because VIP-binding sites were found in rat and rabbit conjunctiva[8] and VIP-like immunoreactivity was shown in nerves of rat limbal blood vessels.[9] Since sites exist in the conjunctiva for VIP, the effect of topical application VIP ($10^{-8}$ M) on goblet cell secretion was determined. In inferior conjunctiva treated with $10^{-8}$ M VIP, the mucin-containing goblet cell density was significantly decreased by 39 % ($P <$ 0.005) (n = 5 - 6) from contralateral controls (Fig. 2). In contrast, goblet cell density in buffer-treated eyes was not significantly different from contralateral control eyes. This VIP stimulation of conjunctival goblet cell mucous secretion is unlike that of intestinal goblet cells in which VIP does not effect secretion.[2] In most tissues VIP is localized in parasympathetic nerves. VIP could be one of the neurotransmitters of parasympathetic nerves that mediates the efferent arm of the reflex arc that stimulate conjunctival goblet cell secretion.

Serotonin has been identified in the nerves of the cornea.[10] Serotonin was shown to enhance phosphoinositol turnover in rabbit cornea and to cause electrolyte and water secretion into the tear film.[11] The mucin-containing goblet cell density in inferior conjunctiva treated with $10^{-8}$ M serotonin was significantly decreased from contralateral eyes by 54 % ($P < 0.02$) (n = 3 - 4) (Fig. 2). The buffer-treated eyes were not significantly different from contralateral control eyes. Interestingly in intestinal goblet cells, unlike the conjunctiva goblet cells, serotonin had no effect on mucous secretion.[2] However, in the conjunctiva serotonin is a possible neurotransmitter that could mediate the efferent arm of the reflex arc to stimulate goblet cell secretion.

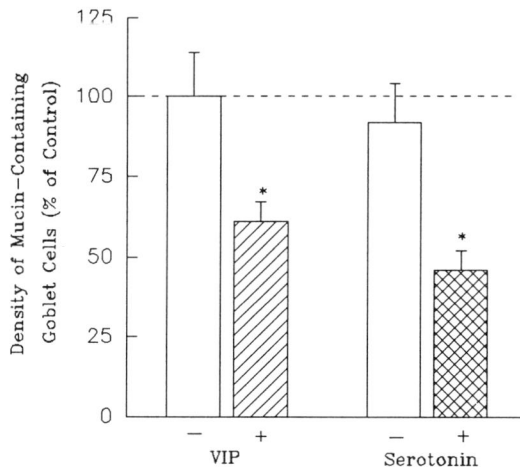

**Figure 2.** Mucin-containing goblet cell density of inferior conjunctiva in buffer-treated control (open bars), $10^{-8}$ M VIP (right-hatched bar), $10^{-8}$ M serotonin (right-hatched bar) treated eyes, and contralteral control eyes (dashed line). Results are expressed as mean $\pm$ SE (n = 3 - 4) ($P < 0.02$).

Since sympathetic nerves innervate the conjunctiva, the adrenergic neurotransmitters epinephrine ($10^{-4}$ M), phenylephrine ($10^{-4}$ M) and dopamine ($10^{-8}$ M) were each applied to the ocular surface. The mucin-containing goblet cell density in inferior conjunctiva treated with $10^{-4}$ M epinephrine, $10^{-4}$ M phenylephrine, or $10^{-8}$ M dopamine was significantly decreased by 46 %, 40 %, and 64 %, respectively ($P < 0.03$) (n = 3 - 6) from contralateral control values (Fig. 3). The buffer-treated eyes were not significantly different from contralateral control eyes. The stimulation of mucous secretion in conjunctival goblet cells with epinephrine, pheylephrine, or dopamine was unlike that of intestinal goblet cells where no effect on secretion was observed.[2] Also, in contrast to the present study, long term exposure of rabbit ocular surface with the vasoconstrictor, phenylephrine do not significantly affect conjunctival goblet cell density.[5] However, Shellans *et al.* does not report on the short-term effect it has goblet cell secretion. The results of the present study suggest that sympathetic nerves could also mediate the efferent arm of the stimulatory pathway for goblet cell secretion.

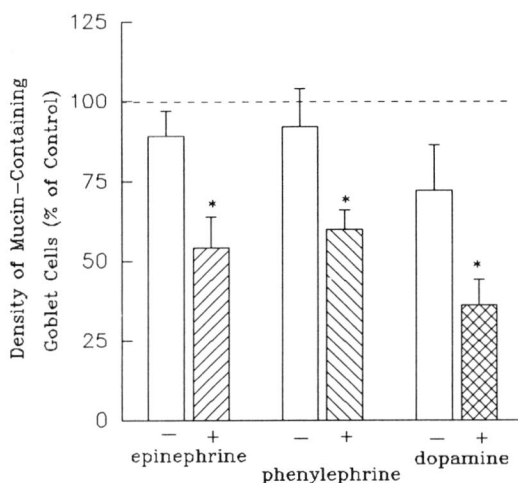

**Figure 3.** Mucin-containing goblet cell density of inferior conjunctiva in buffer-treated control (open bars), $10^{-4}$ M epinephrine (right-hatched bar), $10^{-4}$ M phenylephrine (left-hatched bar), $10^{-8}$ M dopamine (cross-hatched bar) treated eyes, and contralateral control eyes (dashed line). Results are expressed as mean $\pm$ SE (n = 3 - 6) ($P < 0.03$).

We conclude that corneal debridement probably by stimulation of reflex afferent sensory nerves causes conjunctival goblet cell mucous secretion. Topical application of VIP, serotonin, epinephrine, dopamine or phenylephrine stimulates conjunctival goblet cell mucous secretion. Parasympathetic, serotonergic, dopaminergic, and sympathetic nerves could be involved in the efferent arm of the reflex pathway.

ACKNOWLEDGEMENT

Supported by NIH # EY09057 and NIH # EY06472.

REFERENCES

1. Phillips TE, Phillips Tl, and Neutra MR, 1989, Cholinergic responsiveness of goblet cells during intestinal maturation, *Biol Neonate.* 55:197-203.
2. Neutra Mr, O'Malley LJ, and Speican RD, 1982, Regulation of intestinal goblet cell secretion. II. A survey of potential secretagogues, *Am J Physiol.* 242 (*Gastrointest Liver Physiol.* 5):G380-G387.
3. Tokuyama K, Kuo H-P, Rohde JAL, Barnes PJ, and Rogers DF, 1990, Neural control of goblet cell secretion in guinea pig airways, *Am J Physiol.* 259 (*Lung Cell Mol Physiol.* 3):L108-L115.
4. Aragona P, Candela V, Caputi AP, Micali A, Puzzolo D, and Quintieri M, 1987, Effects of a stable analogue of $PGE_2$ (11-deoxy-13,14-didehydro-16(S)-methylester methyl $PGE_2$:FCE 20700) on the secretory processes of conjunctival goblet cells of rabbit, *Exp Eye Res.* 45:647-654.
5. Shellans S, Rich LF, and Louiselle I, 1989, Conjunctival goblet cell response to vasoconstrictor use, *J Ocular Pharm.* 5:217-220.

6. Tseng SCG, Hirst LW, Farazdaghi M, Green WR, 1984, Goblet cell density and vascularization during conjunctival transdifferentiation, *Invest Ophthalmol Vis Sci.* 25:1168-1176.
7. Kessler TL and Dartt DA, 1993, An in vivo method to study rat conjunctival goblet cell mucous secretion, Submitted.
8. Denis P, Dussaillant M, Nordmann J-P, Elena P-P, Saraux H, and Rostene W, 1991, Autoradiographic characterization and localization of vasoactive intestinal peptide binding sites in albino rat and rabbit eyes, *Exp Eye Res.* 52:357-366.
9. Stone RA, 1986, Vasoactive intestinal polypeptide and the ocular innervation, *Invest Ophthalmol Vis Sci.* 27:951-957.
10. Klyce SD, and Crosson CE, 1985, Transport processes across the rabbit corneal epithelium: a review, *Curr Eye Res.* 4:323-331.
11. Akhtar RA, 1987, Effects of norepinephrine and 5-hydroxytryptamine on phosphoinositidde-$PO_4$ turnover in rabbit cornea, *Exp Eye Res.* 44:849-862.

BIOAVAILABILITY OF PREDNISOLONE IN RABBITS: COMPARISON OF A HIGH-VISCOSITY GEL AND AN AQUEOUS SUSPENSION—SINGLE- AND REPEATED APPLICATIONS

Sven Johansen,[1] Eva Rask-Pedersen,[2] and Jan Ulrik Prause[1]

[1] Eye Pathology Institute
University of Copenhagen
Denmark
[2] Leo Pharmaceutical Products
Denmark

**KEY WORDS:** carbomer - aqueous suspension - vehicle - prednisolone acetate - fusidic acid - sulfacetamide sodium  - opthalmic bioavailability - rabbit.

## AIM

To investigate the possible beneficial effect of carbomer gel[1] on the bioavailability of prednisolone acetate versus the bioavailability of prednisolone acetate in aqueous suspension. In addition we investigated the influence of the concomitant presence of fusidic acid and sulfacetamide sodium. Two main questions were given:
Does carbomer increase the concentration of the drug in ocular tissues?
Does it increase the total amount taken up?

## MATERIALS

We examined a synthetic high molecular weight polymer of acrylic acid cross-linked with allylsucrose (acrylic acid polymer, Leogel, Carbopol, carboxypoly-methylene and carboxyvinyl polymer):

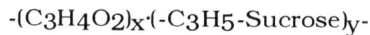

$$-(C_3H_4O_2)_x-(-C_3H_5-Sucrose)_y-$$

### Animal

126 Copenhagen white rabbits (females, 2.5-3.5 kg).

### Topical test preparations

Prednisolone acetate 0.5% in Leogel with fusidic acid 1.0%[2] (leogelpred. 0.5%) Fig. 1. Prednisolone acetate 1.0% in Leogel with fusidic acid 1.0% (leogelpred. 1.0%). Prednisolone acetate 0.5% in an aqueous suspension with sulfacetamide sodium 10% (aquapred. 0.5%).

*Lacrimal Gland, Tear Film, and Dry Eye Syndromes*
Edited by D.A. Sullivan, Plenum Press, New York, 1994

## METHODS

### Study

The study was divided into two parts (see Table 1.), a single application study (IA and IB) and a repeated applications study (II):

In part IA the gel preparation with prednisolone acetate 0.5% in Leogel with fusidic acid, was compared with prednisolone acetate 0.5% in aqueous suspension with sulfacetamide sodium 10% (Metimyd®, Schering).

In part IB the 0.5% prednisolone gel was compared with an 1% prednisolone gel.

In part II, the gel preparation was given twice daily and the aqueous suspension four times daily for one week. The preparations were the same as in part IA.

**Table 1.** Study preparations, sampling times and number of rabbits.

| Study | | Preparation | Sampling time | No. |
|---|---|---|---|---|
| IA Single application | | leogelpred. 0.5% versus aquapred. 0.5% | 0.5, 1, 1.5, 2, 3, 4, 6, 8 & 12 h after application | 54 |
| IB Single application | | leogelpred. 1% | 1, 4, 7 & 10 h after application | 12 |
| II Repeated applications | twice daily for one week | leogelpred. 0.5% versus | 0.5, 1, 1.5, 2, 3, 4, 6, 8, 10 & 12 h after last application | 60 |
| | four times daily for one week | aquapred. 0.5% | | |

### Applications

The gel preparation (50 mg) was placed in the inferior fornix by a Mikroman pipette Model M 250 (49.35 ± 0.20 mg). The aqueous suspension (one drop, 50 mg) was delivered by a plastic container fitted with a pipette.

### Samples

Animals were killed with an overdose of pentobarbital 20%. Samples were taken from conjunctiva (approximately 245 mg), cornea (approximately 16 mg), aqueous- ($\geq 120 \mu$l) and vitreous humour ($\geq 200 \mu$l). All samples were stored at -25° C until assayed.

FUSIDIC ACID            PREDNISOLONE ACETATE

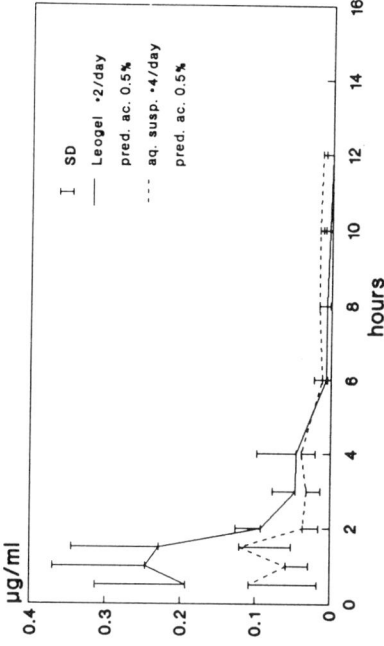

Figure 1.

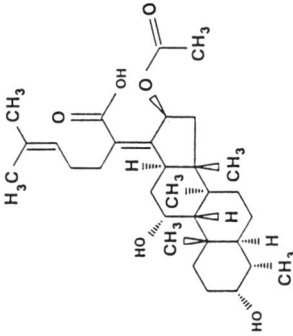

Figure 2. Mean prednisolone concentration in rabbit aqueous humor after single application.

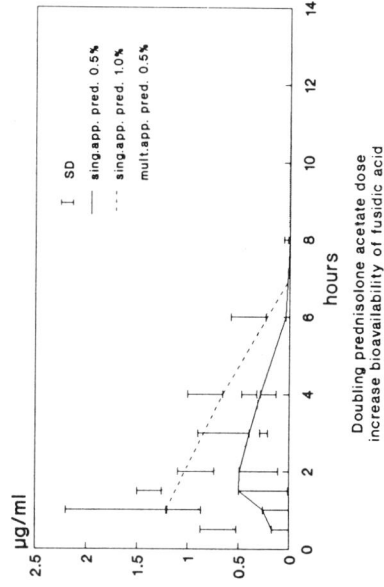

Figure 3. Mean prednisolone concentration in rabbit aqueous humour after repeated applications for one week.

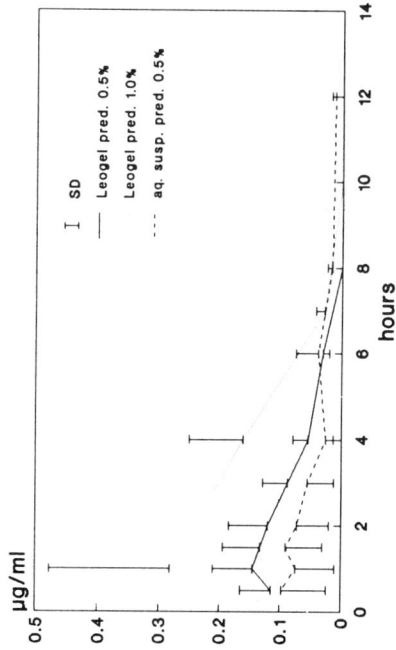

Figure 4. Mean fusidic acid concentration in rabbit aqueous humour after single and repeated applications.

401

## ANALYSES

Prednisolone and sulfacetamide sodium concentrations were measured by HPLC. Fusidic acid was determined in two biological methods: an agar plate disc method (ocular samples), and a turbidimetric method (serum and plasma) - both measuring the inhibitory effect of fusidic acid on bacterial growth.

## STATISTICS

Mean values were compared by the unpaired t - test. Areas under the curves (AUC) were taken as an expression of the bioavailability. AUC's were compared using the u - test.

## RESULT

Cornea, conjunctiva and aqueous humour exhibited the same dose/concentrations curves (Fig.'s 2,3,4). After single application of prednisolone acetate in Leogel, significantly higher bioavailability was found in all three tissues six hours after application compared to application in an aqueous suspension. Repeated applications of the drugs in Leogel significantly increased their concentrations in all three tissues.

Fusidic acid was found in serum after repeated applications, and concentrations of 0.21-0.30 $\mu$g/ml was measured three hours after last application.

## CONCLUSION

PREDNISOLONE: Higher concentrations in cornea, conjunctiva and aqueous humour after both single- and repeated applications in Leogel than in aqueous suspension.

FUSIDIC ACID: Repeated applications in Leogel gave higher concentrations in cornea, conjunctiva and aqueous humour than after a single application.

SULFACETAMIDE SODIUM concentration does not change with type of applications.

## COMMENT

Complete reports on the study including all results and a detailed discussion are under preparation[3,4].

## REFERENCES

1. R.D Schoenwald and J.S Boltralik, A bioavailability comparison in rabbits of two steroid formulations as high viscosity gel and reference preparation.Invest Opthalmol Vis Sci.18:61-66 (1979).
2. K. Benzen et al, Fusidic acid, an immunusuppresive drug with functions similar to cyclosporin A. Cytokine. 6:423-429 (1990).
3. S. Johansen et al, A bioavailability comparison in rabbits after a single application of prednisolone acetate formulated as a high-viscosity gel and as an aqueous suspension. To be published in Acta Ophthalmologica.
4. S. Johansen et al. A bioavailability comparison in rabbits of prednisolone acetate after repeated applications formulated as a high-viscosity gel and as an aqueous suspension. To be published in Pharmacology & Toxicology.

# TEAR FILM - CONTACT LENS INTERACTIONS

Donald R. Korb

Korb Associates
80 Boylston Street
Boston, MA  02116

The relationship between the precorneal tear film and contact lenses has been recognized since the conception of contact lenses by da Vinci, who identified the optical effect of tears in a contact lens system.[1]  With the initiation of the modern age of contact lenses by Feinbloom in 1936[2] and Mullen in 1938,[3] an insertion solution was required to fill the vaulted and sealed space between the posterior lens surface and the cornea.  Since no significant interchange of fresh tears to the retro-lens space occurred, pH, osmolarity, and buffering agents of the insertion solutions were manipulated in unsuccessful attempts to eliminate corneal edema and to increase wearing time.[4]  In 1943, Bier introduced fenestration, allowing the exchange of tear fluid into the retro-lens space and reducing edema, with a concurrent increase in wearing time.[5]

The invention of the corneal lens by Touhy[6] in 1947 and the rapid development of corneal lenses focused attention on the relationship of the tear film and the contact lens.  In 1951, Graham reported: "The primary objective of designing lenses is to permit optimum tear flow to the cornea beneath the lens."[7]  Fluorescein made it possible to observe the interchange of the tears underneath the lens, and its use in the fitting of corneal lenses became routine.[8]  Despite a multitude of lens designs and fitting refinements to promote the exchange of tears behind the non-permeable methylmethacrylate corneal lens (hard lens), the exchange was inadequate, leading to short- and long-term corneal complications.

Contemporary contact lenses are of two primary types: rigid gas permeable lenses and hydrogel lenses.  Rigid gas permeable lenses are an evolutionary development of the hard lens.  A co-polymer of methylmethacrylate and silicon, termed a siloxane-acrylate polymer, was the first rigid lens material designed specifically for contact lens use,[9] and was introduced in 1979.  The most frequently used polymers are siloxane-acrylate, fluoro-acrylate and fluorocarbon.  The diameter of rigid gas permeable lenses is smaller than that of the cornea, thus exposing the peripheral portions of the cornea.  Hydrogel lenses are made from polymers which absorb and bind water into their molecular structure, and were first reported by Wichterle and Lim in 1960.[10]  Hydrogel lenses cover the entire cornea and a small portion of the bulbar conjunctiva.  Contemporary hydrogel polymers vary from 30% to over 85% water of hydration (by

*Lacrimal Gland, Tear Film, and Dry Eye Syndromes*
Edited by D.A. Sullivan, Plenum Press, New York, 1994

weight), may be ionic or nonionic, and are manufactured in various materials and designs throughout the world.

The structure of the tear film on the eye is well established. The classical description utilizing a three-layer model was introduced by Wolff in 1946,[11] while a more complex model of six layers was introduced by Tiffany in 1988.[12] The inner surface of all contemporary contact lenses is in apposition with the tear film, although the exact nature of the tear film in the retro-lens space is subject to question. The characteristics of the tear film on the outer surface of all lenses are also the subject of current investigation. The tear film thickness on the eye is reported to be up to 10 microns, decreasing to 4.5 microns between blinks.[13] The tear film is relatively thin when compared with the thickness of any contact lens, which varies from a minimum of 30 to an average of 60-120 microns, and over 250 microns for lenses of considerable optical power. Thus, the shear mass of any contact lens may compromise the specific functions of the tear film, which include: the flushing action,[14] the prevention of desiccation of the ocular tissue,[15] the lubrication of the ocular and palpebral surfaces,[16] the formation of an optically smooth-curved surface,[17] a vehicle for oxygen and carbon dioxide transport,[16] and the defense of the cornea against trauma, infection or disease.[14]

## TEAR FILM CHARACTERISTICS SUSCEPTIBLE TO ALTERATION BY CONTACT LENSES

The characteristics of the tear film may be influenced by the following factors associated with contact lens wear: the physical presence of a contact lens must alter the tear menisci along the upper and lower eyelids;[24] the tear film is unstable and requires frequent blinking for resurfacing;[16] the wearing of a contact lens may compromise the blink frequency and/or blink amplitude;[18] lid/surface congruity is altered by all contact lenses;[19] the tear volume is limited to seven microliters ($\pm$ two microliters),[20] thus the volume of tear fluid available to physically cover contact lenses or to hydrate hydrogel lenses is limited; the lipid layer on the anterior surface of the contact lens is altered, and is either very thin or absent;[21,22] all contact lenses create a retro-lens tear film, whose interchange with the habitual tear film is restricted; and, all contact lenses prevent the habitual, direct rubbing action of the lids upon the covered ocular surfaces.

## OCULAR SURFACE DESICCATION PHENOMENA

The most common corneal and conjunctival complications associated with contact lens-tear film compromise are ocular surface desiccation phenomena, including: desiccation and staining of the exposed corneal areas with rigid lenses (three and nine o'clock staining), followed by infiltrates if severe and longstanding; desiccation of the cornea occurring with hydrogel lenses despite the hydrogel lenses covering the area; and xerotic type changes of the exposed horizontal bulbar conjunctiva occurring with all types of contact lenses.

The three and nine o'clock staining of the exposed portions of the cornea has many possible causes, including: alteration of the normal lid-ocular congruity, interference with the resurfacing of the tear film;[17] the influence of the lens and edge design;[23] thinning of the tear film at the edge of the lens resulting from the lens meniscus;[24] infrequent or partial blinking and inadequate lens movement;[25] and inadequate tear film or marginally dry eye.[25] While treatment methods are usually able to minimize and/or control this problem, three and nine o'clock staining frequently remains despite all treatment, and when persistent and severe, may require termination of rigid lens wear.

Corneal desiccation with hydrogel lenses, in contrast to rigid lenses, may occur on the corneal areas covered by the lens. The desiccation usually occurs on the inferior and/or central portions of the cornea, rarely on the superior portion. The primary cause is related to lens thickness, the phenomenon being reported with thin hydrogel lenses,[26,27] where lens dehydration desiccates the underlying epithelium.[27] The problem can usually be solved with a thicker hydrogel lens.

Bulbar conjunctival xerotic changes occur with and without contact lens wear. When associated with contact lens wear, they may be considered a variant of corneal three and nine o'clock staining with essentially the same causes, particularly infrequent and/or partial blinking. This may occur with either rigid or hydrogel lenses.

## VISUAL PERFORMANCE AND THE TEAR FILM

The visual performance achieved with hydrogel lenses is highly dependent upon tear film interaction, while the performance of rigid lenses is essentially independent of the tear film. The hydrogel lens must remain hydrated to provide an anterior surface of adequate optical quality for clear vision. Clinical evaluation of the wetting and quality of the anterior surface in situ may be performed with several techniques.[28] Our in vitro studies of optical imaging of hydrogel lenses demonstrate that a clear optical image requires a fully hydrated lens and a wet surface, while rigid lenses require a clean, dry surface. Low-contrast visual acuity was measured with normal and suppressed blinking without lenses, and repeated with rigid and hydrogel lenses. The drying following suppressed blinking resulted in a significant reduction of 4.1 lines of low-contrast visual acuity for hydrogel lenses, but no significant reduction without lenses or with rigid lenses.[29] Thus, optimal visual performance with hydrogel lenses requires adequate blinking and hydration, while rigid lenses require only a clean anterior surface.

## TEAR FILM CHANGES ASSOCIATED WITH CONTACT LENS WEAR

The studies for tear film changes occurring with contact lens wear are extensive, but also frequently conflicting. The initial insertion of a lens on an unadapted eye usually results in increased reflex tear secretion, with definitive changes in concentration of certain tear ingredients and subsequent return to baseline.[30] There is evidence that adapted wearers of rigid and extended wear lenses present elevated tear osmolarity.[31] The lipid layer is compromised by the presence of a contact lens,[21,22] leading to increased evaporation.[32] Mucus production may be increased by non-goblet cells in the tarsal conjunctiva in response to contact lens wear.[33] The relationship between contact lens wear and the pH of the tears has been the subject of many investigations, leading to the conclusion that there are no definitive changes in pH with lens wear.[34] Tear albumen, lysozyme, and lactoferin do not change with long-term wear of either rigid or hydrogel lenses.[31] The changes in lacrimal and serum-derived proteins do not appear to be of clinical relevance at this time. The tear film changes occurring with contact lens wear which are of present relevance to clinical performance and contact lens design include: lipid layer changes, increased osmolarity, increased evaporation, increased mucus, and indirectly the frequency and amplitude of the blink. Further investigation is required to determine if the nature and magnitude of other changes are of significance and clinical relevance.

## CONTACT LENS DEPOSITS

Deposition on contact lenses has been studied extensively, but certain aspects remain ambiguous. Among the primary factors influencing deposition are: lens surface chemistry, ionic character, water content, individual patient tear chemistry, evaporation rate, ocular secretions including those from the palpebral conjunctiva, and the use of adjuvants. Deposits include lipids, proteins, inorganic materials and bacteria, and may result in numerous ocular responses from the reversible phenomena of Trantas' dots and giant papillary conjunctivitis, to severe and potentially compromising infective keratitis. Bacterial attachment remains an area of concern and possible controversy.

## WETTABILITY, LIPID LAYER AND EVAPORATION

The cornea exhibits essentially optimal wetting, defined as the spreading of a liquid over a surface. Rigid lenses have limited wetting characteristics,[35] while hydrogel lenses, which are hydrophilic, demonstrate superior wetting properties. The wettability of both rigid and hydrogel lenses is reported to be improved by the surfacing of mucin on the lens surfaces; however, rigid lenses still exhibit limited wetting.[36,37]

In 1981, Holly suggested that the ideal contact lens/tear film relationship would be as follows: "A well-fitted contact lens has to rest on a continuous aqueous tear layer sandwiched between the lens and the epithelium, and it has to be coated with a continuous tear film complete with a superficial lipid layer. The stability and the continuity of both behind-the-lens (postlens) and in front-of-the-lens (prelens) tear films are important for good wearing performance."[37] However, all contemporary contact lenses are unable to mimic the ocular surface properties, and therefore a comparable tear film on the lens surfaces is unable to form.

The lipid layer on the surface of all contact lenses is compromised as compared to the lipid layer of the cornea without a contact lens. J. P. Guillon in 1982, was the first to report both qualitative and quantitative data for the lipid layer on contact lenses.[21] It is agreed that a lipid layer does not form on rigid lenses.[21,22] There are conflicting reports regarding the presence and/or characteristics of the lipid layer on hydrogel lenses; some claim the complete absence of a lipid layer, while others report it as present but thin,[21,22] dependent on the water content of the lens.[21]

Our investigations confirm that a normal lipid layer is not present with rigid lenses, and that lipid layers on hydrogel lenses are variable, and rarely similar to the lipid layers observed on eyes without lenses. We have found that lipid layers on all hydrogel lenses are very sensitive to humidity.[47] It is usually necessary to create a high humidity environment of over 70%, and frequently 100%, in order to form a normal lipid layer on hydrogel lenses. Certain types of hydrogel polymers allow the formation of better lipid layers than others, a characteristic not necessarily related to water content.

The role of the lipid layer in preventing evaporation is relevant to contact lens wear. If the meibomian glands are obstructed, essentially eliminating the lipid layer, the rate of evaporation dramatically increases by a factor of 10 or more times.[38] Clinical experience indicates that individuals without objective signs of dry eyes or subjective symptoms may experience classical dry eye symptoms while wearing contact lenses. The most plausible explanation, as first reported by Tomlinson, appears to be the increase in evaporation from the eye occurring with all contact lens wear,[32] and the resulting sequelae. Tomlinson has further pointed out that: "The contact lens acts as provocation and turns the marginal into a fully manifest dry eye, in fact, producing a contact lens syndrome of increased tear evaporation."[39] This situation is analogous to many situations for non-contact lens wearers, where symptoms of dryness only occur when the rate of evaporation is significantly increased, as with conditions of low humidity (heated, closed spaces or airplane interiors), excess wind or drafts, or prolonged visual activities at near (reading or video display terminal operation), resulting in less frequent blinking.

Blinking has an obvious relationship to the wetting of the ocular surfaces, the contact lens surfaces, and evaporation. The role of blinking in maintaining the lipid layer and in preventing evaporation has received minimal attention. Woolf in 1946[11] and Linton in 1961[40] reported that blinking promoted meibomian gland secretion. Benedetto et al in 1984 determined that forceful blinking substantially increased tear film thickness.[41] Korb et al in 1993 reported a significant increase in lipid layer thickness with deliberate forceful blinking, and suggested that blinking, in addition to spreading lipid, was important in the maintenance of the lipid layer by augmenting the expression of lipid from the meibomian glands.[42,43] Since contact lens surfaces, rarely if ever, form a complete tear layer, including the lipid layer, in a comparable manner to that of the ocular surfaces, the lids, in passing over the lens surfaces, may experience sensation and blinking may therefore be inhibited. The increase in evaporation occurring with contact lens wear would suggest a compensatory increase in the frequency and efficacy of blinking. A decrease in blinking associated with contact lens wear, as the author believes frequently occurs, would further increase evaporation, compromise the wetting of the eye and contact lens, and may decrease meibomian gland secretion and lipid layer thickness. Thus, attention must be directed to promoting wet anterior lens surfaces and designing contact lenses so as to minimize the tendency to inhibit blinking. Lubricating and wetting formulations (artificial tears) for contact lenses are usually of limited efficacy, probably the result of their further compromising the lipid layer and therefore resulting in increased evaporation.[44] Nevertheless, until an adjuvant is available which will replenish not only the aqueous, but also the lipid layer of the tear film, the present formulations should be recommended for patient trial, since they may provide some relief by hydrating the tear film and/or lens and wetting the lens surfaces.

The effects of environmental humidity on contact lens comfort are universally recognized by clinicians. Comfort frequently decreases markedly with the arrival of the winter season, dry indoor heat, and the accompanying decrease in indoor humidity. This may also occur in warm climates, when air conditioning results in lowered indoor humidity. Andrasko and Schoessler in 1980 were the first to quantify the effects of environmental relative humidity on hydrogel lenses reporting a dramatic decrease of 14 to 19% hydration in a low humidity environment, compared to a 6 to 12% decrease in a high humidity environment.[45] Finnemore et al investigated the effects of 100% humidity for 30 minutes with contact lens wearers experiencing dry eye symptoms. Over 75% of all hydrogel lens wearers reported marked alleviation of all symptoms, while only 15% of rigid gas permeable wearers obtained relief.[46] Korb et al measured the preocular lipid layer thickness prior to and following 20 minutes in goggles which created a 100% humidity environment. The mean thickness doubled from 61 to 126 nm, indicating the importance of environmental humidity to lipid layer thickness and tear film stability.[47] With hydrogel lenses it was found that a 100% humidity environment for 30 minutes would improve the prelens lipid layer thickness in all instances,[48] and would also with certain types of previously unworn lenses create a normal lipid wave pattern and movement. Thus, the lipid layer on the eye and on hydrogel lenses was found to be very sensitive to environmental humidity.

## SUMMARY

Contemporary contact lenses, when considered in perspective, are remarkably effective; the remaining primary challenge, however, appears to be the relationship of the contact lens to the tear film. The number of contact lens wearers in the United States is no longer increasing, perhaps the result of discomfort which may occur initially, preventing contact lens wear, or which may develop over a period of years, resulting in termination of wear. Contemporary lenses can usually provide comfort if tear film integrity can be maintained. The environmental humidity is a critical component in the formation and thickness of the lipid layer for both the preocular and the prelens tear films, thus influencing tear film stability and evaporation. A common clinical observation is the gradual development of discomfort over 10 or more years of contact lens wear by patients who initially are totally asymptomatic. This discomfort, probably the result of compromise to the tear film occurring with age, frequently leads to discontinuation of contact lens wear. Intermittent discontinuation of contact lens wear may also occur as the result of seasonal discomfort when the relative humidity is low, or when the nature of the visual demand, as in extended VDT operation, inhibits blinking.

The fragile tear film and its relation to contact lenses is readily understood by two quotations: "The human tear film is rather unstable, but it is regenerated by frequent blinking," and "When a contact lens is placed in the eye, the lens alters the normal structure of the tear film and affects its rate of evaporation. These changes affect the ocular surface as well as the contact lens itself."[49] Both blinking and evaporation are recognized as critical factors in contact lens wear. Thus, future goals should be to increase the precorneal tear film thickness, and particularly to increase lipid layer thickness and quality,[50] in order to minimize evaporation and maintain ocular surface integrity. New contact lens materials are required which would mimic the ocular surfaces so as to allow a prelens tear film which more nearly replicates the remarkably efficacious preocular tear film, complete with lipid layer. This would decrease the evaporation rate and thus improve contact lens tolerance. New materials should also minimize the thickness of the lens, particularly the edge thickness, since evaporation increases with elevation of the meniscus at the edge of any contact lens. Current evidence suggests that the rigid nature of gas permeable lenses prevents many, if not the majority, of contact lens candidates from achieving optimal comfort. This is probably the result of trauma to the ocular surfaces and/or lids occurring upon blinking. Further, the substantial meniscus about the edge of all rigid lenses, necessitated by the need for edge lift with rigid lenses, is a significant factor in increased evaporation. No obvious resolution for these problems with a rigid lens is apparent. Current hydrogel lenses provide immediate and sustained comfort if the ambient relative humidity approaches 100%; therefore, the primary direction for improvement for hydrogel lenses is the development of materials which will decrease evaporation.

In conclusion, the tear film of many individuals is inadequate to support the increased evaporation rate encountered with contact lens wear. Among the therapeutic possibilities for tear film enhancement are: increased blink efficacy, meibomian gland treatment to improve the quantity and quality of secretions, surgical elevation of the lower lid, and the use of tear film additives, i.e., artificial tears and/or contact lens lubricating and rewetting solutions. Contact lens surfaces must replicate more closely the ocular surfaces in order to limit dehydration and evaporation phenomena and permit a more normal lipid layer. Tear film additives should become more efficacious; the next advancement should more closely replicate the tear film, specifically the lipid layer. Future research in lens materials, designs, and tear additives should consider compatibility with the anthropological design and function of the ocular surfaces and lid actions. Models derived from primates, marine life, and amphibians, where evolution has solved analogous problems may be useful. Regardless of the specific direction, the tear film-contact lens relationship promises to be an area of active research in the near future.

NOTE: Space constraints required the elimination of many references.

## REFERENCES

1. L. da Vinci. Codex of the Eye, Manuscript D (circa 1508). For translation and illustrations, see H.W. Hofstetter and R. Graham, Leonardo and contact lenses, *Amer. J. Optom.* 30(1):41-44 (1953).
2. W. Feinbloom, A plastic contact lens, *in*: "Transcript of 15th Annual American Academy of Optometry Meeting," 10:37-44, Chicago (1936).
3. J.E. Mullen. "Contact Lens," U. S. Patent 2,237,744 (1938).
4. L. Lester Beacher. "Contact Lens Technique," Third ed., pp. 92-105, New York Contact Lens Research Laboratories, New York (1944).
5. N. Bier. "Method for Contact Lens Fenestration," U.K. Patent 592,055 (1943).
6. M.W. Nugent, The corneal lens, a preliminary report, *Ann. West. Med. Surg.* 2(6):241 (1948).
7. R. Graham R, Corneal lenses--a supplementary report, *Amer. J. Optom and Arch. Amer. Acad. Optom.* 29(3):137 (1952).
8. R.W. Lester, Fluorescein and contact lenses, *Contacto* 2(4):91-95 (1958).
9. N.G. Gaylord. "Method for Correcting Visual Defects, Compositions and Articles of Manufacture Useful Therein," U.S. Patent 4,120,570 (1978).
10. O. Wichterle and D. Lim, Hydrophilic gels for biological uses, *Nature (Lond)* 185(4706):117 (1960).
11. E. Wolff, The muco-cutaneous junction of the lid-margin and the distribution of the tear fluid, *Trans. Opthalmol. Soc. U.K.* 66:291-308 (1946).
12. J.M. Tiffany, Tear film stability and contact lens wear, *J. Br. Contact Lens. Assoc.* 11(Meeting Suppl.):35-38 (1988).
13. N. Ehlers, The precorneal film--biomicroscopical, histological and chemical investigations, *Acta Ophthal. (Suppl.)* 81:5-186 (1965).
14. M.R. Allansmith, How the cornea defends itself, *Trans. Ophthal. Soc. U.K.* 98:361-362 (1978).
15. S. Mishima and D.M. Maurice, The oily layer of the tear film and evaporation from the corneal surface, *Exp. Eye Res.* 1:39-45 (1961).
16. F.J. Holly and M.A. Lemp, Tear physiology and dry eyes, *Surv. Ophthalmol.* 22:69-87 (1977).
17. F.J. Holly and M.A. Lemp, Surface chemistry of the tearfilm; implications for dry eye syndromes, contact lenses, and ophthalmic polymers, *C.L. Soc. of Am. J.* 5(1):12-19 (1971).
18. M.D. Sarver, J.L. Nelson, and K.A. Polse, Peripheral corneal staining accompanying contact lens wear, *J. Am. Optom. Assoc.* 40(3):310-313 (1969).
19. D.R. Korb and J.E.Korb, A study of 3 and 9 o'clock staining after unilateral lens removal, *J. Am. Optom. Assoc.* 41:233-236 (1970).
20. M.B. Abelson, N.A. Sotter, et al., Histamine in human tears, *Amer. J. Ophthal.* 83:417-418 (1977).
21. J.P. Guillon, Tear film photography and contact lens wear, *J. Brit. C.L. Assoc.* 5:84-87 (1982).
22. G. Young and N. Efron, Characteristics of the pre-lens tear films during hydrogel contact lens wear, *Ophthalmic Physiol. Opt.* 11:53-58 (1991).
23. T. Holden, et al., The effect of secondary curve liftoff on peripheral corneal desiccation, *Am. J. Optom. Physiol. Opt.* 64:108P (1987).

24. J.E. McDonald and S. Brubaker, Meniscus-induced thinning of tear films, *Am. J. Ophthal.* 72(1):139-146 (1971).

25. D.R. Korb and J.E. Korb, A new concept in contact lens design: Parts 1 and 2, *J. Am. Optom. Assoc.* 41:1-12 (1970).

26. L.N. Kline and T.J. DeLuca, Pitting stain with soft contact lenses--Hydrocurve thin series, *J. Am. Optom. Assoc.* 48:372-376 (1977).

27. B.A. Holden, D.F. Sweeney and R.G. Seger, Epithelial erosions caused by thin high water content lenses, *Clin. Exp. Optom.* 69:103-107 (1986).

28. L.S. Mengher, et al., Non-invasive assessment of tear film stability, *in*: "The Preocular Tear Film in Health, Disease, and Contact Lens Wear," F.J. Holly, ed., pp. 64-75, Dry Eye Institute, Lubbock, TX (1986).

29. G.T. Timberlake, M.G. Doane and J.H. Bertera, Short-term, low-contrast visual acuity reduction associated with in vivo contact lens drying, *Optom. & Vis. Sci.* 69(10):755-760 (1992).

30. G.E. Lowther, R.B. Miller and R.M. Hill, Tear concentrations of sodium and potassium during adaptation to contact lenses: I. Sodium observations, *Am. J. Optom. Arch. Am. Acad. Optom.* 47:266-275 (1970).

31. R.L. Farris, Tear analysis in contact lens wearers, *CLAO J.* 12:106-111 (1986).

32. A. Tomlinson and T.H. Cedarstaff, Tear evaporation from the human eye: The effects of contact lens wear, *J. Br. C.L. Assoc.* 5:141-150 (1982).

33. J.V. Greiner and M.R. Allansmith, Effect of contact lens wear on the conjunctival mucous system, *Ophthalmology* 88:821-832 (1981).

34. D.J. Browning and G.N. Foulks, Tear pH in health, disease, and contact lens wear, *in*: "The Preocular Tear Film in Health, Disease, and Contact Lens Wear, F.J. Holly ed., pp. 954-965, Dry Eye Institute, Lubbock, TX (1986).

35. M. Sarver, et al., Wettability of some gas-permeable hard contact lenses, *Int. C.L. Clin.* 11:479-490 (1984).

36. W.J. Benjamin, M.G. Piccolo and H.A. Toubiana, Wettability: A blink by blink account, *Int. C.L. Clin.* 11:492-498 (1984).

37. F.J. Holly, Tear film physiology in contact lens wear: II. Contact lens - tear film interactions, *Am. J. Optom. Physiol.* 58:331 (1981).

38. S. Mishima and D.M. Maurice, The oily layer of the tear film and evaporation from the corneal surface, *Exp. Eye Res.* 1:39-45 (1961).

39. A. Tomlinson, Complications of Contact Lens Wear, pp. 178, Mosby-Year Book, Inc., St. Louis, (1992).

40. R.G. Linton, D.H. Curnow and W.J. Riley, The Meibomian glands. An investigation into the secretion and some aspects of the physiology, *Br. J. Ophthalmol.* 45:718 (1961).

41. D.A. Benedetto, T.E. Clinch and P.R. Laibson, In vivo observation of tear dynamics using fluorophotometry, *Arch. Ophthalmol.* 102:410-412 (1984).

42. D.R. Korb, et al., Meibomian gland secretion as a function of blinking, *Invest. Ophth. Vis. Sci. (ARVO Suppl.)* 34(4):1473 (1993).

43. D.R. Korb, et al., Meibomian gland secretion as a function of blinking, (*Submitted* 1993).

44. G.R. Trees and A. Tomlinson, Effect of artificial tear solutions and saline on tear film evaporation, *Optom. Vis. Sci.* 67:886-890 (1990).

45. G. Andrasko and J.P. Schoessler, The effect of humidity on the dehydration of soft contact lenses on the eye, *Int. C.L. Clin.* 7(5):30-32 (1980).

46. V.M. Finnemore, et al, The effect of 100% humidity on contact lens symptoms, *In preparation*.

47. D.R. Korb, et al, The effect of 100% humidity on lipid layer thickness, *In preparation*.

48. D.R. Korb, et al, The effect of 100% humidity on the prelens lipid layer for rigid gas permeable and hydrogel contact lenses, *In preparation*.

49. M.J. Refojo, The tear film and contact lenses: the effect of water evaporation from the ocular surface, *Simposio del Societa Optalmologica Italiana*, Rome (1984).

50. D.R. Korb and J.V. Greiner, Increase in tear film lipid layer thickness following treatment of meibomian gland dysfunction, *in*: "Lacrimal Gland, Tear Film and Dry Eye Syndromes: Basic Science and Clinical Relevance," D. Sullivan et al., eds., Plenum Press, New York (in press 1993).

# ELECTROPHORETIC PATTERNS OF HUMAN DENATURED TEAR PROTEINS AND GLYCOPROTEINS FROM NORMAL SUBJECTS AND SOFT CONTACT LENS WEARERS

Véronique Claudon-Eyl and Joël Baguet

Laboratoire Meuse Optique Contact  (MOC)
Centre Hospitalier
Bar le Duc, France

## INTRODUCTION

The biological reactivity of soft contact lenses (SCL) is influenced by a variety of factors. The composition of the tear fluid contacting the SCL is one of the most important factors affecting biocompatibility. Maintenance of adequate tear functions seems to be an important way to control SCL deposits[1]. Two major mechanisms lead to an abnormal protein deposition  on SCL: abnormal protein denaturation and instability of the mucous tear film.

Polymer surfaces are known to denature proteins during adsorption of proteins and the denaturation process is crucial in the adsorption of the proteins on biomaterials[2]. Many investigators reported conformation changes of proteins upon UV exposition, reactive oxygen species (ROS) interactions and thermodynamic reactions at the interface SCL-tear fluid. These involve increased exposure of nonpolar residues of the proteins to the aqueous or hydrophilic medium. In human tear fluid, it was recently demonstrated[3] that tear specific prealbumin (TSP) belongs to the group of hydrophobic molecule transporters called lipocalins. In this article, as suggested by Delaire[3] to normalize the nomenclature, we call TSP "Tear Lipocalin" (TL: followed by their molecular weight) and protein G "TL aggregates" (TLA). Fullard[4] showed the strong tendency of purified TL to denature, dimerize or/and aggregate in TLA 35. The phenomenon is increased in presence of strong denaturant like sodium dodecyl sulfate (SDS)[5]. So, it is possible that TL possesses some particular physico-chemical characteristics to play a main role in SCL spoilage. Changes of protein profiles of human tears of healthy persons as well as persons suffering from various eye diseases have been investigated[6-12]. Very little has been reported on possible changes in tears from patients related to contact lenses wearing[13,14] or deposit formation.

On the other hand, glycoproteins play a key role in the maintenance of tear film stability[6]. There are ample indications that carbohydrate moieties of glycoproteins perform other important biological roles including stabilization of protein conformation and inhibition of protein adsorption on surfaces. Various diseases and contact lenses can induce changes in the composition of conjunctival glycoproteins[15-17]. To our knowledge, a few studies have attempted a characterization of the tear glycoproteins[18-21] and no studies were available on tear fluid glycoproteins profiles during SCL wearing and deposit formation.

The original motivation of the present work was to study basal and reflex tear fluid protein and glycoprotein profiles of apparent healthy SCL wearers (with low or high SCL deposits formation susceptibility) to determine whether the SCL wearers possessed the same patterns as control patients (non SCL wearers). Tear fluid was denatured to avoid intermediate TL aggregates and obtain maximal TLA level. Electrophoretic analysis method was SDS-polyacrylamide gel electrophoresis on minigel.

## MATERIALS AND METHODS

### Tears, Patients, SCL and Types of Deposits

Basal (1 μl) and reflex tears (20 μl) were collected by glass capillaries from volunteers. The patient populations consisted of control patients and SCL wearers with low or high level of deposits. Control subjects (group I) were healthy people without complaint of ocular pain and who never worn contact lenses. "Low-depositors" SCL wearers (group II) had no complaint of ocular pain, discomfort or lens-related problems and had no recent history of ocular disease. The population of apparently healthy "high depositors" SCL wearers (group III) produced a high level and frequency of deposits. The SCL used by the patients were co-polymer of polymethylmethacrylate (PMMA) and N-vinylpyrrolidone (NVP) or polyacrylamide (PA) containing 70 to 78 % water ."High depositors" SCL wearers produced two types of deposits classified according to Le Naour[22]: Type C deposits are called "lens calculi" containing mucoproteins, lipids and calcium; Type D deposits are "crinkled film of grid-like structure" containing proteins.

### SDS - PAGE of Denatured Tear Fluid

Samples were diluted in SDS buffer (10 mM Tris-HCL, pH 8.0 with 2.5% SDS, 1 mM EDTA and 0.005% $NaN_3$) and cold denatured by 6 freeze-thaw cycles (FTC) at -20°C. Electrophoresis was run on precast 4%-15 % gradient minigels with Pharmacia PhastSystem according to the manufacturer's programmed instructions. After electrophoresis, the gels were silver stained.

### Electroblotting and Lectin Affinity Staining

After electophoresed under SDS-PAGE conditions, minigels were transferred electrophoretically onto a 0.45 μm polyvinyldifluoride (PVDF) membrane (Immobilon, Bedford, MA) in the Pharmacia PhastTransfer unit system according to the manufacturer's programmed instructions. Glycoproteins sugar residues were detected by biotinylated lectins and horseradish peroxidase - avidine D system. Lectins used were: Jacalin and Concanavalin A (Con A). Jacalin appears to bind only O-glycosidically linked oligosaccharides preferring the structure galactosyl (β-1,3) N-Acetylgalactosamine. Con A recognizes α-linked mannose residues.

## RESULTS

### SDS-PAGE

Tear fluids from group I (n=8), group II (n=8) and group III (n=20) subjects were analyzed. Representative SDS-PAGE patterns of cold SDS denatured basal human tear proteins were shown on figure 1. Some individual differences were observed in the tear profiles of the subjects :
- TL 23 was irregularly detected
- TLA bands at about 30 kD showed little variations.
- protein patterns of molecular weight (MW) higher than lactoferrin were slightly different.
Qualitatively, SDS PAGE analysis of denatured basal tear fluid from normal subjects or SCL wearers ("low" or "high depositors") showed a largely similar protein profile.

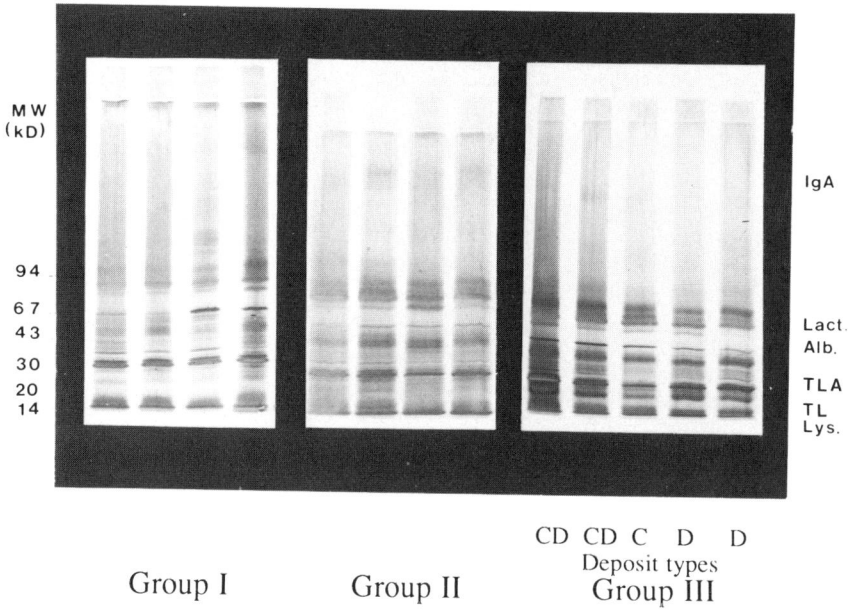

**Figure 1** . Representative SDS-PAGE patterns of cold SDS denatured basal human tear (1μl 1/75 dilution) proteins obtained after silver staining. Group I : control subjects, Group II : low depositors, Group III : high depositors. Type C deposits: "lens calculi" containing mucoproteins, lipids and calcium. Type D deposits: "crinkled film of grid-like structure" containing proteins. Alb. : albumin, Lact. : lactoferrin, Lys. : lysozyme, TL : tear lipocalin, TLA : TL aggregates. Estimated molecular weights markers are indicated on the left of the patterns. (Slight differences in position and in appearance of corresponding bands are due to different gels used in the studies. With 4-15% gradient minigels, lysozyme is partly recovered by TL 17 and not clearly visible).

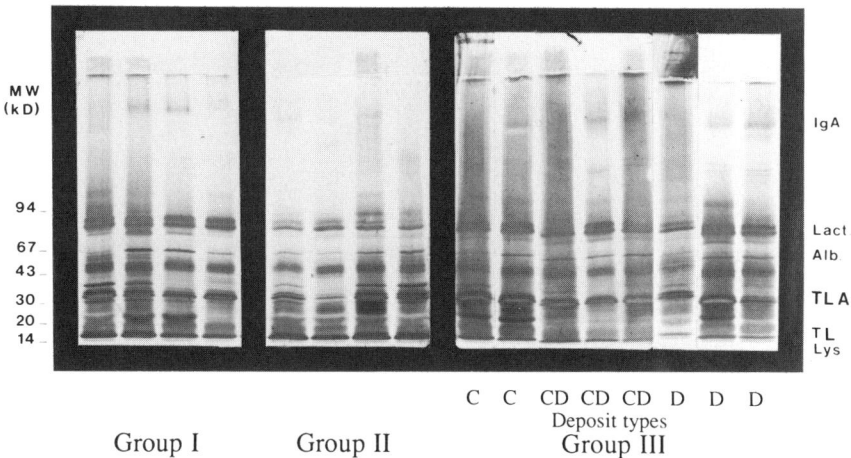

**Figure 2** . Representative SDS-PAGE patterns of cold SDS denatured reflex human tear (1μl 1/25 dilution) proteins obtained after silver staining. Group I : control subjects, Group II : low depositors, Group III : high depositors. Type C deposits: "lens calculi" containing mucoproteins, lipids and calcium. Type D deposits: "crinkled film of grid-like structure" containing proteins. Alb. : albumin, Lact. : lactoferrin, Lys. : lysozyme, TL : tear lipocalin, TLA : TL aggregates. Estimated molecular weights markers are indicated on the left of the patterns. (Slight differences in position and in appearance of corresponding bands are due to different gels used in the studies. With 4-15% gradient minigels, lysozyme is partly recovered by TL 17 and not clearly visible).

Representative SDS-PAGE patterns of cold SDS denatured reflex human tear proteins on minigels 4-15% were shown on figure 2. In reflex tear protein profiles, some individual variations were observed as in basal tear profiles. They were not related to SCL wearing or deposit formation.

## Electroblotting and Lectin Affinity Staining

Tear fluids from group I (n=5), group II (n=5) and group III (n=10) subjects were analyzed. Representative patterns of cold SDS denatured basal and reflex tear glycoproteins stained with Jacalin (Fig. 3) and Con A (Fig. 4) were shown. No significant difference in the glycoprotein profiles of the subjects was detected.

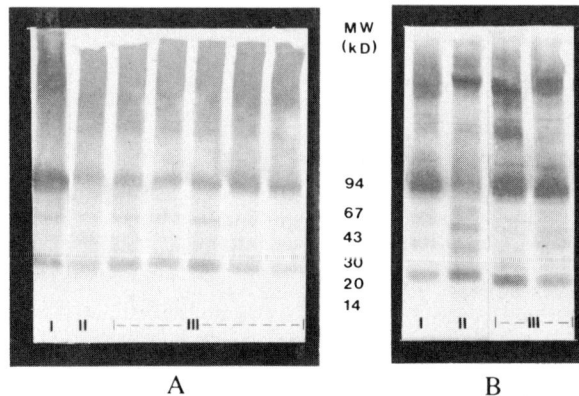

A                    B

**Figure 3** . Representative SDS PAGE patterns of cold SDS denatured human tear glycoproteins obtained after electrotransfer and biotinylated Jacalin staining. **A**: Basal tears (1μl 1/50 dilution), **B**: Reflex tears (1μl 1/10 dilution). *Lanes A1, B1*: control subjects (Group I), *lanes : A2, B2*: low depositors (Group II), *lanes: A 3-7,B 3-4* : high depositors (Group III). Estimated molecular weights markers are indicated. (Lanes were transferred from different gels, which caused slight differences in position and in appearance of corresponding bands).

## CONCLUSION

In this study, protein and glycoprotein electrophoretic patterns of denatured tear fluid from control patients and SCL wearers were determined with techniques with improved sensitivity. Tear fluid was subject to denaturation to avoid intermediate TL aggregation (maximum level of TLA) and to investigate abnormal protein denaturation related to SCL deposits. Despite the very sensitive technique used in this study, it is not possible to show any evident difference between basal or reflex denatured tear proteins and glycoproteins patterns of control patients and those of contact lenses wearers. Similar protein profiles were determined with all patients and some additional proteins could be found irregularly in the three groups of patients. It seems that variations in protein profiles were related to individual physiology and experimental conditions but not to contact lenses wear or susceptibility to form deposits on SCL surfaces. Moreover, this study demonstrated no evident variations of sugar receptor sites of tear glycoproteins for Con A and Jacalin in tear fluid from the subjects of the three different groups. These results indicate that deposit formation on SCL would not be related to abnormal denatured proteins or glycoproteins profiles of tear fluid.

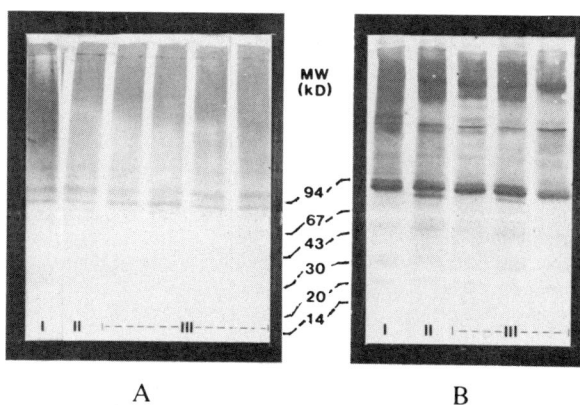

**Figure 4** . Representative SDS PAGE patterns of cold denatured human tear glycoproteins obtained after electrotransfer and biotinylated Con A staining. **A** : Basal tears, **B** : Reflex tears. *Lanes A1, B1*: control subjects (Group I), *lanes A2, B2*: low depositors, *(Group II), lanes: A3-6, B3-5*: high depositors (Group III). Estimated molecular weights markers are indicated. (Lanes A, B were transferred from different gels, which caused slight differences in position and in appearance of corresponding bands).

SCL spoilage is known to be due, firstly to an abnormal protein deposition and secondly to inflammatory and/or immunological reactions and presence of bacteria (with or without signs of infections). Our results indicate that the first step takes place in an apparent normal protein environment and that contact lens wearers seem to have similar denatured tear protein and glycoprotein profiles. However, SDS is a stronger denaturant than polymer surfaces. So, further investigations on tear - polymer surface denaturing interactions may be worthwhile. Tears possess particular properties to inhibit proteins adsorption on polymers. In 1985, Boonstra[23] showed that human tears contain a factor preventing the binding of tear lactoferrin to an acrylate surface. Boot[24] showed that too low an activity of this factor in tears is associated with deposits on contact lenses. The mechanism of coating inhibition and the role of this tear surfactant in the binding of other tear proteins to different types of biomaterials are not well known. In case of "high-depositors" patients, abnormal protein built up on contact lenses could be an abnormal activity of proteases or the deficiency of an hypothetic anti-denaturing factor. It is likely that an interactive system including mucin (hydrophilic residues and hydrophobic domains/ polar-apolar ratio), reactive oxygen species (oxydative molecules), major tear proteins, lipids (cholesterol/phospholipids ratio), complex molecules (lipoproteins and glycolipids), TL (hydrophobic molecule transporters) and epithelial cell activation (proteolytic activity and ROS release, mucin hypersecretion) is involved in the abnormal coalescing protein deposition on contact lenses. SCL spoilage was certainly due to a complex process including thermodynamic and biological mechanisms.

## ACKNOWLEDGEMENTS

Supported by grants from Conseil Général et Comité d'Aménagement, de Promotion et d'Expansion du département de la Meuse (CAPEM), Mairie de Bar le Duc, Hôpital de Bar le Duc, Conseil Régional de Lorraine, Essilor International.

## REFERENCES

1. R.L. Farris, The dry eye : its mechanism and therapy, with evidence that contact lens is a cause, *CLAO J.* 12, 234 (1986).
2. J.D. Andrade and V. Hlady, Protein adsorption and materials biocompatibility : a tutorial review and suggested hypotheses, *In* "Advances in polymer sciences," K Düsek, ed., Springer-Verlag, Berlin Heidelberg, 79:3 (1986).

3. A. Delaire, H. Lassagne, A.M.F. Gachon, New members of the lipocalin family in human tear fluid, *Exp. Eye Res.* 55:645 (1992).

4. R.J. Fullard and D.M. Kissner, Purification of the isoforms of tear specific prealbumin,*Curr. Eye Res.* 10:613 (1991).

5. J. Baguet, V. Claudon-Eyl, A.M. Gachon, Tear protein G originates from denatured tear specific prealbumin as revealed by two-dimensional electrophoresis, *Curr. Eye Res.* 11:1057 (1992).

6. F.J. Holly and B. Hong, Biochemical and surface characteristics of human tear proteins, *Am J. Optom. Phys. Optics*, 59, 43 (1982).

7. A.M. Gachon, J. Richard and B. Dastugue, Human tears: normal protein pattern and individual protein determinations in adults, *Curr. Eye Res.* 2:301 (1982).

8. A.M. Gachon, P. Lambin and B. Dastugue, Human tears: electrophoretic characteristics of specific proteins, *Ophtalmic Res.* 12:277 (1980).

9. A. Kuizenga, N.J. Van Haeringen and A. Kijlstra, SDS minigel    electrophoresis of human tears. Effects of sample treatment on protein pattern, *Inv. Ophthalmol. Vis. Sci.* 32:381 (1991).

10. G. Wollensak, E. Mur, A. Mayr, G. Baier, W. Gottinger, and G. Stoffler, Effective methods for the investigation of human tear film proteins and lipids, *Graefe's Arch. Clin. Exp. Ophthalmol.* 228:78 (1990).

11. R.J. Fullard, Identification of proteins in small tear volumes with and without size exclusion HPLC fractionation, *Curr. Eye Res.* 7:163 (1988).

12. P.K. Coyle, P.A. Sibony and C. Johnson, Electrophoresis combined with immunologic identification of human tear proteins, *Inv. Ophthalmol. Vis. Sci.* 30:1872 (1989).

13. I. Tapaszto, G. Hajos, A. Bujdoso, Z. Tapaszto and B. Tapaszto, Biochemical changes in the human tears of hard and soft contact lens wearers. V- Protein changes in the human tear of hard and soft contact lens wearers, *Contact Lens J.* 17:316 (1990).

14. R.L. Farris, Tear analysis in contact lens wearers, *CLAO J.* 12:234 (1986).

15. P. Versura, M.C. Maltarello, M. Cellini, F. Marinelli, R. Caramazza and R. Laschi, Detection of mucus  glycoconjugates in human conjunctiva by using the lectin-colloidal gold technique in TEM, *Acta Ophthalmologica,* 65:661 (1987).

16. P.A. Wells, C. Desiena-Shaw, B. Rice and C.S. Foster, Detection of ocular mucus in normal human conjunctiva and conjunctiva from patients with cicatricial pemphigoid using lectin probes and histochemical techniques, *Exp. Eye Res.* 46:485 (1988).

17. J.U. Prause, O.A. Jensen, K. Pashides, A. Stovhase and P. Vangsted, Conjunctival cell glycoprotein of healthy persons and of patients with Primary Sjögren's syndrome - Light microscopical investigation using lectin probes, *J. Autoimmun.* 2:494 (1989).

18. C.H. Dohlman, J. Friend, V. Kalevar, D. Yagoda and E. Balazs, The glycoprotein (mucus) content of tears from normal and dry eye patients, *Exp. Eye Res.* 22:359 (1976).

19. P.Halken, T.C. Bog-Hansen and J.U. Prause, Differences between glycoprotein profiles of normal tear fluid from patients with   primary Sjögren's syndrome, *Scand. J. Rheumatology* 61(Suppl.):234 (1986).

20. A. Kuizenga, N.J. Van Haeringen and A. Kijlstra, Identification of lectin binding proteins in human tears, *Inv. Ophthalmol. Vis. Sci.* 32(Suppl.):732 (1991).

21. K. Bjerrum, P. Halken, and J.U. Prause, The normal human tear glycoprotein profile detected with lectin probes, *Exp. Eye Res.* 53: 431 (1991).

22. L. Le Naour and P.R. Day, Observation of deposits on high water content lenses, *Opt. Today*, 28:127 (1987).

23. A. Boonstra, N. Van Haeringen and A. Kijlstra, Human tears inhibit the coating of proteins to solid phase surfaces, *Curr. Eye Res,* 4:1137 (1985).

24. N. Boot, J. Kok and A. Kijlstra, The role of tears in preventing protein deposition on contact  lenses, *Curr. Eye Res.* 8:185 (1989).

# THE REFRACTIVE INDEX OF TEARS IN NORMALS
# AND SOFT LENS WEARERS

Sudi Patel, Lynn Anderson, and Katrina Cairney

Department of Vision Sciences
Glasgow Caledonian University
Cowcaddens Rd., Glasgow
Scotland, U.K, G4 OBA

## INTRODUCTION

The dry eye has tear osmolarity above normal levels. The measurement of tear osmolarity is regarded by some authorities to be the 'gold standard' differentiating the dry eye from the normal[1,2]. Essentially, the osmolarity of a fluid is equivalent to its concentration. the concentration of a fluid directly influences its refractive index (RI). The theoretical relationship between the RI of a fluid and its concentration was examined by Gladstone and Dale[3]. Batsanov[4] has reviewed the attempts by several other groups who have also attempted to equate these 2 parameters. It is possible to estimate the concentration of tear fluid by way of refractometry. Refractometry as a method for estimating the protein concentration in biological fluids was first suggested by Reiss in 1903[5]. The idea was explored by Barer and Joseph[6] and with the development of the hand held refractometer, the protein concentration of tear fluid was estimated by Stegman and Miller[7]. Recently, refractometry has been used to differentiate the dry eye from the normal[8].

Contact lens wearers can develop dry eye like symptoms. However, there has not been any corroborative evidence suggesting the contact lens induced dry eye (CLIDE) is a genuine dry eye. The lens wearer may have reduced tear stability[9], a feature found in genuine dry eyes[10] but this may be short lived[11]. Certainly, no common trait of reduced tear production or osmolarity has been detected except in cases of aphakics who are on extended wear[12,13]. If the soft lens wearer is at risk of developing a genuine dry eye then perhaps, in the asymptomatic case, the tear concentration differs from the norm. Osmolarity (units osmoles/ litre) can be measured by freezing point depression[1] or osmolality (units osmoles/ kgm) can be calculated from the results using a thermocouple hygrometer[14]. Both methods are essentially laboratory methods which are difficult to use in a clinical setting. On the other hand, refractometry could be used in the busy clinic. The purpose of this investigation was to:

I)   Assess the validity of refractometry as a method of tear concentration evaluation.
ii)  Compare the concentration of tears in both asymptomatic soft lens wearers and normals.

## METHODS

**Apparatus:** Tear concentration was estimated using 2 refractometers, a standard bench refractometer (by Beck) with a scale resolution of 0.0001 and a hand held salinity refractometer (model S-10[T.M.] by Atago). The hand held portable refractometer has a scale range from 0 to 10% saline with a resolution of 0.1%.

### Procedure:
I) Calibration
11 saline samples of varying concentration (0 to 2.5%) were made up. Each sample was assigned a random number code. The refractive index of each was measured using both refractometers after pre-adjusting the instruments according to manufacturer's recommendations. The sodium 'D' line (589.3 nm) was used as the light source in both I) and II). The room temperature was 21°c.

II) Subjects
A 1μl thin walled micropipette was used to obtain a tear sample from the lower tear meniscus of the left eye. This sample was measured using the bench refractometer. 2 minutes later, a 2nd sample was taken and measured using the portable refractometer. The lens wearers did not remove their lenses prior to sample taking. 20 asymptomatic soft lens wearers (age range) and 20 age and sex matched normals were entered into the study. Using the calibration curves for the two instruments, the equivalent % saline concentrations of the tear samples (80 in total) were calculated. All subjects were of good general health with no history of ocular conditions which may have affected tear properties. Nervous and hypersecreting subjects were excluded.

## RESULTS

I) Calibration
The least squares calibration curves were as follows:
Bench refractometer,  $y = 1.3332 + 0.0017 x$   (c.c.= 0.987)
S-10 refractometer,   $Y = 0.132 + 0.968 x$   (c.c.= 0.993)
where, y is the refractive index, Y is the scale reading of % saline and x is the actual % saline.

II) Subjects
The average equivalent % saline concentrations, calculated using the appropriate calibration curves, for the 2 samples taken from the 2 groups are listed in Table 1. The mean refractive indices, standard deviation and range of refractive index for the normals was 1.336, +/- 0.001 and 1.334 to 1.338, respectively. These results were identical with thoses found within the lens wearing group. The equivalent saline concentrations for the individual eyes are shown in Figures 1 and 2.

**Table 1.** Mean tear concentrations and results of statistical analyses.

| Refractometer | Mean E.S.C. [1](%) | | t' statistic | Significance (@1%) |
|---|---|---|---|---|
| | Normals | Lens wearers | | |
| Bench | 1.66 (0.58)[a] | 1.73 (0.56)[b] | a and c, -2.75 | Yes |
| S-10 | 2.07 (0.34)[c] | 1.98 (0.21)[d] | b and d,-1.84 | No |

[1]Equivalent Saline Concentration. [2]Standard deviation,(+/- $Ó_{n-1}$), shown in parenthesis after mean value. [3]Application of 't'-test revealed there to be no significance in the apparent difference between normals and lens wearers.

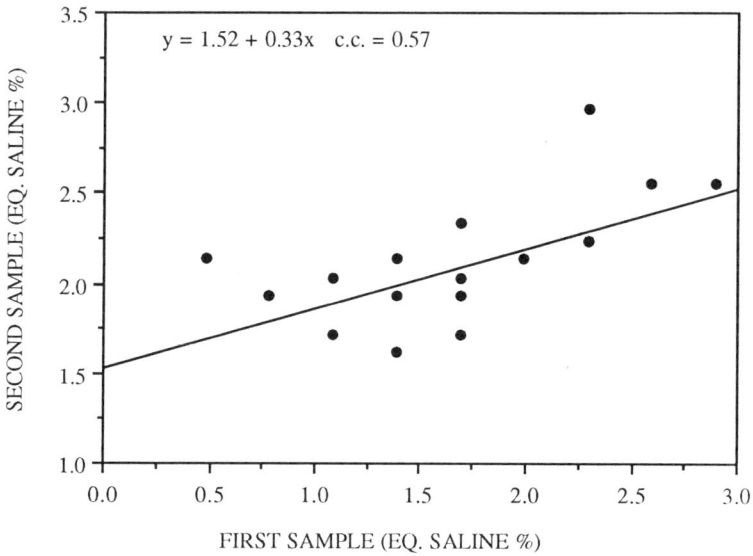

$$y = 1.52 + 0.33x \quad c.c. = 0.57$$

**Figure 1.** Tear concentrations of normal subjects

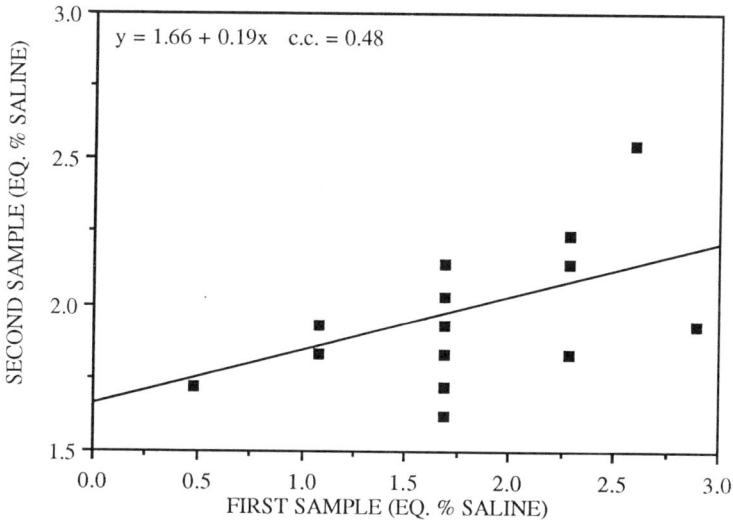

$$y = 1.66 + 0.19x \quad c.c. = 0.48$$

**Figure 2.** Tear concentrations of lens wearers

## DISCUSSION

The high correlation coefficients for both refractometers indicate that either model could be used to measure the concentration of tear samples in a clinical situation. The 95% confidence limit for the S-10 refractometer was +/- 0.06%. This is the likely error in estimating the concentration of a single tear sample in terms of equivalent saline concentration. The refractive indices of the 2 groups were identical and agree well with those found by others [8,16] (1.3357 -1.3374), but they are much lower than the average figure of 1.357 according to Milder[17]. The differences in tear concentration between the normals and soft lens wearers was insignificant. This is in keeping with previous studies where tear osmolarity was the parameter of measurement.

In both groups, the average saline equivalent concentrations of 2.07% and 1.98% are higher than the commonly quoted figures of 0.9% to 1.02%[14]. This must be viewed with caution. The 2.07% saline equivalent is the concentration of NaCl solution which has the same refractive index as the tear sample hence, it is a refractive equivalent and not an osmotic equivalent. The other tear constituents will raise the net refractive index of the tear sample. The refractive index of meibomian oil is 1.482 at ocular surface temperature[18], and tear proteins may have refractive indices of the order 1.53 to 1.60[6]. These and the remaining high molecular weight additions to the basic salt solution of lacrimal fluid will raise the net refractive index of tears.

Within the normal group, there was a significant difference between the first and second tear sample concentrations. The rise in concentration is accountable if we assume that some degree of lacrimation, as a result of anticipation, took place when the first tear sample was obtained and that the lacrimation rescinded by the time the second sample was taken. This has a bearing on all tear sampling. A similar difference was not found in the average data from the soft lens wearing group. Perhaps, the soft lens wearer is adapted to the lens and finger touching the eye and lids and therefore does not lacrimate quite so readily. This does not hold in all cases. Referring to Figures 1 & 2, only within the normals is there a significant correlation between the first and second tear sample concentrations. In the soft lens wearing group, it is not possible to predict the concentration of the second sample from a measurement performed on the first sample. In untrained subjects the investigator is advised to pay more attention and rely on the findings of the second rather than the first sample. Some of the tear samples had a concentration equivalent $< 2.5\%$, the relatively high concentration may be the signs of incipient dry eye. Further work investigating the refractive index of tears and comparing the results with those of other tests are required before making a final decision on the true value of tear refractometry.

## REFERENCES

1. J.P. Gilbard, R.L. Farris and J. Santamaria, Osmolarity of tear microvolumes in KCS, *Arch Ophthalmol.* 96: 677-681 (1978).
2. J.P. Gilbard and R.L. Farris, Tear osmolarity and ocular surface disease in KCS, *Arch Ophthalmol.* 97: 1642-1646 (1979).
3. J.H.Gladstone, Refraction equivalents, *J Chem Soc (U.K).* 23: 101-115 (1870).
4. S.S. Batsanov. "Refractometry and Chemical Structure" Von Nostrand, Princetown (1960).
5. Reiss (1903), quoted in 6).
6. R. Barer and S. Joseph, Refractometry of living cells, *Quarterly J Microscopic Sci.* 95: 399-423 (1954).
7. R. Stegman and D. Miller, A human model of allergic conjunctivitis, *Arch Ophthalmol.* 93: 1354-1358.
8. T.R. Golding and N.A. Brennan, Tear refractive index in dry and normal eyes, *Clin & Exp Optom.* 74: 212 (1991).
9. S. Patel, Constancy of the front surface dessication times for Igel 67 lenses in vivo, *Am J Optom & Physiol Opt.* 64: 167-171 (1987).
10. J. Farrell, D.J. Grierson, S.Patel, R.D. Sturrock, A classification of dry eyes following comparison of tear thinning time with Schirmer tear test. *Acta Ophthalmol.* I70: 357-360 (1992).
11. T.R. Golding, N. Efron, N.A. Brennan, Soft lens lubricants and prelens tear film stability. *Optom & Vis Sci.* 67: 461-465 (1990).
12. R.L. Farris, Tear analysis in contact lens wearers. *CLAOJ.* 12: 106-111 (1986).
13. R.L. Farris, The dry eye: its mechanisms and therapy with evidence that contact lens is a cause. *CLAOJ.* 12: 234-246 (1986).
14. J.E. Terry and R.M. Hill, Human tear osmotic pressure. *Arch Ophthalmol.* 96: 120-122 (1978).
15. F.J. Rohlf and R.R. Sokal. "Statistical Tables" 2nd ed. Freeman, San Francisco (1981).
16. A. von Röth, Uber die tränenflüssigkeit, *Klin Monat sbl Augenh.* 86: 598-602 (1922).
17. B. Milder, The lachrimal apparatus, *in*: "Adler's Physiology of the Eye," R.A. Moses, ed., C.V. Mosby, St. Louis (1987).
18. J.M. Tiffany, Refractive index of meibomian and other lipids, *Curr Eye Res.* 5: 887-889 (1986).

# THE ROLE OF TEAR DEPOSITS ON HYDROGEL CONTACT LENSES INDUCED BACTERIAL KERATITIS: *Pseudomonas aeruginosa* ABHESIVES

Marta Portolés and Miguel F. Refojo

Schepens Eye Research Institute
Department of Ophthalmology
Harvard Medical School
Boston, MA 02114

## INTRODUCTION

Bacterial keratitis is a devastating acute ocular disease that is increasing being found in association with contact lens wear.[1, 2] The extended overnight wear of hydrogel contact lenses is thought to be a major risk factor for ulcerative keratitis among contact lens wearers.[3, 4] Corneal hypoxia and accumulation of carbon dioxide are complications of all extended-wear hydrogel contact lenses.[5] It has been shown that these effects and the tear exchange barrier effect of a hydrogel lens on the cornea surface can alter the physiology and morphology of the cornea, compromising its integrity and health.[6]

The adherence of bacteria to compromised corneal surface is considered to be the first step in the infection process. Bacteria then colonize the corneal surface, produce bacterial enzymes and toxins, and finally, invade the corneal epithelium and stroma.[7-9] When a contact lens is present, it could act as a reservoir from which attached bacteria may be transferred to the compromised cornea and initiate the infectious process.[10]

Shortly after insertion on the eye, a contact lens becomes coated with a thin diffuse tear-deposit layer.[11, 12] There is a prevalent belief that the etiology of contact lens induced *Pseudomonas aeruginosa* keratitis is directly related to the presence and amount of tear-deposits on the lenses.[13, 14] Enzymatic and detergent treatments improve lens tolerance but these treatments fail to eliminate or significantly reduce lens coating.[15, 16]

The role of molecules, similar to certain tear components, in bacterial adherence to contact lenses has been extensively studied. It has been reported that submaxillary bovine mucin adsorbed on soft contact lenses facilitates *in vitro P. aeruginosa* adherence to contact lenses and also increases the incidence of experimental pseudomonal keratitis.[17-19] Others have found that submaxillary bovine mucin and sialic acid in aqueous solution decrease the *P. aeruginosa* adherence to soft contact lenses *in vitro*.[20] Also, egg-white lysozyme, bovine serum albumin, bovine milk lactoferrin and human calostrum immunoglobulin A, adsorbed to the lens surface or in solution were reported to enhance *P.aeruginosa* adherence to soft contact lenses *in vitro*.[17, 20]

From these reports one could predict that new or clean (without tear-deposits) lenses will be less likely to lead to infections than other hydrogel lenses. Unfortunately, laboratory and clinical studies do not support this prediction, e.g. the age of a lens and the amount of protein deposits do not determine the microbial contamination of the lens.[21] A clinical

Supported by a grant from "Colegio Oficial de Biólogos", Spain (MP) and by NIH grant EY 00327 (MFR).

study found that the relative risk for ulcerative keratitis was not related to the age of the lenses (hence, the amount of tear coating).[3] In addition, there is no evidence in the literature to indicate that the incidence of bacterial keratitis is dependent on the type of hydrogel lens used although it is well known that the ionic hydrogel lenses are coated *in vivo* with larger amounts of protein than the non-ionic lenses.[22-24] Furthermore, it has been shown that *Pseudomonas aeruginosa* attach even to new never worn contact lenses [25] and that *Pseudomonas aeruginosa* attachment to low- or high-water, ionic or non-ionic, lenses worn 24 hours by rabbits was significantly reduced compared to the attachment to the unused lenses of the same type. [26] This was confirmed in a recent study that found less *P. aeruginosa* adherence to disposable contact lenses following a week of continuous, day and night, wear in human eyes than to new disposable lenses.[27] Also, after 5 hours of eye human wear, low-water non-ionic hydrogel lenses have less adhered microorganisms than after patient handling before lens insertion.[28] The results on this study demonstrate the efficacy of the ocular defense system and the failure of tear coatings to increase microbial adherence. Therefore, it seems that the film coating of used contact lenses, rather than enhancing bacterial adhesion, has anti-adhesive (abhesive[29]) properties.

In any case, the importance of bacterial adherence to contact lenses and their relevance to contact lens induced bacterial keratitis is still an unsolved problem. Recently it was shown that *in vitro Pseudomonas aeruginosa* adherence to new or worn hydrogel contact lenses and the development of infectious keratitis in a rabbit model were directly related to the concentration of the bacterial inoculum.[30] Our hypothesis is that if bacterial adhesion to the lenses is eliminated or significantly decreased, we might reduce the incidence of infection. Therefore, the object of our research is to identify an inert abhesive that will hinder bacterial adhesion to contact lenses, preventing or at least diminishing the incidence of bacterial keratitis associated to contact lens wear.

Poloxamers are water-soluble non-toxic block copolymers of poly(ethylene oxide) and poly(propylene oxide) (Table 1). By varying the block size, different poloxamers with different properties are obtained. Poloxamers have been used in eye drop solutions as vehicles that gel at body temperature for drug delivery to mucous membranes, and as detergents in cleaning solutions for contact lenses.[31-33] Polyethylene glycols (PEGs) are similar in chemical structure to poly(ethylene oxide) but have different molecular weight. PEGs bound to different polymers have been reported as reducing bacterial adherence[33, 34].

In this study, the effect of solutions of poloxamers and PEGs on *Pseudomonas aeruginosa* adherence to new hydrophilic contact lenses was investigated.

**Table 1.** Characteristics of polymers used in *Pseudomonas aeruginosa* adherence to hydrophilic contact lens assays.

| POLYMER | PHYSICAL FORM | AVERAGE Mw | VALUE * a | b | BRAND NAME |
|---------|---------------|------------|-----------|-----|------------|
| Poloxamer 123 | Liquid | 1,850 | 7 | 21 | PLURONIC® L43 |
| Poloxamer 182 | Liquid | 2,500 | 8 | 30 | PLURONIC® L62 |
| Poloxamer 184 | Liquid | 2,900 | 13 | 30 | PLURONIC® L64 |
| Poloxamer 188 | Solid | 8,350 | 76 | 30 | PLURONIC® F68 |
| Poloxamer 234 | Paste | 4,200 | 22 | 39 | PLURONIC® P84 |
| Poloxamer 238 | Solid | 10,800 | 97 | 39 | PLURONIC® F88 |
| Poloxamer 288 | Solid | 13,500 | 122 | 47 | PLURONIC® F98 |
| Poloxamer 335 | Paste | 6,500 | 38 | 54 | PLURONIC® P105 |
| Poloxamer 338 | Solid | 14,000 | 128 | 54 | PLURONIC® F108 |
| Poloxamer 403 | Paste | 5,750 | 21 | 67 | PLURONIC® P123 |
| Poloxamer 407 | Solid | 12,500 | 98 | 67 | PLURONIC® F127 |
| Polyethylene glycol 6,000 | Solid | 6,000-7,500 | --- | | --- |
| Polyethylene glycol 18,500 | Solid | 18,500 | --- | | --- |

* Graphic formula [36]: $\alpha$-hydro-$\omega$-hydroxypoly(oxyethylene)-poly(oxypropylene)-poly(oxyethylene)

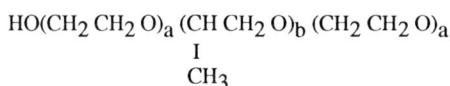

$$HO(CH_2 CH_2 O)_a (CH\ CH_2\ O)_b (CH_2\ CH_2\ O)_a$$
$$|$$
$$CH_3$$

## MATERIAL AND METHODS

*P. aeruginosa* 6294, a clinical isolate from a corneal ulcer, was obtained from Channing Laboratory (Harvard Medical School, Boston, MA, USA). One milliliter aliquots of *P. aeruginosa* suspension were maintained at -70°C in Tryptic Soy Broth with 10% glycerol. For each experiment, 100 μl from one of these *P. aeruginosa* suspensions was grown overnight in a Tryptic Soy Agar (TSA) plate at 37°C. The *P. aeruginosa* inoculum, obtained from the plates, was spectrophotometrically adjusted to a concentration of $10^8$ colonies forming units per milliliter in phosphate buffered saline (cfu/ml PBS) (optical density 0.1 at 590 nm).

New sterile etafilcon A hydrophilic contact lenses (CL, ACUVUE™, Vistakon, Inc., Jacksonville, FL, U.S.A.; ionic, 58% $H_2O$ content, FDA group IV) were used.

Eleven poloxamers (PLURONIC®, BASF Wyandotte Corp., NJ, U.S.A.): 123, 182, 184, 188, 234, 238, 288, 335, 338, 403 and 407, and two PEGs: Mw 18,500 (Polysciences Inc., PA) and Mw 6,000 (J.T. Baker Chemical Co., NJ, U.S.A.), were assayed. A sterile filtered (0.22 μm Millipore filter) 4% solution in PBS was freshly prepared for each polymer. The final concentration for each polymer, after dilution in *P. aeruginosa* suspension was 2%. PBS alone was used as control solution. Polymers physical form, molecular weight, graphic formula and brand name are summarized in Table 1.

CLs were incubated statically for 60 minutes at room temperature in 0.5 ml of the *P. aeruginosa* suspension diluted in either 0.5 ml of 4% poloxamer solution, 4% PEG solution or PBS (control). The lenses were then rinsed in 5 aliquots of 10 ml of PBS (10 dips in each aliquot) to remove non-adherent bacteria. The CLs were then ground and three serial dilutions (1:10) were performed. One milliliter of each dilution was plated by duplicate in TSA. After 36 h incubation at 37°C colonies were counted and the average for the duplicates was calculated.

The results were expressed as colony forming units per lens (cfu/lens). The arithmetic means and standard deviations (x±SD) were determined for poloxamers, PEGs and the control. A percentage of adherence was calculated for each polymer with adherence to control (PBS) representing 100%.

The comparisons between bacterial adhesion to CLs incubated in poloxamers or PEGs solutions and the lenses incubated in PBS (control) were determined by an analysis of variance (ANOVA, Scheffé Test) using a computer statistical package (SPSS-X, Release 3.1 for VAX/VMS).

**Table 2.** Effect of poloxamers and polyethylene glycols compared to the control on the percentage of *Pseudomonas aeruginosa* adherence to new hydrophilic contact lenses.

| GROUP (n) | % ADHERENCE | % INHIBITION |
|---|---|---|
| Control-PBS (30) | 100 | 0 |
| | | |
| PEG 6,000 (3) | 37 | 63 |
| PEG 18,500 (11) | 34 | 66 |
| | | |
| POLOXAMERS: | | |
| 123 (3) | 213 | --- |
| 182 (3) | 613 | --- |
| 184 (3) | 636 | --- |
| 188 (8) | 34 | 66 |
| 234 (3) | 176 | --- |
| 238 (3) | 54 | 46 |
| 288 (5) | 18 | 82 |
| 335 (3) | 6 | 94 |
| 338 (11) | 3 | 97 |
| 403 (7) | 3 | 97 |
| 407 (12) | 2 | 98 |

\* Polymer groups different from control group ($p < 0.05$)

## RESULTS

The results of the effect of 2% poloxamer or PEG solutions in PBS on the *P. aeruginosa* attachment to CLs (x±SD cfu/lens) are summarized in Figure 1. The percentage of attachment for each polymer compared to control (PBS) is presented in Table 2.

Both PEGs (Mw 18,500 and Mw 6,000) diminished *P. aeruginosa* adherence to CLs, by 60%, but this decrease was not statistically significant.

Among the polymers assayed, poloxamers: 123, 182, 184 and 234, increased bacterial adherence. Specifically 182 and 184 increased *P. aeruginosa* attachment more than 500% above the control (p<0.05).

Poloxamers 188, 238 and 288 showed a tendency to decrease *P. aeruginosa* adherence, reducing attachment by 66%, 46% and 82% respectively. Like PEGs, none of these decreases in bacterial attachment were statistically significant.

Poloxamers 335, 338, 403 and 407 inhibited *P. aeruginosa* adherence to CLs by 94% or more, and this was statistically significant (p<0.05).

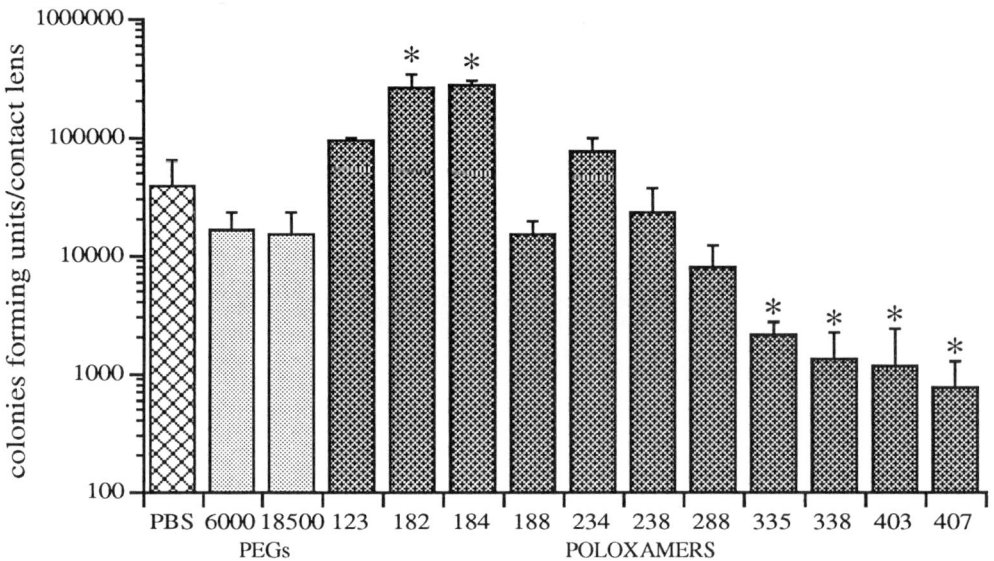

**Figure 1.** Effect of poloxamers and polyethylene glycols on *Pseudomonas aeruginosa* adherence to new hydrophilic contact lenses (* Polymer groups different from control group, p<0.05)

## DISCUSSION

Most studies of bacterial adherence mechanisms during ocular infection are based on the specific interaction between certain adhesins on the bacterial surface and receptors on the epithelial cells. Receptors and adhesins have been described in detail as mediators in *Pseudomonas aeruginosa* adherence to the corneal epithelial cells.[37, 38]. However bacteria can adhere in the absence of specific receptors. Bacteria adhere to many inert surfaces such as glass, plastic or never-worn contact lenses.[25, 27]. That suggests the participation of non-specific interactions such as hydrophobic/hydrophilic interactions between the bacteria and the host.

The results of this investigation showed that poloxamers: 335, 338, 403 y 407 inhibit bacterial adherence to CLs by more than 90% as compared to the control group. The outstanding abhesive effect of these poloxamers is higher than that described for the biological compounds assayed in aqueous solution or adsorbed onto contact lenses, such as submaxillary bovine mucin or sialic acid.[17, 18, 20, 39]

Poloxamers are block copolymers with a central hydrophobic block of poly(propylene oxide) and two lateral hydrophilic blocks of poly(ethylene oxide). The four poloxamers with the highest abhesive effect have the largest hydrophobic block while the ones with a smaller hydrophobic block either have no significant inhibitory effect or even enhance bacterial adherence to the lenses. Therefore, the size of the hydrophobic block could be a critical characteristic on the inhibition of adherence while the size of the hydrophilic blocks seems to be less important.

Preliminary data obtained in our laboratory show that the mechanism of action of poloxamer is on the bacteria, perhaps interacting with the adhesins, rather than on the contact lens surface. The poloxamer could be adsorbed to the bacteria surface by the hydrophobic block exposing the lateral hydrophilic blocks to the aqueous media, creating a highly hydrated layer around the bacteria that inhibits adherence in a non-specific manner. The results suggest that a minimum size of the hydrophobic block could be critical for the poloxamer interaction on the bacterial surface. The fact that aqueous solutions of polyethylene glycol (with the same chemical structure of poly(ethylene oxide) but different molecular weight) do not decrease significantly *P. aeruginosa* adherence to contact lenses also indicates an important role for the hydrophobic block on the inhibition of adherence. Besides, polyethylene glycol chemically bound onto the surface of synthetic polymers has been described to inhibit bacterial, cell and protein adhesion *in vitro*.[34, 35]

Recently, it was reported poloxamers adsorbed onto polystyrene microspheres inhibited both *Staphylococcus epidermidis* adherence and macrophages adherence and phagocytosis to the microspheres.[40-42] In those studies, the anchorage of the poloxamers onto the microsphere by the hydrophobic block and the formation of a hydrophilic layer around the microspheres was proposed as a non-specific mechanism of inhibition of bacterial and cell attachment.

The abhesive mechanism of action of poloxamers is not due to bacteriostatic or bacteriocidal effects. Therefore, topical use of poloxamers will not produce the emergence of bacterial resistance which is one of the major problems associated with the frequent use of antibiotics and chemiotherapics.

Poloxamers have been used as detergents in cleaning solutions for both hard and soft contact lenses.[33] Thus, poloxamers could have a double beneficial effect as bacterial abhesives and as a surface active agents to clean the lenses.

In conclusion, some polymers of the poloxamer series decrease *Pseudomonas aeruginosa* adherence to new hydrophilic contact lenses by 94% or more. Poloxamers are non toxic, high stable, water soluble and lack antimicrobial activity. For these reasons, poloxamers: 335, 338, 403 and 407 could be used to prevent bacterial attachment to contact lenses and, therefore, may be potential candidates to reduce the incidence of contact lens induced keratitis.

## REFERENCES

1. J.K.G. Dart. *Br J Ophthalmol*, 72, 926-930 (1988).
2. E. Alfonso, S. Mandelbaum, M.J. Fox and R.K. Forster. *Am J Ophthalmol*, 101, 429-433 (1986).
3. O.D. Schein, *et al.*. *N Engl J Med*, 321, 773-778 (1989).
4. J.K.G. Dart, F. Stapleton and D. Minassian. *The Lancet*, 338, 650-653 (1991).
5. J.P.G. Bergmanson and L.W.-F. Chu. *Am J Optom Physiol Opt*, 59, 500-506 (1982).
6. B.A. Holden, D.F. Sweeney, A. Vannas, K.T. Nilsson and N. Efron. *Invest Ophthalmol Vis Sci*, 26, 1489-1501 (1985).
7. J.K.G. Dart and D.V. Seal. *Eye*, 2 (Suppl.), S46-S55 (1988).
8. R.A. Hyndiuk. *Trans Am Ophthalmol Soc*, 79, 541-624 (1981).
9. T.J. Liesegang. in *The cornea* (eds. Kaufman, H.E., *et al.*) 217-270 (Churchill Livingstone, New York, Edinburgh, London, Melbourne, 1988).
10. M.M. Slusher, Q.N. Myrvik, J.C. Lewis and A.G. Gristina. *Arch Ophthalmol*, 105, 110-115 (1987).
11. T. Bilbaut, A.M. Gachon and B. Dastugue. *Exp Eye Res*, 43, 153-165 (1986).
12. S.A. Fowler, D.R. Korb and M.R. Allansmith. *CLAO J*, 11, 124-127 (1985).
13. S.I. Butrus and S.A. Klotz. *Curr Eye Res*, 9, 717-724 (1990).

14. M.I. Aswad, T. John, M. Barza, K. Kenyon and J. Baum. *Ophthalmology*, 97, 296-302 (1990).
15. S.A. Fowler and M.R. Allansmith. *Arch Ophthalmol*, 99, 1382-1386 (1981).
16. J. Jung and J. Rapp. *CLAO J.*,19:47 (1993)
17. M.J. Miller, L.A. Wilson and D.G. Ahearn. *J Clin Microbiol*, 26, 513-517 (1988).
18. G.A. Stern and Z.S. Zam. *Cornea*, 5, 41-45 (1986).
19. M. DiGaetano, G.A. Stern and Z.S. Zam. *Cornea*, 5, 155-158 (1986).
20. S.I. Butrus, S.A. Klotz and R.P. Misra. *Ophthalmology*, 94, 1310-1314 (1987).
21. M.F. Mowrey-McKee, *et al.*. *CLAO J*, 18, 87-91 (1992).
22. G.E. Minno, L. Eckel, S. Groemminger, B. Minno and T. Wrzosek. *Optom Vis Sci*, 68, 865-872 (1991).
23. S.T. Lin, R.B. Mandell, C.D. Leahy and J.O. Newell. *CLAO J*, 17, 44-50 (1991).
24. C.D. Leahy, R.B. Mandell and S.T. Lin. *Optom Vis Sci*, 67, 504-511 (1990).
25. J.A. Durán, M.F. Refojo, I.K. Gipson and K.R. Kenyon. *Arch Ophthalmol*, 105, 106-109 (1987).
26. C.A. Lawin-Brüssel, M.F. Refojo, F.-L. Leong and K.R. Kenyon. *Invest Ophthalmol Vis Sci*, 32, 657-662 (1991).
27. S.F. Boles, M.F. Refojo and F.-L. Leong. *Cornea*, 11, 47-52 (1992).
28. M.F. Mowrey-McKee, H.J. Sampson and H.M. Proskin. *Contact Lens Assoc Ophthalmol J*, 18, 240-244 (1992).
29. C.V. Cagle. *Adhesive bonding* 1-306 (McGraw-Hill Book Company, New York, 1968).
30. C.A. Lawin-Brüssel, M.F. Refojo, F.-L. Leong, L. Hanninen and K.R. Kenyon. *Cornea*, 12, 10-18 (1993).
31. G.J. Sherman. US Patent. 4,356,100, (1982)
32. J.Z. Krezanoski. US Patent, 4.188.373, (1980).
33. J.Z. Krezanoski and J.C. Petricciani. US Patent, 3,954,644, (1976).
34. K. Holmberg, *et al.*. 201st Amer Chem Soc Nat Meeting: Division of Colloid and Surface Chemistry, Abstract #304 (Atlanta, Georgia, USA, 1991).
35. J.A. Hubbell and N.P. Desai. 201st Amer Chem Soc Nat Meeting: Division of Colloid and Surface Chemistry, Abstract #255 (Atlanta, Georgia, USA., 1991).
36. M.C. Griffiths, C.A. Fleeger and L.C. Miller. *United States Adopted Names and the United States Pharmacopeial dictionary of drugs names* (The United States Pharmacopeial Convention, Inc., Rockville, Md., 1983).
37. L.D. Hazlett, M. Moon and R.S. Berk. *Infect Immun*, 51, 687 689 (1986)
38. N. Panjwani, *et al.*. *Infect Immun*, 58, 114-118 (1990).
39. S.I. Butrus and S.A. Klotz. *Curr Eye Res*, 5, 745-750 (1986).
40. M.J. Bridgett, M.C. Davies and S.P. Denyer. *Biomaterials*, 13, 411-416 (1992).
41. L. Illum, I.M. Hunneyball and S.S. Davis. *Inter J Pharm*, 29, 53-65 (1986).
42. L. Illum, L.O. Jacobsen, R.H. Müller, E. Mak and S.S. Davis. *Biomaterials*, 8, 113-117 (1987).

# THE CLOSED-EYE CHALLENGE

Brien A. Holden, Kah Ooi Tan, and Robert A. Sack

Cornea and Contact Lens Research Unit, School of Optometry, and
Cooperative Research Centre for Eye Research and Technology
University of New South Wales
Sydney, Australia

Eye closure induces many changes in the precorneal environment. These environmental changes have been implicated as contributing factors in the development of acute and chronic changes in the morphology and physiology of the anterior eye. Recent research has suggested that eye closure in association with hydrogel contact lens extended wear (EW) is a major risk factor for acute ocular inflammation, culture-negative peripheral ulcers and serious corneal infections.

## THE CLOSED-EYE ENVIRONMENT

A question that has puzzled researchers for many years is whether the closed eye is a dry eye. In 1973 David Maurice[1] first asked where the tears go at night. Jules Baum suggested that clinical conditions such as recurrent epithelial erosions and the 'tight' contact lens syndrome may be due in part to a relatively dry eye during sleep.[2] However, attempts by Kurihashi[3] and others to measure the overnight decrease in tear volume have been inconclusive, although these results have suggested that there is a slight but inconsistent reduction in tear volume during sleep. The difficulty in measuring closed-eye tear volume is that after one or two blinks on eye opening the tear volume is fully restored.

Measurements of the biophysical characteristics of the precorneal environment, however, have been more conclusive. Holden and Sweeney measured the oxygen tension and temperature under the closed eyelid.[4] Holden *et al* also recorded accumulation of carbon dioxide following eye closure.[5] These studies have demonstrated that the environmental changes occurring in the closed eye include a rise in temperature, decreased oxygen tension, and accumulation of $CO_2$. A resultant acid shift in the stroma, from a pH of 7.54 in the open eye to 7.39 after eye closure, has been demonstrated by Bonanno and Polse,[6] and it is this stromal acidosis which is thought to cause the endothelial bleb response.[7] In view of the many environmental changes in the closed eye, it is therefore not surprising that prolonged eye closure is associated with many changes in corneal physiology, such as reduced corneal sensitivity,[8] loss of acetylcholine from the epithelium,[9] corneal ulceration in comatose patients,[10] endothelial polymegethism,[11] and contact lens-associated adverse responses.[12]

## CONTACT LENSES AND EYE CLOSURE

Perhaps the most succinct summary of the long-term effects of contact lenses worn under closed-eye conditions comes from the work of Sweeney and colleagues in the CCLRU Long-Term Wearers Study.[13] In this study, the ocular characteristics of 34 patients who had been wearing hydrogel lenses for EW for an average of 6.7 years were compared to 24 control patients. The major contact lens effects found in the study involved changes in epithelial structure and function. Increased numbers of microcysts and vacuoles, thinning of the epithelium, and reduced epithelial oxygen uptake rate all represent significant epithelial compromise, and are likely to be directly related to the increased risk of infection during hydrogel EW.

## CLOSED-EYE TEAR FILM

To investigate other environmental changes associated with eye closure, a number of studies have been conducted to evaluate changes in tear film composition in the closed eye. In a study reported by Sack, Tan and colleagues,[14] tears were collected from the outer canthus and analysed using a variety of techniques. Results indicated that when the eyes are closed, complement is activated and there is a substantial increase in secretory IgA and albumin levels compared to the open eye.

Another major finding from investigations of inflammatory mediators in the tear film is an increase in plasmin activity during sleep. Vannas and colleagues have demonstrated significant increases in plasmin activity immediately on eye opening, which took approximately 15 minutes to reduce to normal low open-eye levels.[15] Studies by Wilson, O'Leary and Holden have also shown that epithelial and PMN cells accumulate in the tear film during eye closure.[16] Some 6,500 PMN cells on average were found in corneal washings from nonlens-wearing subjects on awakening, compared to an average of 19 during the day.

Work by Tan, Sack and colleagues indicates that when the eyes are closed, concentrations of plasmin, secretory IgA and albumin, and epithelial cell numbers, increase steadily in the tear film.[17] Complement appears to be activated 2 to 3 hours after eye closure and PMN cell numbers increase very abruptly after 3 to 5 hours of eye closure, indicating possible recruitment of these cells by chemotactic factors which may be released by the conversion of complement to the active form. Eye closure appears to be the essential trigger for these tear film changes, as eye closure without sleep is also capable of inducing similar changes in tear composition.[18]

## BACTERIA IN THE CLOSED EYE

Our interest in bacterial contamination of contact lenses was stimulated by the observation, from unpublished studies by Grant and LaHood, that when overnight corneal edema is combined with bacterial lens contamination the incidence of acute inflammatory responses such as the acute red eye response increases dramatically. Sharma and colleagues have shown, in a group of nonlens-wearing subjects, that after 8 hours of sleep the incidence of significant numbers of bacteria in the conjunctival sac increases to about 45%, compared to approximately 2% in the open eye.[19] Studies of patient-worn lenses, conducted initially by Grant and colleagues and later by Chuck and Baleriola-Lucas,[20] indicate that bacteria can attach to lenses *in vivo*, and Chuck and colleagues (unpublished data) have also demonstrated that after 8 hours of closed-eye lens wear the number of organisms on the surface of contact lenses increases.

Routine sampling of patient-worn disposable contact lenses has shown that daily wear and extended wear contact lenses seem to accumulate similar levels of bacterial contamination.[20] However, because of the clear association between gram-negative bacterial contamination and acute red eye responses during EW,[21] it would appear that sleeping in contaminated lenses is the key problem.

## THE CLOSED-EYE CHALLENGE

Overall, these studies have demonstrated that eye closure leads to significant changes in the precorneal tear film which allow the proliferation or accumulation of bacteria, and the routine activation of complement and recruitment of PMN cells. We suspect that under certain circumstances systems for control of the sub-acute inflammatory response seen during normal eye closure can be disturbed by excessive numbers of bacteria on the surface of lenses. Sack *et al* have proposed that vitronectin in the closed-eye tear film may act to minimize bacterial toxin stimulation of the inflammatory cycle by down-regulating plasmin and complement activation, and terminating formation of the membrane attack complex, while enhancing the efficiency of phagocytic processing of entrapped micro-organisms.[22] Further basic research to investigate inflammatory pathways in the closed-eye tear film is clearly needed to elucidate these mechanisms.

In summary, it is clear that eye closure induces substantial changes in almost every aspect of the precorneal environment. If prolonged, these changes can lead to alterations in structure and function of the cornea, and the presence of significant bacterial contamination can lead to acute ocular inflammatory and ulcerative episodes. Until these issues are more fully understood and addressed, sleeping with contact lenses will continue to present a significant risk for ocular complications, and it is recommended that extended wear of hydrogel lenses should be avoided.

# REFERENCES

1.  D.M. Maurice, The dynamics and drainage of tears, *Int. Ophthalmol. Clin.* 13:103 (1973).

2.  J. Baum, Clinical implications of basal tear flow, *in:* "The Preocular Tear Film in Health, Disease, and Contact Lens Wear," F.J. Holly, ed., Dry Eye Institute, Lubbock, Texas (1986).

3.  K. Kurihashi, Diagnostic tests of lacrimal function using cotton thread, *in:* "The Preocular Tear Film in Health, Disease, and Contact Lens Wear," F.J. Holly, ed., Dry Eye Institute, Lubbock, Texas (1986).

4.  B.A. Holden, and D.F. Sweeney, The oxygen tension and temperature of the superior palpebral conjunctiva, *Acta Ophthalmol.* 63:100 (1985).

5.  B.A. Holden, R. Ross, and J. Jenkins, Hydrogel contact lenses impede carbon dioxide efflux from the human cornea, *Curr. Eye Res.* 6:1283 (1987).

6.  J.A. Bonanno, and K.A. Polse, Measurement of in vivo human corneal stromal pH: open and closed eyes, *Invest. Ophthalmol. Vis. Sci.* 28:522 (1987).

7.  B.A. Holden, L. Williams, and S.G. Zantos, The etiology of transient endothelial changes in the human cornea, *Invest. Ophthalmol. Vis. Sci.* 26:1354 (1985).

8.  M. Millodot, and D.J. O'Leary, Loss of corneal sensitivity with lid closure in humans, *Exp. Eye Res.* 29:417 (1979).

9.  J.S. Mindel, P.I.A. Szilagyi, J.A. Zadunaisky, T.W. Mittag, and J. Orellana, The effects of blepharorrhaphy induced depression of corneal cholinergic activity, *Exp. Eye Res.* 29: 463 (1979).

10. W.L. Hutton, and R.R. Sexton, Atypical Pseudomonas corneal ulcers in semicomatose patients, *Am. J. Ophthalmol.* 73:37 (1972).

11. J.P. Schoessler, and G.N. Orsborn, A theory of corneal endothelial polymegethism and aging, *Curr. Eye Res.* 6:301 (1987).

12. B.A. Holden, The Glenn A. Fry Award Lecture 1988: The ocular response to contact lens wear, *Optom. Vis. Sci.* 66:717 (1989).

13. D.F. Sweeney, C. Gauthier, R. Terry, M.S. Chong, and B.A. Holden, The effects of long-term contact lens wear on the anterior eye, *Invest. Ophthalmol. Vis. Sci.* 33(suppl):1293 (1992).

14. R.A. Sack, K.O. Tan, and A. Tan, Diurnal tear cycle: evidence for a nocturnal inflammatory constitutive tear fluid, *Invest. Ophthalmol. Vis. Sci.* 33:626 (1992).

15. A. Vannas, D.F. Sweeney, B.A. Holden, E. Sapyska, E.M. Salonen, and A. Vaheri, Tear plasmin activity with contact lens wear, *Curr. Eye Res.* 11:243 (1992).

16. G. Wilson, D.J. O'Leary, and B.A. Holden, Cell content of tears following overnight wear of a contact lens, *Curr. Eye Res.* 8:329 (1989).

17. K.O. Tan, R.A. Sack, and B.A. Holden, Temporal sequence of changes in tear protein composition following eye closure, *Invest. Ophthalmol. Vis. Sci.* 34(suppl):1468 (1993).

18. K.O. Tan, B.A. Holden, and R.A. Sack, Changes in tear protein composition: sleep versus closed eye, *Invest. Ophthalmol. Vis. Sci.* 33(suppl):1288 (1992).

19. B.A. Holden, S. Sharma, M. Reddy, K.R. Lakshmi, C. Fleming, N. Rao, and D.F. Sweeney, Eye closure increases the microbial challenge, *Invest. Ophthalmol. Vis. Sci.* 34(suppl):853 (1993).

20. C. Baleriola-Lucas, B.A. Holden, H. Gardner, C. Gauthier, and M.S. Chong, Habitual bacterial contamination among users of daily and extended wear hydrogel contact lenses, *Optom. Vis. Sci.* 68(suppl):75 (1991).

21. C. Baleriola-Lucas, T. Grant, J. Newton-Howes, D.F. Sweeney, and B.A. Holden, Enumeration and identification of bacteria on hydrogel lenses from asymptomatic patients and those experiencing adverse responses with extended wear, *Invest. Ophthalmol. Vis. Sci.* 32(suppl):739 (1991).

22. R.A. Sack, P.A. Underwood, K.O. Tan, H. Sutherland, and C.A. Morris, Vitronectin: possible contribution to the closed-eye external ocular host-defense mechanism, *Ocular Immunol Inflam*, in press (1993).

# DRY EYE THERAPY: EVALUATION OF CURRENT DIRECTIONS AND CLINICAL TRIALS

Mark B. Abelson[§] and Emma Knight[†]

[§]Department of Immunology, Schepens Eye Research Institute, Boston, MA; Ophthalmic Research Associates, N. Andover, MA; Department of Ophthalmology, Harvard Medical School, Boston, MA
[†]United States Food and Drug Administration

## INTRODUCTION

Beginning in the time of Hippocrates, complaints of dry eye were treated with a number of methods, including bathing the eyes in a variety of lubricating environments. Egg white, goose fat, and warm water were each commonly used to saturate the eyes in an effort to alleviate some or all of the discomfort brought on by a lack of adequate tears. In the 18th century, herbs were boiled in vinegar or wine and prescribed as elixirs to aid in the physiological tear manufacturing process. During this time, human tears were thought to be excrement of the brain itself. In the 19th century, oils and glycerins were used, and the first balanced salt solution was developed. Gelatin was often added to these solutions for viscosity to enhance the therapeutic effect in cases of severe dry eye syndrome. As gelatin proved to be both unstable as well as an excellent medium for microbial growth, methylcellulose was developed and today remains a major constituent of many tear substitutes. Polyvinyl alcohol and polyvinylpyrrolidone are hydrophilic substances which, in general, are less viscous and widely used by dry eye patients.

Today, the search for an ideal tear substitute continues. Artificial tears are currently the only available treatment for dry eye symptoms, and these remain the mainstay of therapy. These products have succeeded in enhancing the comfort of patients, although no formulation has shown significant improvement in the clinical parameters which determine dry eye. Understanding the mechanisms of dry eye and conducting well-controlled clinical trials are essential for the enhancement and development of dry eye therapy.

# DRY EYE SYNDROMES

Dry eye syndromes are common and comprise a heterogeneous group of diseases with many shared characteristics. Otherwise known as sicca syndrome, a dry eye condition exists when the quantity or quality of the tear film is not sufficient to maintain a healthy epithelium. The tear film consists of three layers: lipid, aqueous, and mucin.

The lipid layer is approximately 0.1$um$ thick and is secreted primarily by the meibomian glands, as well as the glands of Zeiss and Moll. The function of this outermost layer of the tear film is to prevent tear evaporation from the surface of the eye. Its polar and non-polar constituents consist of waxy, fatty lipids and cholesterol esters.

The aqueous layer tear film constitutes the middle layer of the tear film. Aqueous layer tear film components are produced by the lacrimal gland and its accessory organs, the glands of Kraus and Wolfring. Inorganic salts, glucose, urea, trace elements, lactoferrin, NAHCO$_3$ ions, IgA, LDH, and proteolytic inhibitors all dissolve in this layer, which also contains the lion's share of the oxygen supply to the cornea. Electrolytes and proteins are secreted with the aqueous layer, creating the plasma-like fluid of "hot eyes." The tear film is normally isotonic, while the tear film of dry eye patients is hypertonic. The tear film composition in reflex, open eyed patients and closed eyed patients varies. Components of reflex tear fluid include lysozyme, lactoferrin, and TSPA. Closed eye tear fluid, on the other hand, has a 40 fold increase in secretory IgA, the principle immunoglobulin present in tears, as well as measurable evidence of C3 activation.

The third layer of the tear film, the mucin layer, is produced by the goblet cells. Although microscopically much thinner than the aqueous layer, the mucin layer is extremely important in providing the hydrophilic coating for the normally hydrophobic corneal epithelium. The mucin layer lies adjacent to the ocular surface and its function is to provide continuity to the tear film by reducing the surface tension and insulating the aqueous layer from polar cell membranes. Through its high molecular weight and greater carbohydrate to protein ratio, mucin enables the tears to completely wet the cornea.

Although research has concentrated on the dwell time of tear substitutes, there is no available information on the interaction of the tear substitute with the epithelium and the tear film. The constituents of the tear film include glucose, electrolytes, metabolic enzymes, bacteriolytic enzymes, and other immunological components. In contrast, most commonly used tear substitutes contain polyvinyl alcohol (1.4-3%), purified water, polymeric vehicles which enhance dwell time, povidone, chlorobutanol, dextrose, carboxymethylcellulose, and a preservative. Other commonly used tear substitute components include white petrolatum, mineral oil, and lanolin in special vehicles such as Durasite® (InSite), Gelrite® (Merck), Lipiden® (Iolab), and Durasorb® (Alcon).[1]

## CLINICAL EVALUATIONS

Rose bengal staining is the clinical gold standard for the diagnosis of dry eye. This dye stains only devitalized or injured conjunctival epithelial cells, mucus, and filaments. Fluorescein staining can be used to assess damage to the corneal epithelial cells. Impression cytology is an experimental procedure which is purported to allow the clinician the ability to quantify the integrity of the epithelial cells. No significant decrease in rose bengal staining has been shown with the use of tear substitutes. Similarly, tear break-up time and non-invasive tear film interferotometry, both of which evaluate the stability of the tear film, are typically unaffected by the use of artificial tears. Measurements of tear secretion, such as Schirmer's and tear meniscus height, are relatively crude and notoriously variable. Use of topical anesthesia with Schirmer's is the oldest available test for dry eye syndrome. The lack of improvement in these tests may be related to other systemic or environmental factors which may confound the data.

## CURRENT DIRECTIONS IN THERAPY

Due to the lack of an efficacious formulation to remedy the common and chronic symptoms of dry eye syndrome, recent efforts have focused on more sensitive measurements and identification of tear film constituents for potentially new parameters to evaluate dry eye. Differential diagnosis of the dry eye patient has also come into play in an effort to identify each patient's individual insufficiency, be it mucoid, aqueous, or lipid in nature. True lipid deficiency is present only in the rare cases of anhydrotic ectodermal dysplasia, where the meibomian glands are congenitally absent. Lipid deficiency is most commonly seen in the changed meibomian gland secretion associated with infection. Bacterial invasion with subsequent lipase secretion leads to the breakdown of oily lipids into free fatty acids, which immediately rupture the otherwise intact tear film.

Aqueous deficient diseases are actually diseases of the main and accessory lacrimal gland. A diverse group of disorders exist which ultimately involve the lacrimal gland itself, such as congenital absence, trauma, auto-immune diseases, inflammation, tumor, and other neurological defects. All of these result in the decreased tear production. Mucin deficient diseases are manifested by any disease which permanently injures the conjunctiva, interfering with the vital function of the goblet cell. The future of therapy for the dry eye patient, in addition to identifying and treating the etiology of their symptoms, lies in the promotion of epithelial growth to ensure tear film stability. Some current ideas are to develop an agent which would bind harmlessly to the epithelium in a continuous layer, acting as a pseudo-mucin layer, or the development of an agent which would stimulate selective gland secretion and/or alter the effect of inflammation of the glands and the resulting loss of tear production.

## EVALUATION OF DRUGS FOR FDA APPROVAL

### Role Of The FDA

New drug applications as well as investigational new drugs for the treatment of dry eye syndrome require intricate research and clinical trials which ultimately involve the Food and Drug Administration (FDA). The Center for Drug Evaluation and Research (CDER) promotes, protects, and enhances the health of the public through drug development and evaluation which primarily serves two purposes: (1) approving drugs for marketing which are effective and have been adequately studied for adverse effects, providing benefits which outweigh any risks, and have directions for use which are complete, honestly communicated, and of high quality standards; and (2) evaluating marketing applications in a timely fashion, with special priority to drugs for serious illnesses which do not have current adequate therapy. CDER shares the goals of identifying new and promising drugs and their expeditious development with the regulatory industry, scientific groups, and the public.

Through collaborative efforts with industry, academia, and other domestic and foreign agencies and organizations, CDER assures that the safety and rights of all patients are adequately protected by monitoring product quality and maintaining product safety. CDER defines a drug product as an article intended for use in the diagnosis, cure, mitigation, treatment, or prevention of disease in man or other animals. Further definition includes any articles recognized in the Official United States Pharmacopeia, Official Homeopathic Pharmacopeia of the United States or the Official National Formulary, or any supplement to each or any of these publications. Drugs are also defined by CDER as any article or component of the article, other than food, intended to affect the structure or function of the body of man or other animals.

Currently approved therapies for dry eye include unit dose, single use preservative-free products as well as preserved, multi-dose tear substitutes. To qualify in the latter group, a product must demonstrate successful completion of preservative efficacy testing to confirm that the product does not support microbial growth for the period of expected use.

## Well-Controlled Clinical Trials

Appropriately designed, well-controlled clinical trials are vital to all areas of research, allowing the researcher to distinguish between the effect of a drug and other influences, such as spontaneous change in the course of a disease, placebo effect, or biased observation. The FDA divides clinical trials during drug development into three distinct stages, each designed to answer a different question. Phase 1 studies are required to demonstrate drug safety and at times drug kinetics. Although efficacy data may be gathered in these studies, it is not required. Phase 2 studies are used to establish a dose-response curve as well as to gather safety and efficacy data. Finally, Phase III studies are larger, adequate and well controlled trials to establish efficacy and demonstrate safety of products. In dry eye they are usually randomized, double masked, placebo-controlled efficacy studies. When appropriate, the outcome of these studies will ultimately determine the indication of the product, as well as the package labeling.

Prior to approval of any medication indicated for a chronic indication, the FDA currently requires at least one trial with 12 months of chronic dosing. A clinical efficacy trial for a dry eye product requires 6 weeks to 3 months of efficacy data in order to provide a primary basis for determining "substantial evidence" in support of the drug's claim for effectiveness.

## Protocol Development

At all stages of protocol development, the FDA encourages interchange and discussion between themselves and the sponsor of the clinical trial. In the evaluation and approval process of a drug for the treatment, cure, or palliation of dry eye syndrome, the FDA looks at certain parameters to ensure the quality of each clinical trial. A typical study includes a heterogenous population of patients representative of those expected to use the product after approval. Demographic factors including age, sex, and race should be considered. Any population excluded in the study will most likely be excluded in the package labeling and insert. The FDA also looks to the sponsor to consider other factors, including escape medication, concomitant drug therapies, and systemic safety concerns (drug effect on cardiac tissue, pulmonary function, or teratogenic effect on a fetus or pregnant woman). The FDA encourages the use of placebo, dose-comparison or active treatment controls. The historical or "no treatment controls" of years past are no longer considered impressive.

## Combination Products

If the sponsor wishes to demonstrate enhanced the effectiveness by combining two products, be it a new entity or a currently approved product, each component of the combination must demonstrate a contribution to the total effect of the drug. In addition, combination therapies should show statistical significance over individual components at the same concentration. The FDA requires justification of the presence of each inactive ingredient included in the formulation.

## Demonstration of Efficacy

In order to evaluate efficacy, the FDA utilizes both statistical and clinical data. All statistics should be carried out in a two-tailed manner to show whether the test product is better or worse than its control. The clinical efficacy parameters for any new drug should show progressive improvement in at least one subjective and one objective sign or symptom of dry eye syndrome, dependent upon the drug's indication and the etiology of the patient's disease. Symptoms of dry eye syndrome are most frequently subjective and described as one of the following: dryness, burning, photophobia, foreign body sensation, blurry vision, discomfort, environmental intolerance, tearing, or mucus formation. Clinical signs of dry eye syndrome are evaluated by discrete, measurable tests which allow the clinician the ability to determine the integrity of the tear film. Objective observations by the clinician include: absent or decreased tear production (measured most frequently by Schirmer's test), redness, corneal and conjunctival staining (measured by fluorescein and rose bengal staining), debris in the tear film, papilla, mucus plaque, filament formation, a dull conjunctival appearance, and tear break-up time. The Schirmer's test is often used, though it is well-accepted that a dry eye may have a normal Schirmer's test. Changes may be meaningful, but lack of change may not be indicative of poor efficacy.

## Conduct of Study

The FDA encourages the use of clinical correlates to these signs and looks for definite, measurable outcomes. Therefore, a dry eye study should include diagnostic results of fluorescein and rose bengal staining, Schirmer's tests, tear break-up time, tear film osmolarity, tear lysozyme levels, along with other cytologic and serologic testing. Examination of the dry eye subject should include, but not be limited to, visual acuity, slit lamp biomicroscopy, pupil size, intraocular pressure, and fundoscopic exam, depending on the individual protocol and medication being studied.

## Analysis of Data

Comfort and other subjective evaluations have historically been measured by a variety of scales. While there is no one acceptable, standardized scale, specificity to recognize change is recommended. The FDA looks for a scale to be both consistent and reliable. The more specific a scale is, the more consistent it will be between investigators and sites. It must be remembered that statistical significance is important, but it is useless without clinical significance. Therefore, all scales may be as sensitive as possible but must maintain clinical significance.

## FUTURE DIRECTIONS

Future dry eye syndrome research should focus on the lacrimal gland and the causes of its inflammation and dysfunction, as well as determination of the components necessary to enhance epithelial healing while maintaining an adequate tear film. Treatment of the individual dry eye patient must aim at differential diagnosis, since each etiology may require a distinct treatment. Thus, each therapy should be geared to the patient's insufficiency. Following close scrutiny of clinical trials and FDA approval of a product based on proven safety and efficacy variables, the tear substitutes of the future may also offer an attractive alternative for the delivery of all ophthalmic agents that is non-toxic to the surface epithelium and comfortable to the patient.

## ACKNOWLEDGEMENT

The mention of commercial products, their sources, or their use in connection with material reported herein is not to be construed as either an actual or implied endorsement of such products by the FDA or the Department of Health and Human Services (HHS).

# ARTIFICIAL TEAR ISSUES

Murray Sibley

Ross Laboratories
Columbus, Ohio 43215

## INTRODUCTION

The designing of artificial tear products has been undergoing constant reevaluation for many decades. Currently available artificial tear formulations differ remarkably with regard to their compositions. Today, significant controversies exist among pharmaceutical scientists in this field. Many formulation issues need to be addressed before agreement can be reached on the desirable attributes of truly efficacious artificial tear formulas.

## DESIGNING ARTIFICIAL TEAR PRODUCTS

This paper will place emphasis on the over-the-counter (OTC) ophthalmic drugs as defined by the FDA Ophthalmic Drug Monograph. This document lists those ingredients considered safe and effective for use in OTC ophthalmic drug products. Many other ingredients are possible but their inclusion in ophthalmic formulas would require New Drug Applications and three to five years for FDA approvals.

From the formulation aspect, the designing of artificial tear products is complex due to several sometimes contradictory requirements such as: (1) the products must be assuredly sterile in manufacturing, delivered sterile to the user and remain safe when opened and used; (2) the ingredients must be stable for two years or more to assure product performance.

## EARLY ARTIFICIAL TEAR CONCEPTS

The earliest approaches to artificial tear formulations were attempts to copy the composition of natural tears. Considerable R&D effort was conducted in analyzing natural tear components. However, the many attempts to produce such a product failed dismally due to problems with stability, sterility and compatibility.

The next approaches to artificial tear formulations were directed toward selecting a single ingredient as the "active" ingredient and building a compatible base around it. The most extensively investigated ingredients in this period were a variety of cellulose compounds - methyl cellulose, hydroxyethyl cellulose, carboxymethyl cellulose, etc. These early formulations were very unsatisfactory due to forming gummy coatings on eyes without providing sufficient relief.


## CURRENT DIRECTIONS

Very few improvements occurred until researchers realized that "key" elements of artificial tears must be defined and combined properly with emphasis on the benefits of the total formulation. Since determining these key elements is complex, it is not surprising that several approaches have been pursued.

Contrary to the beliefs of some, not all artificial tear formulations are the same. Significant compositional differences occur among currently available products - particularly in the way various key elements are combined.


TABLE I

| KEY ARTIFICIAL TEAR ISSUES |
| --- |
| • Selection of active ingredients |
| • Decisions on salt composition / osmolarity |
| • Selection of viscosity agents |
| • Choice of pH / buffering agents /buffering capacity |
| • Inclusion of other ingredients |
| • Exclusion or inclusion of preservatives |

Table I provides a very useful template of significant formulation decisions for developing a truly satisfactory modern artificial tear product. We shall briefly review each item as follows:

### Selection of Active Ingredients

The list of active ingredients allowed by the FDA Monograph is not very lengthy but contains some excellent candidates - namely, polyvinyl alcohol, polyvinyl pyrrolidone, celluloses, ocular emollients (mineral oil, petrolatum), and certain combinations of the above ingredients.

## Salt Composition / Osmolarity

Recent ophthalmic research has re-emphasized the importance of balanced electrolytes in artificial tear formulations. This inclusion of properly balanced amounts of sodium ($Na^+$), potassium ($K^+$), calcium ($Ca^{++}$) and magnesium ($Mg^{++}$) ions are essential.

In addition, one must select the proper osmolarity characteristics based upon desires to have the product be hypotonic, isotonic or hypertonic to natural tears.

## Viscosity Characteristics

A key element of an artificial tear product is its viscosity. A decision must be made regarding: (1) the ideal range of viscosity; and, (2) the choice of viscosity agents to use. Generally the most acceptable agents are one of the members of the cellulose family of: (a) HEC (hydroxyethyl cellulose); (b) HPMC (hydroxypropylmethyl cellulose); or (c) CMC (carboxymethyl cellulose).

## Choice of pH / Buffering Agents / Capacity

Four buffering systems have been regularly used for artificial tear products, as follows:
- borates
- phosphates
- citrates
- bicarbonates

The selection among these buffering systems depends upon the optimization of the pH range desired. One must carefully choose the concentration of buffering agents to assure sufficient capacity to provide a stable pH without over-burdening the natural buffering system of natural tears.

## Inclusion of Other Ingredients

In addition to the several important decision points discussed above, the formulator must consider the possible benefit of ingredients such as glucose, dextran and/or enzymes. Some currently available artificial tear products contain one or more of these ingredients and they may offer benefits.

## Preservatives - Exclusion / Inclusion / Selection

Presently, the most controversial topic in artificial tear formulation is the decision as to the exclusion or inclusion of preservatives. The commonly used preservatives are listed in Table II.

TABLE II

| ARTIFICIAL TEAR PRESERVATIVES |
|---|
| • *Benzalkonium chloride* |
| • *Sorbic Acid* |
| • *Chlorobutanol* |
| • *Parabens* |
| • *Thimerosal* |
| • *Polyquad* |

In recent years, the tendency has been away from the use of preservatives to minimize irritation and sensitivity reactions. Several unit-dose disposable ophthalmic delivery systems have appeared. Due to their relatively high cost, these unit-dose systems have met with limited commercial success. Therefore, an opportunity exists for a low cost, multi-dose delivery system for preservative-free ophthalmic products.

Overall, we must remember that the primary role of artificial tears is to relieve ocular irritation. In our attempts to optimize the many complex decisions leading to truly satisfactory artificial tear formulas, we have investigated special combinations of active ingredients with some success. Of particular interest was the combination of polyvinyl alcohol (PVA) and hydroxyethyl cellulose (HEC).

Trying various different ratios of these ingredients, we determined that a ratio of 3:1 HEC:PVA produced a noticeable synergistic improvement in the film forming ability of these agents, used independently. Additional experiments have suggested similar synergistic benefits with combinations of polyvinyl pyrrolidone (PVP) and HEC.

## The Future of Artificial Tear Products

It is clear that successful artificial tear formulations must provide a balance of essential natural ingredients as well as include appropriate film forming ingredients to provide tear stability. Optimization of the key elements of pH, buffering, buffer capacity, osmolarity and viscosity must be accomplished.

Ultimately, optimum artificial tear formulas must provide long lasting action in a soothing formula.

EVALUATION OF EFFECTS OF A PHYSIOLOGIC ARTIFICIAL TEAR
ON THE CORNEAL EPITHELIAL BARRIER: ELECTRICAL
RESISTANCE AND CARBOXYFLUORESCEIN PERMEABILITY

John L. Ubels[1], K. Keven Williams[2]
Dolores Lopez Bernal[3] and Henry F. Edelhauser[2]

[1]The Eye and Ear Institute and University of Pittsburgh
Pittsburgh, PA

[2]Department of Ophthalmology
Emory University
Atlanta, GA

[3]University of Murcia
Murcia, Spain

INTRODUCTION

Artificial tear solutions are widely used for relief of ocular surface irritation by patients with dry eye syndromes and by persons with ocular discomfort due to climatic conditions, air pollutants and various other causes. Although about 200,000 persons in the United States are classified as having a severe dry eye, nearly 5 million people report using dry eye protectants, resulting in the purchase of over 10 million units of ethical over-the-counter artificial tears worth nearly $80 million dollars per year[1]. This has resulted in a highly competitive market for these products with at least 25 different solutions currently on the market in the United States[2] and new ones appearing frequently.

These tear solutions vary widely in composition. Some, notably hypotonic solutions, contain no electrolytes while sodium chloride and potassium chloride at various concentrations are common components. Additional ions, including calcium, magnesium and zinc, are present in some products and formulations based on Ringer's solution are available. The viscosity agents used in tear formulations are numerous, including hydroxypropyl methylcellulose, carboxymethylcellulose, dextran, polyvinyl alcohol, polyethylene glycol and glycerine. Buffers used in tear solutions include borate, citrate and phosphate. Some products include nutrients such as dextrose and vitamins but these are of little value. Finally, since these formulations are usually packaged in multidose dispensers, a preservative is required, commonly benzalkonium chloride (BAK), sorbic acid, chlorbutanol, polyquaternium 1, and EDTA. The potential toxicity of these preservatives, especially BAK[3,4] has long been recognized leading to the recent trend toward the production of non-preserved solutions packaged in sterile, unidose dispensers. Although some manufacturers continue to use EDTA this chelator causes stinging when applied to the eye and has been eliminated from some products.

In occasional use for intermittent irritation or in mild dry eye conditions most products are probably innocuous and can provide temporary relief of symptoms. Frequent application of an inappropriate tear formulation to a moderate to severe dry eye can, however, potentially exacerbate ocular surface disease. It would seem that the most rational tear for use in such patients would resemble human tears, although given the plethora of product formulations available, it is clear that this concept has often been ignored. While it would be difficult to replace the meibomian lipids or to duplicate the mucin layer, a logical first step in designing a tear solution would be to as closely as possible match the electrolyte composition of human tears. The effect of this approach would be to more effectively replace the tear fluid in aqueous deficiency conditions or, in primary ocular surface diseases which lead to tear film instability, provide by use of an appropriate solution, conditions compatible with recovery of ocular surface function.

There are many reports in the literature on the ionic composition of human tears. Few of these studies systematically analyzed the entire makeup of the tears and, as reviewed by Van Hearingen[5], a relatively wide range of values has been reported. In a previously published study we undertook a comparative study of rabbit and human tear electrolyte composition[6]. The results of the human component of this study are summarized in Table 1. Notable characteristics of human tear electrolytes are a potassium concentration more than three times that in plasma, a bicarbonate concentration about half the level in plasma and the presence of zinc, which is an important enzyme cofactor.

Based on the above study we developed a physiologic artificial tear solution the composition of which is shown in Table 1. This solution is also similar to an electrolyte solution used in studies of corneal epithelial cell sloughing by Wilson and co-workers[7,8]. In these studies they showed that potassium is required to prevent excess cell loss from the epithelium. The physiologically elevated tear levels of potassium have, however, been ignored in the formulation of many tear products. Bicarbonate was included in the physiologic tear formula because it is the natural buffer in tears and, as will be shown, it may be an essential component of an effective artificial tear. The alkaline pH of tears has also been matched in this formulation. Keller et al[9] have shown that acidic conditions increase the permeability of the corneal epithelium which raises questions concerning the relatively low pH (6.6) of tear formulas based on lactated Ringer's solution. Finally, an osmolality of 290 mOsm/kg was chosen to be near that of the tears. Because tears in keratoconjunctivitis sicca are hypertonic[10], low osmolality solutions have been promoted based on the rationale that such solutions might bring the osmolality of tears to normal levels. We have recently observed that these low osmolarity solutions may be of no advantage in the treatment of dry eye[11].

Artificial tear solutions that contain components permitted by the U.S. Food and Drug Administration monograph for these products are considered to be safe and may be marketed without testing. Therefore, beyond comfort studies very few tear products have been tested for efficacy in treatment of dry eye. Various components of tear solutions, especially preservatives, have been studied using methods such as electron

TABLE 1. Composition of Human Tears and Artificial Tear Solutions (mM/I)

| | AQA 5 * | AQA 20 | Phys Tear ‡ | Phys Tear +polymer ‡ | Human** Tear |
|---|---|---|---|---|---|
| $Na^+$ | 144 | 132 | 112.9 | 112.9 | 128.7 |
| $K^+$ | 4.9 | 18 | 17.4 | 17.4 | 17 |
| $Cl^+$ | 103.4 | 103.4 | 132.6 | 132.6 | 141.3 |
| $Ca^{++}$ | 1.4 | 1.4 | 0.36 | 0.36 | 0.32 |
| $Mg^{++}$ | 0.6 | 0.6 | 0.31 | 0.31 | 0.35 |
| $HPO_4^-$ | 0.6 | 0.6 | - | - | - |
| Glucose | 26 | 26 | - | - | - |
| $SO_4^{--}$ | 7.5 | 7.5 | - | - | - |
| $HCO_3^-$ | 20 | 20 | 11.9 | 11.9 | 12.4 |
| Gluconate | 2.8 | 2.8 | - | - | - |
| $Zn^{++}$ | - | - | $11\mu M$ | $11\mu M$ | $11\mu M$ |
| HEPES | 25 | 25 | - | - | - |
| Dextran 70 | - | - | - | 0.1% | - |
| HPMC | - | - | - | 0.3% | - |
| pH | 7.8 | 7.8 | 7.7 | 7.7 | $7.7\pm0.1$ |
| mOsm | 305 | 305 | 290 | 290 | 302 |

* Ref. 18
** Reference 6
‡ Prepared by Alcon Laboratories, Inc. Fort Worth, Texas.

microscopy[3], electrophysiology[12], and cytotoxicity in cell culture to study effects on corneal epithelium[13,14]. Gilbard et al has also tested tear solutions by analysis of conjunctival goblet cells[15]. Many of these studies emphasize the disruption of the corneal epithelial barrier that can occur when it is exposed to preservatives. The corneal epithelium forms a semipermeable barrier between the environment and the inside of the eye by virtue of the tight junctions in the superficial cell layers. Although ion transport by the corneal epithelium may play a minor role in corneal deturgescence, the maintenance of the barrier is probably the primary function of the corneal epithelium. Disruption of the barrier can lead to corneal edema and infection; this is particularly relevant to the problem of dry eye because Gobbels and Spitznas have reported that corneal epithelial permeability of sodium fluorescein is three times higher than normal in dry eye patients[16]. To evaluate the physiologic tear solution described above , we have therefore studied the effects of this solution on the corneal epithelial barrier by measurement of epithelial electrical resistance and by measurement of corneal uptake of 5,6 carboxyfluorescein.

### Effects of the Physiological Artificial Tear Solution on the Electrophysiology of the Corneal Epithelium

It has been shown that the epithelium of the cornea generates a potential difference of 20 plus millivolts negative to the tear side. In addition it has been shown that damage to the outer epithelial layer, which accounts for approximately 60% of corneal resistance, will decrease or eliminate this potential.[17] Therefore, monitoring the electrical properties of the cornea provides a good indication of the integrity of this outer epithelial layer. In 1977, Burnstein and Klyce used an in vitro Ussing chamber to measure the effects of different ophthalmic preparations on the electrical properties of the cornea.[12] They concluded that low concentrations of preservatives in ophthalmic preparations disrupt the barrier and transport properties of the corneal epithelium. The current study utilized a similar method to evaluate the effects of a physiological tear on the electrical properties, potential (Pd), short circuit current (SCC), and resistance (ohms cm$^2$) of the rabbit corneal epithelium.

NZW rabbits weighing 5 kg. were anesthetized and then euthanized by either intravenous or intracardiac overdose with pentobarbital. During the anesthesia period, the eyelids were taped shut to avoid corneal dehydration. After euthanasia, the eyes were enucleated and mounted atraumatically in the Ussing chambers using the vacuum ring mounting technique described by Klyce.[18] The chambers used for this experiment were manufactured by Jim's Instrument Mfg. Inc. (Iowa City, Iowa). Special inserts for the chambers were designed to meet the radius of curvature of the cornea. After removal of the lens iris and ciliary body, the hemichambers were gently reassembled and each side was filled with an AQA 5 Ringers solution as described in table 1. Electrical measurement were made using agar bridges in polyethylene tubing containing 0.9% NaCl connecting the chambers to electrode wells containing calomel half cells. Pd, SCC and resistance were recorded using a WPI DVC-1000 (World Precision Instruments, Inc. Sarasota, Fla.) dual voltage clamp and Kipp & Zonen chart recorders (Holland). Corneas were allowed to equilibrate electrically in the AQA 5 Ringers solution and were then voltage clamped in order to measure initial SCC and resistance. The voltage clamp was then released and the solution on both sides of the cornea was replaced by the solution to be tested. The corneal Pd was allowed to equilibrate then the SCC and ohms cm$^2$ were again measured. After thirty minutes exposure, the test solution was once again replaced with AQA. The potential was allowed to equilibrate a third time and the corneal SCC and ohms cm$^2$ were measured. An example of a typical chart recording is shown in figure 1. The test solutions

444

included in this study were as follows; 1) bicarbonate buffered physiological tear with ion composition similar to natural tears, 2) Bicarbonate buffered tears with the same ion composition as number 1 with polymers Hydroxypropyl methylcellulose and Dextran 70 added, 3) AQA 20 Ringers solution with a $K^+$ concentration (20 mM) to equal that of tears (table 1). Following electrophysiologic measurement, the corneas were carefully removed from the chambers and placed in a fixative solution formulated according to Nichols et al[19] to preserve the mucinous layer of the cornea. The tissues were then routinely processed for TEM and then compared against control corneas for any changes in the mucin layer.

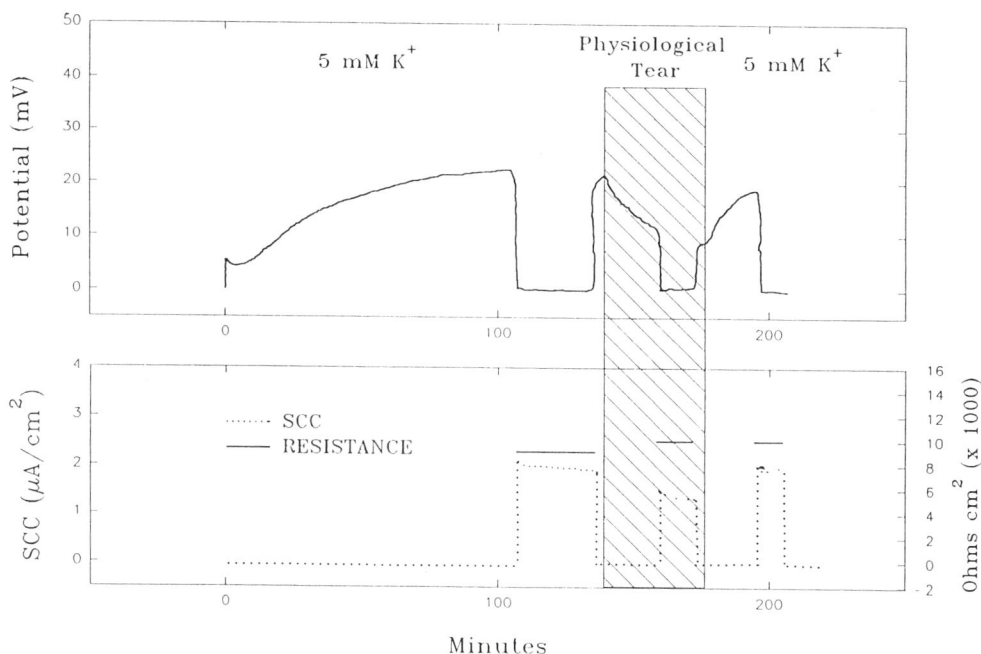

FIGURE 1. Chart recording showing the effects of a physiological tear solution on corneal epithelial potential, short circuit current and resistance.

Results of this study showed that all three tear preparations caused fluctuations in the electrical properties of the corneal epithelium. The initial epithelial potentials ranged from 20 to 40 millivolts when corneas were bathed with the AQA Ringers. Replacement of the Ringers with either of the two artificial tear formulations caused a decrease in the potential while the 20 mM $K^+$ AQA had very little effect. (fig. 2) The short circuit current decreased during exposure to all three preparations (fig. 3). When the tear preparations were replaced with the normal AQA Ringers, the values for both Pd and SCC increased to near baseline levels (these differences were not significant). In all three cases there was a slight but not statistically significant drop off in the resistance (fig. 4) which recovered after the replacement of the tear solution with AQA.

Electron microscopy examination of the corneas exposed to the tear formulations as well the 20 mM $K^+$ AQA showed no difference in the integrity of the mucin layer as compared with the controls.

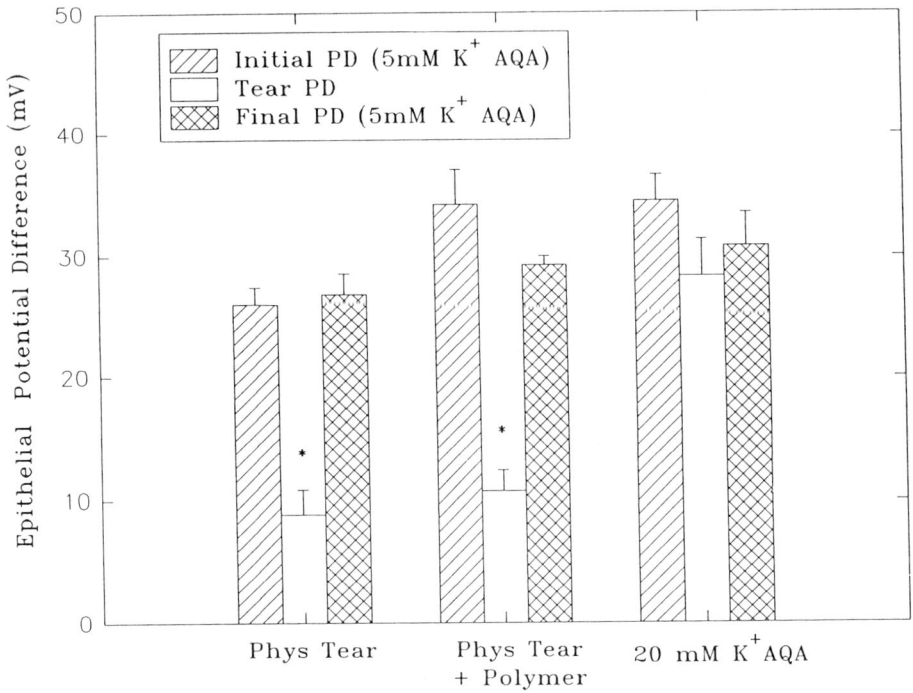

FIGURE 2. Comparison of rabbit corneal epithelial potential difference when exposed to physiological tear solution with polymer, without polymer, and AQA 20. (n=5 ± SEM)
* P < 0.01

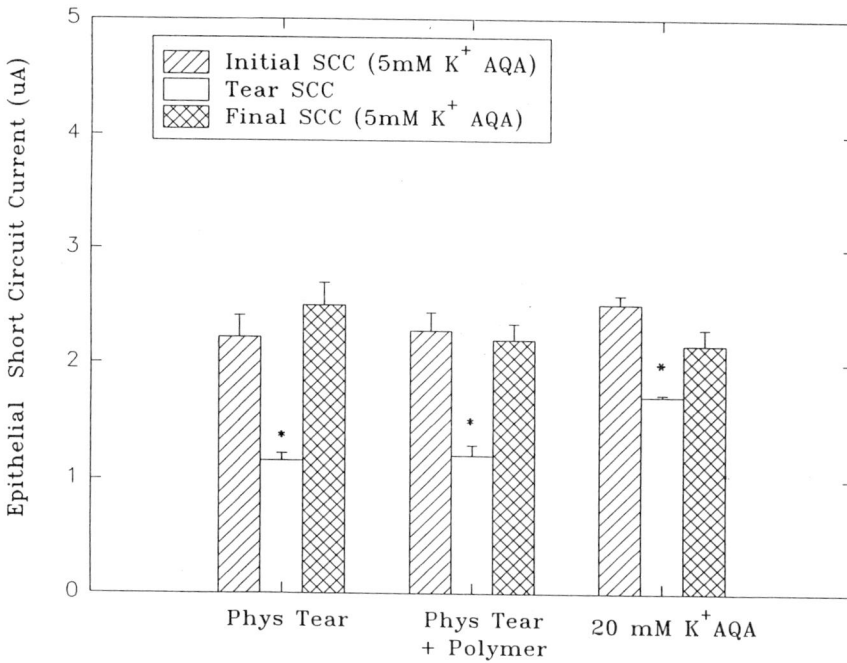

FIGURE 3. Comparison of rabbit corneal epithelial short circuit current when exposed to physiological tear solution with polymer, without polymer, and AQA 20. (n=5 $\pm$ SEM)
$^*P < 0.01$

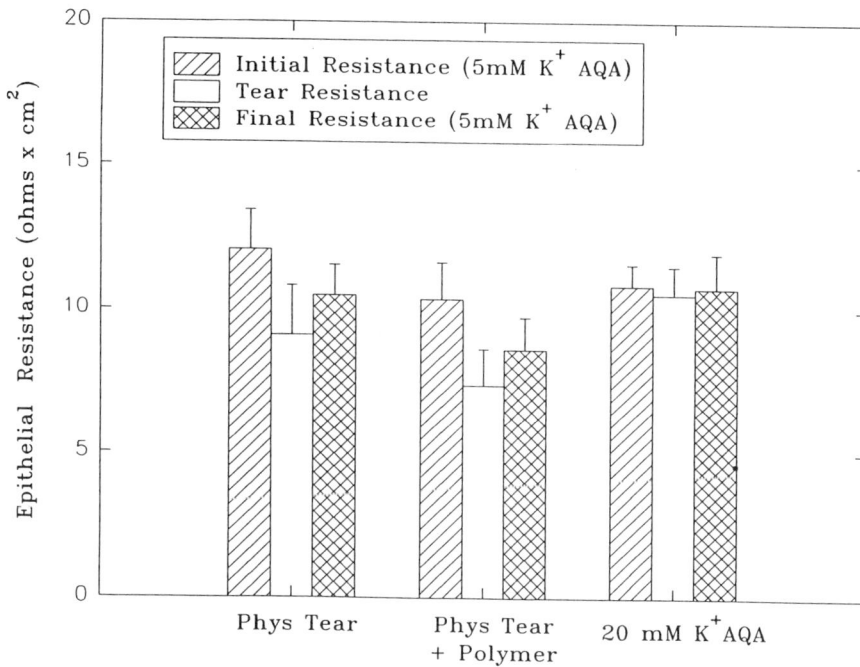

FIGURE 4. Comparison of rabbit corneal epithelial resistance when exposed to physiological tear solution with polymer, without polymer, and AQA 20. (n=5 $\pm$ SEM)

# Effects of the Physiologic Tear Solution on Corneal Uptake of 5,6-Carboxyfluorescein

To test the physiologic tear solution in vivo using an animal model, corneal uptake of 5,6-carboxyfluorescein (CF) was used to measure the integrity of the corneal epithelial barrier. Although sodium fluorescein has been commonly used for measurements of corneal epithelial permeability[12,16] the 5,6-carboxy form of this dye does not penetrate cell membranes and therefore can only pass through pericellular spaces. The methods used for the studies described here have been described in detail elsewhere[11,20]. Briefly, the cornea of an anesthetized rabbit is exposed to an artificial tear solution for 1.5 hr using a conjunctival cup. The contralateral eye is taped shut and serves as a control. The animal is then euthanized and the eyes are enucleated and inverted on a small beaker containing $2.7 \times 10^{-3}$ M CF in balanced salt solution (BSS). After 5 min exposure to CF, with stirring, the corneas are removed from the eyes, weighed and dialyzed in 20 ml BSS for 48 hr. The CF concentration of the dialysate is measured in a fluorometer and total corneal uptake of CF is expressed as nmoles/g.

In a previously published study we investigated the effects of various commercial and experimental artificial tear solutions on normal rabbit corneas[20]. In agreement with previous studies we showed that prolonged exposure to solutions containing

TABLE 2. Composition of Tear Solutions

| Sol | $Na^+$ (mM) | $K^+$ (mM) | Polymers | Other | Osmolarity mosm/kg |
|-----|-----|-----|-----|-----|-----|
| A | 4 | 0 | PVA, PEG | EDTA dextrose | 230 |
| B | 186.5 | 0 | PEG, Dextran 70 Polycarbophil | EDTA | 240 |
| C | 130 | 4 | CMC | $CaCl_2$ lactate | 306 |
| D | 100 | 29 | Glycerin | $CaCl_2$, $MgCl_2$ $ZnCl_2$, Phosphate Citrate | 260 |
| E | 130 | 16 | Dextran 70 HPMC | Borate | 290 |

Na, K and osmolality measured by investigators. Other information from package inserts.

PVA, polyvinyl alcohol; PEG, polyethylene glycol; HPMC, hydroxypropylmethyl cellulose; CMC, carboxymethylcellulose preservatives can damage the cornea, with BAK causing significantly more CF uptake than polyquaternium-1 or thimerosal. It was also demonstrated that non-preserved solutions, including the physiologic tear, do not damage the epithelium. These experiments were conducted using normal eyes, however, artificial tears are often applied to eyes with some degree of ocular surface disease. Experiments were therefore designed to test the ability of several commercial (Table 2) and experimental, non-preserved tear solutions to provide an environment in which the damaged corneal epithelium can recover normal barrier function. To accomplish this rabbit corneas were exposed to 0.01% BAK in BSS for 5 min using the method described above, followed by measurement of CF uptake. Exposure to BAK caused a 6.6-fold increase in CF uptake compared to control (Table 3). If animals were allowed to recover for 1.5 hr before measurement of CF uptake no improvement in barrier function was detected. Artificial tears were then tested by exposing corneas to BAK for 5 min followed by a 1.5 hr exposure to the tear solution using the conjunctival cup. Corneal CF uptake was then measured to determine whether recovery of the barrier had occurred. Treated eyes were compared to paired controls by the paired t-test and treatments were compared to one another by analysis of variance ($p \leq 0.05$).

TABLE 3. Carboxyfluorescein Uptake by Corneas Exposed to Artificial Tear Solutions

| Solution | Treated | Control[†] |
|---|---|---|
| BAK-5 min (0.1% in BSS) | 17.7 ± 2.32[*] | 2.7 ± 1.51 |
| A | 12.5 ± 1.29[*] | 2.0 ± 1.26 |
| B | 30.1 ± 1.28[*#] | 2.9 ± 0.68 |
| C | 10.9 ± 1.11[*#] | 3.9 ± 0.92 |
| D | 8.1 ± 1.97[*#] | 1.31 ± 0.27 |
| E | 6.7 ± 0.75[*#] | 1.4 ± 0.2 |
| PT | 3.7 ± 1.13[#] | 2.6 ± 0.71 |
| PT (no buffer) | 16.6 ± 1.49[*] | 0.73 ± 0.11 |
| PT - borate | 13.2 ± 1.47[*] | 0.94 ± 0.21 |

Values represent nmoles CF/g cornea, $\bar{x}$ ± SE, n=5
[*] Different than paired control.
[#] Different than BAK.
[†] Control values do not differ.
PT, Physiologic Tear
Corneas treated with tear solutions for 1.5 hr following 5 min BAK exposure.
2

Neither solution A or B allowed recovery of the barrier as CF uptake was not different than that of corneas exposed to BAK alone (Table 3). Both of these solutions are hypotonic and contain EDTA which chelates calcium, possibly interfering with tight junction formation. There was a significant reduction in CF uptake during exposure to commercial solutions C, D or E indicating some recovery of barrier function but these values did not reach control levels. The ineffectiveness of solution C may be related to the low pH of this solution (6.6). It should also be noted that solutions D and E both contain elevated levels of potassium.

In contrast to the above results, during exposure to the physiologic tear solution (PT)(Table 1) corneal CF uptake returned to control levels demonstrating that this solution provides conditions compatible with recovery of the epithelial barrier (Table 3). The unique feature of this formulation compared to all other solutions tested is the presence of the bicarbonate buffer. Because of the special packaging requirements that would be involved in the production of a stable product containing bicarbonate it was necessary to determine if the bicarbonate is an essential component of the solution. Therefore two other versions of the physiologic tear were tested, one without buffer and the other in which bicarbonate was replaced with borate. As shown in Table 3, neither of these solutions permitted a decrease in CF uptake to control levels. It is concluded that the physiologic tear solution is superior to the other solutions tested and may prove to be effective in reducing the epithelial permeability of corneas of dry eye patients.

## DISCUSSION

In the case of severe dry eye, persons may instill artificial tears up to several times an hour to avoid eye discomfort. In these patients it is imperative that the formulation of the artificial tear can be well tolerated by the eye. This excludes the use of preservatives as their damaging effects to the sensitive ocular tissues have been confirmed in numerous studies. This study has demonstrated that the use of non preserved tear solutions will not cause an increase in the permeability of the corneal epithelium as demonstrated by the lack of carboxyfluorescein uptake and the maintenance of electrical resistance after exposure to these solutions. It was also demonstrated in this study that tear solutions which are physiological in nature allow full recovery of epithelial barrier function when the cornea has been chemically damaged by BAK. In the case of the non physiologic formulations, less than full recovery in the same amount of time was observed. The results suggest that a physiologic tear containing bicarbonate provides an environment for the corneal surface in which tight junctions may reform resulting in the recovery of the epithelial barrier. The data also show that artificial tear solutions that lack elevated $K^+$ levels and contain EDTA are less effective in providing such an environment. Previously Lopez Bernal and Ubels reported that the epithelium was still resistant to the uptake of ruthenium red stain into the intercellular spaces after exposure to the non preserved tears whereas tear solutions containing preservatives caused an increase in ruthenium red uptake. This shows a good correlation with the carboxyfluorescein uptake data generated in this study.

The electrophysiological studies have provided data to support the importance of a physiological tear solution's ability to maintain the epithelial resistance, potential, and SCC. Both physiological tears decreased the corneal potential and SCC, however the epithelial resistance remained unchanged. The decrease in potential and SCC is most likely related to the 29 mM higher concentration of $Cl^-$ in the physiological tear when compared to the AQA 5. The corneal epithelium is a $Cl^-$ transporting tissue[18] and a lower Pd and SCC may be related to the diffusion of $Cl^-$ and $K^+$ from the test media into the corneal stroma across the endothelium (cornea is bathed on both sides with the test solution). The corneal epithelium also showed a decreased SCC with the AQA 20 (18 mM $K^+$) without a change in Pd or resistance which would be consistent with a $Cl^-$ transporting tissue. The major point of these electrophysiological measurements is that bathing the epithelial surface with a physiological tear containing $HCO_3$ and 17 mM $K^+$ with and without polymer for 30 - 60 minutes maintained the epithelial resistance.

Based on these studies it is apparent that the best artificial tear solution to protect the corneal epithelial tight junctions is one that contains ions similar to natural tears and is buffered with bicarbonate. When experimental corneas are exposed to this solution, the epithelial barrier can be maintained for an extended period of time. The epithelial barrier may also be re-established if the cornea has been damaged with BAK.

## ACKNOWLEDGEMENTS

Supported in part by a departmental Core Grant from NIH P30 EY06360, Alcon Laboratories, Inc. and Research to Prevent Blindness, Inc.

## REFERENCES

1. J.D. Nelson and J.L. Ubels. Dry-Eye Syndromes: A Handbook for Diagnosis and Management. Alcon Laboratories, Inc., Fort Worth (1991).

2. Artificial tear solutions and ocular lubricants, in: "Ophthalmic Drug Facts", J.D. Bartlett, N.R. Ghormley, S.D. Jaanus, J.J. Rowsey and T.J. Zimmerman, ed. Facts and Comparisons, Inc., St. Louis, pp 66-68 (1992).

3. R.R. Pfister and N. Burstein. Invest. Ophthalmol. 15:246 (1975).

4. N.L. Burstein. Invest. Ophthalmol. Vis. Sci. 25:1453 (1984).

5. N.J. Van Haeringen. Surv. Ophthalmol. 26:84 (1981).

6. V. Rismondo, T.B. Osgood, P. Leering, M.G. Hattenhauer, J.L. Ubels and H.F. Edelhauser. CLAO J. 15:222 (1989).

7. W.G. Bachman and G. Wilson. Invest. Ophthalmol. Vis. Sci. 26:1484. (1985).

8. R.J. Fullard and G.S. Wilson. Curr. Eye Res. 5:847 (1986).

9. N. Keller, D. Moore, D. Carper and A. Longwell. Exp. Eye Res. 30:203 (1980).

10. J.P. Gilbard, R.L. Farris and J. Santamaria. Arch. Ophthalmol. 96:677 (1978).

11. D. Lopez Bernal and J.L. Ubels. Cornea 12:115-120 (1993).

12. N.L. Burstein and S.D. Klyce. Invest. Ophthalmol. Vis. Sci. 16:899 (1977).

13. P.S. Imperia, H.M. Lazarus, R.E. Botti, Jr., and J.H. Lass. J. Toxicol.-Cut. Ocular Toxicol. 5:309 (1986).

14. J. Adams, M.J. Wilcox, M.D. Trousdale, D-S Chien, and R.W. Shimigu. Cornea 11:234 (1992).

15. J.P. Gilbard and S.R. Rossi. Ophthalmology 99:600 (1992).

16. M. Gobbels and M. Spitznas. Graefes Arch. Clin. Exp. Ophthalmol. 229:345 (1991).

17. D.M. Maurice. Exptl. Eye Res. 6:138-140 (1967).

18. S.D. Klyce, A.H. Neufeld, and J.N. Zadunaisky. Invest. Ophthalmol. 12:127-139 (1973).

19. B.A. Nichols, M.L. Chiappino, and C.R. Dawson. Invest. Ophthalmol. Vis. Sci. 26:464-473 (1985).

20. D. Lopez Bernal and J.L. Ubels. Curr. Eye Res. 10:645 (1991).

# EVALUATION OF A PHYSIOLOGICAL TEAR SUBSTITUTE IN PATIENTS WITH KERATOCONJUNCTIVITIS SICCA

J. Daniel Nelson,[1]  Margaret M. Drake,[2] James T. Brewer, Jr.[2]
and Michael Tuley[2]

[1]Departments of Ophthalmology at Ramsey Clinic
St. Paul-Ramsey Medical Center
640 Jackson Street, St. Paul MN 55101
The University of Minnesota, Minneapolis, MN
[2]Clinical Sciences, Alcon Laboratories, Inc., Fort Worth TX

## INTRODUCTION

In the absence of any definitive cure for dry eye disorders, artificial tears remain the only therapeutic choice. These lubricating products are marketed for relief of symptoms of dry eye. However, none are currently allowed for use on the basis of their amelioration of objective signs of dry eye.

Formulations which contain viscosity agents such as hydroxymethylcellulose,[1] and hyaluronic acid[2-6] have been reported effective in treating symptoms, and in prolonging fluid residence time, but have not been found to improve objective findings. Carboxymethylcellulose has been reported to increase fluid residence time, and in one comparative study, to provide some objective benefit.[7,8]

Significant improvement in ocular surface damage has been shown with chronic administration of an electrolyte solution in rabbits with experimentally induced dry eye and in preliminary clinical studies.[9,10] It appears that specific ion concentrations were needed for these positive effects to occur. A physiological tear formulation was also reported to ameliorate signs of preservative-induced corneal toxicity in rabbits.[11]

These findings have led to the current investigation of a physiological tear formulation containing the electrolytes found in human tears. A solution which resembles human tears in ionic composition, including the presence of a bicarbonate buffer, should be beneficial in treating patients with aqueous tear deficiency due to keratoconjunctivitis sicca (KCS). The objective of this study was to evaluate and compare the objective and subjective effects of two such formulations, one with bicarbonate and one without, in subjects with moderate to severe dry eye disorders.

## MATERIALS AND METHODS

Diagnostic tests used both prior to, and during treatment with the artificial tear products included: a Schirmer test, rose bengal staining, and conjunctival impression cytology (CIC). The Schirmer test was performed unanesthetized over a five minute period. Rose bengal staining was performed by applying 2 µL of a 1% solution of rose bengal to the inferior marginal tear strip. The eye was examined with green-filtered light and the degree of staining on the temporal and nasal conjunctiva and the cornea scored on a scale of 0 to 3. The scores for each area were summed, providing a total possible score of 9.[12] CIC was performed as previously described.[13] Circular strips of cellulose acetate filter material, 6.2 mm in diameter, were used. Specimens were obtained from the temporal, nasal and superior bulbar conjunctiva and the inferior palpebral conjunctiva. Each specimen was stained, examined microscopically and assigned a grade (0 to 3) based upon the degree of squamous metaplasia present. For each eye, the bulbar grades were summed (total possible score 9) and the palpebral grade recorded separately. Patient symptoms were scored on a 0 (none) to 3 (severe) scale.

For entry into the study, patients were required to have: 1) a Schirmer value of less than 5 mm; 2) a rose bengal score of > 3 out of 9; and 3) objective evidence of ocular surface disease as defined by a bulbar conjunctival impression cytology score of > 3 out of 9, and a palpebral conjunctival impression cytology score of < 1 out of 3. Excluded from entry were patients who were current contact lens wearers, had ocular surgery within the past six months, and who had active ocular inflammation or corneal disease (other than related to KCS). If patients were using systemic medications, they were required to remain unchanged throughout the study. All patients provided written informed consent.

Patients were examined one week prior to study entry, and on Days 0, 7, 28 and 56. Treatment was one to two drops, O.U., every one to two hours as required. Examinations included visual acuity, ocular symptoms, and slit lamp examination. Within-group comparisons were made using the Sign Rank test. A critical p value of $\leq 0.05$ (two-tail) was used.

Two formulations were evaluated, each of which were packaged in unpreserved, unit dose containers, at pH 7.4, containing hydroxypropylmethylcellulose (HPMC). The products contained the following components of human tears at a concentration at or near physiological: calcium, magnesium, sodium, potassium, and zinc. Formulation #1 also contained bicarbonate.

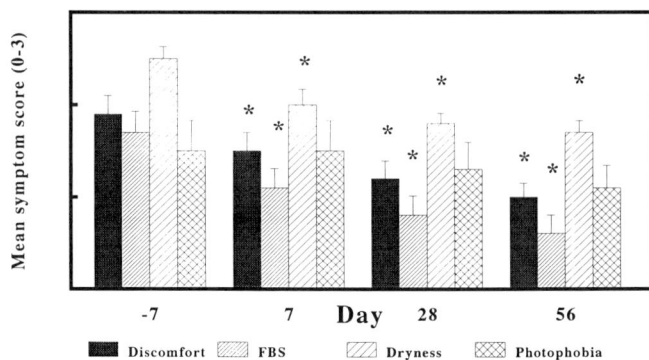

Figure 1. Mean severity (± s.e.m.) for patient symptoms for Formulation #1 (n = 14).

## RESULTS

In Study #1, Formulation #1 was evaluated in 14 female Caucasian patients. The mean age was 52 ± 14 years, and ranged from 26 to 71 years, and the mean duration of disease was ten years. Prior dry eye therapy was unpreserved artificial tear preparations alone, or in combination with ophthalmic ointments or Lacriserts®. No patients were using preserved products prior to study entry.

The effect of treatment on patient symptoms is shown in Figure 1. There was a decrease of up to 1 severity grade in all four solicited symptoms: discomfort, foreign body sensation, dryness and photophobia. For all but photophobia, these decreases were statistically significant.

The effect of treatment on objective measures is shown in Figure 2. The severity of rose bengal staining decreased by approximately one grade at the first follow-up visit, and remained decreased throughout the 56 day treatment period (p = 0.003 to 0.015). At the 56

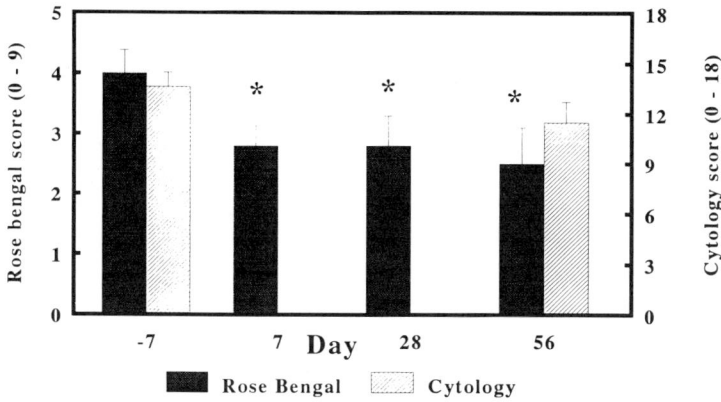

Figure 2. Mean severity (± s.e.m.) for objective measures for Formulation #1.

day examination, the mean impression cytology score was decreased by 2.1 severity grades (p = 0.07).

Thirteen patients were enrolled in Study #2, an evaluation of Formulation #2. The mean age was 52 ± 13 years, and ranged from 34 to 72 years, and the mean duration of disease was 12 years. Seven of these patients also had participated in Study #1, and there was at least a two week washout interval between studies. In three patients there were both signs and symptoms of dry eye severe enough to prevent them from completing the study.

The effect of treatment on patient symptoms is shown in Figure 3. There was a decrease of up to 0.6 severity grades in all four solicited symptoms: discomfort, foreign body sensation, dryness and photophobia. For all but discomfort and photophobia, these decreases were statistically significant.

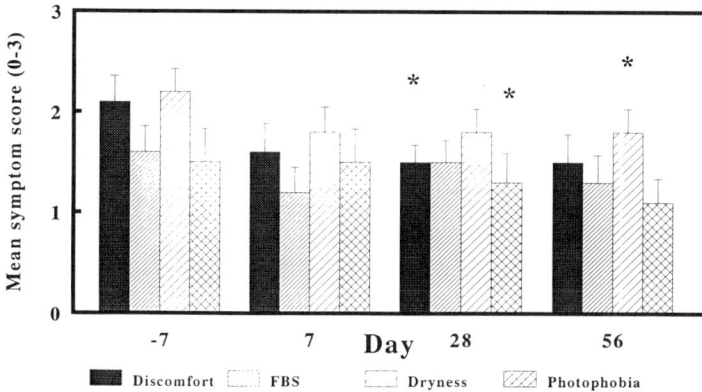

Figure 3. Mean severity (± s.e.m.) of patient symptoms for Formulation #2 (n = 10).

The effect of treatment on objective measures is shown in Figure 4. The severity of rose bengal staining decreased by approximately one grade at the first follow-up visit (p = 0.016), and remained decreased throughout the 56 day treatment period (p = 0.420 to 0.200). At the 56 day examination, the mean impression cytology score increased by 1.1 grades.

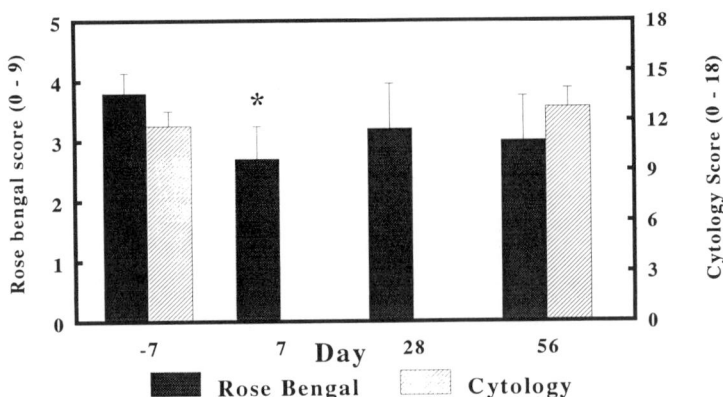

Figure 4. Mean severity (± s.e.m.) for objective measures for Formulation #2.

At baseline, for Formulation #1 and #2, the mean Schirmer score was 5.7 ± 1.8 and 5.3 ± 2.6 seconds, respectively. At Day 56, the mean was 4.8 ± 2.0 and 3.3 ± 1.7 seconds, respectively (p = 0.27 and 0.17, respectively).

## DISCUSSION

The primary objectives of the physician caring for dry eye subjects are to minimize ocular surface desiccation and cell death by maintaining a well lubricated surface and to improve subjective comfort. A multitude of diagnostic tests are available for dry eye such as tear break up time, Schirmer's test, tear meniscus height, and tear lysozyme content. However, CIC and rose bengal staining are the only objective parameters which can be directly related to ocular surface cell damage. If an artificial tear improves either of these test parameters, it would indicate recovery of the ocular surface damage caused by the lack of aqueous tears in these patients.

We found an improvement in the objective signs of CIC and rose bengal staining seen with treatment with an unpreserved physiological tear substitute with bicarbonate (Formulation #1). This effect was of statistical significance for rose bengal staining, and of borderline statistical significance for CIC. This observation suggests improvement in ocular surface disease in these patients with moderate to severe dry eye. The identical formulation without bicarbonate (Formulation #2) showed improvements only in patient symptoms of discomfort and dryness.

Improvement in the ocular surface as measured by CIC has not been seen in other clinical trials involving sodium hyaluronate[6] and topical fibronectin.[14] In a double-masked, parallel study, 1% carboxymethylcellulose provided a decrease in impression cytology score of approximately 11% of full scale.[8] This was approximately the same magnitude as seen with physiological tears in the present study (12% of full scale). In contrast to the 1% carboxymethylcellulose, the physiological tears are much less viscous, and unlikely to interfere with daytime vision.

Studies using a dry eye rabbit model showed that administration of an artificial tear preparation containing bicarbonate resulted in an increase in conjunctival goblet cell densities.[15] Our present report confirms the importance of bicarbonate in alleviating the objective findings of dry eye.

Ocular symptomatology remains a vital component in the evaluation of a tear substitute. Subjects using both formulations reported improvement in symptoms of discomfort and dryness; however, subjects using Formulation #1 showed a significant reduction in foreign body sensation compared to baseline and Formulation #2. The benefit of Formulation #1 is further supported by the dropout of three subjects from the Formulation #2 group due to worsening of symptoms and clinical findings. Patient symptoms of dryness and discomfort were similar for both formulations over the study period. However, foreign body sensation measurements improved over the course of the study in the Formulation #1 group compared to the Formulation #2 group. This would suggest that improvement in foreign body sensation measurements may be related to the recovery of the ocular surface as demonstrated by improvement in rose bengal staining and CIC scores.

The open-label, non-comparative nature of the present studies suffer from the obvious biases inherent in such studies. However, as all patients were using non-preserved artificial tear preparations prior to study entry, the design may be viewed as a crossover study, with comparisons to marketed products. As all of the patients had long-standing dry eye which was under treatment, we suggest it unlikely that there was an increased compliance during the study period.

On the basis of the present study, we suggest further studies and consideration of a physiological tear substitute containing bicarbonate in patients with KCS.

## ACKNOWLEDGMENTS

The authors acknowledge the assistance of Gary D. Novack, Ph.D.

## REFERENCES

1. Versura, P., Maltarello, M.C., Stecher, F., Caramazza, R. and Laschi, R. (1989) Ophthalmologica 198, 152-162.
2. Polack, F.M. and McNiece, M.T. (1982) Cornea 1, 133-136.
3. Stuart, J.C. and Linn, J.G. (1985) Ann Ophthalmol 17, 190-192.
4. DeLuise, V.P. and Peterson, W.S. (1984) Ann Ophthalmol 16, 823-824.
5. Sand, B.B., Marner, K. and Norn, M.S. (1989) Acta Ophthalmol (Copenh) 67, 181-183.
6. Nelson, J.D. and Farris, R.L. (1988) Arch Ophthalmol 106, 484-487.
7. Hawi, A.L., Smith, T.J. and Diogenis, G.A. (1990) Suppl to Invest Ophthalmol Vis Sci 31, 517.
8. Grene, R.B., Lankston, P., Mordaunt, J., Harrold, M., Gwon, A. and Jones, R. (1992) Cornea 11, 294-301.
9. Gilbard, J.P., Rossi, S.R. and Heyda, K.G. (1989) 107, 348-355.
10. Gilbard, J.P. and Rossi, S.R. (1991) Suppl to Invest Ophthalmol Vis Sci 32, 1114.
11. Lopez Bernal, D. and Ubels, J.L. (1992) (In Press).
12. von Bijsterveld, O.P. (1969) Arch Ophthalmol 82, 10-14.
13. Nelson, J.D. (1989) In: *Contact Lenses: Update 5*, Little, Brown and Co., p. 3C.1-3C.7.
14. Nelson, J.D., Gordon, J.F., and The Chiron KCS Study Group (1992) Am J Ophthalmol 114:441-447.
15. Gilbard, J.P. and Rossi, S.R. (1992) Ophthalmology 99, 600-604.

# A SCANNING ELECTRON MICROGRAPHIC COMPARISON OF THE EFFECTS OF TWO PRESERVATIVE-FREE ARTIFICIAL TEAR SOLUTIONS ON THE CORNEAL EPITHELIUM AS COMPARED TO A PHOSPHATE BUFFERED SALINE AND A 0.02% BENZALKONIUM CHLORIDE CONTROL

Kendyl Schaefer[1], Michelle A. George[1], Mark B. Abelson[1,2], and Christopher Garofalo[1]

[1]Department of Immunology
Schepens Eye Research Institute
Boston, MA
[2]Department of Ophthalmology
Harvard Medical School
Boston, MA

## INTRODUCTION

Common ophthalmic preservatives can be divided into five classes: quaternary ammonium compounds (benzalkonium chloride), mercurials (thimerisol), alcohols (chlorobutanol), esters of parahydroxybenzoic acid, and other compounds (polymyxin and chlorhexidine).[1]

While these ophthalmic preservatives possess effective antimicrobial properties, they also have undesired cytotoxic effects on the corneal epithelium. However, benzalkonium chloride (BAC), the most commonly-used antimicrobial preservative in topical multi-use ophthalmic preparations, does not appear to have adverse effects unless used more than 4 to 6 times daily. When used excessively, BAC has been shown to induce corneal epithelial toxicity and retard epithelial healing and regrowth in animal studies.[2-6]

Currently, tear substitutes containing methylcellulose or polyvinyl alcohol are commonly prescribed to supplement tear film in treating keratoconjunctivitis sicca (KCS). Mild cases require instillation several times per day to provide adequate patient comfort, but severe cases of KCS require more frequent use. Many corneal specialists are concerned that preservatives in tear preparations can cause ocular surface damage, such as disruption of the precorneal tear film and damage to the epithelial surface.[7,8] Many patients requiring frequent use of tear supplements are now being treated with preservative-free artificial tear preparations to prevent corneal epithelial toxicity.

The objective of this study was to determine whether preservative-free artificial tear preparations induced corneal epithelial damage in an animal model, and to compare these results with the effects of a solution containing benzalkonium chloride (0.02%).

## MATERIALS AND METHODS

### Formulations

Two preservative free solutions were tested. Sterile, unit-dose containers of HypotearsPF® (Iolab, Inc., Claremont, CA) containing EDTA and Refresh® (Allergan, Inc., Irvine, CA), were purchased as over the counter preparations. A third solution containing 0.02% BAC was prepared in phosphate buffered saline (PBS) while maintaining isotonicity and physiologic pH. This concentration of BAC was selected as it is the highest concentration currently used in topical, multi-use ophthalmic preparations.

### Experimental Procedure

New Zealand White rabbits (2-4 kgs) were randomly assigned to receive either an aqueous tear substitute (n=4 eyes/group), 0.02% BAC (n=5 eyes/group) or PBS control solution (n=2 rabbits). A total of 20 rabbits were used in this study. This protocol was approved by the Animal Care and Use Committee at the Schepens Eye Research Institute.

All rabbits were examined by slit-lamp biomicroscopy to assess baseline epithelial keratopathy. Only rabbits with minimal punctate staining were included in the study. Group 1 (mild usage) received two 40 µl drops of either HypotearsPF®, Refresh® or 0.02% BAC every thirty minutes for two hours (four doses), while Group 2 (exaggerated usage) received two 40 µl drops of either HypotearsPF®, Refresh® or 0.02% BAC every three minutes for one hour (twenty doses). Doses were delivered, one drop at a time, onto the limbus in a randomized, masked manner.

Twenty minutes after receiving the last drop, animals were anesthetized with 0.5 ml of ketamine HCl (100 mg/ml) and 0.5 ml of xylazine HCl (20 mg/ml). Ten minutes later, animals were sacrificed by cardiac puncture with 1.0 ml of sodium pentobarbital (6 grs.). The eye was immediately washed with half-strength Karanovsky's fixative, which was applied frequently to prevent corneal drying and facilitate tissue fixation until enucleation was complete. A notch was cut at the 12 o'clock position on the cornea to orient the tissue. The entire globe was removed and placed in fixative for ten minutes. The cornea was then excised and returned to fresh fixative for 24 hours prior to imbedding procedures for scanning electron microscopy (SEM).

After fixation, a triangular, limbus based wedge of tissue was removed from each cornea and prepared by dehydration in a series of ethanol dilutions ranging in concentration from 70% to 95%, each for twenty minutes. Following dehydration, the tissue was infiltrated with LKB Historesin® (glycol methacrylate, benzoyl peroxide) and allowed to sit overnight. The tissue was imbedded with the LKB Historesin® and cut into 3 µm thin sections and stained with hematoxylin-eosin. These slides were evaluated in a masked fashion to corroborate SEM findings.

The remaining corneal tissue was prepared for SEM by the Morphology Department at the Schepens Eye Research Institute. SEMs were assessed for epithelial damage according to a modification of the rating system proposed by Burstein[9] (Table 1).

**Table 1.** Rating system of relative damage to the corneal epithelium.

| Relative Damage Score | SEM Findings |
|---|---|
| 0 | No loss of microvilli; no plasma membrane wrinkling; normal number of epithelial holes; no cell peeling |
| 1 | Mild diffuse loss of microvilli or loss of peripheral microvilli only; plasma wrinkling <10% of cells; cell peeling <2% of cells; increased number of epithelial holes; increased number of dark cells |
| 2 | Moderate diffuse loss of microvilli; plasma membrane wrinkling >10% and <50% of cells; cell peeling >2% and <25% of cells; loss of hexagonal shape (smoothing) |
| 3 | Plasma membrane wrinkling >50% of cells; diffuse cell peeling >25% of cells; retraction of cell membrane borders |
| 4 | Loss of superficial cell layer; second cell layer intact |
| 5 | Above plus peeling of second cell layer |

The relative damage score (RDS) assigned to each eye was an average (mean value) of scores assigned to the three evaluated areas of the cornea. The statistical evaluation of data subsets was performed using the paired, two-tailed Students' $t$ test.

## RESULTS

### Mild Use

Mild use of PBS resulted in a mean RDS of $1.38 \pm 0.38$ (Table 2), with an increased number of epithelial holes, loss of the normal epithelial hexagonal shape and mild to moderate loss of microvilli (Figure 1). Mild use of BAC eyes had a mean RDS of $1.20 \pm 0.12$ (Table 2). The corneas in this group demonstrated minimal loss of microvilli, increased number of epithelial holes and loss of the normal hexagonal shape (Figure 2).

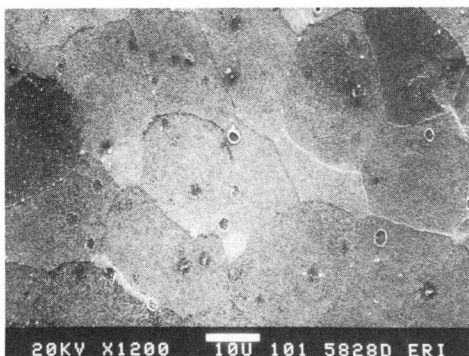

Figure 1. PBS control, mild use: Score = 0.5. These eyes showed an increased number of epithelial holes and dark cells, although there is no plasma membrane wrinkling or cell peeling. The microvilli still appear healthy, with minimal loss confined to the periphery, and the cells have maintained their morphologic integrity.

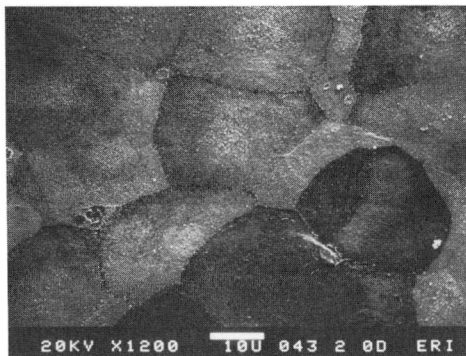

Figure 2. BAC, mild use: Score = 2.0. Eyes from this group showed minimal loss of microvilli, an increased number of epithelial holes and loss of their normal hexagonal shape. This SEM, however, shows a moderate to diffuse loss of microvilli, plasma membrane wrinkling in > 25% of the cells, and cell peeling in > 10% of the cells.

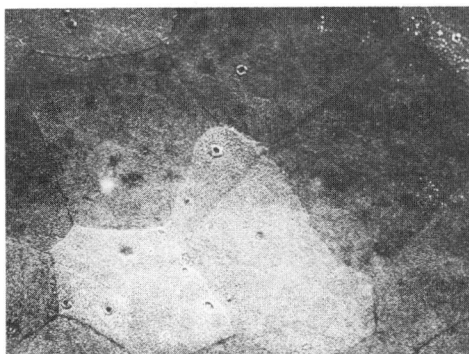

Figure 3. HypotearsPF®, mild use: Score = 0.5 HypotearsPF® treated eyes displayed normal surface epithelium with velvety microvilli, although there was an increase in the number of dark cells present. No alterations in cell morphology are apparent, and there is no cell peeling.

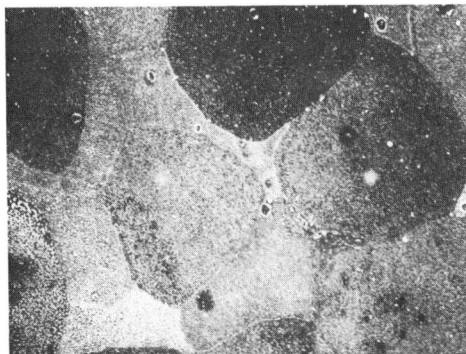

Figure 4. Refresh®, mild usage: Score = 0.5. Eyes treated with Refresh® showed normal epithelial architecture except for an increased in the number of dark cells and epithelial holes. There is a mild and diffuse loss of microvilli as well.

461

Mild usage of HypotearsPF® resulted in a mean RDS of 0.75 ± 0.16 (Table 2), with normal corneal surface epithelial morphology, velvety microvilli, and normal hexagonal shape. However, there was an increase in the number of dark cells present (Figure 3). Mild use of Refresh® resulted in a mean RDS of 1.02 ± 0.23 (Table 2), with a normal corneal epithelial architecture, except for an increase in the number of dark cells and epithelial holes (Figure 4). The mean RDS for both preservative-free preparations was not significantly different from control eyes receiving PBS or from that of eyes treated with 0.02% BAC (Table 2).

## Exaggerated Use

Exaggerated usage of PBS resulted in a mean RDS of 1.26 ± 0.13 (Table 2), with mild peripheral microvilli loss, increased numbers of dark cells and minimal wrinkling of surface cells. The epithelial cells maintained their normal shape and showed no evidence of retraction (Figure 5). Exaggerated use of BAC resulted in a mean RDS of 4.00 ± 0.16 (Table 2) and exhibited diffuse cell peeling and retraction of cell membrane borders. There was loss of the normal fine, velvety appearance attributable to destruction of microvilli and loss of the superficial layer of the corneal epithelium. The cell membrane borders had begun to retract in most areas and there was marked loss of the superficial epithelial layer (Figure 6).

Exaggerated use of HypotearsPF® resulted in a mean RDS of 1.31 ± 0.21 (Table 2), with mild diffuse microvilli loss, loss of the normal hexagonal corneal epithelial shape and an increased number of epithelial holes (Figure 7). Exaggerated use of Refresh® resulted in a mean RDS of 1.35 ± 0.08 (Table 2), with mild diffuse and peripheral microvilli loss and loss of the normal corneal epithelial shape (Figure 8).

Exaggerated use with both HypotearsPF® and Refresh® induced only minimal corneal epithelial damage which was not different from PBS treated control eyes. However, when compared to 0.02% BAC, both HypotearsPF® and Refresh® induced significantly less epithelial damage in the exaggerated usage group (p=0.0001). Light microscopic examination of the tissues under 10x and 40x magnification corroborated the pathologic changes seen in all corneas with SEM (Table 2).

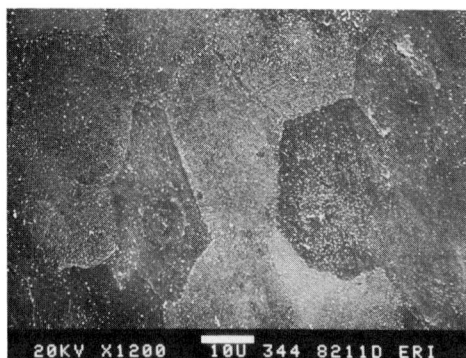

20KV X1200    10U 344 82110 ERI

Figure 5. PBS control, exaggerated use. Score = 0.5. These eyes showed no peripheral microvilli loss, increased numbers of dark cells and minimal wrinkling of surface cells. The cell morphology appears normal, but there seems to be some retraction of cell borders.

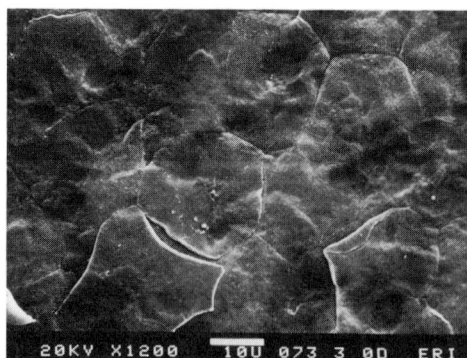

20KV X1200    10U 073 3 0D    ERI

Figure 6. BAC, exaggerated use: Score = 3.5. In general, these corneas exhibited diffuse epithelial peeling and retraction of the cell membrane borders. In this photo, most of the superficial cell layer is gone (noted by the ghosts), but the second cell layer is intact.

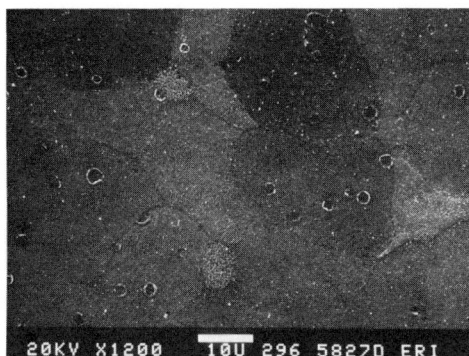

Figure 7. HypotearsPF®, exaggerated use: Score = 1.0. The corneal epithelium displayed mild diffuse microvilli loss, loss of the normal hexagonal epithelial shape and an increased number of epithelial holes. Plasma membrane or cell peeling is not apparent, but there is cell borders.

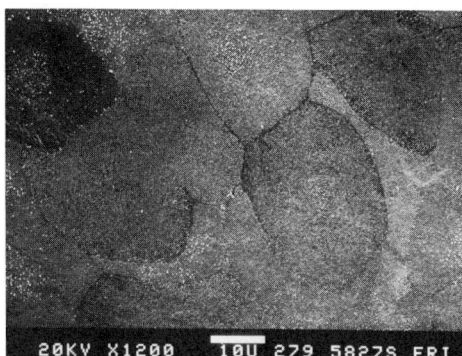

Figure 8. Refresh®, exaggerated use: Score = 1.0. These eyes showed mild diffuse and peripheral microvilli loss and loss of epithelial shape. Plasma membrane wrinkling is beginning. Cell peeling is not obvious, but there is some early retraction of retraction of cell membranes.

**Table 2.** Individual and group mean relative damage scores for rabbit corneal epithelium treated with mild and exaggerated use of HypotearsPF® (n= 4), Refresh® (n=4), benzalkonium chloride (BAC, n=5) and phosphate buffered saline (PBS, n=2).

| | MILD USE | | | |
|---|---|---|---|---|
| **Rabbit** | **HypotearsPF®** | **Refresh®** | **BAC** | **PBS** |
| 1 | 1.08[1] | 0.96 | 1.00 | 1.75 |
| 2 | 0.83 | 1.00 | 1.50 | 1.00 |
| 3 | 0.33 | 0.50 | 1.00 | - |
| 4 | 0.75 | 1.63 | 1.00 | - |
| 5 | - | - | 1.50 | - |
| Mean ± SEM | 0.75 ± 0.16 | 1.02 ± 0.23 | 1.20 ± 0.12 | 1.38 ± 0.38 |

| | EXAGGERATED USE | | | |
|---|---|---|---|---|
| **Rabbit** | **HypotearsPF®** | **Refresh®** | **BAC** | **PBS** |
| 1 | 1.75 | 1.38 | 4.00 | 1.13 |
| 2 | 1.50 | 1.50 | 4.00 | 1.38 |
| 3 | 1.25 | 1.13 | 3.50 | - |
| 4 | 0.75 | 1.40 | 4.50 | - |
| 5 | - | - | 4.00 | - |
| Mean ± SEM | 1.31 ± 0.21* | 1.35 ± 0.08* | 4.00 ± 0.16 | 1.26 ± 0.23* |

[1] Individual scores represent the mean of the three representative areas from the cornea. The grading was performed in increments of 0.5.
* Denotes statistically significant difference from BAC score.

## DISCUSSION

Ophthalmic preservatives in general are a "double edged sword". Clinicians must weigh the possible benefit of sterility against the risk of cytotoxicity to the corneal surface. Presently, BAC is the most common antimicrobial ophthalmic preservative used clinically. Subtle to profound damage to the corneal surface can occur with overuse of BAC, leading to chronic irritation and possible infection. BAC in concentrations as low as 0.005% has produced cellular damage in cat and rabbit corneas.[9]

Recently, Olson and White[10] have resurrected the long-standing belief held by many corneal specialists that KCS may be aggravated by frequent use of preservative containing artificial tear preparations. This study attempted to recreate two common scenarios encountered when treating patients with KCS. Group 1 (mild usage) represents those patients who have mild symptoms and who would be treated with a tear preparation several times a day. Group 2 (exaggerated usage) represents those patients with more severe symptoms characteristic of KCS and who may use tear substitutes hourly. As we have shown in this study, corneal epithelial damage induced by mild dosing with 0.02% BAC was minimal and not unlike that seen in control eyes treated with PBS. However, when 0.02% BAC was used excessively, changes consistent with epithelial toxicity were found. We have previously found toxic effects of BAC on the rabbit corneal epithelium in concentrations as low as 0.001%.[11] Evaluation of preservative-free polyvinyl alcohol tear preparations demonstrated these agents to be non-toxic to the corneal epithelium in both the mild and exaggerated usage protocols. The epithelial changes observed were no different from those seen in control eyes.

## CONCLUSION

The results of this study indicate that exaggerated usage of the preservative-free aqueous tear preparations, HypotearsPF® and Refresh®, did not induce corneal epithelial damage and were less toxic than a solution of 0.02% BAC in a rabbit model. This study supports the belief that preservative-free preparations are safe to use in patients, especially when frequent dosing is required.

## REFERENCES

1. A. Osol, ed. "Remington's Pharmaceutical Sciences," 14th ed. Mack Publishing Company, Easton, PA (1970).
2. R.R. Pfister and N. Burstein, The effects of ophthalmic drugs, vehicles, and preservatives on corneal epithelium: a scanning electron microscope study, *Invest. Ophthalmol. Vis. Sci.* 15:246-259 (1976).
3. A.M. Tonjum, Effects of benzalkonium chloride upon the corneal epithelium studied with scanning electron microscopy, *Acta Ophthalmol.* 53:358-366 (1975).
4. N.L. Burstein and S.D. Klyce, Electrophysiologic and morphologic effects of ophthalmic preparations on rabbit cornea epithelium, *Invest. Ophthalmol. Vis. Sci.* 16:899-911 (1977).
5. N.L. Burstein, Preservative cytotoxic threshold for benzalkonium chloride and chlorhexidine digluconate in cat and rabbit corneas, *Invest. Ophthalmol. Vis Sci.* 19:308-313 (1980).
6. I. Rucker, R. Kettrey, F. Bach, and L. Zeleznick, A safety test for contact lens wetting solutions, *Ann. Ophthalmol.* 4:1000-1006 (1972).
7. M. Gobbels and M. Spitznas, Influence of artificial tears on cornea epithelium in dry eye syndrome, *Graefes Arch. Clin. Exp. Ophthalmol.* 227:139-143 (1989).
8. H.B. Collin and B.E. Grabsch, The effect of ophthalmic preservatives on the healing rate of the rabbit corneal epithelium after keratectomy, *Am. J. Optom. Physiol. Opt.* 59:215-222 (1982).
9. N.L. Burstein, Preservative alteration of corneal permeability in humans and rabbits. *Invest. Ophthalmol. Vis. Sci.* 25:1453-1457 (1984).
10. R.J. Olson and G.L. White, Preservatives in ophthalmic topical medications: a significant cause of disease, (letter) *Cornea* 9:363-364 (1990).
11. K.M. Schaefer, M.A. George, and G.J. Berdy, A scaning electron microscope comparison of the effects of 0.001% sodium silver chloride complex vs 0.001% benzalkonium chloride on the rabbit corneal epithelium, (suppl) *Invest. Ophthalmol. Vis. Sci.* 33:998 (1992).

# A UNIQUE THERAPEUTIC ARTIFICIAL TEAR FORMULATION

Jeffrey P. Gilbard[1,2] and Scott R. Rossi [1]

[1]Cornea Research Unit, Schepens Eye Research Institute
[2]Department of Ophthalmology, Harvard Medical School
Boston, MA 02114

## PURPOSE

The tear film has a unique electrolyte balance distinct from aqueous humor and serum.[1] We postulated that this electrolyte balance was critical for the maintenance of ocular surface health.

As seen in clinical studies of dry eye and in our studies of dry-eye disease in a variety of rabbit models, decreases in conjunctival goblet-cell density are consistently among the earliest changes detectable and measurable in the ocular surface.[2-6] Clinically the cornea remains free of rose Bengal staining in the presence of significant conjunctival staining, and the cornea always stains less than the conjunctiva in all stages of the disease. In rabbit models for dry-eye disease decreases in conjunctival goblet-cell density precede morphological changes in corneal epithelium by one year.

The purpose of the project reviewed here was to develop an artificial tear solution capable of restoring conjunctival goblet-cell density in keratoconjunctivitis sicca (KCS) and, in the process, reversing other less sensitive changes of dry-eye disease.

## RABBIT STUDIES

### 12-Hour Bathing Experiments

We studied the effect of electrolyte solution composition and balance on goblet-cell density in flat mounts of conjunctiva from normal rabbits after 12 hours of continuous bathing. From a group of formulations that tended to preserve goblet cells, we selected a formulation that was optimally capable of maintaining normal conjunctival goblet-cell density after prolonged exposure.[7] A formulation from outside that group (BTS) is compared to our optimal solution (Solution 15) in Table 1.

The data in Table 1 demonstrate that Solution 15 maintains conjunctival goblet-cell density in normal rabbit conjunctiva.[7,9]

**Table 1.** Effect of two "tear-like" electrolyte solutions on goblet-cell density

|  | BTS[8] | | Solution 15[9] | |
| --- | --- | --- | --- | --- |
| NaCl | 116.4 mmol/l | 72%[*] | 99 mmol/l | 63% |
| KCl | 18.7 mmol/l | 11% | 24 mmol/l | 15% |
| NaHCO$_3$ | 25.9 mmol/l | 16% | 32 mmol/l | 20% |
| CaCl$_2$ | 0.4 mmol/l | <1% | 0.8 mmol/l | 1% |
| MgCl$_2$ | 0.6 mmol/l | <1% | 0.6 mmol/l | <1% |
| NaH$_2$PO$_4$ | 0.7 mmol/l | <1% | 1 mmol/l | 1% |
| Goblet Cells Remaining | 60.3% | | 99.4% | |

[*]Percentages for salts represent percent of total salt content.

## Treatment Of A Rabbit Model For KCS

ATF is a hypotonic version of Solution 15 with a demulcent and buffer added. In 12-hour bathing experiments, ATF maintains conjunctival goblet-cell density as well as solution 15.[9] We next studied the effect of four artificial tear solutions, including ATF, on conjunctival goblet-cell density and corneal epithelial glycogen levels in a surgically-induced rabbit model for KCS.[10] When left untreated, this rabbit model demonstrates a predictable decrease in conjunctival goblet-cell density and corneal glycogen levels 20 weeks after surgical creation of disease.

In this experiment the lacrimal gland excretory ducts were closed in the right eye of 40 rabbits. Rabbits were then divided into five groups. KCS eyes of group 1 rabbits received no treatment. KCS eyes of groups 2 through 5 were treated with one of four artificial tear solutions four times a day five days a week for 12 weeks from 8 to 20 weeks postop.

Group 2 rabbits were treated with ATF containing NaCl, KCl, NaHCO$_3$, MgCl$_2$, NaH$_2$PO$_4$, boric acid, sodium borate, demulcent, and H$_2$0. Group 3 rabbits were treated with a hypotonic lubricant solution containing polyvinyl alcohol, polyethylene glycol 400, dextrose, edetate disodium and H$_2$0. Group 4 rabbits were treated with an isotonic lubricant solution containing dextran, hydroxypropyl methylcellulose, polyquaternium-1 (a preservative), edetate disodium, KCL, NaCl, and H$_2$0. Group 5 rabbits were treated with an isotonic lubricant solution containing polyvinyl alcohol, povidone, NaCl, and H$_2$0.

The study design provided two sets of controls: contralateral (left) eyes that were unoperated and untreated, and group 1 rabbits that were operated but left untreated.

Weekly microvolume tear samples were taken for masked measurement of osmolarity, sodium and potassium. At 20 weeks postop (after 12 weeks of treatment) masked measurements of conjunctival goblet-cell density and corneal glycogen were performed.

## Results

ATF treatment of group 2 rabbits produced a significant decrease in elevated tear osmolarity and tear sodium from 9 weeks of treatment on (P<0.05) whereas treatment with the other solutions did not. Conjunctival goblet-cell density and corneal glycogen in KCS eyes of group 2 rabbits was significantly higher than untreated KCS eyes of group 1 rabbits(P<0.01) (Figures 1 and 2). Goblet-cell density and corneal glycogen in treated KCS eyes of group 4 rabbits were significantly lower than operated and untreated group 1 eyes (P<0.05). Conjunctival goblet-cell density and corneal glycogen in group 3 and 5 rabbits remained unchanged relative to untreated rabbits after 12 weeks of treatment.

## Conclusion

These data indicate that ATF increases conjunctival goblet-cell density and corneal glycogen in a rabbit model for KCS. ATF also decreases elevated tear osmolarity and sodium in this rabbit model for KCS.

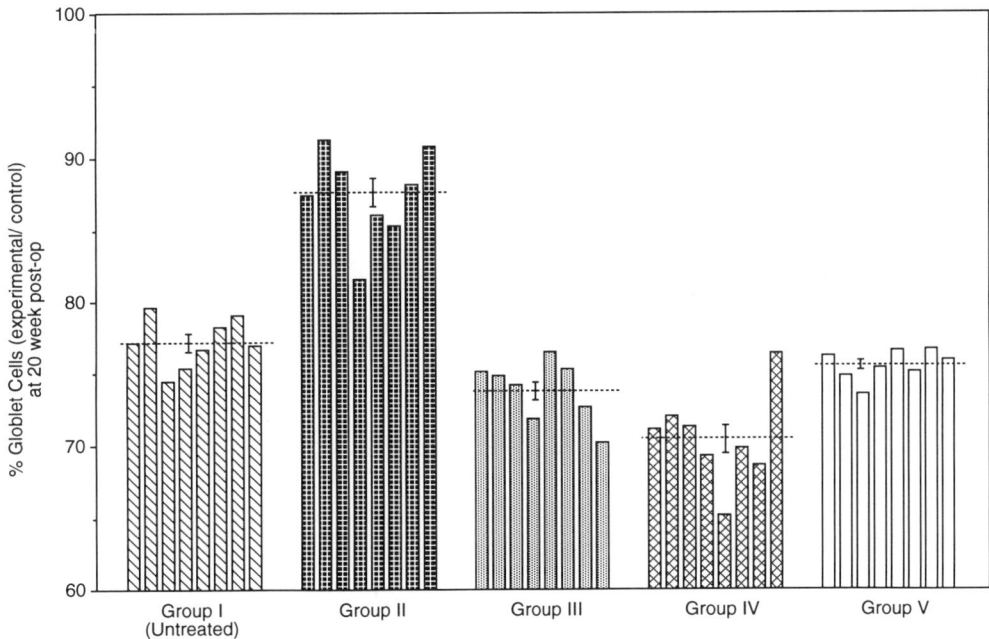

**Figure 1.** The effect of 12 weeks of treatment with 4 different artificial tear solutions on conjunctival goblet-cell density. Published courtesy of *Ophthalmology* (1992;99:600-4).

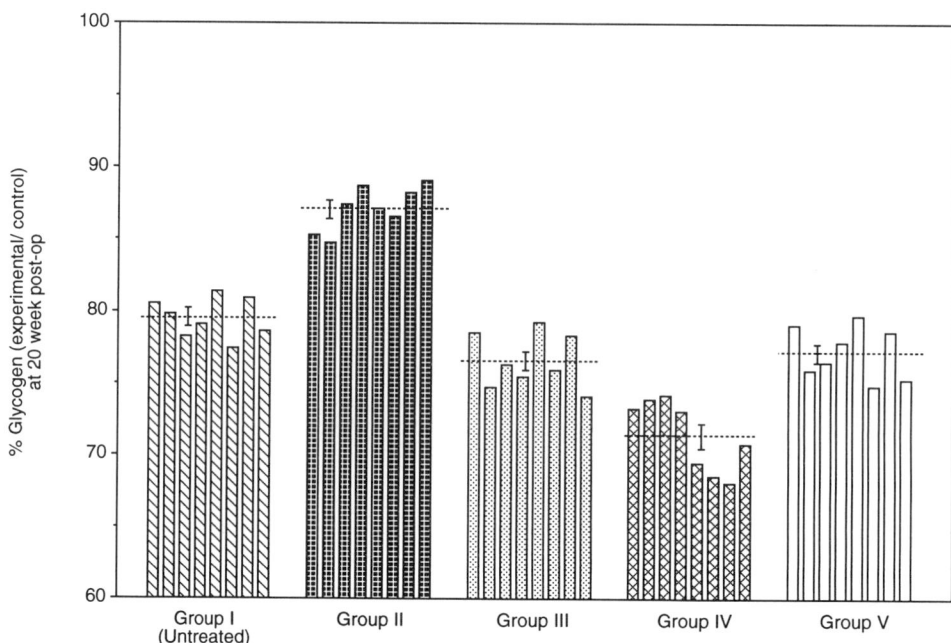

**Figure 2.** The effect of 12 weeks of treatment with 4 different artificial tear solutions on corneal glycogen levels. Published courtesy of *Ophthalmology* (1992;99:600-4).

## CLINICAL STUDIES

### Treatment of Dry-eye Patients

We studied the effect of ATF in a 16-week prospective double-masked "active-agent-controlled" clinical study.[10] Eleven patients participated in the study. Entry into the study required a clinical history and signs typical for dry-eye disease and described in detail elsewhere.[11] Entry also required elevated tear film osmolarity measurements. ATF and a control (a commercially available artificial tear solution) were used in contralateral eyes at least six times a day. At eight weeks treatments were crossed. Tear osmolarity measurements were performed at weeks 0, 4, 8 and 16. Rose Bengal staining was performed and photographed at weeks 0, 8 and 16.

### Results

Treatment with ATF produced a decrease in tear osmolarity similar to that observed in the rabbit studies. Eyes treated with ATF had significantly lower tear osmolarity than eyes treated with control 4 weeks into treatment ($P<0.01$), and when eyes undergoing treatment with ATF and control were examined relative to baseline osmolarity measurements, ATF-treated eyes had a significant decrease in tear osmolarity ($P<0.025$) whereas control-treated eyes did not.

Rose Bengal staining decreased after treatment with ATF but increased after treatment with control. The difference in the change in rose Bengal staining with each treatment was significant ($P<0.05$).

# CONCLUSION

ATF, a preservative-free hypotonic electrolyte-based solution, maintains conjunctival goblet-cell density in normal rabbit conjunctiva, and is the first treatment to restore conjunctival goblet cells and increase corneal glycogen levels in a rabbit model for keratoconjunctivitis sicca. In our clinical study ATF decreased elevated tear film osmolarity; this effect was also observed in our rabbit model for KCS. In the rabbit study we also measured tear sodium levels and demonstrated an associated decrease in elevated tear sodium. We postulate that the decrease in rose Bengal staining observed clinically corresponds to a restoration of conjunctival goblet cells, and an improvement in the health and structure of the conjunctival and corneal epithelial surface.

These studies demonstrate an effective new treatment for dry-eye disease.

# ACKNOWLEDGMENTS

This work was supported in part by grant EY03373 from the NEI. Solution 15 and ATF are protected by U.S. Patent 4,775,531 and additional foreign patents have issued or are pending. Dr. Gilbard is the inventor and the Schepens Eye Research Institute (SERI) is the assignee. SERI has a proprietary interest in this technology and Dr. Gilbard will participate in its commercialization.

# REFERENCES

1. J.P Gilbard, S.R. Rossi, Changes in tear ion concentrations in dry eye disorders, *Invest Ophthalmol Vis Sci.* Supp 33:1287 (1992).
2. J.P. Gilbard, S.R. Rossi, K.L. Gray, A new rabbit model for keratoconjunctivitis sicca, *Invest Ophthalmol Vis Sci.* 28:225 (1987).
3. J.P. Gilbard, S.R. Rossi, K.L. Gray, L.A. Hanninen, K.R. Kenyon, Tear film osmolarity and ocular surface disease in two rabbit models for keratoconjunctivitis sicca, *Invest Ophthalmol Vis Sci.* 29:374 (1988).
4. J.P. Gilbard, S.R. Rossi, K. Gray Heyda, Tear film and ocular surface changes after closure of the meibomian gland orifices in the rabbit, *Ophthalmology.* 96:1180 (1989).
5. J.P. Gilbard, S.R. Rossi, K.L. Gray, L.A. Hanninen, Natural history of disease in a rabbit model for keratoconjunctivitis sicca, *ACTA Ophthalmol.* (Suppl 192)67:95 (1989).
6. J.P. Gilbard, S.R. Rossi, Tear film and ocular surface changes in a rabbit model of neurotrophic keratitis, *Ophthalmology.* 97:308 (1990).
7. J.P. Gilbard, Non-toxic ophthalmic preparatons, U.S. Patent 4,775,531.
8. W.G. Bachman, G. Wilson, Essential ions for maintenance of the corneal epithelial surface, *Invest Ophthalmol Vis Sci.* 26:1484 (1985).
9. J.P. Gilbard, S.R. Rossi, K. Gray Heyda, Ophthalmic solutions, the ocular surface, and a unique therapeutic artificial tear formulation, *Am J Ophthalmol.* 107:348 (1989).
10. J.P. Gilbard, S.R. Rossi, An electrolyte-based solution that increases corneal glycogen and conjunctival goblet-cell density in a rabbit model for keratoconjunctivitis sicca, Ophthalmology. 1992; 99:600-604.
11. J.P. Gilbard, Dry Eye Disorders, *in* "Principles and Practice of Ophthalmology: Clinical Practice," D. Albert and F. Jackobiec eds., W.B. Saunders Company, Philadelphia (in press).

# NON-SJÖGREN DRY EYE: PATHOGENESIS DIAGNOSIS
# AND ANIMAL MODELS

Anthony J. Bron

Nuffield Laboratory of Ophthalmology
University of Oxford
Oxford OX2 6AW, UK
England

## INTRODUCTION

Non-Sjögren dry eye (NSDE), is a term which may be used to cover a group of local and systemic disorders,excluding Sjögren's syndrome giving rise to the symptoms and signs of dry eye.Their causes are given in Tables 1,2, and 3 .

The condition can be defined as a disorder of the ocular surface, especially involving, but not confined to the interpalpebral zone, due to a deficiency of tear components or their distribution. It occurs in the absence of the systemic autoimmune disease which characterises of Sjögren,s syndrome. The different forms have interpalpebral ocular surface damage in common.

Non-Sjögren dry eye includes disorders due to defects of tear aqueous, oil, mucin and protein,and disorders of blinking and of lid apposition in the absence of Sjögren features (The latter include eg. a positive labial salivary biopsy, dry mouth, and serological features including anti Ro and La antibodies, rheumatoid factor, anti-tissue antibodies (Lancet editorial 1992 ).

## TEAR COMPONENT DEFICIENCIES LEADING TO NSDE

NSDE can be caused a by deficiency of any of the major components of the tears.

## AQUEOUS DEFICIENCY

Aqueous deficiency results from a failure in lacrimal gland function, which may be due to congenital absence of the gland, denervation, inflammation and infection, toxicity, trauma, and obstruction due to cicatricial changes at the glandular orifices. Two auto-immune disorders, graft versus host disease and the dry eye of HIV infection are included, as conditions which are borderline between between NSDE and Sjögren dry eye (SDE). (Table 1).

*Lacrimal Gland, Tear Film, and Dry Eye Syndromes*
Edited by D.A. Sullivan, Plenum Press, New York, 1994

The commonest form of aqueous-deficient NSDE is keratoconjunctivitis sicca (KCS), which is an age-related disorder, which, like Sjögren's syndrome, is more common in women than in men (Holm 1949;Bron 1985,Williamson 1979). The lacrimal gland is infiltrated by lymphocytes, which appear to be responsible for the destruction of lacrimal acinar and duct tissue (Nasu et al 1984; Chomette et al 1986).

**Table 1.** Aqueous Deficiency Syndromes.

| UNILATERAL | BILATERAL |
|---|---|
| Paralytic hyposecretion | Keratoconjunctivitis sicca |
| Dacryoadenitis | Sjögren Syndrome |
| Dacryoadenectomy | Sarcoidosis |
| Lacrimal gland and ductular scarring | Congenital Alacrima |
| Anhydrotic ectodermal dysplasia | Multiple Neuromatosis |
| | Riley Day Syndrome |
| | "Cri du Chat" Syndrome |
| | HIV Infection |
| | Graft vs Host Disease |

In the normal lacrimal gland there is an age-related increase in round cell infiltration of the gland and duct tissues, which suggests that KCS may represent an exaggeration of this process (Damato et al 1984;Williamson 1985). This age-related process has not been shown in all studies (Murray et al 1981).

## MUCIN DEFICIENCY

The conjunctival goblet cells are regarded as the major, if not the sole source of tear mucin. Although at various times the lacrimal gland (Allen et al 1972) or the subsurface vesicles of the conjunctiva ( Greiner,Allansmith 1977;Greiner et al 1979, 1980,1985; Dilley and Mackie 1981) have been proposed as additional sources of mucin, adequate evidence for this has not been available. However,recently Gipson has presented evidence to suggest that ocular surface glycocalyx has features of a mucus glycoprotein (Gipson et al 1992;1993). Nonetheless, it may be reasonable currently, to equate qualitative and quantitative losses of goblet cells, with a loss of the tear mucin component. However, since the relationship between conjunctival goblet cell density and the quality and quantity of tear mucin is not known, goblet cell density can only provide a guide to this parameter. Tear mucin is the major source of tear viscosity and hence is the basis of the lubricative function of the tears (Kaura,Tiffany 1984).

**Table 2.** Goblet Cell Deficiency.

Hypovitaminosis A (Xerophthalmia)
Keratoconjunctivitis Sicca
Mucous membrane pemphigoid
Erythema multiforme
Cicatricial conjunctivitis (trachoma)
Chemical and thermal burns
Drug induced  (Practolol)

Mucin lowers the surface tension of the tears (Holly and Lemp 1971 ) and plays a role in stabilising the tear film, probably in conditions of reduced surface wettability. The surface tension effect of mucin, is reflected in the tear break-up time, a test of tear stability. A reduced break-up time, indicates reduced tear stability, and implies an increased tear surface tension. In normal subjects the ocular surface is wettable, and mucin therefore performs no special role in this respect (Cope et al 1986 ).

Table 2 lists the major disorders resulting in goblet cell deficiency. These include xerophthalmia, KCS, various forms of cicatrising conjunctivitis, and toxic causes due to chemical burns.

## TEAR OIL DEFICIENCY

The tear oil derives from the Meibomian glands of the lids. It is layered onto the preocular tear film with each blink. It serves to retard water evaporation from the tear film, but has several other functions (Holly 1980;Tiffany 1987).

**Table 3.** Meibomian Deficiency.

Congenital absence
Dystichiasis
Meibomian gland disease
Retinoid therapy
KCS
Cicatrising conjunctival disease

Causes of tear oil abnormality are listed in Table 3. They include congenital absence of the meibomian glands (Bron and Mengher 1987), dystichiasis, ( a congenital replacement of the oil glands by an extra row of lashes), a group of conditions collectively known as meibomian gland disorder (MGD) (McCulley, Sciallis 1977;McCulley et al 1982;Bron et al 1991), KCS, and toxic causes of meibomian gland disorder including exposure to retinoids (Lambert and Smith 1988) adrenergic drugs (Jester et al 1989), and polychlorinated biphenyls (Ohnishi and Kohno 1979).

## THE AGENTS OF OCULAR SURFACE DAMAGE

These component deficiencies generate agents which, acting alone or in concert,cause the ocular surface damage characteristic of the disease. The damaging agents include hyperosmolarity, mucin deficiency and, the products of inflammatory reaction.There are no doubt other, unknown agents. The pathological changes occurring in the conjunctiva in KCS were described by Abdel-Khalek et al (1978).

### 1.Hyperosmolarity

Hyperosmolarity has been proposed to result from the effects of a normal rate of evaporation from a preocular tear film of reduced volume, which in turn results from a reduced rate of aqueous flow (von Bahr 1941;Mastmann et al 1961;Mishima et al 1971;Gilbard et al 1978;Gilbard,Farris 1979 ). All of these events may arise in KCS, where there is a deficiency of aqueous secretion (van Bijsterveld 1969). Alternatively, hyperosmolarity may result when the period over which the tears evaporate, before replenishment, is extended by a reduced blink rate, as in Parkinsons disease. Imperfect lid/globe congruity, as in atopic keratoconjunctivitis, or an enlarged palpebral aperture as in endocrine exophthalmos (Gilbard et al 1983), may interfere with the re-establishment of the tear film after the blink and lead to hyperosmolarity in this way.A deficiency of tear oil causes increased hyperosmolarity by increasing the water evaporation rate from the preocular film in the experimental model, (Mishima and Maurice 1961;Gilbard et al 1989) and patients with keratoconjunctivitis sicca (Rolando et al 1983).

The levels of hyperosmolarity reached clinically (Gilbard et al 1978),and experimentally (Gilbard et al 1987a;1989) as a result of oil gland disease or lacrimal dysfunction, is associated with ocular surface change and has been demonstrated to impair corneal epithelial cell survival in tissue culture (Gilbard et al 1984). Although it cannot be guaranteed that cell culture provides an ideal model for dry eye surface damage,it seems likely that the levels of hyperosmolarity measured clinically and studied in vitro are far lower than those that pertain at the surface of the eye within the preocular film. In human studies osmolarity measurements have universally been made on the tear meniscus, while clinical surface damage, except in the most severe cases, lies in relation to the interpalpebral preocular film, within the confines of the tear menisci. Since the volume of the interpalpebral film is smaller, and its area vastly greater than that of the menisci, it would be expected that the hyperconcentration of the preocular film would be greatest in the interpalpebral zone. Therefore the meniscal data for hyperosmolarity or cation content is likely to underrepresent the degree of change due to excessive evaporation, although some degree of exchange between the meniscal and inter-palpebral pools would be expected both in vivo and during tear sampling.

## 2.Mucin Deficiency

The acceptance of mucin deficiency as an agent of ocular surface damage in NSDE has a sound theoretical basis, but lacks the experimental evidence available for the hyperosmolar mechanism. One reason for this is the absence of a model for pure goblet cell deficiency. Althougth the vitamin A deficiency model might apparently fulfill this role, it must be noted that xerophthalmia causes not only a loss of goblet cells, but also, probably, a primary abnormality of the ocular surface, since it essential for the incorporation of glycosidic residues in the corneal epithelium (Kiorpes et al 1979). This too might be expected to influence surface wetting.

Mucin is the basis for the rheological properties of the tears, and it is assumed that the significant viscosity of the tears (2.82 mPa sec) provides an important lubricative function during the rapid, high shear movements of lid on globe or vice versa, which occur during the blink or saccade. Therefore loss of goblet cells, and hence mucin, might be expected to cause frictional damage and induce symptoms by this mechanism at some critical concentration.

Holly (1980) has argued for the role of a tear mucin deficiency in reducing tear stability, and it has been demonstrated that a reduced break-up time will occur in mucin-deficient eyes in the absence of an aqueous tear deficiency (Lemp 1973). The implications for ocular surface damage and for symptoms in NSDE are discussed below.

Holly (1980) has demonstrated that the spreading of oil over an aqueous surface is facilitated by the presence of mucin.It follows from this that a major absence of mucin could influence the integrity of the oil film and hence influence evaporation.

## 3.Inflammation

Inflammation is a feature of the dry eye. The conjunctiva is injected, and there are increased proteins of serum origin in the tears. White cells (PMN's) and inflammatory mediators such as superoxide, or prostaglandins may be detected in the tears in keratoconjunctivitis sicca (Bron et al 1985). It is assumed that this inflammatory response is a response of the ocular surface to the damaging agents mentioned above, but it is likely that the products of inflammation themselves contribute to the process of damage.

Very little is written about the natural history of the dry eye, but the subject has been addressed in the rabbit model made aqueous-deficient by lacrimal duct occlusion and extirpation of the nictitans and Harderian gland. Gilbard et al (1988) have shown decreases in corneal epithelial glycogen and conjunctival goblet cell density, in addition to histopathological conjunctival changes, which follow the onset of tear hyperosmolarity. Similar events were recorded after closure of the meibomian gland orifices in the rabbit and there was also staining of the ocular surface with Bengal Rose (Gilbard et al 1989 ). These changes increase with time.

It is likely that a similar series of events occurs in patients with aqueous-deficient or tear oil-deficient eyes. In both instances the primary deficiency leads to tear hyperosmolarity while goblet cell loss may be regarded as a secondary change due to ocular surface damage. Surface damage may then be construed as giving rise to inflammatory events, which might themselves cause further damage.

## DIAGNOSIS OF NSDE

A number of ocular abnormalities have been reported in dry eye, but not all of these have been used for diagnostic purposes. A smaller number of tests have been utilised as diagnostic criteria. Such tests are not qualitatively equivalent. Some are tests of tear component deficiency, some detect the damaging agents and others record features of the ocular surface damage itself. In some respects the quantification of ocular damage is the most important since the damage characterises the disorder and is responsible for the symptoms. Without ocular damage there is no disease, only a pathological change in function. The value of available diagnostic tests may be appraised by using KCS as a model of NSDE.

## DIAGNOSIS OF KERATOCONJUNCTIVITIS SICCA (Table 4)

### Aqueous Deficiency

Schirmer's test is the most commonly used test for dry eye in the clinic, and is a measure of reflex aqueous tear flow. Van Bijsterveld (1969) proposed a diagnostic cut-off of 5.5 mm or below.

A critically reduced tear flow in the presence of a normal rate of surface evaporation is responsible tear hyperosmolarity. Gilbard et al (1978) and Gilbard and Farris (1979), in a series of studies using nanolitre sampling of tears and a depression of freezing point method of measurement, selected a value of 312 mOsM as a diagnostic cut-off to discriminate dry from normal eyes. Rolando et al (1985) presented a graded change in the ability of tears to form fern patterns on drying,as a diagnostic test for dry eye, but although it was proposed as an indicator of mucin deficiency in dry eye, it appears that abnormal ferning can result from changes in electrolytes alone (Kogbe et al 1991). Raised tear sodium and potassium levels have been reported in dry eye (Menger 1990).

A deficiency of tear proteins of lacrimal gland origin has been reported by several authors who have shown a fall in the levels of lysozyme and lactoferrin and to a lesser extent IgA (van Bijsterveld 1969;Stuchell et al 1981). This is discussed further in this symposium. These levels fall with age, and Mackie and Seal (1984;1984) have indicated the importance of using age-specific normative levels to determine cut-off values .

**Table 4.** Diagnostic Test of Dry Eye.

| TEST | CUT OFF | % SENSIT. | % SPECIF. | MEASURE |
|---|---|---|---|---|
| Rose Bengal[1] | 4mm | 58 | 100 | Damage |
| Rose Bengal[2] | 4mm | 95 | 96 | |
| Schirmer[1] | 3mm | 10 | 100 | Flow |
| Schirmer[2] | 6mm | 85 | 83 | |
| Basal Tear Vol[1] | 104 | 59 | 77 | Volume |
| Osmolarity[1] | 312 mOsm/L | 76 | 84 | Evaporation |
| Nibut[3] | 10 sec | 83 | 85 | Stability |
| Lysozyme: mg[1] | 110mg | 67 | 67 | Damage |
| Lysozyme: Diam[2] | 21mm | 99 | 99 | |

[1] Farris et al 1983;      [2] Bijsterveld 1969;      [3] Mengher et al 1986.

## Mucin Deficiency

Recent studies of goblet cell density using conjunctival biopsy (Ralph 1975) or impression cytology (Nelson and Wright 1984) indicate that the density is reduced in KCS. Rolando et al (1990) have reported a greater loss in the upper and lower bulbar conjunctiva than in the horizontal meridian. This is of great interest, since the ocular surface damage demonstrated by staining, is initially interpalpebral,and only affects the vertical meridian (usually upper), in more severe degrees of dry eye. It raises the question as to whether frictional forces operating between the lid and globe in the early stages of the disease may direct the loss, to the vertical meridian. Goblet cell loss occurs in the full dry eye model described by Gilbard et al (1989) in the rabbit. Hyperosmolarity is established within days of creating the model and goblet cell loss develops after about 20-30 weeks. No direct measure of mucin deficiency is available, although such methods are under development (Huang and Tseng 1987). The fluorescein break-up test (FBUT), reported by Norn (1969), and by Holly and Lemp (1970) and Lemp and Hamill (1973), provided a test of tear stability, and it was demonstrated that this could be reduced when tear flow was normal. A correlation between goblet cell density and the break-up time implied that in this situation the BUT was an indirect measure of tear mucin activity. A cut-off value of 10 seconds has been suggested as a diagnostic criterion for dry eye. A non-invasive break-up test of tear stability has been reported (NIBUT), which discriminates KCS patients from normals with a sensitivity of 83 % and specificity of 85 %,using a cut off of 10 seconds (Mengher et al 1985a). The two tests are not equivalent, since the instillation of fluorescein for the FBUT, itself lowers tear stability, so that the FBUT reading is usually lower than the NIBUT reading. The FBUT test may be regarded as a provocative test of tear stability (Mengher et al 1985b ).

## Lid Oil Deficiency

The amount of tear oil on the lid margins has been estimated by the technique of meibometry, which involves lifting tear oil from the lid margin onto a plastic tape, and quantifying the amount of oil present by the change in optical density. (Chew et al 1993 a,b). The amount of oil on the lid margin rises with age in both sexes, although it is lower in women in the post pubertal period up to the fifth decade (1993 b). A proportion of patients with KCS exhibit meibomian gland disease

(McCulley 1988;Chew et al 1992),but it is not yet clear whether this represents the chance association of two common disorders, or whether for instance, MGD is a feature of KCS, perhaps resulting from ocular surface injury in the same way as goblet cell loss. The presence and degree of MGD can be graded clinically, (Mathers et al 1991;Bron et al 1991) and its effect on tear osmolarity and stability has been measured (Chew 1992).

## Ocular Surface Change

A number of significant changes have been noted in the conjunctiva in KCS. Lemp has recorded a reduction in size of the corneal epithelial cells on specular microscopy, accompanied by an increase in area of of the neighbouring conjunctival epithelial cells (Lemp 1987;Lemp et al 1984). Impression cytology has demonstrated a fall in the area ratio of nucleus to cytoplasm (Nelson et al 1984; Rolando et al 1990; Prause and Marner et al 1986), and Marner (1980) and Marner et al (1986) have reported the presence of a rod- or snake-like condensation of chromatin in the upper bulbar conjunctiva in a proportion of KCS patients. The change is not confined to dry eye patients, and is for instance also encountered in contact lens wearers. It has been proposed that it may reflect frictional damage to the upper bulbar conjunctiva resulting from alterred lubrication. The reduction in bulbar conjunctival goblet cell density in NSDE biopsy material and on conjunctival impression cytology has already been noted. It is accompanied by increased keratinisation, and squamous metaplasia (Abdel-Khalek et al 1978).

Interpalpebral staining of the globe with vital dyes has long been accepted as an indicator of ocular surface damage in dry eye. The pattern may be characteristic, and the staining may be graded, to provide a diagnostic criterion. Van Bijsterveld (1969) proposed that staining be graded from 0-3 in each exposed corneal and conjunctival zone (giving a range of 0-9), using Rose Bengal as the vital dye. Using a score of 4 or above as the cut off, a high sensitivity and specificity was acheived (Table 4 ). Because of the low pH of commercial Rose Bengal drops, they are painful to use diagnostically and this is particularly so in aqueous-deficient dry eye patients,who are unable to wash the excess dye from the eye. Patients may experience ocular pain for over 24 hours following the use of the drop. A similar pattern of staining may be observed using fluoresein dye, if a yellow barrier filter (eg Kodak Wratten 12 or 15 ) is used in conjunction with the blue exciting light. Alternatively, lissamine green has been used by Norn (1973,1983) and others as an alternative vital dye with the same staining properties as Bengal Rose.

## Inflammatory Events

Infection is uncommon in KCS, despite a deficiency of antimicrobial agents in the tears (Jannsen and van Bijsterveld 1986) but redness of the globe is a commonplace feature of the disease. There is also an increase in the concentration of plasma-derived proteins in the tears (Mackie and Seal 1981,1984) which has usually been assumed to imply an increase in conjunctival vascular permeability. It has, however, never been established that the increase in tear proteins such as, albumin, ceruloplasmin and lactoferrin is due solely to permeability changes in the conjunctiva, as opposed to the lacrimal gland itself. More important still, it has always been assumed that the leakage of serum proteins into the tears reflects a change in vascular permeability. However, conjunctival capillaries are fenestrated and would be expected to be highly permeable to plasma proteins in their normal state (Raviola 1983). Since the conjunctival surface cells are attached to one another by tight junctions (Hogan et al 1972; Huang Tseng and Kenyon 1989), these surface cells themselves must represent a barrier to the diffusion of plasma proteins, and an increase in conjunctival permeability must be assumed to explain in part the rise in tear proteins of plasma origin which occurs.

In addition to signs of vasodilatation and increased permeability, whether of conjunctival epithelium or of conjunctival vessels other signs of inflammation have been detected in dry eye, such as the presence of PMNs in the tears and of inflammatory mediators such as PGE2. There is also evidence for the presence of superoxide, which could be of white cell origin (Bron 1986). It is a reasonable expectation that the inflammatory events themselves may result in ocular surface damage, and may therefore represent a damage mechanism in their own right.

## OTHER FORMS OF NSDE

Certain other forms of NSDE may be considered,characterized by a major tear component deficieny

In meibomian gland disorder (MGD),there may be both qualitative and quantitative abnormalities in lid oil production,and in some forms there is both obstruction and deficiency. Lid oil gland abnormalities have been quantified by clinical grading (Bron et al 1991), and more recently it has been possible to measure the amount of oil on the lid margin by the technique of meibometry [Chew et al 1993].

Since meibomian gland disease can coexist with KCS, the attribution of dry eye damage to MGD alone requires that KCS is excluded. In a recent study a reduction of tear stability was shown to occur in a group of patients with clinical features of MGD, and a normal Schirmer test (Chew 1992) and a reduction of meibomian oil is capable of increasing tear osmolarity. Clinically the demonstration that meibomian gland disease is not contributing to ocular surface damage in a KCS patient is usually based on the absence of any morphological evidence of gland disease at the slit-lamp. However,when MGD is clinically present, the relative contribution of aqueous and oil deficiencies to ocular surface damage cannot yet be construed. Our experience is, that these disorders compound one another.

Table 2 lists those disorders which are associated with goblet cell loss,and hence are assumed to be associated with a tear mucin deficiency. As noted earlier, there does not appear to be a pure goblet cell deficiency disorder which would allow us to determine the effect of an isolated mucin defect on the ocular surface. [Although xerophthalmia would appear to provide an example,vitamin A deficiency is responsible for a primary abnormality of the surface epithelium (xerosis) which is quite distinct In appearance from that of NSDE itself, and therefore it is not possible to determine an ocular surface change attributable to mucin lack alone.] Notwithstanding this, it should be possible to distinguish a mucin-deficient disorder such as xerophthalmia, from 'pure ' forms of KCS and MGD,by demonstrating ocular surface damage in the presence of tear mucin deficiency,and the absence of signs of aqueous or meibomian deficiency. However, as has been noted, more than one tear component deficiency may occur in a single disorder eg. goblet cell and lid oil deficiency may occur in KCS. In a study of xerophthalmia in Indonesia, Sommer and Emran (1982), and Sommer (1982) found Schirmer's test to be abnormal in 59 %. This must be assumed to be a overestimate, since a cut off value of 15 mm wetting was used in this study but it emphasizes that tests of tear component deficiency alone may not be sufficient to distinguish such disorders and that other labels may be necessary to characterize a particular form of NSDE.In the case under discussion,the presence of Bitot's spots and of a xerotic conjunctiva might serve this function.

A number of other forms of NSDE are listed in Tables 1-3.Their diagnosis depends on the use of criteria of ocular surface damage and of tear component deficiency discussed above. Their characterization as a particular form of NSDE depends on identifying unique features which place them in unique groupings, distinct from of other causes. Thus the diagnosis of the cicatricial causes of NSDE might depend on the demonstration of symblepharon, and their further separation might depend on age, clinical history, and possibly conjunctival immunocytochemistry. The attribution of NSDE to sarcoidosis would demand the diagnosis of sarcoidosis by agreed criteria, although the most stringent requirement for assignment might be the demonstration of sarcoid changes in the lacrimal gland.

**Table 5.** A Proposed Sequence of Tests for NSDE.

1. TEAR SAMPLING (eg. for lysozyme/lactoferrin; for osmolarity)
   then, leave an interval with the eyes undisturbed, for say 10 minutes.

2. [NIBUT] ie. non-invasive break-up time if performed.
   Then instill fluorescein, eg. one drop of unpreserved, Minims (SNP) saline on the end of a Fluoret strip (SNP), the excess shaken off and the residual tapped onto the lower tarsal plate.

3. FBUT ie. fluorescein break-up time.

4. [FLUORESCEIN OCULAR SURFACE STAINING]. If a yellow barrier filter is now used in front of the slit-lamp oculars, with the cobalt blue exciting light, then staining can be graded on the bulbar conjunctiva in addition to the cornea. Use of fluorescein is relatively painless compared to Bengal Rose. Then leave an interval with the eye undisturbed, eg. 10 minutes.

5. SCHIRMER TEST: Schirmer papers (Minims-SNP) are placed at the lateral third of the lids, and the lids closed. Wetting is recorded over a period of 5 minutes.

6. OCULAR SURFACE STAINING AFTER INSTILLATION OF BENGAL ROSE:
   eg. instill 2 drops of Bengal Rose 1% (Minims) into the lower conjunctival sac. Allow the patient to blink a few times while still keeping the lower lid under traction, to prevent overspill. Let the patient close the upper lid gently while using a tissue to mop the overspill as the lid closes. (There will be staining of the lower conjunctiva at the site of insertion of the Schirmer paper, but this does not interfere with the reading of the grade of interpalpebral staining over the globe and cornea).
   Rose Bengal is uncomfortable to instill, particularly in dry eye patients. Lissamine Green has been recommended as an alternative, for staining of the ocular surface, and the 1% solution is regarded as, equivalent to the 1% Bengal Rose solution. (Sautter 1976; Norn 1983; Liotet 1993). A scheme for grading ocular surface damage has been proposed by van Bijsterveld (1969).

## SELECTING DIAGNOSTIC CRITERIA FOR NSDE

Three aspects must be established in the characterisation of NSDE:
1. Diagnosis of dry eye
2. Diagnosis of mechanism
3. Diagnosis of associated systemic disorder.

### 1.Diagnosis of Dry Eye

Diagnostic criteria for a disease are commonly selected on the basis of the sensitivity and specificity data for one or more tests.In the case of dry eye the generation of such data is often confounded in a number of ways.First, in selecting persons to be assessed by the tests of interest it is necessary to distinguish between normals and affecteds.This usually depends on historical information which includes the results of at least some of the tests under assessment.This interferes with the process of assignment as normal or affected.

The size and composition of the disease group studied will also influence the derived sensitivity and specificity.The result will differ if all the disease group are mild compared to if they are all severe. Some but not all studies take this into account. Test-retest repeatability data is rarely presented. It does mean that some caution should be taken before accepting the value of a particular test.

When diagnosis is dependent on a battery of tests then the sequence in which the tests are performed may be important since one test may influence the other.This may demand that tests are done on separate days. For instance, instillation of fluorescein dye into the conjunctival sac for the FBUT test could interfere with a subsequent Schirmer test. The period to wait between tests is not established. A sequence of testing which might be acceptable is shown in Table 5.

**Table 6.** Diagnostic Criteria for Non-Sjögren Dry Eye.

| DISEASE | PRIMARY DEFIC. | 1° DAMAGE AGENT | 1° TESTS | 2° TESTS |
|---|---|---|---|---|
| KCS | Lacrimal secretion | Hyperosmol. RB[1] | Schirmer test[2] Tear lysozyme Tear lactoferrin | BUT[3], osmol. Meibometry Surface tension |
| MGD | Meibomian oil | Hyperosmol. RB[1] | Meibometry | BUT[3], osmol. |
| Vitamin A Deficiency | Tear Mucin | Mucin defic. RB[1] | Tear surface tension Goblet density Vitamin A levels | Bicots' spots |
| Blink Defic. | Film | Hyperosmol. RB[1] | Blink interval | BUT[3], osmol. |
| Surface Abnormality | Ocular surface | Hyperosmol. RB[1] | | |

KEY:   KCS = Keratoconjunctivitis sicca;  MGD = Meibomian Gland Disease;
RB = Rose Bengal Stain: 1. Fluorescein staining observed through yellow filters will provide similar information; 2. Fluorimetry provides an alternative to the Schirmer test.
BUT = Fluorescein Break-up time: 3. The non-invasive test (NIBUT) will give similar information. The NIBUT usually gives a higher value.

For most ophthalmologists the tests available readily available in the clinic are the Schirmer test, the fluorescein BUT, and grading of ocular surface damage using Rose Bengal. For each of these, Sensitivity and Specificity data have been generated by different authors, using different cut-offs and are also available for a number of other tests listed.

There is however a problem in accepting such diagnostic criteria to distinguish a dry from a normal eye, in that there is no gold standard for the disease against which various tests can be assessed. Therefore, usually the tests have been initiated in dry eye patients whose diagnosis has in part been dependent in the use of at least some of the tests which are under investigation. Van Bijsterveld (1990) has addressed this problem by using multivariate analysis to discriminate between dry and non-dry eye.

## 2. Diagnosis of mechanism

Table 6 illustrates how selected forms of NSDE might be characterised by their causative mechanism, using KCS, MGD, vitamin A and blink deficiency (eg. Parkinsonism) as examples. Each disorder shows interpalpebral ocular surface damage, and staining with vital dyes, as a nonspecific feature. The primary damage mechanism is presumed to be hyperosmolarity in all but vitamin A deficiency where a mucin deficiency and probably a glycocalyx abnormality at the ocular surface are thought to be responsible. Each disorder is characterised by loss of particular components of the tear film. The acinar atrophy of KCS is accompanied by reduced tear flow and loss of proteins synthesised by the lacrimal gland. Although it has been noted that there is evidence of goblet cell loss and mucin deficiency in this disorder, and that MGD may be associated, there will be individual patients in whom this is not the case, and who can be regarded as pure forms of the disorder, diagnosed not only with the evidence of loss of lacrimal functions, but also with the evidence of normal goblet cell and meibomian gland function. It appears that tests are now being refined sufficiently to make this characterisation possible.

In the same way, it is possible to identify populations of patients with clinical features of MGD,ocular surface damage, reduced breakup time, but a normal Schirmer test (Chew et al 1993). These are the patients with 'pure' oil gland deficiency. In xerophthalmia it appears that goblet cell loss and primary ocular surface change probably characterize the disorder adequately, although a proportion may have additional reduced aqueous secretion in addition. Information on the state of the oil glands is not available.

### 3.Diagnosis of an associated systemic disorder

The diagnosis of systemic disease as the basis of NSDE is based on the level of clinical information available or sought. The presence of conjunctival scarring or symblepharon will suggest the cicatrising disorders such as mucous membrane pemphigoid, erythema multiforme or trachoma. The presence of uveitis, pulmonary or skin disease might suggest sarcoid, while the onset of dry eye symptoms in a patient who has undergone an organ transplant may indicate the onset of a graft vs "host" reaction. This approach can be amplified to include other disorders in this category.

### Sjögren Syndrome

Sjögren syndrome is an exocrinopathy associated with selected forms of systemic autoimmune disorder including rheumatoid arthritis, polyarteritis nodosa, Wegeners granulomatosis, systemic lupus erythematosis, and systemic sclerosis. The lacrimal and salivary glands undergo invasion and damage by lymphocytes. The diagnostic ocular features of the disorder are the same as those used to diagnose KCS. Diagnosis of the salivary features is based on the presence of symptomatic xerostomia and the demonstration of a reduced basal and systemic salivary flow rate (Manthorpe et al 1986 a,b; Prause et al 1986). A positive biopsy showing extensive lymphocytic infiltration of the minor salivary glands of the buccal mucosa is taken by some to be one of the most important diagnostic features of the disorder (Alarcon-Segovia 1989; Moutsopoulos, Manoussakis 1989).

Serological evidence of systemic autoimmune disease includes the following: 1.Positive Rh F,> 1:160; 2.Positive ANA, > 1:160; 3. Positive anti Ro (SSA) or anti-La (SSB) antibodies. Patients with graft-versus host disorder, or positive HIV, although autoimmune in nature,are generally placed in the category of NSDE.

### ANIMAL MODELS OF DRY EYE

In an attempt to understand both Sjögren and non-Sjögren dry eye, the regulation of tear production in animals, and several models of dry eye have been studied. Extensive studies of the endocrine control of lacrimal secretion have been carried out in rats (Table 7). Orchidectomy increases the volume of secretion, while reducing the content of secretory component and of IgA. (This response may be species-specific since androgen administration increases tear volume in autoimmune NZB/NZW F1 female mice, and normal rabbits ). Testosterone reverses the effects in the rat, while hypophysectomy prevents this reversal (Sullivan et al 1984,1990; Sullivan, Allansmith 1986; 1987). It is relevant that dry

**Table 7.** Control of Lacrimal Function in the Rat.

| | |
|---|---|
| * | Orchiectomy increases volume and reduces IgA/SC secretion [1,2,3] |
| * | Testosterone reverses these effects [1,2,3] |
| * | Hypophysectomy prevents reversal[2] |
| * | Thyroidectomy, adrenalectomy, oestrogen therapy, no effect[3] |

[1] Sullivan et al 1984a;  [2] Sullivan, Allansmith 1985, 1986;  [3] Sullivan et al 1984b.

eye is commoner in female animals than male, in the same way that both Sjögren and non-Sjögren dry eye is commoner in women than men.

Dry eye is common in the West Highland Terrier, and 70 % occurs in female dogs (Sansom and Barnett 1985). Also, 5-amino salicylic acid toxicity,which has been reported in dogs, is more common in females (Barnett and Josephs 1987). In the dry eye syndrome which occurs in the NZB and NZB/NZW F1 mouse, lacrimal and salivary gland infiltration was greater in females (Kessler 1968,Gilbard et al 1987b). However, there are also reports of equal sex incidence of dry eye in animals (Aguirre et al 1971;Plate 1971;Helper 1970,1976.)

**Table 8.** Dry Eye Syndromes in the Dog.

| Local ocular disease | Trauma |
| | Blepharitis/conjunctivitis |
| Neurogenic | |
| Congenital | Pug |
| | Yorkshire terrier |
| Inherited | Minature snauzer |
| | American cocker spaniel |
| | English bulldog |
| | Beagle |
| | West Highland white terrier[1] |
| Drug-induced | Atropine[2] |
| | Topical anaesthesia |
| | General anaesthesia |
| Drug toxicity | Phenazopyridine[3] |
| | Sulphadiazine[4] |
| | Sulphasalazine[5] |
| | 5-ASA[6] |
| Systemic disease | Distemper |
| | Auto-immune[7] |
| | Hypothyroidism |

[1] Sansom and Barnett, 1985; [2] Aguirre et al., 1971; [3] Bryan and Slatter, 1973; [4] Todenhofer, 1969 [5] Aguirre, 1973; [6] Barnett and Joseph, 1987; [7] Kaswan et al., 1983,1984,1985,1989.

Dry eye has been induced in the rabbit by extirpation of the lacrimal gland (Maudgal 1978 ) or by lacrimal duct occlusion (Gilbard et al 1987a), in conjunction with excision of the Harderian gland and nictitans, without whose excision the full syndrome of dry eye does not occur. In the full model, there is increased tear osmolarity, increased corneal epithelial desquamation, decreased epithelial glycogen, and positive staining with Rose Bengal. Goblet cell loss occurs as in the human situation, which is proportional to the degree of tear hyperosmolarity (Gilbard et al 1978, 1979, 1988, 1989; Farris et al 1983).

In the dog too, the nictitans makes an important contribution to the total tear flow. Gelatt et al (1975) has estimated the lacrimal contribution to be 62 %, the nictitans 35 %, and the goblets cells 3 %. Curiously, in the dog, while lacrimal extirpation results in dry eye in 5-23 %, nictitans extirpation causes dry in a higher percentage, 29-57 %. However, combined extirpation causes dry eye in 100% (Helper et al 1970;1976). Barnett and Sanson (1987) have described the clinical features of dry eye in the dog,which include a thickened, corrugated and hyperaemic conjunctiva, an oedematous, vascularised cornea which ultimately ulcerates,and the production of a thick, ropy mucus. A swirling pigment migration from the conjunctiva onto the corneal surface is a characteristic feature. The several causes of dry eye in the dog are summarised in Table 8.

## CONCLUSIONS

Dry eye syndromes consist of a variety of disorders of the ocular surface caused by tear hyperosmolarity and instability. The contribution of selected tear component deficiencies to this state can now be determined more accurately than in the past, so that we are close to being able to differentiate between primary deficiencies of the lacrimal and meibomian gland and the goblet cells. Because of the convergence of the damage mechanism in these conditions, there may be evidence of combined deficiencies in such disorders, when secondary mechanisms come into operation. Keratoconjunctivitis sicca is a good example of this. A standard battery of tests may be performed in any clinic (Schirmer test, BUT and staining with vital dyes), and further tests are available which will characterise theds local mechanism and any systemic association. The tests which distinguish NSDE from Sjögren dry eye have been noted above. A variety of animal models of dry eye exist which provide an opportunity to explore both mechanism and therapy.

## REFERENCES

Abdel-Khalek, L.M.R., Williamson, J. and Lee, W.R., 1978, Morphological changes in the human conjunctival epithelium. II. In keratoconjunctivitis sicca, Br J Ophthalmol. 62: 800.

Aguirre, G.D., Rubin, L.F. and Harvey, C.E., 1971, Keratoconjunctivitis sicca in dogs, J Am Vet Med Assoc. 158: 1566.

Alarcon-Segovia, D., 1989, Editorial. Primary Sjogren's syndrome. Six Characters in search of an author, J Rheumatol. 16: 1177.

Allen, M., Wright, P. and Reid, L., 1972, The human lacrimal gland. A histochemical and organ culture study of the secretory cells, Arch Ophthalmol. 88: 493.

Ariga, H., Edwards, J. and Sullivan, D.A., 1989, Brief Communication. Adrogen control of autoimmune expression in lacrimal glands of MRL/Mp-lpr/lpr mice, Clin Immunol Immunopathol. 53: 499.

Barnett, K.C. and Joseph, E.C., 1987, Keratoconjunctivitis sicca in the dog following 5-aminosalicylic acid administration, Hum Toxicol. 6: 377.

Barnett, K.C. and Sansom, J., 1987, Diagnosis and treatment of keratoconjunctivitis sicca in the dog, Vet Rec. 120: 340.

Baum, J.L., 1976, Keratoconjunctivitis sicca, Trans Am Acad Ophthalmol Otolaryngol. 81: 619.

Bron, A.J., 1985, Prospects for the dry eye. (Duke-Elder Lecture), Trans Ophthalmol Soc UK. 104: 801.

Bron, A.J., 1986, Quantification of external ocular inflammation. In "The Preocular Tear Film in Health, Disease, and Contact Lens Wear", ed. F.J.Holly, Dry Eye Inst, Lubbock, Texas. 1: 776.

Bron, A.J., Benjamin, L. and Snibson, G.R., 1991, Meibomian gland disease. Classification and grading of lid changes, Eye. 5: 395.

Bron, A.J. and Mengher, L.S., 1987, Congenital deficiency of meibomian glands, Br J Ophthalmol. 71: 312.

Bron, A.J. and Mengher, L.S., 1989, The ocular surface in keratoconjunctivitis sicca, Eye. 3: 428.

Bron, A.J., Mengher, L.S. and Davey, C.C., 1985, The normal conjunctiva and its response to inflammation, Trans Ophthalmol Soc UK. 104: 424.

Bryan, G.M. and Slatter, D.H., 1973, Keratoconjunctivitis sicca induced by phenazopyridine in dogs, Arch Ophthalmol. 90: 310.

Chew, C.K.S., Hykin, P.G., Jansweijer, C., Dikstein, S., Tiffany, J.M. and Bron, A.J., 1993, The casual level of meibomian lipids in humans, Curr Eye Res. 12: 255.

Chew, C.K.S., Jansweijer, C., Tiffany, J.M., Dikstein, S. and Bron, A.J., 1993, An instrument for quantifying meibomian lipid on the lid margin: the Meibometer, Curr Eye Res. 12: 247.

Chew, C.K.S., Tiffany, J.M., Dikstein, S. and Bron, A.J., 1992, Lipid levels on the lid margins of patients with meibomian gland dysfunction, ARVO Suppl: Invest Ophthalmol Vis Sci. 33: 950.

Chomette, G., Auriol, M. and Liotet, S., 1986, Ultrastructural study of the lacrimal gland in a case of Sjogren's syndrome, Scand J Reheumatol Suppl. 61: 71.

Cope, C., Dilly, P.N., Kaura, R. and Tiffany, J.M., 1986, Wettability of the corneal surface: a reappraisal, Curr Eye Res. 5: 777.

Damato, B.E., Allan, D., Murray, S.B. and Lee, W.R., 1984, Senile atrophy of the human lacrimal gland: the contribution of chronic inflammatory disease, Br J Ophthalmol. 68: 674.

Dilly, P.N. and Mackie, I.A., 1981, Surface changes in the anaesthetic conjunctiva in man with special reference to the production of mucus from a non-goblet cell source, Br J Ophthalmol. 65: 833.

Editorial 1992: Diagnosis of Sjogren's Syndrome. Lancet 340: 150.

Farris, R.L., Gilbard, J.P., Stuchell, R.N. and Mandel, I.D., 1983, Diagnostic tests in keratoconjunctivitis sicca, CLAO J. 9: 23.

Gelatt, K.N. et al., 1968, Evaluation of tear formation in the dog, using a modification of the Schirmer tear test, J Am Vet Med Assoc. 166: 368.

Gilbard, J.P., Carter, J.B., Sang, D.N., Refojo, M.F., Hanninen, L.A. and Kenyon, K.R., 1984, Morphologic effect of hyperosmolarity on rabbit corneal epithelium, Ophthalmology. 91: 1205.

Gilbard, J.P. and Farris, R.L., 1979, Tear osmolarity and ocular surface disease in keratoconjunctivitis sicca, Arch Ophthalmol. 97: 1642.

Gilbard, J.P. and Farris, R.L., 1983, Ocular surface drying and tear film osmolarity in thyroid eye disease, Acta Ophthalmol. 61: 108.

Gilbard, J.P., Farris, R.L. and Santa Maia, J., 1978, Osmolarity of tear microvolumes in keratoconjunctivitis sicca, Arch Ophthalmol. 96: 677.

Gilbard, J.P., Hanninen, L.A., Rothman, R.C. and Kenyon, K.R., 1987, Lacrimal gland, cornea, and tear film in the NZB/NZW F1 hybrid mouse, Curr Eye Res. 6: 1237.

Gilbard, J.P., Rossi, S.R. and Gray, K.L., 1987, A new rabbit model for keratoconjunctivitis sicca, Invest Ophthalmol Vis Sci. 28: 225.

Gilbard, J.P., Rossi, S.R., Gray, K.L. and Hanninen, L.A., 1989, Natural history of disease in a rabbit model for keratoconjunctivitis sicca. (Supplement 192), Acta Ophthalmol. 67: 95.

Gilbard, J.P., Rossi, S.R., Gray, K.L., Hanninen, L.A. and Kenyon, K.R., 1988, Tear film osmolarity and ocular surface disease in two rabbit models for keratoconjunctivitis sicca, Invest Ophthalmol Vis Sci. 29: 374.

Gilbard, J.P., Rossi, S.R. and Heyda, K.G., 1989, Tear film and ocular surface changes after closure of the meibomian gland orifices in the rabbit, Ophthalmology. 96: 1180.

Gipson, I.K., Spurr-Michaud, S., Kublin, C., Cintron, C. and Tisdale, A., 1993, Characteristics of an ocular surface glycocalyx glycoprotein of the rat. (Abstract 2941), Invest Ophthalmol Vis Sci. 34: 1300.

Gipson, I.K., Yankauckas, M., Spurr-Michaud, S.J., Tisdale, A.S. and Rinehart, W., 1992, Characteristics of a glycoprotein in the ocular surface glycocalyx, Invest Ophthalmol Vis Sci. 33: 218.

Greiner, J.V. and Allansmith, M.R., 1981, Effect of contact lens wear on the conjunctival mucus system, Ophthalmology. 88: 821.

Greiner, J.V., Covington, H.I. and Allansmith, M.R., 1977, Surface morphology of the human upper tarsal conjunctiva, Am J Ophthalmol. 83: 892.

Greiner, J.V., Henriquez, A.S., Wideman, T.A., Covington, H.I. and Allansmith, M.R., 1979, 'Second' mucus secretory system of the human conjunctiva, ARVO Suppl: Invest Ophthalmol Vis Sci. 18: 123.

Greiner, J.V., Kenyon, K.R., Henriquez, A.S., Korb, D.R., Weidman, T.A. and Allansmith, M.R., 1980, Mucus secretory vesicles in conjunctival epithelial cells of wearers of contact lenses, Arch Ophthalmol. 98: 1843.

Greiner, J.V., Weidman, T.A., Korb, D.R. and Allansmith, M.R., 1985, Histochemical analysis of secretory vesicles in non-goblet conjunctival epithelial cells, Acta Ophthalmol. 63: 89.

Helper, L.C., 1970, The effect of lacrimal gland removal on the conjunctiva and cornea of the dog, J Am Vet Med Assoc. 157: 72.

Helper, L.C., 1976, Keratoconjunctivitis sicca in dogs. Symposium: Diseases of Unknown Etiology - Contributions of Animal Studies to their Understanding, Trans Am Acad Ophthalmol Otolaryngol. 81: 624.

Hogan, M.J., Alvarado, J.A. and Weddell, J.E., 1971, Histology of the Human Eye, W B Saunders, Philadelphia. 1: 1.

Holly, F.J., 1980, Tear film physiology, Am J Optom Physiol Optics. 57: 252.

Holly, F.J., 1986, Dry eye and the Sjogren's system, Scand J Rheumatol Suppl. 61: 201.

Holly, F.J. and Lemp, M.A., 1971, Wettability and wetting of corneal epithelium, Exp Eye Res. 11: 239.

Holm, S., 1949, Keratoconjunctivitis sicca and the sicca syndrome. (Supplement 33), Acta Ophthalmol. 27: 1.

Huang, A.J.W. and Tseng, S.C.G., 1987, Development of monoclonal antibodies to rabbit ocular mucin, Invest Ophthalmol Vis Sci. 28: 1483.

Huang, A.J.W., Tseng, S.C.G. and Kenyon, K.R., 1989, Paracellular permeability of corneal and conjunctival epithelia, Invest Ophthalmol Vis Sci. 30: 684.

Janssen, P.T. and Bijesterveld, O.P., 1986, Local antibacterial defense in the sicca syndrome. International Tear Film Symposium. Lubbock, Texas, November 1984, Proceedings (ed) F.J. Holly, . 1: 1.

Janssen, P.T. and van Bijsterveld, O.P., 1983, A simple test for lacrimal gland function: a tear lactoferrin assay by radial immunodiffusion, A von Graefes Arch Klin Exp Ophthalmol. 220: 171.

Jester, J.V., Nicholaides, N., Kiss-Polvolgyi, I. and Smith, R.E., 1989, Meibomian gland dysfunction. II. The role of keratinisation in a rabbit model of MGD, Invest Ophthalmol Vis Sci. 30: 936.

Kaswan, R.L., Martin, C.L. and Chapman, W.L., 1984, Keratoconjunctivitis sicca: histopathologic study of nictitating membrane and lacrimal glands from 28 canine cases, Am J Vet Res. 45: 112.

Kaswan, R.L., Martin, C.L. and Dawe, D.L., 1983, Rheumatoid factor determination of 50 dogs with keratoconjunctivitis sicca, J Am Vet Med Assoc. 183: 1073.

Kaswan, R.L., Martin, C.L. and Dawe, D.L., 1985, Keratoconjunctivitis sicca: immunological evaluation of 62 canine cases, Am J Vet Res. 46: 376.

Kaswan, R.L. and Salisbury, M.A., 1990, A new perspective on canine keratoconjunctivitis sicca. Treatment with ophthalmic cyclosporine, Vet Clin North Am Small Anim Pract. 20: 583.

Kaswan, R.L., Salisbury, M.A. and Ward, D.A., 1989, Spontaneous canine keratoconjunctivitis sicca. A useful model for human keratoconjunctivitis sicca: treatment with cyclosporine eye drops, Arch Ophthalmol. 107: 1210.

Kaura, R. and Tiffany, J.M., 1985, The role of mucous glycoproteins in the tear film: A comparative rheological study of native tears and model mucus solutions including a purified ocular mucous glycoprotein. Int.Tear Film Symposium, Lubbock, Texas, 1974, . 1: 1.

Kern, T.J., Erb, H.N., Schaedler, J.M. and Dougherty, E.P., 1988, Scanning electron microscopy of experimental keratoconjunctivitis sicca in dogs: cornea and bulbar conjunctiva, Vet Pathol. 25: 468.

Kiorpes, T.C., Kim, Y.C.L. and Wolf, G., 1979, Stimulation of the synthesis of specific glycoproteins in corneal epithelium by vitamin A , Exp Eye Res. 28: 23.

Kogbe, O., Liotet, S. and Tiffany, J.M., 1991, Factors responsible for tear ferning, Cornea. 10: 433.

Lambert, R.W. and Smith, R.E., 1988, Pathogenesis of blepharoconjunctivitis complicating 13-cis-retinoic acid (Isotretinoin) therapy in a laboratory model, Invest Ophthalmol Vis Sci. 29: 1559.

Lemp, M.A., 1973, The mucin-deficient dry eye, Int Ophthalmol Clin. 13: 185.

Lemp, M.A., Gold, J.B., Wong, S., Mahmood, M. and Guimaraes, R., 1984, An in vivo study of corneal surface morphologic features in patients with keratoconjunctivitis sicca, Am J Ophthalmol. 98: 426.

Lemp, M.A. and Hamill, J.R., 1973, Factors affecting tear film break up in normal eyes, Arch Ophthalmol. 89: 103.

Lemp, M.A., Holly, F.J., Iwata, S. and Dohlman, C.H., 1970, The precorneal tear film, Arch Ophthalmol. 83: 89.

Lemp, M., 1987, Diagnosis and treatment of tear deficiencies. In "Clinical Ophthalmology", eds. T.D.Duane, E.A.Jaeger (Chapter 14), Harper & Row, NY & Philadelphia. 4: 1.

Mackie, I.A. and Seal, D.V., 1981, The questionably dry eye, Br J Ophthalmol. 65: 2.

Mackie, I.A. and Seal, D.V., 1984, Diagnostic implications of tear protein profiles, Br J Ophthalmol. 68: 321.

Manthorpe, R., Andersen, V., Jensen, O.A., Oxholm, P., Prause, J.U. and Schoidt, M., 1986, Editorial comments to the four sets of criteria for Sjogren's syndrome, Scand J Reheumatol Suppl. 61: 31.

Manthorpe, R., Oxholm, P., Prause, J.U. and Schiodt, M., 1986, The Copenhagen criteria for Sjogren's syndrome, Scand J Reheumatol Suppl. 61: 19.

Marner, K., 1980, Snake-like appearance of nuclear chromatin in conjunctival epithelial cells from patients with keratoconjunctivitis sicca, Acta Ophthalmol. 58: 849.

Marner, K., Manthorpe, R. and Prause, J.U., 1984, Snake-like nuclear chromatin in imprints of conjunctival cells from patients with Sjogren's syndrome. (Pub. Rheumatology service, Hasharon Hospital), Prog Rheumatol. 11: 127.

Mastmann, G.L., Baldes, E.J. and Hendersen, J.W., 1961, The total osmotic pressure of tears in normal and various pathologic conditions, Acta Ophthalmol. 65: 509.

Mathers, W.D., Shields, W.J., Sachdev, M.S., Petroll, W.M. and Jester, J.V., 1991, Meibomian gland dysfunction in chronic blepharitis, Cornea. 10: 277.

Maudgal, P.C., 1978, The epithelial response in keratitis sicca and keratitis herpetica (an experimental and clincal study), Doc Ophthalmol. 45: 223.

McCulley, J.P., 1988, Meibomitis. In "The Cornea", eds. H.E.Kaufman, B.A.Barron, M.B.McDonald, S.R.Waltman, Churchill Livingstone, Lond & Edin. 1: 125.

McCulley, J.P., Dougherty, J.M. and Deneau, D.G., 1982, Classification of chronic blepharitis, Ophthalmology. 89: 1173.

McCulley, J.P. and Sciallis, C.F., 1977, Meibomian keratoconjunctivitis, Am J Ophthalmol. 84: 788.

Mengher, L.S., 1990, Tear assessment in the dry eye. (University of Oxford), D Phil Thesis. 1: 1.

Mengher, L.S., Bron, A.J., Tonge, S.R. and Gilbert, D.J., 1985, A non-invasive instrument for clinical assessment of the pre-corneal tear film stability, Curr Eye Res. 4: 1.

Mengher, L.S., Bron, A.J., Tonge, S.R. and Gilbert, D.J., 1985, Effect of fluorescein instillation on the pre-corneal tear film stability, Curr Eye Res. 4: 9.

Mishima, S., Kubota, Z. and Farris, R.L., 1971, The tear flow dynamics in normal and in keratoconjunctivitis sicca cases, in Ophthalmology Proceedings of the XXI International Congress Mexico, D.F. 8-4, March 1970, ed. M.P.Solanes, Excerpta Medica; Amsterdam, NY, Oxford. 2: 1801.

Mishima, S. and Maurice, D.M., 1961, The oily layer of tear film and evaporation from the corneal surface, Exp Eye Res. 1: 39.

Mountz, J., 1990, Animal models of systemic lupus erythematosus and Sjogren's syndrome, Curr Opin Rheumatol. 2: 740.

Moutsopoulos, H.M. and Manoussakis, M.N., 1989, Immunopathogenesis of Sjogren's syndrome: "facts and fancy", Autoimmunity. 5: 17.

Murray, S.B., Lee, W.R. and Williamson, J., 1981, Ageing changes in the lacrimal gland: A histological study, J Clin Exp Gerontol. 3: 1.

Nasu, M., Matsubara, O. and Yamamoto, H., 1984, Post-mortem prevalence of lymphocytic infiltration of the lacrymal gland: a comparative study in autoimmune and non-autoimmune diseases, J Pathol . 143: 11.

Nelson, S.D. and Wright, J.C., 1984, Conjunctival goblet cell densities in ocular surface disease, Arch Ophthalmol. 102: 1049.

Norm, M.S., 1969, Desiccation of the precorneal film. I. Corneal wetting time, Acta Ophthalmol. 47: 865.

Norn, M.S., 1973, Lissamin green. Vital staining of cornea and conjunctiva, Acta Ophthalmol. 51: 483.

Norn, M.S., 1983, External eye. Methods of examination. (2nd Edition), Scriptor. 1: 1.

Ohnishi, Y. and Kohno, T., 1979, Polychlorinated biphenyls poisoning in monkey eye, Invest Ophthalmol Vis Sci. 18: 981.

Plate, H.R., 1971, Surgical treatment of keratoconjunctivitis sicca in the dog, Veterinaria Mexico. 1: 4.

Prause, J.U., Manthorpe, R., Oxholm, P. and Schiodt, M., 1986, Definition and criteria for Sjogren's syndrome used by the contributors to the first international seminar on Sjogren's syndrome - 1986, Scand J Reheumatol Suppl. 61: 17.

Ralph, R.A., 1975, Conjunctival goblet cell density in normal subjects and in dry eye syndromes, Invest Ophthalmol. 14: 299.

Raviola, G., 1983, Conjunctival and episcleral blood vessels are permeable to blood borne horseradish peroxidase, Invest Ophthalmol Vis Sci. 24: 725.

Rolando, M., Baldi, F. and Calabria, G.A., 1985, Tear mucus ferning in K.C.S. International Tear Film Symposium. Lubbock, Texas, November 1984, Proceedings (ed) F.J.Holly, . 1: 1.

Rolando, M., Refojo, M.F. and Kenyon, K.R., 1983, Increased tear evaporation in eyes with keratoconjunctivitis sicca, Arch Ophthalmol. 101: 557.

Rolando, M., Terragna, F., Giordano, G. and Calabria, G., 1990, Conjunctival surface damage distribution in keratoconjunctivitis sicca. An impression cytology study, Ophthalmologica. 200: 170.

Sansom, J. and Barnett, K.C., 1985, Keratoconjunctivitis sicca in the dog: a review of 200 cases, J Small Anim Pract. 26: 121.

Sauter, J., 1976, Xerophthalmia and measles in Kenya. v. Denderen, Groningen, . 1: 235.

Scherz, W. and Dohlman, C., 1975, Is the lacrimal gland dispensible? Keratoconjunctivitis sicca after lacrimal gland removal, Arch Ophthalmol. 93: 281.

Sommer, A., 1982, Nutritional blindness. Xerophthalmia and keratomalacia, Oxford Univ Press. 1: 1.

Sommer, A. and Emran, N., 1982, Tear production in a vitamin A responsive xerophthalmia, Am J Ophthalmol. 93: 84.

Stuchell, R.N., Farris, R.L. and Mandel, I.D., 1981, Basal and reflex human tear analysis. II. Chemical analysis: lactoferrin and lysozyme, Ophthalmology. 88: 858.

Sullivan, D.A. and Allansmith, M.R., 1986, Hormonal modulation of tear volume in the rat, Exp Eye Res. 42: 131.

Sullivan, D.A. and Allansmith, M.R., 1987, Hormonal influence on the secretory immune system of the eye: endocrine interactions in the control of IgA and secretory component levels in tears of rats, Immunology. 60: 337.

Sullivan, D.A., Bloch, K.J. and Allansmith, M.R., 1984, Hormonal influence on the secretory immune system of the eye: androgen regulation of secretory component levels in rat tears, J Rheumatol. 132: 1130.

Sullivan, D.A., Hann, L.E., Yee, L. and Allansmith, M.R., 1990, Age- and gender-related influence on the lacrimal gland and tears, Acta Ophthalmol. 68: 188.

Takakusaki, I., 1969, Fine structure of the human palpebral conjunctiva with special reference to the pathological changes in vernal catarrh, Arch Histol Jap. 30: 247.

Tiffany, J.M., 1987, The lipid secretion of the meibomian glands, Adv Lipid Res. 22: 1.

Tiffany, J.M., 1990, Measurement of wettability of the corneal epithelium. 2. Contact angle method, Acta Ophthalmol. 68: 182.

Tiffany, J.M., 1990, Measurement of wettability of the corneal epithelium. 1. Particle attachment method, Acta Ophthalmol. 68: 175.

Tiffany, J.M., Winter, N. and Bliss, G., 1989, Tear film stability and tear surface tension, Curr Eye Res. 8: 507.

Todenhofer, H., 1969, Toxische Nebenwirkungen von Sulfadiazin (Debenal, Sulfatidin) bei der Anwendung als Geriatrikum fur Hande, Dtsch Tierarztl Wochenschr. 76: 14.

Tseng, S.C.G., 1985, Staging of conjunctival squamous metaplasia by impression cytology, Ophthalmology. 92: 728.

Tseng, S.C.G., Huang, A.J.W. and Sutter, D., 1987, Purification and characterization of rabbit ocular mucin, Invest Ophthalmol Vis Sci. 28: 1473.

van Bijsterveld, O.P., 1969, Diagnostic tests in the sicca syndrome, Arch Ophthalmol. 82: 10.

van Bijsterveld, O.P., 1990, Diagnosis of keratoconjunctivitis sicca associated with tear gland degeneration. In "The Lacrimal System", eds. O.P.van Bijsterveld, M.A.Lemp, D.Spinelli, Kugler & Ghedini Publ, Amsterdam, Berkeley, Milano. 1: 1.

van Bijsterveld, P.O., 1974, Standardization of the lysozyme test for a commercially available medium. Its use for the diagnosis of the sicca syndrome, Arch Ophthalmol. 91: 432.

Vendramini, A.C.L.M., Soo, C. and Sullivan, D.A., 1991, Testosterone-induced suppression of autoimmune disease in lacrimal tissue of a mouse model (NZB/NZW F1) of Sjogren's syndrome, Invest Ophthalmol Vis Sci. 32: 3002.

von Bahr, G., 1941, Konnte der Flussig keitsabgang durch die cornea von physiologischer Bedentung sein, Acta Ophthalmol. 19: 125.

Watanabe, H., Tisdale, A.S. and Gipson, I.K., 1993, Eyelid opening induces expression of a glycocalyx glycoprotein of rat ocular surface epithelium, Invest Ophthalmol Vis Sci. 34: 327.

Williamson, J., 1985, Modern studies of lacrimal gland and conjunctival histology in the aging population. International Tear Film Symposium. Lubbock, Texas, November 1984, Proceedings (ed) F.J. Holly, . 1: 1.

Williamson, J. and Loudon Brown, R., 1978, The eye in connective tissue disease, Edward Arnold, Lond. 1: 1.

# ABNORMALITIES OF THE STRUCTURE OF THE SUPERFICIAL LIPID LAYER ON THE IN VIVO DRY-EYE TEAR FILM

Marshall G. Doane

Biomedical Physics Unit
Schepens Eye Research Institute
and
Department of Ophthalmology
Harvard Medical School
Boston, MA

## INTRODUCTION

The complexity of the tear film covering the anterior surface of the human cornea has become increasingly evident with recent investigations into its structure and chemical composition. The three-layered structure proposed by Wolff[1], perhaps useful as an approximation, is a considerable simplification of reality. While the oily lipid portion of the tear film usually forms a thin layer on the anterior surface of the tear film, as proposed by Wolff, the distribution of other components, such as mucins, appear to be not as well stratified. For instance, there is increasing evidence that a significant portion the mucin component, in highly-hydrated form, is present throughout the aqueous phase as well as being adherent, in a rather thick coating, to the surface of the epithelial layer of the cornea. This mucin coating is much thicker than that proposed by early investigators, at least 1 micron, and perhaps very much greater. Most of this mucin is probably restricted in its ability to freely enter the aqueous phase of the tear film. Thus, the *fluid* portion of the tear film is now often described as having two layers, rather than three. These consist of the floating lipid phase, secreted primarily by the meibomian glands within the eyelids, and the aqueous phase, a watery solution containing "everything else", the major components being secreted by the main and accessory lacrimal glands.

Abnormalities in the volume, composition, thickness, and other structural factors of the tear film can be expected to cause problems in corneal coverage, wetting action, evaporation characteristics, etc. Such abnormalities can give rise to decreased tear film stability, expressed as shortened tear film breakup times, resulting in the class of symptoms which are grouped into the common description of "dry-eye". Thus, while the insufficient tear volume associated with keratoconjunctivitis sicca appears to be a major cause of dry-eye, it is by no means the only factor. It is the purpose of this article to describe and illustrate a different type of dry-eye modality, which appears to be linked to abnormalities in the superficial lipid layer of the tear film , which may or may not be accompanied by insufficient tear volume.

## DRY-EYE AND MEIBOMIAN GLAND DYSFUNCTION

For some time it has been evident to many investigators that the "dry-eye" malady is not a simple, single manifestation but can encompass a multitude of causes giving rise to similar symptoms, such as burning, dryness, itching, and related discomfort. Certainly insufficiency of tear fluid volume is a common finding, but abnormalities in any of the other major components, or in the blinking mechanism, can also be a contributory or major factor. Thus, meibomian gland dysfunction, resulting in insufficient or compromised lipid secretion, can, in itself, result in a class of dry-eye symptoms. During the past decade, such abnormalities have been described and photographed by several groups.

For instance, Klip et al[2] and Hamano and colleagues[3] have used differential interference microscopy and reflectance microscopy to observe relatively small areas of the tear film under high magnification, noting changes in the colored interference patterns with alterations in the lipid layer, that often correlated with the clinical observation of various dry-eye conditions. Similarly, Josephson[4] has recorded the interference colors of the lipid layer by specular reflectance from the prescleral tear film, using a photographic slit lamp with a camera (still or video) mounted in one eyepiece. Guillon[5] has used a custom-designed device to obtain detailed color photographs of the lipid layer on the precorneal tear film in normal and dry-eye cases. Thus, there is considerable evidence that an abnormal lipid layer on the preocular tear film is associated with many manifestations of dry-eye. The current study further illustrates some of the basic structural changes in the precorneal superficial lipid layer associated with dry-eye, using a specialized video-recording tear film interferometer.

## METHODS: THE TEAR FILM INTERFEROMETER

The instrument used to obtain the interference images and the optical principles of thin-film interferometry have previously been described in detail[6] so only a cursory description will be provided here. The device is constructed specifically for optimizing the observation of thin film interference fringes from the in-vivo tear film. The instrument is mounted upon an optical table having a slit-lamp-type head-rest attached to one end. High-contrast interference fringes can be readily observed that originate from the lipid layer overlying the tear film on normal and dry-eye subjects. A tungsten-halogen white-light source is directed by condensing lenses onto a translucent screen. This screen is imaged by a large lens and diagonal mirror onto the central cornea such that the specular reflectance of the incident light can be collected by a third optical system and brought to a focus upon the detector of either a low-light level color CCD video camera or a silicon-intensified-target (SIT) monochrome video camera .

A key feature of the instrument is that a circular area of approximately 5mm diameter can be observed from virtually anywhere on the anterior surface of the eye, and the dynamics of the lipid spreading and thickness distribution can be recorded in high-resolution video. Appropriate filtering of the light source is made to exclude all radiation outside the visible spectrum, i.e., outside the 400-700 nm range. Light intensities are kept as low as possible, facilitated by the 0.5 lux usable sensitivity of the color video detector, with a minimum of 0.001 lux required when using the monochrome video SIT camera.

In use, the subject's head is positioned by the head-rest, which has a lateral movement to allow centering the illumination beam on the desired eye. The optical table supporting the light source, interferometric system, and camera are moved on ball-bearing ways to the proper distance from the subject's cornea so that in-focus interference patterns are seen on a video monitor. The subject is asked to either blink normally or, when evaluating tear film stability, to suppress blinking until the tear film is seen to undergo thinning and eventual breakup. The entire process is recorded on S-VHS video tape for later review.

## RESULTS

The variations in lipid layer structure and thickness distribution over large areas of the cornea are easily seen using the interferometric technique described above. Importantly, the *dynamics* of motion of the lipid and tear film as a result of the blinking lids are also evident, an action that can only be appreciated by viewing the full-motion video recordings. For instance, in normal eyes the oily lipid can be seen streaming upward from the inferior lid margin following strong blinks when the inferior limbal region is observed. It often requires several subsequent blinks to distribute this lipid in a fairly uniform layer over the tear film.

In normal, non-symptomatic eyes, the lipid layer is near the critical thickness to allow interference colors to be seen. Even in normal eyes however, the thickness can vary substantially at varying locations on the anterior ocular surface. There also can be considerable variation in the thickness distribution of the lipid layer from blink to blink, although in normal eyes it moves quite freely over the underlying aqueous phase of the tear film.

The blink action drastically alters the lipid distribution; as the upper lid descends during the initial phase of a blink, the lipid is forced to "pile up" on itself, since it normally does not enter the "hydrophilic" space beneath the lids. Since the interpalpebral area is rapidly diminishing as the upper lid descends, the lipid layer must necessarily become thicker, a process easily documented by tear film interference patterns. As the upper lid returns to its open, resting position the lipid film is seen to follow it, acting much as if it was attached to the upper lid margin. Thus, no "bare" aqueous surface of the tear film is normally exposed. When the upper lid has come to rest following the blink, the lipid layer is seen to continue its upward motion for another second or two, due to the inertia imparted to it by the lid motion.

Figures 1-6 which follow are photographs taken as the video recordings of the tear film of various subjects are played back on a high-resolution monitor.

## DISCUSSION

As can be seen from Figures 1-6, there are obvious differences between the lipid layer in a normal eye and that in *some* instances where dry-eye symptoms are present. The lipid layer thickness and distribution pattern appear markedly different, and the presumption can be made that some or all of the dry-eye symptoms are caused by these abnormalities.

That abnormal spreading of the lipid on the surface of the aqueous phase of the tear film can cause more rapid drying with associated shortened breakup times does not come as a surprise. We know that a normal, intact lipid layer reduces the rate of evaporation of the underlying aqueous phase of the tear film[7]. Holly[8] has stated that the lipid-air interface has about half the surface tension as would be the case for pure water in direct contact with the air, thus greatly increasing the ability of the aqueous portion of the tear film to spread uniformly over the surface of the cornea (or, more properly, over the surface of the adherent mucin layer coating the epithelium). Also, the accumulation of lipid in thick patches, particularly the non-polar oils, may contaminate the underlying mucin layer, thus making it unwettable. A variation of this kind of scenario has been proposed by Holly[8], and Lemp[9] where lipid migrates downward through the aqueous phase of the tear layer, causing localized areas of non-wetting on the mucin overlying the epithelium.

A more recent theory relating to the inevitable breakup of the tear film in normal and abnormal eyes has been given by Fatt[10]. As he notes, the lids can be held open manually until the tear film breaks. At the break point, the suction from the tear in the lid margin (due to the surface tension difference between the tear fluid in the marginal meniscus and that in the preocular tear film) has thinned the tear film on the cornea to the point where the tensile strength of this thin water layer is less than the suction force created by the menisci in the lid margins. While this scenario does not presume the direct action of lipid or mucin in the thinning and breakup process *per se*, the absence or irregular distribution of lipid on the surface of the aqueous phase can significantly speed up the thinning process due to an increase in evaporation rate. Also, the fact that the overlying lipid layer diminishes by about 50% the surface tension of the tear layer would be expected to lengthen the time that would be required for the surface tension difference (now considerably reduced) between the tear fluid in the marginal menisci and that in the tear film proper to thin the tear film to the point where breakup occurs. Thus, the presence of an intact lipid layer is an important one even in this surface-tension-driven mechanism of tear film breakup.

It should be pointed out that it is not unusual for short-term irregularities in the superficial lipid layer to be seen in normal subjects exhibiting no dry-eye symptomology. It is my experience, however, that in normal subjects such surfacing irregularities *are* transient, and within another blink or two the lipid pattern usually reverts to a normal, thin, relatively uniform distribution pattern. In the cases reported here, however, the lipid layer irregularities were pronounced and persistent.

Of course there are many instances where dry-eye symptoms *are* present, yet the superficial lipid layer appears normal, or nearly so. These facts support the conclusion that dry-eye is a multi-faceted malady, caused by any one, or combination, of several physiological problems. It appears likely that abnormalities in *any* of the major components of the tear film (i.e., lipid, aqueous volume, mucin, etc.) can give rise to the range of symptoms we associate with dry-eye.

In any case, it is certainly true that abnormal lipid spreading is often observed in dry-eye subjects; to what extent these abnormalities *cause* the dry-eye symptoms remains to be shown. It is surely suggestive, however, that there is a definitive cause and effect relationship between the drastically altered lipid distribution patterns often seen, as illustrated

Figure 1.  Typical appearance of normal lipid layer on precorneal tear film. Most of the illuminated area has uniform, minimal color, indicating a smooth, thin lipid layer with relatively little thickness variation.

Figure 2.  Same eye as in Fig. 1, but with fairly uniform thickening of lipid layer due to smaller palpebral fissure width as lid is slightly closed, giving more pronounced overall color.

Figure 3.  Punctate "orange peel" appearance of lipid in subject exhibiting moderate dry eye symptoms. Poor lipid uniformity.

Figure 4.  Irregular surface of tear film, poor lipid uniformity, perhaps with a contribution from contaminated mucin. Moderate dry-eye symptoms.

Figure 5.  Pronounced overall punctate lipid "clumping"; highly irregular lipid coverage. Moderate-to-severe dry-eye symptoms.

Figure 6.  Punctate lipid distribution over superior area of cornea only; very sparse lipid coverage of inferior tear layer. Most inferior meibomian ducts were blocked and dysfunctional. Moderate-to-severe dry-eye symptoms.

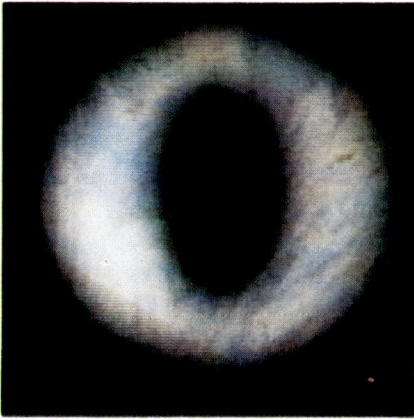

Fig. 1

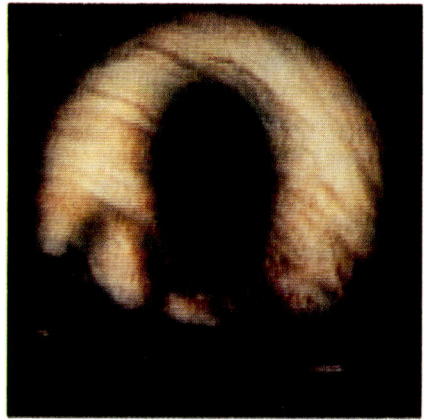

Fig. 2

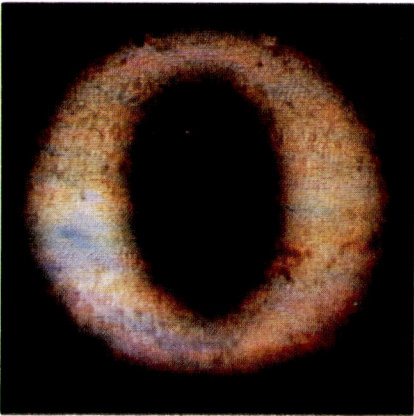

Fig. 3

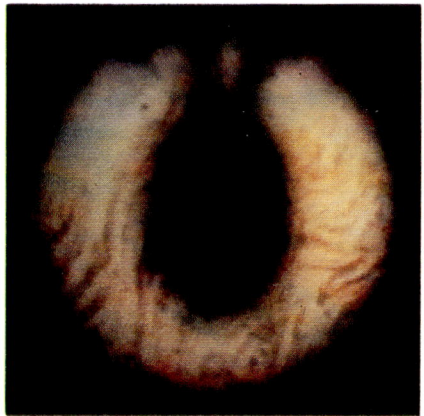

Fig. 4

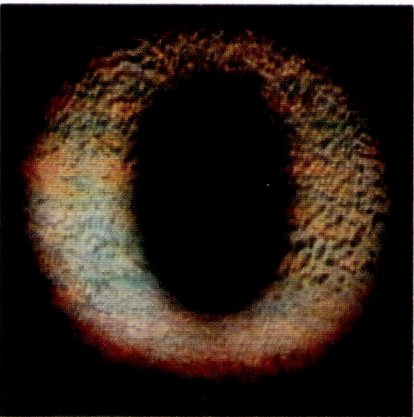

Fig. 5

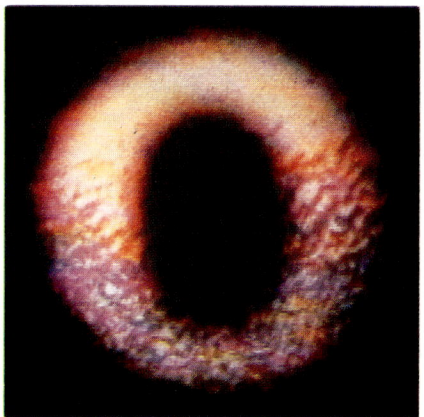

Fig. 6

here, and the dry-eye symptomology accompanying them. As Korb[11] has observed, certain kinds of contact lens intolerance with associated symptoms of ocular dryness and fluorescein staining are also accompanied by lipid irregularities, most notably obstruction of the meibomian gland orifices and/or expression of abnormal, milky lipid. Forceful expression of the lipid to clear the meibomian ducts and treatment of the meibomian glands by daily hot compresses and frequent cleansing of the lid margins using a cotton swab will often alleviate the contact lens intolerance and associated symptoms of dryness and staining, indicating the direct cause-and-effect of lipid abnormalities.

## REFERENCES

1.  E. Wolff."Anatomy of the Eye and Orbit, 4th Edition," Blakiston Co., New York (1954). P. 207.

2.  H. Klipp, E. Schmid, L. Kirchner, and A. Zipf-Pohl, Tear film observation by reflecting microscopy and differential interference contract microscopy, *in*: "The Preocular Tear Film in Health, Disease, and Contact Lens Wear," F.J. Holly, ed., Dry Eye Institute, Lubbock TX (1986). P. 564.

3.  H. Hamano, M. Hori, H. Kawabe, M. Umeno, S. Mitsunaga, Y. Ohnishi, and I. Koma, Change of surface pattern of precorneal tear film due to secretion of meibomian gland, *Folia Ophthalmol Jap.* 31:353 (1980).

4.  J. E. Josephson, Appearance of the preocular tear film lipid layer, *Am. J. Optom. and Physiol. Opt.* 60:883 (1983).

5.  J.P. Guillon, Tear film structure and contact lenses, *in*: "The Preocular Tear Film in Health, Disease, and Contact Lens Wear," F.J. Holly, ed., Dry Eye Institute, Lubbock TX (1986). P. 914.

6.  M.G.Doane, An instrument for in vivo tear film interferometry, *Optom Vis Science* 66:383 (1989).

7.  S. Mishima and D.M. Maurice, The oily layer of the tear film and evaporation from the corneal surface, *Exp. Eye Res.* 1:39 (1961).

8.  F.J. Holly, Formation and rupture of the tear film, *Exp. Eye Res.* 15:515 (1973).

9.  M.A. Lemp and J.R. Hamill, Factors affecting tear film breakup in normal eyes, *Arch. Ophthalmol.* 89:103, (1973).

10. I. Fatt and B.A. Weissman, "Physiology of the Eye. An Introduction to the Vegetative Functions, 2nd Edition," Butterworth-Heinemann, Boston (1992). P. 235.

11. D.R. Korb and A.S. Henriquez, Meibomian gland dysfunction and contact lens intolerance, *J. Am. Optom. Assoc.* 51:252, (1980).

# TEAR OSMOLARITY - A NEW GOLD STANDARD?

R. Linsy Farris

Department of Ophthalmology, College of Physicians and Surgeons of
Columbia University, and The Edward S. Harkness Eye Institute of
Presbyterian Hospital, and Harlem Hospital Medical Center, 506 Lenox
Avenue, New York, NY 10037

## PROPOSAL

The purpose of this presentation is to propose tear osmolarity measurement as a new
gold standard for the diagnosis of keratoconjunctivitis sicca. Agreement on the diagnostic
criteria is needed in order to permit meaningful comparisons of scientific results obtained
from varying patient populations. The method employed is a review of previous studies of
clinical symptoms, signs and diagnostic tests employed in making a diagnosis of
keratoconjunctivitis sicca. Data from previous studies are reviewed and the results of
individual and combinations tests are compared in regard to sensitivity, specificity and overall
efficiency in establishing accurate diagnoses. The results are that the measurement of tear
osmolarity measurement provides the greatest sensitivity, specificity and overall efficiency of
a single test. Adding either the Schirmer test without anesthetic or tear lactoferrin measured
by the Lactoplate™ method in parallel to tear osmolarity measurement did not increase the
sensitivity of diagnotic testing beyond 90% which was obtained by using tear osmolarity
measurement alone. The specificity of such combination diagnostic testing was increased
only from 95% to 100%. The simplicity of tear osmolarity measurement and its established
reliability supports the conclusion that this test is a reasonable candidate for a new
international gold standard in the diagnosis of keratoconjunctivitis sicca.

## INTRODUCTION

The Proceedings of the First International Seminar on Sjögren's syndrome was
published in 1986 as a supplement to the Scandinavian Journal of Rheumatology. Prause,

Manthrope, Oxholm and Schiodt reported significant lack of agreement on the definitions and criteria used for Sjögren's syndrome.[1]  Although ninety-six percent (96%) of the contributors agreed on the simultaneous presence of keratoconjunctivitis sicca (KCS) and xerostomia as the definition of primary Sjögren's syndrome, all required one or more abnormal objective tests for the definition of keratoconjunctivitis sicca.  Only thirteen percent (13%) required subjective symptoms as well.  The number of abstracts requiring differing number of tests were thirty (30) for at least two (2) tests, thirteen (13) for at least one (1) abnormal test, five (5) for at least three (3), and five (5) for at least four (4) abnormal objective tests.

Similarly, all contributors defined xerostomia by the presence of one or more abnormal objective tests, but only twenty-two percent (22%) required the presence of subjective symptoms as well.  Twenty-nine percent (29%) required at least one abnormal objective tests, twenty-one (21) demanded at least two (2) abnormal objective tests while three (3) abstract contributors demanded at least three (3) abnormal tests.

Ninety-four percent (94%) of the participants agree upon the terminology of primary and secondary Sjögren's syndrome with all in agreement that secondary Sjögren's syndrome indicated along with, the keratoconjunctivitis sicca and/or xerostomia, another well defined chronic inflammatory connective tissue disease such as lupus, rheumatoid arthritis or dermato(poly)-myositis.  The survey indicated that more criteria were required for a diagnosis of keratoconjunctivitis sicca than xerostomia.

From this conference, four (4) set of criteria emerged for the diagnosis of Sjögren's Syndrome.[2]  They were called the Copenhagen, Japanese, Greek and California criteria.  The Copenhagen criteria were formulated in 1976 and 1977 and were based upon only objective tests.[3]  The objective tests used for the diagnosis of keratoconjunctivitis sicca and xerostomia required at least two (2) of three (3) tests which must be abnormal for a diagnosis of KCS or xerostomia.  The Greek criteria required only one abnormal test, whereas the Japanese and California also required at least two (2) abnormal tests.  The Schirmer I test and the van Bijesterveld Rose Bengal staining score were the most frequently used tests although the Copenhagen group also used the tear film break up time.  The disadvantage seen for Rose Bengal staining and tear film break up time was the requirement for a slit lamp exam and an ophthalmologist.  The Japanese and California criteria required abnormal fluorescein staining and/or van Bijesterveld Rose Bengal staining score as a separate point.


## STATEMENT OF PURPOSE

As pointed out at this meeting, international discussions are needed to obtain agreement on the criteria for Sjögren's syndrome and its components so that meaningful comparisons of scientific results can be obtained from different patient populations.

The inability to compare results of studies becomes even more pronounced when in addition to the utilization of different tests, different cutoff or referent values are used for an abnormal test result.  Populations for studies are selected in different settings which produces considerable variation in study results.  For example, a rheumatology clinic and an ophthalmologist's office or a dental clinic may examine Sjögren's syndrome patients presenting with a variety of chief complaints. In many cases the severity of symptoms of one organ system may mask or cause the patient to overlook mild symptoms from a disorder in another system. A study investigating the incidence of serum autoantibodies in Sjögren's syndrome reported a ten-fold disparity of serum autoantibodies was detected in the patients with keratoconjunctivitis sicca presenting in an ophthalmologist's office compared to those presenting in a rheumatology clinic.[4]

A "gold standard" for the diagnosis of the disorders of each organ system in Sjögren's syndrome is required which can be used by all researchers. The gold standard is the test or criteria used to unequivocally define the disease.[5] At the present each investigator or geographic region seems to have their own criteria. I would like to review my efforts to determine if tear osmolarity could be a "gold standard" for keratoconjunctivitis sicca.


## HISTORY OF THE TEAR OSMOLARITY TEST

Twenty-four years ago I was invited by Sai Mishima to join him and Zenichi Kubota to measure the osmolarity of microvolumes of tears in normals and dry eye patients in relation to their tear flow as measured by fluorescein dilution.[6] We selected dry eye patients on the basis of history and clinical examination. Compared to a group of normals, the results indicated a distinct separation of the two groups into dry eye patients with elevated tear osmolarity and low flow and normals with only a slightly hypertonic tear film and greater tear flow (Figure 1).

Eight years later, a medical student, Jeff Gilbard, asked to do some summer research. We had a new instrument called a Nanoliter Tear Osmometer purchased from Clifton Technical Physics, Hartford, New York to replace an older, more cumbersome Ramsay-Brown micro-osmometer. We had not been able to standardize and use the instrument. Jeff made the instrument work and compared a group of dry eye patients collected on the basis of symptoms and at least one of the following signs: a deficient inferior marginal tear strip, debris in the tear film, or a viscous appearing tear film. A group of normal subjects and a group of patients with conjunctivitis were used as controls. Distinct separation of the dry eye population was evident and we adopted 312 mOsm/L as the cutoff or abnormal referent value. The tear osmolarity test in this initial study was found to be 94.7% sensitive and 93.7% specific (Figure 2).[7]

My next question was, "How does this compare with other diagnostic tests?" Having been stimulated to investigate tear tests because of the limitations of the Schirmer test, I wanted to compare its results with tear osmolarity, Rose Bengal staining and other clinical tests of tear function. In a group of 28 eyes with KCS diagnosed according to symptoms and at least one of three slit lamp findings, tear osmolarity was positive in all cases but the Schirmer test with anesthetic was positive in only 29% when using cutoff values of 312 Mosm/L for the tear osmolarity test and 5mm of wetting in five minutes for the Schirmer test with anesthetic.[8] Similarly in a group of 23 eyes with KCS, tear osmolarity was positive in all cases but tear film break up time was positive in only 43% using referent values of 312 Mosm/L for the tear osmolarity test and less than ten seconds for the tear film break up time (BUT).[7]

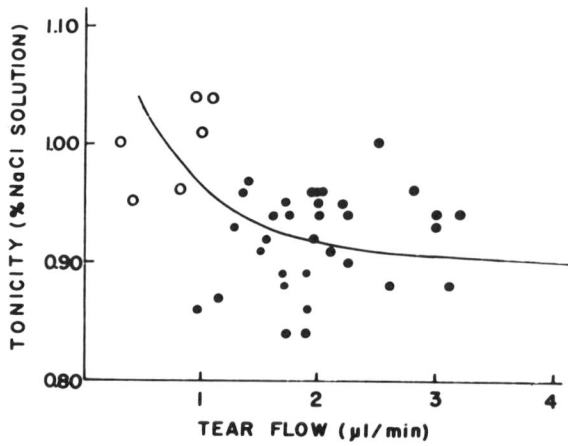

Figure 1. Osmotic pressure of the tears (equivalent of Ns Cl solution) and the rate of tear flow. Closed circles: Normal subjects. Open circles: cases of keratoconjunctivitis sicca. (From Mishima, Kubota and Farris, Excerpta Medical Internaitonal Congress Series No. 222, pp 1801-1805. Copyright 1970 Elsevier North Holland. Reprinted by permission.)

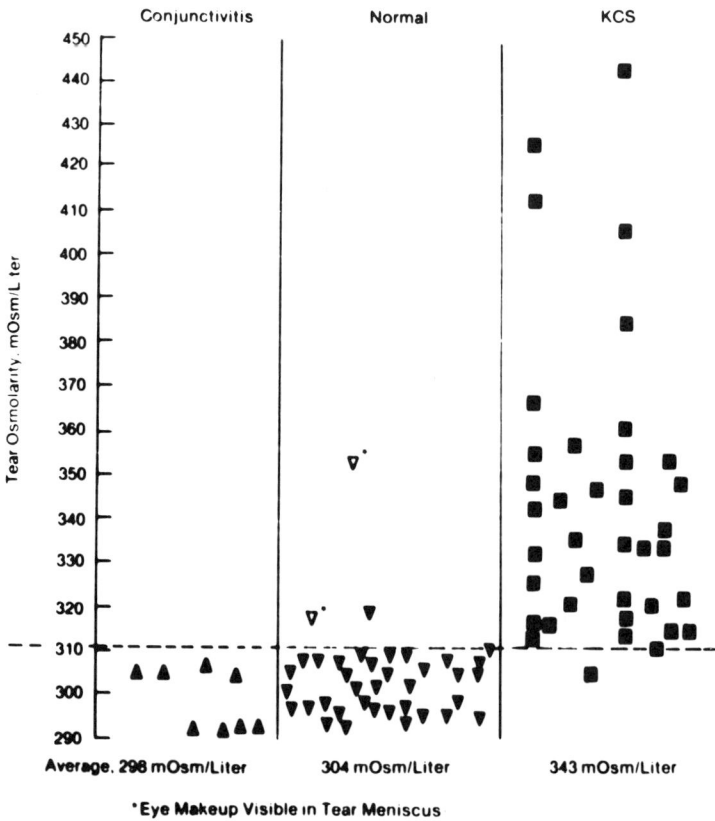

Figure 2. Tear osmolarity in conjunctivitis, normal eyes and KCS. (From Gilbard, Farris and Santamaria, Arch of Ophthalmol, 96:677-681. Copyright 1978 American Medical Association, Chicago, reprinted by permission.)

We then decided to go ahead and use tear osmolarity as a gold standard which would predict in a population of patients with symptoms of a dry eye the subsequent course of the disease and provide a more consistent association with the symptoms and clinical signs of KCS. We were aware that such a decision would subject us to the criticism that we were defining a new disease, but we understood that any new state of the art such as a new diagnostic test would require considerably more research before being accepted.[9] We were fortunate to have studies completed about the same time by Rolando and Refojo who measured increased evaporation rates in external eye diseases.[10] In addition, Jeff Gilbard and Ken Kenyon demonstrated that a hypertonic medium produced changes in epithelial cells growing in culture which resemble changes seen in the epithelial cells of dry eye patients.[11]

We then collected a larger group of patients presenting only with complaints of eye discomfort and performed tear osmolarity tests as well as the tests of basal tear volume, Schirmer without anesthetic, Rose Bengal staining, and lysozyme and lactoferrin concentration in basal and reflex tears, as well as the percentage increase of the concentration of lysozyme and lactoferrin with reflex tearing.[12] Tear osmolarity was 76% sensitive and 84% specific compared to the Schirmer test which was only 10% sensitive but 100% specific when 3mm or less of wetting in five minutes was considered abnormal.[10] Rose Bengal staining using a cutoff value of 3.5 was only 58% sensitive but 100% specific. The lactoferrin concentration of reflex tears provided a specificity of 94% compared to only 67% specificity for the lysozyme concentration in tears. The percentage increase in lactoferrin was 95% sensitive when less than a 100% increase in lactoferrin with reflex tearing was considered abnormal.

## OTHER STUDIES AND COMBINATION TESTING

How do these values compare with previous studies? van Bijesterveld was dismayed with the overlap of normal and abnormal values with the Schirmer test and Rose Bengal staining and found the lysozyme test measured by agar diffusion to be the most sensitive test.[13] Analyzing his data, the sensitivity was 98.8% and the specificity was 98.5% using a diameter limit of 21.5mm of lysis as the cutoff. Rose Bengal staining was 95% sensitive and 96% specific with a referent value of 3.5. The Schirmer test was 58% sensitive and 83% specific with a referent value of 5.5mm of wetting in 5 minutes. The test population was selected on the basis of several practitioners' opinions and was most likely a more severely affected population than patients in our study who had only symptoms of eye discomfort as a criterion for entry.

Goren and Goren have published results of tear test in a more similar population which includes mildly affected as well as more severely affected KCS patients.[14] Patients were selected only on the basis of symptoms. The patients with symptoms were divided into those with minimal symptoms, moderate to severe symptoms and both ocular and systemic symptoms. This study demonstrated the effect of combining tests into a battery of tests which is considered positive when any one of the tests within the group is positive. This is called combination testing by a parallel approach.[15] A or B or both may be positive for the combination to be positive. Goren and Goren used combination testing in this manner i.e. a test batter was considered positive if any test within the group was positive. Even though a

"+" sign was used in their display of results, both tests were not required to be positive for the combination to be considered positive. Contrast this with a series approach in which the positive test from one test are retested with a second test. As a result, A and B must be positive for the combination to be considered positive. The data of Goren and Goren was used to recalculate test results after combining groups of patients with ocular symptoms ranging from mild to severe including those with or without systemic symptoms into one group in order to compare with our studies.[16]

**TABLE 1**    Patients with minimal to severe symptoms and combining those without and with systemic Involvement.

| TESTS | SENSITIVITY (%) | SPECIFICITY (%) |
|---|---|---|
| BUT = A | 52 | 72 |
| Schir = B | 66 | 77 |
| LF = C | 64 | 90 |
| RB = D | 25 | 90 |
| | | |
| A + B | 78 | 56 |
| A + C | 71 | 64 |
| A + D | 57 | 62 |
| B + C | 79 | 70 |
| B + D | 77 | 49 |
| C + D | 70 | 54 |
| A + B + C | 84 | 51 |
| A + B + D | 80 | 49 |
| A + C + D | 74 | 54 |
| B + C + D | 83 | 62 |
| A + B + C + D | 87 | 44 |

Combining Tables 2, 3, and 4 Goren and Goren:  Am J Ophthalmol 106:570,1988.
LF - Lacatoplate,™  But - Tear film breakup time, Schir-Schirmer without anesthetic, RB - Rose Bengal Staining.  + = parallel combination testing, either test positive or both positive are a positive combination test.

The Schirmer test using less than 8mm of wetting as the cutoff for an abnormal value has the highest sensitivity, 66%, and the specificity is 60%.

Lactoferrin concentration in the tears was measured using the Lactoplate™ and was 64% sensitive and 90% specific. Rose Bengal staining was only 25% sensitive and 90% specific using any staining as the cutoff for abnormal. Tear film break up time was 52% sensitive and 72% specific using less than 8 seconds as the cutoff for an abnormal test. Combining the tests so that parallel testing was done, that is one or more tests of the combination must be positive to consider the combination positive, a combination of all four tests was most sensitive at 87%. As would be expected with parallel testing, the combined sensitivity is greater than the individual sensitivities of the contributing tests. The specificity of all four test using parallel testing was only 44%. Parallel testing results in the highest sensitivity but the lowest specificity. Series testing would have provided the opposite yielding lowest sensitivity but highest specificity but series combination testing was not reported. Of the combination of three tests, tear film break-up time, the Schirmer test and lactoferrin combination was the most sensitive at 84% with the lowest sensitivity of a three test parallel combination being tear film break-up, lactoferrin and Rose Bengal at 74%. The most sensitive of the two test parallel combination are Schirmer and lactoferrin at 79%, tear film break-up and Schirmer at 78% and Schirmer and Rose Bengal at 77%. The specificity of these parallel combinations are 69%, 56%, and 48% respectively. Thus overall efficiency appears best with the Schirmer and lactoferrin test combination.

**TABLE 2** Combination tear testing: Tear Osmolarity, Lactoplate,™ and Schirmer

| TESTS | SENSITIVITY % | SPECIFICITY % |
|---|---|---|
| Tear Os = A | 90 | 95 |
| LF = B | 35 | 70 |
| Schir = C | 25 | 90 |
| **Series:** | | |
| A and B | 35 | 100 |
| A and C | 25 | 100 |
| B and C | 20 | 100 |
| A and B and C | 20 | 100 |
| **Parallel:** | | |
| A or B | 90 | 65 |
| A or C | 90 | 85 |
| B or C | 40 | 60 |
| A or B or C | 90 | 55 |

Combining and calculating from Table I and II Lucca et al: CLAO Journal 16:109, 1990.

## RECENT STUDIES

In our last study, we compared the performance of tear osmolarity, lactoplate and Schirmer tests in 20 keratoconjunctivitis sicca patients and 20 age matched controls.[16] The diagnosis of keratoconjunctivitis sicca was made on the basis of history, symptoms, and clinical examination. Abnormal cutoff values for tear osmolarity, lactoferrin and Schirmer were 312 mOsm/L, 0.90 mg/dl, and less that 1mm/min of wetting.

The single test which provided the greatest sensitivity, specificity and overall efficiency was tear osmolarity. The next most sensitive test was the Lactoplate™ test which was 35%

sensitive but 70% specific. The Schirmer test was 90% specific but only 25% sensitive. Adding any one of the other tests to tear osmolarity did not make the combination more sensitive with parallel testing, i.e. A or B positive. Specificity was 100% with series testing using any two of the tests in this fashion, i.e. A or B negative. Series testing maximizes specificity whereas parallel testing maximizes sensitivity.

Impression cytology is a more direct measure of the cellular damage produced by keratoconjunctivitis sicca than tear osmolarity.[17] However, impression cytology reveals several variations of cell structure which may not be the result of a tear film deficiency but the result of ocular surface disease. Tear osmolarity measurement reflects only changes in the aqueous environment of the surface epithelium in a KCS patient with an excessively hypertonic tear film. Nelson [17] has described the sensitivity and specificity of impression cytology as a diagnostic test for dry eye and has stated that impression cytology is superior to tear osmolarity with 100% sensitivity and 87% specificity. However, the referent value appears to require a series combination of goblet cell density less than 350 cell/mm$^2$, mean epithelial cell areas greater than 1000 square micron/cell on the interpalpebral bulbar ocular surface and goblet cell densities greater than 100 cells/mm$^2$ on the inferior palpebral ocular surface in the absence of inflammatory cells. The results of impression cytology in normal controls are not included in the paper. As demonstrated, series combinations are more specific but less sensitive than parallel combinations which leads us to question how the data produced a 100% sensitivity but only 87% specificity. The labor of cytologic examination and clinical judgement required to determine endpoints on each patient gathered from three impression samples seems to disqualify impression cytology when compared to the simplicity of tear osmolarity determination which provides one number from the thawing and disappearance of a final ice crystal through the microscope. I cannot explain why Nelson's studies yielded only 44% sensitivity for tear osmolarity and 75% specificity. It may be important here to explain that osmolarity is the term used more by the physiologist and osmolality is the term used by the chemist with both terms meaning the same for our purposes of tear tests.

## CONCLUSION

In summary, tear osmolarity is a simple test to determine the freezing point of a microsample of basal tears. Studies have shown that it is a highly sensitive and specific test for keratoconjunctivitis sicca. Since the simplicity of the test and it's reliability have become well established, it does appear to be a reasonal candidate for a new international gold standard in the diagnosis of keratoconjunctivitis sicca.

## REFERENCES

1. J.U. Prause, R. Manthrope, P. Oxholm, and M. Schiodt, Definition and Criteria for Sjögren's Syndrome used by the Contributors to the First International Seminar on Sjögren's Syndrome - 1986, *Scand. J. Rheumatology* 61 (Suppl):17-18 (1986).
2. R. Manthorpe, V. Andersen, O.A. Jensen, P. Oxholm, J.U. Prause, and M. Schiodt, Editorial Comments to the Four Sets of Criteria for Sjögren's Syndrome, *Scand. J. Rheumatology* 61 (Suppl):31-35 (1986).

3. R. Manthorpe, P. Oxholm, J.U. Prause, and M. Schiodt, The Copenhagen Criteria for Sjögren's Syndrome, *Scand. J. Rheumatology* 61(Suppl):19-21 (1986).

4. R.L. Farris, R.N. Stuchell, and R. Nisengard, Sjögren's Syndrome and Keratoconjunctivitis Sicca, *Cornea* 10:207-209 (1991).

5. R.K. Riegelman, and R.P. Hirsch, Studying a Study and Testing a Test. Little, Brown and Company, Boston (1989).

6. S. Mishima, Z. Kubota, and R.L. Farris, The Tear Flow Dynamics in Normal and in Keratoconjunctivitis Sicca Cases. In Solanes MP (Ed); Proceedings of Amsterdam. Excerpta Medica International Congress Series No. 222, pp 1801-1805 (1987).

7. J.P. Gilbard, R.L. Farris, and J. Santamaria, Osmolarity of Tear Microvolumes in Keratoconjunctivis Sicca, *Arch. Ophthalmol.* 96:677-681 (1978).

8. J.P. Gilbard, and R.L. Farris, Tear Osmolarity and Ocular Surface Disease in Keratoconjunctivitis Sicca. *Arch. Ophthalmol.* 97:1642-1646 (1979).

9. R.L. Farris, Tear Osmolarity Variation in the Dry Eye, *Tr. Am. Ophthal. Soc.* 84:250-268 (1986).

10. M. Rolondo, M.F. Refojo, and K.R. Kenyon, Increased Evaporation in Eye with Keratoocnjunctivitis Sicca, *Arch. Ophthalmol.* 101:557-558 (1983).

11. J.P. Gilbard and K.R. Kenyon, Tear Diluents in the Treatment of Keratoconjunctivitis Sicca, *Arch. Ophthalmol.* 29:6446-650 (1985).

12. R.L. Farris, J.P. Gilbard, R.N. Stuchell, and IID. Mandel, Diagnostic Test in Keratoconjunctivitis Sicca, *CLAO J.* 9:23-28 (1983).

13. O.P. van Bijsterveld, Diagnostic Tests in the Sicca Syndrome, *Arch. Ophthalmol.* 82:10-14 (1969).

14. M.B. Goren, and S.B. Goren, Diagnostic Tests in Patients with Symptoms of Keratoconjunctivitis Sicca, *Am. J. Ophthalmol.* 106:570-574 (1988).

15. R.S. Galen, and S.R. Gambino, Beyond Normality - The Predictive Value and Efficiency of Medical Diagnosis. John Wiley and Sons (1975).

16. J.A. Lucca, J.N. Nunez, and R.L. Farris, A Comparison of Diagnostic Tests for Keratoconjunctivitis Sicca: Lactoplate, Schirmer, and Tear Osmolarity, *CLAO J.* 16:109-112 (1983).

17. J.D. Nelson and J.C. Wright, Impression Cytology of the Ocular Surface in Keratoconjunctivitis Sicca. In Holly FJ; Dry Eye Institute, Inc. Lubbock, Texas, pp 140-156 (1986).

# MICROPACHOMETRIC DIFFERENTIATION OF DRY EYE SYNDROMES

Hans-Walter Roth[1,2] and Donald Lewis MacKeen[2,3]

[1]Institut fuer wissenschaftliche Kontaktoptik, Ulm, D-7900, F.R.G.
[2]Center for Sight, Georgetown University Medical Center, Washington D.C.,20007, U.S.A.
[3]MacKeen Consultants Ltd, Bethesda, MD 20816, U.S.A.

## INTRODUCTION

More than 100 years ago Helmholtz[1] noted that any alteration of corneal physiology results in a change in corneal thickness. These changes are dependent not only on aging, refractive errors and the multitude of endogenous and exogenous factors but also on the status of the preocular tear film.[2,3]

Micropachometric investigation of the central cornea have shown that the thickness of the normal cornea changes continuously during the waking hours. Maximum values are found in the morning, these values decrease during the day and are minimal in the evening.[3,4]

A careful computer analysis of data collected in a long term study indicated that hypoxia could not be implicated as the only factor for overnight swelling of the cornea[5]. This was shown by causing experimental hypoxia by either corneal or scleral lens wear.

Central and peripheral corneal thickness measurements of normal and diseased eyes indicated that a second factor associated with thinning or swelling of the cornea was a quantitative and/or qualitative disturbance of the tear flow. Various disturbances of tears may result in identical pachometric reactions of the cornea. Many authors have noted the dependence of the central corneal thickness on the volume and quality of tears.[5,6]

Previous methods of investigating corneal thickness were problematic. The majority of previous ultrasound systems did not have sufficient accuracy to detect the small changes in corneal thickness that might accompany disturbances in tear volume or composition.

The use of micropachometry eliminated the sources of inaccuracy and enabled measurement of corneal thickness in the living eye with an accuracy of three microns.[5,7-9] Micropachometric measurements of 1360 normal subjects aged 25 to 45 established that the average corneal thickness of emmetropic European eyes is 545 ± 20 uM.[7]

## PURPOSE OF THE INVESTIGATION

The purpose of this investigation was to utilize micropachometry to determine if there were a correlation between central corneal thickness (CCT) changes in dry eye patients and the results of any of a series of clinically appropriate dry eye tests.

## INVESTIGATIONAL DETAILS

In the present study measurements were made of 1,470 eyes that showed symptoms of dry eyes; the statuses and apparent causations were determined in order to provide appropriate treatments. Measurements were made prior to initiation of therapy, when therapy was halted follow up measurements were made one week or later.

In all subjects the following tests were done between 3 and 5 o'clock in the afternoon to coincide with the minimum values of corneal thicknesses. This was based on previous findings of corneal thickness changes during the waking hours. The following tests were made:

1) Measurements were made of the meniscus of the lower lids; this was done by means of a Zeiss slit lamp at 30X ; the eyepiece was fitted with a calibrated reticle.
2) BUT tests were done with the slit lamp, illumination was done with 5 light sources using red-free light of 445 nm without the addition of an indicator substance such as fluorescein.
3) Schirmer I Test
4) Schirmer II Test (after topical anesthetic)
5) Measurement of tear conductivity. The electric resistance of strips wetted in Schirmer I tests was measured (microohms/mm$^2$). The osmolality can be calculated from the corrected resistance value.
6) Photometric assessment after Rose Bengal (0.5 %) instillation. This was done by photographing the stained corneal surface with a Hasselblad slit lamp camera, the 6 x 6 cm slides were quantitated photometrically. The values are expressed as a range of values from 1 to 9.
7) Micropachometric method of Roth.[3]

All values were obtained with the help of an electronic data program and statistically evaluated with a computer program.

## RESULTS

### Central Corneal Thickness: Relationship to Schirmer I and II Values

As shown in Figure 1, dry-eyed patients with Schirmer I test values greater than 15-20 mm (per 5 minutes) had normal central corneal thickness (CCT) values. Those with values of less than 5 mm had decreased thicknesses. In contrast corneas with values of 2 mm or less had greater thicknesses, approximately 60 to 70 microns greater than the average normal CCT of 545 microns[7].

The curves in Figures 1 and 2 are grossly similar; however, the curve from Schirmer II data appears 'delayed', presumably because of the anesthesia.

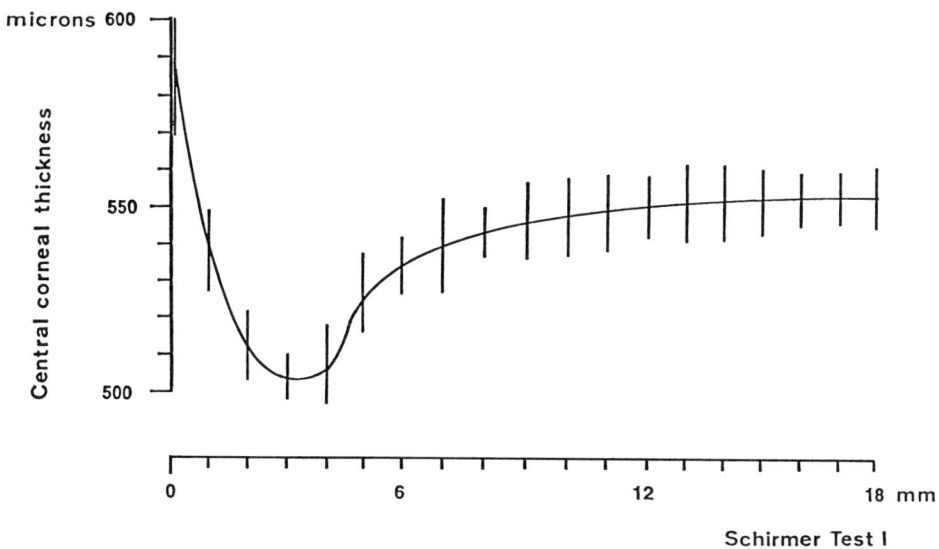

**Figure 1.** Central Corneal Thickness of Dry-Eyed Patients vs Schirmer I values (av. ± 1SD).

## Central Corneal Thickness: Relationship With BUT

Figure 3 shows the relationship between CCT and BUT values in dry-eyed patients. The CCT values of corneas with BUT 10 seconds or greater were in a normal range. The CCT decreased when the BUT was lower (10 sec.to 3 sec.). This indicated a relative decrease in tear film stability associated with CCT thinning. Significant increases of CCT were noted in eyes with BUT one second or less; this probably resulted from severe surface damage from drying and accompanying corneal decompensation.Examination of Figs. 1 to 3 show that markedly decreased Schirmer or BUT values can be associated with seemingly normal CCT thicknesses, this is a condition termed pseudonormal corneal thickness.

## Central Corneal Thickness: Relationship With Meniscus Test

As shown in Fig. 4, attempts to correlate the width of the lower meniscus with corneal thickness were relatively unproductive in contrast to the Schirmer and the BUT tests. We found no significant changes in corneal thickness in eyes with meniscal widths of 0.5 mm and greater. Widths less than 0.5 mm were accompanied by either increases or decreases in corneal thicknesses.

## Central Corneal Thickness: Relationship With Rose Bengal Staining

As shown in Fig. 5 staining resulting from the instillation of 0.5% Rose Bengal showed a correlation between the degree of staining and corneal thickness changes; the greater the corneal swelling, the greater the corneal staining.

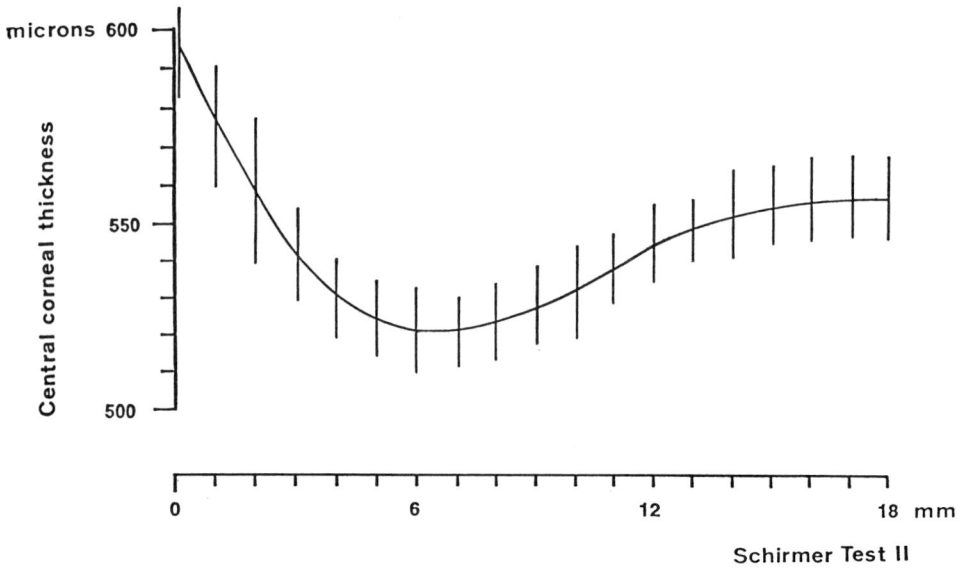

**Figure 2.** Central Corneal Thickness of Dry-Eyed Patients vs Schirmer II values (av. ± 1SD).

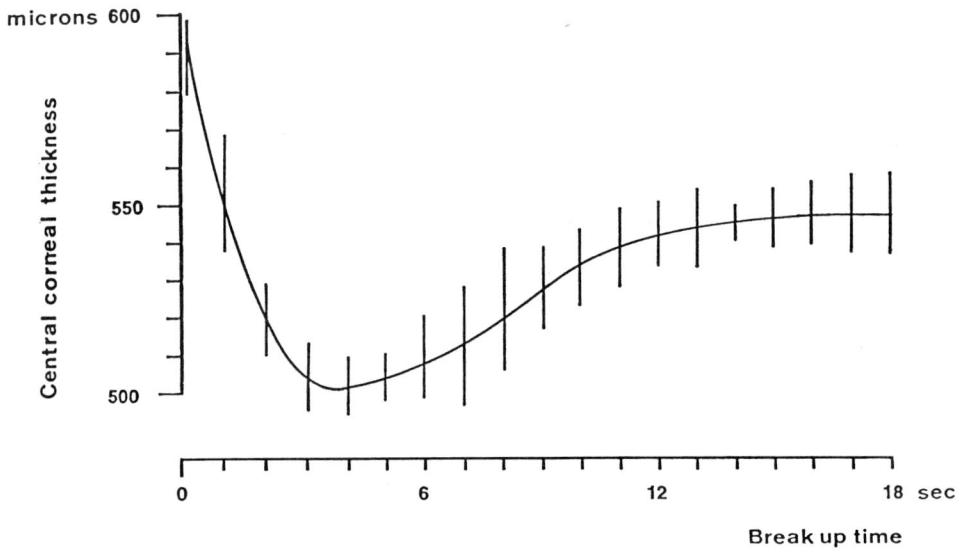

**Figure 3.** Central Corneal Thickness of Dry Eyed Patients vs BUT Values in seconds (av.± 1SD).

## Central Corneal Thickness:    Relationship With Electrical Resistivity of Tear Samples

Figure 6 shows the relationship between corneal thickness and the electric resistance of tears. There is an increase in CCT associated with a resistivity greater than 8 ohms/sq.mm. Osmotic strength is inversely related to resistance and can be readily calculated. Therefore, it is not surprising that tears with decreased osmolality, i.e. hypotonic, are associated with corneal edema.  When the osmolality was maximal, i.e. <2 ohms/sq.mm., there was a significant decrease in CCT. Reportedly there is a correlation between hyperosmolality and epithelial changes in dry eye states.

## DISCUSSION

These more sensitive measurements of corneal thickness have shown correlations with tear volume, preocular tear film stability, corneal staining and tear resistance.

One factor that initiated these investigations were the reports of Holden[10] that corneal swelling following contact lens wear is an inverse function of lens Dk. Bijsterfeld[6] has shown that corneal thickness can either increase or decrease as a result of changes in tear secretion. Roth[7] reported finding that there are other factors than hypoxia, e.g., tear composition changes involved in corneal swelling. Honegger[11] stated that both the pH and the osmotic pressure of tears affect the corneal thickness. Wilson et al.[13] reported on the corneal swelling that occurs when ions such as potassium, calcium and magnesium are absent from artificial tears.  Green and MacKeen[14] reported on the relationship between tear potassium concentration and corneal thickness.

It was interesting to compare the findings of various commonly used dry eye clinical tests with the corresponding CCT values.  For example, when the tear flow (Schirmer value) or BUT were less than normal the cornea was thinned, presumably due to evaporative loss. However, when these values were greatly decreased,e.g., BUT approached zero, the CCT increased markedly. Presumably, the stroma swelled after the injured epithelium decompensated.

The Rose Bengal test showed a curvilinear correlation between corneal staining and increased CCT. This is in contrast to the Schirmer test results where there was no relation between increase in central corneal thickness and millimeters of wetting when there was no staining of the corneal surface. We found that when there was an optically detectable defect of the corneal epithelium, a classic sign of severe dry eye, the cornea was decompensated and swollen.

Resistance measurements using tear-wetted Schirmer strips, even without conversion to the corresponding ionic strength, offer the clinician a simple, quick method of assessing this important aspect of tears. The present findings show the correlation between electrolyte concentration and changes in corneal thickness.

The resistivity of tears has been measured using capillary tubes, cuvettes and strips from Schirmer I tests for sampling; we found that resistivity of the wetted portion of Schirmer strips made this measurement practical for clinical use. We did not find any clinically usefulness in measuring the lower meniscal width in dry eye patients to detect changes in CCT.

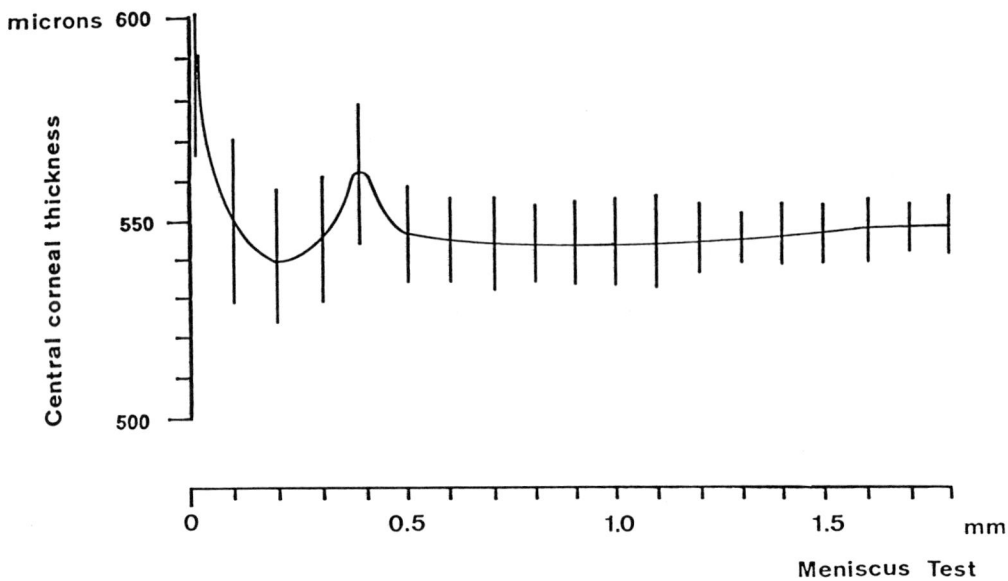

**Figure 4.** Central Corneal Thickness of Dry Eyed Patients vs Width of the Lower Meniscus (av.± 1SD).

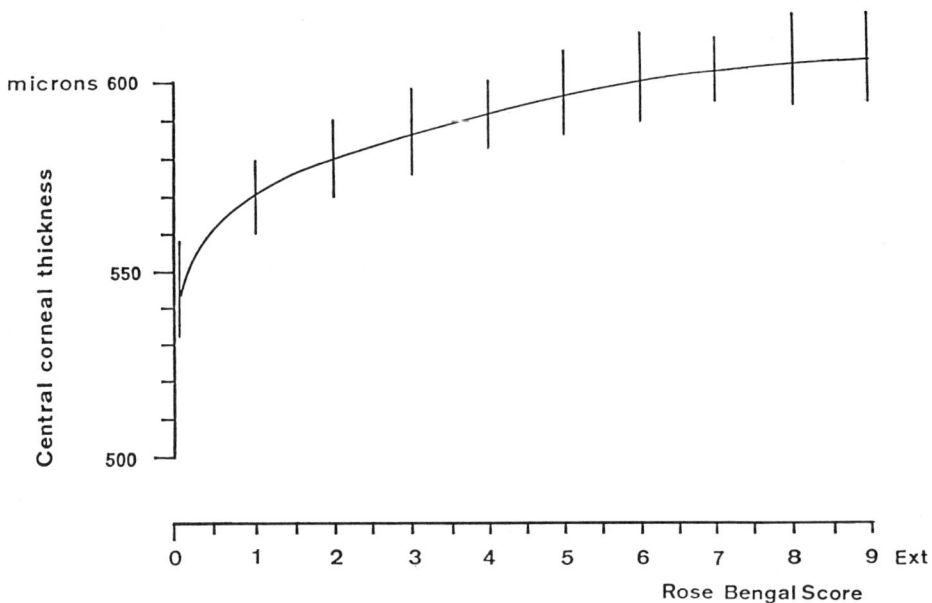

**Figure 5.** Central Corneal Thickness of Dry Eyed Patients vs Rose Bengal Staining (Range 1 to 9, av. ± 1SD).

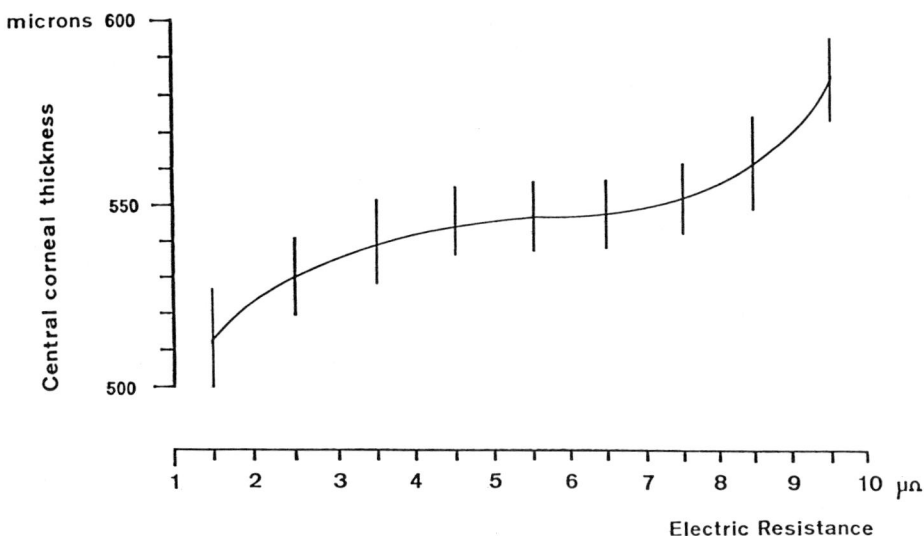

**Figure 6.** Central Corneal Thickness of Dry Eyed Patients vs Electric Resistance (av. microOhms ± 1SD).

On the basis of these investigations one can differentiate the severity and status of the dry eye into one of three classes based on corneal thickness: normal, thinned or swollen.

As an aside, previous research in this institute (Roth[5]) has shown that type B keratoconus shows an identical pachometric progression, suggesting that tear dysfunction is an important common factor. The initial thinning of the cornea is followed by an epithelial defect with resultant thickening.

Finally, although not unexpected, these findings reaffirm the necessity of assessing several different variables in making a meaningful diagnosis.

## REFERENCES

1. H. von Helmholtz, "Handbuch der physiologischen Optik", Voss, Hamburg (1866).
2. D.L. MacKeen, and H-W. Roth, Successful maintenance of dry eyed contact lens wearers with the punctum flow controller, VIIth International Medical Contact Lens Symposium, Singapore Mar 15. (1990).
3. H-W. Roth, Bedeutung der Ultraschallpachymetrie beim Kontaktlinsentragen, Z. prakt. Augenheilk 125-128 (1991).
4. G. von Bahr, Corneal thickness, its measurement and changes, J. Ophthalmol. 42:251 (1956)
5. H-W. Roth, Micropachometric investigation of rigid contact lens-induced corneal dystrophies, CLAO Journal, in press (1993).
6. O.P. van Bijsterfeld, and J. Baardman, Hornhautdickenmessung bei Patienten mit Ceratoconjunctivitis sicca, Klin. Mbl. Augenheilk 197:240 (1990).
7. H.W. Roth, Mikropachometrische Untersuchungen der Hornhautdicke beim Tragen harter Kontaktlinsen, Contactologia 13 D:169 (1991).
8. H-W.Roth, Mikropachymetrische Differenzierung unterschiedlicher Keratokonustypen, Contactologia 14 D: 17 (1992).
9. H-W. Roth, Micropachometric quantification of the healthy, diseased and contact-lens wearing eye, Cornea, in press (1993).
10. B. Holden, Corneal swelling response to contact lens worn under extended wear conditions, Invest. Ophthalmol. Vis. Sci. 24:218 (1983).

11. H. Honegger,Untersuchungen ueber die lokale Osmotherapie der Hornhaut, *Klin. Mbl. Augenheilk* 141:582 (1962).

12. H-W. Roth, Untersuchungen ueber die Verformung der zentralen Hornhautvorderflaeche nach Tragen harter und weicher Linsen, *Fortschr. Ophthalmol.* 86:185 (1990).

13. G. Wilson, W.G. Bachman, and P.L. Call, A nutritional role for tears, in "The Preocular Tear Film in Health, Disease and Contact Lens Wear", F.J. Holly, D.W. Lamberts and D.L. MacKeen, eds., Dry Eye Institute, Lubbock, TX (1986).

14. K. Green, and D.L. MacKeen, Tear Potassium contributes to maintenance of corneal thickness, *Ophthalmic Res.* 24:99 (1992).

# DECREASED TEAR SECRETION IN CHERNOBYL CHILDREN: EXTERNAL EYE DISORDERS IN CHILDREN SUBJECTED TO LONG-TERM LOW-DOSE RADIATION

Dorit Gamus[1], Zeev Weschler[2], Simon Greenberg[2] and Amalia Romano[1]

[1] Goldschleger Eye Research Institute, Sheba Medical Center, Sackler School of Medicine, Tel Aviv University, Tel Aviv
[2] Hadassah University Hospital, Jerusalem

## INTRODUCTION

Long term effect of radiation exposure are not completely understood. Evaluation of late radiation-induced damage is especially important in the case of the Chernobyl disaster in 1986, which is considered to be one of the worst in our century.

More than 130,000 people including 40,000 children were resettled from the 30 km radius zone, while the population from the neighbouring areas continued to live in the presence of low dose radiation due to environmental contamination. Since children are the most vulnerable part of the affected population, medical follow-up and treatment are essential for their future development and general well-being.

The objectives of our study were:
1. Detection of possible radiation induced changes in the eyes of children exposed to different radiation doses.
2. Evaluation of other systemic causes of morbidity among these children.

## METHODS

For the last two years 328 children, aged 6-16 years, who arrived to Israel from the affected zones were subjected to detailed medical examination.

During the period of medical evaluation all children stayed in the Chabad Village, in the central zone of Israel, and were exposed to similar climatic environment and nutritional care.

General medical examination included:
1. Anamnesis of current and previous medical history
2. Systemic pediatric check-up
3. Laboratory tests: blood count; blood chemistry; ultrasound of thyroid gland (when clinical examination revealed thyroid gland enlargement).

The ophthalmic examination was performed in 294 children and included:
1. a. Visual acuity and refraction; ocular movements;

*Lacrimal Gland, Tear Film, and Dry Eye Syndromes*
Edited by D.A. Sullivan, Plenum Press, New York, 1994

b. Slit-lamp examination of the anterior and posterior segments;
c. Ophthalmoscopic fundus examination.
2. Functional tests:
a. Color vision test by pseudoisochromatic plates;
b. Modified Schirmer I test[1]. Deficient tear secretion was defined as <6mm of wetting of Schirmer strips.


## RESULTS

The results of all causes of morbidity among the examined children are summarized in Table 1.

Table 1. Chernobyl children: General morbidity

| | | |
|---|---|---|
| Ophthalmic Disorders* | 115 | 39.1% |
| Deficient tear secretion | 46 | 15.6% |
| Blepharoconjunctivitis | 53 | 18.0% |
| Endocrine Disorders** | 127 | 38.7% |
| Goiter, unspecified | 116 | 35.4% |
| Disorders of Respiratory System** | 75 | 22.9% |
| Psycological Disorders** | 39 | 12.5% |
| Disorders of Genitourinary System** | 30 | 9.1% |
| Disorders of Digestive System** | 23 | 7.0% |

```
 * Out of 294 children
** Out of 328 children
```

The most frequent ophthalmic findings were external eye disorders: external eye infections (acute and chronic blepharoconjunctivis) and decreased lacrimation. (Table 1). The vast majority of cases with Schirmer test <6mm were bilateral. None of the examined children wore contact lenses (which are one of major causes of dry eyes[2].
When we analyzed the occurrence of dry eyes according to the distance of children's residence from the disaster area, we found decreased lacrimation in 18.8% among 218 children who arrived from zones most affected by radiation, and only 6.6% among children from more distant areas (Table 2). While this trend was present in both sexes, significant correlation (by Chi-Square) was demonstrated only among girls (P=0.05).
No radiation damage to retina or to lens was detected.

Table 2. Gender and dry eye distribution according to the distance from Chernobyl.

| | Near Zones | Distant Areas | Total |
|---|---|---|---|
| Females | 132 | 52 | 184 |
| Males | 86 | 24 | 110 |
| Total | 218 | 76 | 294 |
| Schirmer test <6 mm | 41 (18.8%) | 5 (6.6%) | 46 (15.5%) |

## DISCUSSION

Nuclear disaster in Chernobyl, 1986 is unprecedented in its enormity and our knowlege about late consequences of such catastrophes is limited. About 50 million Curie of various radionuclides were spread out into the environment. Among these, the $I^{131}$, $Cs^{137}$, $Sr^{90}$ and Plutonium have the highest impact on human health[3,4]. The assessment of radiation damage to the population was difficult because exact values of radiation exposure were never evaluated, and the distribution of the contamination was not homogenous.

It is well-known that ionizing radiation may cause cataracts. In Hiroshima and Nagasaki atomic bomb survivors high incidence of posterior lenticular cataracts was found[5]. We could not detect such damages in our study group five years after the explosure in Chernobyl nuclear plant .

Summarizing the results revealed that the most prominent findings among the examined children were decreased lacrimation and external ocular infections (Table 1). As the focus of our study was a screening procedure, which consisted of multiple parameters (described in "Methods"), the evaluation of possible deficiency of each tear layer was not performed at that stage.

Although the efficiency of single Schirmer test value in the diagnosis of dry eyes is debated[1,7,10], one of the reports that argue against the validity of single Schirmer test measure[1], showed nonetheless, that while modified Schirmer I test of <6mm was found in 3% of normal eyes, keratonconjunctivitis sicca (KCS) patients produced 0-6mm of wetting of the filter strip. We choose therefore to use this method at cut off point <6mm for determination of decreased tear secretion.

The high incidence of decreased tear secretion in children is an unusual finding: the decline in Shirmer test values are attributed to much older age groups[7]. Almost equally high incidence of acute and chronic blepharoconjunctivitis (Table 1), which are known to be associated conditions of dry eyes[1,8,9], further emphasizies the validity of Schirmer test results, detected by us. It is too early to establish whether our findings represent true "dry eyes", or are secondary to external eye infections. Nonetheless, our results suggest that the ability of eyes of the examined children to resist infections may be compromized.

Decreased or abscent lacrimal secretion following therapeutic X-Ray irradiation has been previously described[11,12]. Radiation damage may result in the defeciency of all three components of tear film[6,11,12], which sometimes appear as a late onset effect[11,12].

While Iodine [131] excretion through sallivary glands is an established phenomenon[13], we found that $I^{131}$ can be also detected in tears of patients who receive this isotope for therapeutic reasons (unpublished results). The lacrimal gland may therefore be affected by the resultant local gamma radiation. The external eye disorders could result either from single high dose exposure, or reflect the effect of long term low-dose radiation by contaminated dust that might have affected conjunctival mucosae.

A significant difference in the Schirmer test values among girls with various radiation exposure was demonstrated in our study group. The inability to confirm such statistically significant correlation in males can be attributed to a much smaller number of the examined boys (Table 2).

In conclusion, our findings suggest a possible correlation between the occurrence of dry eyes and the distance from the Chernobyl plant.

We intend to continue the follow-up of these children in order to determine whether the described conditions are temporary or are of permanent nature. If the finding of dry eyes will persist, tear composition will be evaluated in order to further understand the pathophysiology of radiation damage to the external mucosae, and to provide proper treatment to these children.

The possibility that lacrimal functions may serve as an additional tool in the assessment of radiation damage should be further studied.

## REFERENCES

1. I.A. Mackie and D.V. Seal, The questionably dry eyes, Brit J Ophthalmol 65:2 (1981).
2. R.L. Farris, The dry eye: Its mechanism and therapy, with evidence that contact lens is a cause, CLAO 12(4):234 (1986).
3. E.E. Pochin, Basis for detection and measurement of deposed radionuclides in: "Diagnosis and Treatment of Deposited Radionuclides", H.A. Kornberg and W.D. Norwood, eds., Excerpta Medica Foundation, Amsterdam, (1968).
4. National Council on Radiation Protection and Measurement. Identification of uncertainties associated with model predictions: Predicting the transport, bioaccumulation, and uptake by man of radionuclides released to the environment. NCRP Publications, Washington DC, NCRP Report No.76, (1984).
5. M. Otake and W.J. Schull, Radiation-relate posterior lenticular opacities in Hiroshima and Nagasaki atomic bomb survivors based on DS86 dosimetry system, Radiat Res 121:3 (1990).
6. M.A. Lemp and B. Chacko, Diagnosis and trearment of tear deficiencies, in: "Clinical Ophthalmology", T.D. Duane ed., J.B. Lippincott Company, Philadelphia (1991).
7. R.W. Stricland, J.T. Tesar, B.H. Berne, B.R. Hobbs, D.M. Lewis and R.C. Welton, The frequency of sicca syndrome in an eldrely female population, J Rheumatol 14:766 (1987).
8. M.A. Lemp, Recent developments in dry eye management, Ophthalmology 94:1299 (1987).
9. A.J. Bron, Prospects for the dry eyes, Trans Ophthalmol Soc UK 104: 801 (1985).
10. A. Mackie and D.V. Seal, Diagnostic implications of tear proteins, Brit J Ophthalmol 68:321 (1984).
11. L.A. Karp, B.W. Streeten and D.G. Cogan, Radiation- induced atrophy of the Meibomian glands, Arch Ophthalmol 97:303 (1979).
12. P.A. Macfaul and M.A. Bedford, Ocular complications after therapeutic irradiation, Brit J Ophthalmol 54:237 (1970).
13. R. Chisin, A.M. Noyek, O. Israel, I.J. Witterick, D. Front and J.C. Kirsh, Contribution of nuclear medicine to the diagnosis and management of extracranial head and neck diseases (excluding thyroid and parathyroid), Isr J Med Sci 28:254

# LACRIMAL DYSFUNCTION IN PEDIATRIC ACQUIRED IMMUNODEFICIENCY SYNDROME

Renato A. Neves, Denise de Freitas, Elcio Sato,
Carlos Oliveira, and Rubens Belfort, Jr.

Department of Ophthalmology
Escola Paulista de Medicina
Rua Botucatu 822
S. Paulo, SP Brazil 04023

## INTRODUCTION

In the United States, the Acquired Immunodeficiency Syndrome (AIDS) is the second leading cause of death among men of 25 to 44 years old. By 1987 AIDS had also become the ninth leading cause of death for children between the ages of 1 and 4 . It is estimated that one in 75 men and one in 700 women between the ages of 15 and 49 years are infected with human immunodeficiency virus (HIV). Homosexual/bisexual men still account for most cases, but the groups whose numbers are increasing most rapidly include patients exposed through intravenous drug use or heterosexual contact and children infected by perinatal transmission (CDC, 1991).

In Brazil, 17,796 cases of HIV infection have been reported since 1980, with a 40% average annual growth rate. In relation to pediatric AIDS, by January 1992, 753 children under 15 years of age were reported with AIDS. The World Health Organization has estimated that 30 million adults and 10 million children throughout the world will be infected with HIV by the year 2,000 (Neves, 1993).

Ophthalmic disorders from adult patients with AIDS are associated with disease of microvasculature, opportunistic infections, neoplasm, and neuro-ophthalmic disorders associated with intracranial infections and neoplasm (Holland, 1992).

Also, studies have hypothesized that a Sjögren's syndrome-like illness may be a manifestation in HIV-infected patients (Lucca, 1990).

Very little is known about the ocular findings in pediatric AIDS. We have followed children with AIDS in order to elucidate some of the many questions involving this subject.

## PATIENTS AND METHODS

From January 1990 through March 1992, 64 children (29 males and 35 females) under 12 years old were seen in the Uveitis Center, Escola Paulista de Medicina. According to the CDC criteria, 18 children were diagnosed in the P0 form (indeterminate infection in infants under 19 months born from infected mothers); 10 in the P1 form (asymptomatic infection with positive test over 15 months) and 35 in the P2 form (symptomatic infection).

The ocular examination included examination of the anterior segment of the eye by slit lamp, examination of the posterior pole by indirect ophthalmoscopy, rose bengal staining of the cornea and conjunctiva and/or Schirmer test under topical anesthesia.

A Rose Bengal test was performed in all children and a Schirmer test was performed only in children older than 4. Lacrimal dysfunction was diagnosed when a typical corneal and conjunctival staining were present and/or Schirmer test was below 5 mm. Ten healthy children under the age of 12 were tested for lacrimal gland dysfunction with Rose Bengal and Schirmer test.

## RESULTS

Ocular findings were divided into strabismus (10 patients), optic nerve atrophy (4 patients), amblyopia (2 patients), cytomegalovirus retinitis (2 patients), toxoplasmic retinochoroiditis (2 patients), herpes keratitis (1 patient), and lacrimal dysfunction (36 patients). No retinal hemorrhage or Kaposi sarcoma were observed.

In relation to the lacrimal gland dysfunction, 36 children reveled typical rose bengal staining and 17 revealed abnormal Schirmer test (4, zero; 8, 1-2 mm; 3, 3-4 mm; 2, 5mm) (Table 1). All control children presented Schirmer test above 9 mm (range 34 to 9 mm, average 14.4 mm).

## DISCUSSION

Only a few reports on ocular findings in pediatric AIDS have been published and the different incidence of ocular involvement in the different studies may be explained by different age groups, phases of the disease, ways of transmission and countries. All papers report a lower incidence of ocular manifestation in pediatric AIDS when compared to the disease in adults (13-57%) (Blini, 1989; Kestelyn, 1990). In our study, ocular disease was present in 40.6% of patients excluding lacrimal dysfunction and 60.9% including this finding.

The P0 group (19 patients) revealed neurophthalmologic findings (optic disc atrophy, amblyopia, and strabismus) probably related to the mother's disease. Blini (1989), in Italy observed 11% of patients with similar findings and Kestelyn in Kigali, Africa, has correlated these findings to probable alcohol and drug abuse of the mother. Encephalopathy and neuromotor retardation has been a frequent finding of pediatric HIV infection (Deheny, 1989). CMV retinitis and toxoplasmic chorioretinitis do not seem so frequent in pediatric AIDS. In our series the two CMV retinitis cases were treated with DHPG for 3 weeks and has remained healed for 9 months. One case of toxoplasmosis was found in a previous examined normal eye. At the time of the diagnosis of ocular toxoplasmosis, the patient had specific circulating IgM titers of 1/256 for toxoplasmosis and concurrent central nervous system involvement. The other case of toxoplasmosis had CNS involvement, negative specific IgM and IgG titers of 1/512 for Toxoplasmosis. Several studies have reported cases of Sjogren's syndrome-like illness occurring in patients with HIV infection (Ulirsch,1987; Couderc, 1987; Clerck, 1988; Calabrese, 1989; Green, 1989; Talal, 1990; Lucca, 1990). The precise causes of the sicca syndrome are still unknown. The involvement of CMV or EBV as an etiologic agent in Sjogren's syndrome has been suggested. Kestelyn (1990) has reported 20% of lacrimal dysfunction in his cases of pediatric AIDS and has attributed that also to EBV and not only to HIV infection. In Brazil the EBV infection is frequent among children and antibody positive serum under 8 years of age is the rule. Although Schirmer test and Rose Bengal staining have their limitations, mainly in children, verbal complaint of burning eye sensation and red eye symptoms were frequent in our 36 children with lacrimal dysfunction.

**Table 1.** Ocular manifestations in pediatric AIDS patients

| Group | PO | P1 | P2 | TOTAL |
|---|---|---|---|---|
| Strabismus | 4 | 1 | 5 | 10 |
| Optic atrophy | 3 | 0 | 1 | 4 |
| Amblyopia | 2 | 0 | 0 | 2 |
| Cotton Wool spots | 0 | 0 | 3 | 3 |
| Vasculitis | 0 | 0 | 2 | 2 |
| CMV retinitis | 0 | 0 | 2 | 2 |
| Toxoplasmosis | 0 | 0 | 2 | 2 |
| Hemorrhage | 0 | 0 | 0 | 0 |
| Keratitis | 0 | 0 | 1 | 1 |
| Kaposi sarcoma | 0 | 0 | 0 | 0 |
| Lacrimal dysfunction. | 7 | 1 | 28 | 36 |
| | | | | |
| Total | | | | |
| Without Lacrimal dysfunction | 9 | 1 | 16 | 26 |
| With Lacrimal dysfunction. | 16 | 2 | 36 | 39 |

Although ocular involvement in children suffering from AIDS seems to be less frequent than the one seen in adults, it appears to be quite different and also important. A closer and early follow-up must be considered in order to prevent further ocular complications.


## REFERENCES

M. Blini, G. Bertoni, M. Chiama, E Massironi, A. Plebani, B. D'Arminio and A. Monforte, Ocular involvement in children with HIV infection, Abstracts of the 5th International Conference on AIDS, Montreal, 258 (1989).

L. H. Calabrese, The rheumatic manifestations of infection with the human immunodeficiency virus, Semin. Arthritis Rheum. 18:225 (1989).

Centers for Disease Control, Human immunodeficiency virus/AIDS Surveillance Report. 1:18 (1991).

L.S. Clerk, M.M. Coutennye, M.E. deBroe, and W.J. Stevens, Acquired Immunodeficiency syndrome mimicking Sjögren syndrome and systhemic lupus eritematosus, Arthritis Rheum. 31:272 (1988).

L.J. Couderc, M.F. DAgay, F. Danon, Sicca complex and infection with human immunodefficiency virus, Arch Intern Med. 147:898 (1987).

PJ. Dehhenny, R Warman, J.T. Flynn, G.B. Scott and M.T. Mastrucci, Ocular manifestations in pediatric patients with acquired immunodeficiency syndrome, Arch. Ophthalmol. 107:978 (1989).

M.D. De Smet, B. Rubin, R. Belfort Jr., P.A. Pizzo, S. Mellow and R.B. Nussenblatt, Retinal manifestations of AIDS in pediatric patients, Ophthalmology 97 :153 (1990).

J.E. Green, S.H. Hinrichs, J. Vogel and G. Jay, Exocrinopathy resembling Sjogren Syndrome in HTLV1 tax trangenic mice, Nature 341:72 (1989).

G. N. Holland, Acquired immunodeficiency syndrome and ophthalmology: The first decade, Am. J. Ophthalmol. 114:86 (1992).

P. Kestelyn, Ocular Problems in AIDS, Int. Ophthalmol. 14:165, (1990).

J.A. Lucca, R.L. Farrys, L. Bielory, A.R. Caputo, Keratoconjuntivitis sicca in male patients infected with the human immunodefficiency virus type I, Ophthalmology 97:1008 (1990).

R.A. Neves, C.F. Oliveira, R. Belfort Jr, Ocular findings in pediatric acquired immunodefficiency syndrome in: "Recent Developments in Uveitis". Dernouchamp J.P. Kugler, Amsterdam, (1993).

N. Talal, M.J. Dauphinee, H. Dang, S.S. Alexander, D.J. Hart, R.F. Garry, Detection of serum antibodies to retroviral proteins in patients with primary sjogren syndrome (autoimmune exocrinopathy), Arthritis Rheum. 33:774 (1990).

R.C. Ulirsch, E.S. Jaffe, Sjogren syndrome-like illness associated with the acquired immunodefficiency syndrome-related complex, Hum. Pathol. 18:1063 (1987).

# KERATOCONJUNCTIVITIS SICCA IN HIV-1 INFECTED FEMALE PATIENTS

John A. Lucca, John S. Kung, and R. Linsy Farris

Department of Ophthalmology
Columbia University
New York, NY

## INTRODUCTION

AIDS is caused by infection with the human immunodeficiency virus type-1 (HIV-1) presently affecting over 1.5 million people in the United States. While affecting less than 1% of the general population, keratoconjunctivitis sicca (KCS) occurs in greater than 20% of males infected with HIV.[1] In our preliminary survey of HIV infected females, we found 17% had clinical signs and symptoms compatible with KCS.[2]

## METHODS

On clinical examination of the HIV-infected females (n=59), 10 complained of excessive dryness of the eyes with varying degrees of photosensitivity. These individuals all had signs of dry eye which included a deficient inferior marginal tear strip, excess debris in the tear film, and a viscous appearing tear film. All patients had at least one Center of Disease Control risk factor.[3] All of the patients were between the ages of 23-42 years old, non-contact lens wearers, and were without a history of ocular medication for a period of at least six months. All denied a history of previous globe surgery. None of the 10 KCS patients complained of a mild dry mouth. Clinical evaluation showed no signs of severe disease in these individuals.

The Schirmer test was performed without anesthetic on the 10 suspect patients, as well as 10 control patients. Control patients consisted of 10 age-matched HIV negative females not complaining of dry eyes and with a negative clinical exam. The Schirmer test strips were place simultaneously in both inferior conjunctival sacs of each patient and allowed to remain for five minutes. We considered less than 11 mm of wetting suspect for a tear production deficiency.

Tear osmolarity was determined on samples collected using a fine-tipped pipette. Approximately 0.04 ml of tears were quickly collected by capillary action from the inferior marginal tear strip of each patient. The tear samples were immediately blown into a column of oil contained within a capillary tube and stored under refrigeration. In this study, no sample was stored for more than 6 hours. The samples were then aspirated from the oil and loaded into six separate wells of a Biological Cryostat Nanoliter Osmometer (Clifton Technical Physics, Inc.). A standard was run prior to each test sample. The reading of the last well to thaw was taken as equivalent to the cryostat standard (Advanced Instruments). A value of 312 mOsm/L and greater was suggestive of KCS.

## RESULTS

The mean of the Schirmer test performed in the suspect group of 10 patients was 5.6 ± 4.1 (MEAN ± SD) mm of wetting with a range of 1 - 12 mm. The same test done on the HIV negative control group produced a mean of 15.6 ± 7.8 mm with a range of 11 - 20 mm.

The results of the tear osmolarity tests done on the patients with signs and symptoms of KCS showed a mean of 318.8 ± 9.5 mOsm/L with range of 304 - 335 mOsm/L. Eighteen of the 20 suspect eyes tested positive for KCS. Tear osmolarity of the HIV negative control patients was 304.1 ± 5.6 mOsm/L with a range of 299-312 mOsm. One of the 20 control eyes tested positive for KCS.

## DISCUSSION

KCS occurring as a single entity without associated xerostomia or connective tissue disorder had not been reported prior to our recent study of HIV positive males where greater than 20% had KCS.[1] We now have found a similar proportion of KCS positive females in an HIV positive population.

Persistent generalized lymphadenopathy was seen in 7 of the 10 KCS patients. Although this suggests a possible correlation between generalized lymphadenopathy and KCS, i.e. lymphocyte activation leading to the localized release of fibrotic cytokines, we could neither rule out nor confirm a statistically significant correlation.

Out of the 59 patients, 30 had a previously diagnosed opportunistic infection, under the Center of Disease Control's criteria,[3] before the study began. This included 5 of the patients with KCS. No statistically significant correlation could be found between those carrying a diagnosis of AIDS and those patients with KCS.

We may conclude from the data that KCS occurs at a greater rate in females affected with the HIV than it does in the general population. We also recommend that young females, as well as males, presenting with KCS, in which all previously reported causative factors are ruled out, be further evaluated for HIV infection.

## ACKNOWLEDGMENTS

Supported in part by grants from the Sigma Xi Scientific Research Society, Research to Prevent Blindness, Inc., and the Lions Eye Research Foundation of New Jersey.

## REFERENCES

1. J.A. Lucca, R.L Farris, L Bielory, and A.R. Capto, Keratoconjunctivitis Sicca in male patients infected with human immunodeficiency virus type I, *Ophthalmology* 97:1008, 1990.
2. J.A. Lucca and R.L Farris, Keratoconjunctivitis sicca in HIV-positive female individuals, *Invest. Ophthalmol. Vis. Sci.* 33 (suppl):1288, 1992.
3. Public Health Service Coolfront Report, PHS plan for prevention and control of AIDS. *Pub. Health Rep.* 101:341, 1986.

# THE INCREASE OF THE BLINK INTERVAL IN OPHTHALMIC PROCEDURES

Hiroko Takano[1], Etsuko Takamura[1], Keiko Yoshino[1], and Kazuo Tsubota[2]

[1]Department of Ophthalmology, Tokyo Women's Medical College, Tokyo, Japan
[2]Department of Ophthalmology, Tokyo Dental College, Chiba, Japan

## INTRODUCTION

It has been previously reported[1-3] that the blink interval increases while using a visual display terminal and that this response could lead to a dry eye condition. During ophthalmologic examinations and surgical procedures, some clinicians experience an ocular burning sensation or visual blurriness. This may be due to the development of dry eye conditions. In this study, we measured the blink interval of ophthalmologists during surgical and clinical procedures.

## MATERIALS AND METHODS

Sixteen ophthalmologists between 25 to 65 years of age were selected for this study. All subjects had normal tear secretion as shown by a Schirmer test, and had no corneal fluorescence staining nor conjunctival and corneal rose bengal staining. The mean Schirmer test value was $29.5 \pm 6.2$ (mean $\pm$ SD) mm.

A small amount (2.0 µl) of a 1% solution of fluorescein and rose bengal was placed in the inferior conjunctival sac. After blinking several times, subjects would close their eyes, then open them. The time from the opening of the eyelids to the appearance of the first randomly distributed dry spot on the cornea is the tear break up time (BUT). The mean BUT was $6.5 \pm 2.5$ seconds.

The mesured time that the eye lid was open was defined as the blink interval. The blink interval was measured during non-work periods and while doing routine ophthalmic examinations and surgical procedures, such as biomicroscopy, tonometry, fundus examination, photocoagulation and cataract surgery (Fig.1). Blink intervals were measured under the following conditions; room temperature was 26–29°C, room humidity was 11–40% and the velocity of the wind within the room was 0.00–0.03 m/s.

## RESULTS

The blink interval was 2.1 ± 0.4 seconds during non–working times. During routine ophthalmic examinations, the blink interval increased to 3.7 ± 2.6 seconds for biomicro-scopies, to 4.3 ± 2.2 seconds for tonometries and to 4.8 ± 2.0 seconds for fundus examinations. Surgical procedures increased the blink interval drastically. Thus, the interval rose to 10.5 ± 4.2 seconds during photocoagulations and to 30.5 ± 27.9 seconds during cataract surgeries (Fig.1). The blink intervals during photocoagulation and cataract surgery were longer than the blink interval during non–working time (p<0.01). Following these procedures, in all subjects, there was no indication of injury to the cornea as shown by fluorescence staining tests.

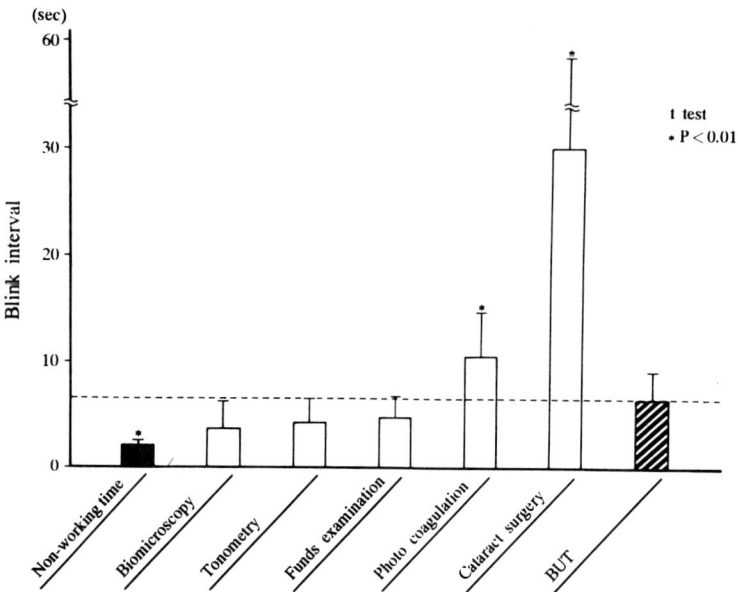

**Figure 1.** The blink interval in ophthalmic procedures.

## DISCUSSION

The blink interval was prolonged during ophthalmic examinations, especially during times of microscopic work. The blink interval during surgery was longer than BUT, whereas the blink interval during routine ophthalmic examinations was less than the BUT. These results suggest that certain ophthalmic procedures cause elongation of the blink interval which may extend the normal BUT and lead to a dry eye condition[4]. In this study, all subjects had normal tear secretions and therefore corneal injury did not occur. However, if the subjects had been dry eye patients, corneal injury may have occurred after surgical procedures. Thus, it may be necessary for ophthalmologists with dry eye to use a moisture eye chamber during surgery to avoid further complications and possible injury to the cornea[5,6].

## REFERENCES

1. K. Tsubota, and K. Nakamori, Dry eyes and video display terminals, *N. Eng. J. Med.* 328:584 (1993).
2. Y.Yaginuma, H. Yamada,and H. Nagai, Study of the relationship between lacrimation and blink in VDT work, *Ergonomics* 33:799 (1990).
3. J. Foreman, San Fransisco passes oridinance regulating VDT use, *Arch. Ophthalmol.* 109:477 (1991).
4. M. Luckiesh , and F. Moss, Relation between blink frequency and break-up time ? *Acta Ophthalmol.* 65:19 (1987).
5. K. Tsubota,and M. Yamada, Tear evaporation from ocuular surface, *Invest.Ophthalmol. Vis. Sci.* 33:2942 (1992).
6. K. Tsubota, The effect of wearing spectacles on the humidity of the eye, *Am. J. Ophthalmol.* 108:92 (1989).

# CHANGES IN TEAR ION CONCENTRATIONS IN DRY-EYE DISORDERS

Jeffrey P. Gilbard[1,2] and Scott R. Rossi[1]

[1]Cornea Research Unit
Schepens Eye Research Institute
[2]Department of Ophthalmology
Harvard Medical School
Boston, MA

## PURPOSE

Meibomian gland dysfunction and autoimmune lacrimal gland disease both increase tear film osmolarity, the former by an increase in tear film evaporation, and the later by a decrease in tear secretion.[1] In both cases the surface disease that results is known as keratoconjunctivitis sicca (KCS) and is characterized by decreased conjunctival goblet-cell density and corneal glycogen levels, specific epithelial abnormalities, and rose Bengal staining.[2-9]

For many years investigators were unable to demonstrate elevated tear film osmolarity in KCS because of limitations in the techniques used to collect and study tear fluid.[10] These techniques required the collection of large volumes of tear fluid, ocular contact and stimulation of tear secretion. Since tear film osmolarity decreases with increased tear flow rate, investigators documented increased osmolarity only after a technique was developed to non-invasively collect and study tear samples as small as 0.1 μl.[11-13]

There are many published studies that report the results of human tear electrolyte measurements.[14] Like the early studies of tear osmolarity, these studies required relatively long collection times and stimulation of tear secretion to collect samples of adequate size. Since tear electrolyte concentrations change as a function of lacrimal gland secretion rate[15] these studies were limited in their ability to accurately measure electrolyte concentrations in the undisturbed tear film. Long collection times create an additional artifact with regard to the measurement of bicarbonate since disassociated $CO_2$ gas rapidly escapes into surrounding air, reducing the amount of bicarbonate in the sample.

*Lacrimal Gland, Tear Film, and Dry Eye Syndromes*
Edited by D.A. Sullivan, Plenum Press, New York, 1994

With the increasing understanding of the importance of tear electrolytes in the maintenance of the ocular surface,[16-19] there has been increased interest in the electrolyte composition of the normal undisturbed tear film. The purpose of this study was to measure tear electrolytes in the normal tear film, and to study how they change in dry eye disease.

## PATIENTS AND METHODS

We measured osmolarity, sodium, potassium, bicarbonate, calcium and magnesium in microvolume tear samples collected from normal subjects and patients with meibomian gland dysfunction, lacrimal gland disease, or both. Patients were assigned to one of three groups using a clinical method based on history and examination described in detail elsewhere.[20]

Patients with ocular irritation were sorted prospectively into three groups based on history, examination and rose Bengal staining: 1) meibomian gland dysfunction (n=9), 2) lacrimal gland disease (n=19) and 3) lacrimal gland disease and meibomian gland dysfunction (n=3).

Samples for tear osmolarity and electrolyte measurement were obtained from a control group of 23 normal subjects (18 women, and 5 men) and from 31 dry-eye patients (27 women, and 4 men). Normal subjects were asymptomatic and had no history of dry-eye complaints.

Tear samples approximately 0.1 μl in size were collected from the inferior marginal tear strip with special L-shaped micropipettes that entered the strip without globe or lid contact.[11] Samples were saved for measurement as previously described.[13] Tear osmolarity was measured by freezing-point depression using a Clifton Nanolitre Osmometer (Clifton Technical Physics, Hartford, NY).[12] Atomic absorption spectrophotometry (Smith Heiftje 22, Thermo Jarrell Ash, Franklin, MA) was used to measure tear sodium ($Na^+$), potassium ($K^+$), calcium ($Ca^{++}$), and magnesium ($Mg^{++}$) as previously described.[21] Tear bicarbonate ($HCO_3^-$) levels were measured using a picapnotherm (model GV-1, World Precision Instruments, New Haven, CT) as previously described.[22,23]

All data are expressed as mean ± (SEM) and statistical analysis was done using the Student's t-test.

## RESULTS

In the 23 normal controls tear osmolarity measured in mOsm/L was 304.4 ± 0.4 and ranged between 299 and 309. Tear electrolytes measured in mmol/L were: $Na^+$, 133.2 ± 0.2; $K^+$, 24.0 ± 0.2; $HCO_3^-$, 32.8 ± 0.2; $Ca^{++}$, 0.80 ± 0.04; and $Mg^{++}$, 0.61 ± 0.03.

In the meibomian gland dysfunction patients mean tear osmolarity and tear $Na^+$, $K^+$, $HCO_3^-$, $Ca^{++}$, and $Mg^{++}$ concentrations increased uniformly by approximately 2%. Five eyes from four patients in the meibomian gland dysfunction group had tear osmolarity less than 310 mOsm/L. In these eyes, osmolarity was 305.6 ± 1.5 mOsm/L; tear electrolyte concentrations were not significantly different from normal controls.

In the lacrimal gland disease patients tear osmolarity and tear $K^+$, $HCO_3^-$, $Ca^+$, and $Mg^{++}$ increased by about 3.5%. Mean tear $Na^+$ increased disproportionately by 6.5%.

Among patients with the diagnosis of lacrimal gland disease there were five eyes with osmolarity measurements less than 310 mOsm/L. In these patients osmolarity measured $308.0 \pm 0.4$ mOsm/L and mean tear $Na^+$ increased 3.8% relative to normal ($P < 0.05$). The other electrolytes did not differ from controls.

Patients in the meibomian gland dysfunction/lacrimal gland disease group also showed a disproportionate increase in tear $Na^+$.

## CONCLUSIONS

We report tear film electrolyte concentrations in rapidly-collected unstimulated tear samples from both normals and patients with the most commonly encountered varieties of dry-eye disease.

In our previous work with a rabbit model, we found that meibomian gland dysfunction alone is sufficient to increase tear film osmolarity.[7] The current clinical data provide additional evidence that meibomian gland dysfunction alone, independent of any lacrimal gland disease, can increase tear film osmolarity and produce dry-eye disease.

Meibomian gland dysfunction increases tear electrolytes uniformly in proportion with tear osmolarity, consistent with an evaporative effect. In contrast, with lacrimal gland disease there is a disproportionate increase in tear $Na^+$. This increase in tear $Na^+$ is consistent with increased $Na^+$ in the primary lacrimal secretion at low flow rates.

Botelho and Martinez measured tear $Na^+$, $K^+$, and $Cl^-$ concentrations in fluid collected from the lacrimal gland excretory duct in the rabbit at various flow rates.[15] They found that at low flow rates $Na^+$ and $Cl^-$ concentrations increased, independent of evaporation. The increase in tear $Na^+$ seen in patients with very early lacrimal gland disease and normal tear osmolarity, and the disproportionate increase in tear $Na^+$ observed in patients with lacrimal gland disease and elevated tear osmolarity, is most likely due to the phenomenon originally described by Botelho and Martinez.

In the rabbit lacrimal gland fluid osmolarity increases as flow rate declines independent of evaporation.[24] The disproportionate increase in tear $Na^+$ in patients with lacrimal gland disease makes it likely that tear osmolarity increases in these patients in part due to higher lacrimal gland fluid osmolarity at low flow rates.

Elevated tear film osmolarity was not required for entry into the dry-eye patient groups. This gave us the ability to study changes in tear electrolyte concentrations that may precede increases in tear film osmolarity. It is notable that there were five eyes diagnosed as having lacrimal gland disease that had normal tear osmolarity but elevated tear $Na^+$. This suggests that in early lacrimal gland disease increased tear $Na^+$ precedes increased tear osmolarity. It may be that initial associated increases in lacrimal gland fluid osmolarity are concealed in the tear film by the osmotically driven transport of water across the cornea and conjunctiva.[25] Such osmotically driven water transport can lower and normalize tear osmolarity but not tear

Na$^+$. In early lacrimal gland disease, ocular surface disease may be due to changes in the secreted lacrimal gland fluid, as opposed to an increase in the influence of tear film evaporation secondary to decreased tear film turnover or decreased volume. These data also suggest that tear volume need not be decreased in early lacrimal gland disease.

We previously demonstrated that conjunctival goblet-cell density is sensitive to changes in the electrolyte composition of bathing solutions.[18,19] We discovered an optimum electrolyte balance for maintaining conjunctival goblet cells, and this balance matches the electrolyte balance that we now report in the normal tear film. It is also significant that with lacrimal gland disease tear Na$^+$ increases to over 140 mmol/L, a concentration that we found incompatible with the maintenance of normal conjunctival goblet-cell density.[18] We now hypothesize that in these patients disproportionate increases in tear Na$^+$, along with increases in tear osmolarity, contribute to the development of surface disease, specifically the depletion of conjunctival goblet cells.

## ACKNOWLEDGMENTS

This work was supported in part by grant EY03373 from the NEI, and the Massachusetts Lions Eye Research Fund, Inc.

## REFERENCES

1. J.P Gilbard, Rossi SR, K.L. Gray, Mechanisms for increased tear film osmolarity, *in* "The Cornea: Transactions of the World Congress on the Cornea III,*"* H.D. Cavanagh, ed., Raven Press, New York (1988).
2. R.A. Ralph, Conjunctival goblet cell density in normal subjects and in dry eye syndromes, *Invest Ophthalmol Vis Sci.* 14:299 (1975).
3. J.D. Nelson, V.R. Havener, J.D. Cameron, Cellulose acetate impressions of the ocular surface. Dry eye states, *Arch Ophthalmol.* 101:1869 (1983).
4. J.D. Nelson, J.C. Wright, Conjunctival goblet cell densities in ocular surface disease. *Arch Ophthalmol.* 102:1049 (1984).
5. L.M.R. Abdel-Khalek, J. Williamson, W.R. Lee, Morphological changes in the human conjunctival epithelium. II. In keratoconjunctivitis sicca, *Br J Ophthalmol.* 62:800 (1978).
6. J.P. Gilbard, S. Rossi, K. Gray, A new rabbit model for keratoconjunctivitis sicca, *Invest Ophthalmol Vis Sci.* 28:225 (1987).
7. J.P. Gilbard, S.R. Rossi, K. Gray Heyda, Tear film and ocular surface changes after closure of the meibomian gland orifices in the rabbit, *Ophthalmology.* 96:1180 (1989).
8. J.P. Gilbard, S.R. Rossi, K.L. Gray, L.A. Hanninen, K.R. Kenyon, Tear film osmolarity and ocular surface disease in two rabbit models for keratoconjunctivitis sicca, *Invest Ophthalmol Vis Sci.* 29:374 (1988).
9. J.P. Gilbard, S.R. Rossi, K.L. Gray, L.A. Hanninen, Natural history of disease in a rabbit model for keratoconjunctivitis sicca, *Acta Ophthalmol.* (Suppl 192) 67:95 (1989).
10. G.J. Mastman, E.J. Baldes, J.W. Henderson, The total osmotic pressure of tears in normal and various pathologic conditions, *Arch Ophthalmol.* 65:509 (1961).
11. J.P. Gilbard, R.L. Farris, J. Santamaria, Osmolarity of tear microvolumes in keratoconjunctivitis sicca, *Arch Ophthalmol.* 96:677 (1978).
12. J.P. Gilbard, R.L. Farris, Ocular surface drying and tear film osmolarity in thyroid eye disease, *Acta Ophthalmol.* 61:108 (1983).

13. J.P. Gilbard, K.L. Gray, S.R. Rossi, Improved technique for storage of tear microvolumes, *Invest Ophthalmol Vis Sci.* 28:401 (1987).
14. N.J. Van Haeringen, Clinical biochemistry of tears, *Surv Ophthalmol.* 26:84 (1981).
15. S.Y. Botelho, E.V. Martinez, Electrolytes in lacrimal gland fluid and in tears at various flow rates in the abbit, *Am J Physiol.* 225:606 (1973).
16. W.G. Bachman, G. Wilson, Essential ions for maintenance of the corneal epithelial surface, *Invest. Ophthalmol Vis Sci.* 26:1484 (1985).
17. R. J. Fullard, G.S. Wilson, Investigation of sloughed corneal epithelial cells collected by non-invasive irrigation of the corneal surface, *Curr Eye Res.* 5:847 (1986).
18. J.P. Gilbard, Non-toxic ophthalmic preparations, US Patent 4,775,531. Oct. 4, 1988.
19. J.P. Gilbard, S.R. Rossi, K. Gray Heyda, Ophthalmic solutions, the ocular surface, and a unique therapeutic artificial tear formulation, *Am J Ophthalmol.* 107:348 (1989).
20. J.P. Gilbard, Dry Eye Disorders, *in* "Principles and Practice of Ophthalmology: Clinical Practice," D. Albert and F. Jackobiec eds., W.B. Saunders Company, Philadelphia (in press).
21. J.P. Gilbard, S.R. Rossi, Tear film and ocular surface changes in a rabbit model of neurotrophic keratitis, *Ophthalmology.* 97:308 (1990).
22. G.G. Vurek, D.G. Warnock, R. Corsey, Measurement of picomole amounts of carbon dioxide by calorimetry, *Anal Chem.* 47:765 (1975).
23. R.L. Bowman, G.G. Vurek, Analysis of nanoliter biological samples. *Anal Chem.* 56:391A (1984).
24. J.P. Gilbard, D.A. Dartt, Changes in rabbit lacrimal gland fluid osmolarity with flow rate, *Invest Ophthalmol Vis Sci.* 23:804 (1982).
25. D. Maurice, The tonicity of an eye drop and its dilution by tears, *Exp Eye Res.* 11:30 (1971).

# QUANTITATIVE CELLULAR EVALUATION OF CONJUNCTIVAL SQUAMOUS METAPLASIA IN THE DRY EYE PATIENT

Etsuko Takamura,[1] Hiroko Takano,[1] Keiko Yoshino,[1] Kazumi Negoro,[1] Kazuo Tsubota [2] and Tadao Kobayashi[3]

[1]Department of Ophthalmology, Tokyo Women's Medical College, Tokyo, Japan; [2]Department of Ophthalmology, Tokyo Dental College, Chiba, Japan; and [3]Department of Cytopathology, Saiseikai Shiga Hospital, Shiga, Japan

## INTRODUCTION

Cytology can be used to evaluate ocular surface changes in patients with dry eye. For example, impression cytology of the conjunctiva has been utilized as a diagnostic tool to examine for possible alterations in both the number of goblet cells and appearance of squamous cells[1,2]. However, although this cytologic technique detects goblet cells and keratinized cells effectively, it is sometimes difficult to observe their morphologic details and to evaluate them quantitatively, because of the overlapping of cells. For the objective assessment of dry eye conditions, it is necessary to accurately visualize and quantitate ocular surface alterations at the cellular level.

The recently developed technique of conjunctival cytology using a special brush has several advantages over impression cytology[3-6]. This brush procedure permits the collection of a whole layer of cells, not just the superficial abnormal epithelium, thereby allowing the analysis of many cells and the calculation of the ratio of keratinized cells to normal cells. Thus, the brush cytology technique appears to provide a simple and accurate method for the evaluation of cellular changes on the ocular surface in dry eye.

The present study was undertaken to demonstrate the use of brush cytology in the quantitative evaluation of conjunctival squamous metaplasia in dry eye patients.

## MATERIALS AND METHODS

A total of twenty-seven dry eye patients, including 16 with Sjögren's syndrome (SS) and 11 with non-SS dry eye, were recruited. The SS patients included 16 females, ranging in age from 26 to 76 years (median age = 54.6 years), all of whom were diagnosed by Fox's criteria[7]. The patients with non-SS included 2 males and 9 females, ranging in age from 35 to 66 years (median age = 53.6 years). To serve as a normal control population, 6 females (30 to 49 years old, median age = 51.5 years) without ocular complaints and with normal ocular examinations were also recruited.

Brush cytology was performed with the modified method previously reported[3]. The eyes were topically anesthetized (0.4% proparacaine) and both the temporal bulbar and the lower tarsal conjunctiva were scraped 7 times using a Cytobrush-S® (Medscand) (Figure 1). The smears with all cellular material were spread on a glass slide by means of cytocentrifugation. Then, the cells were stained with a modified Papanicolaou method. On each slide, 300 cells were counted and the ratio of the keratinized to normal cells was calculated.

In order to determine the number of goblet cells per slide, periodic-acid Schiff (PAS) staining was carried out on Papanicolaou-destained slides and the total number of PAS-positive cells ( i.e. goblet cells) were counted.

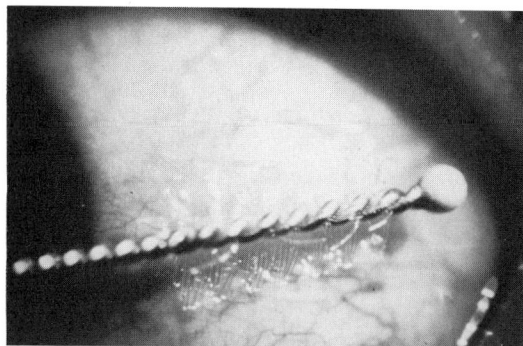

Figure 1. Sample collection from the temporal bulbar using a Cytobrush-S®(Medscand ).

## RESULTS

Conjunctival cells from normal controls showed fine nuclear chromatin and polyhedral cytoplasm. In contrast, conjunctival smears from SS patients showed a clump of keratinized cells with marked cytoplasmic degeneration; some cells were extremely elongated (Figure 2).

In dry eye patients with SS the percentage of keratinized cells in the bulbar and tarsal conjunctiva were significantly ($p<0.05$) higher than that of the non–SS patients. These cells often intermingled with polymorphonuclear leukocytes and lymphocytes. The ratio of keratinized cells in both dry eye patients groups (i.e. SS and non-SS) was significantly ($p<0.01$) higher than that of the age-matched controls.

The number of goblet cells, which showed a typical cellular appearance (Figure 3), was determined after staining conjunctival smears with PAS (Figure 4). The mean number of goblet cells in the bulbar conjunctival samples of SS and non-SS patients was $4.7 \pm 5.8$ and $8.6 \pm 14.8$, respectively, whereas in the lower tarsal conjunctiva these cells numbered $6.7 \pm 7.7$ and $5.3 \pm 7.6$, respectively.

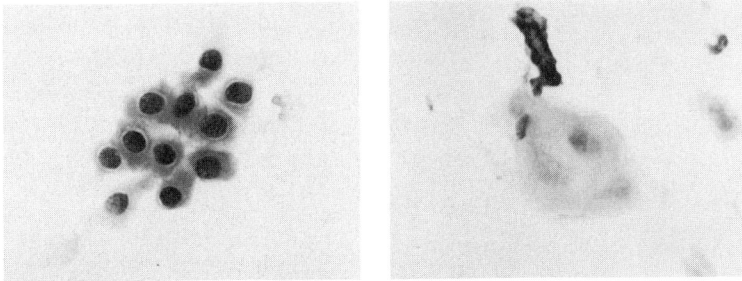

**Figure 2. Left**, Group of conjunctival epithelial cells from the normal control group showing slight nuclear enlargement, fine chromatin and polyhedral cytoplasm. **Right**, Smear from Sjögren's syndrome patients showing clumps of keratinized cells with stained orangenophillic cytoplasm and ill-defined cell margins (Papanicolaou stain, x 400).

Table 1. The percentage of keratinized cells in the conjunctival epithelium.

| Conjunctival sample | Keratinized cells (%) | | |
|---|---|---|---|
| | Sjögren's syndrome | Non-Sjögren's Syndrome | P Value |
| Temporal bulbar | 7.2 ± 3.8 | 4.4 ± 2.5 | < 0.05 |
| Lower tarsal | 6.7 ± 3.6 | 4.4 ± 2.0 | < 0.05 |

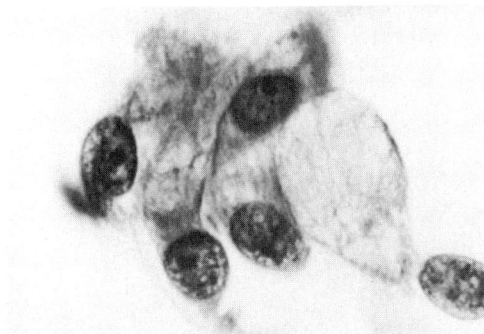

**Figure 3.** Conjunctival smear showing typical appearance of goblet cells intermingled with non-secretory conjunctival cells. Note the vacuoles which distend the cytoplasm (Papanicolaou stain, x 1000).

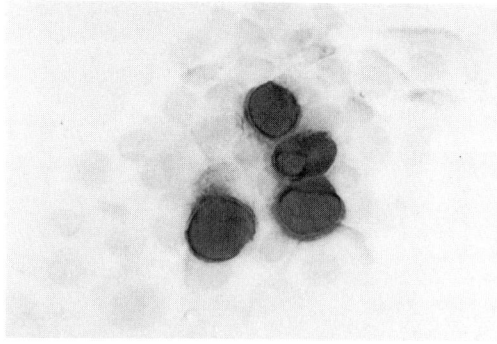

**Figure 4.** Clumps of conjunctival cells showing several cells with diffuse staining of the cytoplasm for mucin (Periodic–acid Schiff reaction, x 1000).

## DISCUSSION

Using the brush cytology procedure, the quantitative cellular evaluation of squamous metaplasia in dry eye patients can be performed. This method permits a definite diagnosis based on the morphology of the conjunctival epithelial cells. Typically, a clinical examination often fails to differentiate SS dry eyes and non-SS dry eyes. However, the cytological information provided by the brush technique allows differentiation between these diseases.

It is possible that this brush method may also be utilized in the future for the enumeration of goblet cells in conjunctival samples. Indeed, this application would have definite diagnostic value. However, in the present study, the limited number of samples and the variation in goblet cell counts prevented any meaningful statistical analysis.

Overall, the brush cytology procedure is useful for the cellular evaluation and classification of conjunctival epithelial changes in dry eye conditions. It is also possible that this technique could be utilized to determine the efficacy of dry eye treatment.

## REFERENCES

1. J.D. Nelson, V.R. Havener, and J.D. Cameron, Cellulose acetate impressions of the ocular surface, dry eye states, *Arch. Ophthalmol.* 101:1869 (1983).
2. S.C.G. Tseng, Staging of conjunctival squamous metaplasia by impression cytology, *Ophthalmology* 92:728 (1985).
3. K. Tsubota, K. Kajiwara, S. Ugajin and T. Hasegawa, Conjunctival Brush Cytology, *Acta Cytol.* 34:233 (1990).
4. E. Takamura, Y. Uchida, and K. Tsubota, Inflammatory cells in the conjunctival epithelium in aqueous tear deficiency, *Jpn. J. Clin. Ophthalmol.* 45:1195 (1991).
5. K. Tsubota, E. Takamura, T. Hasegawa, and T.Kobayashi, Detection by brush cytology of mast cells and eosinophils in allergic and vernal conjunctivitis, *Cornea* 10:525 (1991).
6. T. Kobayashi, S. Sato, K. Tsubota, and E. Takamura, Cytological evaluation of adenoviral follicular conjunctivitis by cytobrush, *Ophthalmologica* 202:156 (1991).
7. R. Fox, C. Robinson, J. Curd, F. Kozin, and F. Howell, Sjögren's syndrome: proposed criteria for classification, *Arth. Rheum.* 29:577 (1986).

# TOPICAL ANESTHETIC INDUCED PAIN AS A DIAGNOSTIC TOOL FOR KERATO CONJUNCTIVITIS SICCA (KCS)

Hans H. Stolze, Andrea Volprecht, and Ute Welter

Augenklinik der RWTH
D-5100 Aachen
Germany

## INTRODUCTION

Most patients with KCS suffer from heavy ocular discomfort, whereas considerable morphological damage of the ocular surface is only found in rare cases. So the question arises, if increased sensitivity plays a major role in KCS. None of the KCS- tests available gives information on subjective sensitivity. In the test described here the induction of pain by a defined trauma of the tear film and ocular surface is used.

To use a defined trauma seems to be a good method for a KCS-test since the provocation of pain by situations stressing the tear film belongs to the typical anamnestic history of dry eye syndrome.

For stressing the tear film we use the application of a drop of local anesthetic, which is routinely done when measuring intraocular pressure by applanation technique. So the information can be easily obtained by just asking the patient for intensity and duration of feeling pain.

## PATIENTS AND METHODS

We investigated 321 normals and 77 KCS-patients defined by the typical complaints and at least two positive KCS-tests: BUT (<10 sec), Schirmer (< 5 mm), tear-meniscus (<0.2 mm) and rose bengal staining(staging>3, van Bijsterveld,1969). We applied one drop of commercially available local anesthetic with fluorescein conserved with phenylmercuriborat (Oxybuprocain, Thilorbin$^R$, Alcon, pH 4.9). After an initial latency of about three seconds many patients claimed to have a burning feeling. They were asked to quantify their pain into none (0), slight (1), medium (2) or strong (3). The duration was measured.

In supplementory investigations we could show that the mean scatter range of the duration of pain in the same patient at different days was 17.5 %.

## RESULTS

### Duration of time

There was no correlation to age in the range tested (18 - 91 years) or to sex. The mean duration of pain in normals is 7.64 seconds (standard deviation (SD) 8.25 s). In KCS-patients the mean duration is 16.23 s (SD 12.82 ). This difference is highly significant (p < 0.001). If the limit is set at 12.0 s, the sensitivity ( = probability to prove a KCS-patient as pathological) is 64 %. The specificity ( = probability to prove a normal person as normal) is 69 %.

*Lacrimal Gland, Tear Film, and Dry Eye Syndromes*
Edited by D.A. Sullivan, Plenum Press, New York, 1994

As can be seen from figure 1, almost half of the normals do not feel any pain. A quarter of them feel pain only below 10 seconds or between 10 and 20 seconds. A longer duration is only found in less then 10 %. In KCS-patients however no pain or very short pain below 10 seconds is only found in 18 % each. Most patients (52 %) feel pain for 11 to 30 seconds, 12 % feel pain even longer.

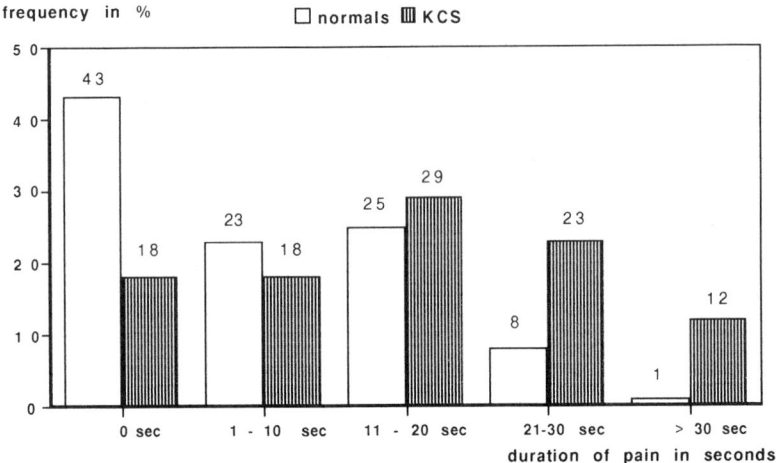

Figure 1. Duration of pain in normals and KCS-patients. Time is measured in seconds, frequency in %.

## Intensity of Pain

The average intensity of pain in normals is 0.63 (SD 0.60). In KCS-patients it is increased to 1.34 (SD 0.91). This difference in intensity of pain is highly significant too (p< 0.001).

In figure 2 the distribution of intensity in the different groups is shown. In normals almost half of the persons tested have no or only slight pain. Pain of medium degree was described in 5 %. None of the 321 normals had claimed strong pain. In KCS-patients 41 % had medium or strong pain. Only 18 % had no pain.

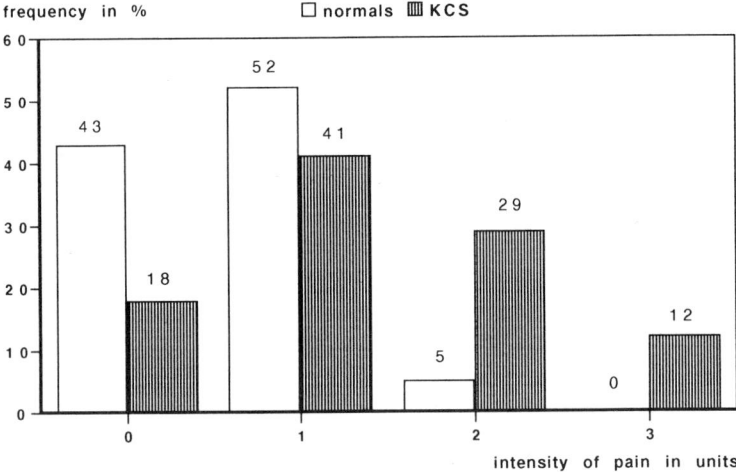

Figure 2. Intensity of pain in normals and KCS-patients. Intensity is scaled from 1 to 3, frequency in %.

## DISCUSSION

Initial pain is a well known phenomenon in local anesthesia. It is often discussed to be caused by the acid pH of local anesthetics. The interesting finding of Höh (1991) who found a strong correlation between pain, octanol-water-distribution-coefficient and local anesthetic side effect but not of the pH in local betablocking eyedrops strongly suggests that the induction of initial pain is a direct specific side effect of local anesthetics.

The efficacy of a local anesthetic eye drop to produce pain depends on individual sensitivity as well as on different factors of the tear film and ocular surface: tear volume, wash out by reflex secretion, buffer capacity of tears, integrity of mucus layer and epithelium. Most of the KCS-tests only give an information on a single element of the impaired tear film like tear volume, tear quality and the state of the epithelium. As Snyder and Fullard (1991) pointed out "this type of approach has unfortunately promoted the idea that dry eye is a specific condition of singular cause."

Furthermore most of these tests only give information at a moment where no stress is applied to the tear film. However one of the most important functions of the tear film is its ability to keep in a stable condition despite a wide range of different environmental influences. In the test described here the tear film is impaired by an acid drop of a local anesthetic. This well standardized stress is able to give an information on the regulatory capacity of the tear film. We think that this is of high diagnostic value since Behrens-Baumann (1986) demonstrated that KCS correlates well with ocular complaints provoked by situations like central heating, air conditioning and cigarette smoke.

The value of different tests for a specified diagnosis is expressed by their sensitivity and specifity. As mentioned above this new test should be compared to other KCS-tests that are based not only on a single element of the tear film and in addition give information on the adaptative capacity of the tear film under environmental stress. These informations can be given by the anamnesis and the break up time (BUT). Also increased tear osmolarity and epithelial defects may be considered here because they are not only a single primary cause of KCS but different aetiologies may finally impair these parameters.

McMonnies (1987) has reported a sensitivity of 98 % and a specificity of 97 % for his questionnaire demonstrating the important information obtained by an exact anamnesis. Sensitivity of BUT is 52 %, its specificity 78 % (Albach et al., 1993). In the more sensitive version of non invasive break up time (NIBUT) it reaches a sensitivity of 83 % and a specificity of 85 % (Mengher et al., 1986). Tear osmolarity has 76 % sensitivity and 84 % specificity (Farris et al., 1983). Since an expensive apparatus is needed in both tests they are not widely used yet. Rose Bengal staining of the epithelium has a sensitivity of 95 % and a specificity of 96 % according to Van Bijsterveld (1969). Farris et al. (1983) reported 58 % sensitivity and 100 % specificity.

In our test we found 64 % sensitivity and 69 % specificity. If you use these numbers to show that a new test outperforms older diagnostic tests you have to agree on the "gold standard test" that defines KCS (Farris, 1992). WHO defines a disease by the impairment of feeling well. So anamnestic ocular pain caused by tear film abnormalities would ideally fulfill the condition of a gold standard. To make sure that reported ocular pain is caused by the tear film, there should be a reported provocation by environmental influences. Classical KCS-tests are not suitable because in many cases a mismatch between symptoms and clinical signs of dry eye is found (Snyder and Fullard, 1991). Since all the values of sensitivity and specificity of the different tests base on different standards one should be very careful to use them to find out if a test is more valuable than the other.

In contrast to all other KCS-tests this new test gives additional information on subjective sensitivity and the influence of expecting eye pain produced by a standardized trauma in the often anxious KCS-patients. So the lower values compared to the classical tests of KCS just indicate that it is different information. Only if all values were calculated on the basis of the same golden standard for KCS, which has to include the subjective disturbance of the patient, sensitivity and specificity could define the diagnostic value of the tests.

For a complete assessment of an individual KCS-patient the accurate anamnesis and the classical KCS-tests are important. But also information on the subjective sensitivity towards a tear film stress should be known, which can easily be obtained by the test described here. It will help to clear many cases where a mismatch between symptoms and clinical signs of dry eye is found . Furthermore it may also be of therapeutic importance to avoid potentially irritating eye drops in the very sensitive patients.

Further investigation has to be done to determine the exact mechanism of pain induction, to prove the specifity for KCS compared to other corneal diseases and to look for a correlation to tests of specific tear film parameters like tear volume, quality and the condition of the epithelium and corneal sensitivity.

## SUMMARY

Duration and intensity of pain induced by instillation of an acid local anesthetic eyedrop were compared in 321 normals and 77 KCS-patients. The medium duration of pain was increased in KCS-patients (16.23 s ) compared to normals (7.64 s). Also the main intensity of pain was increased (KCS: 1.34, normals 0.63). Both differences were highly significant (< 0.001). The sensitivity of duration (limit 12 s) was 64 % and the specificity 69 %.

Except BUT this very simple test is the only one that uses a well defined stress to the tear film to test its capacity for adaptation. Furthermore it is the first KCS-test, that includes the individual sensitivity, which is an important parameter for assessment of KCS-patients.

## REFERENCES

Albach,K.A.,Lauer,M..,and Stolze,H.H., 1993 , Die Wertigkeit verschiedener Tests zur Diagnose der Keratoconjunctivitis sicca (KCS) bei Patienten mit rheumatoider Arthritis, Der Ophthalmologe,90,in press

Behrens-Baumann,W., 1986, Häufigkeit der verschiedenen subjektiven Symptome beim "trockenen Auge", Fortschr. Ophthalmol.,83,118

Farris,R.L., 1992, Tear osmolarity-a new gold standard?,Int.Conf.on the lacrimal gland,tear film and dry eye syndromes:Basic science and clinical relevance, Nov.14.-17., Southampton Parish, Bermuda

Farris,R.L.,Gilbard,J.P.,Stuchell,R.N.and Mandel,D.,1983, Diagnosetests bei keratoconjunctivitis sicca,Contactologia 5D,133

Höh,H.,and Nastainczyk,W., 1991, ß-Blocker und Hornhautsensibilität, Fortschr. Ophthalmol.,88,515

McMonnies,C.W., and Ho,A., 1987, Patient history in screening for dry eye conditions, J.Am.Optom.Assoc., 58,296

Mengher,L.S.,Bron,A.J.,Tonge,S.R.,and Gilbert,D.J.,1986,Non-invasive assessment of the tear film stability, in: "The Precorneal Tear Film in Health,Disease and Contast Lens Wear,"Holly,F.J.,ed, Dry Eye Institute,Lubbock,Tx.

Snyder,C., and Fullard,R.J., 1991, Clinical profiles of non dry eye patients and correlations with tear protein levels,Int. Ophthalmol.,15,383

Van Bijsterveld,O.P., (1969) ,Diagnostic tests in the sicca syndrome, Arch. Ophthalmol.,82,10

# LACRIMATION KINETICS AS DETERMINED BY A SCHIRMER-TYPE TECHNIQUE

Frank J. Holly

Dry Eye Institute, Inc.
P.O. Box 98069
Lubbock, Texas 79499

## INTRODUCTION

The easiest, least expensive, and therefore most popular of lacrimation tests is the so-called Schirmer-I test.[12] This classic test consists of placing the 5 mm long, bent end-piece of an unbonded, porous paper strip, usually 35 mm long and 5 mm wide, in the lower fornix of the subject at one third of the palpebral distance from the temporal canthus. The presence of the rounded strip end in the fornix results in a minor trigeminal nerve irritation inducing reflex tearing. The tears are absorbed by the strip and their amount secreted in a given time interval is evidenced by a certain wetted length of the paper strip. Traditionally the time interval is five minutes and the wetted length is given in millimeters. This test is used either without any anesthesia or with topical anesthesia ostensibly to prevent excessive lacrimation.

## THE SCHIRMER LACRIMATION TEST

Schirmer published his classical paper in 1903.[12] He described three types of test; the first has endured the test of time and is still widely used in the clinic. In the Schirmer-I test, no additional stimulation is imposed on the patient. In the Schirmer-II test, a camel hair brush was inserted in one of the nostrils and twirled to induce additional lacrimation. The second test has not become popular, although in recent times investigators employed nasal stimulation to collect tears.[2,15] The third test required staring into the sun and was soon abandoned.

During the ninety years since the publication of Schirmer's work, numerous articles about his test have appeared in print. Over the years various paper types have been tried, such as blotter paper of high absorbency,[12] litmus paper,[13] cigarette paper,[14] and filter paper of various types.[3,5-8] All the *in vitro* tests, however, were conducted under the conditions of unlimited fluid supply, where the paper and fluid characteris-tics determined the rate of wetting, thus bearing little relevance to the Schirmer test where the supply rate of tears to the strip is the controlling factor.[10]

Halberg and Berens[8] developed a standardized Schirmer kit using No. 589 Black Ribbon filter paper made of cotton fibers and having a high alpha-cellulose content. They reported that the 5 mm wide paper strip had a specific absorbency of about 0.5 microliter of water per one millimeter length.

## KINETIC ANALYSIS OF THE SCHIRMER LACRIMATION TEST

The driving force of liquid flow in porous paper strips is derived from surface forces, i.e., from the tendency of the fluid to wet the cellulose fibers through the pore structure of the paper strip. If the fluid is supplied to the strip in excessive quantities, the wetting rate of the paper strip will be identical to the penetration rate of a fluid into a single horizontal capillary.[9] For a given fluid and paper type, the fluid absorption rate is thus equal to the capillary coefficient[4,9] consisting of parameters such as surface tension, viscosity, contact angle of wetting, and effective pore radius. It has been shown that the capillary characteristics of the Black Ribbon No. 589 filter paper are identical to a clean glass capillary of 1.2 micrometers in internal diameter.[9]

The wetting rate or the velocity of the fluid front in the paper strip (in millimeters per minute) will yield the fluid uptake rate (in microliters per minute) when multiplied by the volume of the fluid (in microliters) necessary to wet a one millimeter segment of the paper strip, V. Under unlimited fluid supply rate, there is a maximal fluid uptake rate for a given liquid by the paper strip that will depend on the magnitude of V and decreases with time.

Tear secretion rate measurements by the Schirmer strip can be meaningful only if the secretion rate does not exceed this maximal uptake rate. When the water (or tear) supply to the filter paper strip is limited to constant values below the maximal uptake rate[9] the wetting rate is again found to be proportional to the supply rate provided that the water evaporation from the strip surface was prevented. Evaporation lowers the apparent wetting rates and the relative error thus caused increases with increased wetting length due to the increased area of evaporation.[9]

In the Schirmer-I test certain conditions have to be fulfilled so that the wetting rate of the paper strip is related to the tear secretion rate in a predictable manner. These conditions are:
1. Tear secretion rate should be equal to the strip absorption rate.
2. Specific wetting volume should be reasonably constant along the strip.
3. All the absorbed fluid should stay within the strip (no evaporation).

It was shown[9] that the Schirmer test can be modified to fulfill these conditions and thus yield quantitative data on the tear secretion rate during the lacrimation response to the insertion of the strip.

## KINETICS OF LACRIMATION INDUCED BY THE SCHIRMER STRIP

When lipid-extracted porous paper strips are used and the evaporation is prevented the tear secretion rate is reflected in the rate of wetting of the paper strip. Therefore, by analyzing the time dependence of the wetting length increase, quantitative infor-mation can be obtained from the tear secretion rate and its time dependence during lacrimation.[9,10]

## Kinetic Parameters

It was found in several hundred determinations[10] that the time dependence of the tear secretion rate during lacrimation is always of the same type. After strip insertion, lacrimation starts with a high secretion rate, which then exponentially decreases to a lower, constant tear secretion rate. Thus such a lacrimation cycle can be described by three kinetic parameters; the initial tear secretion rate, $F_i$, the final tear secretion rate, $F_f$, and the secretion decay coefficient, $k$. In a mathematical form,

$$F = F_f + (F_i - F_f)e^{-kt} \qquad (1)$$

From the wetted length increase as a function of time these three parameters can be readily determined.[10]

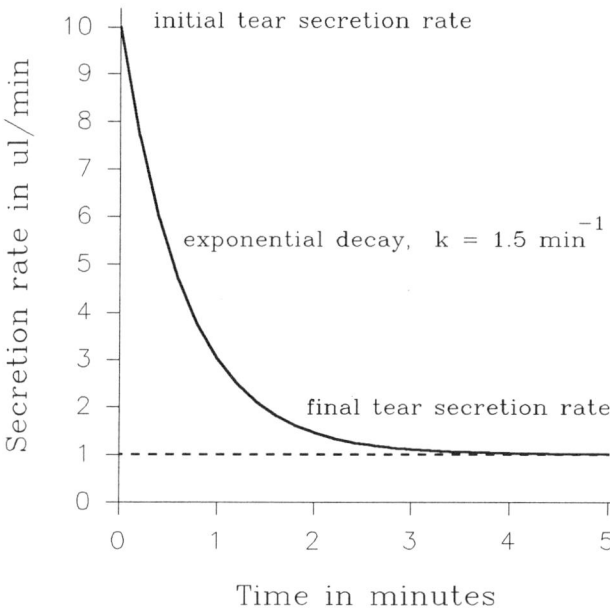

**Figure 1.** The variation of tear secretion rate with time during a lacrimation cycle.

## Multi-cycle Pattern

The strip wetting curve obtained over a time interval of five or ten minutes may result from several lacrimation cycles. Every lacrimation cycle is characterized by the three kinetic parameters defined above, and in general, these parameters are different for each cycle. The most complex wetting data obtained consisted of four lacrimation cycles in 7.5 minutes. It was found without exception that the change to a higher tear secretion rate, signifying the start of a new cycle is always practically instantaneous, while the decay of the secretion rate during the cycle is relatively slow ($1 < k < 3$) and *always exponential.* To facilitate comparison among lacrimation patterns, the multi-cycle patterns can be reduced to a one-cycle pattern.[9,10]

It is important to note that the conventional Schirmer test result will be larger for a multi-cycle than a uni-cycle lacrimation pattern. For example, assuming a typical set of kinetic parameters for a lacrimation cycle in a normal eye to be $F_i = 8$ $\mu\ell$/min, $F_f = 0.8$ $\mu\ell$/min, and $k = 1.7$ min$^{-1}$, the Schirmer test result would be 13 mm. If the kinetic parameters remained the same but the lacrimation pattern would consist of four cycles lasting 1.25 min each, the Schirmer test result would become 30 mm.

## EFFECT OF TOPICAL ANESTHESIA

In the unanesthetized eyes the frequency distribution of the kinetic parameters is quite skewed toward the higher values.[9] In the anesthetized eyes these excessively high values are absent making the frequency distribution of the kinetic parameters less skewed.

As a direct consequence of such skewed frequency distribution, the mean or average values are considerably higher than the median values, and much higher than the mode (most probable) values. Since extreme values are not obtained under topical anesthesia, the mean kinetic parameters obtained for anesthetized eyes are much lower than those for the unanesthetized eyes. The median values are less different and the mode values are almost indistinguishable between unanesthetized and anesthetized eyes (Table 1).[10]

**Table 1.** Statistical values for tear secretion parameters obtained in the absence and presence of tropical anesthesia in normals.

| STATISTICAL VALUES | INITIAL RATE[*] | | FINAL RATE[*] | | DECAY COEFF.[**] | |
|---|---|---|---|---|---|---|
| | without | with | without | with | without | with |
| Mean | 42.4 | 21.8 | 6.2 | 1.4 | 11.6 | 3.98 |
| Median | 12.7 | 9.7 | 2.0 | 1.4 | 1.96 | 1.84 |
| Mode | 3.5 | 4.1 | 0.60 | 0.53 | 0.75 | 0.52 |

[*] in microliter/minute;          [**] in reciprocal minute

## LACRIMATION RESPONSE OF DRY EYE PATIENTS

Selected groups of sicca patients and age-matched controls were examined[1] using the same technique employed previously for normals,[10] and the eyes were anesthetized by topical means. The results obtained showed that the mean, median, and mode values were much closer together, indicating that the frequency distributions of the secretion kinetic parameters are less skewed for the older population than was found for younger normals.

The study further showed that the only final tear secretion rate and the secretion decay coefficient were significantly different in the dry eyes from normal (Table 2). The final tear secretion rate was lower, while the secretion decay coefficient was higher in the sicca patients. Hence, apparently the average dry eye patients are capable of suddenly increasing their tear secretion rate, just as normals do, upon trigeminal nerve stimulus. However, their lacrimation rate diminishes faster and to a lower final value.

In a typical comparison, where the initial tear secretion would remain the same, if the final tear secretion rate would decrease 21% and the secretion decay coefficient would increase by 41%, the Schirmer test value would diminish by about 24% e.g. from 24 mm to 18 mm (assuming no evaporation).

Table 2. Statistical comparison of kinetic parameters of lacrimation in dry eye patients and normals.

| KINETIC PARAMETERS | STATISTICAL QUANTITY | DRY EYE | NORMAL | SIGNIFICANCE OF DIFFERENCE |
|---|---|---|---|---|
| Initial tear secretion rate ($\mu\ell$/min) | mean | 7.8 | 8.6 | p > 0.25 (not significant) |
| | median | 5.7 | 7.4 | |
| | mode | 5.6 | 7.5 | |
| Final tear secretion rate ($\mu\ell$/min) | mean | 0.40 | 0.94 | P < 0.01 (highly significant) |
| | median | 0.31 | 0.65 | |
| | mode | 0.30 | 0.52 | |
| Secretion Decay Coefficient ($min^{-1}$) | mean | 2.45 | 1.86 | p < 0.05 (significant) |
| | median | 2.38 | 1.73 | |
| | mode | 1.90 | 1.83 | |

The lacrimation pattern in dry eye patients was found to be much simpler than in the age-matched controls. In the sicca patients the one-cycle pattern was observed in 60% of the eyes, while only 6% of the control eyes exhibited such a simple pattern.

The reduction from a four-cycle lacrimation pattern to a unicycle lacrimation pattern would decrease the Schirmer test value about 63%, i.e. from 24 mm to 9 mm, even if the kinetic parameters would remain the same.

## CONCLUSIONS

The popular Schirmer test can be modified to obtain the time dependence of the tear secretion rate during lacrimation response induced by the insertion of the Schirmer strip provided that the evaporation of the tears during the measurement is prevented.

The tear secretion rate during lacrimation was found to be a simple exponential function, so that it could be quantitatively described by three kinetic parameters: the initial tear secretion rate, the final tear secretion rate, and the secretion decay coefficient. Topical anesthesia of the eye appears to eliminate the high values obtained in certain, responsive individuals, but the most probable values of the kinetic parameters do not seem to be much affected.

In dry eye patients, the initial tear secretion rate does not seem to be affected. However, the final secretion rate, more closely related to the basal tear secretion rate, is considerably lower and the secretion decay coefficient is considerably higher in dry eye patients than normals. Both of these changes tend to decrease the Schirmer test value by a factor of 3/4, while the reduction in the number of cycles in the lacrimation diminishes the Schirmer test to 1/3rd of the original value in the absence of evaporation.

The conventional Schirmer test result yields a lower wetting length due to evaporation of tears from the strip surface. Since the evaporative loss would tend to decrease the difference between the test values obtained for dry eye patients and normals by about 10%. However, the two factors described above would override the evaporative effect and would enhance the difference between dry eye patients and normals. This is why the classical Schirmer test is still of value in detecting tear secretion abnormalities in dry eye patients despite all the shortcomings of the test.

## REFERENCES

1. W.E. Beebe, E.D. Esquivel, and F.J. Holly, Comparison of lacrimation in dry eye patients and normals, *Curr. Eye Res.* 7:419 (1988).
2. A. Berta, Collection of tear samples with or without stimulation, *Am. J. Ophthalmol.* 96:116 (1983).
3. G.M. Bruce, Keratoconjunctivis sicca, *Arch. Ophthalmol.* 26 :945 (1941).
4. J.T. Davies and E.K. Rideal, **Interfacial Phenomena,** 2nd ed. Academic Press, New York-London (1963) p. 419.
5. A. de Roetth, Sr., Lacrimation in normal eyes, *Arch. Ophthalmol.* 49:184 (1953).
6. A. de Roetth, Sr., On the hypofunction of the lacrimal gland, *Am. J. Ophthalmol.* 26:20 (1941).
7. G. Eisner, Der Einfluss der Papierwahl auf die Resultate des Schirmer'schen Testes, *Ophthalmologica* 141:314 (1961).
8. G.P. Halberg and C. Berens, Standardized Schirmer tear test kit, *Am. J. Ophthalmol.* 51:840 (1961).
9. F.J. Holly, D.W. Lamberts, and E.D. Equivel, Kinetics of capillary tear flow in the Schirmer strip, *Curr. Eye Res.* 2:57 (1982).
10. F.J. Holly, S.J. LauKaitis, E.B. Esquivel, Kinetics of lacrimal secretion in normal human subjects, *Curr. Eye Res.* 3:897 (1984).
12. O. Schirmer, Studien zur Physiologie und Pathologie der Tränen absonderung und Tränen abführ, *Graefes. Arch. Ophthalmol.* 56:197 (1903).
13. H. Sjögren, Zur Kenntnis der Keratoconjunctivitis sicca, *Acta Ophthalmol. (Copenhagen), Suppl 2,* p. 1 (1933).
14. S.A. Spector, Chronic keratoconjunctivitis, chronic pharyngitis, and chronic arthritis due to ovarian insufficiency, *Klin. Med.* 9:876 (1931).
15. R.N. Stuchell, J.J. Feldman, R.L. Farris, and I.D. Mandel, The effect of collective technique on tear composition, *Invest. Ophthalmol. Vis. Sci.,* 25:374 (1984).

# A PRECISE METHOD OF USING ROSE BENGAL IN THE EVALUATION OF DRY EYE AND THE DETECTION OF CHANGES IN ITS SEVERITY

Michelle A. George,[1,2] Mark B. Abelson,[1-3] Kendyl Schaefer,[1,2] Myca Mooshian,[1,2] and Dana Weintraub[1]

[1]Department of Immunology
Schepens Eye Research Institute, Boston, MA 02114
[2]Ophthalmic Research Associates, N. Andover, MA
[3]Department of Ophthalmology
Harvard Medical School, Boston, MA 02115

## INTRODUCTION

Rose bengal (tetrachloro-tetraiodo fluorescein sodium) is a vital dye used to diagnose disorders of the external eye, particularly dry eye syndrome. Historically, it has been written that rose bengal possesses a propensity for dead or degenerate cells, but will also stain mucus. Differentiation between degenerate cells and mucus has been achieved by the addition of alcian blue, allowing the degenerate cells to remain red in coloring while the mucus turns blue[1]. Recent work performed by Feenstra and Tseng[2] has suggested that rose bengal is not a vital dye, but actually absorbed by all cells in a rapid (less than 1 minute), dose dependent manner. Feenstra and Tseng further asserted that rose bengal may actually be cytotoxic, and that this effect could be augmented by exposure to light, perhaps explaining why patients complain of increased discomfort as time passes following instillation of the dye. It would appear from these results that rose bengal is not specific for degenerated cells, but instead may expose breaks in the tear film, particularly the mucin layer, exposing the underlying epithelial cells to dye absorption. While further investigation is still necessary to determine the exact mechanism of action of rose bengal, it is clear that rose bengal staining is, nevertheless, a reflection of a dry eye state. A review of the clinical use of rose bengal in diagnosing dry eye has lead to the development of a precise system in which the changes in the severity of rose bengal staining in dry eye can be mapped. As these changes have been correlated with symptomatology, they can thereby be used to evaluate the effect of specific therapies and therapeutic regimens on the severity of the condition.

The objective of this study was to explore the use of rose bengal as a parameter by which to measure changes in dry eye surface disease and to see if changes in staining patterns could be induced.

## MATERIALS AND METHODS

18 subjects were evaluated in this randomized, single masked study. Only those subjects with a clear history of bilateral dry eye as assessed by prior staining patterns, Schirmer's test scores and tear break-up time, were entered into the study. Subjects were

given a baseline exam, wherein their recent history was taken and their baseline staining patterns were recorded. In a single masked fashion, subjects were randomly assigned to instill Hypotears® PF artificial tear solution in one eye while the contralateral eye remained untreated. Subjects were instructed to instill the artificial tear solution into the assigned eye hourly while awake for a period between 24 and 36 hours, until their next visit on the following day. The importance of correct administration of the drop into the appropriate eye was emphasized. Subjects were told to record instillation times of the drops and if the drops were instilled in the wrong eye at any time. At Visit 2, Day 2, the subjects were asked if they preferred the comfort of one eye over another. Both eyes were then stained with rose bengal.

Two methods of recording the rose bengal staining patterns were piloted. The overall staining pattern was first separated into three areas: conjunctival staining, limbal staining, and staining of the central cornea. Initially, the severity of the staining in each area was assigned a subjective grade ranging from 0 to 3, in 0.5 unit increments, and this value was recorded. Subsequently, it was found that actually counting the number of dots in each area gave a much more accurate description of the condition of the eye. Each eye was evaluated and graded as to have improved (mildly, moderately, greatly), deteriorated (mildly, moderately, greatly), or had no change. Those subjects' comments which were clear enough upon which to base a judgment, were evaluated in the same manner.

## RESULTS

Statistically, eyes treated with Hypotears® PF were significantly more comfortable than untreated eyes after one day of intense therapy in the paired and non paired comparisons using a robust analysis (Table 1, p = 0.029 and p = 0.0001 respectively).

### TABLE 1 COMFORT EVALUATION (n=8)

|  | TREATED | | UNTREATED | |
|---|---|---|---|---|
|  | # SUBJECTS | % | # SUBJECTS | % |
| GREATLY IMPROVED | 0 | - | 0 | - |
| MODERATELY IMPROVED | 0 | - | 0 | - |
| MILDLY IMPROVED | 5 | 62% | 1 | 12% |
| NO CHANGE | 1 | 12% | 6 | 75% |
| MILDLY DETERIORATED | 2 | 25% | 1 | 12% |
| MODERATELY DETERIORATED | 0 | - | 0 | - |
| GREATLY DETERIORATED | 0 | - | 0 | - |

Improvements in rose bengal staining patterns in the treated group over the control group were noted in all areas. Statistically, the treated group was shown to have significantly less staining than the control group in the paired comparison in the area of limbal staining(p = 0.04).

### TABLE 2 LIMBAL STAINING (n=18)

|  | TREATED | | UNTREATED | |
|---|---|---|---|---|
|  | # SUBJECTS | % | # SUBJECTS | % |
| GREATLY IMPROVED | 0 | - | 0 | - |
| MODERATELY IMPROVED | 4 | 22% | 1 | 5% |
| MILDLY IMPROVED | 5 | 28% | 3 | 17% |
| NO CHANGE | 7 | 39% | 9 | 50% |
| MILDLY DETERIORATED | 2 | 11% | 2 | 11% |
| MODERATELY DETERIORATED | 0 | - | 3 | 17% |
| GREATLY DETERIORATED | 0 | - | 0 | - |

While, statistically significant differences could not be shown for comparisons of conjunctival and central staining, the overall distribution suggests a general trend of greater improvement in the treated group (Tables 3 and 4).

### TABLE 3 CONJUNCTIVAL STAINING (n=18)

|  | TREATED | | UNTREATED | |
|---|---|---|---|---|
|  | # SUBJECTS | % | # SUBJECTS | % |
| GREATLY IMPROVED | 0 | - | 0 | - |
| MODERATELY IMPROVED | 2 | 11% | 0 | - |
| MILDLY IMPROVED | 3 | 17% | 3 | 17% |
| NO CHANGE | 8 | 44% | 13 | 72% |
| MILDLY DETERIORATED | 4 | 22% | 0 | - |
| MODERATELY DETERIORATED | 1 | 5% | 1 | 5% |
| GREATLY DETERIORATED | 0 | - | 1 | 5% |

### TABLE 4 CENTRAL CORNEAL STAINING (n=17)

|  | TREATED | | UNTREATED | |
|---|---|---|---|---|
|  | # SUBJECTS | % | # SUBJECTS | % |
| GREATLY IMPROVED | 0 | - | 0 | - |
| MODERATELY IMPROVED | 0 | - | 1 | 6% |
| MILDLY IMPROVED | 2 | 12% | 0 | - |
| NO CHANGE | 13 | 76% | 11 | 65% |
| MILDLY DETERIORATED | 2 | 12% | 4 | 24% |
| MODERATELY DETERIORATED | 0 | - | 1 | 6% |
| GREATLY DETERIORATED | 0 | - | 0 | - |

## DISCUSSION

Differences in staining patterns are not as obvious as an absolute improvement in the treated eye versus absolute deterioration in the control eye. In some cases, for instance, the staining in both eyes increased, but to a lesser degree in the treated eye. Therefore, to gather valuable information from these results, the change or mean difference in scores between treated and control eyes was considered. The distribution of raw scores and percentages suggest that the rose bengal staining in the treated eyes is generally improved over that in the control eyes. The statistic significance found only at the limbus may be indicative of a raised tear meniscus height due to the highly frequent dosing regimen allowing for a more dramatic change to occur.

Realizing that one is unlikely to induce statistically significant changes in the integrity of the ocular surface within 1 day of treatment raises the question of the effects of longer treatment periods. Analysis of data from a seven-week placebo controlled study confirmed that the greatest percentage of change in rose bengal staining is observed in the inferior conjunctiva. A 44 - 50% decrease in staining was observed in groups stratified as having severe dry eye as defined by age >56 years, positive diagnosis of Sjogren's Syndrome, poor comfort scores, or exhibiting medial and lateral conjunctival rose bengal staining scores which were > 2 (moderate). A 33 - 57% decrease in staining was observed in patients who did not meet the above criteria to be classified as severe. Regardless of the stratifications according to treatment group or of treatment group assignment (active or placebo control/Refresh™), a decrease in rose bengal staining was observed in 42 - 78% of patients. It should be noted that in patients with < 2 rose bengal staining in the lateral and medial conjunctiva, no decrease in rose bengal staining was observed.

# CONCLUSION

The significant differences found, coupled with the trends seen in the distribution of scores, suggest that easily observed, clinically significant results could be shown using this model. In addition, counting the number of dots stained, as opposed to assigning a subjective grade to the staining pattern, enables detection of precise differences between eyes. A review of prior experience with rose bengal in the diagnosis of dry eye and modifications made in developing this standardized model suggests that this method could serve as a sensitive and valuable tool in the evaluation of the dry eye condition and potential agents for its treatment. Feenstra and Tseng's[2] findings that rose bengal may actually be staining breaks in the tear film or mucin barrier, could indicate that this model may be ideally suited for evaluating dry eye conditions resulting from mucin deficiencies but, further work must be done to confirm their analyses. Particularly, these findings were performed on cell culture deprived of medium. This work should perhaps be repeated in a Boyd's chamber model with an intact and functional cornea or an appropriate *in vivo* experiment. Furthermore, clinical observations have revealed that dry eye conditions resulting from other dysfunctions in the tear system, such as meibomianitis, are also reflected by rose bengal staining. As it is not clear what effects lipid or aqueous deficiencies have on the continuity of the mucin layer, it remains difficult to conclusively state that rose bengal only invades breaks in the mucin layer.

While additional work is being performed to elucidate the exact mechanism of action of rose bengal, a precise system has been developed to map the severity of rose bengal staining in dry eye. Our initial findings have been corroborated by changes in rose bengal staining in the placebo (Refresh™) controlled study. Future studies may entail better control over external conditions. Environmental parameters could be better controlled, performing trials in the winter, when the humidity is lowest and/or requiring subjects to remain in one environment for the duration of the study. In addition, in accordance with the clinical observation that different etiologies of dry eye seem to induce characteristic patterns of symptomatology in individuals, the examination times could be controlled so that they coincide with the time of day when symptomatology is worst. Finally, by assembling a population whose dry eye has a similar etiology and then exacerbating their condition with an external stimulus (or even internal, as seen with the use of antihistamines), we can influence the severity of their dry eye and exert a difference in rose bengal staining patterns that has a higher consistency. Rose bengal staining may offer the unique opportunity of acting as a double parameter in the assessment of the dry eye condition: evaluation of the staining pattern could serve as an objective parameter, while the patient's response to the staining procedure and analysis of comfort, in a diary format, could serve as a subjective parameter.

# REFERENCES

1. M.S. Norn. "External Eye: Methods of Examination," Scriptor, Copenhagen, Denmark (1974).
2. R.P. Feenstra, and S.C. Tseng. What is actually stained by rose bengal? Arch. Ophthalmol. 110:984 (1992).

# DRY EYE SYNDROMES: TREATMENT AND CLINICAL TRIALS

Michael A. Lemp

Clinical Professor of Ophthalmology
Georgetown University
and
University Ophthalmic Consultants of Washington
4910 Massachusetts Avenue, NW
Suite 210
Washington, D.C. 20016

The evaluation of treatments in medicine has progressed at an accelerated pace in the last half century. No development has been more central to this progress than the advent of the randomized controlled clinical trial. Sir Austin Bradford Hill, who was the father of this scientific genre was judged by the President of the Royal College of Physicians to have made a contribution to medicine "as important and valuable as the discovery of penicillin."[1]

It is instructive to consider why the introduction of scientific principles to the design and application of clinical trials is appropriate and how this relates to the subject at hand.

The history of medicine can be viewed as a series of informed opinions. Opinions are, however, subject to considerable bias and are based on experiences with individual patients. It has been pointed out that medical training emphasizes the uniqueness of patients and the study of their distinctive characteristics.[2] Mainland has stated that "the training of a doctor as a doctor is in some ways the reverse of an investigator's training."[3] Thomas Lewis noted that "Self confidence is by general consent, one of the essentials to the practice of medicine, for it breeds confidence, faith and hope. Diffidence, by equally general consent, is an essential quality in investigation for it breeds inquiry. A natural companion of confidence is an easy and uncritical acceptance of statements of fact and of hypothesis. The companion of diffidence is skepticism."[4] Clinicians are presumably interested in dealing with individuals and have a bias for action and results and are not "detached" in identifying with their patients' desire for good results. These factors tend to lead to overly optimistic judgements on the efficacy of treatments. This has lead to the introduction of statistical methods to the design, conduct and interpretation of clinical treatment studies and the use of paramedical personnel as clinical coordinators.

The late eminent Harvard biostatistician, Hugo Muench, developed a series of postulates and laws that are at once amusing, insightful and pertinent[5] (Figure 1). The postulates suggest that there is much talk, unsupported by scientifically validated evidence, that far too much work of poor design and interpretation is published and that careful review of the scientific literature will reveal that most "new" discoveries have antecedents and indeed, often identical, earlier findings. His laws have been borne out by extensive clinical investigative experience.

<u>Muench's Aphorisms</u>

Postulates
1. Everyone talks too much
2. Everyone writes too much
3. Nobody pays any attention

Laws
1. No full-scale study confirms the lead provided by a pilot study.
2. Results can always be improved by omitting controls.
3. In order to be realistic, the number of cases promised in any clinical study must be divided by a factor of at least ten.

**Figure 1.** Muench's aphorisms, from Biometric notes no. 4, Office of Biometry and Epidemiology, NEI, NIH, DHEW, 1974

Clinical trials are of value when the difference between a new and an old treatment is not clear, the disease follows a chronic variable and erratic course, and a large number of known or unknown factors may influence the course of the disease and the outcome of the treatment[6] these conditions characterize the dry eye states.

What, then, are the characteristics of a proper clinical investigation of a new therapeutic modality? Figure 2 lists the major characteristics.

## Ethical Considerations

The recruitment of a patient into a clinical trial involving a treatment which may or may not be beneficial to the patient involves informed consent with full disclosure of possible risks and an explanation of the rationale of treatment. Individuals must never be denied clearly appropriate treatment even if the trial protocols are, thereby, disrupted.

Characteristics of a Clinical Trial

Ethical consideration
Protocol and study design
Control groups
Randomization
Masking
Adequate patient numbers
Biostatistical data analysis
Complete patient followup

**Figure 2.** Characteristics of a clinical trial.

## Protocol and Study Design

Complete and meticulous study of design and protocol development, including decisions such as to design a parallel or crossover trial and faithful adherence to the protocol involves considerable effort and time. Indeed, manuals of procedure are often tedious but absolutely essential to the successful conduct of any clinical trial.

## Control Groups

It has been stated that positive results can be enhanced by the lack of a control. Contemporaneous control groups eliminate or reduce investigative bias and the bias of time which is possible when historical controls are used.

## Randomization

Assignment of patients to specific treatments must be done in a randomized fashion. Randomization removes, or at least ameliorates the potential conflict that clinical investigators have in resolving their obligations both to the patient and to science and are at the very core of successful study design.

## Masking

The best study design is that of a so-called double blind or double masked study. They are far superior to single masked studies but these are often difficult to conduct. "A placebo should not only look, smell and taste like the active drug but should have the same side effects".[7] The differences in the appearance of treatments, e.g. gels vs drops, particularly in the field of dry eyes, present problems in this regard.

## Adequate Patient Numbers

As Muench's third law specifies, most clinical investigators are overly optimistic about the number of patients they will be able to recruit into a study. Adequate numbers are essential to fairly evaluate the treatments. The greater the difference between two treatments being evaluated, the smaller the number of subjects will be required to demonstrate that difference. If the endpoints to be measured are definitive, i.e. death, a smaller number of subjects is necessary to demonstrated differences between two treatments. As we shall see, somewhat indefinite endpoints of dry eye studies create problems in this regard.

## Biostatistical Data Analysis

Sophisticated data analysis methodologies have been developed. Independent analysis of the data by individuals not conducting the investigation is considered ideal.

## Complete Patient Followup

Patients must not only be recruited but also maintained in the study in order to conclude a valid assessment.

All of these considerations lead to what Fredrickson has called the "indispensable ordeal" of clinical trials.[8] Virtually all published clinical trials in the treatment of dry eyes fail to meet the previously discussed characteristics of a clinical trial on one or more counts. There are currently no NEI funded clinical therapeutic trials in dry eye. The condition itself presents certain problems in study design, conduct and interpretation.

The term "dry eye" is a rubric to describe a variety of conditions of diverse pathogenesis which affect the pre-ocular tear film and/or the ocular surface.[9] Investigations over the last two decades have clarified, but at the same time complicated our understanding of these conditions. We have progressed from a relatively simplistic concept of dry eyes being simply a lack of tears to a much more sophisticated understanding of diverse pathogenetic factors involving the lacrimal glands and the ocular surface that ultimately lead to ocular surface disease. Some of these factors include dysfunction of the secretory activity of the main and accessory lacrimal glands, inflammatory disease of the lacrimal glands and/or the ocular surface itself, dysfunctional changes in the action of the lids, morphologic changes in the corneal and conjunctival epithelium, functional changes in the secretion of the meibomian glands, the effect of systemic hormonal changes on the ocular surface and/or tear production, and finally the effect of environmental influences on the tear film and ocular surface. This complicated schema presents considerable problems and diagnosis all important in standardizing the recruitment of clinical patients for clinical studies. Diagnostic criteria, for example, range all the way from "I know it when I see it" to the utilization of specific diagnostic tests such as Schirmer test, tear film osmolarity, tear lactoferrin levels, tear lysozyme levels and conjunctival morphologic changes as measured by impression cytology. The lack of a consensus on appropriate diagnostic criteria and even a classification of dry eye states severely hampers the design of studies and the consequent interpretation of results. It would seem important to sort out the different factors present in each patient prior to enrolling patients in the study designed to achieve meaningful results.

Sjögren's syndrome, a systemic disease with a high prevalence of dry eyes, has certain characteristics not necessarily present in other cases of keratoconjunctivitis sicca, i.e. ocular surface inflammatory disease and other evidence of autoimmune phenomena of the ocular surface e.g. rheumatoid nodules, scleritis, and corneal ulceration. Should these patients be lumped together with other patients with keratoconjunctivitis sicca but without these definite autoimmune signs?

Clinical study design is greatly simplified if there are definitive end-points to be measured. In dry eye states, what constitutes "improvement"? Is it an increase in aqueous tear secretion? Is it, rather, an improvement in signs of ocular surface disease? Or is it, rather, a decrease in subjective symptomatology? A variety of answers to these questions are apparent in the diverse clinical trials which have been reported in the literature. In the case of ocular surface disease, it is not even known how long it would be expected to take for clinical signs of improvement to appear.

There is a paucity of studies employing objective criteria for measuring changes in treatment. In terms of aqueous tear secretion, the gold standard is fluorescein dilution. This is a difficult test to standardize and requires expensive, sophisticated equipment only available at a few centers; very few studies have used this methodology to assess improvement. Other parameters utilized have included tear film osmolarity, changes in break-up time of the tear film and changes in conjunctival morphology as evidenced by impression cytology. Since the pathogenesis of dry eye states is diverse, it would seem important to choose end-points carefully based on hypotheses about the actions of treatment. Drugs designed to increase the secretory activity of the lacrimal glands are probably best judged by determining changes in aqueous secretion. Treatments designed to directly affect the ocular surface, on the other hand, would be best assessed using objective criteria of ocular surface morphology. Rose bengal staining is a gross indication of ocular surface disease. When assessing aqueous tear secretion, it would of course be important to assess the ocular surface and likewise important to assess tear secretion when addressing treatment of the ocular surface but these end-points would be secondary ones.

Alternative methods of assessing putative improvement have included subjective criteria. Figure 3 lists some of the objective and subjective symptoms that have been used to assess improvement. Semi-quantitative schemes based on questionnaires have been employed. As new treatment modalities emerge, the possibility of modulating other parameters of dry eyes states, such as meibomian gland function will appear. Appropriate end-points for assessing the effects of treatment in these conditions will need to be developed.

As pointed out earlier, results can always be improved with the absence of controls. Because of the diverse pathogenesis and lack of any consensus for grading disease truly effective non-disease controls are almost impossible to incorporate into dry eye clinical study design. As a consequence, patients have been used as their own controls. This involves either using two patient groups, each with different treatment, or using the same patients and using the cross-over design. Most of these studies involve an initial wash-out. This, in itself, presents problems because most patients to be recruited into a study are dependent on continued treatment and are unwilling to go without treatment for a period of time. This has led to the development of "minimal treatment," for example, with artificial tears alone for a wash-out period. An additional treatment is then added in the first phase of a crossover study. Since the sequence of events in renewal of the ocular surface is still incompletely understood, the time necessary between the treatments in crossover studies is very poorly understood. These factors confound the design and interpretation of results.

| Objective | Subjective |
|---|---|
| Schirmer test | Comfort |
| Rose bengal staining | Blurriness |
| Impression cytology | Stickiness |
| Tear break up time | Burning |
| Tear osmolarity | Stinging |
| Tear lysozyme levels | Foreign body sensation |
| Specular microscopy of surface | Dryness |
| | Photophobia |
| | Itching |
| | Redness |

**Figure 3.** Endpoints in clinical trials.

If one seeks to prove treatment differences, one can enhance the likelihood of seeing those differences by employing one treatment containing an ingredient which is known to have adverse effects. An example of this is the comparison of preservative-containing artificial tear preparations with one containing no preservative. This does not really test the other ingredients of the artificial tear.

These considerations are but a few of the important obstacles to the truly scientific design of clinical trials which are likely to lead to real advances in the management of this group of conditions. The task of designing improved clinical trials is a daunting one but one which I think must be pursued. To this end, I propose an academic-clinical-practice-industry-governmental effort to develop a consensus to 1) To define diagnostic criteria 2) To determine time intervals for clinical objective changes to be discerned 3) To agree on an approach to the diffuse components of the problem e.g. tear production, ocular surface disease, associated drying problems 4) to develop a standardized format for designing clinical trials including subjective patient evaluation with psychometric expertise. This is not to suggest a rigid format that must be adhered to in all studies emerge from these efforts. Rather, I believe that this is an ideal area in which a collaborative effort in study design could be a first step that could yield considerable improvement in our ability to evaluate new treatment modalities which will be emerging from the laboratories in the coming years.

The careful scientific design of high quality clinical trials is indeed, as Fredrickson has pointed out, an "indispensable ordeal".[8] These protocols must, as Bearman has pointed out, be "excruciatingly specific".[10] While the task is arduous, it is imperative. Fredrickson further described the characteristics of such efforts, "They lack glamour; they strain our resources and patience and they protract to excruciating limits the moment of truth. Still, they are among the most challenging tests of our skills. I have no doubt that when the problem is well chosen the study is appropriately designed and all the populations concerned made aware of the route and the goal, the reward can be commensurate with the effort. If in major medical dilemmas, the alternative is to pay the cost of perpetual uncertainty, have we really any choice?"

## ACKNOWLEDGEMENT

I would like to express my gratitude to Fred Ederer of the National Eye Institute and the Emmes Corporation for his assistance in providing me the materials to help me prepare this presentation. I have drawn heavily on his expertise.

## REFERENCES

1. H. Atkins, Conduct of a controlled clinical trial, *Brit. Med. J.* 2:377 (1966).
2. F. Ederer, The randomized clinical trial, *in:* "Clinical Practice and Economic," C. J. Phillips and J. N. Wolfe, eds., Pitman Press. Bath, Avon (1977).
3 D. Maitland, Clinical trials and health care administrators, Note 42, Notes on Biometry in Medical Research, Washington, D.C., Va Monoge 10-1 (Suppl U) (1969).
4. R. Lewis, Research in medicine: its position and needs, *Br. Med. J.* 3/15/30: 479.
5. J.W. Bearman, R.B. Loewenson, and W.H. Gullen, Muench's postulates, laws, and corollaries: a biometricians' views on clinical studies, *in:* "Biometrics Note No 4," Office of Biometry and Epidemiology, NEI, NIH, DHEW. Bethesda, MD.
6. C. Kupfer, Foreword. Evaluating new approaches to the treatment of eye and vision disorders, *in:* NIH Publication No 90-2910 (1990).
7. F. Ederer, Patient bias, investigator bias and the double-masked procedure in clinical trials, *Amer J. Med.* 58: 295 (1975)
8. D.S. Fredrickson, The field trial; some thoughts on the indispensable ordeal. *Bull NY Acad. Med.* 44:985 (1968).
9. M.A. Lemp, Basic principles and classification of dry eye disorders, *in:* "The Dry Eye," M.A. Lemp and R. Marquardt, eds., Springer-Verlag, New York (1992).
10. J.E. Bearman, Writing the protocol for a clinical trial, *Amer. J. Ophthalmol.* 79: 775 (1975).

# KERATOPROSTHESIS IN END-STAGE DRY EYE

Claes Dohlman and Marshall Doane

Massachusetts Eye and Ear Infirmary
Schepens Eye Research Institute
Harvard Medical School
243 Charles Street
Boston, MA 02114

In very dry eyes with opaque corneas, regular keratoplasty is rarely successful. In these cases a keratoprosthesis (window of artificial material) seems a logical possibility. Progress in this field has been slow, however. The difficulties have been less related to the materials used, or even to the designs, but rather to the frequent occurrence of tissue necrosis around the device, resulting in leak, infection and even extrusion. Formation of a retroprosthesis membrane as well as late secondary glaucoma or retinal detachment have also diminished the long-term usefulness of this approach. (For a recent review of the field of keratoprostheses, see reference[1]).

The authors have used a collarbutton-shaped keratoprosthesis since the early 1960's. Thus, 36 cases were reported between 1965 and 1975[2]. During the last three years another 11 operations have been performed. Of these, 5 have been done in end-stage dry eyes and they will be reported here.

## METHODS AND RESULTS

The basic model of our keratoprosthesis is shown in Fig. 1. The material is medical grade polymethylmethacrylate. It consists of an anterior plate, 7mm in diameter, connected to a stem with screw-threads, 3mm. in diameter. The separate backplate is 10.5mm in diameter, has a laser ridge and has several holes. Total antero-posterior length is 2.7mm. In the operating room a frozen donor cornea is trephined with a 9.0mm trephine and, centrally, with a 3.0mm trephine. The resulting doughnut-shaped graft is then pushed over the stem from behind and the backplate is screwed on.

The patient (under general anesthesia) first has a wide conjunctival flap mobilized. Next, the cornea is opened with an 8.5mm trephine. The iris is usually removed as well as the lens, whether cataractous or not. The posterior lens capsule is left intact. The graft-prosthesis combination is then sutured in place like a regular graft with 10-0 nylon. Finally, the conjunctival flap is brought over the graft with the prosthesis and anchored. After about two months the conjunctiva over the flap is opened.

This standard procedure was used only in one of our dry eye patients (D.H.). In the remaining four patients, the keratoprosthesis was modified by adding a 1or 2 mm long nub to the front plate. In these cases the prosthesis was covered by lid skin rather than conjunctiva[3] and after two months an opening was made for the nub to protrude.

The present status of vision in the five cases is shown in table I. None of the cases were free from complications:

Figure 1. Basic model of the collarbutton-shaped keratoprosthesis.

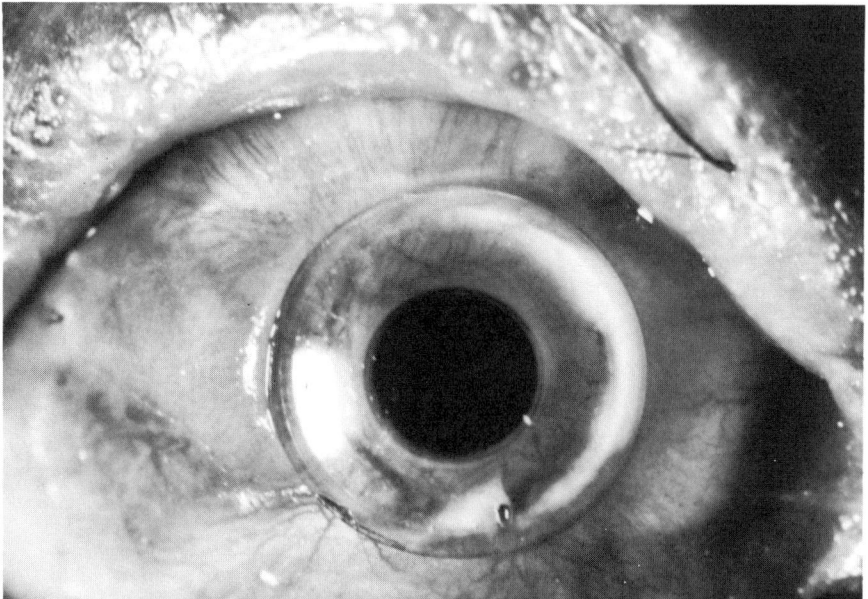

Figure 2. Keratoprosthesis in patient with Stevens-Johnson syndrome.

Table I.

| Pt | Diagnosis | Preop VA | Best VA | Present VA | Follow-up |
|----|-----------|----------|---------|------------|-----------|
| SS | Pemphigoid | HM4' | 20/25 | - | 27 months |
| HN | Pemphigoid | HM | 20/25 | 20/25 | 23 months |
| SL | Pemphigoid | LP | CF 4' | LP | 21 months |
| DH | Stevens-Johnson | HM | 20/40-2 | 20/70+ | 18 months |
| FC | Pemphigoid | HM | 20/25+ | 20/25+ | 6 months |

SS (pemphigoid). Had 20/25 vision for 9 months. Between two visits she developed a retraction of the lid skin, ulceration, leak and infection. The eye had to be eviscerated.

SL (pemphigoid). Had a well-placed prosthesis but was found to have massive macular degeneration with vision of finger counting. After five months the anterior part of the prosthesis became unscrewed and repair was only partially successful.

HN (pemphigoid). Had some overgrowth of skin over the prosthesis nub, requiring minor revisions. Now stable.

DH (Stevens-Johnson). Had stormy course with unscrewing of the front part of the prosthesis. After repair has settled down on topical medication of Provera (Fig.2).

FC (pemphigoid). Had threatening retraction of lid skin. Skin flap from behind the ear was used to augment the lid skin. After opening, a small plate was screwed on the front nub to protect the skin wound edge from evaporation and retraction. The situation seems stable (Fig. 3).

## DISCUSSION

Keratoprosthesis is clearly a high-risk procedure, especially in the extremely dry eye. However, the risks may no longer be unacceptable considering that there is simply no other known way to restore vision. During our work certain principles have emerged which we feel are important in achieving success:

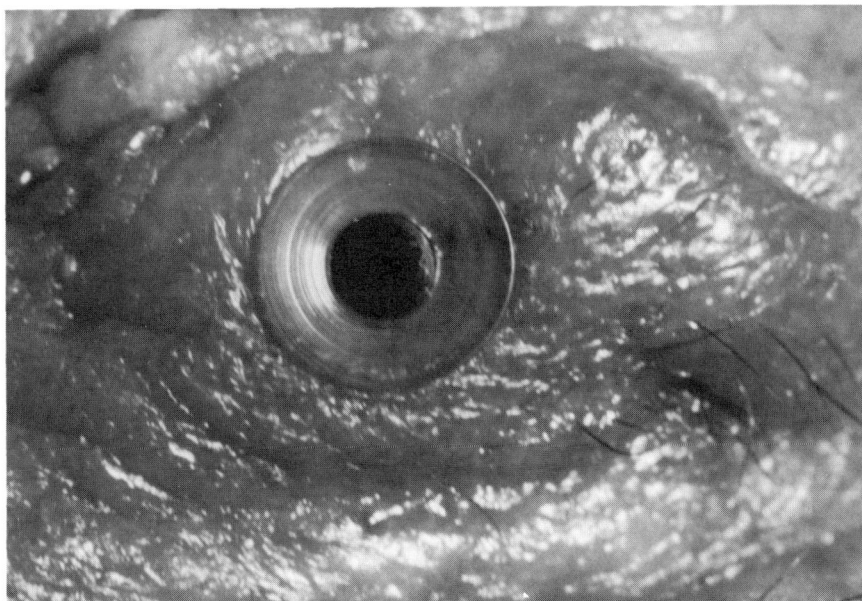

Figure 3. Keratoprosthesis in patient with pemphigoid.

1) A collarbutton design appears more practical than the more commonly used long stem with a skirt design. Repair is easier and, since the collarbutton stem is shorter, the field is wider and alignment with the macula less critical.

2) It seems to be very important to cover the keratoprosthesis completely with tissue (conjunctiva or lid skin) for months before opening up to allow vision.[2] Healing around the device has then become more complete and inflammation has subsided, making subsequent necrosis of the supporting graft less likely.

3) After opening of the tissue covering the prosthesis, topical application of drugs suppressing the proteolytic enzymes that cause tissue necrosis appears highly beneficial. Medroxyprogesterone ("Provera", Upjohn) reduces synthesis of collagenase[4] and we use it as a 1% suspension two to four times daily. Tetracycline ("Achromycin", Lederle) is a direct collagenase inhibitor[5] and is useful as a 1% suspension in a similar regimen. Necrosis and ulceration of the tissue holding the keratoprosthesis have decreased sharply.

4) In through-the-lid keratoprosthesis, it is very important to protect the skin wound from evaporation and exposure after opening. Anything that protects the wound around the prosthesis is beneficial. The patient in Fig. 3 has an anterior plate which gently touches the skin and prevents evaporation.

## SUMMARY

Five cases with end-stage dry eye (four with pemphigoid, one with Stevens-Johnson syndrome) have been operated with a keratoprosthesis and have been followed for 6 to 22 months. Two patients have been failures, three still have good vision.

## REFERENCES

1. C.H. Dohlman. Keratoprostheses, in: "Principles and Practice of Ophthalmology," D.M. Albert and F.A. Jakobiec eds., WB Saunders Company, Philadelphia (in press).
2. C.H. Dohlman, H.A. Schneider and M.G. Doane, Prosthokeratoplasty, Am J Ophthalmol. 77:694, (1974).
3. H. Cardona and A.G. DeVoe, Prosthokeratoplasty, Trans Am Acad Ophthalmol Otolaryngol 83:271, (1977).
4. D.A. Newcombe, J. Gross, Prevention by medroxyprogesterone of perforation in the alkali-burned rabbit cornea: inhibition of collagenolytic activity, Invest Ophthalmol Visual Sci 16:21, (1977).
5. J.A. Seedor, H.D. Perry, T.F. McNamara, L.M. Golub, D.F. Buxton and D.S. Guthrie, Systemic tetracycline treatment of alkali-induced corneal ulceration in rabbits, Arch Ophthalmol 105:268, (1987).

# AMYLASE IN MARE LACRIMALE IN PATIENTS WITH SUBMANDIBULAR SALIVARY GLAND TRANSPLANTATION TO TIIE LACRIMAL BASIN

Juan Murube, Manuel G. Marcos and Reynaldo Javate

Department of Ophthalmology, Hospital Ramon y Cajal, and
University of Alcalá, Moralzarzal 43, E-28034, Madrid, Spain

## INTRODUCTION

Transplantation of salivary gland tissue to the subconjunctival space of the lids has increased the volume of resident tear fluid in patients with dry eyes. However, frequently neither heterotransplants nor autotransplants survive in those with auto-immune diseases. Many methods have been employed with limited success to assess the functional status of the transplanted tissue. These include symptomatology and objective findings such as Schirmer values and staining of epithelial lesions with fluorescein or rose bengal (1,2). All these methods have limitations in evaluating the secretory activity of the transplanted salivary tissue. Biopsies cannot be repeated often because of the paucity of tissue. Assessment of the quantity and quality of the fluid in the meniscus is occasionally useful, but in dubious cases is difficult to evaluate. Even in successful cases Schirmer values increase only 1-3 mm because the transplanted gland cannot respond to reflex stimulation; moreover the inherent inaccuracy of this test does not give much reliability. The value of subjective feelings of dryness or wetness is variable.

In an attempt to add another test to follow the activity of the transplanted salivary tissue, we have measured amylase concentrations in the fluid entering the conjunctival sac; one can assume that high levels of amylase indicate viability of the transplanted salivary glands.

## SUBJECTS, MATERIALS AND METHODS

We tested six patients with keratoconjunctivitis sicca who underwent submandibular salivary transplants to the subconjunctival tissue. All patients were given routine ocular and physical examinations. All had symptoms of dry eye in various stages from mild to severe. None had pancreatitis or gastrointestinal disease. Three had Sjögren's syndrome, one had pemphigoid and two had facial palsy due to acoustic neuroma. Five underwent a transplant of submandibular glands to the more affected eye and one to both eyes.

The technique of transplantation has been described elsewhere (3,4). One to 2 ml of submandibular gland (SMG) was taken and divided into slices of 2 mm or less in thickness which were placed and sutured under the superior and inferior lids, between the septal conjunctiva and the palpebral muscles/orbital septum.

Amylase activity levels were determined in saliva, and in the secretion of both eyes (transplanted and fellow non-transplanted). at various intervals after surgery. Saliva was obtained by having the patient spit at least 30 ml into a sterile Eppendorf tube. Fluid was collected with a micropipette from the temporal canthal meniscus (cisterna lacrimalis) until at least 20 µl were obtained; it was transferred to a sterile Eppendorf tube. The tubes were kept at room temperature for 10 minutes and centrifuged for 2.5 minutes at 10,000 x g. The amylase activity of the supernatant was determined spectrophotometrically at 405 nm (Hitachi 4020) using a commercially available kit (Boehringer,Mannheim) with PNP maltoheptaose as the substrate.

## RESULTS

**Case 1.** Female, born 1945, Sjögren II Syndrome with bilateral KCS (xerophthalmia, xerostomia, rheumatoid arthritis). Preoperative Schirmer tests were OD (right eye) 9, 8, 3 mm and OS (left eye) 12, 16, 12. Each Schirmer test was repeated 3 times without intervals between measurements; each result is reported, the data were not averaged. On May 28, 1991 an autotransplant of 1.2 ml of SMG was transplanted to the OD. Ten months later the condition of the OD had improved subjectively and objectively (fluorescein and rose bengal tests). Schirmer tests were OD 5, 6, 13 and OS 10, 11, 15. Amylase measurements (in U/l) were 153,700, 1,454 and 313 in the saliva, OD and OS respectively.

**Case 2.** Male, born 1927. Sjögren's II with bilateral KCS. Preoperative Schirmer test was OU 0, 0, 0 . On July 5, 1990 a bilateral SMG autotransplant (1.3 ml OU) was done. The eyes subsequently developed a slight moistness. One year later the Schirmer tests were OD 4, 2, 3 and OS 1, 0, 0 . The amylase activity (in U/l) was 310,000 and 7,821 in the saliva and the tear fluid (OD) respectively. There was insufficient fluid collectible from the OS for analysis. Two years later the Schirmer tests were OD 3, 1, 2 and OS 3, 0, 1; the amylase activity was 261,000; 4, 530 and 9, 110 in the saliva, OD and OS respectively.

**Case 3.** Female born 1919. Bilateral KCS due to pemphigoid. In the preoperative examination Schirmer tests were OD 8, 7, 7 and OS 12, 15, 9. -Impression cytology showed few goblet cells. On May 28, 1991 1 ml of salivary gland was autotransplanted to the OD. Five months later the Schirmer tests were OD 12, 7, 9 and OS 12, 12, 8.; SMG tissue was detected on biopsy of the OD. Amylase activity was 109,000; 1,935 and 500 in the saliva, OD and OS, respectively. The condition of the OD deteriorated in the ensuing five months and one year postoperatively the Schirmer tests were OD 13, 10, 16 and OS 14, 12, 10 . Biopsy of the site of the transplant site failed to show acinar tissue and the amylase activity in U/l was 61,268; 1,210 and 871 in the saliva, OD and OS respectively.

Case 4. Female born 1925. Bilateral KCS due to Sjögren's syndrome plus lesion of the left N VII in the pontocerebellar fossa after extirpation of an acoustic neuroma. Preoperative Schirmer tests were OD 16, 18, 10 and OS 0, 2, 5. Jones tests were OD 1, 1, 1 and OS 0, 0, 0. On July 2, 1991 an autotransplant was done to the OS. Seven months later the clinical picture of the OS had improved dramatically. Schirmer tests were OD 12, 8, 11 and OS 3, 2, 1, and Jones tests were OD 2, 1, 2 and OS 3, 0, 2. Amylase activity in saliva, OD and OS were respectively 492,700, 1,900 and 3,954.

Case 5. Female, born 1924; KCS due to acoustic neuroma on the right side. Schirmer tests were OD 4, 4, 1 and OS 10, 12, 11. On October 18, 1990 an autotransplant of 1.5 ml of SMG was performed. Subjective and objective symptoms improved; the last Schirmer tests were OD 12, 13, 6 and OS 16, 10, 18. On November 8, 1990 the amylase activity was 130 and 557 in the OD and OS respectively. On November 22, 1990 the values were 270 and 36; on December 10, 1990 the values were 649 and 946; on January 15, 1991 the values were 3,318 and 23; on February 8th 3,643 (salivary iso-amylase 3,488 and 155) and 800 (salivary iso-amylase 800 and pancreatic iso-amylase 0); on May 27, 1991 the values were 6,686 and 615.

Case 6. Female born 1929. Bilateral KCS due to Sjögren I syndrome. Preoperative Schirmer tests were OD 3, 3, 2 and OS 5, 1, 4. On November 19, 1990 an allotransplant of 1.6 ml of SMG was performed to the OD. Postoperative treatment consisted of corticoids, cyclosporin A and azathioprin. No clinical improvement was noted. Rejection of the transplant was confirmed by biopsy. Three months after surgery values of amylase in U/l were 170,000; 400 and 280 in saliva, OD and OS respectively; one year after surgery these values were 21,000, 380 and 400.

The results of these cases are summarized in Table 1.

**Table 1.** Transplantation of submaxillary salivary gland tissue in KCS

| Case Number | Etiology | Transplant Volume (ml) | Result | Last Amylase Activity (U/l) | |
|---|---|---|---|---|---|
| | | | | Transplanted Eye | Nontransplanted Eye |
| 1 | SS II | 1.2 | Success | 1,454 | 313 |
| 2 | SS II | 1.3 | Success | 4,530 | --- |
| | SS II | 1.3 | Success | 9,110 | --- |
| 3 | Pemphigoid | 1.0 | Failure | 1,210 | 871 |
| 4 | SS I + VIIN OS | 1.2 | Success | 3,954 | 1,900 |
| 5 | VIIN Palsy | 1.5 | Success | 6,686 | 615 |
| 6 | SS I | 1.6[**] | Failure | 380 | 400 |

SS = Sjögren's syndrome; All autotransplants except one allotransplant, as noted[**]; Amylase activity in International Units per liter (U/l).

## DISCUSSION

Amylase activity has been reported in tears since the early 60's (5-7); normal values considered to be 48.6 Wohlgemuth units (7); 376 Somogyi units per gram of protein (8) or 200-300 International Units per liter (U/l) (9).

The normal activity of amylase is 100,000 - 500,000 U/l in saliva and 320 U/l in serum. Lacrimal amylase is produced totally in the lacrimal gland (10,11) from zymogen granules in the acinar cells as occurs in the parotid gland (12). In autopsy tissues (13) total activity of amylase was highest in parotid glands (1,710 ± 897 U/g wet weight of tissue; followed by submandibular gland 605 +/- 354 U/g and pancreas (258 ± 137 U/g); other tissues (intestinal tract, lung, skeletal muscle, testes, ovary, kidney, liver, gall bladder, spleen, thyroid gland) have less than 5 U/g. The small activity in the cornea (14) might represent contamination from lacrimal fluid. According to Watson et al. (8) lacrimal amylase decreases in cases of malnutrition to mean values of 322, 252 and 199 U Somogyi in moderate, medium and severe cases respectively. General surgery has been reported to produce amylasemia (15). However, according to Liotet (7) intra and extraocular diseases do not significantly alter amylase activity in tears; the mean values in conjunctivitis, glaucoma, retinal detachment and cataract are 44.2, 44.7, 53.4 and 66.2 U Wohlgemuth respectively. In cases 1, 2, 4 and 5 (cases 1, 2 and 4 were Sjögren's Syndrome; cases 4 and 5 were acoustic neuroma) signs and symptoms improved, i.e., less grittiness, increased Schirmer values and lessened ocular surface dryness. There was a parallel increase in amylase activity in fluid sampled from the meniscus at each lateral canthus; the mean value increased to 5,147 U/l while the value in the fellow control eye was 947 U/l.

Salivary glands and thyroid contain more than 90% S-isoamylase; pancreas, skeletal muscle, intestinal tissue, spleen and testes contain more than 90% P-isoamylase (13). In the present series of patients, iso-enzymes were assayed only in case 5, and it was 96% S iso-amylase.

In cases 3 and 6 (the first a pemphigoid, the second a Sjögren's syndrome which underwent the only allotransplant) both failed and the mean values (U/l) of the amylase were 795 and 635 respectively in the transplanted and the non-transplanted eyes.

Experience with other cases not included in this series suggests that pemphigoid cases are not good candidates for autotransplants with the current status of management. Although the transplant survives initially, the eye returns to the dry state after an initial period of improvement: later the transplant is destroyed, possibly because of occlusion of the outflow channels by the pemphigoid.

The increase in amylase activity associated with a decrease in the Schirmer values may reflect a decrease in lacrimal gland fluid and a concomitant increase in transplanted salivary gland output.

Case 6 was the only allotransplant in this series. In salivary allotransplants the immune rejection of the graft ordinarily can only be detected by biopsy, a maneuver that cannot be repeated many times because of the paucity of tissue.

The Schirmer test is of minimal value in evaluating the functional status of the transplanted tissue. The test measures mainly reflex secretion but transplanted tissue is not innervated. Nonetheless, Schirmer testing is an additional measure than can be added to other objective and subjective signs.

In all six patients there was an evident correlation between the objective and subjective evolution of the signs and symptoms of the eye with the transplanted tissue and the levels of amylase in the secretion of that eye.

## SUMMARY

Six patients with dry eyes of different etiologies underwent transplants of 1.0 to 1.6 ml of submandibular salivary gland tissue, five of them to one eye and one to both eyes. In the four cases in which the transplants survived, the amylase activity in tear fluid sampled from the cisterna lacrimalis (temporal canthal meniscus) had a mean value of 5,147 U/l in contrast with the mean value 943 U/l of the fellow control eyes . In the two eyes in which the transplants failed to survive, the average value was 635 U/l .

The small sample size does not enable calculation of statistical significance to the results but suggests that salivary amylase determinations in tear fluid would facilitate assessment of the functional status of the transplanted salivary tissue.

## ACKNOWLEDGMENT

This work was supported by the A.I.E.T.I. Foundation (Madrid, Spain)

## REFERENCES

1. J. Murube del Castillo, and C. Rodrigo, Eye parameters for the diagnosis of xerophthalmos, Clin. Exp. Rheumatol. (Pisa), 7:145 (1989).
2. J. Murube del Castillo, and I. Murube Jiménez, Transplantation of sublingual salivary gland to the lacrimal basin in patients with dry eye. Histopathological postoperative study, in: "The Lacrimal System," O.P. van Bijsterveld, M.A. Lemp, and D. Spinelli, eds., Kugler & Ghedini, Amsterdam, p.63 (1991).
3. J. Murube del Castillo, Transplantation of salivary gland to the lacrimal basin. Scand. J. Rheumatol. Suppl. 61:264 (1986).
4. J. Murube del Castillo,, Tratamiento del xeroftalmos con trasplantes de glándula salival, Arch. Soc. Españ. Oftalmol. 54:151 (1988).
5. E.A. Mylius, Amylase in tears. A histochemical study, Acta Path. Microbiol. Scand. Suppl. 148:143 (1961).
6. T. Mizukawa, T. Otorio, and M. Iga, Histochemistry of the lacrimal gland in humans, Jap. J. Ophthalmol. 6:17 (1962).
7. S. Liotet, Pouvoir amylasique des larmes humaines, Ann. Ocul. (Paris) 200:526 (1967).
8. R.R. Watson, M.A. Reyes, and D.N. McMurray, Influence of malnutrition on the concentration of IgA, lysozyme, amylase, and aminopeptidase in children's tears, Proc. Soc. Exp. Med. Biol. 157:215 (1978).
9. N.J. van Haeringen, and E. Glasius, Enzymatic studies in lacrimal secretion, Exp. Eye Res. 19:135 (1974).
10. N.J. van Haeringen, F. Ensink, and E. Glasius, Amylase in human tear fluid: Origin and characteristics, compared with salivary and urinary amylases, Exp. Eye Res. 21:395 (1975).
11. N.J. van Haeringen, and E. Glasius, The origin of some enzymes in tear fluid, determined by comparative investigation with two collection methods, Exp. Eye Res. 22:267 (1976).

12. F.W. Kraus, and J. Mestecky, Immunohistochemical localization of amylase, lysozyme and immunoglobulins in the human parotid gland, Arch. Oral Biol. 16:781 (1971).

13. R.O Whitten, W.L. Chandler, M.G.E. Thomas, K.J. Klayson, and J.S. Fine, Survey of alpha-amylase activity and isoamylases in autopsy tissue, Clin. Chem. 34:1552 (1988).

14. M. Jonadet, Activités béta-glucuronidasique et amylasique au niveau de la cornée, Rev. Pathol. Comp. Hyg. Gen. 67:453 (1964).

15. R. Morrissey, J.E. Berj, L. Fridhandler, and D. Pelot, The nature and significance of hyperamylasemia following operation, Ann. Surg. 180:67 (1974).

# A SUBJECTIVE APPROACH TO THE TREATMENT OF DRY EYE SYNDROME

Robert S. Herrick

Corneal Disease Specialist
Los Angeles, California
Director of Research and Development
Lacrimedics, Inc.
Rialto, California

## INTRODUCTION

This International Meeting has dealt mainly with scientific studies of the lacrimal gland. I'd like to discuss the treatment of Dry Eye Syndrome, a product of the lacrimal glands, as well as the treatment of the secondary symptoms of Dry Eye Syndrome involving the nose, throat and sinus, by following a subjective diagnosis and treatment plan. Several speakers have referenced the need for objective clinical studies in the diagnosis and treatment of lacrimal gland dysfunction. I agree that we need this highly scientific approach, but strongly believe in applying practical experience to the study of the diagnosis and treatment of lacrimal gland dysfunction, specifically in the area of Dry Eye Syndrome. This involves:

1. Obtaining an adequate and comprehensive patient history, using a "Symptoms Checklist"
2. Initial examination specifically looking for landmarks of Dry Eye Syndrome.
3. Dependable, diagnostic testing to see if the patient experiences temporary symptomatic improvement regarding the cross-reference of subjective complaints.
4. Simple, practical, effective treatment -- only if the patient has responded favorably to the testing and has had significant improvement in the cross-reference of the unstimulated tear.
5. Consistent reversibility -- in case there is resultant epiphora from too much retention of the unstimulated tear.

# ADEQUATE HISTORY

A comprehensive patient history is the key to the successful and accurate diagnosis of Dry Eye Syndrome. A "Symptoms Checklist" (Figure 1) is used to chart the patients primary, eye-related symptoms and secondary symptoms involving the nose, throat and sinus. This "Symptoms Checklist" becomes a very practical tool to assisting in the proper diagnosis of Dry Eye Syndrome and related disorders of the nose, throat and sinus. The majority of patients check between five and twelve symptoms. It is this multiple combination of symptomatic complaints that greatly strengthens the credibility of this subjective approach to diagnosing and subsequent treatment of Dry Eye Syndrome.

# INITIAL EXAMINATION

Following a review of the "Symptoms Checklist", a comprehensive examination to identify evidence of Ocular Surface Disease. Clinical signs include:

1. Pinguecula
2. Pterygium
3. Blepharitis
4. Contact Lens Discomfort, decrease in wearing time
5. Excessive protein deposition on lenses

I do not generally put any drops in the eye during the initial exam as this helps insure that nothing impacts the effect of the Lacrimal Efficiency Test to be performed next.

# DEPENDABLE TESTING: LACRIMAL EFFICIENCY TEST™

After a careful review of the patients primary and secondary symptoms and an examination for specific clinical symptoms, we need to test whether lacrimal occlusion with non-dissolvable lacrimal plugs will provide long-term symptomatic relief. In 1983, I developed collagen implants for use in children, since the collagen implant did not require anesthetic injections like the Temporary Stitch Test. The Lacrimal Efficiency Test utilizes these collagen implants (Figure 2). With this test, sterile, dissolvable collagen implants (.3mm in length) are inserted through all four puncta down into the horizontal canaliculus. This results in a 60 - 80% increase in surface tear retention on the eye. The collagen implants will dissolve in four-to-seven days.

Once the implants have been inserted, the patient is scheduled for a two-week follow-up. When the patient returns, the "Symptoms Checklist" is revisited to determine whether the patient has experienced relief of the symptoms originally checked. In nearly 90% of the cases, the patients experience tremendous improvement of their primary and secondary symptoms, and request long-term treatment with non-dissolvable lacrimal plugs. In fact, many patients request an earlier follow-up as the collagen implants dissolve and symptoms return, so as to avoid any further suffering.

| Do you ever experience the following eye symptoms? | Left (✓) | Right (✓) | How Long? | General Symptoms | Yes (✓) | No (✓) |
|---|---|---|---|---|---|---|
| Dryness of the eye | | | | Sinus congestion | | |
| Mucous discharge | | | | Nasal congestion | | |
| Redness | | | | Runny nose | | |
| Sandy or gritty feeling | | | | Post-nasal drip | | |
| Itching | | | | Cough-chronic | | |
| Burning | | | | Bronchitis-chronic | | |
| Foreign body sensation | | | | Head allergy symptoms | | |
| Constant tearing | | | | Seasonal allergies | | |
| Occasional tearing | | | | Hay fever symptoms | | |
| Watery eyes | | | | Frequent cold symptoms | | |
| Light sensitivity | | | | Middle ear congestion | | |
| Eye pain or soreness | | | | Sneezing | | |
| Chronic infection of eye or lid | | | | Dry throat, mouth | | |
| Sties, chalazion | | | | Headaches | | |
| Fluctuating visual acuity | | | | Asthma symptoms | | |
| "Tired" eyes | | | | Muscle pain | | |
| Contact lens discomfort | | | | Joint pain | | |
| Contact lens solution sensitivity | | | | | | |

**Figure 1.** Symptoms listed to the left of the checklist are "Primary" dry eye symptoms, pertaining directly to the eyes. The items listed under "General Symptoms" relate to secondary dry eye symptoms involving the nose, throat and sinus.

# TREATMENT

Following positive symptomatic relief, long-term treatment of Dry Eye Syndrome is accomplished through the placement of non-dissolvable lacrimal plugs into the horizontal canaliculi. The long-term treatment procedure involves placement of a Herrick Lacrimal Plug into the lower punctum of each eye (Figure 3). Collagen implants are again placed into the upper punctum of each eye to maximize the symptomatic relief. The patient is asked to track their symptomatic relief over the next two-weeks. If the patient returns and has experienced maximum symptomatic relief during the period in which the collagen implants were present, placement of the non-dissolvable Herrick Lacrimal Plug into the upper puncta is indicated.

**Table 1.**  Through this subjective approach to the diagnosis of Dry Eye Syndrome, the following conditions have been shown to improve:

| | |
|---|---|
| Chronic conjunctivitis | Recurrent conjunctivitis |
| Allergic conjunctivitis | Blepharoconjuntivitis |
| Rosacea keratoconjuntivitis | Herpes Simplex Keratitis |
| Phylctenularkeratoconjuntivitis | Meibomian keraconjuntivitis |
| Vernal conjunctivitls | Chronic blepharitis |
| Pinguecula | Pterygium |
| Recurrent chalazion | Recurrent acute lid abscesses |
| Recurrent anterior uveitis | Lid twitch |
| Marginal corneal ulcer | Central corneal ulcer |
| Filamentary keratitis | Recurrent corneal erosion |
| Corneal vascularization | Keratitis secondary to ectropion |

**Table 2.** In addition to the primary eye-related symptoms and conditions positively treated through lacrimal occlusion with lacrimal plugs, the following secondary symptoms involving the nose, throat and sinus have seen dramatic improvement.

| | |
|---|---|
| Hay Fever | Sinusitis |
| Postnasal drip | Chronic cough |
| Bronchitis | Rhinorrhea |
| Asthma | Recurrent or chronic otitis medai |
| Frontal headaches (secondary to sinus congestion) | |

# COMPLICATIONS OF TREATMENT

Epiphora is a contraindication of long-term treatment with lacrimal plugs. It's important however, when epiphora occurs, to determine the type of tears being exhibited. If there is too much Constant tear, then the patient has been overtreated and the canaliculus must be reopened to stop the epiphora. This is easily accomplished using a 25 gauge cannula .3mm in diameter (allowing for a water tight seal between the cannula hub and the

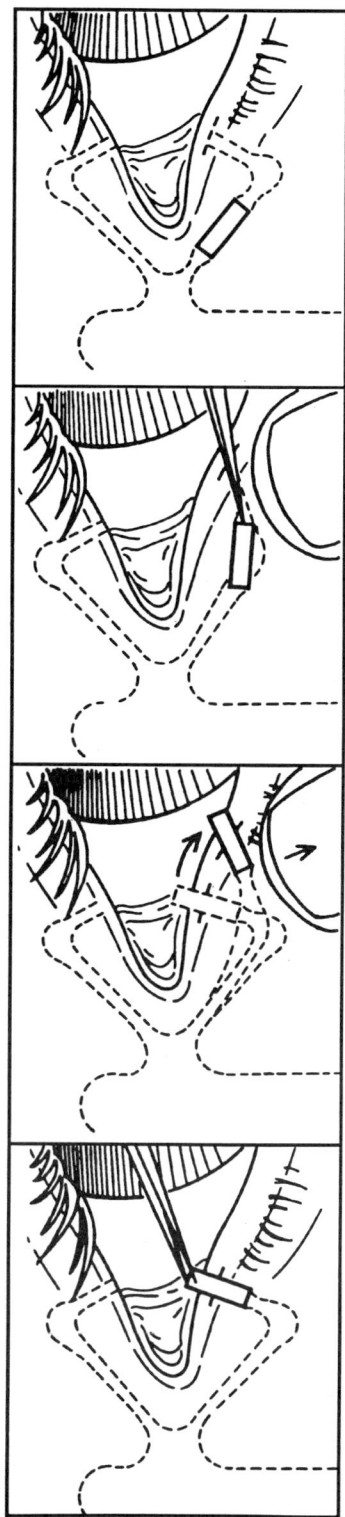

**Figure 2.** Collagen implants are inserted down into the horizontal canaliculus through the punctum. This increases the volume of natural, infection-fighting tears on the surface of the eyes.

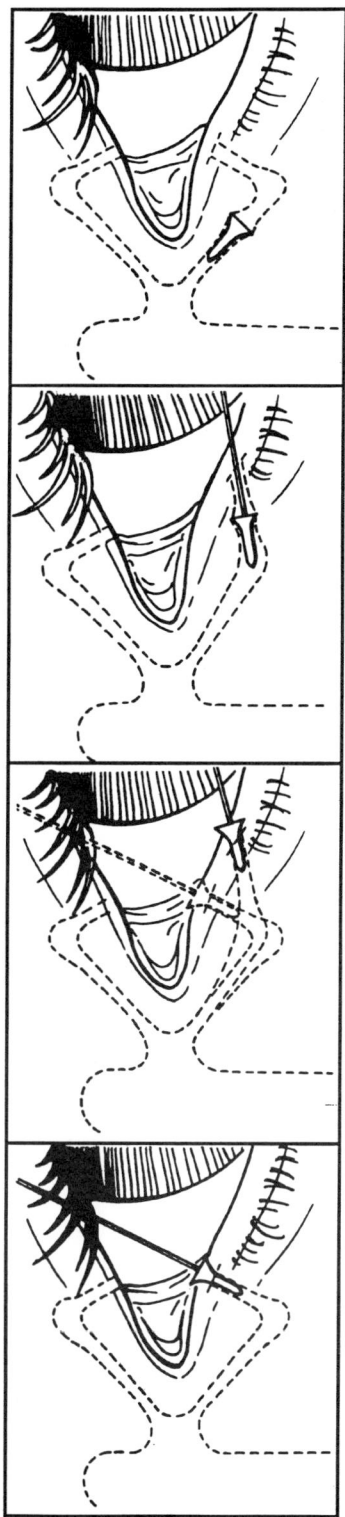

**Figure 3.** Herrick Lacrimal Plugs differ from punctal plugs used for lacrimal occlusion in that they are intracanalicular plugs which are inserted down into the horizontal canaliculus versus resting on the punctum.

punctum) filled with sterile unpreserved saline solution. The pressurized saline readily pushes the lacrimal plug through the lacrimal drainage system (canaliculus) and out through the nose or throat at no discomfort to the patient. If the tearing (epiphora) is of the Reflex variety, the patient must receive additional treatment for the irritation and the secondary stimulation of the Reflex tear.

## AVAILABLE TREATMENT ALTERNATIVES

Beyond lacrimal plugs, there exists a few other notable methods of treatment for the treatment of Dry Eye Syndrome. Artificial Tears are often prescribed for non-acute symptoms. Unfortunately, these drops only provide temporary symptomatic relief, and in fact, wash away the natural, infection-fighting properties of the natural tears. In addition, studies have proven that up to 80% of an eye drop is lost to rapid drainage within 15 - 30 seconds after installation.[1]

Over the years, argon laser canaliculoplasty and electrocautery have been used to help increase the natural lubricating pre-corneal tear film, despite the fact that each has its shortcomings. Argon Laser Canaliculoplasty requires a significant investment in argon laser technology and training, making it financially inaccessible for many patients, while electrocautery is not easily reversible, and may cause thermal damage which may permanently alter the punctum.

**Table 3.** Other proven benefits of lacrimal occlusion with lacrimal plugs include:

---

1. Increase contact lens comfort and wearing time.

2. Increase absorption and efficacy of topically applied medications, eye drops and lubricants.

3. Increased efficacy of corneal grafts.

---

## CONCLUSION

Lacrimology, and more specifically, the diagnosis and treatment of dry eye syndrome and its related disorders is truly an exciting opportunity for eye care practitioners. Properly practiced, the patient most often responds with dramatic improvement to their dry eye and related symptoms through a change in their tear dynamics. In summary, the diagnosis and treatment with this simple, practical and effective method can lead to increased patient satisfaction and practice growth.

## REFERENCE

1. M.D. Huang, and M.D. Lee, Amer. J. Ophthalmology 107:151 (1989).

# SIMULATION OF LACRIMAL GLAND OUTPUT: A TEAR JET
# FOR REPLACING EYE MOISTURE IN SJÖGREN'S SYNDROME

James H. Bertera

Schepens Eye Research Institute
and
Department of Ophthalmology, Harvard Medical School
Boston, MA  02114

## INTRODUCTION

In this paper a completely new method, called a tear jet, is presented for continuously adding moisture to the human eye. The tear jet functions as an artificial lacrimal gland, external to the eye, mounted on eyeglass frames. The essential part of the tear jet system is a micro pump taken from computer printing technology, which has the small size and output volume required for adding moisture to very dry eyes. The symptoms of ocular dryness can range from irritation and burning to intense discomfort and tissue damage. The symptoms can be eliminated by adding moisture to the eyes to supplement or replace insufficient or absent tears.

There are an estimated 6 million severely dry eye patients in the United States alone (Alexander, 1992), and another one million suffer dry eye symptoms as a result of work related visual tasks. It is generally agreed that frequently adding moisture to a dry eye is the mainstay of effective therapy, although there is debate about the exact formula, the thickness, and the neural and chemical control of the human tears and the tear layer.

The tear film provides the cornea with an optically high quality surface and acts as a lubricant to the lids during blinking. Lysozyme and beta lysin concentration in tears are also known to be bactericidal (Holly and Lemp, 1977). Dry eye conditions are associated with a variety of diseases as well as visually intensive work such as visual display terminal operation or visual inspection, and environmental conditions such as wind and dry air. Severe dry eye, associated with Sjögren's syndrome, predominantly affects women (9:1). A dry eye or keratoconjunctivitis sicca (KCS) is said to exist when the quantity or quality of the pre-corneal tear film is insufficient to ensure the well being of the ocular epithelial surface (Lamberts, 1987, p. 387). Sjögren's syndrome, in which KCS is a major feature along with xerostomia, may be similar in autoimmune response to rheumatoid arthritis, multiple sclerosis, and other autoimmune diseases.

*Lacrimal Gland, Tear Film, and Dry Eye Syndromes*
Edited by D.A. Sullivan, Plenum Press, New York, 1994

The volume input possible with the micro pump method described here is sufficient to replace the entire tear volume, overcome the tear drainage system, and evaporation losses. The drainage system for the tears, through the puncta and down to the nose, limits the value of typical drop applications because the fluid is not retained in the eye for very long. The tear jet input rate can match or exceed the normal volume requirements for human tears which is a tear flow rate of 1 μl per minute. The total steady state volume of tears is about 7 μl and the turnover rate is therefore about 16% per minute (Lamberts, 1987, p. 38).

## Advantages of Pico Droplet Application

The tear jet application for dry eyes is a secondary application for highly developed computer printing technology, and, therefore, has the advantage of years of development work and investment as a starting point. Adding moisture continuously and directly to dry eyes may be a convenient, timely, and affordable treatment for a wider range of patients than any one drug, gene, or surgical treatment, and with fewer side effects. The miniaturization made possible by the small size requirements for pumps in computer printing may convey ease of use, cosmetic acceptability, and low cost, which are required for long term treatment of severely dry eye patients.

Several advantages relate to the bio-compatibility of the droplet size. The jet pumps used for the tear jet produce droplets of about 200 Pico liters in volume. These Pico droplets may be small enough to avoid washing away the beneficial tear film structures that results when a large drop floods into the eye from a typical dropper bottle (Lamberts, 1987, p. 38). Pico liter droplets are so small that there is almost no sensation when they land on the eye. The sensation is more of a temperature change since the droplets are below body temperature (about 73 degrees Fahrenheit) when they reach the eye. The 200 Pico liter droplet size means there is no blink or startle response as there would be with larger droplets. Small droplets are also a better simulation of the normal functioning of the tear or lacrimal gland which emits fairly continuously at about 1 μl per minute. With this steady and even addition of fluid there should be no parch and flood as with periodic administration and no overflow from the lids or build up to distort vision (epiphora). Droplets landing on the skin during a blink or prolonged downward gaze evaporate very quickly owing to their very large ratio of surface area to volume.

Adjustments in the tear jet input rate for the wetting solutions may be used to alter tear osmolality towards a normal value in dry eye patients. Attempts to dilute the higher salt concentration in dry eyes, whether a primary defect in tear production or a result of evaporation (Gilbard, Farris, and Santamaria, 1978), by manually adding hypo-osmotic drops has not succeeded, since the tears return to hyper-osmolality within 60 to 90 seconds (Holly and Lamberts, 1981). Thus, a continuous method like a tear jet is required to adjust and maintain correct tear osmolality in some patients.

## Earlier Attempts at Continuous Irrigation

Most recently in an ARVO abstract, a belt mounted, battery driven, droplet pump method was described by Vo (1992). Vo used five microliter droplets launched periodically (every 5 minutes) from a peristaltic type pump worn on the waist. Tubing carried the fluid up from the belt and through the eyeglass frames. The tubing end was placed near the eye but not in contact with it. This 5 μl droplet is much larger and less frequent than the normal tear output. This pump and battery may be the best that can be expected from conventional pump technology, but they have no potential for miniaturization sufficient to be fully head mounted. The tubing carrying the fluid from the waist level pump to eyeglass frames must be replaced for sterility.

The earlier tubing methods were not practical or tolerable for general use but they satisfied some of the moisture needs for highly motivated and attentive users. An older eye-contacting method was developed by a dry eye sufferer and had no moving parts (Flynn and Shulmeister, 1967). A brief ooze of fluid was passed to the eye by a tube which rested directly on the lower eyelid. The tube was fed with fluid from a reservoir mounted on eyeglass frames. The drop oozed out when the user bobbed his head forward causing fluid to move forward inside the tube. Of course, contact on the lid with the tube would present a foreign body sensation to some users. This method was further elaborated with the use of a belt mounted pump (Dohlman, Doane, and Reshmi, 1971).

A surgical approach was tried by Ralph, Doane, and Dohlman (1975), in which a tube from a reservoir and ooze pump was passed under the skin of the neck, the side of the face, and the lower lid. The fluid was pumped in small drops through the tube and passed up to the cornea during blinking. Re-routing some of the salivary flow into the eye has also been tried. While some patients may tolerate such procedures because of the severity of their dry eye and the prospect of loosing vision, there are obvious drawbacks and limitations to a surgical approach.

In the tear jet, the output pump and the reservoir are one unit without tubing. Alternative methods for continuous irrigation described above all require a long run of tubing. All tubing is an element requiring sterilization. In addition, the end of a lid borne tube is easily displaced by blinking or facial muscle movements. The long lengths of tubing needed in belt mounted pumps generate flow resistance so that saline or non-viscous solutions are the only formulation choices. The tube-connected reservoirs are easily infected adding further threat to an already compromised dry eye (Wright, 1985).

## METHODS AND RESULTS

A single subject used the tear jet for four days of continuous irrigation of 10 hours per day. The tear jet system consists of the jet substrate, a reservoir, and the battery pulser. A micro jet pump can form droplets in the 200 Pico liter range and eject them at 100 Hz. A suitable and available pump is the resistive thermal expansion pump used by Hewlett-Packard Company (Allen, Meyer, and Knight, 1985) as the principle component in the ink jet printer cartridge. A 10 μs current applied to a 50 μm square film resistor rapidly heats 3% of a confined liquid and the evolving steam bubble ejects the 200 Pico liter droplet of fluid from a small orifice at an adjustable velocity of from 0.5 to 10 m/s (See Figure 1). The droplets can travel up to 50 mm. The human tear volume of 1μl/minute and a droplet size of 200 Pico liters requires about 100 droplets/s. Continuous use for a severe dry eye condition might require 14 use hours per day and a droplet yield per day of 360,000 per hour X 14 hours or about 5M droplets. The 100 Hz rate is lower than the 2000 Hz upper limit on the available ink jet substrate pump and preliminary tests show a more than adequate droplet yield of 10-15 M.

The problem of accidental touch by the tear jet assembly was addressed by insulating all the edges and possible contact points in plastic to form a buffer between the pump and eye. An accidental touch by the device may be slightly more likely than a touch by eyeglass frame components themselves, which is rare. The sterility of the unit was ensured by a single use reservoir and pump combination.

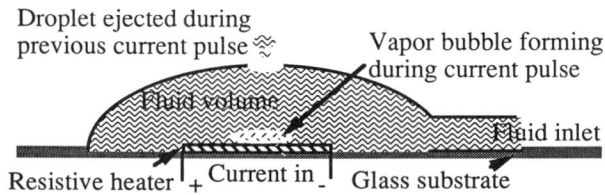

Figure 1. A single thermal expansion pump is shown in cross section. A cavity with an output orifice is formed on the surface of a glass substrate. A 10 μs current pulse applied to a resistive heater element, 50 μm square, deposited as a layer on the glass substrate, vaporizes a small amount of the fluid volume and the expansion forces the fluid from the orifice. A train of current pulses, e.g. 100 Hz, creates a train of droplets which can be directed to the ocular surface.

Two Hewlett-Packard ink jet cartridges were emptied, washed, sterilized, filled with saline, and mounted on a stereotaxic carrier attached to eyeglass frames (See Figure 2). While this demonstration method is unsuitable for use on patients, and occluded most of the visual field, it shows that the jets can be mounted on eyeglass frames and that the droplet impact had no immediate harmful effect on one subject over several hours of use. An electronic controller and battery were housed in a separate package in the shirt pocket, connected to the eyeglass frame mounted tear jet with a flexible wire cable (Figure 2). The electronic control and batteries were held in a shirt pocket sized box to avoid unnecessary prototype construction for this initial test.

Fluid flow of accumulated droplets on the ocular surface and mixing and spreading with the indigenous tears was verified by adding a fluorescein solution to the tear jet reservoir. The droplets mixed with the tears and moved, from a nasal conjunctiva landing position across the eye and onto the cornea.

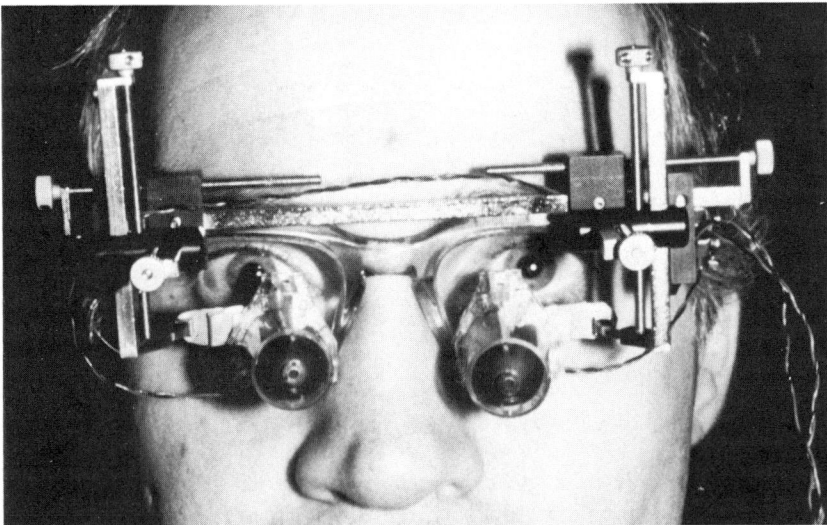

Figure 2. Preliminary test prototype tear jet and reservoir mounted on stereotaxic eyeglass frames. The droplet train is aimed inward at the eye; droplet landing zones were tested on the nasal and temporal bulbar conjunctiva and directly on the cornea. Droplet accumulation on the cornea between blinks caused blurring. Wire cables (exiting at right) allowed shirt pocket mounting for batteries and reduced prototype construction cost for these initial tests. Final version of tear jet includes eyeglass frame mounting of all system components, cf Figure 3.

## FURTHER IMPROVEMENTS

Prospects are very good for miniaturization. The largest and heaviest component of the system is the batteries. Estimating from the available resistive thermal expansion pumps, battery weight would be at least 15 g using off-the-shelf batteries. Lightweight metal frame eyeglasses with plastic lenses weigh about 40-50 g. Longer lived batteries, supplying several days of continuous operation, can weigh 90 g. The prospects are good for high current, lighter weight batteries, now becoming available. The alternative to eyeglass frame mounted batteries is the use of a belt or shirt pocket mounted battery and flexible wire cable, but a tubing connection from head to belt eventually results in the most user complaints, even in patients in great need of a supplemental tear system (Dohlman, et al, 1971).

Efficacy should be demonstrated in both statistical and clinical improvement using subjective and objective measures. In future tests with dry eye patients, an analog dimension will be used for self rating by the patient for dryness, irritation, burning, photophobia, visual acuity, tearing and mucus, wateriness, and running nose (if there is increased nasal drainage). Objective measures will include photographs to rate redness and gross pathology.

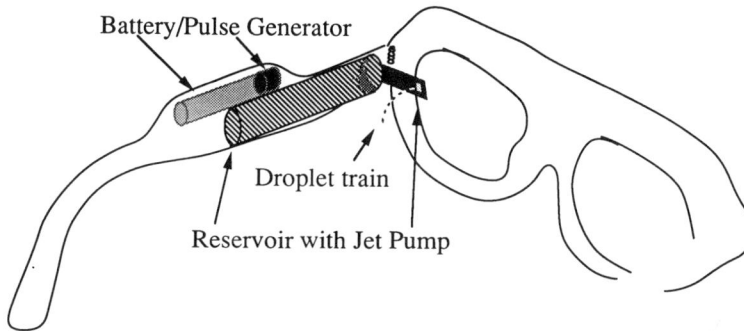

Figure 3. Proposed tear jet prototype showing jet pump, reservoir, batteries, and pulse former mounted on temple of eyeglass frames.

Slit lamp exams will be used to monitor corneal and conjunctival health. Corneal and conjunctival staining, e.g. rose bengal, are mainstays of evaluation, although there is some debate as to the exact meaning of the staining process. Conjunctival appearance of dullness and papillae are also observable signs.

Since the aqueous portion of the tears are composed of many elements besides water (ions of salts, enzymes, proteins, etc.), all artificial tear or wetting solutions, are only low fidelity simulations of the naturally occurring fluids on the normal eye. The relationships between tear physiology and dry eyes has been reviewed by Holly and Lemp (1977). Evidence suggests that a simplified saline solution used in a continuous instiller, like a tear jet, might require the addition of balanced salts in order to prevent toxic effects on ocular tissues. Simple saline solution applied to the eye can be toxic to the corneal epithelium in rabbits (Gilbard, Scott, Rossi, and Heyda, 1989). Preservatives solutions can also impair ocular surface functions, especially after long term use (Göbels and Spitznas, 1992; Gilbard, Rossi, and Heyda, 1989). While some forms of balanced salt solution are commercially available, the least disruptive to ocular function appears to be the Gilbard solution 15.

## ACKNOWLEDGMENTS

This work was supported in part by grant DAMD17-90-Z-0014 from the Army Medical Research Acquisition Agency and the Schepens Eye Research Institute Development Fund.

## REFERENCES

Allen, RR, Meyer, JD, and Knight, WR, Thermodynamics and hydrodynamics of thermal ink jets. Hewlett-Packard Journal, 1985, 21-26.

Alexander, E, Sjögren's syndrome underrecognized. Moisture Seekers Newsletter, The Sjögren's Syndrome Foundation, 1992, 9, 4.

Bertera, JH, Simulation of lacrimal gland output: A tear jet system for projecting fluids onto the ocular surface. International Conference on the lacrimal gland, Bermuda, 1992. p. 23.

Dohlman, CH, Doane, M, and Reshmi, CS, Mobile infusion pumps for continuous delivery of fluid and therapeutic agents to the eye. Annals of Opthalmology, 1971, 3, 126.

Flynn, F, and Schulmeister, A, Keratoconjunctivitis sicca and new techniques in its management, Medical Journal of Australia, 1967, 1, 34.

Gilbard, JP, Farris, RL, and Santamaria, J, II Osmolarity of tear microvolumes in keratoconjunctivitis sicca. Archives of Ophthalmology, 1978, 96, 677.

Göbels, M and Spitznas, M, Corneal epithelial permeability of dry eyes before and after treatment with artificial tears. Ophthalmology, 1992, 99.

Holly, FJ and Lamberts, DW, Effect of non-isotonic solutions on tear film osmolality. Investigative Ophthalmol and Visual Science, 1981, 20, 236.

Holly, FJ and Lemp, MA, Tear physiology and dry eyes. Survey of Ophthalmology, 1977, 22, 69-87.

Lamberts, DW, Keratoconjunctivitis sicca. In The Cornea. Smolin, G and Thoft, RA (Eds.), Little, Brown, and Company, Boston, 1987.

Ralph, RA, Doane, MG, and Dohlman, CH, Clinical experience with a mobile ocular infusion pump. Archives of Ophthalmology, 1975, 93, 1039-1043.

Rolando M, Refojo MF and Kenyon KR, Increased evaporation in eyes with keratoconjunctivitis sicca. Archives of Ophthalmology, 1983, 101, 557-558.

Talal, N, Sjögren's syndrome: Close to cause and cure? International Conference on the Lacrimal Gland, Bermuda, 1992, p. 129.

Vo, Van Toi, Automatic administration of eye medication by injection of droplets. Investigative Ophthalmology and Visual Science, Supplement, 1992, 1012.

Wright, P, Other forms of treatment of dry eyes. Transactions of the Ophthalmology Society of the United Kingdom, 1985, 104, 497-98.

# CHARACTERISTICS OF A CANINE MODEL OF KCS: EFFECTIVE TREATMENT WITH TOPICAL CYCLOSPORINE

Renee Kaswan

Departments of Small Animal Medicine and
Veterinary Physiology/Pharmacology
College of Veterinary Medicine
University of Georgia
Athens, GA 30602

## INTRODUCTION

For the past 12 years, we've studied keratoconjunctivitis sicca (KCS) in dogs to develop a therapeutic intervention to benefit both veterinary and human KCS patients. The spectrum of ocular surface pathology in canine KCS can support a battery of assessment criteria for therapeutic evaluation. Although there are multiple causes of canine KCS, the vast majority appear to be immune mediated.

Topical application of cyclosporine (CsA) leads to a gradual recovery of lacrimation in 80% of affected dogs. Conjunctivitis and keratitis are reduced with long term application. The resolution of surface pathology is largely attributed to cyclosporine's anti-inflammatory activity because improvement occurs with or without increased lacrimation. Current hypotheses on the mechanism of action of cyclosporine on lacrimation include 1) modulation of lymphocyte cytokine production in the lacrimal gland, 2) decreased recruitment of autoreactive lymphocytes from the conjunctiva to the lacrimal gland, and 3) a direct neurohormonal effect of cyclosporine mediated through prolactin receptors identified on lacrimal acinar epithelium.

## CHARACTERIZATION OF KCS IN DOGS

### Clinical Signs

The ocular lesions seen in dogs are much more severe than those seen in people. The increased severity may be attributed to delayed diagnosis in dogs related to the patient's inability to report discomfort as an early sign, and to their inability to self medicate at a frequency compatible with comfort. Lagophthalmos and exophthalmos, common in many breeds, further exacerbate the effects of an insufficient tear film.[1]

**Conjunctivitis.** Clinical signs of conjunctivitis of the globe, lids and third eyelid include; mucoid to mucopurulent discharge, hyperemia, chemosis, and hypertrophy. Opportunistic bacteria can be cultured from the ocular discharge, but their abundance is secondary to the effects of aqueous deficiency. Conjunctivitis is the consistent component of KCS, whereas keratitis occurs only in more advanced cases and in exophthalmic or lagophthalmic breeds. Conjunctivitis is irritative, but, typically, not painful.

**Pain.** Discomfort in dogs with KCS is highly variable; often the apparent discomfort is inconsistent with the apparent clinical signs. Dogs respond to conjunctivitis by pawing or rubbing the eyes against rugs and furniture, but blepharospasm is unusual unless corneal ulcers have also occurred. Corneal sensation was evaluated by anesthesiometer and found to vary with head conformation (brachiocephalics have less sensitivity than doliocephalics),[2] consistent with differences in the density of corneal nerves. Corneal sensation is much less sensitive in dogs than persons, and is often further diminished in chronic KCS, therefore, even ulcers may not be associated with signs of discomfort.

**Keratitis.** In severe cases of KCS, the corneal epithelium becomes keratinized, vascularized, and hypertrophic, as much as 30 cells thick.[3] Corneal hypertrophy can become so extreme as to preclude lid closure, and lagophthalmos compounds the effects of tear deficiency. An undulated corneal surface can occur from extreme hypertrophy with inflammation induced subepithelial stromal edema. Concurrent with epithelial and subepithelial vascularization, dystrophic superficial or subepithelial precipitates can include lipids, calcium, and/or pigment.

**Pigmentary Keratitis.** In exophthalmic breeds, and in breeds with periocular pigmentation such as the Chinese pug, miniature schnauzer,and dachshund, pigmentary keratitis can be a devastating consequence of KCS. Free pigment granules, melanin in macrophages, and melanocytes can be deposited beneath the corneal epithelium.

**Blindness.** Is frequently the presenting complaint in dogs with KCS. Advanced pigmentation and corneal scarring cause sight loss. Superficial hypertrophy and, to a lesser degree, subepithelial fibroplasia are reversible with antiinflammatory treatment (corticosteroids and/or cyclosporine), but pigmentary keratitis is relatively recalcitrant.

**Corneal Ulcers.** Ulcers caused by dessication generally occur in the central cornea, the area exposed within the lid fissure. Large deep melting ulcers also occur occasionaly; rapid resolution has been seen with cyclosporine ophthalmic (unreported clinical observation).

# Diagnosis

Normal wetting in dogs is 15 to 25 mm/min on non-anesthetized Schirmer tear test (STT).[4] KCS cases typically wet less than 10 mm/min, with the majority of symptomatic cases wetting less than 5 mm/min on repeated trials. A diagnosis of KCS is made when decreased STT values occur with mucopurulent conjunctivitis, corneal inflammation, ulceration, or pigment deposition in the corneal surface.[1,5]

# Incidence

Review of the incidence of KCS in dogs presented to 20 veterinary teaching hospitals during a period of 24 years, revealed a progressive increase in the number of recorded cases from an initial incidence of 0.04% in 1965 to an incidence of >1.5% in 1988. The absolute incidence of KCS in the general canine population is unknown.[6,7]

**Breeds** with a high relative incidence (RI) of KCS are: English bulldog (RI=20), West Highland white terrier (RI=18), lhasa apso (RI=17), Chinese pug (RI=8), American cocker spaniel (RI=7), Pekingnese (RI=6), Yorkshire terrier (RI=5), shih tzu (RI=4), miniature schnauzer (RI=4), and Boston terrier (RI=4).[1]

**Sex and age** predisposition to KCS have been reported for female and aged dogs.[6,7] Using the Veterinary Medical Data Program (VMDP), we evaluated the interaction of gender, neutered status, and age, in a population of > 1 million dogs. The incidence of KCS increased with age in all sex groups. There were a larger number of older female dogs with KCS compared to aged male dogs, however, when the sex groups were further subdivided, to intact female, female spay, intact male, and male castrate, the gender predisposition disappeared. Neutered animals of either gender are predisposed to develop KCS, and females are neutered 3.4 times as frequently as males. The incidence of KCS increased with age at a higher rate in neutered dogs age 10-15 (p<.000001) compared to intact dogs of either sex.[6]

# Etiology/Immune Mediated

Although the cause of KCS in the dog can often not be determined, circumstantial evidence, suggests autoimmune processes account for the majority of KCS cases.

**Histopathology.** The dog has 2 major lacrimal glands, with similar anatomy.[8] In a series of 49 nictitating membrane glands or orbital lacrimal glands removed from dogs with severe KCS, no infectious agent was seen on either light or electron microscopy.[9] The most common lesion was multifocal mononuclear cell infiltration with varying degrees of fibrosis. The proportion of T-cell and B-cell subsets in dog lacrimal glands has yet to be assessed because specific canine T-cell markers were not available. Paradoxically, many glands had few focal inflammatory lesions and the majority of glands had large areas of apparently nonfunctional acini. Lesions were classified into 4 stages of progression. In stage 0 lesions were absent on light microscopy and degenerative acinar changes such as loss of secretory granules, and nuclear degeneration were seen on electron microscopy.[a] In stage 1, inflammatory lesions appeared to begin as small multifocal mononuclear infiltrates and progess to stage 2; large, confluent areas of mononuclear infiltrates. Lesions classified as stage 3 had diffuse, coalescing mononuclear inflammatory lesions, fibrosis and atrophy of the acinar elements.[9] Contrary to the conventional belief in human medicine, that lacrimal glands in KCS patients have progressive, irreversible atrophy, we contemplated that glands with stage 0-2 lesions could regain function following acinar repair and regeneration, if the inflammatory response could be aborted.

**Autoantibodies.** In a series of 50 dogs, rheumatoid factor was positive in 34% of dogs,[10] antinucleolar antibodies were positive in 40%, and anti-glandular antibodies were demonstrated in occaisional dogs affected by KCS.[11] Serum electrophoresis showed elevated gammaglobulins in 90% of canine KCS cases. The patient population for these studies included approximately 50% outpatient ophthalmology patients and 50% hospitalized internal medicine patients. When these tests were repeated in a group of 35 patients presented specifically for ophthalmic disorders, the incidence of autoantibodies was decreased.

**Concurrent autoimmune disorders.** Canine KCS occurs as an isolated disorder without associated immune mediated diseases, and in association with previously recognized autoimmune disorders including; juevenile diabetes mellitus, systemic lupus erythematosis (SLE), pemphigus foliaceous, rheumatoid arthritis, hypothyroidism, polymyositis, polyarthritis, and glomerulonephritis. Concurrent glandular disorders include xerostomia, parotid sialoadenitis, seborrhea, atopy, and hyperadrenal corticoidism.[11] In a colony of dogs with SLE, 2 dogs were found to have severe KCS, xerostomia, autoimmune thyroiditis, a nonspecific chronic GI disorder, vaginal dryness, as well as multiple serum autoantibodies. This was the first description of dogs having a polyglandular autoimmune exocrinopathy, similar to Sjogren's syndrome.[12]

## Other Etiologies of Canine KCS

**Distemper virus** can cause acute lacrimal adenitis. Unlike most cases of KCS which are chronic and progressive, KCS due to distemper usually occurs precipitously, and resolves spontaneously if the animal recovers systemically.[13] In one case of distemper induced KCS, biopsy of the third eyelid lacrimal gland revealed diffuse intense neutrophillic inflammation. The dog had an STT of 0 mm/min at the time of biopsy but spontaneously regained normal tear production suggesting that the severe diffuse acute inflammation had resolved and the acini regenerated or recovered function.[13]

**Congenital alacrima** occurs occasionally as an extreme xerosis. It is often unilateral, and is most commonly seen in small breeds.

**Drug induced.** Atropine,[14] phenazopyridine,[15] sulfadiazine,[48] and salicylazosulfa-pyridine often cause transient KCS.[16] Some cases did not resolve following discontinuation of treatment.[17]

**Neurologic xerosis** is seen in some cases of facial trauma, inner ear infections and brainstem disease.[18] The parasympathetic lacrimal nerve, which courses first with the facial nerve and later with the trigeminal, can be lost with injury to either.

**Obstruction** of the lacrimal ductules due to chemosis or conjunctival cicatrization is a possible cause of transient or permanent sicca.[3]

**Vitamin A deficiency** is an unlikely cause of canine KCS.[3,19]

**Conventional therapies** of canine KCS include: artificial tears, topical mucolytic (acetylcysteine), intermittent topical antibiotics, systemic and/or topical cholinergic stimulation (pilocarpine), intermittent topical corticosteroids, and parotid duct transposition.[5]

## OPHTHALMIC CYCLOSPORINE: THERAPEUTIC MANAGEMENT FOR KCS

Cyclosporine is a non-cytotoxic immunosuppressant used primarily for organ transplantation. Its primary effect is to inhibit T-helper cell activity while sparing T-suppressor cell activity, thereby shifting the balance of T-cell immune regulation toward immune tolerance. To avoid toxicity and expense, we tested cyclosporine in an ophthalmic preparation, at doses < 1/100th of the canine oral therapeutic dose.

Initially, two distinct therapeutic effects of BID cyclosporine ophthalmic were observed: (1) increased tearing in most cases; and (2) improved ocular surface lesions.

Major improvements over conventional treatment are anticipated for the cyclosporine treated patient. Cyclosporine may treat the underlying cause of KCS, and therefore may prevent the progressive deterioration of the lacrimal glands and otherwise inevitable disease progression. Dogs resume production of natural tears, apparently with the composite physiologic tear proteins preserved.[20]

### Mechanism of Action of Cyclosporine

Initially cyclosporine was not advocated for autoimmune diseases because it was felt it could only prevent development, but not reverse, an ongoing immune response. The T helper cell secretes many cytokines which are either essential requirements or strong amplifiers in the activation of macrophages, production of antibodies and B cell proliferative responses, activation and proliferation of helper and cytotoxic T cells, and proliferation of fibroblasts. The effects of T helper cell inhibition on immune reactions are much more pervasive than initially anticipated.[21] The antigen stimulus to T cells is required continuously to maintain an active autoimmune response.[22] Interruption of the continuous afferent response to self antigens could prevent interleukin 2, gamma interferon and other cytokines from being synthesized following T cell exposure.[21] Without a constant source of amplifying soluble factors, suppressor T cells are unopposed in re-establishing self tolerance. Based upon numerous clinical studies, cyclosporine is now considered a promising agent for management of a variety of autoimmune disorders.

Cyclosporine, FK506 and rapamycin are non-cytotoxic immunosuppressants which share a common mechanism of action, but have specific (not interchangable)[23,24] nuclear protein substrates. They act upon a newly recognized family of enzymes, petidy prolyl cis-trans isomerases (also called rotamases), which activate transcription in T cells.[25] Cyclosporine specifically binds the rotamase, cyclophilin a 17 Kd protein, while FK506 specifically binds FK506 binding protein, and cannot bind cyclophilin.[23,24] The intrinsic substrate for cyclophilin is prolactin[26,27] (see lacrimomimetic effects). Cyclophilin is found in highest concentration in T lymphocytes, but has also been identified in all tissues examined to date.[28] Although the majority of cyclosporine's activity is attributed to suppression of the T-cell, it also binds cyclophilin in basophils and inhibits their release of histamine when challenged with anti-IgE.[29] The dose dependent side effects of cyclosporine on epithelial cells including hair growth and gingival hyperplasia may be direct effects on epithelial cyclophilin.

## Cyclosporine Distribution in the Eye

Cyclosporine is a highly lipophilic drug and is absorbed into the cornea in high levels.[30] The purported therapeutic serum trough concentration for immunosuppression is 0.2 ug/ml.[31] Corneal levels following topical administration peak at 4 ug/gm; 20 fold higher than minimum effective serum levels.

## Anti-Inflammatory Effects of Cyclosporine on the Ocular Surface

Improvement of the corneal and conjunctival lesions in dogs with KCS can be dramatic, with or without improvement in lacrimation,[32-37] suggesting that ocular surface disease is caused in large part by secondary inflammaton, as opposed to dessication.[38,39] Conjunctival hyperplasia, leucoplakia corneal granulation-like tissue, and mucopurulent ocular discharge, improved in most of the cyclosporine treated KCS dogs, within 2-3 weeks. Corneal vascularization and pigmentation also resolved in most dogs, however, these lesions resolved much more gradually; improvement was first apparent by 3 months, and continued resolution occured beyond 12 months.

Cyclosporine shows promise to be of benefit in a spectrum of chronic ocular surface inflammatory disorders as an alternative to topical corticosteroids. The apparent advantages of topical cyclosporine compared to topical corticosteroids are that cyclosporine does not appear to promote cataractogenesis, glaucoma or corneal collagenase activation. It appeared to be effective and well tolerated in uncontrolled trials in severe corneal ulcers[40] rheumatoid corneal ulcers,[41] necrotizing scleritis,[42] and ligneous conjunctivitis.[43] Limited controlled trials in human vernal keratoconjunctivitis[44,45] and canine chronic superficial keratitis (pannus) indicated dramatic antiinflammatory benefits.[46]

An application for a New Animal Drug Approval (NADA) for ophthalmic cyclosporine

(Optimmune[R], Schering-Plough Animal Health Inc.) for the treatment of canine KCS, is scheduled for submission to the FDA Center for Veterinary Medicine for August, 1993. All data collected for the NADA are confidential and are therefore not included in this review.

## The Lacrimomimetic Effects of Cyclosporine

In an experimentally induced mouse model of Sjogren's syndrome, systemically administered cyclosporine prevented and reversed the lymphocytic infiltration of the lacrimal glands.[47] A progressive improvement in lacrimal function is seen in dogs beyond 1 year's treatment.[36] These observations suggest to us that regeneration of lacrimal gland tissue is occuring when chronic inflammation is controlled.

Cyclosporine penetrates to the lacrimal gland following topical administration[30,b] When first administered to KCS-affected dogs, an average delay of 2-3 weeks occurs before improved lacrimation is seen.[32,34,37] This delay is consistent with an immunosuppressive mechanism, and supports the basic hypothesis, that cyclosporine would suppress autoreactive T helper cells in the conjunctiva and lacrimal gland and abort local autoimmunity. Presuming the lacrimomimetic effect of cyclosporine is T cell dependent, there are 2 sites of activity to be considered: 1- Cyclosporine acts directly on T cells in the lacrimal gland or 2-Cyclosporine suppresses conjunctival associated lymphocytes; these purportedly are bathed in protein rich lacrimal fluid then,[20] recycle to the lacrimal gland to stoke the inflammation.

There is a growing body of evidence that cyclosporine also affects tearing by a hormonal mechanism. When normal beagles were treated with topical cyclsporine or a placebo control in a double-blind cross over study, tearing progressively increased in the cyclosporine treated dogs but not in the placebo control dogs. Since these beagles presumably did not have lacrimal inflammation, suppression of autoimmunity would not explain increased tearing in this trial.[32] Epiphora is observed in renal transplant patients as a side effect of systemic cyclosporine.[c]

Although inducing a lacrimal response in KCS dogs requires 2-3 weeks of treatment with ophthalmic cyclosporine (Table 1), a precipitous drop in tearing occurs within 12-24 hours of stopped treatment. When cyclosporine is reinstituted, tearing rebounds to maximal levels in 3 hours.[34] This rapid recapture of tearing capacity supports the hormonal effect hypothesis. Using an in vitro assay of lacrimal secretion, we tested the effects of cyclosporine on lacrimal fluid secretion. Lacrimal acini were isolated from rat lacrimal glands and exposed in vitro to therapeutic amounts of cyclosporine (800 ng/ml). A 42% increase in peroxidase (a marker for lacrimal secretion in rodents) was consistently seen.[34]

Once we suspected a hormonal activity of cyclosporine, we re-evaluated the biochemical effects of cyclosporine which might be responsible for hormonal regulation of lacrimation. The cytosolic receptor for cyclosporine, cyclophilin is the natural ligand for prolactin.[26,27]

Prolactin has been identified in lacrimal acini and tear fluid,[48,49] and prolactin receptors occur on membranes of lacrimal acini and then prolactin is endocytosed into the lacrimal acinar cell cytoplasm.[50] Dihydrotestosterone (DHT) and prolactin act synergistically to reverse the hypophysectomy-induced acute regression of female rat lacrimal glands.[51] Prolactin has a biphasic effect in that higher doses act to suppress the stimulatory effects of DHT (see Warren et al this issue).[52]

We hypothesize that in immune mediated KCS, secretion of local lymphokines includes the prolactin-like protein.[50] Like testosterone, the prolactin-like protein plays a regulatory role in lacrimal acini regeneration, secretory granule production, granule secretion, and membrane trafficking of Na/K pump receptors.[51,52] We hypothesize that cyclosporine binds regulatory prolactin receptors in lacrimal tissue. We propose that the 2-3 week average delay until first effect of cyclosporine relates to a regeneration of cells and secretory granules in degenerate lacrimal acini; while the brief (3 hour) interval to regained secretion following treatment interruption, relates to secretory membrane effects of cyclosporine.

Table 1. Summary of reported clinical trials using cyclosporine ophthalmic to treat KCS in dogs.

| Patients | Duration RX | Pretreatment STT mm/min | Cyclosporine STT Increase with Rx | Control STT Change With Placebo |
|---|---|---|---|---|
| 32 client owned dogs 34 | 0 mo<br>1 mo.<br>2 mo.<br>3 mo<br>6 mo<br>12 mo<br>withdrawal 2 wk | 3.6 m | +5.0mm<br>+5.6mm<br>+5.6mm<br>+6.7mm<br>+11.1mm | +1.6mm[#] |
| 60 client 36 owned dogs 100 eyes | 12 mo. | 3.1 | +9.6mm[$] | uncontrolled trial |
| 36 client 32 owned dogs | 1-26 wk<br>ave = 10.5 wk | 3.6 OD<br>2.3 OS | +9.0[^]<br>+8.9 | uncontrolled trial |
| 42 client 37 owned dogs 22 CsA 20 vehicle | 4 wk | 3.8<br>2.9 | +6.5mm[@] | +0.1 mm |
| 18 laboratory beagles 35 age 15-17 | 10 Wk | < 10 (4 eyes)<br>> 10 (14 eyes)<br>< 10 (6 eyes)<br>> 10 (12 eyes) | +17mm[*]<br>+5.5mm | +2.5mm/min<br>+5.0mm/min |

[^]P < .0005 compared to pretreatment; [*]P < .01 compared to placebo, 9 dogs CsA, 9 dogs placebo; [@]P < .05 compared to placebo; [$]P value not given; 32 dogs began this study, 30 dogs completed 6 mo, 19 dogs completed 12 mo.; [#]2 wk placebo given after 12 mo. CsA; NR - not reported.

At this time, it is impossible to determine if the lacrimal secretogogue effects of hormones such as prolactin and testosterone, occur indirectly, by effecting the secretion of lymphokines, or if these hormones have a direct growth regulatory effect upon lacrimal acini and ducts.[53] Similarly, it has not been determined if the immunosuppressant drug, cyclosporine, acts by competing for local hormone receptors (most likely prolactin receptors) on lacrimal acini and ducts, or if cyclosporine affects local lymphokine secretion (including the prolactin-like protein), or if both activities contribute to the lacrimomimetic effect.

## Types of KCS Responsive to Cyclosporine

The primary determinant of whether or not cyclosporine will increase tearing in a dog with KCS seems to be the stage of KCS at which it is first instituted. If cyclosporine ophthalmic therapy was instituted in dogs that had not yet reached end-stage lacrimal disease, tearing increased by at least 5 mm/min, within 3 mo., in 87% to 100% [32-37] of treated eyes. If KCS was endstage, STT = 0-1 mm/min, when cyclosporine was first introduced, improved lacrimation was seen in only 59%[34], 60%[36], 71%,[37] or 53%[32] of treated eyes in 3 clinical trials. When biopsies were taken of nonresponsive dogs, diffuse inflammation and fibrosis characterized the non-responsive glands.[9]

It is uncertain if dogs with neurologic lesions as the cause of KCS will respond to cyclosporine. However, the utility of cyclosporine for KCS is not restricted solely to immune mediated KCS because dogs with KCS caused by sulphonimide toxicity respond very well to ophthalmic cyclosporine, with increased tear production.[32,34,36]

## Criticism of Cyclosporine Use in KCS

Some critics have argued that cyclosporine could cause tearing by an irritative mechanism. However, the typical KCS patient cannot increase tearing in response even to ammonia irritation. To argue that ocular irritation could be therapeutic for KCS is contrary to the common clinical observation that environmental irritants exacerbate KCS.

A second objection to ophthalmic cyclosporine used to treat KCS has been the presumption that topically applied cyclosporine would not likely penetrate to the lacrimal gland. The lacrimal glands are covered by conjunctiva, a tissue which is highly permeable to lipophillic drugs such as cyclosporine. Although therapeutic agents are not routinely prescribed for topical therapy of the lacrimal gland, topical drugs have well established side effects on lacrimation. Topical atropine and epinephrine lead to iatrogenic KCS, and topical pilocarpine has been advocated for the treatment of KCS. While local penetration of topically applied drugs to the lacrimal gland is a novel concept, which has not been a topic of controlled investigation, the assumption that ophthalmic drugs penetrate to the lacrimal gland is implicit to the recognized xerosis side effect of topical atropine.

## Current Status of Ophthalmic Cyclosporine for Commercial Use

A saturated 2% cyclosporine ointment containing chlorbutanol was tested by Sandoz Pharmaceuticals Inc. in normal human volunteers. Subjects had significant ocular irritation with both the placebo and cyclosporine ointment.[55] A multicenter clinical trial of ophthalmic cyclosporine 1% for the treatment of high risk keratoplasty was completed in Oct. 1992. A pilot trial of 1% cyclosporine ointment with chlorbutanol for the treatment of severe Sjogren's syndrome in human subjects found trends of improved clinical signs and lacrimation in cyclosporine vs placebo control subjects.[56] Anectdotal reports of human patients taking between 0.2-1.0% cyclosporine ophthalmic have suggested topical cyclosporine may also have lacrimomimetic effects in man (personal communications T. Mauger and others). Ocular irritation in patients taking both placebo or cyclosporine was high, but appeared to improve with time.[56] A multicenter clinical trial of 3 doses of cyclosporine for treatment of KCS in people was terminated prior to completion due to a high level of drop outs due to ocular irritation. Intolerance in the KCS study compared to the corneal transplantation study likely has several factors, including; intolerance to the chlorbutanol preservative in KCS patients, inherent ocular sensitivity of KCS patients to any irritant, and an inordinantly high concentration of cyclosporine. The improved tolerance with the length of time using cyclosporine in Sjogren's syndrome patients[56] may suggest that the therapeutic effects of cyclosporine lead to a reduction in ocular irritability. Some KCS patients avoided ocular irritation by using a gradual introduction to ophthalmic cyclosporine[d]. In 1993, ophthalmic cyclosporine was sublicensed to Allergan Pharmaceuticals Inc. for reformulation and development for use to treat human KCS.

## FOOTNOTES

a. The EM result is a personal communication from C.L. Martin
b. Unpublished results, M. D'Souza, R. Kaswan, D. Zhou.
c. Personal communication Keith Green
d. Personal communication T. Mauger

## REFERENCES

1. R.L. Kaswan, C.L. Martin. In: Kirk RW ed. Veterinary Therapy VIII. Philadelphia: WB Saunders, 1983; 550.
2. P.M. Barrett, R.H. Scagliotti, R.E. Merideth, et al. Absolute corneal sensitivity & corneal trigeminal nerve anatomy in normal dogs. Prog Vet Comp Ophthalmol 1991; 1(4):245-254.
3. G.D. Aguirre, L.F. Rubin, C.E. Harvey. Keratoconjunctivitis sicca in dogs. J Am Vet Med Assoc 1971; 158:1566-79.
4. L.F. Rubin, R.K. Lynch, W.S. Stockman. Clinical estimation of lacrimal function in dogs. JAVMA 1965; 147:946.
5. R.L. Kaswan, M.A. Salisbury. A new perspective on canine keratoconjunctivitis sicca, treatment with ophthalmic cyclosporine. Vet Clin N Amer: Sm Anim Pract-Ophthalmol 1990; 20(3):583-613.
6. R.L. Kaswan, M.A. Salisbury, C.D. Lothrup. Interaction of age and gender on occurrence of canine keratoconjunctivitis sicca. Prog Vet Comp Ophthalmol. 1991; 1(2)93-97.
7. J. Sansom, K.C. Barnett. Keratoconjunctivitis sicca in the dog: A review of two hundred cases. J Small Anim Pract 1985; 26:121-31.
8. C.L. Martin, J. Munnell, R. Kaswan. Normal ultrastructure and histochemical characteristics of canine lacrimal glands. Am J Vet Res 1988; 49(9):1566-1572.
9. R.L. Kaswan, C.L. Martin, W.L. Chapman. Keratoconjunctivitis sicca: Histopathologic study of nictitating membrane and lacrimal glands from 28 dogs. J Vet Res 1984; 45(1):112-118.
10. R.L. Kaswan, C.L. Martin, D.L. Dawe. Rheumatoid factor determination in 50 dogs with keratoconjunctivitis sicca. JAVMA 1983; 183(10):1073-1075.
11. R.L. Kaswan, C.L. Martin, D.L. Dawe. Keratoconjunctiitis sicca: Immunological evaluation of 62 canine cases. Am J Vet Res 46:376-83.
12. F.W. Quimby, R.S. Schwartz, T. Positt et al. A disorder in dogs resembling Sjogrens syndrome. Clin Immuno Immunopathol 1979; 12:471-476.
13. C.L. Martin, R.L. Kaswan. Distemper-associated kekratoconjunctivitis sicca. JAAHA 1985; 21344-359.
14. J.W. Ludders, J.E. Heavner. Effect of atropine on tear formation in anesthetized dogs. J Am Vet Med Assoc 1979; 175-585-586.

15. D.H. Slatter. Keratoconjunctivitis sicca in the dog produced by oral phenazopyridine hydrochloride. J Small Anim Pract 1973; 14:749-771.

16. R.V. Morgan, A. Bachrach. Keratoconjunctivitis sicca associated with sulfonamide therapy in dogs. J Am Vet Med Assoc 1985;180:432-434.

17. S. Berger, R. Scagliotti. Quantitative study of the effects of Tribrissen[R] on canine tear production. Vet Pathol 1992; 29(5):479.

18. T. Kern, H.N. Erb. Facial neuropathy in dogs and cats: 95 cases (1975-1983). J Am Vet Med Assoc 1987; 191:1604-1609.

19. D.H. Slatter. Disorders of the lacrimal system. part 1. Deficiency of precorneal tear film. Compend Contin Educ 1980; 10:801-807.

20. R.L. Kaswan, D.H. Zhou, R.J. Fullard. Components in normal dog tears and tears from dogs with KCS treated with cyclosporine. In Press, Invest Ophthalmol Vis Sci Supp. 1993.

21. A.D. Hess, A.H. Esa, P.M. Colonbone. Mechanisms of action of cyclosporine: Effect on cells of the immune system and on subcellular events in T-cell activation. Transplantation Proc 1988; 20:29-40.

22. P.D. Hodgkin, A.J. Hapel, R.M. Johnson et al. Blocking of delivery of the antigen-mediated signal to the nucleus of T cells by cyclosporine. Transplantation 1987; 43:685-691.

23. R.K. Harrison, R.L. Stein. Substrate specificities of the peptidyl proyl cis-trans isomerase activities of cyclophilin and FK-506 binding protein: Evidence of the existence of a family of distinct enzymes. Biochemistry 1990; 29(16):3813-3816.

24. M.K. Rosen, R.F. Standaert, A. Falat, et al. Inhibition of FKBP rotamase activity by immunosuppresent FK506: Twisted amide surrogate. Science 1990; 248:863-866.

25. E.A. Emmel, C.L. Verweij, D.B. Durand, et al. Cyclosporin A specifically inhibits function of nuclear proteins involved in T cell activation. Science 1989; 246:1617-1620.

26. D.H. Russell, R. Kibler, L. Matrisian, et al. Prolactin receptors on human T and B lymphocutes: Antagonism of prolactin binding by cyclosporine. J Immunol 1985; 134:3027-37.

27. D.F. Larsen. Mechanism of action: Antagonism of the prolactin receptor. Prog. Allergy 1986; 32:222-238.

28. W.H. Marks, M.W. Harding, R. Handschumache, et al. The immunochemical distribution of cyclophilin in normal mammalian tissues. Transplantation. 1991; 52(2):340-345.

29. R. Cirillo, M. Tiggiani, L. Siri, et al. Cyclosporin A rapidly inhibits mediator release from human basophils presumably by interacting with cyclophilin. J Immunol 1990; 144: 3891-3897.

30. R.L. Kaswan. Intraocular penetration of topically applliled cyclosporine. Transplantation Proc 1988; 20:650-655.

31. P.A. Keown. Optimizing cyclosporine therapy: Dose, levels and monitoring. Transplantation Proc 1988; 20:382-389.

32. R.L. Kaswan, M.A. Salisbury, D.A. Ward. Spontaneous canine keratoconjunctivitis sicca. A useful model for human keratoconjunctivitis sicca: Treatment with cyclosporine eye drops. Arch Ophthalmol 1989; 107:1210-16.

33. M.A. Salsibury, R.L. Kaswan, D.A. Ward, et al. Topical application of cyclosporine in the management of keratoconjunctivitis sicca in dogs. J Am Anim Hosp Assoc 1990; 26:269-274.

34. R.L. Kaswan, M.A. Salisbury, A.K. Mircheff, Gierow J.P. Lacrimomimetic effects of cyclosporine. Invest Ophthalmol Vis Sci Supp 1990; 31:46.

35. J.F. Baer, R.L. Buschbom, R.E. Weller, et al. Efficacy of topical cyclosporine versus corn oil vehicle in the treatment of keratoconjunctivitis sicca in the dog. Proc Am Col Vet Ophthalmol 1989; 72-80.

36. R.V. Morgan, K.L. Abrams. Topical administration of cyclosporine for treatment of keratoconjunctivitis sicca in dogs. JAVMA 1991; 199(8):1043-1046.

37. D.K. Olivero, M.G. Davidson, R.V. English, et al. Clinical evaluation of 1% cyclosporine for topical treatment of keratoconjunctivitis sicca in dogs. JAVMA 1991; 199(8): 1039-1042.

38. S.C. Pflugfelder, K.R. Wilhemus, M.S. Osato, et al. The autoimmune nature of aqueous tear deficiency. Ophthalmology 1986; 93:1513-17.

39. T. Hikichi, A. Yoshida, K. Tsubota. Lymphocytic infiltration of the conjunctiva and the salivary gland in Sjogren's syndrome. Arch Ophthalmol 1993; 111:21.

40. M. Zierhut, H.J. Thiel, E.G. Weidle, et al. Topical treatment of severe corneal ulcers with cyclosporin A. Graefe's Arch Clin Exp Ophthalmol 1989; 227:30-35.

41. J.T. Liegner, R.W. Yee, J.H. Wild. Topical cyclosporine therapy for ulcerative keratitis associated with rheumatoid arthritis. Am J Ophthalmol 1990; 109:610-612.

42. F. Hoffmann, M. Wiederhold. Local treatment of necrotizing scleritis with cyclosporine A. Cornea 1985/1986; 4:3-7.

43. E.J. Holland, C.C. Chan, T. Kuwabara, et al. Immunohistologic findings and results of treatment with cyclosporine in ligneous conjunctivitis. Am J Ophthalmol 1989; 107:160-166.

44. D. BenEzra, J. Pe'er, M. Brodsky, E. Cohen. Cyclosporine eyedrops for the treatment of severe vernal keratoconjunctivitis. Am J Ophthalmol 1986; 101:278-282.

45. J.H. Bleik, K.F. Tabbara. Topical cyclosporine in vernal keratoconjunctivitis. Ophthalmology 1991; 98(11):1679-1684.

46. P.A. Jackson, R.L. Kaswan, R.E. Merideth, et al. Chronic superficial keratitis in dogs: A placebo controlled trial of topical cyclosporine treatment 1991; 1(4):269-275.

47. D.H. Zhou, S.H. Liu. Successful treatment of experimental autoimmune dacryoadenitis with cyclosporin A. In Ophthalmol Vis Sci Supp 1992; 33(4):845.

48. W.H. Frey, J.D. Nelson, M.L. Frick, et al. Prolactin immunoreactivity in human tears and lacrimal gland: Possible implications for tear production. In HOlly FJ (ed): The Preocular Tear Film in Health and Disease, and Contact Lens Wear. Lubbock, TX Dry Eye Institute, 1986, pp 798-807.

49. D.W. Warren, R.L. Kaswan, R.L. Wood, et al. Prolactin binding and effects on peroxidase release in rat exorbital lacrimal glands. Invest Ophthalmol Supp 1990; 31-540.

50. A.K. Mircheff, D.W. Warren, R.L. Wood, et al. Prolactin localization, binding, and effects on peroxidase release in rat exorbital lacrimal gland. Invest Ophthalmol Vis Sci 1992; 33(3):641-650.

51. A. Azzarolo, K. Bierrum, A.K. Mircheff, et al. Dihydrotestosterone and prolactin reverse lacrimal gland regression after hypophysectomy of female rats. Invest Ophthalmol Vis Sci Supp 1992; 33(4).

52. D.W. Warren, A.M. Azzarolo, L. Becker, et al. Effects and interaction of dihydrotestosterone and prolactin of lacrimal gland function in hypophysectomized female rats. Intern Conf Lacrimal Gland. Bermuda, Nov. 14, 1992; 142.

53. H. Ariga, J. Edwards, D.A. Sullivan. Androgen control of autoimmune expression in lacrimal glands of MRL/Mp-lpr/lpr mice. Clin Immunol Immunopathol 1989; 53: 499-508.

54. A. Bachrach, R. Scagliotti. Topical cyclosporine for KCS. Adv SAM 1988; 1:1-2.

55. S. Solch, P.I. Nadler, M.H. Silverman. Safety and tolerability of 2% Cyclosporine (Sandimmune[R]) ophthalmic ointment in normal volunteers. J Ocular Pharmacol 1991; 7(4):301-312.

56. Laibovitz RA, Solch S, Andriano K, O'Connell M, Siverman MH. Pilot trial of cyclosporine 1% ophthalmic ointment in the treatment of ketoconjunctivitis sicca. Cornea 12(4);315-323:1993.

# TREATMENT OF SEVERE EYE DRYNESS AND PROBLEMATIC EYE LESIONS WITH ENRICHED BOVINE COLOSTRUM LACTOSERUM

C. Chaumeil, S. Liotet, and O. Kogbe

Centre Hospitalier National d'Ophtalmologie des Quinze-Vingts
28 rue de charenton, Paris, France 75012

## INTRODUCTION

In dry eye states, reduced tear film leads to mechanical stress and metabolic asphyxia, which cause serious dystrophic lesions. This consequence of reduced tear secretion is due not only to the lack of an aqueous tear component, but also to a decrease in many other glandular secretory products, including enzymes, immunoglobulins, nutrients, anti-microbial substances, and surface active agents. These changes result in alterations in the tear film pH and osmolarity, as well as ocular surface metabolism, which all add up to cause epithelial cell breakdown and an inability to recover by regeneration.

To manage dry eye syndromes, especially keratoconjunctivitis sicca (KCS), much effort has been directed towards the development of tear replacements, in order to offset the decrease in secretion by the major and accessory lacrimal glands. Such a tear substitute, if it contained some of the missing tear constituents as well as water and ions, could help reduce the ocular surface damage associated with dry eye.

For many years, attention has been drawn to colostrum as a natural, rich source of these essential nutritional components (Table 1). Colostrum lactoserum is prepared from bovine colostrum obtained during the first 3 days after delivery. This preparation is rich in oligo-elements, amino-acids, antibacterial agents like lysozyme and lactotransferrin, cellular growth factors and vitamins.

This preparation might be used in dry eye syndromes to replace, both in volume and function, the diminshed or absent tears. Moreover, given its nutritive action and growth stimulating factors, colostrum lactoserum could be used to hasten scarring in corneal ulcers from various etiologies, especially from metaherpetic ulcers.

The purpose of this study was to examine whether colostrum lactoserum treatment of patients with keratitis or ulcers elicits regenerative and cicatrizing effects on corneo-conjunctival epithelia. The efficacy of treatment was evaluated by different parameters, including the Rose Bengal score, corneal flourescein test and a record of the subjective sensations of improvement.

**Table 1.** Components contained within colostrum lactoserum.

| |
|---|
| - β lactoglobulin |
| - lactalbumin |
| - albumin |
| - IgG and IgA immunoglobulins (20% of protein content) |
| - IgA specific helper factor (αHF), which induces IgA synthesis |
| - lactoferrin, iron-chelator |
| - lysozyme |
| - amino acids |
| - mineral salts |
| - oligo-elements |
| - many water and fat soluble vitamins, including a high level of pantothenic acid and biotin, both of which play an important part in the development of dermal tissues |
| - epithelial growth factor and insulin (100 X serum level) both stimulate growth |
| - prolactin (7 X serum level) with reported activity on lacrimal gland secretion |
| - a powerful antibacterial system: lactoperoxidase-thiocyanate with other less important antibacterials, such as the xanthine oxidase |
| - an activity on cells multiplication: this preparation has been tested by Mme Adolphe (Laboratoire de pharmacologie cellulaire de l'Ecole pratique des Hautes Etudes, Paris) on fibroblasts L929 in culture, a stimulation of the cells multiplication was observed with concentrations of 5, 10, 20 and 40% lactoserum. |

Information was obtained from articles by Tomasi (1972), Pittard and Bill (1979), Ogra and Faden (1981), and Hanson (1982).

## MATERIAL AND METHODS

An open study was carried out to test the benefit of a colostrum lactoserum preparation in the treatment of individuals who had problematic cases of keratitis or ulcers, and whose clinical condition was not helped by currently available therapy. The study involved observing the regenerative and cicatrizing effects of colostrum lactoserum treatment on corneal epithelia and the conjunctiva in 30 patients with keratitis or ulcers. These ocular pathologies were due to dry eye syndromes (n = 11), viral infection (n = 6), and other various causes (n = 13), including post-traumatic (n = 5), Recklinghausen disease (n = 1), Stevens-Johnson syndrome (n = 1), Goldenhar disease (n = 1), degenerative dystrophy (n = 1), follicular conjunctivitis (n = 1), corneal ulcer secondary to a metabolic disorder (n = 1), or unknown etiology (n = 2).

A special preparation of lactoserum from bovine colostrum was formulated into eye drops. This lactoserum was prepared by treating bovine colostrum, after collection within the first 5 days following delivery, to coagulation under rennet after removal of all its

Table 2. Effect of colostrum treatment on patients (n = 6) with viral superficial punctate keratitis (KPS) or ulcer.

| | Clinical Condition | KPS | Ulcer | Evolution Following Previous Treatment | Further treatment | Results |
|---|---|---|---|---|---|---|
| Mme R | Keratoconj, viral ulcer, subepithelial nodules | | yes | persistent problem | colostrum /4 weeks | improvement |
| Mr C | conj. papillofollicular, false membrane | *** | yes | slight improvement, then false membrane | colostrumX6, Rifamycin®, hyaluronidase | healed |
| Mr O | recurrent KPS, dendritic ulcer/monopht. | *** | | KPS, no ulcer | colostrumX3 / 15 days | healed |
| Mr D | KPS, amoeboid ulcer then KPS | *** | | persistent KPS | colX4, Chibrocad.®, Timoptol®, Zovirax ®/20 days | reduced KPS |
| Mr P | herpetic ulcer | | yes | metaherpetic ulcer / corticoid | colX4, Rifamycin®X4 / 20 days | improvement, dendritic scars |
| Mr A | KPS, contact lens wear | *** | | improvement | colX6, Ophtaglobuline®, Rifamycin® | healed |

Table 3. Effect of colostrum treatment on patients (n = 11) with dry eye symptoms.

| | Etiology, Symptoms | Duration | Dryness | Further Treatment | Results |
|---|---|---|---|---|---|
| Mme B | Sjögren's syndrome | 9 years | yes + KPS | colostrum / 3 months | improvement, slight KPS / RE |
| Mme F | KCS, KPS / RLE | 8 years | yes | colostrum, Gellarnes®, vit B12/30 days | seems better, KPS remains impt./ palpebral fissure area |
| Mme A | KCS, diffuse KPS / RLE | 5 years | yes | colostrumX3 and X6 | improvement less dense KPS |
| Mme R | KCS, KPS / RLE | 10 years | yes | colostrum | improvement, persist. KPS |
| Mme R | KCS, KPS / RLE | 4 years | yes | colostrumX4 / 5 months | no KPS, Rose Bengal negative |
| Mme B | KCS, KPS / RLE | 11 years | yes | colostrum, vit B12, vit A | improvement +++ reduced staining with RB, reduced KPS |
| Mme D | KCS, KPS 1/3 inf. / RLE | 3.5 years | yes | colostrum, vit A | improved Schirmer, KPS unchanged |
| Mr K | KCS, monopht. Schirmer=4, RB=3+, BUT=34 | 15 months | yes | colostrum | improvement, Schirmer=13, BUT=5 |
| Mme F | KCS, trophic keratitis post-op. / RE | 5 months | yes +keratitis | colostrumX6 | improvement, relapse |
| Mme D | episcleritis KPS / LE, hyperaemia / RE | 1 month | yes + KPS | colostrum, Rifamycin®, Solucort® | improvement |
| Mme B | KCS, Rhematoid polyarthritis, KPS / RLE | 5 years | yes | colostrum | unchanged |

Table 4. Effect of colostrum treatment on patients (n = 13) with KPS or ulcer from different etiologies.

| | Etiology | Dryness | KPS | Ulcer | Evolution Following Previous Treatment | Further Treatment | Results |
|---|---|---|---|---|---|---|---|
| Mme B | contact lens, post cataract torpid ulcer | | | yes | ulcer since 1 year | colostrum | improvement, deep scars |
| Mme B | Eye dryness, GLV, renal failure tt / EDTA | yes | yes | yes | recurrent | colostrumx6 months | no ulcer, persist. KPS |
| Mme H | chronic ulcer / Stevens Jonhson | | | yes torp. | improvement | colostrum+Tifomycine® / 30 days | slight improvement |
| Mme L | detergent splash | | | yes | slow improvement | colostrumx6,Rifamycin®, padding / 5 days | healed |
| Mme E | trauma / camping gaz 3 years ago | | | yes RLE | recurrent ulcer | colostrumx3, padding | great improvement |
| Mme K | post-op./cataract | | yes superf | | dragging | colostrum | good, no KPS |
| Mr A | post-op. keratitis treat/Timoptol® | | important | | aggravation | colostrum / 1 month | healed |
| Mme M | post-op.[cataract] keratitis | | yes | | dragging | colostrum / 20 days | excellent, healed |
| Mr B | Exposure ulcer post-ptosis operation / Recklinghausen | | yes | yes | Gellarnes®,Antibiotic | colostrumx6, Rifa. padding | improvement |
| Mme T | Goldenhar, palpebral fissure ulcer | yes | yes | yes | | colostrum, alltern. vit A, Lacrygel®, vit B12 | improvement, recurrent KPS |
| Mr V | ??, corneal dysplasia treated / corneal abrasion | | yes | | | colostrum / 1 week | good, scarring |
| Mme T | ??, filamentous keratitis | yes | yes | | artificial tears | colostrum | transient improvement |
| Mr C | contact lenses, OT treated / Timoptol® | yes | yes | | chronic KPS / antibiotics, vitA, Keratyl® | colostrum, Chibroxine® | improvement then lowered VA and KPS recurrent following each treat. interruption |

cream. As most of the fat soluble vitamins would have been removed with the cream and by caseine coagulation, the residual lactoserum was enriched by adding vitamin A.

## RESULTS

Application of colostrum lactoserum to the eyes of patients with virus-induced superficial punctate keratitis (KPS) or ulcer resulted in improvement in all cases (Table 2). Similarly, clinical improvement was observed in 10 out of 11 patients with dry eye symptoms after administration of the colostrum lactoserum for varying time periods (Table 3). In the dry eye group, though, a relapse followed improvement because of the persistence of the etiological problem (i.e. post operative scar). In addition, colostrum lactoserum treatment had no effect in one patient with rheumatoid polyarthritis and KPS.

Colostrum lactoserum therapy was also quite effective in patients with KPS or ulcer from different etiologies (Table 4). The only therapeutic failure was observed in a case of filamentous keratitis, wherein the etiology of disease was neither established nor effectively treated. In this latter situation, improvement following colostrum lactoserum administration was only transient, because the cause of the ulcer could not be or had not been treated.

## CONCLUSION

Our results show that the use of colostrum lactoserum in the treatment of KPS and corneal ulcers of viral origin, including metaherpetic ulcers, gave good functional and clinical results. Similar beneficial results were also observed in patients with post-traumatic ulcers.

For ulcers present in corneal and conjuctival infectious states, colostrum brought rapid improvement. This effect, though, was only transient when the cause of the ulcer had not been treated. The same transient improvement was observed in palpebral fissure ulcers provoked by exposure, because the etiological problem remained unresolved.

For the dry eye syndrome cases, 10 out of 11 patients reported an improvement in their sensation of dryness, and had an improvement in clinical signs.

This was not a double-blind study, and the results need to be confirmed by evaluating a larger number of patients.

## REFERENCES

Hanson, L.A., 1982, The mammary gland as an immunological organ, *Immunol. Today* 3:168.
Ogra P.L., and Faden H., 1981, Breast milk as an immunologic vehicle for transport of immunocompetence, *in:* "Textbook of Gastroenterology and Nutrition in Infancy," Lebenthal, ed., Raven Press, New York, p. 355.
Pittard W.B., and Bill K., 1979, Immunoregulation by breast milk cells, *Cell. Immunol.* 42:437.
Tomasi T.B., 1972, Secretory immunoglobulins, *New Engl. J. Med.* 287:500.

# THE EFFICIENCY OF SHORT-WAVE DIATHERMY
# AND LASER STIMULATION OF THE LACRIMAL GLAND
# IN THE TREATMENT OF DRY EYE SYNDROME

Tadeusz Kęcik, Iwona Świtka-Więcławska,
Ligia Portacha, and Joanna Ciszewska

Department of Ophthalmology
Medical University
Warsaw - Poland

## INTRODUCTION

The decrease of tear flow connected with tear film dysfunction leads to changes known as the dry eye syndrome, which is the cause of many undesirable symptoms. The patients complain of burning, photophobia and a feeling of dry eye. Clinical signs of reduction of tear flow may be seen as chronic conjunctivitis. Conjunctivitis is often followed by a defect of corneal epithelium and symptoms of keratitis filiformis, which may lead to severe complications such as corneal ulcerations.

The diagnosis of dry eye syndrome is based on the decreased tear production (Schirmer test), decreased tear break up time and biochemical changes in the examined tears. The treatment of this syndrome presents a problem. Topical drops which moisten the cornea need frequent and constant administration. Mucolitics and cholinergics give temporary relief but may cause undesirable side effects. Application of both of those methods is not always effective.

In search of other modes of therapies, laser stimulation and short-wave diathermy to the lacrimal glands have been applied in the Department of Ophthalmology in Warsaw.

## METHODS

10 patients ( 20 eyes ) were treated. In order to compare results of both methods the same patients were treated with the two therapies.

The interval between laser stimulation and short-wave diathermy ranged from 3 to 6 months. All patients had a decrease in tear secretion ranging from 1 to 6mm and symptoms such as dry eyes, prickling, burning, and foreign body sensation. None of these patients presented with pathological changes in the cornea.

All patients have been treated before with moistening drops (e.g. methyl-cellulose, artificial tears) and with systemic drugs (e.g. prostygmin and flegamin). After each course of treatment a Schirmer test I and break up time were performed.

For laser stimulation, the red laser (633nm) was used. Treatment consisted of 10 daily courses. The laser beam was directed through closed eye lids to the surface of 3cm of the upper temple of the orbit. Exposure time was -5min, the power 0.1-0.15W.

For short-wave diathermy, 10 courses of treatment were applied to the same patients. Electrods were placed about 5cm from the lacrimal gland region. The power given ranged from 30-40V, the time exposure 15min.

## RESULTS

The results are presented on figures (fig. 1, 2). The analysis of the Schirmer test performed before and directly after treatment showed that tear secretion varied.

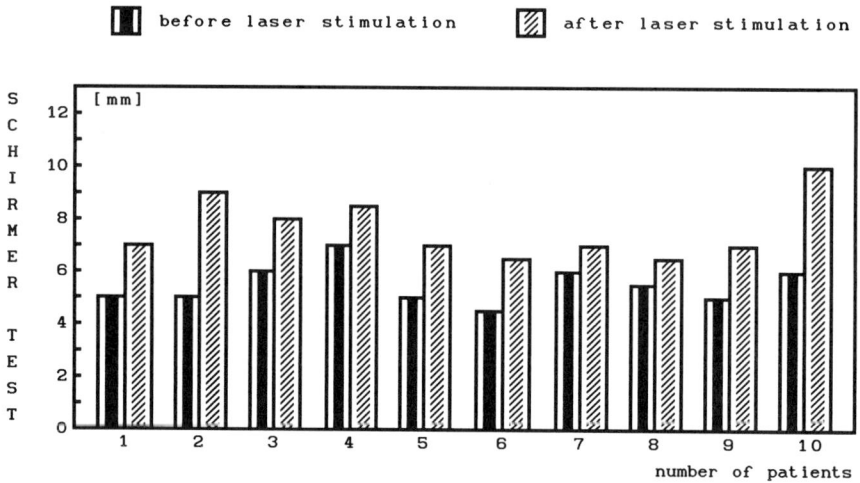

Figure 1. The average tear secretion

However, the comparison of tear secretion before and 10 days after laser stimulation showed an increase in tear flow from 2 to 6mm (mean 3mm) and in case of short-wave diathermy from 2 to 7mm (mean 4mm). The break up time increased from 3.7 to 10.6 after laser stimulation and from 3.2 to 10.3 after diathermy.
After completion of treatment with both methods all patients had a subjective improvement. The symptoms subsided, the feeling of wet eye returned. The efficiency of therapies lasted from 3 weeks to 6 months.

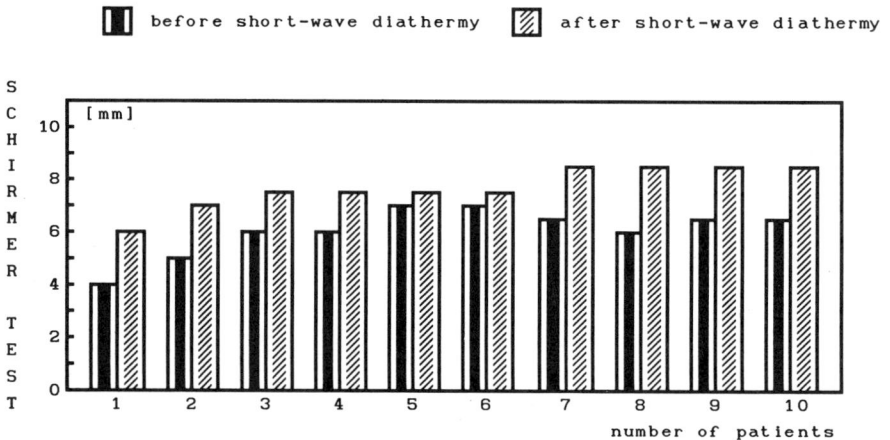

Figure 2. The average tear secretion

## DISCUSSION

The stimulating effects of low dose laser radiation especially it's red and infrared spectrum is useful in many diseases.[8,10] The low and moderate power of laser radiation causes specific chemical and metabolic reactions in the cells known as biostimulation. The tissue absorption of such energy causes biochemical, bioelectric and bioenergic effects.

The effective biochemical stimulation is probably based on the process of restoring the biological balance of the cells by increasing ATP in the mitochondria, increasing metabolism and number of mitotic figures. The bioelectric effect causes normalisation of the cell membrane potential. A normal cell has more negative charges and it's potential ranges from -60 to -80mV. In pathologic changes, due to penetration of positive ions (sodium) into the cells, the cell's potential decreases. To reverse this process the cells need energy which can be obtained from the hydrolysis of ATP, which takes place during laser stimulation. The bioenergetic effect is a factor stimulating nutrition and cell growth. It also regulates cellular processes. These effects, called primary, found in irradiated tissues, generate secondary effects presents also in surrounding tissues. These are analgetic, antiinflamatory and biostimulating effects.[1,2,3,4,9] Working through the biostimulating mechanism the laser energy improves blood circulation, nutrition and cell regeneration. In addition, by hyperpolarizing the cell membrane, the low dose energy changes the sensitivity threshold of the nerve fibers (sensory).[2,5] It is difficult to define the dose of energy which directly reaches the lacrimal gland. Passing through skin, and bone structures of the orbit the beam is reflected, scattered and partially absorbed. The increase in tear secretion directly after stimulation could be caused not only by the laser beam on lacrimal glands but also through a thermal effect. The increase of tear production 10 days after treatment indicates the effect of the laser beam on the glands. The biological effect of short-wave diathermy is based on the action of heat. It causes blood vessel dilatation, increased blood circulation and cell metabolism.[7,10] The increase in tear secretion directly after diathermy is caused by improved blood perfusion and after 10 days by improved metabolism of the lacrimal glands.

## CONCLUSIONS

1. The clinical results obtained with both methods are satisfactory but temporary.
2. Tear secretion increases faster after short-wave diathermy.
3. Short-wave diathermy or laser stimulation of lacrimal gland can be used as an adjunct to classic therapy in dry eye syndrome.

## REFERENCES

1. G. J. Abstern, and S. Jofte, Lasers in medicine: 14-22(1985)
2. V. P. Bersnev, I. V. Yakovenko, The effects of low-power laser irradiation on partially injured nerves. International Conference, Mediolan June(1992)
3. B. S. Briskin, A. K. Polonsky, I. M. Alien, Surgical diseases laser therapy -International Conference, Mediolan, June, 1992
4. J. Colls, Laser therapy today-International Conference, Barcelona(1986)
5. D. W. Jegorow, F. Liens, Materiały konferencji "Primienienie mietodow i sriedstw łaziernoj techniki w biologii i medicinie": 24-25(1981)
6. L. A. Linnik, Łaziernaja tierapia w oftalmologii. Oftalmologiczeskij Zurnal 8,280: 451-454(1985)
7. T. Mikka, Fizykoterapia, PZWL: 222-232(1973)

8. G.A.Peyman,M.Raichand,R.Zeimer,Ocular effects of various laser wave-
   lengths,Survey of Ophthalmology 5,28:391-404(1984)
9. L.Pokora,Lasery w stomatologii:39-43(1992)
10. P.Richand,J.L.Boulnois,The use of laser beams in medical therapy,
    Minerva Medica,77:1675-1682(1983)
11. G.Straburzyński,Fizjoterapia,PZWL:308-322(1988)

# EFFECT OF COLLAGEN PUNCTAL OCCLUSION ON TEAR STABILITY AND VOLUME

Sudi Patel[1] and David Grierson[2]

[1]Department of Vision Sciences
Glasgow Caledonian University
Cowcaddens Rd, Glasgow
Scotland U.K. G4 OBA
[2]Eye Department
Royal Infirmary, Alexandra Parade, Glasgow
Scotland U.K. G31 2ER

## INTRODUCTION

When the patient has dry eye symptoms, the clinician can apply several techniques to, evaluate tear status, the likely cause of symptoms, decide on appropriate therapy and monitor the progress and reaction of the eye. When the dry eye is the result of insufficient lacrimal output coupled with an abnormally high evaporation rate, the clinician can attempt to retain moisture at the ocular surface by blocking the lacrimal canaliculi. In this endeavour, occlusion of the inferior and/or superior punctum lacrimale has been applied for many years [1,2]. Permanent occlusion can be produced by cautery. However, reversible occlusion is preferable and this can be achieved by inserting either a plastic or silicone plug of appropriate design. More useful are intracanalicular implants manufactured out of processed collagen which have the property of self-dissolution. They can be used as temporary 'plugs' which will help the clinician decide whether or not a more permanent occlusion is the better option. In the literature there are several anecdotal reports praising the punctal plug and reporting on patient satisfaction. There are very few studies reporting on objective changes in the tear film properties as a consequence of punctal or canalicular occlusion.

In this study our intention was to monitor the stability of the tear film and volume of the lower tear meniscus prior to, and during a 7 day period of temporary intracanalicular occlusion in a small group of dry eye subjects.

## METHODS

Tear stability was measured using the non-invasive method of Tear Thinning Time (TTT.) measurement, using a standard Bausch and Lomb keratometer[3]. Tear volume was inferred by measuring the height of the lower tear meniscus (TMH). This was estimated by photographing the lower tear meniscus using a standard slit lamp. After fixing and developing, the film was magnified using a C.C.T.V. magnifier. The height of the lower tear meniscus was measured at the junction between the lower eyelid and the 6 o' clock point at the corneal limbus. The true magnification of the TMH was 50 times. The system was used after initial calibration.

Three 'dry eye' subjects (6 eyes) were enrolled. All had at least 3 of the following: Schirmer result less than 10mm/5mins, mucin filaments, +ve subjective symptoms (McMonnies questionnaire[4]), Rose Bengal staining > 3.5 (Van Bijsterveld system).

Subjects were asked to cease any existing therapy for 48hrs prior to the investigation. TTT and TMH was measured prior to plug insertion. After insertion, measurements were repeated at 2 hrs, 24 hrs and 7 days. The 0.4mm diameter intracanalicular collagen implants (supplied by Eagle Vision, Memphis, TE) were inserted into both upper and lower puncta of each eye.

## RESULTS

The results obtained from the 6 eyes are shown graphically in Figures 1 to 6.

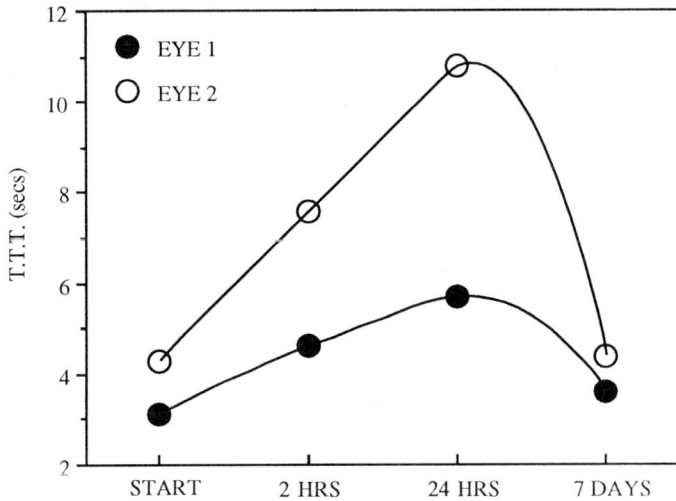

**Figure 1.** Tear stability changes (TTT) in eyes 1 & 2

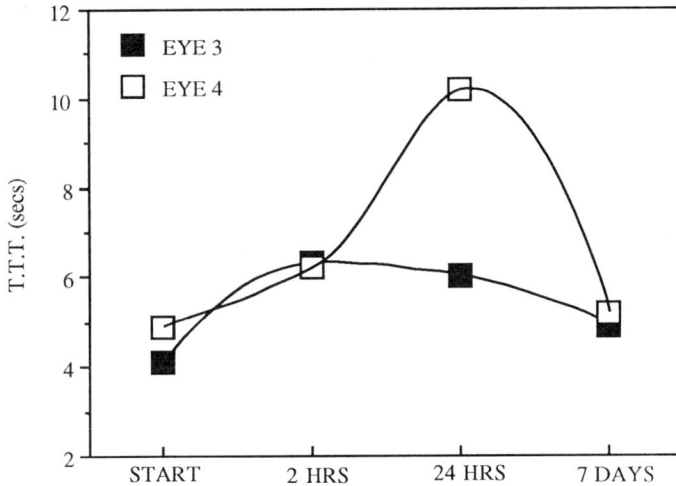

**Figure 2.** Tear stability (TTT) changes in eyes 3 & 4

606

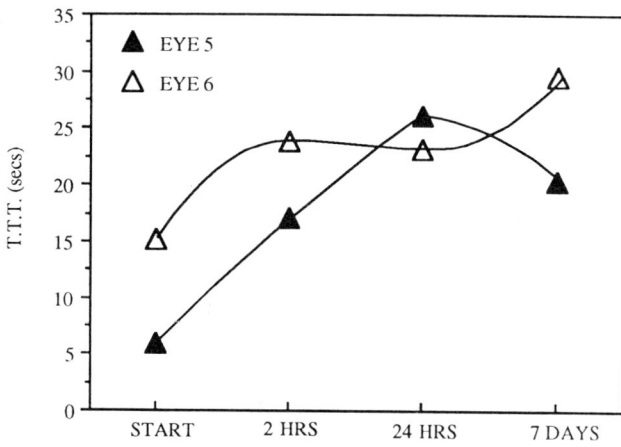

**Figure 3.** Tear stability (TTT) changes in eyes 5 & 6

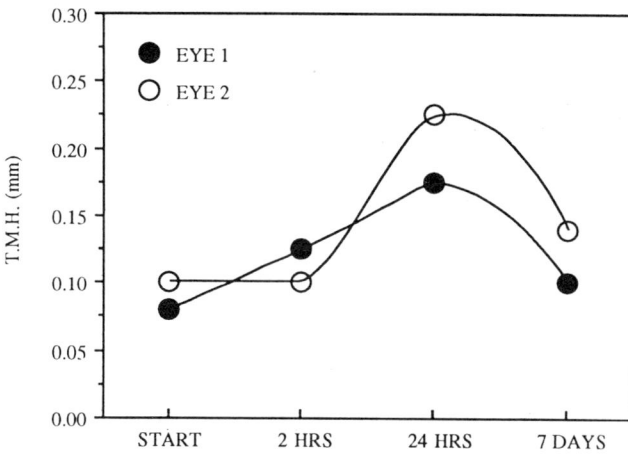

**Figure 4.** Tear meniscus height (TMH) changes in eyes 1 & 2

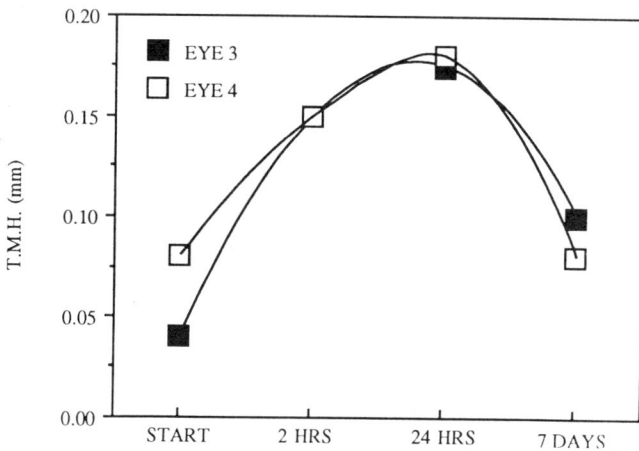

**Figure 5.** Tear meniscus height changes in eyes 3 & 4

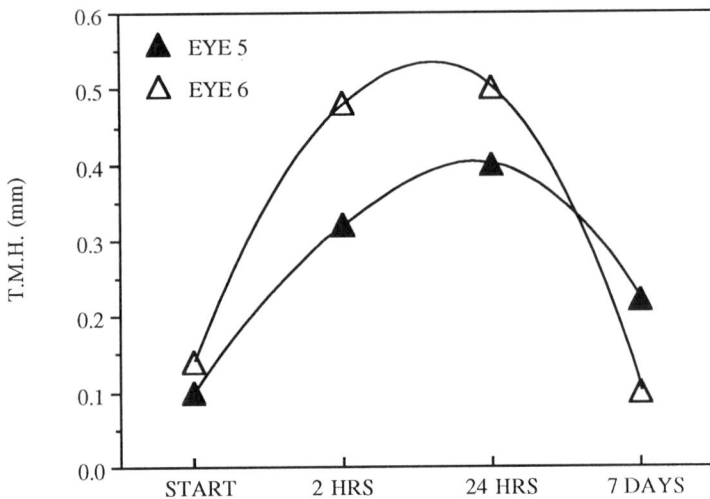

**Figure 6.** Tear meniscus height (TMH) changes in eyes 5 & 6

## DISCUSSION

Viewing Figures 1 to 6, clearly, punctal occlusion increases the stability of the pre-corneal tear film and increases the height of the lower tear meniscus. Initial mean TTT values were all below the dry eye population average of 7secs[5], except in Eye 6. In all cases an immediate improvement in TTT was detected after plug insertion reaching a peak at 24 hours. If measurements had been taken on a daily basis over the period, we could have been more specific with regard to the time gap between insertion and peak effect. On the 7th day, TTT returned to base level in all eyes except 5 and 6. By the 7th day the implant had dissolved in all eyes except 5 and 6 in which the overall performances are suggestive of a borderline, variable dry eye which may not benefit from a more permanent occlusion. The averaged standard deviation in TTT measurement was +/- 4 secs.

TTT rise parallels TMH rise suggesting a relationship between TTT and TMH. A relationship of this type has not been detected within normals[6]. In all cases initial TMH was below the normal population mean[7], however, after 24 hours all eyes had a TMH within the normal range. All subjects commented on a relief from symptoms and felt that during the period, they did not require 'topping up' with drops.

## CONCLUSION

We conclude, tentatively, this form of treatment produces an objective improvement in tear properties in dry eyes. Temporary occlusion can be used to assess the value of more permanent occlusion and differentiate the borderline from the more definite dry eye. Certainly there is a need for more controlled study using larger subject numbers before making a final decision on the value of this treatment.

## REFERENCES

1. C.H. Dohlman, Punctal occlusion in KCS. *Ophthalmology.* 85: 1277-1281 (1978).
2. R.M. Willis, R. Folberg, J.H. Krachmer, E.J. Holland, Treatment of aqueous deficient dry eye with removable punctum plugs, *Ophthalmology.* 94: 514-518 (1987).
3. S. Patel, J.C.Farrell, Age related changes in precorneal tear film stability.*Am.J Optom & Physiol. Opt.* 66: 175-178 (1989).
4. C.W. McMonnies, Key questions in a dry eye history. *J.Am. Optom. Assoc.* 57: 512-517 (1986).
5. J.C. Farrell, D.J. Grierson, S. Patel, R.D. Sturrock, A classification for dry eyes following comparison of tear thinning time with Schirmer test. *Acta Ophthalmol.* 70: 357-360 (1992).
6. S.Patel, M.J.A. Port, Tear characteristics of the VDU operator.*Optom.& Vis. Sci.* 68: 798-800 (1991).
7. M.J.A. Port, T.S. Asaria, Assessment of human tear volume. *J. Brit. Cont. Lens. Assoc.* 13:76-82 (1990).

# SJÖGREN'S SYNDROME: IMMUNOLOGIC AND NEUROENDOCRINE MECHANISMS

Robert I. Fox[1] and Ichiro Saito[2]

[1]Department of Rheumatology
Scripps Clinic and Medical Foundation
10666 North Torrey Pines Road
La Jolla, California 92037

[2]Medical Research Institute
Tokyo Medical and Dental University
Tokyo, Japan

## OVERVIEW

Sjögren's syndrome (SS), an autoimmune disorder characterized by decreased lacrimal and salivary gland function, can exist as either a primary condition (1° SS) or in association with other autoimmune disorders (i.e. secondary SS, 2° SS). These patients have lymphocytic infiltrates of their salivary and lacrimal glands, resulting in glandular destruction and interference with the neural regulation of glandular function. The lymphocytes infiltrating the salivary glands are predominantly CD4+ T-cells, while those infiltrating the lacrimal gland exhibit a high proportion of B-cells. Cytokines produced in the SS glands include interleukin-1, interleukin-2, interleukin-6, interleukin-10, interferon-$\gamma$, and tumor necrosis factor-$\alpha$. Both CD4+ T-cells and glandular epithelial cells are sources of transcription of these cytokine mRNA's, indicating that both lymphocytes and "target cells" are co-participants in the inflammatory process.

SS patients have a characteristic pattern of autoantibodies including rheumatoid factor (IgM anti-IgG Fc) and antibodies against nuclear antigens including the ribonuclear proteins

*Lacrimal Gland, Tear Film, and Dry Eye Syndromes*
Edited by D.A. Sullivan, Plenum Press, New York, 1994

SS-A (Ro) and SS B (La). Although the role of these antibodies in salivary and lacrimal glandular destruction remains unclear, it is likely that extraglandular manifestations such as vasculitis are closely linked to immune complex formation involving autoantibodies and complement activation.

Both genetic and environmental factors play a role in pathogenesis. Genetic factors include both major histocompatibility complex class (MHC) II antigens and non-MHC genes. In Caucasians, the extended haplotype HLA-DR3-DR52a-DQA4-DQB3 is found in increased frequency in 1° SS patients and in the subset of systemic lupus erythematosus (SLE) patients with 2° SS. However, different MHC alleles are found in Chinese and Japanese 1° SS patients. The environmental factors responsible for SS remain unknown, but indirect roles for Epstein-Barr Virus (EBV) and perhaps for retroviruses have been suggested. Animal models for SS have been proposed but still lack many of the pathogenetic features found in human disease. Recently, implantation of human salivary gland under the kidney capsule of immunodeficient (SCID) mice may provide a model for studying SS in a pre-clinical model.

## DEFINITION OF SJÖGREN'S SYNDROME

There remains a great deal of debate about the criteria for diagnosis of SS. Symptoms of dryness are very common in the general population, particularly in older patients, and the diagnosis of SS should be reserved for those with salivary/lacrimal gland dysfunction as the result of a systemic autoimmune process. The criteria used at our clinic for the diagnosis of SS listed are listed Table 1; they emphasize the need for objective demonstration of dry eyes and dry mouth on physical examination, the presence of characteristic autoantibodies (such as anti-nuclear antibodies {ANA} and rheumatoid factor {RF}), and lymphocytic infiltrates on minor salivary gland biopsy.

SS is divided in to 1° SS and 2° SS on clinical grounds, since the latter condition is associated with well defined diseases such as rheumatoid arthritis (RA), SLE, primary biliary cirrhosis (PBC), polymyositis or systemic sclerosis (SSC). The clinical distinction between 1° SS and the subset of SLE with 2° SS is quite difficult, since both conditions share clinical, immunogenetic and laboratory features. Exclusions to the diagnosis of 1° SS include pre-existent lymphoma, chronic infections including tuberculosis and HIV infection, sarcoidosis, metabolic conditions (i.e. fatty infiltration associated with hypercholesterolemia, hypertriglyceridemia, or chronic alcoholism), amyloidosis, and graft-versus host disease after bone marrow transplantation.

The differential diagnosis of the patient with dryness includes a wide spectrum of infectious, metabolic, inflammatory and behavioral (i.e. anxiety, depression) conditions. The absence of autoantibodies (ANA, SS-A, RF) in a patient with salivary gland swelling should make the clinician consider a disease process other than SS, particularly a tumor or retroviral infection. In the patient lacking autoantibodies and lacking salivary gland swelling, a minor salivary gland biopsy may help determine whether a diffuse infiltrative process plays a role in the sicca symptoms. In many older patients, decreased salivary and lacrimal gland function may occur in the absence of lymphoid infiltrates; this presumably reflects an imbalance in the normal regulation of glandular secretion under the control of the autonomic neuro-endocrine system. Factors including increasing age-related glandular atrophy, mucus inspissation of the glandular ducts, anxiety and depression (that alter cholinergic input to the gland) and a wide spectrum of medications with anti-cholinergic activity can all contribute to symptomatic dryness.

Table 1. Criteria for diagnosis of primary and secondary Sjögren's syndrome

---

Primary Sjögren's syndrome

    Symptoms and objective signs of ocular dryness

        Schirmer test less than 8 mm wetting per 5 minutes
        Positive rose bengal or fluorescein staining of cornea and conjunctiva to demonstrate
            keratoconjunctivitis sicca
    Symptoms and objective signs of dry mouth
        Decreased parotid flow rate using Lashley cups or other methods
        Abnormal biopsy of minor salivary gland (focus score of ≥ 2 based on average of 4 evaluable lobules
    Evidence of a systemic autoimmune disorder
        Elevated rheumatoid factor ≥ 1:320
        Elevated antinuclear antibody ≥ 1:320
        Presence of anti-SS-A (Ro) or anti-SS-B (La) antibodies

Secondary Sjögren's syndrome
    Characteristic signs and symptoms of SS (described above) plus clinical features sufficient to allow a
        diagnosis of RA, SLE, polymyositis or scleroderma

Exclusions: sarcoidosis, pre-existent lymphoma, acquired immunodeficiency disease and other known causes
        of keratitis sicca or salivary gland enlargement

---

# GENETIC FACTORS IN SS

Determination of genetic factors in SS may be important for several reasons. First the alleles of the class II region of the MHC encode cell surface proteins (i.e. the HLA-DR, -DQ and -DP α and β chain molecules) that "present" endogenous and exogenous peptide antigens to CD4+ lymphocytes [1]. Antigenic peptides physically associate with the α and β chains of these class II MHC molecules and this complex is recognized by CD4+ T-cells. The ability of specific MHC molecules to bind (or not bind) a specific autoantigenic peptides is likely to be the structural basis for the MHC association with autoimmune disease. Second, MHC class II alleles are induced on the salivary (SG) and lacrimal gland (LG) epithelial cells in SS patients, in contrast to the absence of these antigens on normal glandular epithelial cells. Thus, the induction of class II MHC proteins on the epithelial cells allows for direct "communication" between the CD4+ T-cells and the target organs (i.e. lacrimal and salivary gland epithelial cells) in SS. Third, recognition of genes (MHC and non MHC) associated with SS may provide a more rationale basis to subgroup patients who share a common etiopathogenesis and facilitate the search for the environmental co-factors that may differ in each subgroup.

In Caucasian 1° SS patients, the extended haplotype (HLA-DR3-DR52a-DQA4-DQB3) was strongly associated with clinical sicca symptoms and with the presence of autoantibodies against the SS-A and SS-B ribonuclear proteins [2]. However in Japanese and Chinese 1° SS patients, the presence of sicca symptoms and autoantibodies against these nuclear antigens was associated with a different extended haplotypes [3]. These results demonstrate that no single MHC gene is required for pathogenesis in 1° SS. Comparison of the DNA sequence of MHC alleles associated with SS in different ethnic groups indicated a common sequence of HLA-DQB; the common amino acid sequence was in the third hypervariable region in a location that would be predicted to bind antigen. However, further studies will be required to determine whether this region is critical in binding a pathogenetic peptide or whether this shared sequence is merely coincidental.

In contrast to 1° SS, Caucasion patients with RA and sicca symptoms (i.e. 2° SS plus RA) have a different genetic predisposition (i.e. HLA-DR4), a different pattern of autoantibodies (generally anti-SS A and anti-SS B antibodies are absent) and different clinical picture (i. e., eye dryness in excess of mouth dryness). Similarly, Caucasian patients with scleroderma and sicca symptoms ( i.e. 2° SS plus PSS) exhibit a different HLA class II association, a different pattern of autoantibodies (i.e. anti-topoisomerase and anti-centromere proteins cen A and cen B), and different histologic features in their minor SG biopsy (i.e. scarring of the glands in contrast to the persistent lymphoid infiltrates in 1° SS biopsy). Finally, patients with sicca symptoms and biliary cirrhosis (i.e. 2° SS plus PBC) differ from 1° SS in their pattern of autoantibodies (i.e. the presence of anti-mitochondrial antibodies ), and HLA-class II associations. The relative lack of liver involvement in most 1° SS patients suggests the need to look for hepatitis viruses or drug toxicity when abnormal liver functions are noted in 1° SS patients. Although some of these patients with 2° SS express the same antibody profiles as found in 1° SS patients, this relatively infrequent situation probably represents the co-existence of two relatively uncommon conditions in the same patient.

Family studies in patients with SS (and in studies of large kindreds with other autoimmune diseases) have indicated the importance of genes other than MHC. In animal models of autoimmunity where genetic factors have been most thoroughly studied, the occurrence of autoimmunity is multigenic. In both man and mouse, at least 4 or 5 distinct genetic loci must be involved to give the observed incidence of disease inheritance. These "risk factor" genes include sex linked immune response genes, hormone response genes, genes governing fluctuation in the pituitary-adrenal axis, the immunoglobulin and T-cell antigen variable regions, and genes controlling preprogrammed cell death (apoptosis). Each of these genes (when present) serves as an "acceleration" factor and emphasizes that development of autoimmunity is a multi-step phenomena. None of the genes serves as a "necessary" or "sufficient" factor for disease pathogenesis, but combinations of genes predispose to pathogenesis. Thus, it is not surprising that a variety of different MHC and non-MHC genes each serve to increase the risk of developing SS when the individual is challenged by a particular environmental antigen.

## HISTOLOGY AND IMMUNOHISTOLOGY OF SJÖGREN'S SYNDROME

The characteristic abnormality on biopsy of lacrimal or salivary glands is the presence of lymphoid infiltrates that begin around the capillary blood vessel in the center of the glandular lobule. The central vessel changes from an ordinary capillary into a "high endothelial venule" (HEV) with increased columnar shaped endothelial cells [4]. This process is associated with altered histochemical staining , induction of specific cell surface adhesive molecules that facilitate entry of lymphoid cells into the glands (such as ICAMS, VCAMS, and selectins) and transcription of proto-oncogenes such as c-myc [5]. An important immunohistologic feature of the SS salivary gland is the presence of HLA-class II antigens on the glandular epithelial cells [12,13], since these antigens are absent on normal acinar cells and are induced by proinflammatory cytokines such as interferon-$\gamma$ and tumor necrosis factor-$\alpha$. Thus, the pathogenetic process involves not only the infiltration of lymphocytes into the glands, but also activation of specific genes in the blood vessels and epithelial cells that may contribute to pathogenesis.

The majority of lymphoid cells in the salivary biopsy are CD4+ T-cells, although a smaller proportion of CD 8+ T-cells are detectable [14]. These T-cells express the ab antigen receptor and cell surface antigens associated with mature (CD45-) "memory" cells [15]. Recent studies suggest that a preferential use of specific variable region segments of the variable region of the antigen receptor $\beta$ chain may be utilized in salivary gland T-cells of Japanese SS patients [16]. Of interest, the tissue CD4+ T-cells have an increased expression of cell surface markers associated with adhesive receptors ICAM-1 and LFA-1 [17]. Surprisingly few (if any) natural killer cells or monocytes were detected in the salivary glands even though these cells were frequently present in the same patient's blood [18].

At the level of routine microscopy, lymphocytic infiltrates involve a majority of salivary gland lobules and all methods to evaluate lip biopsies must report the average score based on at least 4 evaluable lobules [6]. The evaluation of a single lobule with dense lymphoid infitrates (while ignoring that the majority of other lobules are spared lymphocytic infiltration) is a common cause in the misdiagnosis of SS. Electron microscopic studies of the SS gland demonstrate the absence of immune complexes at the basement membranes of the vascular endothelial cells and around glandular epithelial cells [7]. Also, CD4+ T-cells are in direct contact with the glandular epithelial cells and may be directly mediating the destruction of the glandular cells (Figure 1). Although "cytotoxic" cells are generally thought to exhibit CD8+ phenotype, activated CD4+ T-cells can be induced to exhibit "killer function" and are recognized by their content of granzyme A and perforin (two enzymes that mediate the killing process). Recently, we (Alpert, Weissman and Fox, manuscript in preparation) and others (Saito and Miyasaka, personal communication) have demonstrated lymphocytes containing granzyme A in salivary gland biopsies, suggesting that activated CD4+ T-cells containing this enzyme may destroy epithelial cells expressing a specific antigen on the cell surface presented by class II molecules.

The histologic changes in the SS salivary gland differ subtly from those in the same patient's lacrimal glands. Key differences include the predominance of small "germinal center" like clusters of B-cells in the lacrimal gland [9] while these are not seen in minor salivary gland biopsies [10]. The function of germinal centers in the SS lacrimal gland is likely to be similar to the function of germinal structures in normal lymph nodes, namely regions where immature B-cells undergo antigen driven cell activation, isotypic switching of their immunoglobulin heavy chain, and somatic mutations of their variable regions. In order to form "germinal centers" in an SS lacrimal gland, follicular type dendritic cells (derived from bone marrow monocytic precursors) must migrate from their bone marrow origin, traverse the blood stream, exit via the central HEV of the lacrimal gland and provide an architectural scaffolding for subsequent B-cell lymphoid infiltrates [11]. In contrast, the SS minor salivary gland biopsy does not show "germinal centers" but rather has predominately CD4+ cells in a pattern more similar to the "interfollicular" T-cell zone of normal lymph nodes. Thus, "T-cell" interfollicular dendritic cells (that are antigenically and functionally distinct from follicular dendritic cells) provide the "architectural" scaffolding for the subsequent lymphoid infiltrates in the SS salivary gland [10].

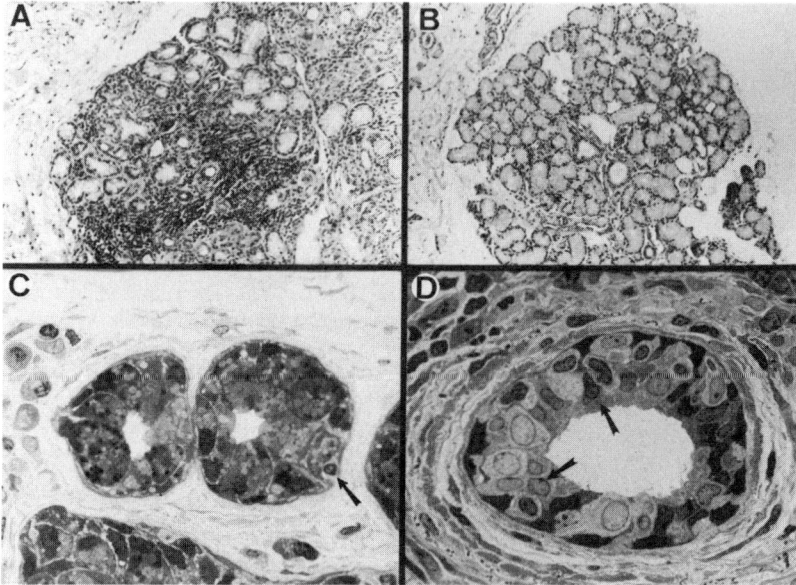

**Figure 1.** Salivary gland biopsy in Sjögren's syndrome. Panels A and B show labial salivary gland biopsies from a patient with primary Sjögren's syndrome and from a patient lacking autoimmune disease, respectively, at low power magnification. Panels C and D show higher magnification of the Sjögren's biopsy to demonstrate increasing numbers of lymphocytes (arrows) adjacent to the glandular and ductal epithelial cells.

Functional studies have shown that salivary gland lymphocytes can produce autoantibodies in vitro after mitogen stimulation [14,19], although little IgG antibody was produced in the absence of mitogenic stimulation [14,20]. When the CD4+ cells are eluted directly from the salivary gland biopsies, the predominant cytokine mRNAs that can be detected are IL-2, IL-10 and interferon-γ (Kang, Fox - manuscript in preparation). Surprisingly little IL-4 or IL-5 mRNA was detected when the cells were first eluted, although these mRNAs could be stimulated by mitogens in vitro; these results are consistent with previous findings of relatively little IgG secretion in the absence of mitogen stimulation in vitro (Kang, Fox-manuscript in preparation).

*In vitro* functional studies on lymphocytes eluted from the SS salivary gland indicate the ability to produce IL-2 [21] and IFN-γ [18]. It is paradoxical that peripheral blood lymphocytes from the same SS patients may be relatively deficient in IL-2 production, when the SS salivary glands can produce IL-2 in large amounts. When mixtures of SS salivary gland and peripheral blood lymphocytes from the same patient were examined, the peripheral blood lymphocytes "suppressed" the IL-2 production by the salivary gland lymphocytes [21]. One interpretation of this data is that cells responsible for down regulating the immune response (including NK like cells) are unable to effectively "home" to the inflamed salivary gland and thus help control the autoimmune response [21]. The production of IFN-γ may play an important role in the induction of HLA-DR or DQ molecules on the salivary gland epithelial cells, a process that allows the target organ (i.e. the epithelial cell) to present antigenic peptides to the CD4+ T-cell in the gland [18].

It is also of interest that SS patients lack anti-salivary gland antibodies [8]. This is in contrast to other autoimmune diseases such as thyroiditis, myasthenia gravis and type I diabetes mellitus where antibodies against the target organ antigens can be demonstrated in sera and in the target organ. Although the role of autoantibodies in glandular destruction remains unclear, it is likely that autoantibodies and complement activation play a role in some extraglandular manifestations such as vasculitic rashes [7].

## NEUROENDOCRINE INNERVATION OF THE SALIVARY GLAND

Neural innervation influences salivary gland function by regulating the rate and types of secretion and represents a major (but poorly understood) factor in the maintenance of glandular integrity. In addition, neural innervation regulates pain perception, vascular blood flow and contributes to inflammation. In rats, treatment with anti-cholinergic drugs leads not only to decreased volume of secretions but also to a surprising induction of new protein synthesis. Due to the complexity of neurovascular and neuroendocrine interactions, it has been difficult to clearly assess the role of neural innervation in SS. However, an interesting feature of salivary gland biopsies from SS patients is that the lobules often show residual glandular epithelial cells that appear grossly intact, even in patients whose eyes and mouth are remarkably dry [14]. An example is shown in Figure 1, frame A, where a biopsy from a SS patient reveals a dense lymphoid infiltrate in the center of the lobule but apparently normal acinar structures at the periphery of the lobule. This implies that dryness is not solely a result of glandular destruction, but that the residual acinar cells are not being fully utilized. Since glandular secretion is under the control of neural stimulation, an understanding of this process may lead to new therapeutic approaches to salvage the function of the residual acinar cells.

Neurovascular innervation enters the glandular lobule in the central region where the high endothelial venules are located; this is the same location where the earliest lymphoid infiltrates develop in the SS salivary gland [14]. Recent immunohistologic studies of normal and SS salivary glands by Konttinen et al.[22,23] have studied the distribution of protein PGP 9.5 (a cytoplasmic, noncytoskeletal marker of neurons), synaptophysin (a glycoprotein present in presynaptic vesicles), CPON (C-flanking peptide of neuropeptide Y found in postganglionic sympathetic fibers), CGRP ( calcitonin gene-related peptide found in sensory fibers) and neurotransmitters including substance P and VIP. These workers found significant alterations in the distribution of these neuroendocrine marker in the SS salivary gland including: a) Synaptophysin (a marker present where the neural synapse innervates the acinar cell) were abundantly present on normal salivary acinar cells but absent in SS glands where the glandular cells had atrophied (23); and b)  CPON, GCRP, and substance P (markers of neural innervation on small blood vessels) were present throughout normal salivary gland acini and intralobular ducts,  but not within, the clusters of lymphocytes in SS biopsies.  These results suggest that alterations of neural innervation may influence both blood flow to the gland as well as glandular secretion.  Further, these neuropeptides have been shown to have important synergistic interactions with cytokines such as IL-1 or TNF-α to influence both the proinflammatory response of the cytokines as well as to influence neuroendocrine functions of neuropeptides on vascular endothelial cells. Thus, the neural innervation in SS biopsies may influence glandular secretory rates and induction of cellular enzymes, vascular permeability and adhesive molecules, response to inflammatory cytokines produced by lymphocytes,  and modulate responses to hormones (discussed below).

SS, like other autoimmune diseases, occurs predominantly in women and thus sex hormone related genes must play a very important role.  It is known that castration of female animals delays or abrogates subsequent autoimmune disease and that androgen administration delays or minimizes autoimmunity in many animal models. Also, it is known that salivary glands are rich sources of growth hormones (including bFGF, insulin like growth factor, nerve growth factor) and that the synthesis of these growth factors is under androgenic control.  However, the specific role of androgens and estrogens in pathogenesis has remained relatively unclear.  Although minor immune response genes have mapped to the sex chromosomes, it is felt that sex hormone responsive genes (including prolactin and hormones relating to control of pituitary-adrenal axis) on other chromosomes may play a more important role in predisposition to autoimmunity.  Recent studies have indicated that transcription of pro-inflammatory cytokines such as interferon-γ is influenced by the binding of estrogens to cell steroid receptors (24)·  Sullivan and co-workers demonstrated that the secretion of secretory component by lacrimal gland acinar cells stimulated by cytokines (i.e. interleukin-1 and TNF-α) or by endocrine agonists (i.e. isoproterenol) is strongly influenced by the presence of androgens in the culture media (25,26). Together with the neuropeptides described above, these results  demonstrate the complex interactions of immunologic, endocrine and neural factors on glandular secretion.  Slight modifications of each of these factors by the inflammatory cytokines produced by lymphocytes in the SS biopsy may lead to significant alterations in glandular function.

Finally, it is known that emotional factors can influence the balance of cholinergic/sympathetic nerve stimulation to the lacrimal and salivary gland.  It has been suggested that anxiety and depression may influence (or be the cause of) imbalance of endogenous enkephalin, VIP, serotonin and/or substance P production in the brain.  This imbalance of the neuroendocrine system may influence salivary gland function, since neuropeptides such as enkephalins (which are known to exist in neural fibers innervating the salivary gland and can be found in saliva) can presynaptically inhibit cholinergic transmission in sympathetic ganglia.  The cholinergic postganglionic parasympathetic neurons contain vasoactive intestinal peptide (VIP), which can potentiate the acetylcholine response in terms of saliva volume (27). Thus, release of neuropeptides such as encephalins and substance P may influence the volume and protein content of saliva. A common pathway that may link cytokine, endocrine and neuropeptide activities may be their influence on the intracellular levels of cAMP (26,27). In summary, it is likely that inflammatory processes as well as behavioral factors (mediated by the autonomic nervous system) play a crucial role in the net neural signal delivered to the lacrimal and salivary gland.

## ENVIRONMENTAL FACTORS IN 1° SS

Studies on autoimmune disease in identical twins have shown a greatly increased concordance rate (i.e. if one twin has the disease, then the second twin has approximately 20% chance of developing the autoimmune disease). Although this observation indicates the strong role of genetics as a risk factor for pathogenesis, it is important to point out that the majority of twin pairs do not show concordance of disease and emphasizes the importance of "nongenetic" factors in pathogenesis. Such nongenetic factors may include exposure to exogenous agents (i.e. environmental antigens) and the occurrence of "random" recombination events occurring at different genetic loci, i.e. the recombination of immunoglobulin and T-cell antigen receptor variable and constant regions. It is likely that both of these factors contribute to the discordance of autoimmune disease among twins and the relatively low proportion of individuals with a specific genotype that actually develop SS.

The environmental co-factor(s) responsible for SS remain unknown. Indirect evidence has supported a role for herpes virus (esp. EBV) since this virus maintains latency in the salivary and/or lacrimal gland in normals after primary infection [28, 29]. Increased levels of EBV DNA and antibody responses to EBV encoded proteins have been found in a subset of SS patients in the United States [30-33], Europe [34], China [35], and Japan [36]. Immunohistology and *in situ* hybridization have suggested that the EBV genome is located in epithelial cells and to a lesser extent in B-cells [34]. However, the long latent period between initial EBV infection (usually before age 10) and onset of 1° SS (usually after age 25 yrs) make it difficult to attribute to EBV a primary pathogenetic role [29]. It is possible that EBV becomes reactivated as a consequence of cytokines released in the SS epithelial cell microenvironment and thus is a secondary event (rather than a primary cause of the autoimmune response). However, the strong T-cell responses against EBV encoded antigens in all normals suggest that the reactivation of EBV must be considered a candidate in disease perpetuation, even if it is not a primary cause [29].

Indirect evidence has also suggested a potential role for retroviruses [37]. In one study, a subpopulation of SS patients had antibodies to HIV retroviral protein p30 [38]. These studies were interpreted to show a role for infection by an exogenous retrovirus, in analogy to the salivary gland syndrome that occurs in some AIDS patients [39]. However, the significance of these studies remains unclear. Low affinity cross-reactions with a wide variety of proteins (including p30) are nonspecific and may not imply exposure to a retrovirus [40]. In two SS patients, a type C (intracisternal) virus was isolated after co-cultivation with SS tissue [41]. However, co-cultivation experiments do not prove that the intra cisternal particle arose from the SS sample and the proportion of SS patients infected by an exogenous retrovirus may be very small. The current debate over the potential role of exogenous retroviral infection in SS is reminiscent of the previous suggestions of an association of multiple sclerosis (MS) with retroviral infection. Subsequent studies showed that the vast majority of MS patients did not have an exogenous retroviral infection, and that the cases initially diagnoses as MS were actually the syndrome of tropical spastic paraparesis, a condition that is viral associated [42]. In view of the very great anxiety among SS patients (and their families) about the implications of retroviral infection, it is very important that the speculations regarding retroviral etiologies in SS be carefully phrased in public announcement to avoid "hysteria" in the patient population.

In addition to exogenous retroviral infections, it is known that the human genome contains fragments of retroviral like genes [43]. Most of these gene fragments are "defective" although at least one human endogenous retroviral like gene (HRES) has a nucleic acid sequence predicted to encode a HTLV-1 like gag protein. An increased antibody response against the predicted structure of HRES protein was found in a minority of SS patient's serum [44]. Shattles et al., [45] found that a monoclonal antibody against a HTLV-1 gag protein (p19) reacted with epithelial cells in about 30% of SS patients but not with normal salivary glands. This HTLV-1 antigen could be induced with IFN-g and was felt to be distinct from HRES [44]. Since the patients lacked antibodies to HTLV-1, they propose tissue specific induction of a distinct endogenous retroviral protein [45] may play a role in autoimmune responses against the salivary gland in SS. The significance of induction of endogenous retroviral proteins in human autoimmune disease is intriguing but remains unclear [46]. In animal models such as the NZB X NZW mouse, retroviral proteins (most notably the envelope proteins) are induced in the liver similar to other acute phase reactants [47]. Antibody responses against these proteins is felt to be a secondary phenomena in pathogenesis, although the resulting immune complexes do contribute to glomerulonephritis [47]. More recently, the significance of T-cell responses against retroviral proteins in murine SLE models has been investigated with emphasis on viral encoded "superantigens" [48]. Although still an intriguing hypothesis, recent genetic (i.e. crossbreeding) studies in autoimmune animals have dissociated superantigen responses from autoimmunity [43].

Finally, it is known that in endemic areas of Japan, HTLV-1 infection may be associated with SS-like symptoms and the occurrence of a "tax" gene may serve to transactivate important other genes. In a subpopulation of HIV infected individuals a SS-like condition may develop [39]. However, the histology and immunohistology of these glandular swelling (i. e. predominance of CD8+ T-cells and collapsing" lymphoid follicles seen in "ARC" like conditions), as well as differences in autoantibody profile (i.e. absence anti-SS A and anti-SS B antibodies), and distinct HLA-DR predisposition (i.e. HLA-DR5) distinguish the HIV associated condition from primary SS.

In summary, evidence for viral proteins derived from exogenous infection (i.e. EBV) and from activation of endogenous retroviral sequences (i.e. HRES or other HTLV-1 like sequences) have been found in salivary gland epithelial cells of SS patients by immunohistologic methods. However, it remains unclear whether these viral proteins play a primary role in pathogenesis or a secondary role (i.e. are targets for a polyclonal immune response and are reactivated as a consequence of increased immune activity in the gland). Even if they are secondary events, they may play a role in disease perpetuation. Thus in the case of human SS, it will be difficult to determine whether such responses are a primary or a secondary event in pathogenesis.

## ANIMAL MODELS OF SS

During the past decade, a number of animal models have been shown to possess lymphocytic infiltrates of the lacrimal and salivary glands. These include the widely used models of SLE such as the MRL/lpr and NZB X NZW mice [47]. In addition, mice undergoing chronic graft versus host disease also develop features of sialoadenitis. More recently, transgenic mice with expression of tax gene have shown lymphoid infiltrates. Although a full discussion of these animal models is beyond the scope of this chapter, each of these models has intriguing insights about mechanisms of pathogenesis. However, all of the models have fundamental differences from the immunopathologic features of human SS. One additional approach may be the immunodeficient SCID mouse, that can accept lymphocytes and tissue grafts from human sources including SS patients [49].

## SUMMARY

SS patients are characterized by decreased volume of lacrimal and salivary secretions. The dryness results from a combination of destroyed glandular elements as well as by interference with the neuro-endocrine innervation of the residual glands. Specific genetic factors (i.e. HLA class II alleles) have been associated with increased risk of SS in Caucasian (US), Chinese and Japanese populations. However, different class II MHC alleles are risk factors in each population. The environmental factors that precipitate SS remain unknown. Future understanding of the mechanisms of destruction of the salivary and lacrimal glands may provide a more rationale approach to therapy.

## ACKNOWLEDGEMENTS

Supported by grants from the National Institutes of Health (MO1RR00833), the Thornton and Ramsell Foundations, and the Scripps-Stedham Fund.

## REFERENCES

1. G.T. Nepom, and H. Erlich, MHC class-II molecules and autoimmunity, *Annu Rev Immunol.* 9:493 (1991).
2. H.M. Fei, H.-I. Kang, S. Scharf, H. Erlich, C. Peebles and R.I. Fox, Specific HLA-DQA and HLA-DRB1 alleles confer susceptibility to Sjögren's syndrome and autoantibody SS-B production, *J. Clin. Lab. Analysis* 5:382 (1991).
3. H.I. Kang, H. Fei and R.I. Fox, Comparison of genetic factors in Chinese, Japanese and Caucasoid patients with Sjöjren's syndrome, *Arthritis & Rheum.* 34:S41 (1991).
4. A. Freemont, C. Jones, P. Bromley and P. Andrews, Changes in vascular endothelium related to lymphocyte collections in diseased synovium, *Arth. Rheum.* 26:1427 (1983).
5. F.N. Skopouli, E.E. Kousvelari, P. Mertz, E.S. Jaffe et al., c-myc mRNA expression in minor salivary glands of patients with Sjögren's syndrome, *J. Rheumatol.* 19:693 (1992).

6. T.E. Daniels, Labial salivary gland biopsy in Sjögren's syndrome, *Arthritis Rheum.* 27:147 (1984).
7. R.I. Fox, F.V. Howell, R.C. Bone and P. Michelson, Primary Sjögren's syndrome: Clinical and immunopathologic features, *Seminars in Arthritis and Rheumatism* 14:77 (1984).
8. R.N.M. MacSween, R.B. Goudie, J.R. Anderson, E. Armstrong, M.A. Murray, D.K. Mason, M.K. Jasani, J.A. Boyle, W.W. Buchanan and J. Williamson, Occurrence of antibody to salivary duct epithelium in Sjögren's disease, rheumatoid arthritis, and other arthritides, *Ann. Rheum. Dis.* 26:402 (1967).
9. S.C. Pflugfelder, K.R. Wilhelmus, M.S. Osato, A.Y. Matoba and R.L. Font, The autoimmune nature of aqueous tear deficiency, *Ophthalmology* 93:1513 (1986).
10. T.C. Adamson III, R.I. Fox, D.M. Frisman and F.V. Howell, Immunohistologic analysis of lymphoid infiltrates in primary Sjögren's syndrome using monoclonal antibodies, *J. Immunol.* 130:203 (1983).
11. I. Weissman, R. Warnke, E. Butcher et al., The lymphoid system. Its normal architecture and potential for understanding the system through the study of lymphoproliferative diseases, *Hum. Pathol.* 9:25 (1978).
12. R.I. Fox, T. Bumol, R. Fantozzi, R. Bone and R. Schreiber, Expression of histocompatibility antigen HLA-DR by salivary gland epithelial cells in Sjögren's syndrome, *Arthritis Rheum.* 29:1105 (1986).
13. G. Lindahl, E. Hedfors, L. Kloreskog and U. Forsum, Epithelial HLA-DR expression and T-cell subsets in salivary glands in Sjögren's syndrome., *Clinical Exp. Immunol.* 61:475 (1985).
14. R.I. Fox, T.C. Adamson III, S. Fong, C.A. Robinson, E.L. Morgan, J.A. Robb and F.V. Howell, Lymphocyte phenotype and function of pseudolymphomas associated with Sjögren's syndrome, *J. Clin. Invest.* 72:52 (1983).
15. F.N. Skopouli, P.C. Fox, V. Galanopoulou et al., T cell subpopulations in the labial minor salivary gland histopathologic lesion of Sjögren's syndrome, *J. Rheumatol.* 18:210 (1991).
16. F. Yonaha, T. Sumida, T. Maeda, H. Tomioka et al., Restricted junctional usage of T cell receptor V beta 2 and B beta 13 genes, which are overrepresented on infiltrating T cells in the lips of patients with Sjögren's syndrome, *Arthritis Rheum.* 35:1362 (1992).
17. Y. Ichikawa, H. Shimizu, M. Hoshida, M. Takaya and S. Arimori, Accessory molecules expressed on the peripheral blood or synovial fluid T lymphocytes from patients with Sjögren's syndrome or rheumatoid arthritis, *Clin. Exp. Rheumatol.* 10:447 (1992).
18. R.I. Fox, T.E. Hugli, L.L. Lanier, E.L. Morgan and F. Howell, Salivary gland lymphocytes in primary Sjögren's syndrome lack lymphocyte subsets defined by Leu 7 and Leu 11 antigens, *J. Immunol.* 135:207 (1985).
19. N. Talal, R. Asofsky and P. Lightbody, Immunoglobulin synthesis by salivary gland lymphoid cells in Sjögren's syndrome, *J. Clin. Invest.* 19:19 (1979).
20. R.I. Fox, T.E. Hugli, L.L. Lanier, E.L. Morgan and F. Howell, Salivary gland lymphocytes in primary Sjögren's syndrome lack lymphocyte subsets defined by Leu 7 and Leu 11 antigens, *J. Immunol.* 135:207 (1985).
21. R.I. Fox, A.N. Theofilopoulos and A. Altman, Production of interleukin 2 (IL 2) by salivary gland lymphocytes in Sjögren's syndrome. Detection of reactive cells by using antibody directed to synthetic peptides of IL 2, *J. Immunol.* 135:3109 (1985).
22. Y.T. Konttinen, T. Sorsa, M. Hukkanen, M. Segerberg et al., Topology of innervation of labial salivary glands by protein gene product 9.5 and synaptophysin immunoreactive nerves in patients with Sjögren's syndrome, *J Rheumatol.* 19:30 (1992).
23. Y.T. Konttinen, M. Hukkanen, P. Kemppinen, M. Segerberg et al., Peptide-containing nerves in labial salivary glands in Sjögren's syndrome, *Arthritis Rheum.* 35:815 (1992).
24. H.S. Fox, B.L. Bond and T.G. Parslow, Estrogen regulates the IFN-gamma promoter, *J. Immunol.* 146:4362 (1991).
25. D.A. Sullivan, Ocular mucosal immunity, in: "Mucosal Immunology," P.L. Ogra, J. Mestecky, M.E. Lamm, W. Strober, J. McGhee, and J. Bienenstock, eds., Academic Press, Orlando, FL, in press (1993).
26. R.S. Kelleher, L.E. Hann, J.A. Edwards and D.A. Sullivan, Endocrine, neural and immun e control of secretory component output by lacrimal gland acinar cells, *J. Immunol.* 146:3405 (1991).
27. B. Lindh and T. Hokfelt, Structural and functional aspects of acetylcholine peptide coexistence in the autonomic nervous system, *Prog. Brain Res.* 84:175 (1990).
28. R.I. Fox, T. Chilton, S. Scott, L. Benton, F.V. Howell and J.H. Vaughan, Potential role of Epstein-Barr virus in Sjögren's syndrome, *Rheum. Dis. Clinics of North America* 13:275 (1987).

29. R.I. Fox, M. Luppi, H.-I. Kang and P. Pisa., Potential Role of EBV Reactivation in Sjogren's, Syndrome, "Springer Seminar in Immunopathology," Springer-Verlag, Heidelberg (1991).

30. I. Saito, B. Servenius, T. Compton and R.I. Fox, Detection of Epstein-Barr virus DNA by polymerase chain reaction in blood and tissue biopsies from patients with Sjögren's syndrome, *J. Exp. Med.* 169:2191 (1989).

31. R.I. Fox, G. Pearson and J.H. Vaughan, Detection of Epstein-Barr virus associated antigens and DNA in salivary gland biopsies from patients with Sjögren's syndrome, *J. Immunol.* 137:3162 (1986).

32. R.I. Fox, S. Scott, R. Houghton, A. Whalley, J. Geltofsky, J.H. Vaughan and R. Smith, Synthetic peptide derived from the Epstein-Barr virus encoded early diffuse antigen (EA-D) reactive with human antibodies, *J. Clin. Lab. Anal.* 1:140 (1987).

33. S.C. Pflugfelder, S. . Tseng, J.S. Pepose, M.A. Fletcher, N. Klimas and W. Feuer, Epstein-Barr virus infection and immunologic dysfunction in patients with aqueous tear deficiency, *Ophthalmology* 97:313 (1990).

34. X. Mariette, J. Gozlan, D. Clerc, M. Bisson and F. Morinet, Detection of Epstein-Barr virus DNA by in situ hybridization and polymerase chain reaction in salivary gland biopsy specimens from patients with Sjögren's syndrome, *Am. J. Med.* 90:286 (1991).

35. J.L. Yang, Z.G. He and N.Z. Zhang, Associations between the renal tubular acidosis of primary Sjögren's syndrome and the infection of Epstein-Barr virus: a preliminary study, *Chung Hua Nei Ko Tsa Chih* 30:151 (1991).

36. N. Inoue, S. Harada, N. Miyasdaka, A. Oya and K. Yanagi, Analysis of antibody titers to Epstein-Barr virus nuclear antigens in sera of aptients with Sjögren's syndrome and with rheumatoid arthritis, *J. Infect. Dis.* 164:22 (1991).

37. N. Talal, E. Flescher and H. Dang, Evidence for possible retroviral involvement in autoimmune disease, *Annals of Allergy.* 69:221 (1992).

38. N. Talal et al., Detection of serum antibodies to retroviral proteins in patients with primary Sjögren's syndrome (autoimmune exocrinopathy), *Arthritis and Rheum.* 33:77 (1990).

39. S. Itescu, L.J. Brancato, J. Buxbaum, P.K. Gregersen, C.C. Rizk, T.S. Croxson, G.E. Solomon and R. Winchester, A diffuse infiltrative CD8 lymphocytosis syndrome in human immunodeficiency virus (HIV) infection: a host immune response associated with HLA-DR5, *Ann. Intern. Med.* 112:3 (1990).

40. J.F. Meilof, H. Arentsen, A.A. Kruize, R.J. Hene and e. al., Sjögren's syndrome and retroviral infection [letter], *Arthritis Rheum.* 35:1403 (1992).

41. R.F. Garry, C.D. Fermin, D.J. Hart, S.S. Alexander, L.A. donehower and L.Z. Hong, Detection of a human intracisternal A-type retroviral particle antigenically related to HIV, *Science.* 250:1127 (1990).

42. D.E. McFarlin and H. Koprowski, Neurological disorders associated with HTLV-1, *Curr. Top. Microbiol. Immunol.* 160:100 (1990).

43. J.A. Gonzalo, I.M. de Alboran and G. Kroemer, Dissociation of autoaggression and self-superantigen reactivity [editorial], *Scand. J. Immunol.* 37:1 (1993).

44. S.M. Brookes, Y.A. Pandolfino, T.J. Mitchell, P.J. Venables et al., The immune response to and expression of cross-reactive retroviral gag sequences in autoimmune disease, *Br. J. Rheumatol.* 31:735 (1992).

45. W.G. Shattles, S.M. Brookes, P.J. Venables, D.A. Clark and R.N. Maini, Expression of antigen reactive with a monoclonal antibody to HTLV-1 P19 in salivary glands in Sjögren's syndrome, *Clin. Exp. Immunol.* 89:46 (1992).

46. A.M. Krieg, M.F. Gourley and A. Perl, Endogenous retroviruses: potential etiologic agents in autoimmunity, *FASEB J.* 6:2537 (1992).

47. A.D. Steinberg, A.M. Krieg, M.F. Gourley and D.M. Klinman, Theoretical and experimental approaches to generalized autoimmunity, *Immunol. Rev.* 118:129 (1990).

48. T. Chatila and R.S. Geha, Superantigens, *Curr. Opin. Immunol.* 4:74 (1992).

49. P. Pisa, M.J. Cannon, E.K. Pisa, N.R. Cooper and R.I. Fox, Epstein-Barr virus induced lymphoproliferative tumors in severe combined immunodeficient mice are oligoclonal, *Blood.* 79:173 (1992).

# MURINE MODELS OF SJÖGREN'S SYNDROME

Douglas A. Jabs and Robert A. Prendergast

The Wilmer Ophthalmological Institute
Department of Ophthalmology
The Johns Hopkins University School of Medicine
Baltimore, MD 21205

## INTRODUCTION

Sjögren's syndrome is characterized by dry eyes and a dry mouth, due to mononuclear inflammatory cell infiltration into the lacrimal and salivary glands, resulting in glandular dysfunction. While its etiology is unknown, Sjögren's syndrome is felt to be an autoimmune disorder, as evidenced by its association with autoantibodies and with other systemic connective tissue disorders. Sjögren's syndrome may occur by itself, in which case it is termed primary Sjögren's syndrome, or an association with another, defined, connective tissue disease, in which case it is termed secondary Sjögren's syndrome. Those connective tissue disorders most commonly associated with secondary Sjögren's syndrome include rheumatoid arthritis, systemic lupus erythematosus, and scleroderma. Serologically, patients with Sjögren's often have rheumatoid factor, antinuclear antibodies (ANA), and antibodies to Ro (SS-A) and La (SS-B).[1]

Several murine models of autoimmunity have been described, including the MRL/Mp-*lpr/lpr* (subsequently abbreviated MRL/lpr), MRL/Mp-+/+ (MRL/+), and (NZB/NZW) $F_1$ hybrid (NZB/W) mice. These different mice share many common immunologic and clinical features, including polyclonal B cell activation, hypergammaglobulinemia, autoantibody formation, and glomerulonephritis.[2] All three mice develop lacrimal and salivary gland inflammatory lesions, which are a model for human Sjögren's syndrome.[3-9] MRL/lpr and MRL/+ mice are congenic substrains, which differ only by a single autosomal recessive gene, the *lpr* gene. The *lpr* gene leads to massive lymphadenopathy in MRL/lpr mice and markedly accelerates the autoimmune disease present;[2] MRL/lpr mice develop an acute glomerulonephritis, arthritis, and vasculitis, with a life span limited to six months, while MRL/+ mice develop a less fulminant multisystem autoimmune disease with chronic glomerulonephritis and generally survive to two years of age.[2] NZB/W mice develop an autoimmune disease with glomerulonephritis at approximately six months of age and survive to nine to twelve months of age.[2]

Several lines of evidence suggest that MRL/lpr and NZB/W mice may have different intrinsic immunologic abnormalities. Neonatal thymectomy or treatment with monoclonal anti-T cell antibodies (anti-Thy 1) results in amelioration of the disease in

*Lacrimal Gland, Tear Film, and Dry Eye Syndromes*
Edited by D.A. Sullivan, Plenum Press, New York, 1994

MRL/lpr mice,[10-12] while neither neonatal thymectomy nor treatment with monoclonal anti-Thy 1 alters the disease course in NZB/W mice.[10-12] These results suggest that T cells are primarily responsible for the disease in MRL/lpr mice, while B cells are important for the disease in NZB/W mice. Furthermore, MRL/lpr lymphocytes spontaneously produce a B cell differentiation factor *in vitro*, suggesting that the polyclonal B cell activation in MRL/lpr mice is a T cell-driven process.[13-14] While these results indicate the importance of T cells in MRL/lpr mice and B cells in NZB/W mice, the successful use of anti-CD4 monoclonal antibody to treat NZB/W mice[15] indicates a contribution of helper T cells to their autoimmune disease. The discrepancy between anti-T cell therapy, which eliminates T cells of all subsets, and anti-CD4 therapy, which selectively inhibits CD4+ T cells, for the treatment of NZB/W mice remains unexplained.

## IMMUNOHISTOLOGY OF MURINE MODELS OF SJÖGREN'S SYNDROME

MRL/lpr, MRL/+ and NZB/W mice all develop inflammatory infiltrates of the lacrimal and salivary glands and are models for the human disorder Sjögren's syndrome (Figure 1). These infiltrates begin as multiple focal, and later confluent, areas of mononuclear inflammatory cells. Such infiltrates are not seen in control strains, such as BALB/c or C3H. These lesions develop in concert with other features of the systemic autoimmunity and are present in a fully-developed fashion in mature mice of each strain. While the lesions in each strain appear similar at the cytologic level, we have investigated the immunocytochemical profiles of these three strains and found differences among them.[7]

Frozen sections of lacrimal gland tissue were prepared from 13 MRL/lpr mice ages five to six months, 14 MRL/+ mice ages six to twelve months, and 18 NZB/W mice ages nine to twelve months, animals with fully-developed disease. Sections were stained using a panel of monoclonal antibodies to cell surface markers and the avidin-biotin-peroxidase-complex (ABC) technique.[16] The monoclonal antibodies used are outlined in Table 1. The percentage of mononuclear inflammatory cells staining positively with a given monoclonal antibody was enumerated using a 10 x 10 grid net micrometer disc, covering an area of 0.16 mm$^2$ with a 25x objective.[7]

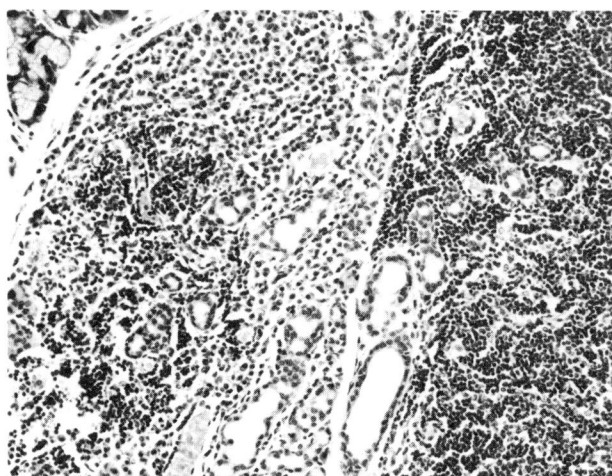

**Figure 1.** Lacrimal gland inflammation in a 5-month-old MRL/lpr mouse. There is extensive replacement of the lobule by mononuclear inflammatory cells (H&E, original magnification x 200. Reprinted with permission from Jabs, Enger, and Prendergast.[9])

**Table 1.** Monoclonal antibodies to cell surface markers.

| Cell type/epitope | Antibody | Clone | Source |
|---|---|---|---|
| T cells | anti-Thy 1.2 | | BD[17,18] |
| CD4 | anti-L3T4 | GK1.5 | ATCC or BD[19,20] |
| CD8 | anti-Lyt 2 | | BD[18,21] |
| B cells | anti-sIg | F(ab')₂ anti-mouse IgG + IgM | Tago |
| | pan-B cell | RA3-LC2/1 | ATCC[22,23] |
| Macrophages | anti-Mac3 | M3/84.6.34 | ATCC[24] |

sIg = surface immunoglobulin; BD = Becton Dickinson (Mountain View, CA); ATCC = American Tissue Culture Collection (Rockville, MD), Tago = Tago, Inc., (Burlingame, CA)

The results of the immunohistologic staining of the lacrimal gland lesions in autoimmune mice are outlined in Table 2.

**Table 2.** Immunohistologic staining of lacrimal gland lesions in autoimmune mice

| | Percentage of mononuclear cells staining with antibody to (mean ± standard deviation): | | | | |
|---|---|---|---|---|---|
| Mouse | Thy 1.2 | CD4+ | CD8+ | B cells | Macrophages |
| MRL/lpr | 85 ± 5 | 63 ± 9 | 14 ± 4 | 10 ± 2 | 3 ± 1 |
| MRL/+ | 78 ± 10 | 49 ± 9 | 30 ± 8 | 13 ± 11 | 2 ± 1 |
| NZB/W | 57 ± 6 | 47 ± 7 | 6 ± 3 | 33 ± 5 | 7 ± 3 |

Modified with permission from Jabs and Prendergast.[7]

While MRL/lpr and MRL/+ mice had a similar percentage of T cells present (mean 85% versus 78%, respectively), NZB/W mice had significantly fewer T cells (57%, student-Newman-Keuls procedure,[25] $p < 0.05$). Furthermore, MRL/lpr mice had significantly more CD4+ T cells (63%) than did MRL/+ mice (49%, $p < 0.05$) or NZB/W mice (47%, $p < 0.05$). Conversely, MRL/+ mice had a significantly greater percentage of CD8+ T cells (30%) than did either MRL/lpr mice (14%, $p < 0.05$) or NZB/W (6%, $p < 0.05$). The difference in the percentage of CD8+ cells between MRL/lpr and NZB/W mice was also significant ($p < 0.05$). B cells were significantly more common in NZB/W mice (33%) than in either MRL/lpr (10%, $p < 0.05$) or MRL/+ (13%, $p < 0.05$) mice.[7]

These results suggest that the immunocytologic profile of the mononuclear inflammatory cells infiltrating the lacrimal gland lesions of autoimmune mice parallels the intrinsic immunologic defects. Specifically, MRL/lpr mice, who appear to have a primary defect in their T cells, have lesions composed largely of CD4+ T cells. Conversely, NZB/W mice have a much greater percentage of B cells present, consistent with an apparent intrinsic abnormality of B cells. Furthermore, MRL/+ mice, who differ from MRL/lpr mice only by the *lpr* gene, but have a much less aggressive autoimmune disease, have a different immunocytochemical profile, in that they have a lower percentage of CD4+ T cells and a higher percentage of CD8+ T cells in the lacrimal gland lesions.

Finally, the massive lymphadenopathy seen in MRL/lpr mice is composed largely of "double-negative" T cells.[26-28] These T cells are Thy 1.2+ but are CD4- and CD8-. Simultaneous staining of lymph node sections confirmed the presence of double-negative T cells in these lymph nodes and demonstrated that the predominance of CD4+ T cells seen in the lacrimal gland was not due to variations in technique.[7] Other authors have reported similar results for the immunohistologic analysis of the salivary gland infiltrates present in MRL/lpr and NZB/W mice.[29,30]

The immunohistology in the lacrimal and salivary glands of MRL/lpr mice are most similar to the minor salivary gland biopsy results from patients with Sjögren's syndrome.[31,32] However, lacrimal gland biopsy results have suggested a relatively greater percentage of B cells in the lacrimal gland lesions[33] than in the minor salivary gland lesions in patients with Sjögren's, results more similar to those seen in NZB/W mice.

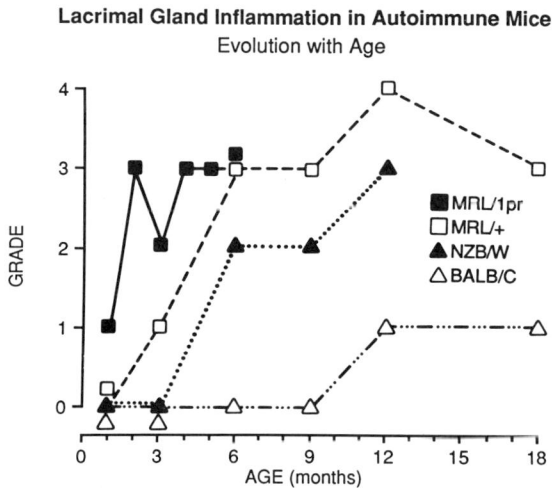

**Figure 2.** Median lacrimal gland focus scale score over time. (Reprinted with permission from Jabs, Enger, and Prendergast.[9])

## EVOLUTION OF THE LACRIMAL GLAND INFLAMMATORY RESPONSE

Because the immune response is dynamic in nature, autoimmune mice at various ages were analyzed to detail the evolution of the lacrimal gland inflammation. Lacrimal gland sections were examined to establish when the lacrimal gland lesions first occurred, and to examine any change in B cell or T cell populations within the gland throughout the course of disease. Groups of approximately 10 mice were killed for each age analyzed: MRL/lpr mice were sacrificed from ages 1 to 6 months; MRL/+ mice at 1, 3, 6, 9, 12, and 18 months of age; NZB/W mice at 1, 3, 6, 9, and 12 months of age; and control BALB/c mice at ages 1 to 6, 9, 12, and 18 months of age. Lacrimal gland sections were scored using a modified focus score scale, in which the histologic sections are scored for increasing severity from 0 to 4 based upon the presence or absence of foci of 50 or more mononuclear inflammatory cells.[9]

The median lacrimal gland grade for each age group is shown in Figure 2.[9] MRL/lpr mice developed lacrimal gland inflammation early with scattered inflammatory cells present at one month of age, and fully developed lacrimal gland inflammatory lesions were present by four to five months of age. MRL/+ mice had no inflammation at one month of age, showed minimal inflammatory lesions by three months of age, had extensive grade 3 lesions by nine months of age, and fully developed lesions by one year. NZB/W mice had no inflammation at one or three months of age, developed inflammatory lesions at six months of age, and had fully developed lesions by nine to twelve months of age. Control BALB/c mice did not develop extensive foci mononuclear inflammatory cells and had only scattered inflammatory cells at advanced ages. At all ages greater than one month, autoimmune mice had significantly more disease than did control BALB/c mice (p < 0.05).[9]

Immunohistology was performed on lacrimal gland lesions for each strain at the various time points using the previously described techniques.[9] For MRL/lpr mice, animals were analyzed at 1 to 2 months, 5 months, and 6 months; for MRL/+ mice, 3 months, 6 months, 12 months, and 18 months; and for NZB/W mice, 6 months, 9 months, and 12 months. Results of the immunohistologic analyses of MRL/lpr lacrimal gland lesions are shown in Table 3, of MRL/+ lacrimal gland lesions in Table 4, and of NZB/W lacrimal gland lesions in Table 5.

**Table 3.** Immunohistologic analysis of MRL/lpr lacrimal gland lesions

| | Percentage of mononuclear cells staining with antibody to (mean ± standard deviation): | | | |
|---|---|---|---|---|
| Age (mo) | Thy 1.2 | CD4 | CD8 | B cells |
| 1-2 | 87 ± 5 | 68 ± 8 | 22 ± 7 | 9 ± 7 |
| 5 | 84 ± 4 | 60 ± 9 | 14 ± 5 | 10 ± 2 |
| 6 | 83 ± 6 | 73 ± 12 | 12 ± 4 | 12 ± 2 |
| p-value | 0.08 | 0.59 | <0.001 | 0.13 |

P-value based on test for significant trend over time.[25] Modified with permission from Jabs, Enger, and Prendergast.[9]

**Table 4.** Immunohistologic analysis of MRL/+ lacrimal gland lesions

| | Percentage of mononuclear cells staining with antibody to (mean ± standard deviation): | | | |
|---|---|---|---|---|
| Age (mo) | Thy 1.2 | CD4 | CD8 | B cells |
| 3 | 88 ± 4 | 56 ± 9 | 32 ± 8 | 8 ± 6 |
| 6 | 81 ± 10 | 51 ± 7 | 31 ± 9 | 11 ± 7 |
| 12 | 74 ± 8 | 46 ± 10 | 28 ± 6 | 14 ± 14 |
| 18 | 80 ± 4 | 40 ± 5 | 30 ± 8 | 28 ± 5 |
| p-value | 0.02 | 0.001 | 0.56 | 0.001 |

P-value based on test for significant trend over time.[24] Modified with permission from Jabs, Enger, and Prendergast.[9]

**Table 5.** Immunohistologic analysis of NZB/W lacrimal gland lesions

| Age (mo) | Percentage of mononuclear cells staining with antibody to: (mean ± standard deviation): | | | |
|---|---|---|---|---|
| | Thy 1.2 | CD4 | CD8 | B cells |
| 6 | 83 ± 5 | 49 ± 5 | 13 ± 5 | 23 ± 4 |
| 9 | 58 ± 6 | 45 ± 9 | 5 ± 3 | 33 ± 5 |
| 12 | 54 ± 2 | 44 ± 1 | 8 ± 3 | 34 ± 3 |
| p-value | 0.001 | 0.22 | 0.01 | 0.001 |

P-value based on test for significant trend over time.[25] Modified with permission from Jabs, Enger, and Prendergast.[9]

For each autoimmune mouse, changes in the immunocytochemical profile over time were clearly seen. For MRL/lpr mice there was a progressive decline in the percentage of CD8+ cells over time. Conversely, for MRL/+ mice there was a higher and unchanging percentage of CD8+ T cells with a progressive decline in CD4+ T cells. Coupled with this change, there was a progressive increase in B cells, suggesting recruitment of B cells into the inflammatory infiltrate over time. In NZB/W mice, there was also a significant decline in CD8+ T cells and an increase in the percentage of B cells over time. Thus, the autoimmune process in these three murine models of Sjögren's syndrome is a dynamic and evolving process with strain and gene-related changes in T cell subsets.

## SUMMARY AND CONCLUSIONS

Autoimmune MRL/lpr, MRL/+, and NZB/W mice all develop lacrimal gland inflammatory lesions, which consist of focal mononuclear inflammatory cell infiltrates. Each strain has a different immunocytochemical profile, which appears to be related to the underlying immunologic defects present in that mouse. The appearance of these lesions parallels the evolution of the systemic autoimmune disease. The lesions are dynamic over time with the early appearance of CD4+ T cells (helper T cells) for each strain. Subsequently, there is an accumulation of B cells over time in MRL/+ and NZB/W mice. In the two more rapidly evolving mouse models, MRL/lpr and NZB/W, there is a progressive decline in the percentage of CD8+ cells. Conversely, in the slowly evolving MRL/+ lacrimal gland lesions, there is a persistent and unchanging percentage of CD8+ T cells (suppressor/cytotoxic T cells). Autoimmune mice provide models for the human disorder Sjögren's syndrome and a mechanism for better understanding the immunopathogenesis of autoimmune lacrimal gland disease.

## ACKNOWLEDGEMENTS

This work was supported by grants EY05912 and EY01765 from the National Institutes of Health, Bethesda, MD. Dr. Jabs is a Research to Prevent Blindness Olga Keith Weiss Scholar.

# REFERENCES

1. D.A. Jabs, Ocular manifestations of the rheumatic diseases. *In* W. Tasman and E.A. Jaeger (eds): *Duane's Clinical Ophthalmology*, volume 5. Philadelphia, J.P. Lippincott Co, 1992. Chapter 26, pp 1-39.

2. A.N. Theofilopoulos and F.J. Dixon, Murine models of systemic lupus erythematosus, *Adv Immunol* 37:269 (1985).

3. H.S. Kessler, M. Cubberly, and W. Manski, Eye changes in autoimmune NZB and NZBxNZW mice, *Arch Ophthalmol* 85:211 (1971).

4. R.W. Hoffman, M.A. Alspaugh, K.S. Waggie, J.B. Durham, and S.E. Walker, Sjögren's syndrome in MRL/l and MRL/n mice, *Arthritis Rheum* 27:157 (1984).

5. D.A. Jabs, E.L. Alexander, and W.R. Green, Ocular inflammation in autoimmune MRL/Mp mice, *Invest Ophthalmol Vis Sci* 26:1223 (1985).

6. D.A. Jabs and R.A. Prendergast, Reactive lymphocytes in lacrimal gland and renal vasculitic lesions of autoimmune MRL/lpr mice express L3T4, *J Exp Med* 166:1198 (1987).

7. D.A. Jabs and R.A. Prendergast, Murine models of Sjögren's syndrome: immunohistologic analysis of different strains, *Invest Ophthalmol Vis Sci* 29:1437 (1988).

8. J.T. Gilbard, L.A. Hanninen, R.C. Rothman, and K.R. Kenyon, Lacrimal gland, cornea, and tear film in NZB x NZW F1 hybrid mouse, *Current Eye Research* 6:1237 (1987).

9. D.A. Jabs, C. Enger, R.A. Prendergast, Murine models of Sjögren's syndrome. evolution of the lacrimal gland inflammatory lesions, *Invest Ophthalmol Vis Sci* 32:371-380 (1991).

10. A.D. Steinberg, J.B. Roths, E.D. Murphy, R.T. Steinberg, and E.S. Raveche, Effects of thymectomy or androgen administration upon the autoimmune disease of MRL/Mp-*lpr/lpr* mice, *J Immunol* 125:871 (1980).

11. L. Hang, A.N. Theofilopoulos, R.S. Balderas, S.J. Francis, and F.J. Dixon, The effect of thymectomy on lupus-prone mice, *J Immunol* 132:1809 (1984).

12. D. Wofsy, J.A. Ledbetter, P.L. Hendler, and W.E. Seaman, Treatment of murine lupus with monoclonal anti-T cell antibody, *J Immunol* 134:852 (1985).

13. G.J. Prud'homme, C.L. Park, T.M. Fieser, R. Kofler, F.J. Dixon, and A.N. Theofilopoulos, Identification of a B cell differentiation factor(s) spontaneously produced by proliferating T cells in murine lupus strains of the *lpr/lpr* genotype, *J Exp Med* 157:730 (1983).

14. G.J. Prud'homme, R.S. Balderas, F.J. Dixon, and A.N. Theofilopoulos, B cell dependence on and response to accessory signals in murine lupus strains, *J Exp Med* 157:1815-1827 (1983).

15. D. Wofsy and W.E. Seaman, Successful treatment of autoimmunity in NZB/NZW F₁ mice with monoclonal antibody to L3T4, *J Exp Med* 161:378 (1985).

16. S. Hsu, L. Raine, and H. Fanger, Use of avidin-biotin-peroxidase (ABC) techniques: a comparison between ABC and unlabeled antibody (PAP) procedures, *J Histochem Cytochem* 29:577 (1981).

17. J.A. Ledbetter and L.A. Herzenberg, Xenogeneic monoclonal antibodies to mouse lymphoid differentiation antigens, *Immunol Rev* 47:63, 1979.

18. J.A. Ledbetter, R.V. Rouse, H.S. Micklem, and L.A. Herzenberg, T cell subsets defined by expression of Lyt-1,2,3 and Thy-1 antigens, *J Exp Med* 152:280, 1980.

19. D.P. Dialynas, Z.S. Quan, K.A. Wall, J. Quintans, M.R. Loken, M. Pierres, and F.W. Fitch, Characterization of murine T cell surface molecule, designated L3T4, identified by monoclonal antibody GK-1.5: similarity of L3T4 to the human Leu-3/T4 molecule, *J Immunol* 131:2445 (1983).

20. D.P. Dialynas, D.B. Wilde, P. Marrack, A. Pierres, K.A. Wall, W. Havran, G.O. Hen, M.R. Loken, M. Pierres, J. Kappler, and F.W. Fitch, Characterization of the murine antigenic determinant, designated L3T4 A, recognized by monoclonal antibody GK-1.5: expression of L3T4 A by functional T cell clones appears to correlate primarily with class II MHC antigen reactivity, *Immunol Rev* 74:29 (1983).

21. J.A. Ledbetter, R.L. Evans, M.I. Lipinski, C. Cunningham-Rundles, R.A. Good, and L.A. Herzenberg, Evolutionary conservation of surface molecules that distinguish T lymphocyte helper/inducer and T cytotoxic/suppressor subpopulations in mouse and man, *J Exp Med* 153:310 (1981).

22. R.L. Coffman and I.L. Weissman, A monoclonal antibody that recognizes B cells and B cell precursors in mice, *J Exp Med* 153:269 (1981).

23. V. Potter, O.N. Witte, R. Coffman, and D. Baltimore, Abelson murine leukemia virus-induced tumors elicit antibodies against a host cell protein, P 50. *J Virol* 36:547 (1980).

24. T.A. Springer, Monoclonal antibody analysis of complex biological systems, *J Biol Chem* 256:3833 (1981).

25. G.W. Snedecor and W.G. Cochran, *Statistical Methods*. Seventh edition. Ames, Iowa State University Press (1980).

26. A.N. Theofilopoulos, R.A. Eisenberg, M. Bourdon, J.S. Crowel, Jr, and F.J. Dixon, Distribution of lymphocytes identified by surface markers in murine strains with systemic lupus erythematosus-like syndromes, *J Exp Med* 149:516 (1979).

27. D.E. Lewis, J.V. Giorgi, and N.L. Warner, Flow cytometry analysis of T cells and continuous T cell lines from autoimmune MRL/1 mice, *Nature* 289:298 (1981).

28. D. Wofsy, R.R. Hardy, and W.E. Seaman, The proliferating cells in autoimmune MRL/*lpr* mice lack L3T4, an antigen on "helper" T cells that is involved in the response to class II major histocompatibility antigens, *J Immunol* 132:2686 (1984).

29. R. Jonsson, A. Tarkowski, K. Backman, L. Klareskog, Immunohistochemical characterization of sialadenitis in NZB x NZW F1 mice, *Clin Immunol Immunopathol* 42:93-101 (1987).

30. R. Jonsson, A. Tarkowski, K. Backman, R. Holmdahl, and L. Klareskog, Sialadenitis in the MRL-1 mouse: morphological and immunohistochemical characterization of resident and infiltrating cells, *Immunology* 60:611-616 (1987).

31. T.C. Adamson III, R.I. Fox, D.M. Frisman, and F.V. Howell, Immunohistologic analysis of lymphoid infiltrates in primary Sjögren's syndrome using monoclonal antibodies, *J Immunol* 130:203 (1983).

32. R.I. Fox, S.A. Carstens, S. Fong, C.A. Robinson, F. Howell, and J.H. Vaughn, Use of monoclonal antibodies to analyze peripheral blood and salivary gland lymphocyte subsets in Sjögren's syndrome, *Arthritis Rheum* 25:419 (1982).

33. F. Akata, S.C. Pflugfelder, S.F. Lee, R.L. Font, and J.S. Pepose, Immunocytologic features of lacrimal gland biopsies in Sjögren's syndrome, ARVO Abstracts, *Invest Ophthalmol Vis Sci* (suppl) 30:386 (1989).

# UTILIZATION OF THE NON-OBESE DIABETIC (NOD) MOUSE AS AN ANIMAL MODEL FOR THE STUDY OF SECONDARY SJÖGREN'S SYNDROME

M.G. Humphreys-Beher, Y. Hu, Y. Nakagawa, P-L. Wang, and K.R. Purushotham

Department of Oral Biology
University of Florida
P.O. Box 100424
Gainesville, FL 32610

## INTRODUCTION

Sjögren's syndrome (S.S.) in the human patient population is an autoimmune inflammatory disease presenting clinical symptoms of xerophthalmia and xerostomia[1]. This condition predominantly affects women. Most diagnoses of S.S. is made in association with autoimmune connective tissue diseases such as rheumatoid arthritis or systemic lupus erythematosus.[2] However, it can also be obseved as an isolated phenomenon, described as primary S.S.

While the availability of human tissue for the study of S.S. has been limited, a number of animal models with lymphocytic attack of the lacrimal and salivary glands have been reported.[3-5] Several are associated with the spontaneous systemic lupus disease developed by the New Zealand Black, the New Zealand Black/New Zealand White hybrid and the MRL/l, MRL/n strains of inbred mice.[3-5] Another system for the study of lymphocyte infiltration into exocrine glands is that described in the histopathological observations of the graft-versus-host mouse.[6] To date, however, none of these mice develop a corresponding salivary or lacrimal gland dysfunction critical in the diagnosis and pathology of S.S.[7-9]

Type 1 insulin dependent diabetes in man results from the autoimmune destruction of the ß-cells of the islet of Langerhans found in the pancreas.[10] Immune system activation also results in a general lymphocytic attack on a number of exocrine tissues, which includes the salivary glands. In fact, 30% of patients with diabetes reports symptoms of xerostomia.[11] In this regard, several animal models have been developed for the study of diabetes.[12] The non-obese diabetic (or NOD) inbred mouse closely resembles the human disease including the infiltration of lymphocytes into the salivary glands.[13, 14] We have now examined this autoimmune-disease prone strain for S.S. pathology.[14] In this paper, we report that the NOD mouse demonstrates both xerophthalmia and xerostomia in response to autoimmune system activation. Further, these animals develop hypergammaglobulimenia with the presence of antinuclear antibodies (ANA). The onset of salivary gland dysfunction is age dependent, with onset of hyposalivation occurring between 15-18 weeks of age.

## METHODOLOGY

Non-obese diabetic (NOD) mice were obtained from the Department of Pathology, University of Florida. All animals used in this study were age-and sex-matched with BALB/c

*Lacrimal Gland, Tear Film, and Dry Eye Syndromes*
Edited by D.A. Sullivan, Plenum Press, New York, 1994

controls. Time course of salivary gland dysfunction was assessed starting at 6 weeks of age through 24 weeks and assayed at three week intervals. Collection of saliva by isoproterenol and pilocarpine stimulation was performed as described previously.

Biochemical analysis of saliva composition by SDS polyacrylamide gels, amylase enzyme assay and total protein determination were carried out as described by Hu et al.[14] Antinuclear antibody staining was performed using a Sigma Chemical Co. test kit which provided rat liver tissue fixed on glass slides and a human primary antiserum containing ANA as a positive control. Test sera from NOD and BALB/c animals was collected as described by Humphreys-Beher et al.[15] Visualization of the staining patterns of the sera was obtained by using and FITC-conjugated goat antimouse second antibody. The slides were evaluated using a Zeiss light microscope. Results were recorded using Kodak 3200 speed Kodachrome print film. Shirmer's test for tear formation was performed as described by Hoffman et al.[7] using Whatman 1 M filter paper cut into 0.5 x 3.0 mm strips subsequently placed in the medial canthus of each eye for 2 min periods. The strips were then removed and the soaked area measured at 10X magnification using the internal ocular micrometer on a dissecting microscope.

## RESULTS

The salivary glands and lacrimal glands of NOD mice were evaluated for their ability to form liquid secretions by stimulation of saliva flow by isoproterenol and pilocarpine treatment or filter paper wetting capacity, respectively. As summarized in Table 1, NOD diabetic and nondiabetic animals showed a reduced capacity to synthesize and secrete saliva fluid and protein constituents. Saliva flow was reduced by 75% and 83% for male and female NOD mice respectively when compared to control animals. Total protein correspondingly declined by 70% and 81% for male and female NOD diabetic when compared to controls. The saliva specific proteins amylase (a product of the parotid acinar cells) and EGF (a product of submandibular gland ductal cells) were assayed to specifically correlate these changes in overall protein synthesis. While EGF showed a decline of 500- and 18-fold for male and female diabetic mice relative to BALB/c mice respectively, amylase levels appeared to recover approximately 50% from the low levels of enzyme activity found in the non-diabetic mice ($p<0.01$).

**Table 1.** Salivary gland function measured at 6 months of age.

| Animal | Sex | Flow Rate[2] | Protein Conc.[3] | Amylase Act.[4] | EGF Concen. | Serum IgG[6] | Schirmer Test[7] |
|--------|-----|------------|------------------|-----------------|-------------|--------------|------------------|
| BALB/c[1] | M | 280 ± 15 | 715 ± 10 | 600 ± 65 | 925 ± 50 | 0.8 ± 0.1 | 2.7 ± 0.3 |
| | F | 375 ± 55 | 1185 ± 120 | 580 ± 20 | 34 ± 10 | 1.7 ± 0.3 | 3.6 ± 0.3 |
| NOD | M | 140 ± 50 | 300 ± 150 | 225 ± 25 | 264 ± 36 | 1.7 ± 0.3 | 3.1 ± 0.3 |
| non-diabetic | F | 73 ± 21 | 300 ± 100 | 230 ± 35 | 52 ± 17 | 2.4 ± 0.3 | 2.6 ± 0.4 |
| NOD | M | 70 ± 16 | 225 ± 50 | 425 ±15 | 1.7 ± 1.0 | 2.4 ± 0.1 | 1.8 ± 0.5 |
| diabetic | F | 66 ± 15 | 230 ± 50 | 390 ± 30 | 1.9 ± 1.3 | 2.4 ± 0.2 | 2.3 ± 0.5 |

[1] n = 6. All values expressed as mean ± standard deviation.

[2] Flow rate expressed as ul/15 min.

[3] Protein concentration expressed as µg/15 min.

[4] Amylase activity expressed as mg starch split/min/mg saliva protein.

[5] EGF concentration expressed as µg/ml saliva.

[6] Values expressed as mg/ml IgG.

[7] Assay of Whatman 1M filter paper wetting over a 20 min period. Values expressed as mm wetted.

A corresponding decline in tear formation was observed in NOD non-diabetic female and male and female diabetic mice (Table 1). Filter paper wetting rates were the same for male control and male non-diabetic NOD mice. The rate of filter wetting in male diabetic NOD mice was reduced by 33% (p < 0.01) when compared to BALB/c animals. For female diabetic mice, filter paper wetting was reduced by 36% (p < 0.05). The decline in saliva flow rates were only observed in the NOD strain of mice. A sampling of a total of six inbred mice which included MRL and New Zealand Black and White strains, which are commonly utilized as models for S.S., revealed the decline in salivary gland function to only be associated with the NOD mouse (Table 2).

**Table 2.** Salivary flow rates for several inbred murine strains developing autoimmune disease.

| Strain[1] | Stimulated Flow[2] |
|---|---|
| NZB | $225 \pm 15$ |
| NZW | $238 \pm 18$ |
| C57/B1 | $264 \pm 20$ |
| NOD | $53 \pm 5$ |
| MRL/lpr | $230 \pm 25$ |
| BALB/c | $245 \pm 16$ |

[1] n = 4. All animals were 6 mo. of age and female sex-matched.
[2] Saliva flow was stimulated as described previously expressed as µl/15 min/100g body wt using a combined injection of isoproterenol and pilocarpine. Values expressed mean ± standard deviation from 2 independent collection times of 15 min duration.

Changes in the salivary gland functional status were examined for age related onset for the parameters outlined in Table 1. Saliva flow rate, declining total protein concentration, and declining amylase activity were all observed to begin changing between 15 and 18 weeks of age in female mice (p < 0.01; Figure 1). It is at this time (15 weeks) that lymphocytic foci begin to become detectable in the submandibular gland (data not shown).

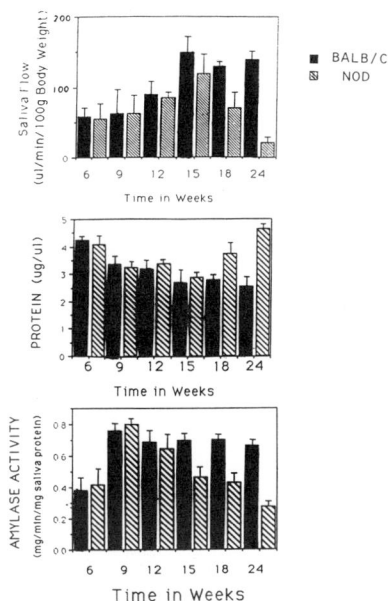

**Figure 1.** Histogram of age-related changes in salivary gland function in BALB/c and NOD mice. Panel A, saliva flow rates following stimulation with a combined injection of isoproterenol and pilocarpine in female mice. Panel B, protein levels in the saliva of mice at different ages. Panel C, alpha amylase activity in saliva of female mice. Solid bars represent BALB/c controls; light hatched bars represent non-diabetic NOD mice.

When proteins in saliva were analyzed by SDS polyacrylamide gel, the most obvious change was the observation of the mobility shift in the glycoprotein at 33,000 Da of NOD mice (Figure 2). Staining for total protein with Coomassie blue showed the decline in amylase protein at 58,000 Da which corresponded with the loss of enzyme activity described above.

Finally, sera from NOD and BALB/c female mice were evaluated for the presence of antinuclear antibodies by fluorescent staining of fixed rat liver sections. ELISA assay of serum IgG levels were found to be increased in all NOD groups examined (Table 1). When NOD sera from female diabetic mice were evaluated for antinuclear autoantibody staining, two patterns were observed (Figure 3). In one sample as indicated in panel C using rat liver slices, a perinuclear pattern was detected while in panel D a uniform speckled nuclear staining was observed. This particular serum had a titer of 1:1600 against HEp-2 cells as well as reacting against the nuclear components of 3T3 fibroblasts.[15] BALB/c serum served as a negative control as shown in Figure 3, panel A, while a human serum containing ANA reactive antibodies, shown in panel B, gave a uniform highly fluorescent nuclear staining pattern.

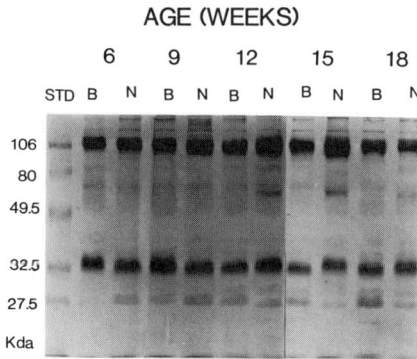

**Figure 2.** SDS polyacrylamide gel separation of total saliva protein stained for glycoproteins by periodic acid-Schiff reagent. Saliva sample is a representative from each age group listed in the figure. Note the shift in mobility of a major glycoprotein of $M_r = 33000$ Da at 15 and 18 weeks. Prestained molecular weight standards are Phosphorylase B, 106000 Da; Bovine serum albumin, 80000 Da; Ovalbumin, 48500 Da; Carbonic anhydrase, 32500 Da; Soybean trypsin inhibitors, 27500 Da. Thirgy-five µg of protein were loaded per well.

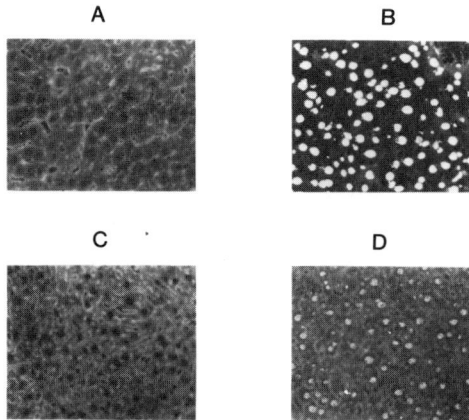

**Figure 3.** Detection of antinuclear antibodies in the sera of NOD mice. Rat liver sections (Sigma Chemical Co.) were reacted with a 1:40 dilution of BALB/c or NOD sera in PBS FITC-conjugated goat antimouse kappa and lambda light chain were subsequently reacted and the fluorescence observed using a Zeiss light microscope equipped with a 35 mm camera. Panel A, reaction with BALB/c serum; panel B, reaction with a positive control human serum containing antinuclear antibodies; panel C, NOD serum showing perinuclear staining patterns; panel D, NOD serum demonstrating speckled nuclear staining pattern. X = 150.

# DISCUSSION

To date, a number of animals models for S.S. have been described based upon the observation of foci of lymphocytic infiltrates of the salivary and lacrimal glands.[4-7] However, the adequacy of these animals in truly representing the etiology of the disease has not been addressed. While showing similar histopathology, mouse models evaluated have so far failed to provide evidence of functional loss of these exocrine tissues as observed in the human disease state. Thus, their true value in evaluating new therapies is questionable as is the pathological consequences of the observations of immune cells subtypes in the specific glandular lesions.

In this paper we have described a new animal model for the study of secondary Sjögren's syndrome associated with the development of autoimmune-induced Type 1 insulin-dependent diabetes. The sexual dimorphism of diabetes onset with females affected more often (4:1) than males is typical of the prevalence of S.S. in human female population over that observed for men.[1] The NOD mouse develops lymphocytic infiltrates of the salivary glands beginning at 15 weeks of age. It is at this time salivary gland dysfunction begins as well, as evidenced by the loss of stimulated flow rates, declining protein synthesis and secretion. The decline in gland function appears to reach a maximum by 24 weeks of age. Concomitant with the decline in oral function is the observed decline in ocular function as presented in the reduced Schirmer test in the NOD mice. Finally, in the NOD mouse S.S. pathology, we have presented evidence for hypergammaglobulinemia which is accompanied by the appearance of autoantibody towards nuclear antigens. The observation of specific ANA S.S.-A/Ro and S.S.-B/La have been reported to aid in the diagnosis of S.S. in the human disease.[16] Therefore we propose, based on the histopathological and biochemical observations presented to date, that the NOD mouse represents a significant advance it the development of an acceptable animal model reflecting not just the histopathology of S.S. but the biological consequences on salivary and lacrimal gland function. Thus, the NOD mouse can be expected to provide new insights into the pathology of this devastating illness as well as provide a system for the evaluation of new and existing drug therapy available to alleviate the negative sequelae.

# ACKNOWLEDGEMENTS

The authors would like to thank Messrs. Micah Kerr and Terry Locey for technical assistance. Additional acknowledgement is given Ms. Marilyn Lietz for preparation of this manuscript. This work was supported by NIDR grant DE 08778 and DE 00291 to MHB. Dr. Hu is supported by a graduate assistantship from the Department of Oral Biology.

# REFERENCES

1. K.J. Bloch, W.W. Buchanan, M.J. Wohl, and J.J. Bunin, Sjögren's syndrome: a clinical, pathological, and serological study of 62 cases, *Medicine* 44:187 (1965).
2. M.A. Alspaugh and K. Whaley , "Sjögren's Syndrome: Textbook of Rheumatology,"E.D. Harris Jr.,S. Ruddy and C.B. Sledge, eds., W.B. Saunder Co., Philadelphia (1981).
3. I. Schwartz, Lacrimal and salivary gland inflammation in the C3H/lpr autoimmune strain mouse: A potential mode for Sjögrens syndrome, *Otolaryngol. Head Neck Surg.* 106:394 (1992).
4. H.S. Kessler, A laboratory model for Sjögrens syndrome, *Am. J. Pathol,* 52:671 (1968).
5. R. Jonsson, A. Tarkowski, K. Bäckman, R. Holmdahl, and L. Klareskog, Sialadenitis in the MRL-1 mouse: morphological and immunohistochemical characterization of resident and infiltrating cells, *Immunol.* 60:611 (1987).
6. I. Sørensen, A.P. Ussing, J.U. Prause, J. Blom, S. Larsen and J.V. Spärck, Histological changes in exocrine glands of murine transplantation chimeras. I: The development of Sjögrens syndrome-like changes secondary to GVH induced lupus syndrome, *Autoimmunity* 11:261 (1992).

7.  R.W. Hoffman, M.A. Alspaugh, K.S. Waggie, J.B. Durham and S.E. Walker, Sjögrens syndrome in MRL/l and MRL/n mice, *Arthritis Rheum.* 27:157 (1984).
8.  A. Wolff, J. Scott, K. Woods and P.C. Fox, An investigation of parotid gland function and histopathology in autoimmune disease-prone mice of different age groups, *J. Oral Pathol. Med.* 20:486 (1991).
9.  A.P. Ussing, J.W. Prause, I. Sørensen, S. Larsen and J.V. Spärck, Histological changes in exocrine glands of murine transplantation chimeras. II: Sjögren's syndrome-like exocrinopathy in mice without lupus nephritis. A model of primary Sjögren's syndrome, *Autoimmunity* 11:273 (1992).
10. G. Eisenbarth, Insulin dependent diabetes mellitus: a chronic autoimmune disease, *N. Engl. J. Med.* 314:1360 (1986).
11. L.M. Screebny, A. Yu, A. Green, and A. Valdini, Xerostomia in diabetes mellitus, *Diabetes Care* 15:900 (1992).
12. P.A. Gottlieb, A.A. Rossini and J.P. Mordes, Approaches to prevention and treatment of IDDM in animal models, *Diabetes Care* 11:29 (Suppl. 1) (1988).
13. H. Asamoto, M. Oishi, Y. Akagawa, and Y. Tochino, Histologic and immunologic changes in the thymus ad other organs in NOD mice, *in:* "Insulitis and Type 1 Diabetes," T. Seiichiro, Y. Tochino, and K. Noraka, Eds., Academic Press, Tokyo (1986).
14. Y. Hu, Y. Nakagawa, K.R. Purushotham and M.G. Humphreys-Beher, Functional changes in salivary glands of autoimmue disease-prone NOD mice, *Am. J. Physiol.* 263:E607.
15. M.G. Humphreys-Beher, L. Brinkley, K.R. Purushotham, P.-L. Wang, Y. Nakagawa, D. Dusek and E.K.L.Chan, Characterization of antinuclear antibodies present in the serum from non-obese diabetic (NOD) mice, *A. J. Physiol.* submitted.
16. E.K.L. Chan and L.E.C. Andrade, Antinuclear antibodies in Sjögrens syndrome, *in:* "Rheumatic Disease Clinics of North America," R.I. Fox, ed., W.B. Saunders Co., Philadelphia (1992).

# EXPRESSION OF GRANZYME A AND PERFORIN IN LACRIMAL GLAND OF SJÖGREN'S SYNDROME

Kazuo Tsubota[1,2], Ichiro Saito[2], and Nobuyuki Miyasaka[2]

[1]Department of Ophthalmology, Tokyo Dental College, Chiba, Japan
[2]Department of Virology and Immunology, Medical Research Institute
Tokyo Medical and Dental University, Tokyo, Japan

## INTRODUCTION

Tissue destruction of the lacrimal and salivary glands is the cardinal feature of Sjögren's syndrome. Although lymphocyte infiltration accompanies this destruction, the mechanism of tissue destruction has not been elucidated. Recently, it was suggested that granzyme A and perforin are associated with immune-mediated cytolysis in such autoimmune diseases as rheumatoid arthritis, where the up-regulation of these compounds in synovial fluid lymphocytes causes tissue destruction.[1] We have investigated whether granzyme A and perforin are associated with Sjögren's syndrome.

## MATERIALS AND METHODS

### Patients

Six primary Sjögren's syndrome patients with severe keratoconjunctivitis sicca, all women ranging in age from 31 to 62 yrs, underwent lacrimal gland biopsy which showed extensive lymphocytic infiltration and gland destruction. As controls, lacrimal gland biopsy tissues without glandular destruction from six non-Sjögren dry eye patients who had no autoimmune background and normal lacrimal glands from six age-and sex-matched autopsy cases were used.

## RT-PCR Amplification

Total RNA preparation was performed using a modified guanidium isothiocyanate procedure. The PCR was performed using the synthesized cDNA. The sequences of the primers and probes were specific, as confirmed by a computer-assisted search of updated versions of GenBank. The following primers were used: The primers sequences for perforin are 5'-ATCCTTCTCCTGCTGCTG-3' and 5'-CTGTAGGGCATTTTCACAG-3', generating a 214-bp PCR product. The internal probe for perforin is 5'-AGTGGACA CACAAAGGTT-3'. The primers for granzyme A are 5'-ACTCCTCATTCAAGACCCTA-3' and 5'-GTGGCTGGCTCATAGGATGG-3' amplifying a 252-bp region of the gene. The internal probe for granzyme A is 5'-ACCAGAGCTGTGCAGCCCCTCAGGCTACCTAG CAACAAGGCCCAGGTG-3'. The primer sequences of β-actin are 5'-CCTTCCTGGGCA TGGAGTCCTG-3' and 5'-GGAGCAATGATCTTGATCTTC-3' generating a 202-PCR product, used as internal control for quantification. The probe for β-actin is 5'-AAAGACC TGTACGCCAACA-3'. The cycle was repeated 35 times by DNA thermal cycler. Specificity of the amplified bands was validated by their predicted size and hybridized with the specific $^{32}$P-labeled internal probe. The autoradiographs were analyzed with laser densitometer.

## Immunohistochemistry

The expression of granzyme A and perforin were also investigated by immunohistochemical staining. Acetone-fixed sections on gelatin-coated slides were incubated with mouse monoclonal antibodies directed against human pore-forming protein (PFP) (courtesy of Dr. Yagita, Juntendo University, Tokyo, Japan), and granzyme A (T cell sciences, Cambridge, MA).

## RESULTS AND DISCUSSION

Reverse transcriptase polymerase chain reaction was employed to detect granzyme A and perforin mRNA, which were present in all 6 lacrimal glands of Sjögren's syndrome but in none of the controls (Fig. 1). Immunohistochemical staining also showed the localized expression of both substances in the infiltrating CD4 [+] T cells (Figures 2 and 3).

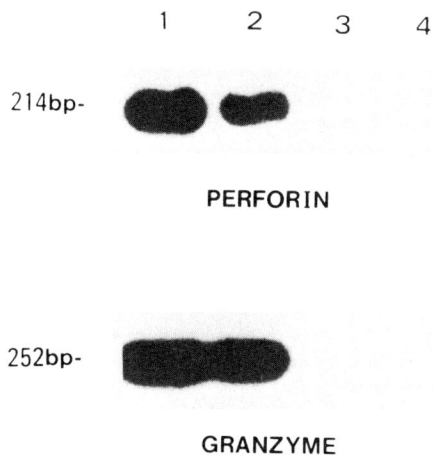

**Figure 1.** Southern blot analysis of amplified products by polymerase chain reaction of granzyme A and perforin in lacrimal gland biopsies from Sjögren's syndrome patients (Lanes 1 and 2), a non-Sjögren's dry eye patient (Lane 3) and a normal healthy control (Lane 4). All autoradiographs were exposed for 4 h. The sizes of the amplification products were 252 bp for granzyme A and 214 for perforin.

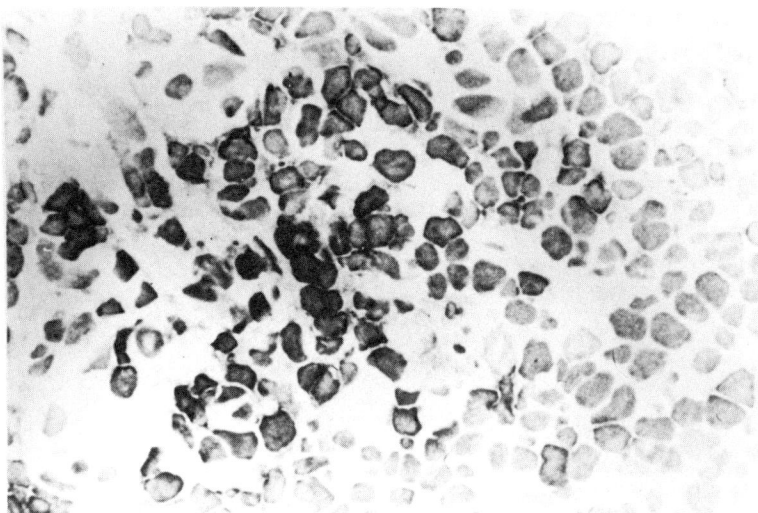

**Figure 2.** Immunohistochemical staining of granzyme A in the infiltrating lymphocytes. The localization to the lymphocytes is apparent.

The data does not prove that granzyme A and perforin are directly involved in cell-mediated killing of the lacrimal gland. However, since granzyme A and perforin are upregulated by the known signal to induce cytotoxic function,[2] and can be produced by CD4 + T cells,[1] it appears that these compounds are associated with a killing role. The immunosuppressive effects of cyclosporin A on T cells are well known, which can in part

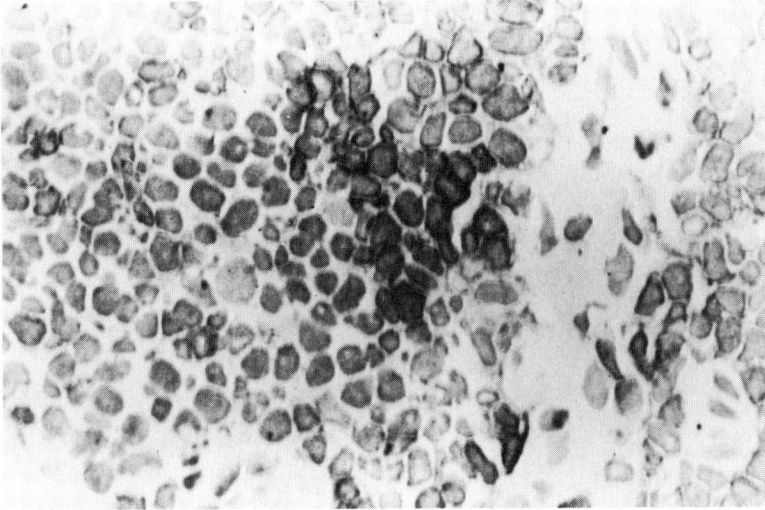

**Figure 3.** Immunohistochemical staining of perforin, which is localized in the infiltrating lymphocytes.

be explained by its suppressive activity of granzyme A and perforin.[2] Thus, cyclosporin A can be a potent agent to prevent the destruction of the lacrimal gland in Sjögren's syndrome through the pathogenetic mechanism proposed in this study.

## REFERENCES

1. G. Griffiths, S. Alpert, E. Lambert, J. McGuire, and I. Weissman, Perforin and granzyme A expression idendifying cytolytic lymphocytes in rheumatoid arthritis, *Proc. Natl. Acad. Sci .USA* 89:549 (1992).
2. C. Liu, S. Rafii, A. Granelli-Piperno, J. Trapani, and J. Young, Perforin and serine esterase gene expression in stimulated human T cells, *J. Exp. Med.* 170:2105 (1989).

# EPSTEIN-BARR VIRUS AND THE LACRIMAL GLAND PATHOLOGY OF SJÖGREN'S SYNDROME

Stephen C. Pflugfelder[1], Cecelia A. Crouse[1,2], and Sally S. Atherton[1,2]

[1]Department of Ophthalmology (Bascom Palmer Eye Institute)
[2]Department of Microbiology and Immunology
University of Miami School of Medicine
Miami, Florida

## INTRODUCTION

Sjogren's Syndrome (SS) is an immunologic disease occurring predominantly in women that is associated with severe dry eyes and disabling ocular irritation. SS has been reported to be the most common immunologic disease next to rheumatoid arthritis, and has an estimated prevalence of 1.2 million patients in the United States (based on a reported prevalence of 0.4% in Sweden[1]).

A B-lymphocyte proliferation develops in the lacrimal glands of SS patients that results in their dysfunction and eventual destruction.[2] Although the cause of the lacrimal gland (LG) lymphoproliferation in SS has not been firmly established, there is increasing evidence suggesting that Epstein-Barr virus (EBV) may play an important role in the LG pathology of SS. EBV is a human herpesvirus that is capable of infecting the lacrimal[3] and salivary glands.[4,5,6,7,8] There have been multiple case reports of primary SS developing immediately after serologically-confirmed infectious mononucleosis.[9,10,11] The current knowledge regarding the role of EBV in the LG pathology of primary SS is presented herein. The LG immunopathology of SS and the biology of (EBV) will be reviewed initially since this information is useful in understanding EBV infection in normal and SS LGs.

## LACRIMAL GLAND PATHOLOGY OF SJOGREN'S SYNDROME

The human LG consists of tubuloacinar structures composed of morphologically and immunohistochemically distinct epithelial cell components.[12] The LG also contains an immunoarchitecture that reflects its role as a component of the secretory immune system.[13] Lymphoid follicles containing central B-cells and peripheral CD4 T-cells are typically found in the center of LG lobules.[2,13] Numerous lymphocytes consisting predominantly of IgA-secreting plasma cells and CD8 T-lymphocytes reside in the interstitium of the LG, surrounding secretory acini.[2,13] The lymphoproliferation in SS LGs consist predominantly of B-lymphocytes, may be clonal, and typically starts in the center of the LG lobule.[2,14] As the B-lymphoproliferation progresses, it spreads in a centrifugal fashion toward the periphery of the LG lobule surrounding epithelial ducts and islands, and eventually replacing secretory acini.[14] Large B-cells expressing CD23 (blast-

2 antigen) and ICAM 1 are found in the center of the B-lymphoproliferation, and smaller mature B-cells are found in the periphery of the B-lymphoproliferation in SS LGs.[15] Unlike normal LGs, <5% of the cells in the lymphoproliferation in SS LGs are plasma cells.[14]

## BIOLOGY OF EPSTEIN-BARR VIRUS

EBV is a human herpesvirus capable of infecting B-lymphocytes and mucosal epithelia in vivo.[16]   One consequence of EBV infection of B-cells is stimulation of intrinsic growth pathways by the EBV antigens, latent membrane protein (LMP) and EBV nuclear antigen 2 (EBNA 2).[16] EBV growth transformed B cells typically express CD21 (the EBV receptor), CD23 (blast-2 antigen), and the adhesion molecules ICAM 1, LFA-1, and LFA-3.[16] Uncontrolled EBV-induced lymphoproliferation in humans is prevented by cytotoxic T-lymphocytes (CTLs) that recognize and lyse EBV-infected cells.[17] EBV-induced B-lymphoproliferative disorders may develop in patients receiving intensive immunosuppressive therapy to prevent immunologic rejection of transplanted organs, or in patients with HIV infection.[18,19] In contrast to B-lymphocytes, EBV infection of epithelial cells often results in cell lysis and release of infectious virus.[16,18].

Similar to other herpesviruses, EBV can persist in a latent non-pathogenic state for the lifetime of an infected individual.[16]   A number of sites of EBV persistence have been identified, and thus far all of these sites are components of the mucosal associated lymphoid tissue (MALT) and include the major and minor salivary glands,[4-8] oropharyngeal epithelia,[20] cervical epithelia,[21] LG,[3] and corneal epithelium.[22]

## EBV INFECTION IN NORMAL AND SJOGREN'S SYNDROME LACRIMAL GLANDS

Direct and indirect evidence has been reported that EBV plays a pathogenic role in the LG pathology of primary SS. In 1990, Pflugfelder and associates reported that primary SS patients had significant elevations of serum antibodies to EBV viral capsid and early antigens compared to patients with non-SS aqueous tear deficiency and normal controls.[23] These results suggested that primary SS patients have a chronic persistent EBV infection that is a risk factor for their disease. Subsequently, Pflugfelder and associates reported the results of studies evaluating peripheral blood mononuclear (PBMN) cells, LG biopsies, and tear specimens from EBV seropositive controls and primary SS patients for the presence of EBV genomes using polymerase chain reaction (PCR).[24] EBV DNA sequences were amplified by PCR in 50% of SS PBMN cell specimens and 80% of SS LG and tear specimens. In contrast, EBV genomic sequences were detected in 32% of normal human LGs, but in none of the PBMN cell specimens from normal controls. Tsubota et al reported the results of similar studies evaluating lacrimal and salivary gland biopsies from normal controls and primary SS patients for the presence of EBV genomes by PCR in 1991.[25] They detected the presence of EBV genomes in 100% of LG biopsies from SS patients, and only 40% of LG biopsies from normal controls.  They also detected the presence of EBV genomes in the majority of salivary gland biopsies from SS patients; however, quantitative analysis of the number of EBV genomes indicated there was a 10-fold greater number of EBV genomes in LG than salivary gland biopsies from primary SS patients. Taken together, these studies indicate that EBV may persist in a small percentage of normal LGs, and that EBV genomes are found in the majority of LGs from primary SS patients suggesting that EBV may be a risk factor in the pathogenesis of the LG disease of SS.  Reported studies using PCR to detect EBV genomes in normal and SS LG biopsies did not indicate the infected cell type(s) within the LG, nor did they determine if the amplified EBV DNA sequences are from latent EBV genomes or replicating virus.

Recent studies performed by Pflugfelder and associates have evaluated LG biopsies from normal controls and primary SS patients to identify the cellular sites of EBV infection using in situ DNA hybridization, determine whether there are differences in the genotype of EBV infecting normal and SS LGs, and evaluate EBV genome expression in normal and SS LGs using a panel of monoclonal antibodies specific for EBV latent and lytic infection cycle antigens.[15]

Hybridization signals for EBV DNA were observed in intralobular ductal epithelia in 3/14 (21%) of normal human LGs. EBV DNA-positive ducts were observed in 12 to 33% of lobules in positive histologic sections. No hybridization signals were observed in lymphocytes or secretory acini in normal LGs. In SS LG biopsies, EBV-positive mononuclear cells were occasionally observed in the LG interstitium adjacent to residual acini; however, hybridization signals in acinar epithelial cells were similar to the background. The majority of cells located in areas of B-lymphoproliferation (identified by staining serial sections with anti-B-cell antibodies) showed positive hybridization signals. Additionally, intralobular ductal epithelia and epithelial islands (identified with cytokeratin antibodies in serial sections) located in areas of B-lymphoproliferation showed strong signals in SS LGs.

Sections cut from paraffin-embedded LG biopsies from 14 normal and eight SS LGs were amplified for three different regions of the EBV region using PCR: 1) the Bam HI W region that is reiterated approximately 12 times in the EBV genomes thus providing a greater likelihood for detection than single copy genes; 2) a polymorphic region in the EBV nuclear antigen 2 (EBNA-2) gene that differs in type 1 (EBNA-2A) and type 2 (EBNA-2B) virus strains; 3) the Bam HI WYH segment that spans a deletion of the EBNA-2 gene similar to that found in the non-transforming P3HR-1 EBV strain and recently detected in mucosal secretions from healthy and HIV-infected adult donors.[26] The Bam HI W internal repeat region was amplified in 5 of 14 (36%) normal human LGs, and 7 of 8 (88%) SS LGs. None of the 5 normal LG specimens positive for EBV DNA using the Bam HI W primers showed amplification products using the EBNA-2A primers; however, all 5 showed deletions of the region of the genome encoding EBNA-2 with the Bam WYH primers. Only type 1, but not EBNA-2 deleted EBV genomes were found in SS LGs. These results indicate that there is an abnormal persistent infection with EBV-1 strains in SS LGs, and there appears to be a predilection for persistence of EBNA-2-deleted strains in normal LGs.

A specific staining pattern for EBV-specific antigens was not observed in any of the 14 normal LG biopsies evaluated. In contrast, EBV-associated antigens were observed in all SS LG biopsies evaluated. EBNA-2 and LMP expression was observed in large CD23-positive lymphocytes located in the center of areas of B-lymphoproliferation. Early antigen restricted component (EA-R)-positive cells were scattered throughout areas of B-lymphoproliferation, and EA-R staining was also occasionally observed in epithelial cells in SS LGs. Viral capsid antigen staining was not observed in mononuclear cells, but epithelia in areas of lymphoproliferation were often VCA positive in SS LGs suggesting that these cells may be lytically infected. Staining with BZ-1, a monoclonal antibody specific for an EBV protein associated with a shift from latent to lytic infection, was occasionally observed in ductal epithelia and mononuclear cells in SS LG biopsies.

## POSSIBLE MECHANISM OF EBV-INDUCED LACRIMAL GLAND PATHOLOGY OF PRIMARY SJOGREN'S SYNDROME

The results of experiments performed in our laboratory suggest that EBV may persist in the normal human LG in a latent non-pathologic state. The cellular site and state of genome expression in normal human LGs persistently infected with EBV appears to be similar to that reported to occur in normal salivary glands.[4,7,8]

In contrast, the results of studies in our laboratory using in situ DNA hybridization and immunohistochemical techniques to evaluate SS LGs for EBV infection indicate that there is a much more extensive infection of ductal epithelia than normal LGs, as well as infection of mononuclear cells in areas of B-lymphoproliferation. EBV antigens were detected in both lymphocytes and epithelial cells in SS LGs; however, the patterns of antigen expression differed in these two cell types. EBV antigens associated with growth transformation of B-cells, LMP and EBNA-2, were detected in mononuclear cells in areas of B-lymphoproliferation. B-cells in SS LGs expressing EBV latent infection cycle antigens also expressed ICAM-1, CD-23, and CD-21, the typical repertoire of antigens up-regulated by EBV following growth transformation of B-cells. Based on these findings, it appears that the EBV infection of B-lymphocytes in SS LGs may be responsible for the B-lymphoproliferation observed in these glands.

In contrast, epithelial cells located in areas of lymphoproliferation in SS LGs strongly express early (EA-R) and late (VCA) EBV lytic-cycle antigens. These findings suggest that a lytic EBV infection may occur in epithelial cells in SS LGs. Because we have previously reported the detection of EBV genomes in the majority (80%) of tear specimens obtained from primary SS patients, it is possible that EBV-infected ductal epithelia may may be the source of the virus shed into the tears.

Similar to other EBV-associated neoplasias, a lymphoepithelial pathology is frequently observed in SS LG biopsies.[27] The lymphoepithelial pathology in SS LGs differs from nasopharyngeal carcinoma (NPC) in that the lymphoproliferation surrounding epithelia in SS LGs consist predominantly of B-lymphocytes, whereas, T-cells typically surround epithelia in NPC.[28] In SS LG lobules with mild inflammation, the B-lymphoproliferation is observed surrounding ducts in the center of the lobule and normal appearing acini are still present in the peripheral lobule. In more severely affected glands, the lymphoproliferation replaces all secretory acini and the ducts in areas of B-lymphoproliferation have an abnormal morphology and pattern of cytokeratin expression.

PCR genotype analysis indicated that the majority of EBV-positive SS LGs are infected with type 1 EBV. Type 1 EBV strains efficiently transform B-lymphocytes into continuous cell lines and the detection of this strain of virus in SS LGs is consistent with the B-lymphoproliferation observed in these LGs. This contrasts with normal LGs from which we were unable to amplify EBV type 1 specific sequences, and were found to be infected exclusively by EBV with EBNA-2 deletions typical of non-transforming type 2 EBV strains. Although the sample size in our study is small and additional studies are needed to confirm our results, the difference in virus strain between normal and SS LGs may be important in the pathogenesis of LG destruction in SS.

The predominant EBV-specific CTLs in humans are human leukocyte antigen (HLA) class I restricted (CD8) T-cells.[17] EBV-specific CTLs have been reported to efficiently lyse HLA-restricted B-cells infected with type 1 EBV strains, and poorly recognize cells infected with type 2 EBV strains.[29,30] CD8 T-cells are the predominant population surrounding acini and proximal ducts in normal LGs.[13] One potential role of CD8 cells in the LG may be to recognize and destroy cells within the LG infected with type 1 EBV. These cells could include EBV-infected ductal epithelia or B-cells which continuously traffic into the gland. LG cells infected with type 2 EBV strains may be able to elude recognition by resident CD-8 CTLs. This hypotheses may explain the fact that only EBNA-2-deleted type 2 EBV DNA was found in normal LGs.

The higher frequency of EBV infection in the blood and LGs of SS patients may result from the inability of CTLs from SS patients to recognize and destroy cells infected with certain strains of type 1 EBV. Misko and associates recently studied paternal EBV-specific CTL activity against EBV-infected lymphoblastoid cell lines (LCLs) established by infecting peripheral blood B-cells obtained from five different children in his family with cither the B95-8 or the BL 74 EBV strains.[31] The paternal HLA type was A1, 11; B51, 8; DR3, 7. Paternal EBV-specific CTLs efficiently lysed haploidentical EBV LCLs infected with the B95-8 strains expressing the HLA A11, B51, DR7 paternal haplotype, but failed to lyse haploidentical LCLs infected with the B95-8 strain expressing the HLA A1, B8, DR3 paternal haplotype. LCLs expressing either of the paternal HLA haplotypes infected with the BL74 strains were efficiently lysed by paternal LCLs. The authors found that the failure to lyse HLA B8-restricted LCLs infected with the B95-8 strains was not due to T-cell dysfunction, and they concluded that the failure to lyse was probably due to an inability of the HLA B8 antigen to present the immunodominant B95-8 epitope to HLA class I restricted CTLs. B95-8 LCLs coated with the BL74 immunodominant peptide were efficiently lysed by the paternal CTLs. Interestingly, the HLA B8, DR3, DW52A, DQW2 haplotype is strongly associated with primary SS (relative risk of 8).[32] As suggested by Misko and associates, the HLA B8 haplotype association in SS patients may be one of the principle risk factors for their abnormal EBV infection. Alternatively, the EBV-induced lacrimal gland B-lymphoproliferation in SS may be related to other cellular immune derangements previously reported to occur in SS patients with severe dry eyes.

The results of the studies reported herein strongly suggest that EBV plays an important role in the B-lymphoproliferation and epithelial pathology occurring in the LGs of primary SS patients. The persistent EBV-1 infection in SS LGs may be related to a genetic inability to lyse LG cells infected with certain EBV-1 strains. These results indicate that the development of LG pathology in primary SS depends on a number of risk factors including HLA type, and type of infecting EBV strain. Additionally, it is likely that female gender is another important risk factors for developing SS. The results of these studies have definite therapeutic implications for primary SS. Since there appears to be a lytic infection in LG ductal epithelial cells, antiviral therapy may prove effective in limiting virus induced LG destruction. Other therapies could be targeted at the risk factors for the disease, and could include stimulation of the defective immune surveillance, or administration of androgen hormones since they appear to provide protection against development of primary SS in males.

## ACKNOWLEDGEMENTS

This investigation was supported in part by Public Health Service Research Grants EY08711 (SCP), EY06012 (SSA), Core Grant EY02180 and Training Grant 32 EY07129 (CAC) National Institutes of Health, National Eye Institute, Bethesda, MD 20205

## References

[1] Manthorpe R, Axell T, Hansen B et al.,1991,Prevalence of primary Sjogren's syndrome in patients with multiple sclerosis. Clin Exp Rheumatol. 9:326 (Abstract).

[2] Pepose JS, Akata RF, Pflugfelder SC, Voight W.,1990, Mononuclear cell phenotypes and immunoglobulin gene rearrangements in lacrimal gland biopsies from patients with Sjogren's syndrome. Ophthalmology. 97:1599-1605.

[3] Crouse CA , Pflugfelder SC, Cleary T, et al.,1990, Detection of Epstein-Barr virus genomes in normal human lacrimal glands. J Clin Microbiol. 28:1026-1032.

[4] Wolf H, Haus M, Wilmes E.,1984, Persistence of Epstein-Barr virus in the parotid gland. J Virol.51:795-798.

[5] Saito I, Servenius B, Compton T, Fox RI.,1989, Detection of Epstein-Barr virus DNA by polymerase chain reaction in blood and tissue biopsies from patients with Sjogren's syndrome. J Exp Med. 169:2191-8.

[6] Venables PJW, Teo CG, Baboonian C, Griffin BE, Hughes RA.,1989, Persistence of Epstein-Barr virus in salivary gland biopsies from healthy individuals and patients with Sjogren's syndrome. Clin Exp Immunol. 75:359-364.

[7] Mariette X, Golzan J, Clerc D, Bisson M, Morinet F.,1991,Detection of Epstein-Barr virus DNA by in situ hybridization and polymerase chain reaction in salivary gland biopsy specimens from patients with Sjogren's syndrome. Am J Med. 90:286-294.

[8]Deacon EM, Matthews JB, Potts AJC, Hamburger J, Bevan JS, Young LS.,1991,Detection of Epstein-Barr virus antigens and DNA in minor salivary glands using immunohistochemistry and polymerase chain reaction: possible relationship to Sjogren's syndrome. J Pathol. 163:351-360.

[9] Whittingham S, McNeilage J, Mackay IR.,1985,Primary Sjogren's syndrome after infectious mononucleosis. Ann Intern Med. 102:490-3.

[10] Pflugfelder SC, Roussel TJ, Culbertson WW.,1987,Primary Sjogren's syndrome after infectious mononucleosis (letter) JAMA. 257:1049-50.

[11]Gaston JSH. Rowe M, Bacon P.,1990, Sjogren's syndrome after infection by Epstein-Barr virus. J Rheumatol.17:558-61.

[12] Yen M, Pflugfelder SC, Crouse CA, Atherton SS.,1992, Cytoskeletal antigen expression in ocular mucosal associated lymphoid tissue. Invest Ophthalmol Vis Sci., 33:3235-3241.

[13] Wieczorek R, Jakobiec FA, Sacks EH, Knowles DM.,1988, The immunoarchitecture of the normal human lacrimal gland: relevancy for understanding pathologic conditions. Ophthalmology. 95:100-9.

[14] Monroy D, Pflugfelder SC.,1993, Immunohistochemical analysis of B-lymphoproliferation and epithelial pathology in Sjogren's syndrome lacrimal glands. Invest Ophthalmol vis Sci [ARVO Abstracts; (in press)].

[15] Pflugfelder SC, Crouse CA, Monroy D, et al.,1993, Epstein-Barr virus and the lacrimal gland pathology of Sjogren's syndrome. Am J Pathol (in press)

[16] Rickinson AB. On the biology of Epstein-Barr virus persistence: A reappraisal, in: Immunobiology and Propylaxis of Herpesvirus Infections, C. Lopez, ed. Plenum Press. New York, 1990, pp 137-146.

[17]Moss DJ, Misko IS, Sculley TB, et al.,1991, Immune regulation of Epstein-Barr virus (EBV); EBV nuclear antigen as a target for EBV-specific T cell lysis. Springer Semin Immunopathol. 13:147-156.

[18] Pagano JS.,1992,Epstein-Barr virus: culprit or consort. N Engl J Med . 24:1751-1753.

[19] Randhawa PS, Jaffe R, Demetris AJ, et al.,1992, Expression of Epstein-Barr virus-

encoded small RNA (by the EBER1 gene) in liver specimens from transplant recipients with post-transplantation lymphoproliferative disease. N Engl J Med. 327:1710-1714.

20 Sixbey JW, Nedrud JG, Raab-Traub N, et al.,1984, Epstein-Barr virus replication in oropharyngeal epithelial cells. N Engl J Med . 310:1225-1230.

21 Sixbey JW, Lemon SM, Pagano JS.,1986, A second site for Epstein-Barr virus shedding: the uterine cervix. Lancet. 2:1122-1124.

22 Crouse CA, Pflugfelder SC, Pereira I et al.,1990, Detection of herpes virus genomes in normal and diseased corneal epithelium. Current Eye Res. 9:569-581.

23 Pflugfelder SC, Tseng SCG, Pepose JS et al.,1990, Chronic Epstein-Barr viral infection and immunologic dysfunction in patients with aqueous tear deficiency. Ophthalmology. 97:313-323.

24 Pflugfelder SC, Crouse C, Pereira I, Atherton SS.,1990, Amplification of Epstein-Barr virus genomic sequences in blood cells, lacrimal glands, and tears from primary Sjogren's syndrome patients. Ophthalmology.97:976-984.

25 Tsubota K, Fujishima H, Toda I, Nishimura S, Saito I, Moro I.,1991, Increased level of Epstein-Barr virus DNA in lacrimal and salivary glands of patients with sjogren's syndrome. Invest Ophthalmol Vis Sci (ARVO abstracts) 32:807.

26 Sixbey JW, Shirley P, Sloas M, Raab-Traub N., Israele V.,1991, A transformation-incompetent, nuclear antigen 2-deleted Epstein-Barr virus associated with replicative infection. J Infect Dis.163:1008-1015.

27 Font RL. Yanoff M, Zimmerman LE.,1967, Benign lymphoepithelial lesion of the lacrimal gland and its relationship to Sjogren's syndrome. Am J Clin Pathol 48:365-76.

28 Weiss LM, Gaffey MJ, Shibata D.,1991,Lymphoepithelioma-like carcinoma and its relationship to Epstein-Barr virus (Editorial). Am J Clin Pathol. 96:156-158.

29 Murray RJ, Young LS, Calender A, Gregory CD, Rowe M, Lenoir GM, Rickinson AB.,1988, Different patterns of Epstein-Barr virus gene expression and of cytotoxic T-cell recognition in B-cell lines infected with transforming (B95-8) or nontransforming (P3HR1) virus strains. J Virol. 62:894-901.

30 Moss DJ, Misko LS, Burrows SR, Burman K, McCarthy R, Sculley TB.,1988,Cytotoxic T-cell clones discriminate between A- and B-type Epstein-Barr virus transformants. Nature. 331:719-721.

31 Misko IS, Schmidt C, Honeyman M, Soszynski TD, Sculley TB, Burrows SR, Moss DJ, Burman K.,1992, Failure of Epstein-Barr virus-specific cytotoxic T lymphocytes to lyse B cells transformed with the B95-8 strain is mapped to an epitope that associates with the HLA B8 antigen. Cin Exp Immunol. 87:65-71.

32 Arnett FC, Bias WB, Reveille JD.,1989, Genetic studies in Sjogren's syndrome and systemic lupus crythematosis. J Autoimmunity. 2:403-413.

# SJÖGREN'S SYNDROME (SS) AND EPSTEIN-BARR VIRUS (EBV) REACTIVATION

Ikuko Toda[1,2], Masafumi Ono[2], Hiroshi Fujishima[1,2], and Kazuo Tsubota[1,2]

[1]Department of Ophthalmology, Tokyo Dental College, 5-11-13, Sugano, Ichikawa, Chiba, Japan, 272
[2]Department of Ophthalmology, Keio University School of Medicine

## INTRODUCTION

Although the etiology of Sjögren's syndrome (SS) is unknown, many researchers have suggested an association between SS and immunologic dysfunction[1], genetic factors[2], or environmental factors. In fact, complex interactions of these factors may contribute to the development of this disorder. Recently, Epstein-Barr virus (EBV) DNA has been detected in the salivary gland[3], lacrimal gland[4] and conjunctiva[5] in SS patients, suggesting that reactivation of EBV contributes to the pathogenesis of SS. In addition, a reduction in cell mediated immunity (e.g. natural killing [NK] activity), which is important for the suppression of EBV infection, has been observed in SS patients[1]. This evidence has led to the hypothesis that EBV reactivation induces the polyclonal proliferation of B cells and the production of various cytokines, thereby resulting in the destruction of lacrimal and salivary glands through T cell activation.

In support of this hypothesis, Pflugfelder et al.[6] and Yamaoka et al.[7] reported an elevation in serum antibody titers to EBV in SS patients. However, the number of patients in these studies was small, and the possible relationship between EBV reactivation and non-SS dry eye was not explored. Moreover, since EBV is an ubiquitous virus that has infected most Japanese people,[7] the pattern of an individual's antibody response to different EBV antigens may have far more diagnostic value in detecting EBV reactivation than the analysis of a single titer to a given viral antigen.

As an additional consideration, many kinds of autoantibodies and immune complexes may be detected in the sera of SS patients (e.g. due to polyclonal B cell activation), but patients with non-SS dry eye may also have immunological abnormalities and low titers of autoantibodies. These autoantibody-positive dry eye (ADE) patients may well have a preliminary stage or mild type of SS.

Therefore, to better clarify the contribution of environmental (e.g. EBV) and immunological (e.g. autoantibodies) factors to the development of dry eye, and to distinguish possible disease-associated EBV reactivation patterns, we measured and compared the level of antibodies to three different EBV antigens in sera of patients with or without (i.e. ADE, simple dry eye [SDE], and control) SS.

## MATERIALS AND METHODS

### Three Immunologically Different Types of Dry Eye

Two hundreds and eighty seven dry eye patients (27 males and 260 females) with an average age of $53.8 \pm 13.3$ years were included in this study. The diagnosis of dry eye was based upon our previously reported criteria,[8] and patients were divided into three groups, according to their immunological status, as follows: (1) SS (n = 62, 2 males, 60 females; $53.6 \pm 11.9$ years old), which was diagnosed by using a modification of Fox's criteria[9], as well as an examination of lacrimal gland biopsies for lymphocytic infiltration; (2) ADE (n = 68, 5 males, 63 females; $55.1 \pm 14.0$ years old), which was associated with the presence of serum autoantibodies, including antinuclear factor (titer > 1:40) or positive rheumatoid factor (titer > 1:40); and (3) simple dry eye (SDE) (n = 157, 20 males, 137 females; $53.5 \pm 13.5$ years old), a condition in which patients showed no evidence of any serum autoantibodies. For control purposes, 47 healthy individuals (37 males, 10 females; $29.1 \pm 4.3$ years old), who had neither dry eye symptoms nor serum autoantibodies, were recruited.

### Detection of Serum Autoantibodies and Antibody Titers to EBV

To assess the autoimmune status of patients, antinuclear antibodies, rheumatoid factor, SS-A, and SS-B were measured in sera, as previously described[10, 11]. To determine EBV reactivation patterns, 3 types of serum antibodies to EBV were quantitated by using reported methods[12]. Briefly, antibody to EBV nuclear antigen (anti-EBNA) was detected by anti-complement immunofluorescence, whereas antibodies to EBV viral capsid antigen (anti-VCA-IgG) and EBV early antigen (anti-EA-IgG) were detected by indirect immunofluorescence. The CATMOD (categorical data modeling; analysis of weighted-least-squares) procedure (SAS Corporation Inc.) and chi square test were used to statistically evaluate the data.

## RESULTS

Striking differences existed among the 3 groups of dry eye patients with respect to their serum antibody profile to EBV antigens. Thus, the serum concentration of anti-EBNA antibodies was significantly higher in SS patients than in individuals with ADE (P<0.05) or

SDE (P<0.0001), or in controls (P<0.0001). Similarly, anti-EA-IgG antibody levels in sera of SS patients was significantly (P<0.001) higher than those of the other groups. In contrast, no significant differences in serum anti-EBNA or anti-EA-IgG antibody concentrations were found between ADE, SDE, and the controls. With regard to anti-VCA-IgG antibody titers, these were significantly higher in sera of SS patients, as compared to those of SDE patients (P<0.01) and control (P<0.05), but not to levels in the ADE group (Figure 1).

In the control population, 95% of individuals had anti-EA-IgG antibody titers of less than 1:10 or anti-VCA-IgG antibody levels lower than 1:320. Thus, we defined the EBV reactivated pattern to be anti-EA-IgG>1:10 and anti-VCA-IgG>1:320. Given these parameters, the EBV reactivated pattern was found in 17.7% of SS, 4.4% of ADE and 1.9% of SDE patients, and 0% of the controls. (Figure 2). This pattern was observed with significantly (p<0.05) higher frequency only in SS patients.

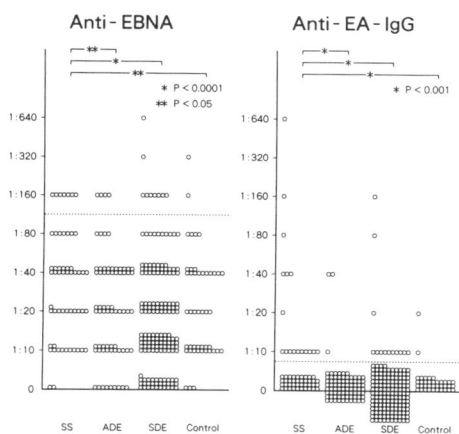

**Figure 1.** Antibody titers to EBV.

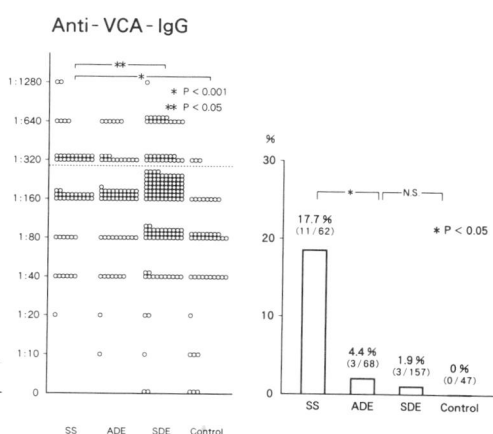

**Figure 2.** EBV Reactivated pattern.

## DISCUSSION

Three types of serum antibodies to EBV were significantly elevated in dry eye patients with SS, as compared to non-SS dry eye patients and controls. Furthermore, the EBV reactivation pattern, which was defined by an increased level of both anti-EA-IgG and anti-VCA-IgG serum antibodies, was seen in 17.7% of SS patients, and that was significantly higher than in ADE, SDE, and the controls. Since EA and VCA are synthesized only during EBV replication, this finding strongly suggests EBV reactivation in SS.

In SS, hypergammaglobulinemia is often observed, which suggests polyclonal B cell activation and the possibility that the elevation of antibody titers to EBV may result from non-specific B cell up-regulation. Pflugfelder et al reported that the antibody levels to other herpes viruses in SS were not elevated[6], which indicates that the increase in antibody titers to EBV in SS is specific. However, the serum antibody titer does not always reflect EBV reactivation, which may occur at selected sites. This may be why some SS patients do not show an elevation in antibody levels. The increased antibody titers to EBV were relatively lower in our SS patients compared to the other reports[6, 7]. This finding is probably due to the fact that most of our SS patients were in chronic stage of disease, wherein EBV replication may be compromised. We believe that EBV may play some role in triggering the pathogenesis of SS. Nevertheless, there may also be other mechanisms for maintaining SS state.

We categorized dry eye patients into three types according to their immunological status and considered that ADE might be a preliminary state of SS, but that SDE was not. However, there were no differences between ADE and SDE in antibody titers to EBV, suggesting that the mechanism of dry eye in ADE is different from that of SS, at least insofar as EBV is concerned. A long term follow-up of these data is still required to determine whether ADE ever converts to SS.

## REFERENCES

1. N. Miyasaka, W. Seaman, A. Bankshi, B. Sauvezie, V. Strand, R. Pope, and N. Talal, Natural killing activity in Sjögren's syndrome, *Arth. Rhuem.* 26:954 (1983).
2. J. Pepose, R. Akata, S. Pflugfelder, and W. Voigt, Mononuclear cell phenotypes and immunogloblin gene rearrangements in lacrimal gland biopsies from patients with Sjögren's syndrome, *Ophthalmology* 97:1599 (1990).
3. I. Saito, B. Servenius, T. Comton, and R. Fox, Detection of Epstein-Barr virus DNA by polymerase chain reaction in blood and tissue biopsies from patients with Sjögren's syndrome, J. Exp. Med. 169:2191 (1989).
4. S. Pflugfelder, C. Crouse, I. Pereira, and S. Atherton, Amplification of Epstein-Barr virus genomic sequences in blood cells, lacrimal glands, and tears from primary Sjögren's syndrome patients, *Ophthalmology* 97:976 (1990).
5. K. Tsubota, H. Fujishima, I. Toda, S. Katagiri, Y. Kawashima, S. Nishimura, I. Kudo, I. Saito, and I. Moro, Increased level of Epstein-Barr virus DNA in lacrimal gland of Sjögren's syndrome patients, *Invest. Ophthalmol. Vis. Sci.* submitted (1993).
6. S. Pflugfelder, S. Tseng, J. Pepose, A. Fletcher, N. Klimas, and W. Feuer, Epstein-Barr virus infection and immunologic dysfunction in patients with aqueous tear deficiency, *Ophthalmology* 97:313 (1990).
7. K. Yamaoka, N. Miyasaka, and K. Yamamoto, Possible involvement of Epstein-Barr virus in polyclonal B cell activation in Sjögren's syndrome, *Arth. Rheum.* 31:1014 (1988).
8. I. Toda, H. Fujishima, and K. Tsubota, Ocular fatigue is a major symptom of dry eye. *Acta Ophthalmol.* in press (1993).
9. R. Fox, C. Robinson, J. Curd, F. Kozin, and F. Howell, Sjögren's syndrome: proposed criteria for classification, *Arth. Rheum.* 29:577 (1986).
10. G. Lockitch, A. Halstead, G. Quigley, and C. MacCallum, Age-and sex-specific pediatric intervals: Study deseign and method illustrated by measurement of serum proteins with the bearing LN nephelometer, *Clin. Chem.* 34:1618 (1988).
11. O. Senju, Y. Takagi, R. Uzawa, Y. Iwasaki, T. Suzuki, K. Gomi, and T. Ishii, A new immunoquantitative method by latex aggrgation-application for the determination of serum C-creative protein (CRP) and its clinical significance, *J. Clin. Lab. Immunol.* 85:99 (1986).
12. W. Henle, G. Henle, and C. Horowitz, Infectious mononucleosis and Epstein-Barr virus-associated malignancies. in: "Diagnostic Procedures for Viral, Rickettsial and Chlamydial Infections, E. Lennette, and N. Schmidt, eds. Washington: American Public Health Association, p.441 (1979).

# HYPOTHESIS FOR AUTOANTIGEN PRESENTATION AND T CELL ACTIVATION

Austin K. Mircheff, J. Peter Gierow, Richard L. Wood, Ronald H. Akashi, and Florence M. Hofman

Departments of Physiology and Biophysics, Ophthalmology, Anatomy and Cell Biology, and Pathology
University of Southern California
School of Medicine
Los Angeles, CA 90033

## INTRODUCTION

Investigators have appreciated for many years that autoimmune processes lead to the periductal fibrosis and acinar atrophy characteristic of Sjögren's Syndrome. Some observations can be interpreted as suggesting that chronic, low-grade autoimmune phenomena account for qualitatively similar, but quantitatively less severe, changes regarded as age-related lacrimal gland atrophy (Damato et al., 1984). We wish here to review several lines of clinical and basic research which have produced new insights into cellular and molecular processes that might play critical roles in the initiation of such autoimmune phenomena.

Bottazzo and coworkers have noted that the targets of autoimmune processes frequently express Class II MHC molecules (1986). These molecules are normally expressed by macrophages and other specialized ("professional") antigen-presenting cells but not by other cell types, and, at the time of the initial observations, their role in antigen presentation was beginning to be uncovered. Accordingly, it was proposed that when a cell is induced to express Class II molecules, it gains the ability to present its own surface proteins as antigens. A number of groups have subsequently confirmed that different Class II$^+$ cell types, given appropriate antigenic peptides and accessory signals, are, in fact, capable of activating Class II molecule-restricted CD4$^+$ lymphocytes. However, the theory has to be modified to accommodate newer concepts of antigen generation and presentation, and questions remain concerning the signals responsible for inducing Class II molecule expression; the pathways by which autoantigens are processed and associate with Class II molecules; and the accessory signals which determine that antigen presentation results in immunity rather then tolerance. In partial answer to these questions, we propose that there are circumstances in which lacrimal gland acinar cells mimic key functions of the professional antigen-presenting cells.

## SPECIALIZED ANTIGEN-PRESENTING CELLS

The process of Class II-mediated antigen presentation normally begins when a macrophage or other professional antigen-presenting cell engulfs a foreign particle. The internalized material is carried to an endosomal compartment containing proteases that generate a mixture of peptide fragments. Certain fragments bind to Class II molecules passing through the endosomal compartment en route to the plasma membranes. Upon arrival at the cell surface, the complexes are recognized by receptors on CD4$^+$ T cells. The professional antigen-presenting cells also release the accessory signals, including IL-1 and IL-2, necessary for T cell activation. However, effective accessory signals may emanate from nearby cells, such as vascular endothelial cells, rather than from the antigen-presenting cells themselves (e.g., Geppert and Lipsky, 1987).

## ANALOGOUS EVENTS IN ACINAR CELLS

That the lacrimal glands contain immunogenic proteins has been amply demonstrated by Liu and coworkers, who have shown that at least two different proteins, one from acinar cells, the other from duct cells, can induce experimental autoimmune dacryoadenitis (Liu and Zhou, 1992). These proteins must be included among the likely suspects in the events leading to clinical autoimmunity. That lymphocytic infiltrates develop suggests that the antigens, or proteolytic fragments derived therefrom, are always present at certain levels in the lacrimal interstitium and that one or more factors normally prevent them from triggering autoimmune reactions. For example, helper and suppressor responses may counterbalance each other, or the combined frequencies of antigen-presenting cells and reactive CD4$^+$ cells may be so low that interactions leading to T cell activation are too rare to sustain a T cell response. Injection of antigen may alter either sort of balance by expanding clones of autoreactive CD4$^+$ cells elsewhere in the body.

Several observations confirm that acinar cells can be induced to express Class II molecules. Class II$^+$ acinar cells are infrequent in the lacrimal glands of young rats and rabbits, but acinar cells from both species readily become Class II$^+$ when placed in primary culture, and this response is enhanced by the cholinergic agonist, carbachol. Cadaver donor lacrimal glands exhibit a range of frequencies of Class II$^+$ epithelial cells, from virtually undetectable to virtually 100% (Mircheff et al., 1991).

Studies reviewed elsewhere in this volume (Mircheff et al., Gierow et al.) reveal an extensive recycling traffic between the basal-lateral membranes and early and late endosomal compartments of lacrimal acinar cells. Recent observations confirm that Class II molecules participate in this traffic. In one experiment, cultured rabbit acinar cells were chilled to 4°, incubated with the anti-Class II monoclonal antibody, 2C4, washed, incubated with ferritin-conjugated goat anti-mouse IgG, and again washed. If the cells were fixed, 2C4 could be seen in clusters at the plasma membrane. If, instead, cells were warmed to 37° and incubated for as little as 15 min prior to fixation, a substantial amount of the ferritin label could be seen to have been internalized. Additional experiments suggest that Class II molecule internalization may reflect turnover, rather than recycling. When cells were incubated with MAB 2C4 for 12 hr at 37°, then fixed, sectioned, and stained with ferritin-conjugated goat anti-mouse IgG, the accumulation of 2C4 in intracellular compartments predicted for equilibrium labeling of a recycling pool was not observed. On the other hand, a latent pool of Class II molecules could be detected by post-embedding labeling with MAB 2C4.

Most cells synthesize cathepsins to degrade cellular proteins in lysosomes. We recently detected cathepsin B immunoreactivity within both secretory vesicles and lysosomes of acinar cells (R.L. Wood, K.-H. Park, J.P. Gierow, and A.K. Mircheff, unpublished). We predicted that they should also be found in various compartments of the intracellular membrane assembly and recycling pathways, at least in the Golgi complex and in the *trans*-reticular Golgi network, but also, by analogy with lysosomal enzyme trafficking in other cell types, in endosomal compartments (Ludwig et al, 1991). Preliminary subcellular fractionation analyses confirm this prediction. Cathepsin B-like catalytic activity is detectible throughout density gradients such as those depicted in the chapter by Mircheff et al. in this volume. It is concentrated in *density window II*, which appears to contain a mixture of early and late endosomal membranes, and in *density window III*, which appears to contain a mixture of Golgi membranes and membranes derived either from a late endosomal compartment or from the *trans*-reticular Golgi complex.

Our working hypothesis for intracellular traffic of newly synthesized cathepsin and Class II molecules is as follows: Common populations of transport vesicles carry them through the Golgi complex to the *trans*-reticular Golgi network, where they are partially segregated, with cathepsins preferentially concentrated in forming secretory vesicles and lysosomes, and Class II molecules concentrated in transport vesicles which carry them, either directly or sequentially, via the late and early endosomes, to the basal-lateral membrane. However, some cathepsin molecules enter the endocytic compartments, perhaps by diffusing into the fluid phase enclosed by forming transport vesicles. Recycling traffic between the endosomes and the *trans*-reticular Golgi network may recapture some cathepsin molecules for translocation to secretory vesicles and lysosomes. Others may be secreted into the interstitium.

Endosomes typically have a mildly acidic pH and so should permit some cathepsin-mediated proteolytic processing to occur. As Class II molecules enter the endosomes they would be expected to release invariant chains and bind proteolytic fragments. It is plausible that some such fragments may be derived from the acinar cell antigens. The antigen's normal subcellular localization has not yet been determined. However, like virtually any cellular protein, it is probably entering the endosomal system at all times via autophagy (Tooze et al., 1990).

Even if the hypothesis outlined above is correct, autoantigen processing and presentation would lead to CD4[+] cell activation only if accompanied by appropriate accessory signals. We can speculate on what these signals might be. Lymphocytes are known to possess receptors for many of the neurotransmitters and biologically active peptides present in the lacrimal gland. These include catecholamines, released from sympathetic nerve endings; enkephalins and VIP, released from parasympathetic nerve endings; and epidermal growth factor, released from ductal epithelial cells. Another peptide, prolactin, is of particular interest because of its established role in immune responses (e.g., Buskila et al., 1991). Acinar cells accumulate prolactin-like proteins, apparently through a combination of local synthesis and uptake from the circulation (Wood et al., this volume), and as posited above for cathepsins, some prolactin may leak into the interstitium. Particular combinations of these mediators might transmit effective accessory signals to the autoreactive CD4[+] lymphocytes. In this context, it is noteworthy that VIP enhances the proliferation of mitogen-stimulated lymphocytes from mouse lacrimal glands (Malaty et al., 1991).

## ACCOUNTING FOR SOME OBSERVATIONS

The hypothesis we propose suggests several possible explanations for the fact that the B8 and DR3 Class II molecule haplotypes occur 3-times more frequently among Sjögren's Syndrome patients than in the normal population. For example, these haplotypes may be the ones whose expression is most readily induced in the lacrimal gland, they may have particular affinities for the lacrimal gland antigens, or they may fail to induce adequate tolerance. That not all individuals expressing B8 or DR3 develop Sjögren's Syndrome may be because there is a degree of randomness at each step of the process, including the events leading acinar cells to become Class II$^+$, the generation of immunogenic fragments and their binding to Class II molecules, the particular milieu of accessory signals, and the occurrence of autoreactive CD4$^+$ cells.

We can also envision ways that gender might influence this model. As reviewed by Warren et al. in this volume, hormonal changes of the female life cycle may impair the lacrimal gland's capacity to produce fluid. The feedback system maintaining the precorneal tear film should attempt to compensate by increasing the level of secretomotor stimulation to the gland. Since carbachol enhances acinar cell Class II molecule expression *in vitro*, it seems possible that increased stimulation might increase the numbers of Class II$^+$ acinar cells. Sustained or excessive stimulation could also alter intracellular membrane trafficking in ways that increase the concentrations of cathepsins in the endosomal compartments. Supramaximal stimulation causes aqueous vacuoles to form in the acinar cell's apical cytoplasm. As reviewed elsewhere (Mircheff et al., 1991), some workers have proposed that such vacuoles result from aberrant fusion of secretory vesicles with lysosomes. However, fusion with the endosomal system also seems possible, and, in support of this suggestion, we have found that stimulation-induced vacuoles accumulate the fluid phase marker, horseradish peroxidase (R.L. Wood, J.P. Gierow, and A.K. Mircheff, unpublished).

Viral infections might also influence the initiation of local autoimmunity. For example, interferon-γ and other cytokines released during acute inflammatory responses may induce acinar cells to begin expressing Class II molecules. Alternatively, stably infected viruses might trigger production of cytokines which provide the accessory signals necessary for autoantigen presentation to culminate in T cell activation.

Finally, we should note that age-related atrophy may reflect the cumulative effects of a few acinar cells activating small numbers of autoreactive T cells without triggering autocatalytic responses that progress to overt autoimmune disease.

## REFERENCES

Bottazzo, G.F., Todd, I., Mirakian, R., Belfiore, A., and Pujol-Borrell, J., 1986, Organ-specific autoimmunity: A 1986 overview, *Immunol. Rev.* 94:137.

Buskila, D., Sukenik, S., and Schoenfeld, Y., 1991, The possible role of prolactin in autoimmunity, *Am. J. Reproduc. Immunol.* 26:118.

Damato, B.E., Allan, D., Murray, S.B., and Lee, W.R., 1984, Senile atrophy of the lacrimal gland: The contribution of chronic inflammatory disease, *Br. J. Ophthalmol.* 68:674.

Geppert, T.D., and Lipsky, P.E., 1987, Dissection of defective antigen presentation by interferon-γ treated fibroblasts, *J. Immunol.* 138:385.

Liu, S.H., and Zhou, D.H., 1992, Experimental autoimmune dacryoadenitis: Purification and characterization of a lacrimal gland antigen, *Invest. Ophthalmol. Vis. Sci.* 33:2029.

Ludwig. T., Griffiths, G., and Hoflack, B., 1991, Distribution of newly synthesized lysosomal proteins in the endocytic pathway of normal rat kidney cells, *J. Cell Biol.* 115:1561.

Malaty, R., Khalaf, S., Griffin. S., Thompson, H.W., and Beuerman, R.W., 1991, Neuropeptide modulation of oncogene expression and proliferation of lymphocytes, *Invest. Ophthalmol. Vis. Sci.* 32s:680.

Mircheff, A.K., Gierow, J.P., Lambert, R.W., Lee, L.M., Akashi, R.H., and Hofman, F.M., 1991, Class II antigen expression by lacrimal epithelial cells. An updated working hypothesis for antigen presentation by epithelial cells, *Invest. Ophthalmol. Vis. Sci.* 32:2302.

Tooze, J., Hollinshead, M., Ludwig, T., Hoflack, B., and Kern, H., 1990, In exocrine pancreas, the basolateral endocytic pathway converges with the autophagic pathway immediately after the early endosome, *J. Cell Biol.* 111:329.

# GINGIVAL IMPRESSION: A NEW BIOLOGICAL TEST FOR XEROSTOMIA IN DIAGNOSIS OF SJÖGREN'S SYNDROME

S. Liotet[1], M.J. Wattiaux[2], and Y. Morin[1]

[1]Centre Hospitalier National d'Ophtalmologie des Quinze Vingts, 28 rue de Charenton, Paris, France 75012
75012 Paris, France
[2]Hôpital St. Antoine, 184 rue de Faubourg St. Antoine, Paris, France 75012

## INTRODUCTION

Considering how difficult it is to diagnose Sjögren's syndrome (SS) on purely clinical grounds, a European prospective multicenter study, supported by the Epidemiology Committee of the EEC, recently defined a new set of diagnostic criteria (to be published in Arthritis and Rheumatism).

The symptomatic expression of keratoconjunctivitis sicca or of dry mouth only becomes evident when Sjögren's syndrome is sufficiently severe. The reliability of various clinical tests for keratoconjunctivitis sicca suffer due to false positive results. Because of the imperfect correlation between the anatomical abnormalities and the degree of the symptoms expressed, numerous tests have been devised in an attempt to overcome these difficulties. These tests include tear osmolarity[1], tear protein electrophoresis[2], tear lysozyme or lactoferrin assay[3], salivary gland biopsy[4], and salivary scintigraphy[5].

One of the most commonly used tests is conjunctival impression cytology[6]. This procedure is equivalent to a microbiopsy and shows modifications of goblet cells and of epithelial cells. In severe sicca syndrome, the test also reveals quite pathognomonic features with the disappearance of goblet cells, parakeratosis and nuclear abnormalities consisting of condensation of chromatin, which is described as "snake like chromatin." This easy impression technique prompted us to apply it to different buccal sites to study possible cytologic modifications in the dry mouths of sicca patients.

*Lacrimal Gland, Tear Film, and Dry Eye Syndromes*
Edited by D.A. Sullivan, Plenum Press, New York, 1994

## MATERIALS AND METHODS

We examined 80 cases of SS, which were diagnosed according to the above mentioned international criteria. Given this information, 61 of these patients were recruited into this study. The patient population included 58 females and 3 males, aged between 28 and 72 years, who had either primary (n = 55) or secondary SS (n = 6). In addition, certain patients also presented with hyperthyroidism (n = 3), systemic lupus erythematosus ( n = 2) or polymyositis ( n = 1), or suffered from a depressive state (n = 4) for which they may have been taking tranquilizers. This SS group was compared to a control group of 23 healthy subjects, 14 women and 9 men, aged between 21 and 62, all free from any mouth disease.

Sampling was performed with a cellulose acetate paper ("GSWP Millipore®" filter), whose rough surface was applied on three different places on the mouth epithelium, arbitrarily selected for their humidity: (a) the jugal epithelium, 2 cm in front of the parotid duct opening; (b) the upper surface of the tongue, half-way from midline and edge; and (c) the medial external surface of the lower gum. The cellulose acetate paper was gently applied to the epithelium, immediately removed and fixed in 10% formaldehyde. Since these epithelia lacked goblet cells, no special staining was required, and our samples were stained with the commonly used Harris-Shorr technique.

The statistic evaluation used either the Chi square or Fischer test.

## RESULTS

### Normal Subjects

The impression samples from normal subjects contained numerous desquamated polyhedric epithelial cells, isolated or in plaques, with a nuclear-cytoplasmic ratio of 1/7 to 1/10. These cells had upturned edges, an acidophilic or basophilic cytoplasm, and sometimes appeared together with polynuclear cells and bacterial flora of various importance (Figure 1). The tongue samples showed less desquamation than the other two sites, but looked "dirtier" (i.e. abundant in mucus and bacterial flora) than those from the gum, which appeared the "cleanest."

### Sjögren's Syndrome Subjects

In this group, a number of impression samples showed nuclear abnormalities, with a dense rod-like chromatin condensation in the middle zone of the nucleus. This condensation sometimes took a snake-like form, identical to the "snake-like chromatin" of conjunctival cells (Figure 2). These abnormalities are more easily detected in basophilic cells than in eosinophilic cells. Fragmentation of nuclei is occasionally noted. Such modifications are more frequent on the gum epithelium, rare on the tongue and never found in the jugal epithelium.

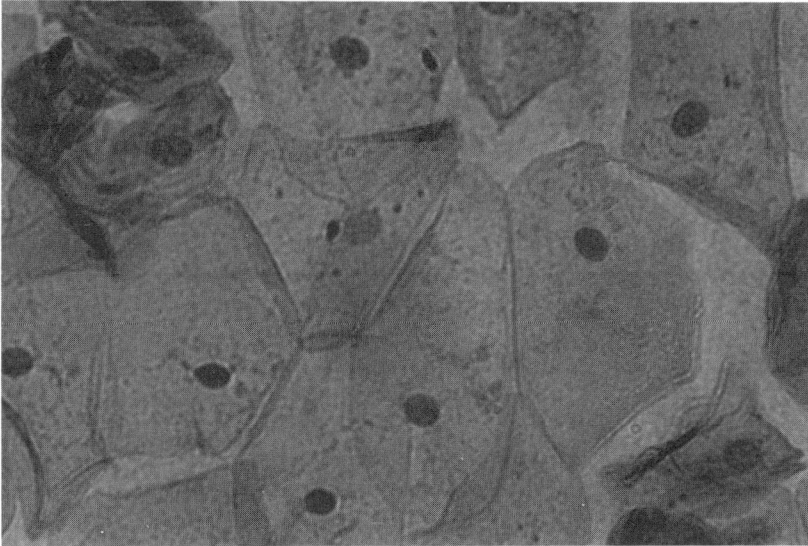

**Figure 1.** Impression sample from the mouth of a normal subject.

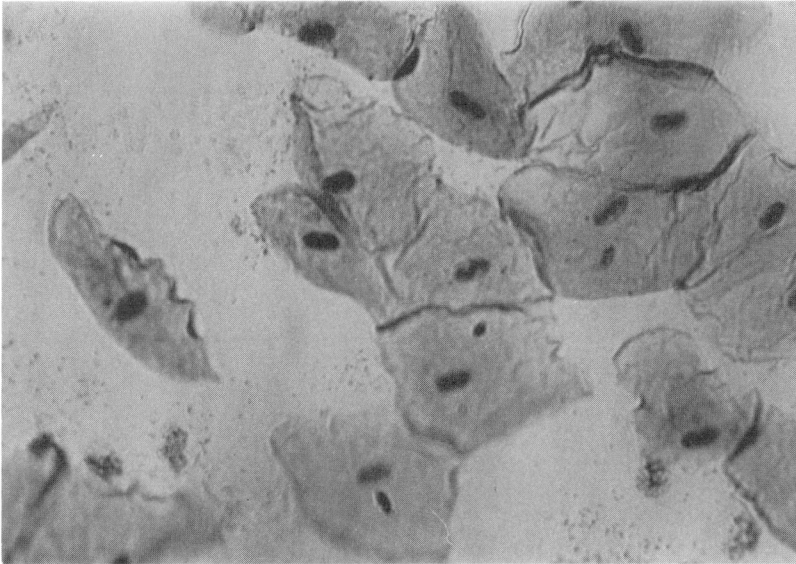

**Figure 2.** Impression sample from the mouth of a patient with Sjögren's syndrome.

Herein, we report only our findings on the gum epithelium samples. The result of the impression was considered abnormal when more than 1% of the cells showed the above mentioned modifications.

When comparing the control group of 23 normal subjects, free of any abnormality, with 57 patients suffering from SS, we found that 38 gingival impressions showed abnormalities. It appears that the difference between these two groups is highly significant (p<0.001), with a specificity of 1 and a sensitivity of 0.67.

The subgroup of patients suffering from mouth dryness consisted of 44 patients, and the gingival impression proved positive in 28 (and negative in 16), as opposed to 9 positive results (and 4 negative) in 13 patients without mouth dryness. The chi Square test was not significant. There was no real difference between the patients with or without mouth dryness (specificity was 0.31 and sensitivity was 0.64).

Of the subgroup without mouth dryness, 8 patients also suffered from keratoconjunctivitis sicca, and the gingival impression proved positive in 5 cases, and negative in 3. Of those without keratoconjunctivitis sicca, gingival impression proved positive in 3 and negative in 2. The difference was not significant, specificity was 0.40 and sensitivity 0.63.

34 patients suffered from both eye and mouth dryness. Among these individuals the conjunctival impression was positive in 10 and the gum impression in 22. The difference was not significant in these 2 results.

6 patients were without eye and mouth dryness; among them 1 gingival impression showed positive and no conjunctival impression did. Specificity is 1 for the gingival impression and 0.33 for the conjunctival impression with sensitivity being respectively 0.22 and 0.65.


## COMMENTS

According to the statistical analysis, the gingival impression is specific for Sjögren's syndrome, but not for mouth dryness. Sensitivity in all cases was around 0.64. Nevertheless, such a statement is based purely on subjective assessment, which can alter the validity of those statistical results.

This reflects the already mentioned imperfect correlation between the anatomical findings and the degree of the symptoms expressed, as well as the fact that the lesions are not uniformly distributed, so that, at least at the beginning of the disease, some samples can come from normal areas.


## CONCLUSION

The gingival impression, derived directly from the conjunctival impression technique, is an easy, simple, non-invasive and quick test which may show objective alterations, whereas the conjunctival impressions of the same patients may be normal. The most characteristic alterations are nuclear, with a rod-like chromatin condensation, sometimes taking the same "snake-like" form as seen in conjunctival cells.

However, this test cannot pretend to outclass a technique such as salivary gland biopsy, and it should be considered only as a complementary test.

# REFERENCES

1. J.P. Gilbard, 1986, Tear film osmolarity and keratoconjunctivitis sicca, *in:* "The preocular tear film in health disease and contact lens wear," F.J. Holly, ed., Dry Eye Institue, Lubbock, Texas, p.137.
2. S. Liotet, H. Hammard, A. Berranger, and M. Arrata M., 1980, Etude des protéines lacrymales au cours des syndromes secs, *Arch. Ophthalmol.* 82:10.
3. D.M. Chisolm, and D.K. Mason, 1968, Labial salivary gland biopsy in Sjögren's disease, *J. Clin. Pathol.* 21:656.
4. T.E. Daniels, M.R. Powell, R.A. Sylvester, and N. Talal, 1979, An evaluation of salivary scintigraphy in Sjögren's syndrome, *Arthritis Rheum.* 22:809.
5. P.M. Egbert, S. Lauber, and D.M. Maurice, 1977, A simple conjunctival biopsy, Am. J. Ophthalmol. 58:849.

# SJÖGREN'S SYNDROME DIAGNOSIS: A COMPARISON OF CONJUNCTIVAL AND GINGIVAL IMPRESSIONS AND SALIVARY GLAND BIOPSY

S. Liotet[1], O. Kogbe[1] and M.J. Wattiaux[2]

[1]C.H.N.O. des Quinze-Vingts, 28 rue de Charenton, Paris, France 75012
[2]Hôpital St. Antoine, 184 rue de Faubourg St. Antoine, Paris, France 75012

## INTRODUCTION

The diagnosis of Sjögren's syndrome is difficult and typically based on the presence of some clinical signs and symptoms, as well as on biological tests chosen from a recommended list. The results of these analyses and tests, though, are often variable and inconsistent, which creates a problem in the diagnosis of this dry eye and dry mouth syndrome. Even complementary tests, which explore the functional defects of exocrine glands (e.g. salivary gland biopsy), or effects of dryness on the epithelium (e.g. conjunctival and gingival impression cytology), also give variable results.

The purpose of this study was to determine which of the above-mentioned complementary tests might have more diagnostic value. For this reason, we recruited a series of patients who had been intensively examined in a special clinic, and evaluated the sensitivity and specificity of salivary gland biopsy, and conjunctival and gingival impression cytology.

## MATERIALS AND METHODS

Patients attending a special clinic for Sjögren's syndrome had gingival and conjunctival impressions taken by cellulose acetate paper (Millipore GS pore size 0.22 μm). These samples were processed as described for conjunctival impressions: fixation in formol and dehydration in alcohols. Harris Shorr's staining was used for gingival impressions and cresyl violet for conjunctival impressions. Clearing was in xylene before covering in balme and examination by light microscopy. Patients also had an accessory salivary gland biopsy from the inferior lip.


## RESULTS

### Conjunctival Impression

Out of a subgroup of 43 patients suffering from keratoconjunctivitis sicca (KCS; according to a Schirmer test), 15 had abnormal conjunctival impression, as opposed to 1 positive result out of 18 patients without KCS. The difference is significant ($P>0.05$). Specificity equaled 0.94 (Figure 1), whereas sensitivity was only 0.35.

In 30 patients with both dry eyes and mouth, conjunctival impression had a specificity of 0.30 and a sensitivity of 1 (Figure 2), when compared to results of those without complaints.

### Gingival Impression

Analysis of gingival impression results shows that there is no significant difference between patients with dry and no dry mouth/eye: 28 out of 44 patients with dry mouth had positive gingival impressions, whereas similar findings were obtained in 9 out of 13 patients without complaints of dry mouth. This test's specificity was 0.31 (Figure 1) and sensitivity was 0.64 (Figure 2).

In 30 patients with both dry eyes and mouth, the gingival impression specificity and sensitivity were 0.25 (Figure 1) and 0.66 (Figure 2), respectively, when compared to results from individuals without complaints.

### Salivary Gland Biopsy

In 30 patients with both dry eyes and mouth, 23 has positive salivary gland biopsies, while 5 out of 6 patients without signs had positive salivary gland biopsies. The difference is not significant, with Fischer P unilateral equal to 0.59. Sensitivity is 0.77, but specificity is only 0.17.

Among 39 patients with dry mouth, 29 were positive, but 11 out of 13 patients without dry mouth were also positive. The difference is not significant, and the specificity is 0.15 (Figure 1) and the sensitivity is 0.79 (Figure 2).

In 37 patients with dry eyes, the results were positive in 28 compared to 11 positive out of 15 without dry eyes. The difference is not significant (Fischer P unilateral = 0.47) and the specificity is only 0.27 (Figure 1) and sensitivity is 0.78 (Figure 2).


## DISCUSSION

The tentative diagnostic criteria for Sjögren's syndrome resulting from the European multicenter study suggest 4 positive out of 6 items, which were shown to have a good

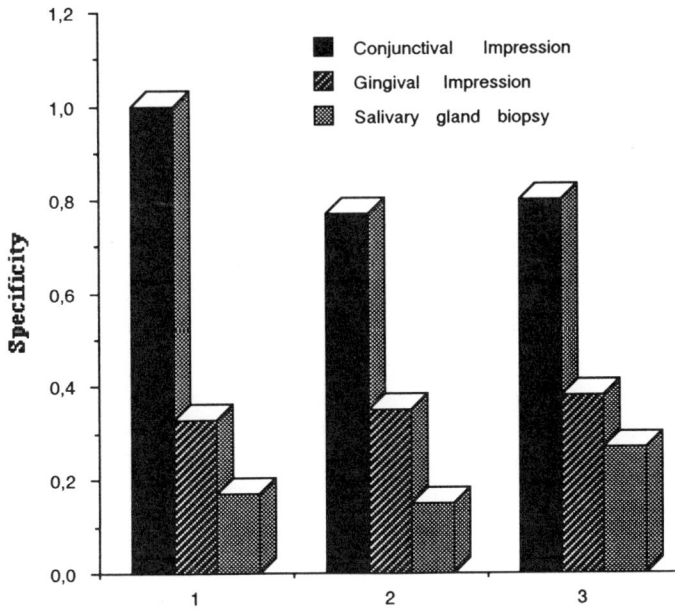

1 : both dry eye & mouth, 2 : dry eyes only, 3 : dry mouth only

**Figure 1.** Specificity of conjunctival and gingival impressions and salivary gland biopsy for the diagnosis of Sjögren's syndrome.

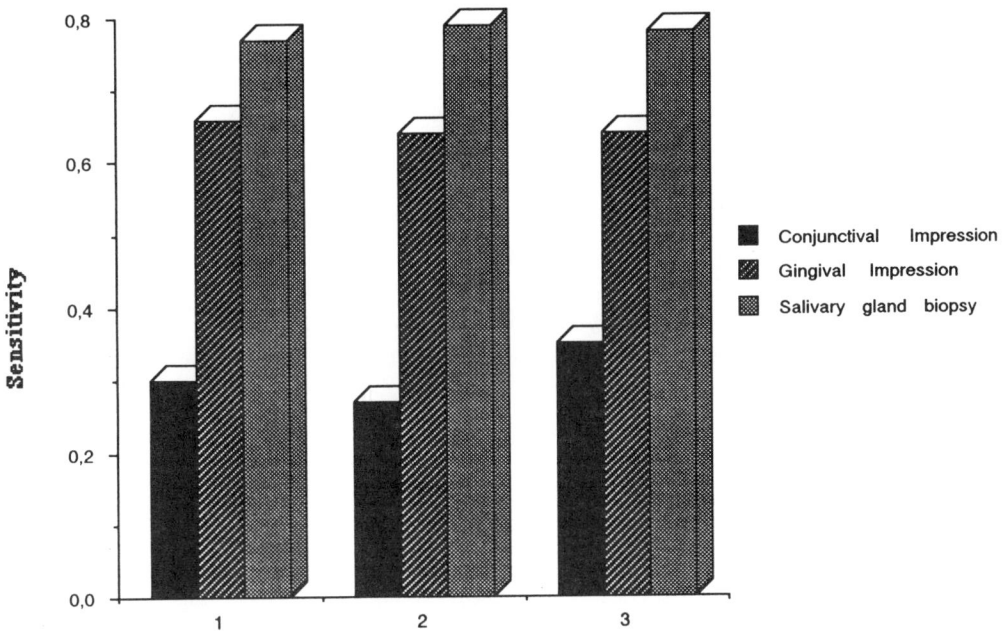

1 : both dry eye & mouth, 2 : dry eyes only, 3 : dry mouth only

**Figure 2.** Sensitivity of conjunctival and gingival impressions and salivary gland biopsy for the diagnosis of Sjögren's syndrome.

sensitivity and specificity for Sjögren's syndrome. Of these 6 items, the first 3 can be easily carried out by all doctors no matter where they practice, as these depend on skills of a good history extracted from the patients. The other three require elaborate tests which are available in specialized medical centers.

Our finding of a simple test which gives nearly as good a result as salivary gland biopsy needs to be looked at by many practitioners, as it could reduce both time and expenses.

In dry eyes, reduced tear volume, reduced tear protein content of lactoferrin and lysozyme, and increased osmolarity are some of the factors known to occur with metaplastic alterations of the ocular surface, as revealed by conjunctival impressions. In Sjögren's syndrome, when patients complain of having dry mouth, we propose testing for abnormal gingival impressions, in addition to testing for reduced saliva volume. This could be done instead of salivary gland biopsy, as we found nearly the same extent of abnormal gingival impressions as salivary gland abnormalities in this group of patients. It would be a pleasant alternative, especially when the patient refuses to have an accessory salivary gland biopsy.

We examined gingival impressions of these patients to test the suggestion that the manifestations of Sjögren's syndrome result from the combined effect of a generalized glandular dysfunction and a local epithelial disease (Frost-Larsen et al., 1980; Oxholm et al., 1987; Prause et al., 1989). In reviewing earlier works, we observed that Nelson and Wright compared results of impression cytology with tear film osmolalitites in the diagnosis of keraconjunctivitis sicca. They found in a population with a 44% prevalence of KCS, 87% positive conjunctival impression cytology to 75% raised osmolalities. Thus, conjunctival impressions yielded a sensitivity of 0.38, while by osmolalities a sensitivity of 0.33 was obtained in their cases of Sjögren's Syndrome. Prause and Marner in commenting on their results stated that positive conjunctival impression results did not exceed 50%. Our results of 0.37 (0.37-0.40) agrees with the findings of both groups. Our finding of a sensitivity of 0.66 in positive gingival impressions suggests that gingival epithelium is affected in more of these cases than their conjunctival epithelium.

To our knowledge, this represents the first documented use of gingival impressions in Sjögren's syndrome. That this very simple, non-traumatic, non-invasive technique revealed a change in the epithelial cells of the gums of Sjögren's syndrome patients to nearly the same extent as revealed by their salivary gland biopsy is remarkable.

We recommend its use especially because it is technically much easier to carry out and less traumatic to the patients than accessory salivary gland biopsy. The findings would certainly expand our work and satisfy any patient who refuses to have a salivary gland biopsy.


## CONCLUSION

The results further confirm that there is an important discrepancy between the clinical signs and the anatomic modifications in Sjögren's syndrome patients, such that one cannot be superimposed on the other.

# SUMMARY

This study reviewed 43 patients with Sjögren's syndrome. All of the patients complained of either dry eye and/or dry mouth. Their clinical records were compared with results of three laboratory diagnostic tests to determine the most sensitive test.

Out of 30 patients who complained of both dry eye and dry mouth, 10 had positive conjunctival impressions, 20 had positive gingival impressions and 24 had positive salivary gland biopsies.

Of 5 patients who suffered from only dry mouth, 2 had positive conjunctival impressions, 3 had positive gingival impressions and 3 had positive salivary gland biopsies.

Of 7 patients with only dry eye complaints, there were 3 positive conjunctival impressions, 5 positive gingival impressions, and 6 positive salivary gland biopsies.

Out of 6 healthy controls without complaints of dry eyes/mouth, there were no positive conjunctival impressions, 4 positive gingival impressions, and 5 positive salivary gland biopsies.

Gingival impression appears to be nearly as sensitive as salivary gland biopsy in this group of patients suffering from Sjögren's Syndrome.

# REFERENCES

Frost-Larsen, K., Isager, H., Manthorpe, R., and Prause, J.U., 1980, Sjögren's Syndrome, *Ann. Ophthalmolol.* 12:836.

Nelson, J.D., and Wright, J.C., 1986, Impression cytology of the ocular surface in keratoconjunc-tivitis sicca, *in:* "The preocular tear film in health disease and contact lens wear," F.J. Holly, ed., Dry Eye Institute, Lubbock, Texas, p.140.

Oxholm, P., Oxholm, A., and Prause, J.U., 1987, Immunohistochemical characterization of intraepidermal in vivo IgG deposits in patients with primary Sjögren's syndrome. *Acta Pathol. Scand. Sect. A.* 95:239.

Prause, J.U., and Marner, K., 1986, Snake-like nuclear chromatin in imprints of conjunctival cells from patients with Sjögren's Syndrome. *in:* "The preocular tear film in health disease and contact lens wear," F.J. Holly, ed., Dry Eye Institute, Lubbock, Texas, p.158.

Prause, J.U., Jensen, O.A., Pachides, K., Stochase, A., and Vangsted, P., 1989, Conjunctival cell glycoprotein patterns of healthy persons and of patients with 1° SS. *J. Autoimmunity* 2: 495.

Vitali, C., Bombardieri, S., Moutsopoulos, H.M., Balestrieri, G., Bencivelli, W., Bernstein, R.M., Coll Daroca, J., De Vita, S., Drosos, A.A., Hatron, P.Y., Hay, E., Isenberg, D.A., Janin, A., Kalden, J.R., Kater, L., Konttinen, Y.T., Maddison, P.J., Maini, R.N., Manthorpe, R., Meyer, O., Ostuni, P., Pennec, Y., Prause, J.U., Richards, A., Sauvezie B., Schiodt, M., Sciuto, M., Scully, C., Schoenfeld, Y., Skopouli, F.N., Smolen, J.S., Snaith, M.L., Tincani, A., Tishler, M., Todesco, S., Valesini, G., Venables, P.J.W., Wattiaux, M.J., and Youinou, P., 1993, Diagnostic criteria for Sjögren's Syndrome. Results of a European prospective multicentre study, in press.

# CORRELATION OF LABIAL SALIVARY GLAND BIOPSIES WITH OCULAR AND ORAL FINDINGS IN PATIENTS WITH SJOGREN'S SYNDROME

John P. Whitcher, Troy E. Daniels, and Mark D. Sherman

The Francis I. Proctor Foundation, School of Dentistry, and Sjogren's Syndrome Clinic
University of California, San Francisco, California USA

## PURPOSE

Sjogren's syndrome is an infrequent but extremely symptomatic condition which occurs primarily in postmenopausal females. The classic triad of dry eyes, dry mouth, and an associated autoimmune disease is not always clinically apparent and the diagnosis of the syndrome becomes problematic [1]. Clinical diagnosis usually depends on a strong positive history for keratoconjunctivitis sicca (KCS) and on typical signs including Rose Bengal staining, fluorescein staining, and tear breakup time (TBUT) [2]. Ocular findings may be correlated with oral symptoms of xerostomia and documented by decreased parotid flow rate (PFR) [3], as well as by rheumatologic evaluation and correspondingly abnormal serologic tests (SS-A, SS-B, ANA, and Rheumatoid Factor) [4]. Even though several ocular tests, including tear osmolarity, cellulose acetate impression cytology and tear fluorescein dilution tests have been proposed as the "gold standard" for diagnosing Sjogren's syndrome, we propose that labial salivary gland (LSG) biopsy should be considered as the basic standard by which all other diagnostic criteria are compared.

## METHODS

Fifty individuals were randomly selected from the Sjogren's Clinic at the University of California Medical Center, San Francisco, who had positive ocular finds consistent with the diagnosis of keratoconjunctivitis sicca (characteristic Rose Bengal staining). A careful history was taken from each patient regarding ocular and oral symptoms as well as the existence of an associated autoimmune. A complete oral examination was performed by a member of the staff in the School of Dentistry and a PFR was documented. A complete ocular examination was performed as well, documenting ocular clinical findings in addition to Rose Bengal staining. Unanesthetized Schirmer tests, TBUT, and tear lysozyme were all measured. Labial salivary gland biopsies were performed on all patients in a standardized fashion excising four lobes of the LSG, and examining 4 sq. mm of the biopsied tissue stained with H&E under high powered microscopy. Lymphocytic infiltration of the gland was graded using a scheme modified after Chisolm and Mason: where 0 = no infiltration, 1 = mild background infiltration, 2 = moderate diffuse infiltration, 3 = one focus of 50 or more lymphocytes for 4 sq. mm of tissue, and 4 = 2 or more foci of 50 or more lymphocytes for 4 sq. mm of tissue.

## RESULTS

The 50 individuals who were randomly selected from the patients seen in the Sjogren's Clinic were picked exclusively on the basis of positive Rose Bengal staining in the characteristic KCS pattern. The average patient age was 53, with females affected greater than males four to one (Table 1). The history of a dry mouth was given by 54% and confirmed by PFR in 70%. An associated autoimmune disease was seen in 24% affected, two-thirds of whom had rheumatoid arthritis.

TABLE 1
KERATOCONJUNCTIVITIS SICCA
(50 PATIENTS)

| | |
|---|---|
| Average Age | 53 |
| Female to Male Ratio | 4 to 1 |
| Xerostomia by History | 54% |
| Xerostomia by PFR | 70% |
| Associated Autoimmune Disease | 24% |
| Rheumatoid Arthritis | 16% |
| Other Autoimmune Disease | 8% |

The most common ocular symptom in the 50 patients (Table 2) was foreign body sensation (68%), followed by excessive secretion (66%), burning (62%), redness (62%), photophobia (58%), blurred vision (56%), itching (52%), pain (52%), and inability to tear (44%).

TABLE 2
SYMPTOMS IN KERATOCONJUNCTIVITIS SICCA
(50 PATIENTS)

| | |
|---|---|
| Foreign Body Sensation | 68% |
| Excessive Secretion | 66% |
| Burning | 62% |
| Redness | 62% |
| Photophobia | 58% |
| Blurred Vision | 56% |
| Itching | 52% |
| Pain | 52% |
| Inability to Tear | 44% |

Ocular clinical findings included (Table 3) Rose Bengal staining (100%), fluorescein corneal staining (96%), scanty tear meniscus (94%), viscous precorneal tear film (74%) and conjunctival infiltration and redness (70%).

Clinical and laboratory findings (Table 4) included an abnormal TBUT (less than 10 seconds) in 90%, an abnormal unanesthetized Schirmer test (less than 15 mm in five minutes) in 90%, an abnormal tear lysozyme (reduced or absent) in 64%, a decreased PFR (less than 5 ml per gland per 10 minutes) in 70%, and a positive LSG biopsy in 100%.

TABLE 3
## CLINICAL FINDINGS IN KERATOCONJUNCTIVITIS SICCA
### (50 PATIENTS)

| | |
|---|---|
| Rose Bengal Staining | 100% |
| Corneal Staining | 94% |
| Conjunctival Staining | 96% |
| Fluorescein Corneal Staining | 96% |
| Scanty Tear Meniscus | 94% |
| Viscous Precorneal Tear Film | 74% |
| Conjunctival Infiltration and Redness | 70% |

TABLE 4
## CLINICAL AND LABORATORY FINDINGS IN KERATOCONJUNCTIVITIS SICCA
### (50 PATIENTS)

| | |
|---|---|
| Abnormal TBUT (less than 10 seconds) | 90% |
| Abnormal Schirmer (less than 15 mm) | 90% |
| Abnormal Lysozyme (reduced or absent) | 64% |
| Decreased PFR (less than 5 ml/glands/10 minutes) | 70% |
| Positive LSG Biopsies (lymphocytic infiltration) | 100% |

When the 50 patients were classified according to severity of ocular disease into severe KCS (corneal filaments), moderate KCS (confluent conjunctival Rose Bengal staining without corneal filaments) and mild KCS (patchy and sparse conjunctival staining with Rose Bengal) and correlated with TBUT values, Schirmer tests, tear lysozyme, and LSG biopsies (Table 5), the LSG biopsy scores were found to be the most sensitive indicator of the presence of Sjogren's syndrome as confirmed by the other associated parameters. The mean biopsy score was 4.0 for severe KCS, 3.3 for moderate KCS, and 3.1 for mild disease.

TABLE 5
## CLINICAL AND LABORATORY CORRELATIONS IN KERATOCONJUNCTIVITIS SICCA
### (50 PATIENTS)

| Clinical Classification (No. Patients) | Tear Breakup Time (Avg. in Secs.) | Schirmer Test (Avg. in mm) | Tear Lysozyme (Mean Value) | Labial Salivary Gland Biopsies (Avg. Grade) |
|---|---|---|---|---|
| Severe (3) | 3 seconds | 4 mm | Absent | 4.0 |
| Moderate (16) | 4 seconds | 7 mm | Reduced | 3.3 |
| Mild (31) | 7 seconds | 9 mm | Low Normal | 3.1 |

## DISCUSSION

Biopsies of the labial salivary glands, when graded for lymphocytic infiltration according to the criteria of Chisolm and Mason, appeared to be an extremely sensitive method for confirming the diagnosis of Sjogren's syndrome. Even though typical Rose Bengal staining of the conjunctiva and cornea in patients with KCS has been used as the clinical standard in ophthalmology for making the diagnosis of Sjogren's syndrome, there may be both false positive and negative results.    Other clinical parameters should be measured including fluorescein staining, TBUT, tear lysozyme, Schirmer testing, and PFR.   Biopsy of the LSG appears to be a more sensitive indicator of the true pathology, however, than any of the other more conventional tests.  We feel that LSG biopsy should be considered as the "gold standard" for the diagnosis of Sjogren's syndrome.  We recommend LSG biopsy in all patients with the putative diagnosis of Sjogren's syndrome.

**REFERENCES**

1.     R.I. Fox, Sjogren's syndrome: proposed criteria for classification, <u>Arthritis and Rheumatism</u>. 29(5):577-585 (1986).

2.     J.P. Whitcher, Diagnosis of the dry eye, <u>International Ophthalmology Clinics</u>. 27:7-23 (1987).

3.     T.E. Daniels, Labial salivary gland biopsy in Sjogren's syndrome: assessment as a diagnostic criterion in 362 suspected cases, <u>Arthritis and Rheumatism</u>. 27:147-56 (1984)

4.     C. Scully, Sjogren's Syndrome: Review of immunopathogenesis, clinical and laboratory features and management in relation to dentistry, <u>Oral Surgery</u>. 62:510-23 (1989).

# A COMPARATIVE STUDY BETWEEN TEAR PROTEIN ELECTROPHORESIS AND ACCESSORY SALIVARY GLAND BIOPSY IN SJÖGREN'S SYNDROME

F. Carré[1], S. Liotet[1], and M.J. Wattiaux[2]

[1]C.H.N.O. des Quinze Vingts, 28 rue de Charenton, Paris, France 75012
[2]Hôpital St. Antoine, 184 rue du Fg St Antoine, Paris, France 75012

## INTRODUCTION

Sjögren's syndrome is a disease that affects the exocrine glands, and especially salivary and lacrimal glands. Clinical symptoms of dry mouth and dry eyes often lead patients to consult specialists or clinicians.

The most reliable test for the positive diagnosis of Sjögren's syndrome is the accessory salivary gland biopsy; but it could be negative because the lesions are multifocal.

Tear protein electrophoresis reflects the lacrimal gland's secretory activity: a decrease or disappearance of one or several of the three main peaks may indicate even a moderate dysfunction of the gland.

We propose to compare the results of tear protein electrophoresis with those of salivary gland biopsy. This analysis will demonstrate whether electrophoresis is as reliable a test as the biopsy, which technique is invasive and not always accepted by patients.

## MATERIALS AND METHODS

The patient population included 54 patients (52 female and 2 male) with Sjögrens's syndrome. These patients ranged in age from 28 to 82 years old, with an average age of 59 years. During clinical examinations, all patients were asked about possible sensations of dry eye and dry mouth. In addition, all patients had: (1) one or several tear collections, to permit the electrophoretic analysis (on cellulose acetate with a barbital buffer, pH 8.6) of tear proteins; and (2) one accessory salivary glands biopsy (on inferior lip).

# RESULTS

Our evaluation of patients with Sjögren's syndrome demonstrated that:

      (1) 37 patients (67.7%) complained of dry eye sensation;

      (2) 40 patients (74%) complained of dry mouth sensation;

      (3) 41 patients (76%) had a pathological electrophoretic pattern of tear proteins;

      (4) 41 patients (76%) had pathological accessory salivary gland biopsy.

The numerical distribution of the different classes of tear electrophoretic patterns and the various stages of salivary gland pathology is shown in figure 1. The classes correspond to tear electrophoretic profiles having a normal pattern (class 0; Figure 2), or decreasing amounts of protein in one or all of the main peaks (classes 1 to 4; Figures 3 to 5), which represent fast migration protein, lactotransferrin and lysozyme. If the dry eye state had advanced such that no tears could be obtained, then this condition was designated as class 5. With regard to the stages of salivary gland pathology, these were based upon the extent of inflammation and fibrosis. Accordingly, biopsies were graded as histologically normal (stage 0; Figure 6), containing discrete (stage 1; Figure 7), moderate (stage 2; Figure 8), or intense (stage 3; Figure 9) mononuclear cell infiltrates, or fibrotic and involuted (stage 4; Figure 10).

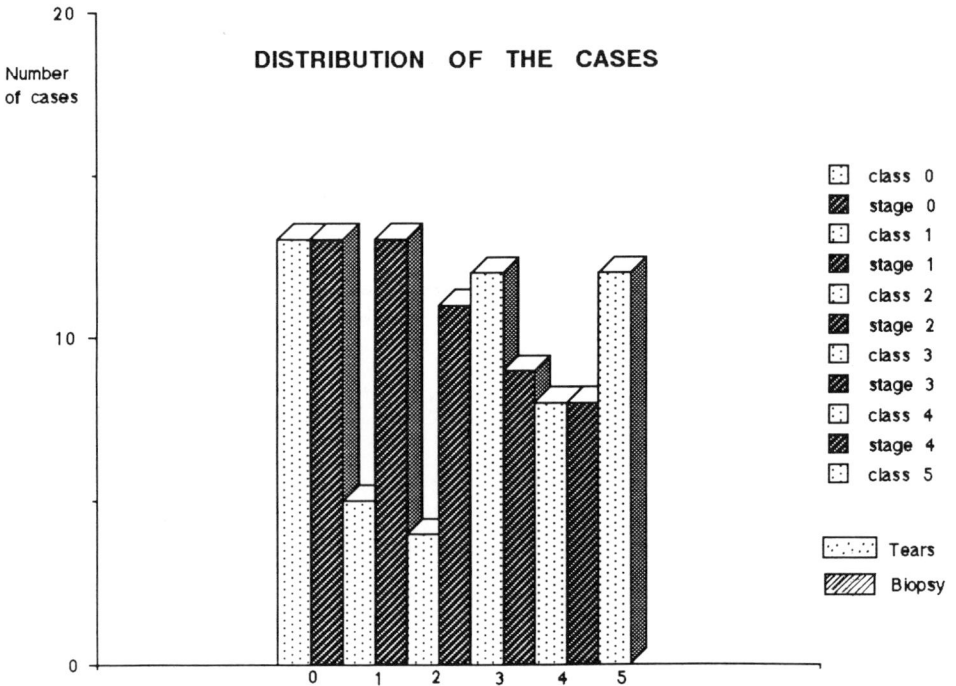

Figure 1. Distribution of the number of different classes and stages shown by the patient population.

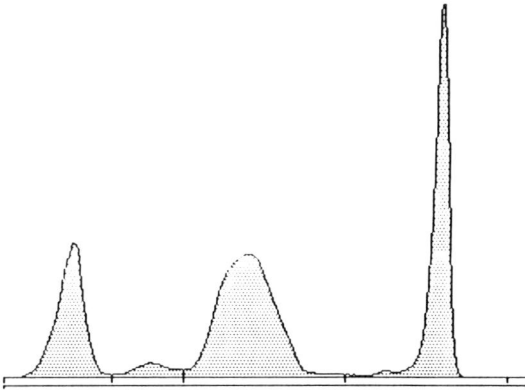

LARMES                              21/05/1991 5
  11  FCH: 11

OD

PROTIDES :   7.2 G/L

        NOM    0/0   CONC   NORMALES %  G/L

        PMR    21.3   1 5   25-35 1.48-2.58
         2      3 3   0 2    4-10 0.12-0.28
        LTF    41.7   3 1   35-45 1.89-3.12
        LYS    33.7   2 4   25-35 1.50-2.60

Figure 2. Normal tear protein electophoretic pattern, corresponding to class 0.

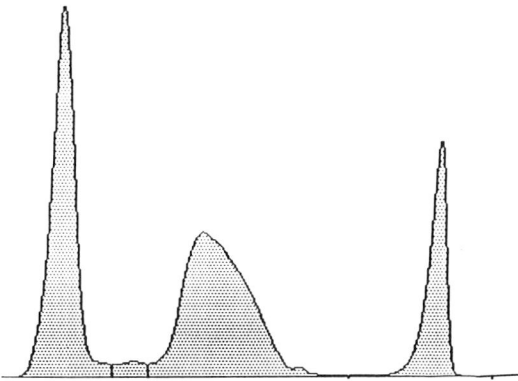

LARMES                              09 04 1991
   7   ECH: 7

08.04.91  OD

PROTIDES :   9.7 G/L

        NOM    0/0   CONC   NORMALES %  G/L

        PMR    37.8   3.7   25-35 1.48-2.58
         2      2.0   0.2    4-10 0.12-0.28
        LTF    42.2   4.1 · 35-45 1.89-3.12
        LYS    18.0   1.7   25-35 1.50-2.60

Figure 3. Class 1 pattern of tear protein electrophoresis.

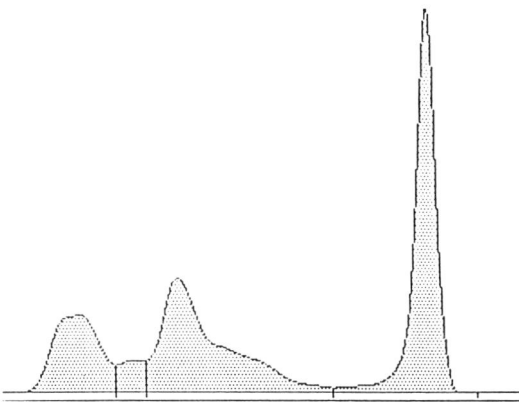

LARMES                              28 DEC  1989
   4   ECH: 4

DG 27.12.89

PROTIDES :   4.8 G/L

        NOM    0/0   CONC   NORMALES %  G/L

        PMR    19.1   0.9   25-35 1.48-2.58
         2      4.0   0.2    4-10 0.12-0.28
        LTF    34.7   1.7   35-45 1.89-3.12
        LYS    42.2   2.0   25-35 1.50-2.60

Figure 4. Class 2 pattern of tear protein electrophoresis.

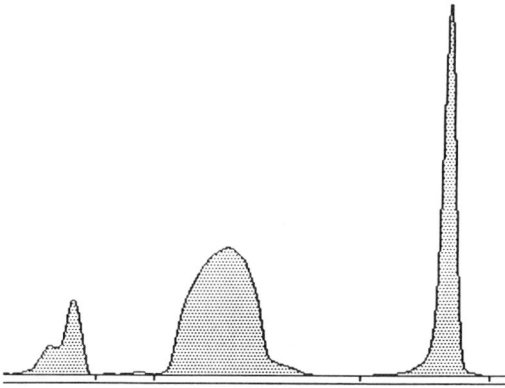

Figure 5. Class 4 pattern of tear protein electrophoresis.

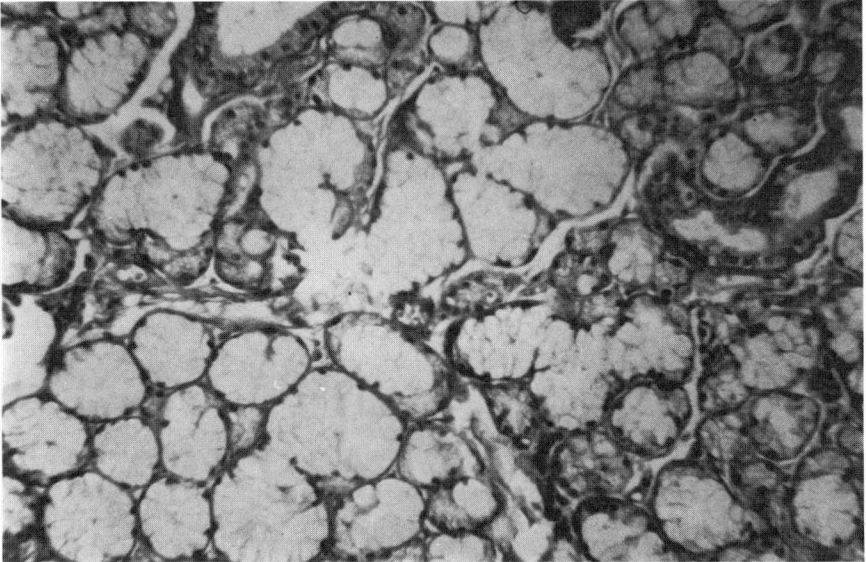

Figure 6. Histological appearance of a normal salivary gland (stage 0) (HES x 200).

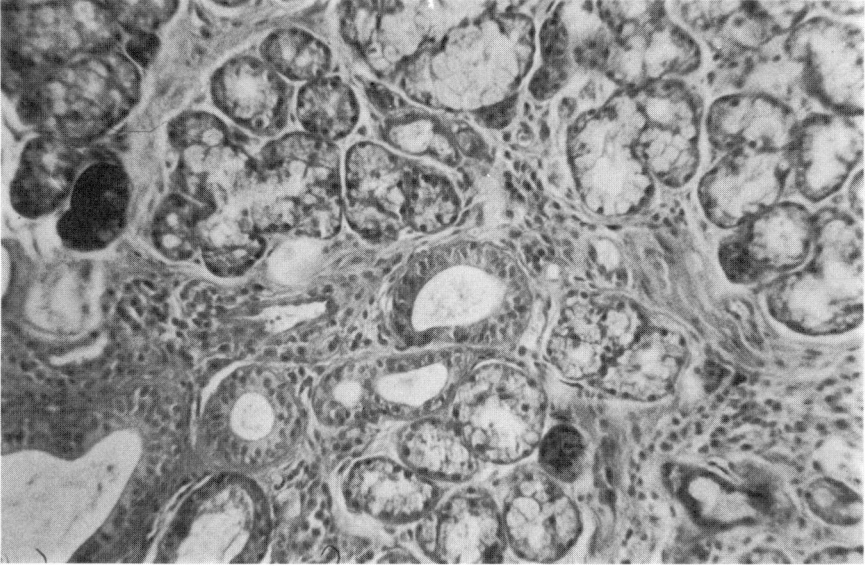

Figure 7. Salivary gland with discrete mononuclear cell infiltrate (stage 1) (HES x 200).

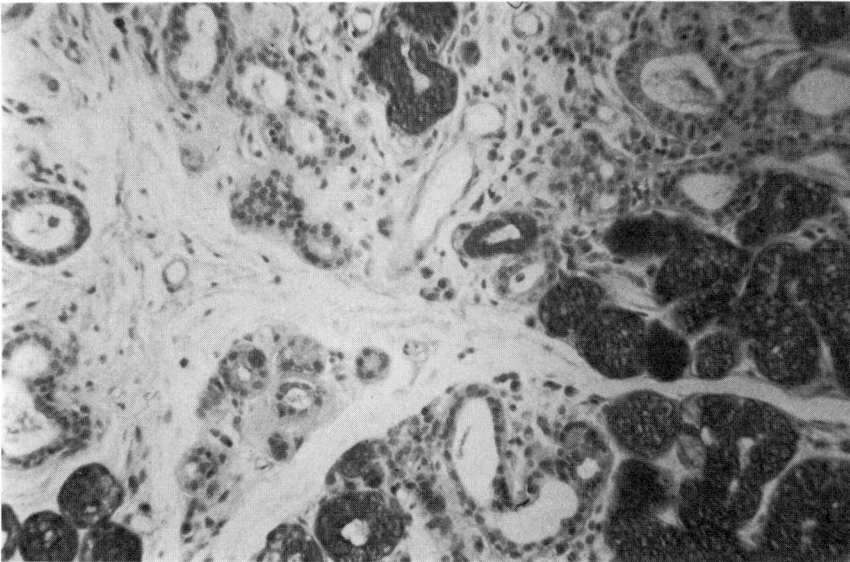

Figure 8. Salivary gland with moderate mononuclear cell infiltrate (stage 2) (PAS x 200).

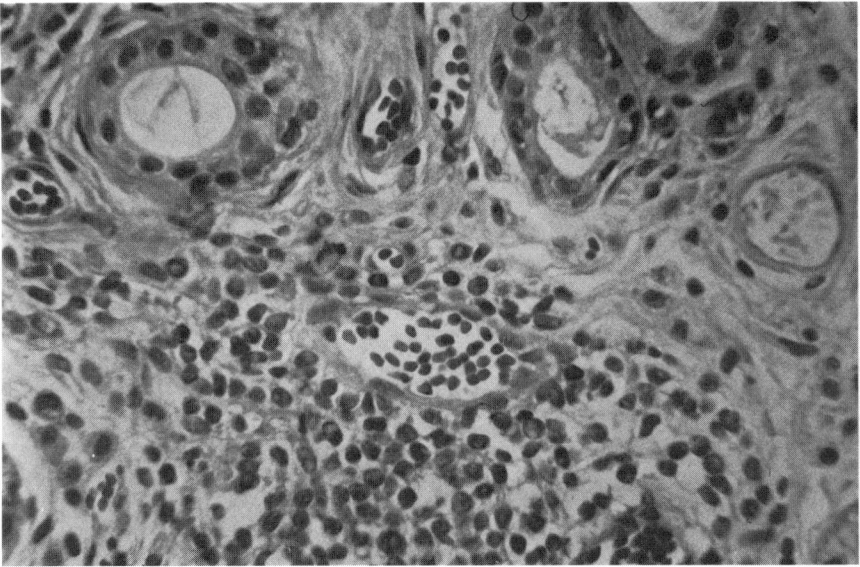

Figure 9. Salivary gland with intense mononuclear cell infiltrate (stage 3) (HES x 400).

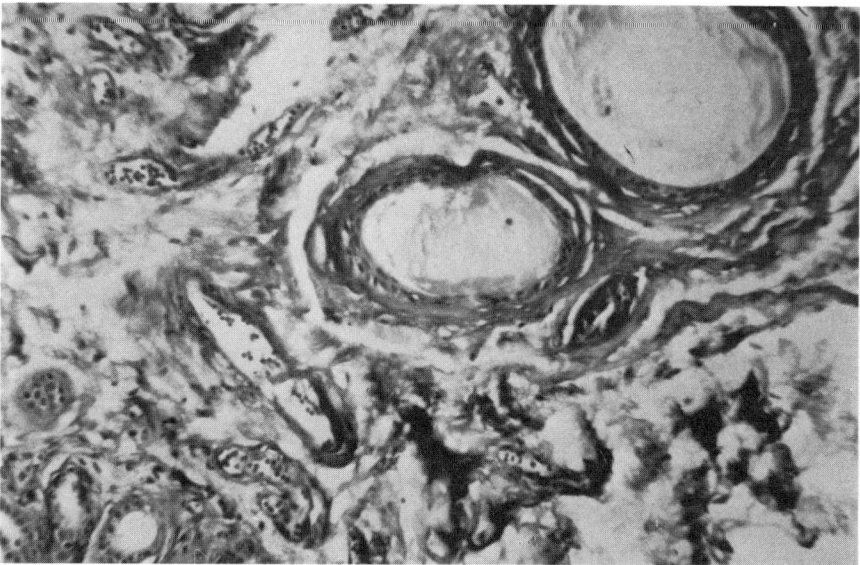

Figure 10. Salivary gland with fibrotic and involuted areas (stage 4) (Trichrome x 200).

## DISCUSSION

In our population of patients with Sjögren's syndrome, 36/54 individuals had both a pathological pattern of tear protein electrophoresis and a pathological salivary gland biopsy. According to the modified X2 test, the test of Fischer, the difference is significant (P<0.05) and the correlation is good. Although neither test is specific for Sjögren's syndrome, both are equally sensitive for diagnostic purposes.

676

Further evaluation of our data from the 54 patients showed that:

    (1) 28 had both dry eyes and a pathological tear electrophoresis pattern;

    (2) 32 had both dry mouth and a pathological tear electrophoresis pattern;

    (3) 29 had both dry eyes and a pathological salivary gland biopsy;

    (4) 32 had both dry mouth and a pathological salivary gland biopsy;

According to the test of Fischer, the difference is not significant for those four studies, and there is no correlation between those patterns. Probably it is because the sensation of dry eye or dry mouth is subjective, and several symptoms should be associated to make the diagnosis of primary Sjögren's syndrome.

## CONCLUSION

The strong correlation existing between the pattern of tear protein electrophoresis and the appearance of a accessory salivary gland biopsy indicates that electrophoresis is a very good test for the diagnosis of primary Sjögren's syndrome. Protein electophoresis is as sensitive, but less invasive, than the biopsy. In contrast, the clinical symptoms of dry eye or dry mouth are subjective and do not appear to have significant diagnostic value by themselves.

To confirm and extend these findings, a larger population of Sjögren's syndrome patients should be studied to determine whether tear protein electrophoresis can replace salivary gland biopsy as a diagnostic tool.

Overall, tear protein electrophoresis is easily reproducible, reliable, and sensitive, and involves far less invasive procedures than the biopsy, which may give some false negative results when the lesions are multifocal. This electrophoretic method could also replace salivary gland biopsy in order to monitor disease progression. If so, the patients would surely appreciate it!

# SJÖGREN'S SYNDROME - CLOSE TO CAUSE AND CURE?

Norman Talal

Clinical Immunology Section, Audie L. Murphy Memorial Veterans Hospital, and
the Department of Medicine, The University of Texas Health Science Center at San
Antonio, San Antonio, TX 78284-7874

## INTRODUCTION

Sjögren's syndrome (SS) is a chronic autoimmune disease associated with the production of rheumatoid factor and other autoantibodies (1). Because it is characterized by lymphocytic and plasma cell infiltration and destruction of salivary and lacrimal glands, SS gives rise to the characteristic symptoms of dry mouth and dry eyes. The term "autoimmune exocrinopathy" has been applied to this disease. The lymphocytic infiltration can extend to other, more-vital organs, so SS is both a systemic (like SLE) and a localized (like autoimmune thyroiditis) autoimmune disorder.

SS is particularly important among the autoimmune diseases for two reasons: first, perhaps one to two million persons in the United States are afflicted with SS (the majority undiagnosed); second, SS is a benign autoimmune disease that may terminate as a malignant lymphoma. Thus, SS offers potential insight into the mechanisms whereby immunologic dysregulation may predispose to a malignant transformation of B cells already involved in an autoimmune process.

## IMMUNOLOGIC FINDINGS

Polyclonal hypergammaglobulinemia is seen in the majority of SS patients (Table 1). Rheumatoid factor and anti-nuclear antibodies are present in 75-90% and 60-70% of patients respectively. The presence of multiple serum autoantibodies is a characteristic feature of SS. Antibodies to Ro(SSA) and La(SSB) occur in the vast majority of primary SS patients and in some SLE patients. These autoantibodies are intimately associated with genetic factors in SS and SLE. Monoclonal immunoglobulins, including homogeneous, light-chain bands, are frequently detected in the blood and urine by sensitive electrophoretic techniques.

Many SS patients have abnormally low levels of interleukin-2. There is a decrease in peripheral blood T cells in about one-third of patients. Abnormalities in T cell function may be present, particularly in patients with systemic features. These patients often have alterations in T cell subsets and natural killer (NK) cell activity.

Recent studies have focused on an unusual B-cell subset that contains the cell surface marker, CD5, that is normally present on T cells. We found CD5-positive cells to be increased in 68 percent of patients with primary SS.

**Table 1.** Immunologic findings in the peripheral blood in Sjögren's syndrome.

- ° Polyclonal hypergammaglobulinemia
  - ° Rheumatoid factor and antinuclear antibodies
  - ° Anti-Ro(SSA) and anti-La(SSB)
  - ° Monoclonal immunoglobulins
  - ° Deficient Il-2 production
  - ° Impaired T-cell function
  - ° Decreased NK cell function
  - ° Increased CD5-positive B cells

Immunologic findings in the salivary glands in SS are presented in Table 2. There are numerous DR-positive glandular epithelial cells in close proximity to the lymphoid infiltrates in which activated T-helper cells are the most prominent population represented. There are fewer B cells, which includes some expressing CD5. This seems particularly important in light both of the subsequent development of B-cell lymphomas in 5 to 10 percent of patients and of the production of monoclonal immunoglobulins in the salivary glands (2).

NK cells play an important role in several immunoregulatory functions, including natural host-defense against malignancy. In SS patients there is defective NK-cell function in the blood and in the salivary gland lesions. These findings suggest that the salivary gland in SS may serve as an initial nidus for lymphoma development.

The presence of activated T-helper cells and hyperactive B cells without local NK-cell defense may be crucial to the malignant transformation that results in lymphoma development.

## LYMPHOMA AND MACROGLOBULINEMIA

In most patients, significant lymphoproliferation remains confined to salivary and lacrimal tissue and has a chronic, benign course of stable or progressive xerostomia and xerophthalmia. In 5 percent of patients, however, even after two decdes of benign disease, there may be extension of lymphoproliferation to extraglandular sites such as lung, kidney, lymph nodes, skin, gastrointestinal tract, and bone marrow. Persistent or massive parotid gland enlargement may also suggest lymphoma.

**Table 2.** Immunologic findings in the salivary glands in Sjögren's syndrome.

- ° Many DR-glandular epithelial cells
- ° Many T-helper cells expressing activation markers
- ° Fewer activated B cells
- ° Production of monoclonal immunoglobulins
- ° Absence of NK cells

When such extraglandular lymphoproliferation occurs, diagnosis is often difficult, and -- clinically and histologically--the disease may simulate frankly malignant lymphoproliferative disorders such as Waldenstrom's macroglobulinemia or non-Hodgkin's lymphoma. Because the subsequent clinical course is frequently that of lymphoma, ending fatally, it is important that the extraglandular lymphoproliferation in SS be recognized early and appropriate therapy be instituted promptly.

The extraglandular lymphoid infiltrates are of two general types. They may be highly pleomorphic and include small and large lymphocytes, plasma cells, and large reticulum cells. In a lymph node, the cells may distort the normal architecture and may extend beyond the capsule; the distinction between benign and malignant lesions is thereby made difficult. The term "pseudolymphoma" has been applied when the lesions show tumor-like aggregates of lymphoid cells but fail to meet histological criteria for malignancy.

In pseudolymphoma, the site of extraglandular lymphoproliferation determines the clinical presentation. Striking regional lymphadenopathy may be the predominant clinical feature. On the other hand, lymphoid infiltration may be selectively excessive in a distant organ such as a kidney or lung. These organs may become functionally impaired, giving rise to renal abnormalities, such as renal tubular acidosis, or pulmonary insufficiency.

## PATHOGENETIC FACTORS

Autoimmune diseases are called multifactorial because we have not yet identified a single etiologic cause. However, failing to find a cause of SS does not imply that one does not exist. A major feature of SS is the genetic predisposition reflected in a tendency to occur in families. Primary SS is one of several autoimmune diseases associated with HLA-B8 and HLA-DR3. The presence of Ro(SSA) and La/SSB antibodies is strongly correlated with HLA-DR3 and -DR2 in primary SS and SLE. The presence of both Ro(SS-A) and La/SSB together in the same patient is most consistent with primary SS and the presence of HLA determinants, B8, DR3, and DRw52. By contrast, the presence of Ro(SSA) without La/SSB is more consistent with SLE, as are a younger age of disease onset, the presence of HLA-DR2 and DQw1, and quantitatively less anti-Ro(SSA). DQ1/DQ2 heterozygosity is found in patients with the highest concentrations of anti-Ro(SSA). Several spontaneously autoimmune mouse strains, particularly the MRL mice, have features of SS and arthritis, further evidence of a genetic predisposition.

Numerous factors (including genetic, infectious, endocrine, and psychoneuroimmunologic) contribute to the pathogenesis of autoimmune diseases. Little is known about the role of infectious agents, aside from the important phenomenon of shared epitopes and molecular mimicry. Interest in viruses and autoimmunity began over two decades ago with the observation that some SS patients, after many years of benign disease, go on to develop lymphoid malignancies. The first virus implicated in SS was the Epstein-Barr virus (3). A role for EBV still remains an attractive possiblity. EBV could be a cofactor along with retroviruses in disease pathogenesis. Since the discovery of the AIDS virus, retroviruses have created great interest as possible causative agents in autoimmune rheumatic diseases, multiple sclerosis, and chronic fatigue syndrome.

The evidence implicating retroviruses in SS is based on both laboratory as well as clinical findings. Salivary gland infiltrates and parotid swelling resembling SS develop in patients infected with the HIV virus. HIV-associated salivary gland disease can occur in adults, children or after transfusion and may be seen with either acquired immune deficiency syndrome (AIDS)-related complex or AIDS itself. HIV-related disease must now be added to the differential diagnosis of any patient presenting with parotid swelling. Xerostomia is present in almost all of these patients. Salivary flow rates may be reduced. Dry eyes and arthralgias may also be present. Generalized lymphadenopathy, lymphocytic pulmonary infiltrates and CNS symptoms can occur as well as antinuclear antibody or rheumatoid factor. It is easy then, to see how the clinician might think of an autoimmune disease like SS or SLE rather than infection with the HIV virus.

The endogenous retroviruses have recently emerged as possible contributors to the etiology of human autoimmune diseases. We suspect that autoimmune diseases such as SLE and SS are disorders in which endogenous retroviral sequences are expressed and act in a pathologic manner (4). A considerable portion of the mammalian genome is made up of these retroviral sequences, the function of which is unknown. A recent series of studies have clearly linked expression of endogenous retroviruses in mice to Mls (minor lymphocyte stimulating) genes. The Mls determinants act like superantigens and simultaneously associate with MHC class II molecules and all T cell receptors encoded by one or more $V_\beta$ gene segments. Positive or negative selection of T cells in the thymus by these determinants controls T cell repertoire selection. Tolerance is established through self-deletion of T cells by these regulatory elements.

MRL/*lpr* mice that develop autoimmunity, extensive lymphoproliferation, and lupuslike syndrome have a defect in the *FAS* antigen gene. The *Fas* antigen is a cell surface protein, possibly a receptor, that helps to mediate apoptosis. Apoptosis is defined as programmed cell death. A lack of *Fas* antigen in the thymus may provide an explanation for the double-negative lymphoid cells that persist in the peripheral lymphoid organs of *lpr* mice, for the lack of tolerance induction by normal processes of negative selection, and for the appearance of autoimmunity.

*Fas* antigen is also expressed in activated B cells. Transgenic mice expressing anti-RBC antibodies develop an autoimmune hemolytic anemia associated with failure of apoptosis and retention of Ly-1$^+$ (CD5$^+$ or B1) B cells. An abnormal extension of the functional life span of CD5$^+$ or B1 cells in patients could explain the retention of germline V genes and immunodominant lupus-associated idiotypes in families and individuals who are susceptible to systemic lupus erythematosus (SLE) and Sjögren's syndrome. These idiotypes, important in the development of the neonatal immune system, could interfere with normal processes of immunoregulation in the adult patient.

The lymphocytic proliferative and destructive processes that result in dry eyes should be sensitive to treatment with new and highly specific immunotherapies now being developed or undergoing early clinical trials in RA, multiple sclerosis and other autoimmune diseases. These new therapeutics approaches often hope for are ultimately superior treatment, possibly even a cure for Sjögren's Syndrome.

## ACKNOWLEDGMENTS

These studies were supported (in part) by the General Medical Research Service of the Veterans Administration and by USPHS grant no. 1R01 DE09311-01.

## REFERENCES

1. N. Talal, H.M. Moutsopoulos, and S.S. Kassan, eds. "Sjögren's Syndrome - Clinical and Immunological Aspects," Springer-Verlag, Heidelberg (1987).
2. N. Talal, Sjögren's syndrome and connective tissue disease with other immunologic disorders, in: "Arthritis and Allied Conditions" (ed 12), Koopman & McCarty D, ed., Lea & Febiger, Philadelphia, PA, pp. 1197-1207 (1989).
3. E. Flescher, and N. Talal, Do viruses contribute to the development of Sjögren's syndrome?, *Am. Jour. Med.*, 90:283-285 (1991).
4. N. Talal, E. Flescher, and H. Dang, Are endogenous retroviruses involved in human autoimmune disease?, *J. Autoimmun*, 5:61-66 (1992).

# ANDROGEN-INDUCED SUPPRESSION OF AUTOIMMUNE DISEASE IN LACRIMAL GLANDS OF MOUSE MODELS OF SJÖGREN'S SYNDROME

David A. Sullivan, Hiroko Ariga, Ana C. Vendramini, Flavio J. Rocha, Masafumi Ono and Elcio H. Sato

Department of Ophthalmology, Harvard Medical School and
Immunology Unit, Schepens Eye Research Institute
20 Staniford Street, Boston, MA 02114

## INTRODUCTION

Almost 2,000 years ago, Claudius Galen, the Greek physician and writer, proposed that the mental status of an individual may significantly influence one's susceptibility to disease.[1] This postulate serves as an historic landmark in the rapidly growing field of neuroendocrinimmunology, which was established through the recognition that the nervous, endocrine and immune systems control each other through bidirectional channels of communication, that employ both similar signals and receptors.[2-7] At present, over 50 neurotransmitters, hormones and secretagogues are known that exert a profound impact on cellular, humoral and mucosal immunity.[2-7] However, the exact nature of these interactions is extremely dependent upon the specific signal, target cell, and local microenvironment.[8] Thus, depending upon the tissue, neuroendocrine action may result in stimulation, inhibition, or no effect, on immune expression.[8] As an additional consideration, antigenic exposure to the immune system may lead to the generation of numerous lymphocytic cytokines (e.g. lymphokines, neuropeptides, hormones), that directly regulate neural and endocrine function.[2-7] In consequence, an extensive, triangular interrelationship exists among the neural, endocrine and immune systems that acts to promote homeostasis and health.

The impact of these intersystem interactions is well exemplified by the endocrine modulation of autoimmune disease. For example, estrogens have been implicated in the etiology and/or progression of many autoimmune syndromes, whereas androgens have been shown to often suppress autoimmune sequelae.[9-12] In fact, androgen therapy has been used to effectively decrease autoimmune expression in animals models of systemic lupus erythematosus (SLE), thyroiditis, polyarthritis and myasthenia gravis, as well as the human condition of idiopathic thrombocytopenic purpura.[9,10] Given these effects of estrogens and androgens, researchers have proposed that the differential actions of these steroids may account for the distinct sexual dichotomy frequently observed in the incidence and severity of a number of autoimmune disorders.[9-12] However, it should also be noted that: [a] estrogens do not always exacerbate autoimmune disease: in certain disorders, these sex steroids may ameliorate pathological manifestations; [b] androgens do not invariably depress autoimmune expression: in specific diseases, androgens may accelerate both morbidity and mortality; and [c] the extent of endocrine influence on autoimmune disorders may be very restricted: for instance, androgens may selectively reduce some, but exert no influence on other, signs and symptoms of a given disease.[9-12] This latter observation underscores the concept of site-selectivity in the hormonal regulation of the immune system.[8]

Recently, research has indicated that the endocrine system may also play a role in the onset, development, as well as potential treatment, of the immune-associated, lacrimal gland defects in Sjögren's syndrome.[13-15] This syndrome is a multifaceted autoimmune disease, that occurs almost exclusively in females, and is accompanied by an insidious lymphocytic infiltration into the main and accessory lacrimal glands, an immune-mediated destruction of acinar and ductal tissues and keratoconjunctivitis sicca.[16,17] The etiology of Sjögren's syndrome is unknown, but may involve the interplay of multiple factors, including those of genetic, neural, viral, environmental and endocrine origin.[9,11,12,16-19] In this last instance, particular attention has been focused upon the detrimental impact of both estrogens[13,14] and prolactin:[20] these hormones appear to be involved in the pathogenesis, acceleration and/or amplification of Sjögren's syndrome (and/or SLE). In contrast, androgens may provide a protective influence: systemic androgen administration to animals or humans after the onset of Sjögren's syndrome may result in a significant suppression of autoimmune sequelae in the lacrimal gland, and/or an apparent reduction in ocular symptoms.[15] The relevant research describing this androgen action, as well as an assessment of the possible mechanisms underlying this hormone effect, are briefly summarized in the following sections.

## RESULTS

### Androgen Influence on Lacrimal Gland Immunopathology in the MRL/Mp-lpr/lpr and NZB/NZW F1 Mouse Models of Sjögren's Syndrome

Several years ago, our laboratory began to examine whether androgen therapy might prevent the progression of, or reverse, autoimmune disease in lacrimal glands after the onset of Sjögren's syndrome. Towards that end, we utilized adult, female MRL/Mp-lpr/lpr (MRL/lpr) mice, which are an animal model of both Sjögren's syndrome and SLE.[21,22] Lacrimal tissues of these mice, as in humans, contain multifocal and extensive lymphocytic infiltrates in perivascular and periductal areas, marked glandular disruption and apparent fibrosis.[21,23] Our experimental approach involved the systemic treatment of age-matched female MRL/lpr mice, after the onset of disease, with placebo compounds or physiological (i.e for an adult male) amounts of testosterone for 17 or 34 days. Our results[24] showed that androgen, but not placebo, administration induced a dramatic, time-dependent decline in the extent of inflammation in lacrimal tissue: after 34 days of hormone exposure, the magnitude of lymphocytic infiltration had undergone a significant, 12-fold decrease. This testosterone action involved a significant reduction in both the number and mean areas of focal infiltrates. Overall, these findings demonstrated that androgen therapy may suppress autoimmune expression in lacrimal glands of the MRL/lpr mouse model of Sjögren's syndrome.[24]

To extend the above results, we also explored whether androgen treatment might diminish autoimmune disease in lacrimal glands of another animal model (female NZB/NZW F1 [F1] mouse) of Sjögren's syndrome. As in humans, lacrimal tissues of this mouse strain, which harbors a fundamental B cell defect, contain dense, lymphocytic aggregates,[21,25] that display a predominance of B and helper T cells.[26] Furthermore, this murine disease is accompanied by a focal disruption of acinar and ductal tissues and apparent ocular surface dryness.[21,25] In contrast, immune dysfunction in MRL/lpr mice appears to have a different pathogenesis, involving a basic, immunoregulatory disorder of T cells.[27] Accordingly, we systemically administered vehicle or varying concentrations of testosterone to age-matched female F1 mice for up to 51 days after the onset of disease. Our findings[28] showed that androgen therapy caused a profound, time-dependent reduction in lymphocyte infiltration in lacrimal tissue: following 51 days of androgen treatment, the extent of lymphocyte accumulation had been decreased by a 46-fold amount, compared to that in placebo-treated tissues. Moreover, this testosterone action, which involved significant diminutions in the density of focal infiltrates, the size of individual foci and the absolute quantity of lymphocyte infiltration per lacrimal section, was also associated with a dose-dependent rise in tear volume.[28] Thus, irrespective of the underlying immune etiology, androgens appear to inhibit the progression of autoimmune disease in lacrimal glands of mouse models of Sjögren's syndrome.

## Impact of Androgen Treatment on Lymphocyte Distribution and Ia Expression in Lacrimal Glands of MRL/lpr Mice

The mechanism by which testosterone suppresses lacrimal gland inflammation in autoimmune mouse models of Sjögren's syndrome could theoretically involve a selective, hormone-induced decrease in the frequency of certain T, B and/or Ia-positive lymphocyte populations. To evaluate this possibility, MRL/lpr females were treated systemically with vehicle or testosterone for 2.5-5 weeks, and lacrimal glands were then analyzed by immunoperoxidase procedures to enumerate total (Thy 1.2), suppressor/cytotoxic (Lyt 2) and helper (L3T4) T cells, total B cells (surface IgM), B220-positive cells, as well as lymphocyte Ia expression. Our findings showed that androgen administration caused a precipitous, but almost uniform, reduction in the total number of T cells, helper T cells, suppressor/cytotoxic T cells, Ia-positive lymphocytes, B cells and B220-positive cells.[29] This result indicates that testosterone action on lacrimal gland immunopathology may target the entire lymphoid aggregate, rather than a specific lymphocyte subclass.

## Comparative Effects of Androgens, Non-Androgenic Steroids and Immuno-suppressive Compounds on Lacrimal Gland Autoimmune Disease in the MRL/lpr mouse model of Sjögren's syndrome

It is possible that testosterone's immunosuppressive impact on autoimmune expression in lacrimal tissue may not be unique, but rather duplicated by the administration of other steroid hormones or anti-inflammatory compounds. To test this possibility, we systemically treated female MRL/lpr mice with vehicle, steroids or immunosuppressive agents for 3 weeks after disease onset, and then examined lacrimal tissues for the extent of lymphocyte infiltration. For comparison, we also examined the influence of these different pharmacologic compounds on the volume of tears, the magnitude of lymphocyte infiltration in the submandibular gland, and the extent of mucosal and peripheral lymphadenopathy. Our results demonstrated that the ameliorative effect of testosterone on inflammatory lesions in lacrimal tissue was reproduced by exposure to an anabolic androgen (i.e. 19-nortestosterone) or to cyclophosphamide, but not by treatment with 17β-estradiol, danazol, an experimental, non-androgenic steroid, cyclosporine A or dexamethasone.[30] We also found that: [a] administration of androgens, cyclophosphamide or dexamethasone significantly decreased the magnitude of inflammation in submandibular glands; [b] treatment with cyclophosphamide, but not androgens or other pharmaceutical compounds, significantly diminished the size of mesenteric lymphatic and splenic tissues; and [c] exposure to 17β-estradiol, the experimental steroid or dexamethasone caused a significant decline in tear volume. It should be noted that the processes involved in the androgen suppression of lacrimal and salivary gland disease appear to be different.[30] Moreover, the capacity of androgens to reduce infiltration in lacrimal, but not peripheral lymphatic, tissues, represents another example of the site-specificity of endocrine-immune interactions.

### Androgen Influence on Lacrimal Function in MRL/lpr and F1 Mice

Given the effect of androgens on lacrimal glands of autoimmune mice, we also explored whether hormone action might be paralleled by an increase in lacrimal gland function (e.g. secretion of IgA antibodies and/or total protein). Therefore, female MRL/lpr and F1 mice were systemically treated with vehicle or physiological levels of testosterone for varying time intervals (17 to 51 days) after the onset of disease. Our findings showed that androgen exposure induced a significant rise in the concentration and total amount of tear IgA and tear protein, relative to levels in pretreatment or placebo controls.[31] Moreover, in other studies, we observed that testosterone administration stimulated the accumulation of tear IgA and protein levels in non-autoimmune, female BALB/c mice.[31] Thus, androgens appear to enhance lacrimal function in autoimmune mice and modulate the murine ocular secretory immune system. In additional experiments, we have also found that: [a] testosterone's regulation of tear IgA and protein content in MRL/lpr mice may be duplicated by treatment with other potent androgens or anabolic analogues, but not by exposure to estrogens or immunosuppressive compounds (e.g. cyclosporine A, cyclophosphamide, dexamethasone); and [b] the androgen-induced increase in tear IgA levels may involve different mechanisms than those underlying the hormone-related suppression of lacrimal infiltration in MRL/lpr mice.[31]

## DISCUSSION AND ACKNOWLEDGMENTS

These studies demonstrate that systemic androgen treatment dramatically curtails the severity of autoimmune disease in lacrimal glands of female mouse models of Sjögren's syndrome. Moreover, androgen action appears to induce an increase in the functional activity (e.g. protein secretion) of lacrimal tissue in these autoimmune murine strains. However, whether androgen exposure might also suppress immunopathological lesions in lacrimal tissue of patients with Sjögren's syndrome, and/or stimulate lacrimal gland function in these individuals, remains to be determined. In this regard, though, several observations are of particular interest, in that: [a] male and female patients with systemic lupus erythematosus, who may also suffer from secondary Sjögren's syndrome, have significantly reduced serum concentrations of androgens;[32,33] in fact, the high estrogen and low androgen levels in these patients may be a significant predisposing factor to disease progression;[12,34] and [b] three clinical studies, albeit uncontrolled, reported that systemic androgen therapy of patients with Sjögren's syndrome leads to an apparent improvement in ocular health[35-37] and 4- to 10-fold rise in tear flow.[37] Thus, it may be that targeted androgen delivery to the lacrimal gland, as compared to systemic administration (e.g. which would be contraindicated for women[38]), may provide a potential therapy for the ocular manifestations of Sjögren's syndrome.

The precise mechanism(s) involved in this immunosuppressive effect of androgens in lacrimal tissue remains to be definitively clarified. In the past, various hypotheses have been proposed to account for the androgen modulation of systemic immunity and/or autoimmune disease. These postulated explanations focused principally upon initial androgen interaction with the thymus,[39-42] hypothalamic-pituitary axis,[41,42] bone marrow[43] and/or spleen,[44,45] although additional interpretations emphasized the effect of androgens on lymphocytes,[41,46] Ia expression,[40] immune complex formation[47] or clearance,[48,49] the central nervous system[50] or genetic factors.[51] Yet, these proposals do not seem to explain the androgen-induced reversal of autoimmune expression in lacrimal tissue.[15]

Rather, androgen action on lacrimal autoimmunity appears to be a unique, tissue-specific effect,[15] that is initiated through androgen binding to receptors in lacrimal gland epithelial cells.[52] In addition, we hypothesize that this androgen interaction then causes the altered expression and/or activity of cytokines, proto-oncogenes and/or adhesion molecules in lacrimal tissue, resulting in the suppression of immunopathological lesions and an improvement in glandular function. In support of this hypothesis, androgens have been shown to: [a] modulate the structure, function and immune expression of lacrimal tissue in a variety of species;[8] [b] associate with specific, high affinity binding sites in the lacrimal gland of male and female animals[53,54] (note: these receptors, which may be up-regulated by androgen treatment, are located almost exclusively in lacrimal epithelial cells[52,55]; in contrast, androgen binding sites do not appear to exist in mature lymphocytes[10,56,57]); [c] directly regulate lacrimal epithelial cell activity;[58] and [d] control the output of immunosuppressive cytokines in other tissues.[59] With regard to cytokines, these peptides, which are produced by a wide variety of cells (e.g. immune, epithelial, endothelial, neural),[6,60-62] have been termed the "mediators of autoimmune disease,"[63] and modulation of their synthesis by sex steroids may well account for the pronounced sexual dimorphism in many autoimmune diseases.[64] It should also be noted that autoimmune disorders invariably present an imbalanced production and release of cytokines,[23,58,64,65] leading, for example, to the subversion of tolerance to specific antigens,[65,66] activation of effector functions of T and B cells,[65,66] stimulation of proto-oncogene and intercellular adhesion molecule expression,[60,67] promotion of the inflammatory process[65,67,68] and destruction of target cells (e.g. epithelial).[65,69] Moreover, cytokines (e.g. IL-1β) may directly suppress transmitter release by adrenergic and cholinergic nerves,[70] which action, if occuring in autoimmune lacrimal tissue, could result in a striking decrease in glandular function and tear secretion.

However, little information is available concerning the array of cytokines typically present in the lacrimal gland, or the cellular origin and function of these peptides, or their potential role in the mediation of lacrimal autoimmune disease, such as in Sjögren's syndrome. Furthermore, no data appear to exist on the hormonal regulation of cytokines in lacrimal tissue in either health or disease. Given that an abberrant expression of cytokines may significantly contribute to the etiology and/or generation of inflammatory eye diseases (e.g. uveitis, ACAID[71]), and that cytokine production is increased in salivary epithelial cells of Sjögren's syndrome patients,[72,73] it is clear that further research is required to clarify the role of cytokines, and their possible modulation by androgens, in autoimmune lacrimal tissue. Such clarification may have great potential for the development of specific therapies for the treatment of lacrimal autoimmunity in Sjögren's syndrome. [This research was supported by NIH grant EY05612 and a grant from the Mass. Lions' Research Fund]

## REFERENCES

1. N.R.S. Hall, and A.L. Goldstein, Thymosin modulation of the immune system, *in*: "The Neuro-Immune-Endocrine Connection," C.W. Cotman, R.E. Brinton, A. Galaburda, B. McEwen, and D.M. Schneider, eds., Raven Press, New York, pp. 59-69 (1987).
2. S. Freier, ed., "The Neuroendocrine Immune Network," CRC Press, Boca Raton, FL (1989).
3. R. Ader, D. Felten, and N. Cohen, eds., "Psychoneuroimmunology," Acad. Press, San Diego, CA (1991).
4. S. D'Orisio, and A. Panerai, eds., "Neuropeptides and Immunopeptides: Messengers in a Neuroimmune Axis," *Ann. N.Y. Acad. Sci.* vol. 594 (1990).
5. R.H. Stead, M.H. Perdue, H. Cooke, D. Powell, and K. Barrett, eds., "Neuro-Immuno-Physiology of the Gastrointestinal Mucosa," *Ann. N.Y. Acad. Sci.* vol. 664 (1992).
6. E.J. Goetzl, and S.P. Sreedharan, Mediators of communication and adaptation in the neuroendocrine and immune systems, *FASEB J.* 6:2646 (1992).
7. N. Fabris, B.D. Jancovic, B.M. Markovic, and N.H. Spector, eds., "Ontogenetic and Phylogenetic Mechanisms of Neuroimmunomodulation," *Ann. N.Y. Acad. Sci.* vol. 650 (1992).
8. D.A. Sullivan, Hormonal influence on the secretory immune system of the eye, *in*: "The Neuroendocrine-Immune Network," S. Freier, ed., CRC Press, Boca Raton, FL, p199-238 (1990).
9. S.A. Ahmed, W.J. Penhale, and N. Talal, Sex hormones, immune responses and autoimmune diseases, *Am. J. Pathol.* 121:531 (1985).
10. S.A. Ahmed, and N. Talal, Sex hormones and the immune system, *Bailliere's Clin. Rheum.* 4:13 (1990).
11. S.A. Ahmed, and N. Talal, Importance of sex hormones in systemic lupus erythematosus, *in*: "Dubois' Lupus Erythematosus," D. Wallace, and B. Hahn, eds., Lea & Febiger, Philadelphia, p 148-156 (1993).
12. F. Homo-Delarche, F. Fitzpatrick , N. Christeff, E.A. Nunez, J.F. Bach, and M. Dardenne, Sex steroids, glucocorticoids, stress and autoimmunity, *J. Ster. Biochem. Mol. Biol.* 40:619 (1991).
13. H. Carlsten, A. Tarkowski, R. Holmdahl, and L.A. Nilsson, Oestrogen is a potent disease accelerator in SLE-prone MRL lpr/lpr mice, *Clin. exp. Immunol.* 80:467 (1990).
14. S.A. Ahmed, T.B. Aufdemorte, J.R. Chen, A.I. Montoya, D. Olive, and N. Talal, Estrogen induces the development of autoantibodies and promotes salivary gland lymphoid infiltrates in normal mice, *J. Autoimmunity* 2:543 (1989).
15. D.A. Sullivan, and E.H. Sato, Potential therapeutic approach for the hormonal treatment of lacrimal gland dysfunction in Sjögren's syndrome, *Clin. Immunol. Immunopath.* 64:9 (1992).
16. R.I. Fox, ed., "Sjögren's Syndrome," *Rheum. Dis. Clin. N.A.* vol. 18 (3) (1992).
17. N. Talal, H.M. Moutsopoulos, and S.S. Kassan, eds., "Sjögren's Syndrome. Clinical and Immunological Aspects," Springer Verlag, Berlin (1987).
18. H. Carlsten, R. Holmdahl, and A. Tarkowski, Analysis of the genetic encoding of oestradiol suppression of delayed-type hypersensitivity in (NZB/NZW) F1 mice, *Immunol.* 73:186 (1991).

19. R.I. Fox, M. Luppi, H.I. Kang, and P. Pisa, Reactivation of Epstein-Barr virus in Sjögren's syndrome, *Springer Semin. Immunopathol.* 13:217 (1991).

20. R. McMurray, D. Keisler, K. Kanuckel, S. Izui, and S.E. Walker, Prolactin influences autoimmune disease activity in the female B/W mouse. *J. Immunol.* 147:3780 (1991).

21. R.W. Hoffman, M.A. Alspaugh, K.S. Waggie, J.B. Durham, J.B., and S.E. Walker, Sjögren's syndrome in MRL/l and MRL/n mice. *Arthritis Rheum.* 27:157 (1984).

22. J.D. Mountz, W.C. Gause, and R. Jonsson, Murine models for systemic lupus erythematosus and Sjögren's syndrome, *Curr. Opin. Rheumatol.* 3:738 (1991).

23. D.A. Jabs, E.L. Alexander, and W.R. Green, Ocular inflammation in autoimmune MRL/Mp mice, *Invest. Ophthalmol. Vis. Sci.* 26:1223 (1985).

24. H. Ariga, J. Edwards, and D.A. Sullivan, Androgen control of autoimmune expression in lacrimal glands of MRL/Mp-lpr/lpr mice, *Clin. Immunol. Immunopath.* 53:499 (1989).

25. H.S. Kessler, A laboratory model for Sjögren's syndrome, *Am. J. Pathol.* 52:671 (1968).

26. D.A. Jabs, and R.A. Prendergast, Murine models of Sjögren's syndrome. *Invest. Ophthalmol. Vis. Sci.* 29:1437 (1988).

27. B. Lieberum, and K.U. Hartmann, Successive changes of the cellular composition in lymphoid organs of MRL-Mp/lpr/lpr mice during the development of lymphoproliferative disease as investigated in cryosections, *Clin. Immunol. Immunopathol.* 46:421 (1988).

28. A.C. Vendramini, C.H. Soo, and D.A. Sullivan, Testosterone-induced suppression of autoimmune disease in lacrimal tissue of a mouse model (NZB/NZW F1) of Sjögren's syndrome, *Invest. Ophthalmol. Vis. Sci.* 32:3002 (1991).

29. E.H. Sato, H. Ariga H, and D.A. Sullivan, Impact of androgen therapy in Sjögren's syndrome: Hormonal influence on lymphocyte populations and Ia expression in lacrimal glands of MRL/Mp-lpr/lpr mice, *Invest. Ophthalmol. Vis. Sci.* 33:2537 (1992).

30. E.H. Sato, and D.A. Sullivan, Comparative influence of steroid hormones and immunosuppressive agents on autoimmune expression in lacrimal glands of a female mouse model (MRL/Mp-lpr/lpr) of Sjögren's syndrome, submitted (1993).

31. D.A. Sullivan, J. Edwards, C. Soo, A.C. Vendramini, H. Ariga, and E.H. Sato, article submitted (1993).

32. R.G. Lahita, H.L. Bradlow, E. Ginzler, S. Pang, and M. New, Low plasma androgens in women with systemic lupus erythematosus, *Arth. Rheum.* 30:241 (1987).

33. C. Lavalle, E. Loyo, R. Paniagua, J.A. Bermudez, J. Herrera, A. Graef, D. Gonzalez-Barcena, and A. Fraga, Correlation study between prolactin and androgens in male patients with Systemic Lupus Erythematosus, *J. Rheum.* 14:268 (1987).

34. R.G. Lahita, The importance of estrogens in SLE, *Clin. Immunol. Immunopath.* 63:17 (1992).

35. R. Bruckner, Uber einem erfolgreich mit perandren behandelten fall von Sjogren'schem symptomen komplex, *Ophthalmologica* 110:37 (1945).

36. M. Appelmans, La Keratoconjonctivite seche de Gougerot-Sjogren, *Arch. 'Ophtalmologie* 81:577 (1948).

37. A. Bizzarro, G. Valentini, G. Di Marinto, A. Daponte, A. De Bellis, and G. Iacono, Influence of testosterone therapy on clinincal and immunological features of autoimmune diseases associated with Klinefelter's syndrome, *J. Clin. End. Metab.* 64:32 (1987).

38. R. Lahita, Sex hormones, Sjögren's syndrome and the immune response, *The Moisture Seekers Newsletter* 8:1 (1991).

39. J. Comsa, H. Leonhardt, and H. Wekerle H, Hormonal coordination of the immune response, *Rev. Physiol. Biochem. Pharmacol.* 92:115 (1982).

40. N. Talal, and S.A. Ahmed, Sex hormones and autoimmune disease, *Int. J. Immunotherapy* 3:65 (1987).

41. C.J. Grossman, Are there underlying immune-neuroendocrine interactions responsible for immunological sexual dimorphism?, *Prog. NeuroEndocrinImmunology* 3:75 (1990).

42. E.J. Goldsteyn, and M.J.L. Fritzler, The role of the thymus-hypothalamus-pituitary-gonadal axis in normal immune processes and autoimmunity, *J. Rheumatol.* 14:982 (1987).

43. A.H.W.M. Shuurs, and H.A.M. Verheul, Effect of gender and sex steroids on the immune response, *J. Steroid Biochem.* 35:157 (1990).

44. Olsen NJ, Watson MB and Kovacs WJ: Studies of immunological function in mice with defective androgen action, Immunology 73: 52, 1991.

45. Y. Weinstein, and Z. Berkovich, Testosterone effect on bone marrow, thymus and suppressor T cells in the (NZB/NZW) F1 mice: its relevance to autoimmunity, *J. Immunol.* 126:998 (1981).

46. N. Talal, H. Dang, S.A. Ahmed, E. Kraig, and M. Fischbach, Interleukin 2, T cell receptor and sex hormone studies in autoimmune mice, *J. Rheum.* (suppl. 13) 14:21 (1987).

47. K.A. Melez, W.A. Boegel, and A.D. Steinberg, Therapeutic studies in New Zealand mice. VII. Successful androgen treatment of NZB/NZW F1 females of different ages, *Arthritis Rheum.* 23:41 (1980).

48. J.R. Roubinian, N. Talal, J.S. Greenspan, J.R. Goodman, and P.K. Siiteri, Effect of castration and sex hormone treatment on survival, anti-nucleic acid antibodies, and glomerulonephritis in NZB/NZW F1 mice, *J. Exp. Med.* 147:1568 (1978).

49. J.B. Allen, D. Blatter, G.B. Calandra, and R.L. Wilder, Sex hormonal effects on the severity of streptococcal cell wall-induced polyarthritis in the rat, *Arthritis Rheum.* 26:560 (1983).

50. N. Talal, S.A. Ahmed, and M. Dauphinee, Hormonal approaches to immunotherapy of autoimmune diseases, *N.Y. Acad. Sci.* 475:320 (1986).

51. H. Carlsten, R. Holmdahl, A. Tarkowski, and L.A. Nilsson, Oestradiol- and testosterone-mediated effects on the immune system in normal and autoimmune mice are genetically linked and inherited as dominant traits, *Immunology* 68:209 (1989).

52. M. Ono, F.J. Rocha and D.A. Sullivan, Cellular distribution and hormonal control of androgen receptors in lacrimal tissue of the MRL/Mp-lpr/lpr mouse model of Sjögren's syndrome, submitted (1993).

53. M. Ota, S. Kyakumoto, and T. Nemoto, Demonstration and characterization of cytosol androgen receptor in rat exorbital lacrimal gland, *Biochem. Internat.* 10:129 (1985).

54. D.A. Sullivan, J.A. Edwards, and R.S. Kelleher, Analysis of androgen binding sites in the rat lacrimal gland, submitted (1993).

55. F.J. Rocha, L.A. Wickham, J.D.O. Pena, J. Gao, M. Ono, R.W. Lambert, R.S. Kelleher, and D.A. Sullivan, Influence of gender and the endocrine environment on the distribution of androgen receptors in the lacrimal gland, submitted (1993).

56. L. Danel, M. Menouni, J. Cohen, J. Magaud, G. Lenoir, J. Revillard, and S. Saez, Distribution of andro-gen and estrogen receptors among lymphoid and haemopoietic cell lines, *Leukemia Res.* 9:1373 (1985).

57. H. Takeda, G. Chodak, S. Mutchnik, T. Nakamoto, and C. Chang, Immunohistochemical localization of androgen receptors with mono- and polyclonal antibodies to androgen receptor, *J. Endocr.* 126:17 (1990).

58. D.A. Sullivan, R.S. Kelleher, J.P. Vaerman, and L.E. Hann, Androgen regulation of secretory component synthesis by lacrimal gland acinar cells in vitro, *J. Immunol.* 145:4238 (1990).

59. E. Nagy, I. Berczi, and E. Sabbadini, Endocrine control of an immunoregulatory cytokine of the submandibular gland, Hans Selye Symp. in Neuroendocrinology, Budapest, Hungary, p. 51 (1992).

60. W.L. Sibbitt, Oncogenes, growth factors and autoimmune diseases, *Antican. Res.* 11:97 (1991).

61. A. Mantovan, F. Bussolino, E. Dejana, Cytokine regulation of endothelial cell function, *FASEB J.* 6:2591 (1992).

62. S.A. Robertson, M. Brannstrom, and R.F. Seamark, Cytokines in rodent reproduction and the cytokine-endocrine interaction, *Curr. Opin. Rheumatol.* 4:585 (1992).

63. J.D. Mountz, and C. Edwards, Murine models of autoimmune disease, *Curr. Opin. Rheum.* 4:621 (1992).

64. N. Sarvetnick, and H.S. Fox, Interferon-gamma and the sexual dimorphism of autoimmunity, *Mol. Biol. Med.* 7:323 (1990).

65. G. Kroemer, and A. Martinez, Cytokines and autoimmune diseases, *Clin. Immunol. Immunopath.* 61:275 (1991).

66. A.D. Steinberg, Concepts of pathogenesis of systemic lupus erythematosus. Clin. Immunol. Immunopathol. 63:19 (1992).

67. R.P. Wuthrich, A.M. Jevnikar, F. Takei, L.H. Glimcher, and V.E. Kelley, Intercellular adhesion molecule-1 (ICAM-1) expression is upregulated in autoimmune murine lupus nephritis, *A. J. Path.* 136:441 (1990).

68. G.S. Firestein, Cytokines in autoimmune diseases, *Clin. Mol. Asp. Autoimmune. Dis.* 8:129 (1992).

69. R.L. Deem, F. Shanahan, and S.R. Targan, Triggered human mucosal T cells release tumour necrosis factor-α and interferon-γ which kill human colonic epithelial cells, *Clin. exp. Immunol.* 83:79 (1991).

70. S.M. Collins, S.M. Hurst , C. Main, E. Stanley, I. Khan, P. Blennerhassett, and M. Swain, Effect of inflammation of enteric nerves. Cytokine-induced changes in neurotransmitter content and release, *Ann. N.Y. Acad. Sci.* 664:415 (1992).

71. D. Wakefield, and A. Lloyd, The role of cytokines in the pathogenesis of inflammatory eye disease, *Cytokine* 4:1 (1992).

72. F.N. Skopouli, E. Kousvelari, P. Mertz, E.S. Jaffe, P.C. Fox, and H.M. Moutsopoulos, c-myc mRNA expression in minor salivary glands of patients with Sjögren's syndrome, *J. Rheumatol.* 19:693 (1992).

73. R.I. Fox, and H.I. Kang, Pathogenesis of Sjögren's syndrome, *Rheum. Dis. Clin. N.A.* 18:517 (1992).

# EFFECT OF SODIUM SUCROSE-SULFATE ON THE OCULAR SURFACE OF PATIENTS WITH KERATOCONJUNCTIVITIS SICCA IN SJÖGREN'S SYNDROME

Jan U. Prause, Kirsten Bjerrum and Sven Johansen

Department of Ophthalmology, Rigshospitalet and
Eye Pathology Institute, University of Copenhagen
DK 2100 Copenhagen, Denmark

## INTRODUCTION

The dry eye state in autoimmune exocrinopathy, as in patients with primary Sjögren's syndrome (1°SS), is not only due to the lack of water. It is caused by a cascade of events: lack of tear enzymes, immunoglobulins, hormones, nutrients, changes in osmolarity, pH .. (Frost-Larsen et al., 1980). This leads, together with a change in the epithelial cells of the ocular surface, to epithelial destruction and ulceration. Many attempts have been made to solve this problem. We have found some analogy between the lesions of the dry eye surface and the lesions of the gastric mucosal ulcers, and have performed an open study (Prause, 1991) and a masked controlled study on the beneficial effect of sodium sucrose-sulfate (SSS) on keratoconjunctivitis sicca (KCS) in patients with 1°SS. Sodium sucrose-sulfate is the water soluble salt of sucralfate (Caralfate®, Ulcogant® or Antepsin®). From studies, mainly of the effect of sucralfate, the drug is known to Improve the healing of mucosal ulcers, stimulate the production of prostaglandins, bind and probably protect growth factors (Sasaki et al., 1983; Nagashima et al., 1980). It adheres to mucosal epithelia acting as a mucin and stimulates the secretion of mucins.

## MATERIAL AND METHODS

### Test Solutions

**Sodium sucrose-sulfate** was given as a 2% solution in a viscous vehicle of oxipropyl methylcellulose 0.25%, made isotonic with sodium chloride and preserved with phenyl mercurinitrate 0.001%.

**The Placebo** solution was the same vehicle including the preservative as used for the SSS solution. The two solutions were presented to the patients in identical brown glass bottles fitted with a dripping device.

The patients received two bottles/week, and use the drops at least five times/day.

## Open Study

We examined 22 patients (median age 59 years, range: 43 - 70 years) all with 1°SS according to the Copenhagen criteria (Manthorpe et al.1986). The test persons did not receive any systemic or topical steroids or non-steroidal anti-inflammatory drugs during the trial. The patients entered the study at their routine controls at the Sjögren clinic after having given their informed consent. They were reexamined every 1 - 3 months for more than six months. All persons were given the SSS solution at entrance. At each examination we performed Break-up time (BUT)(Anonymos, 1989), Schirmer-1-test (S1T)(Anonymos, 1989) and Rose-Bengal score (RBS)(Anonymos, 1989) in addition to a slit-lamp examination.

Differences between test results at entrance and test results obtained at various times during the trial were evaluated using the Wilcoxon-Pratt test. A level of $P < 0.05$ was considered significant.

## Double Masked Controlled Cross-Over Study

Design: We had two treatment periods of one months, separated by one month wash-out period. The patients used their ordinary tear substitute before the study, and during the wash-out period. We tested the SSS solution against the plain tear substitute in a double masked fashion.

Thirty patients with 1°SS (29 females and one male, median age 69.5 years, range 44 - 86 years) according to the Copenhagen criteria completed the study. All test persons gave their informed consent.

We performed the clinical tests at the beginning and at the end of each treatment period. In addition to the tests performed in the open study we also performed imprint biopsy from the conjunctiva of both eyes near the limbus at the 12 o'clock position (Marner et al.1984). The morphology of the obtained conjunctival cells and their number were evaluated microscopically on PAS stained imprints. The number of cells showing snake-like chromatin was scored 0 = no snakes, 1 = less than 25% of cells show snake-like chromatin, 2 = more than 25% of cells show snakes. All patients expressed their subjective effect of the drops on a visual analog scale at each examination, and in addition they completed a diary.

Treatment effects were expressed as differences between test results at the beginning of each test period and test results obtained at the end of the period. Differences between the various test results as well as between treatment effects were evaluated using the Wilcoxon-Pratt test. A level of $P < 0.05$ was considered significant.

## RESULTS

### Open study

No differences between entrance values and any examination values were observed concerning S1T. BUT values improved during the first 1 - 4 months ($P < 0.05$), but this effect was lost in the subsequent periods. The RBS score dropped significantly during the entire study. The main improvement (drop in score) appeared during the first three months of treatment ($P < 0.0007$), and was followed by a small further improvement ($P < 0.02$) (Fig. 1).

No side effects were observed during the open study, however, three patients would have preferred a more viscous solution.

## Double Masked Controlled Cross-Over Study

At entrance the patients had low BUT (median 4.3 s, range 0 - 40 s), low S1T (median 3.5 mm/5 min, range 0 - 25 mm/5 min) and high RBS (median 3.8, range 0 - 8), and expressed a marked feeling of dryness on the visual analog scale (median 80% of maximal dryness, range 49% - 99%).

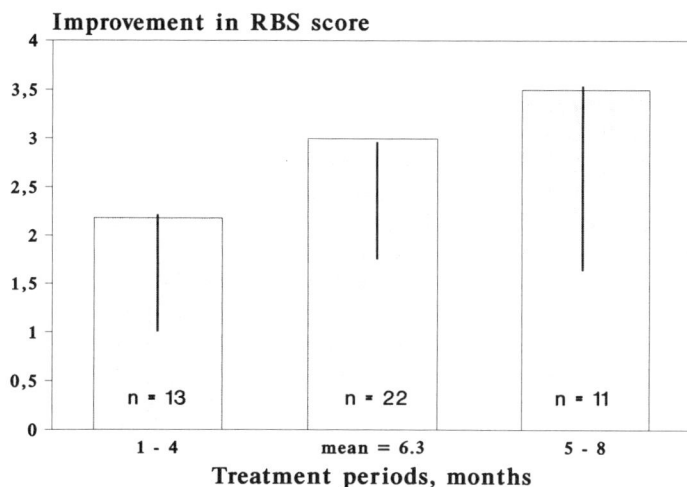

**Figure 1.** Improvement in Rose-Bengal score during treatment with sodium sucrose-sulfate in an open study expressed as mean difference between entrance score and score at examination. Bars indicate SD.

BUT did not improve during treatment with placebo, but improved during treatment with SSS. Both when testing start and end values of treatment and when testing treatment effect against effect of placebo a significant improvement was found ($P < 0.05$). Neither SSS nor placebo affected the S1T. During treatment with placebo a slight positive effect could be observed ($P < 0.05$). The same could not be shown for SSS, however, when testing the treatment effects of placebo against the treatment effects of SSS in a paired fashion, SSS was significantly better than placebo ($P < 0.05$). The imprints showed significant changes in the morphology of the conjunctival cells induced by SSS. The most prominent changes was found in the number of cells exhibiting the snake-like chromatin (Fig. 2), statistically the changes were significant both as a treatment effect and when comparing with the effect of the placebo ($P < 0.05$).

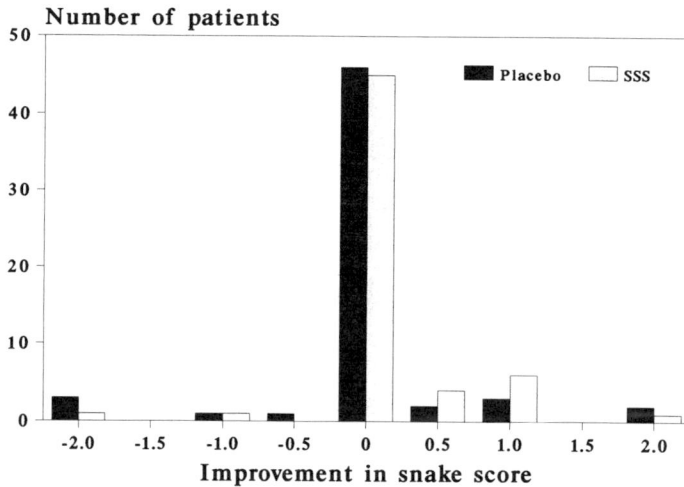

**Figure 2.** After treatment with topical sodium sucrose-sulfate a small but significant reduction in number of conjunctival cells showing snake-like chromatin was observed in conjunctival imprint biopsies.

No patients experienced side effects induced by the SSS. Subjectively the patients indicated effect of both placebo and SSS ($P < 0.05$) when comparing with entrance values i.e. the status produced by their normal tear substitute. A slight trend towards the preference for SSS could be noted from the diaries, but this could not be verified statistically when comparing the scores on the visual analog scales.

## DISCUSSION

Keratoconjunctivitis sicca in 1°SS is the result of a generalized exocrinopathy involving also the tear glands (Frost-Larsen et al., 1980). However the disease process seems also to involve the epithelia of the ocular surface (Oxholm et al., 1987; Prause et al., 1989). Despite the systemic nature of the disease, only a few systemic drugs have proven any effect (Prause et al., 1984; Manthorpe et al., 1984; Oxholm et al., 1986; Manthorpe and Prause, 1986; Prause et al., 1986). Most therapeutic measures have been directed towards the lack of the aqueous component of the tear fluid, hoping that the auto-reparative capacity of the ocular surfaces would restore normal function (Manthorpe and Prause, 1986).

Healing of mucosal ulcers with the use of synthetic mucins with protective, anti-inflammatory- and possible growth stimulating potentials has been performed for many years in other regions of the body. Sodium pentosan polysulfate, a synthetic, sulfonated glycosaminoglycan, stabilizes the mucosal coat of the urinary bladder. Unfortunately this drug had no effect on the surface condition of patients with 1°SS when tested in a controlled manner (Prause et al., 1986).

For more than 20 years the aluminium sucrose-sulfate has been used with success in the treatment of gastric ulcers (Steiner et al., 1982). The drug seems to form a protective plaque over the damaged mucosa, preventing the destructive action of the gastric juice (Sasaki et al., 1983; Nagashima et al., 1980; Ramano et al., 1990; Knight et al., 1988; Steiner et al., 1982). An anti-inflammatory action of the drug seems to be mediated via the prostaglandin system (Payno et al., 1989; Ramano et al., 1990). The healing of the mucosal surface may also be favoured by the ability of the drug to transport and protect various growth factors (Nexø and Poulsen, 1987; Konturek et al., 1989). All these effects may also act in the tear fluid.

In the present studies significant improvements in the conditions of the ocular mucosa were found. Most directly in the improvement in RBS and supported by the fall in snake score. These findings may be the cause of a healing effect of SSS upon the ocular surface, since a concomitant effect on tear production was not found in any of the two studies. We conclude that SSS has a beneficial effect upon the ocular surface condition of patients suffering from 1°SS.

## REFERENCES

Anonymos, 1989, Suggested tests for Sjögren's syndrome, *Clin Exp Rheumatol* 7:100.

Frost-Larsen, K., Isager, H., Manthorpe, R. and Prause, J.U., 1980, Sjögren's syndrome, *Ann Ophthalmol* 12:836.

Knight, L.C., Maurer, A.H., Kollmann, M., Ammar, I.A., Fisher, R.S. and Malmud, L.S., 1988, Selenium-75-labeled sucralfate: comparison with other radiolabels and initial clinical studies, *Am J Physiol Imaging* 3:10.

Konturek, S.J., Brozozowski, T., Bielanski, W., Warzecha, Z. and Drozdowiecz, D., 1989, Epidermal growth factor in the gastroprotective and ulcer-healing actions of sucralfate in rats, *Am J Med* 86:32.

Manthorpe, R., Petersen, S. Hagen and Prause, J.U., 1984, Primary Sjögren's syndrome treated with Efamol/Efavit. A double-blind cross-over investigation, *Rheumatol Int* 4:165.

Manthorpe, R., Oxholm, P., Prause, J.U. and Schiødt, M., 1986, The copenhagen criteria for Sjögren's syndrome, *Scand J Rheumatol* Suppl 61:19.

Manthorpe, R. and Prause, J.U., 1986, Treatment of Sjögren's syndrome: An overview, *Scand J Rheumatol* Suppl 61:237.

Marner, K., Manthorpe, R. and Prause, J.U., 1984, Snake-like nuclear chromatine in imprints of conjunctival cells from patients with Sjögren's syndrome, *in*: Progress in rheumatology, I. Machtey, ed., Rheumatology service, Hasharon Hospital, Petah-Tiqva.

Nagashima, R., Hinohara, Y. and Hirano, T., 1980, Selective binding of sucralfate to ulcer lesion. Experiments in rats recieving 14C sucralfate, *Drug Res* 30:88.

Nexø, E. and Poulsen, S.S., 1987, Does epidermal growth factor play a role in the action of sucralfate?, *Scand J Gastroent* 127:45.

Oxholm, P., Manthorpe, R., Prause, J.U. and Horrobin, D., 1986, Patients with primary Sjögren's syndrome treated for two months with Evening Primrose Oil, *Scand J Rheumatol* 15:103.

Oxholm, P., Oxholm, A. and Prause, J.U., 1987, Immunohistochemical characterization of intraepidermal in vivo IgG deposits in patients with primary Sjögren's syndrome, *Acta Pathol Scand Sect A* 95:239.

Payno, A., Lopez-Novoa, J.M. and Rodriguez-Puyol, D., 1989, Prostanoid production in post-gastrectomy gastritis. Influence of sucralfate, *Am J Med* 86:17.

Prause, J.U., Frost-Larsen, K., Høj, L., Isager, H. and Manthorpe, R., 1984, Lacrimal and salivary secretion in Sjögrens syndrome. The effect of systemic treatment with bromhexine, *Acta Ophthalmol* 62:489.

Prause, J.U., Krogsaa, B. and Lose, B., 1986, Treatment of keratoconjunctivitis sicca with elmiron, *Scand J Rheumatol* Suppl 61:259.

Prause, J.U., Jensen, O.A., Paschides, K., Støvhase, A. and Vangsted, P., 1989, Conjunctival cell glycoprotein patters of healthy persons and of patients with 1°SS, *J Autoimmunity* 2:495.

Prause, J.U., 1991, Beneficial effects of sucrase sodium sulphate on the ocular surface of patients with severe KCS in primary Sjögren's syndrome, *Acta Ophthalmol* 69:417.

Ramano, M., Razandi, M. and Ivey, K.J., 1990, Effect of sucralfate and its components on taurocholate-induced damage to rat gastric mucosal cells in tissue culture, *Dig Dis Sci* 35:467.

Sasaki, H., Hinohara, Y., Tsunoda, Y. and Nagashima, R., 1983, Binding of sucralfate to duodenal ulcer in man, *Scand J Gastroent* 83:13.

Steiner, K., Buhring, H.-U., Faro, H.-P., Garbe, A. and Nowak, H., 1982, Sucralfate: Pharmacokinetics, metabolism and selective binding to experimental gastric and duodenal ulcers in animals, *Arseimittelforschung* 32:512.

# COMPARATIVE EFFICACY OF ANDROGEN ANALOGUES IN SUPPRESSING LACRIMAL GLAND INFLAMMATION IN A MOUSE MODEL (MRL/lpr) OF SJÖGREN'S SYNDROME

Flavio Jaime Rocha, Elcio H. Sato, Benjamin D. Sullivan and David A. Sullivan

Department of Ophthalmology, Harvard Medical School and
Immunology Unit, Schepens Eye Research Institute
20 Staniford Street, Boston, MA 02114

## INTRODUCTION

Recent research from our laboratory has demonstrated that testosterone therapy induces a precipitous decrease in autoimmune expression in lacrimal glands of female mouse models (MRL/Mp-lpr/lpr [MRL/lpr) and NZB/NZW F1) of Sjögren's syndrome.[1-4] Thus, systemic administration of testosterone to autoimmune mice after the onset of disease may: [a] significantly reduce the mean area of focal lymphocytic infiltrates, the absolute number of immune foci, and the overall extent of lymphoid infiltration in lacrimal tissues;[1-4] and [b] remove any apparent evidence of acinar or ductal cell destruction in infiltrate-free regions.[1,2] In addition, testosterone treatment appears to stimulate the functional activity (e.g. protein secretion) of lacrimal glands in both mouse models of Sjögren's syndrome.[5] These combined effects of testosterone seem to be somewhat unique, in that exposure of female MRL/lpr mice to other pharmaceutical compounds, including 17β-estradiol, dexamethasone, danazol, cyclosporine A, or an experimental non-androgenic steroid, does not result in either immunosuppression in, or enhanced function of, lacrimal tissue.[4]

These findings suggest that testosterone treatment could theoretically provide a potential therapy for lacrimal gland defects in Sjögren's syndrome,[6] a disorder that occurs primarily in women.[7] Indeed, several clinical studies, albeit uncontrolled, support this contention, by reporting that systemic androgen administration alleviates dry eye symptoms in such patients.[8-10] However, systemic treatment with testosterone would appear to be contraindicated: such generalized testosterone exposure could well lead to many undesirable side effects, including virilization, menstrual irregularities (e.g. amenorrhea), hepatic dysfunction, edema, hematologic abnormalities, behavioral changes and/or metabolic alterations.[11] In contrast, androgen therapy might possibly be safe and effective if: [a] non-virilizing androgen analogues were identified that duplicate the ameliorative effects of testosterone on autoimmune disease in lacrimal tissue; and/or [b] androgen treatment was targeted directly to the lacrimal gland, given that the testosterone-induced immunosuppression may be mediated by epithelial cells in this tissue.[6]

To begin to explore these possibilities, we focused in the present investigation on determining whether various anabolic or modified analogues of androgens might reproduce, or surpass, the effect of testosterone on autoimmune disease in the lacrimal gland. As part of this study, we also examined whether the androgen-related reduction of lacrimal inflammation is reversible upon cessation of hormone treatment.

## MATERIALS AND METHODS

Adult, female MRL/lpr mice were obtained from The Jackson Laboratory (Bar Harbor) and housed in constant temperature rooms with fixed light/dark intervals of 12 hours duration. After the onset of autoimmune disease, animals were administered subcutaneous implants of placebo (cholesterol, methyl cellulose, lactose)- or androgen (10 mg)-containing pellets (Innovative Research of America, Toledo, Ohio) in the subscapular region. These pellets were designed for a slow, but continuous, release of hormone over a 21 day period and were reimplanted every 3 weeks. Fourteen different androgens were included in this study[12] and were originally purchased from Steraloids or Innovative Research before packaging into the pelleted form. These hormones were representative of the major structural subclasses of androgens, including: [a] androgenic compounds with unusual structural features (e.g. oxandrolone); [b] testosterone derivatives (e.g. methyltestosterone); [c] 4, 5α-dihydrotestosterone derivatives (e.g. oxymethelone); [d] 17β-hydroxy-5α-androstane derivatives containing a ring A unsaturation, excluding testosterone derivatives (e.g. 2, (5α)-androsten-17β-ol); [e] 19-nortestosterone derivatives (e.g. 19-nortestosterone propionate); and [f] adrenal cortical androgens (e.g. dehydroepiandrosterone, an androgen precursor). In addition, relative to standards (typically testosterone), these androgens included compounds displaying: [a] augmented androgenic (i.e. virilizing) activity coupled with an even larger increase in anabolic activity (e.g. fluoxymesterone); [b] enhanced anabolic action with unchanged androgenic effects (e.g. oxymetholone, dihydrotestosterone); [c] decreased androgenic ability with unchanged anabolic activity (e.g. 19-nortestosterone propionate); and [d] decreased androgenic capacity paralleled by increased anabolic activity (e.g. oxandrolone, stanozolol).

Immediately before (pretreatment), or after, the experimental time course, lacrimal glands were removed, fixed in 10% buffered formalin, dehydrated, embedded in Historesin (Reichert-Jung), cut into 3 μm sections and stained with hematoxylin and eosin (Fisher). Sections were examined with a Zeiss Videoplan 2 image analysis system to quantitatively calculate the percentage of lymphocyte infiltration.[12] This measurement was determined by dividing the sum of all focal infiltrate areas/section by the entire section area and multiplying the resulting number by 100.

Statistical analysis of the data was conducted by utilizing Student's unpaired, two-tailed t test.

**Table 1.** Effect of selected androgen analogues or precursors on lymphoid infiltration in lacrimal tissue of female MRL/lpr mice

| Treatment | Number of Tissue Sections | Percentage Infiltration |
|---|---|---|
| Pretreatment | 28 | 7.50 ± 1.07 |
| Placebo | 32 | 11.20 ± 1.11 |
| Testosterone | 32 | 2.97 ± 0.52 * |
| Oxandrolone | 28 | 2.47 ± 0.29 * |
| 5α-androstan-2α-methyl-17β-ol-3-one | 28 | 1.45 ± 0.28 *† |
| Methyldihydrotestosterone | 40 | 1.35 ± 0.21 *† |
| Dehydroepiandrosterone | 36 | 8.55 ± 0.70 |

Lacrimal glands were collected from female MRL/lpr mice either before (pretreatment, n = 7 mice) or following (n = 7-10 mice/group) 6 weeks of systemic treatment with placebo or androgen compounds. After histological processing, lacrimal gland sections (4/tissue) were evaluated for the extent of lymphocyte infiltration, as described in the Materials and Methods. * Significantly ($p < 0.0005$) less than value of pretreatment and placebo controls; † Significantly ($p < 0.05$) lower than value of testosterone-treated group. Data from reference (12).

## RESULTS

### Influence of Systemic Treatment With Androgen Analogues on the Extent of Lymphocyte Infiltration in Lacrimal Glands of Female MRL/lpr Mice [12]

To examine the comparative efficacy of anabolic and/or modified androgen compounds in suppressing immunopathological lesions in lacrimal tissues of female MRL/lpr mice, animals (n = 7 to 10/group) were treated with vehicle or one of a variety of androgens for a 6 week period. After experimental treatment, lacrimal glands were obtained and processed for morphometric analysis. Our results demonstrated that androgens from almost all subclasses, including compounds with enhanced anabolic, and reduced androgenic, activity, were effective in significantly attenuating the extent of lacrimal gland inflammation, compared to that in tissues of pretreatment or placebo controls.[12] Thus, as shown in Table 1, the immunosuppressive action of certain androgen analogues (e.g. 5α-androstan-2α-methyl-17β-ol-3-one) significantly exceeded that of testosterone, whereas other hormones had almost no effect on lacrimal lymphocyte infiltration (e.g. dehydroepiandrosterone).[12]

# Reversibility of the Androgen-Induced Suppression of Inflammation in Lacrimal Glands of Female MRL/lpr Mice [1][2]

Given that continuous testosterone treatment results in a significant suppression of lacrimal gland inflammation in female MRL/lpr mice, studies were also performed to determine whether this hormone effect might be reversible upon cessation of androgen therapy. These experiments, which involved female MRL/lpr mice (n = 9-11/group) and have been described in detail,[12] showed that stopping testosterone exposure (i.e. after 6 weeks) for almost 2 weeks led to a slight, but significant ($p < 0.05$), increase in the percentage of lymphocyte infiltration in lacrimal tissue, compared to that found in glands of chronically treated mice.

## DISCUSSION AND ACKNOWLEDGMENTS

In summary, these studies show that anabolic and/or modified androgens with minimal virilizing activity may significantly decrease autoimmune disease in lacrimal glands of the female MRL/lpr model of Sjögren's syndrome. However, this hormone action may possibly be reversible upon discontinuation of androgen treatment. To extend these findings, our current research is focusing upon the processes involved in the androgen-induced suppression of lacrimal gland inflammation. Ideally, information gained from these experiments may assist in the development of therapeutic strategies to effectively treat lacrimal dysfunction in patients with Sjögren's syndrome.

This research was supported by grants from NIH (EY05612) and the Massachusetts Lions' Research Fund.

## REFERENCES

1. H. Ariga, J. Edwards, and D.A. Sullivan, *Clin. Immunol. Immunopath.* 53:499 (1989).
2. A.C. Vendramini, C.H. Soo, and D.A. Sullivan, *Invest. Ophthalmol. Vis. Sci.* 32:3002 (1991).
3. E.H. Sato, H. Ariga H, and D.A. Sullivan, *Invest. Ophthalmol. Vis. Sci.* 33:2537 (1992).
4. E.H. Sato, and D.A. Sullivan, article submitted (1993).
5. D.A. Sullivan, J. Edwards, C. Soo, A.C. Vendramini, H. Ariga, and E.H. Sato, article submitted (1993).
6. D.A. Sullivan, and E.H. Sato, *Clin. Immunol. Immunopath.* 64:9 (1992).
7. R.I. Fox, ed., "Sjögren's Syndrome," *Rheum. Dis. Clin. N.A.* vol. 18 (3) (1992).
8. R. Bruckner, *Ophthalmologica* 110:37 (1945).
9. M. Appelmans, *Arch. 'Ophtalmologie* 81:577 (1948).
10. A. Bizzarro,G. Valentini, G. Di Marinto, A. Daponte, A. De Bellis, and G. Iacono, *J. Clin. End. Metab.* 64:32 (1987).
11. J.D. Wilson, and D.W. Foster, eds., "Williams Textbook of Endocrinology, Seventh Edition," W.B. Saunders Company, Philadelphia (1985).
12. F.J. Rocha, E.H. Sato, and D.A. Sullivan, article submitted (1993).

# NATIONAL SJOGREN'S SYNDROME ASSOCIATION

Betsy Latiff

National Sjogren's Syndrome Association
3201 West Evans Drive
Phoenix, Arizona 85083

Dear friends, there are differences all around us. We think differently. We act differently. Different things are important to us. We even look different. We can compare our differences all day and never get to know one another. At this point, you may be wondering why I chose to speak about differences? You certainly hear enough about this from the hollow rhetoric of bureaucrats. Why mention differences at a gathering of people that are interested in relieving pain and suffering?

My reason is this. We can allow our differences to impede us or we can look to our common ground to enrich us. We have that choice.

Look around at the world. Once a nation was divided by differences. Once differences erected a wall. For years, people were kept from one another. They could not share discoveries. They could not share resources. They couldn't even relate to one another as caring human beings. But, a vision of a common goal grew bigger than the differences. The desire for freedom tore the wall down. Economic strength is being built. Freedom is being expressed. Families are being united. Healing is taking place! East is meeting West and our quality of life will improve as a result. Our "us and them" attitude is slowly turning to "we."

Imagine if you were injured in a serious accident. Just before the emergency room physician was going to save your life, he was requested to verify that he was politically, socially, and religiously compatible with you. You're probably thinking, that sounds ridiculous. Well, we do just that when we dwell on our differences rather than support our common goals.

Doctors and their patients tend to behave likewise. At times, the physician intentionally or unintentionally gears down his explanations or isolates himself from the painful experiences that his patient is sharing with him.

*Lacrimal Gland, Tear Film, and Dry Eye Syndromes*
Edited by D.A. Sullivan, Plenum Press, New York, 1994

Patients also try and hide their true feelings from their physicians and say "he'll never understand what I'm going through."

We limit what both can accomplish. Patients of Sjogren's Syndrome face incredible disbelief and suspicion from many health care professionals. They reach points of desperation because they feel that no one is listening. Doctors can be unaware of the distress and mental anguish that well meaning reassurances can cause when they are superficial.

Yet, I have seen what can happen when informed doctors treat Sjogren's patients. Both doctor and patient realize that together they can strive to seek a path to healing. Through their mutual cooperation, they have faced their common fear and are gaining ground.

My own story began in isolation and fear. Isolation from even my family because I thought that that couldn't possible understand what I was going through. Fear of anything that might upset the delicate balance of health. Only when I began to break down my walls was I able to make real progress. Only when I could convince my physicians that the enemy was Sjogren's and not me, did doors begin to open in drug and medical fields.

I have experienced the "cut glass in the eye pain" with torn and ulcerated corneas, temporary blindness, pericarditis, persistent oral and vaginal yeast infections, kidney flare-ups chronic sinusitis, and my most recent, interstitial cystitis. Last year, I was treated with D.M.S.O./steroids weekly for many months. It is difficult to express the anxiety felt before each treatment. Can you empathize with the knowledge of chemicals being induced into a raw bladder? I started having nightmares before appointments. The F.D. A. saved me, by opening the doors to get the drug, Elmiron, by a compassionate I.N.D. There are many helpful drugs out there for our relief -- many of you have wonderful research discoveries soon to become available for the Sjogren patient. Because of my suffering, and because of your interest, we are making progress.

As a member of the National Sjogren Syndrome Association, our mission is to be an authentic resource to those who suffer from this dreaded disease and to assist the medical community, in whatever way we can, in providing the necessary care for Sjogren's patients. We are determined to bring awareness to the medical community, the afflicted and the community at large.

We provide the resources needed to cope with the disease, we are a referral service assisting the patient on connections with a whole host and variety of caregivers. We support research, praying that we will one day discover the cures for this formidable disease, and we are a support group enabling the individual to find comfort in the midst of deepening anxieties. You see, we do care, not just about the disease, but also about our neighbors, those who suffer from Sjogren's.

We have working chapters of NSSA throughout the United States and in many other parts of the world. Those of us in leadership maintain contact with the local chapters. We provide these local chapters with information about our involvement with the disease and also about what is happening on the international scene in medicine, treatment programs, and government regulatory activities. While our progress, has been steady, I am sure you will agree, there is much work to be done.

I believe suffering produces perseverance and character. My purpose and hope being here will be satisfied if you leave and act on tearing down the walls of differences. I want you to know that this is real. I want you to hear the truth about what can happen if we work together.

National Sjogren's Syndrome Association (NSSA) sponsors support groups throughout the world. It publishes the Sjogren's Digest, Patient Education Series and a patient guide, Learning to Live with Sjogren's Syndrome, as well as a patient/health professional unique educational video for our Outreach Program.

The National Sjogren's Syndrome Association is an international, non-profit, all-volunteer organization dedicated to providing emotional support to patients and their families and educational information to both patients and health professionals worldwide. NSSA is also committed to encouraging research which will identify the cause of Sjogren's Syndrome and how it can be cured.

For further information about NSSA, contact: Barbara Henry, President, 3201 West Evans Drive, Phoenix, Arizona 85023, U.S.A., or call 1-800-395-NSSA.

# SJÖGREN'S SYNDROME FOUNDATION INC.

Elaine K. Harris

Sjogren's Syndrome Foundation Inc.
382 Main Street
Port Washington, NY 11050

## GOALS

The Sjogren's Syndrome Foundation (SSF), which had its birth in December, 1983, at the Long Island Jewish Medical Center (LIJMC), New Hyde Park, NY, came into this world as "The Moisture Seekers". In 1985, when the Foundation was formally incorporated, the Board of Directors voted to use the name of the disease as the official name of the organization for more recognizable identification. *The Moisture Seekers* name was retained for our monthly newsletter.

The purposes of SSF are to:
- Educate patients and their families about SS and help them cope with it.
- Increase medical awareness and knowledge about Sjögren's syndrome.
- Increase public awareness and knowledge about Sjögren's syndrome.
- Support and stimulate research to find better treatments and develop a cure for Sjögren's syndrome.

In order to accomplish this, the SSF:
- Publishes a 12-page monthly newsletter (11,000 monthly distribution to patients and professionals); is translated into Danish, Dutch, French, Hebrew, Japanese and Spanish.
- Sponsors local support groups.
- Sponsors educational symposia and regional conferences.
- Publishes *The Sjogren's Syndrome Handbook* (over 14000 copies sold).
- Awards research fellowships and summer student fellowships.
- Maintains library including back issues of newsletter; cassette tapes of symposia; selected video tapes of symposia; speakers' bureau. A subject index covering the contents of all back issues of *The Moisture Seekers* will be available Fall 1993.
- Works with professionals and pharmaceutical companies to develop new and/or improved therapeutic modalities; announcement of clinical trials appears in *The Moisture Seekers.*

## PROGRAMS AND ACTIVITIES

### Sjögren's Syndrome Symposia

"Living with Sjögren's, Day-to-Day with a Chronic Disease", the subtitle of our first annual symposium, aptly described the focus of the program, namely to bring to patients and interested health professionals the latest news concerning treatment and practical ways of dealing with the various aspects of the disease. Currently we present two to three major all day symposia annually, for both patients and professionals; professionals attending receive continuing education credits from their respective accrediting organizations. The programs feature recognized authorities who address the many different aspects of SS. Time is allowed for audience questions and "roundtable responses" from the participating specialists. The symposia presentations are reported in *The Moisture Seekers* and tapes of the programs are also available, as well as selected videos.

### Young Moisture Seekers (YMS) – SS Patients Under 40

It has become quite apparent that SS builds up slowly and insidiously, that although the destruction may not become obvious until a woman reaches middle age, the deterioration actually started much earlier in her life. It is imperative that studies be undertaken to determine when and why the lymphocytic infiltration of the exocrine glands begins so that prophylactic measures can be undertaken to prevent or at least impede the destruction that eventually occurs. If there are 4 million SS patients in the US, 3,600,000 of whom are women (SS ratio, 9 women to 1 man), studies to determine just when SS begins and professional education to teach early identification techniques are a necessity so that these patients can be identified early enough to begin treatments, thus helping to prevent corneal abrasion and to preserve natural teeth. These same young women frequently experience what one could call "premature vaginal dryness", resulting in irritation, infection and painful intercourse. Early identification of young women with SS is important so that SS patients contemplating having a family are aware of the possible problems that may occur during their pregnancy.

To respond to the special needs of these younger SS patients, in 1992 the SSF created a special support group, the Young Moisture Seekers. Special meetings of this group, with medical speakers, are held in conjunction with our semi-annual symposia. These patients also have a special hot line.

### The Moisture Seekers Newsletter

Each issue of *The Moisture Seekers* (11 issues per year) carries informative articles related to various aspects of SS, articles that provide information to patients as well as to professionals who are not well acquainted with the "many faces" of SS, particularly as this applies to the most current information on etiology and treatment in the areas other than their particular specialty.

## Support Groups

Local support groups and Contact Leaders (members who respond to local requests for information) attempt to increase local public and medical awareness about SS. SSF Support Groups require a leader or steering committee and a medical advisor. The support groups hold educational meetings with local medical speakers, or host special meetings when a visiting SS expert is coming to their area. The groups' activities must conform with the Foundation's standards and by-laws and the IRS regulations for not-for-profit foundations. Many of our groups are doing outstanding work in educating the public as well as their members. Our high standards have resulted in recognition by many "networking" organizations. The National Institute of Dental Research (NIDR), the Arthritis Foundation, health columnists, the American Academy of Ophthalmology (AAO), the Office of Public Health Information Clearinghouse (OPHIC) of the National Institutes of Health (NIH) each refer SS patients to our Foundation for information and help on SS.

## Medical Interest and Awareness

As the number of diagnosed SS patients increases, so too does the interest of the medical community in responding to their need for knowledgeable and caring medical care. The Foundation is actively involved in encouraging and assisting medical centers to set up SS teams for the treatment of Sjögren's syndrome patients. Recently developed centers include SS teams at the Hospital for Joint Diseases in New York City and the Toronto General Hospital in Toronto, Canada.

## International Affiliates

Several of our overseas medical friends who I have met at professional conferences are working with their local patients to start organizations similar to ours in their own countries.

## SS Statistics

Abbey Meyers, Executive Director of the National Organization for Rare Disorders states, "Sjögren's syndrome is not really a rare disorder; it is simply massively undiagnosed). According to the Center for Disease Control in Atlanta, Georgia, Sjögren's syndrome doesn't even exist. They have no statistics on its incidence. However, we hope this deplorable situation will soon be remedied. Dr. Stanley Pillemer of NIAMS, National Institutes of Health is heading an inter-institute project (NEI and NIDR are also involved) to conduct an epidemiologic study on SS.

## Need For Patient Literature

When I was finally diagnosed as having Sjögren's syndrome (after a year of having gone from doctors to doctor with my medical complaints), I became aware of the acute need for patient literature and information about this hard-to-diagnose, poorly understood, massively ill-treated disease. Our Foundation responds to inquiries for information with a packet consisting of a leaflet, *"Do You have?"* about SS, a *Patient History Questionnaire* (lists the many apparently discrete manifestations of SS which many people would not think of mentioning to the examining doctor), developed of course with doctor help, a copy of our newsletter, *The Moisture Seekers.*, and a flyer about the handbook. We have a membership package of articles which are sent out upon receipt of an application.

When doctors request information, we respond with a letter outlining the highlights of our programs and services, a set of the membership materials, and several scientific professional articles. We have found that although the doctors themselves do not always become paying members of the Foundation, they do keep our materials to show their SS patients and encourage them to join the Foundation.

Our handbook, *Sjogren's Syndrome: An authoritative guide for the patient* represents the first time that a book on Sjögren's syndrome has been written by medical specialists expressly for patients and their families. This handbook is another example of the cooperation that exists between doctors and our Foundation. Each of the 26 authors and the three medical editors - physicians, dentists, psychologists, nutritionists, etc.- has contributed his/her services. Many doctors keep a copy in their desks to show new patients.

The SSF exhibits our literature and services at professional conferences, such as the American Academy of Ophthalmology, the American College of Rheumatology; and the American Dental Association annual conferences. Connective tissue diseases such as scleroderma, lupus, and Raynaud's syndrome, not only affect SS patients, but have several common manifestations. Networking among the organizations representing these diseases is important. We share patients, information, and skills.

**Financial Support of Research**

Beginning with the academic year 1991-2, the SSF has awarded an annual $20,000 fellowship for research on SS. For the 1993-4 academic year, the Foundation will also award Summer Student Fellowships in the amount of $2,000 each to encourage early interest in SS. In addition, the Foundation presents a plaque at the Academy of Rheumatology and also the International Association of Dental Research for the best abstract submitted on SS.

The Sjogren's Syndrome Foundation has turned Sjögren's syndrome from a disease that hardly anyone had even heard about into one that is receiving increasing attention and help from medical professionals.

# LACRIMEDICS, INC.

Lacrimedics, Inc. is proud to have been a sponsor of the International Conference on the Lacrimal Gland, Tear Film and Dry Eye Syndromes: Basic Science and Clinical Relevance. As one of the leading suppliers of products and services for lacrimology, Lacrimedics continues to promote increased awareness for the benefits of the Herrick Lacrimal Plug for long term treatment of dry eye syndromes and related disorders. It is the hope of Lacrimedics, Inc. that practioners around the world will begin to implement this very simple, practical treatment method as part of their primary care offering.

## HERRICK LACRIMAL PLUGS™

The Herrick Lacrimal Plug is considered by many as the next generation of lacrimal plug. Its' differentiating feature is that it is inserted past the punctum, down into the horizontal canaliculus. It does not rest on the punctum or touch the eye at any time. In addition, the Herrick Lacrimal Plug generally requires no anesthetic or dilation for insertion. To date, more than 100,000 patients have been successfully treated with the Herrick Lacrimal Plug.

## LACRIMAL EFFICIENCY TEST™

To help determine the efficacy of long-term lacrimal occlusion with the Herrick Lacrimal Plug, Lacrimedics, Inc. offers the Lacrimal Efficiency Test. Dissolvable collagen implants are inserted through all four puncta, down into the horizontal canaliculus for an evaluation period of four to seven days. If the patient experiences positive symptomatic relief, Herrick Lacrimal Plugs are indicated.

## THE BENEFITS OF LACRIMAL OCCLUSION

Beyond the long-term treatment of dry eye, lacrimal occlusion with lacrimal plugs can increase contact lens comfort and wearing time; increase the efficacy of topically-applied ocular medications; and help to decrease systemic side-effects caused by ingestion.

## DEVELOP A LACRIMOLOGY SUB-SPECIALTY

Lacrimedics, Inc. believes a lacrimology sub-specialty, utilizing Herrick Lacrimal Plugs for lacrimal occlusion, can greatly benefit your patients and your practice. To help you promote the benefits of lacrimal occlusion to your patient base, Lacrimedics, Inc. offers a variety of patient communications materials specifically designed to educate your patients about the symptoms of dry eye and about the long-term treatment methods available. Practice training aids as well as physician and patient educational videos are also available.

# CONTRIBUTORS

# INDEX

contamination of contact lenses, 429
infection, 351-353
keratitis, 421-425
lid during asymptomatic HIV, 339, 340, 343
lysis, 303, 305
ocular surface, 240, 244-246, 584
Barium, 89
Basement membrane of lacrimal gland, 37-43
Basophils, 223, 588
β-endorphin, 129-131
Benoxinate hydrochloride, 93
Benzamil, 107
Bicarbonate, 83, 84
Biostimulation, 603
Blepharitis, 235, 236, 572
Blepharoconjunctivitis, 514, 515
Blink interval and influence of ophthalmic
   procedures, 525-527
Blinking, 233, 239, 240, 242-245, 264, 266,
   267, 270, 279-281, 490, 491
B-N-acetylhexosaminidase, 363, 364
Bone marrow transplantation, 610
Bromhexine, 385, 386, 390
   derivatives, 141-146
8-Bromoadenosine-3':5'-cyclic monophosphate 5,
   176, 178, 179, 220
8-Bromoguanosine-3':5'-cyclic monophosphate,
   178, 179
Bronchial-associated lymphoid tissue, 161
Burns, chemical and thermal, 472

C-flanking peptide of neuropeptide Y, 616
c-fos mRNA, 225-230
c-myc, 613
C-reactive protein, 371-374
Calciosome, 116
Calcitonin gene-related peptide, 4, 7, 13, 68-72,
   616
Calcium, 18, 89, 90, 115-118, 121-124, 134,
   135, 137, 138, 147, 181, 223, 509,
   584
   calcium channel, 18, 116, 134, 135,138, 152
   cell secretion and regulation, 4, 5, 7, 8
   current, 151-155
Calcium/calmodulin pathway, 113
Calmodulin, 4
Calmodulin-dependent protein kinase activity, 124
*Candida albicans*, 356
Carbachol, 14, 32, 33, 35, 40, 64, 67, 68, 82-85,
   90, 111-114, 122, 124, 134, 135,
   138, 142, 148, 154, 155, 176-179,
   219, 223, 226-228, 652, 654
Carbon dust, 265
Carbonic anhydrase, 31-35, 91
Castration, 101, 616

orchiectomy, 158-160, 175, 190-192, 220,
   222, 481
ovariectomy, 102, 158, 159, 175, 190-192,
   220, 222
Cataract, 515, 568
Catecholamine, 653
Cathepsin B, 652-654
CD21+ antigen, 642, 643
CD44+ molecule, 163
Cell membrane recycling, 81, 82, 85, 111, 114
Chelerythrine hydrochloride, 122, 124
Chernobyl, 513, 515
Chlamydial infection, 169, 170
Chloride, 79, 83, 84, 100, 105, 109, 111
   channel, 79, 88-90
Cholera toxin, 162, 176-179, 220
Choline, 149, 150
Cholinergic agonists, 121, 147, 148, 150
Cholinergic nerves, 687
Chronic fatigue syndrome, 681
Chronic rhinitis sicca, 386
Circadian rhythm, 93-97
$Cl^-/HCO_3^-$ antiporter, 83
Clinical trials, 433-436, 553-559
   biostatistical data analysis, 556
   control groups, 555
   ethical considerations, 554
   masking considerations, 555
   patient followup, 556
   patient population, 555
   phases, protocol development, efficacy,
      conduct and analysis, 433-436
   protocol and study design, 555
   randomization, 555
Closed eye, 240, 244, 427-430
   acetylcholine, 428
   carbon dioxide levels, 427
   complement activation, 345
   contact lens adverse responses, 428
   corneal sensitivity, 428
   corneal ulceration, 428
   endothelial blebs, 427
   endothelial polymegathism, 428
   oxygen tension, 427
   plasminogen, 345
   polymorphonuclear leukocytes, 345
   secretory IgA, 345
   stromal acidosis, 427
   tear film, 302, 304, 305
   temperature and environmental ocular surface
      changes, 427
   vitronectin protection against inflammatory
      damage, 345-349
Collagen, 22
   implants, 572, 574

Oxymetazoline, 386

Paracrine stimulation, 3
Parakeratitis, 655
Parasympathetic innervation, 2-4, 7, 13, 67, 68,
        71, 393, 396, 397
Parkinson's disease, 474, 480
Parotid gland
    acinar epithelial cell culture, 178, 179
    acinar epithelial cells and sialodacryoadenitits
        infection, 194, 196
    flow rate, 667-670
Patch clamping, 87, 135, 152
Peanut agglutinin, 183
Pemphigoid, 472, 481, 564, 566-568
Pemphigus, 586
Percelan, 38, 39, 42
Perforin, 613
    mRNA and protein, 637-640
Peroxidase, 13, 31-33, 35, 37, 40, 41, 42, 68,
        148, 149, 309, 327, 328, 330, 589
Petrolatum, 432
PGP-5, 615
Phenazopyridine, 587
Phenyl mercurinitrate, 691
Phenylephrine, 3, 32, 33, 35, 148-150, 393, 394,
        396, 397
4β-Phorbol-12,13-dibutyrate, 113, 121, 123, 124,
        148
Phorbol ester, 118, 121-124, 147, 149
4β-Phorbol-12-myristate-13-acetate, 121, 123,
        124
Phosphatidylcholine, 148-150
Phosphatidylinositol 4,5-bisphosphate, 4, 121,
        147
Phosphatidylserine, 122, 123
Phospholipase $A_2$, 149, 150
Phospholipase C, 4, 115, 117, 118, 121
Phospholipase D, 4, 122, 124
Phospholipid, 121, 123, 124
Pilocarpine, 5, 278, 587, 631
Pinguecula 572
Plasma cell, 133, 134, 138, 642
    Harderian gland, 11, 13, 16, 17, 151, 154, 155
        maxi-K-channels, 64
        muscarinic receptors, 62, 63, 134
    lacrimal gland, 22
        IgA, 161, 185-188, 301, 302
Plasmin    304, 345, 348
Plasminogen activator, 304
Platelet-derived growth factor mRNA, 216
Platelet-derived growth factor-α, 213, 216
    receptor, 213, 216
Platelet-derived growth factor-β, 213, 216
    receptor, 213, 216

p-Nitrophenyl N-acetyl B-D-glucosaminide, 363
p-Nitrophenyl penta-N-acetyl B-chitopentaoside,
        363, 364
Poloxamer, 422-425
Polyarteritis nodosa, 481
Polyarthritis, 586, 684
Polychlorinated biphenyls, 473
Polyethylene glycol, 268, 269, 422-424
Polymerase chain reaction, 213, 642, 644
Polymorphonuclear leukocytes, 223, 240, 345,
        348, 474, 477, 536
Polymyositis, 586, 610, 655
Polyphosphomannan, 182
Potassium, 79, 441, 442, 445, 467, 509
    channel, 79, 88-91, 137, 138, 152, 154, 155
    current, 152, 155
Practolol, 472
Prednisolone and formulation for topical
        application, 399-402
Primary biliary cirrhosis, 610, 612
Progesterone, 159
Progestins, 219
Prolactin, 75-77, 85, 99, 100, 102-104, 219, 583,
        588, 589, 616, 653, 684
    mRNA, 104
    receptor, 589, 590
Proparacaine, 143
Prostaglandin, 475, 477, 691, 694
    $E_2$, 176, 219, 223, 393
Protein G, 303, 335-338, 411
Protein kinase
    C, 4, 118, 147-150
    calcium/calmodulin, 4
    cyclic AMP-dependent, 4, 138
    isoforms, 121-124
Proto-oncogenes, 613, 687
Pseudolymphoma, 681
*Pseudomonas aeruginosa*, 260, 352, 353, 359-
        362, 421-425
*Pseudomonas pyocyanea*, 356
Pterygium, 572
Pulmonary insufficiency, 681

Quinidine, 88

Radiation, 513-515
Rapamycin, 588
Rat cytomegalovirus, 189-192
Rat lacrimal cord and parenchyma accompanying
        major extraglandular ducts, 53-56
Rat von Ebner's gland protein, 208, 209
Recklinghausen disease, 596
Redoxon$^{TM}$, 286
Refractive index, 417-420
Refractometry, 417-420

Epstein-Barr virus reactivation, 647-650
genetic factors, 610-612, 617
retrovirus, 610, 617, 618
primary vs. secondary, 609-612
tear proteins, 303, 305
treatment
androgen-induced suppression of lacrimal
gland inflammation, 683-690, 697-
700
azothioprin, 567
bromhexine, 390
corticosteroids, 567, 686
cyclosporine, 567, 591, 592, 686
danazol, 686
17β-estradiol, 686
sodium sucrose sulfate, 691-696
tear jet, 577-582
Sjogren's Syndrome and Patient Education Series,
703
Sjogren's Syndrome Foundation, 705-708
Snake-like chromatin, 655, 656, 658, 692, 693,
695
Sodium, 79, 82-85, 88-90, 100, 111, 467, 469
Sodium hyaluronate, 268, 269
Sodium pentosan polysulfate, 694
Sodium sucrose-sulfate, 691-695
Specular reflection microscopy, 294, 295
Sphingosine, 122-124
Spiperone, 143
Squamous metaplasia, 292, 454, 477, 535, 538
*Staphylococcus aureus*, 356
*Staphylococcus epidermoidis*, 356
Staurosporine, 122-124
Steven's Johnson syndrome, 564, 596
Strabismus, 518
*Streptococcus pyogenes*, 356
Submandibular gland -
lymphocyte infiltration androgen suppression,
686
neural and endocrine control of secretory
component production, 178, 179
sialodacryoadenititis infection, 194, 196
Substance P, 4, 7, 13, 25-27, 29, 30, 57, 68-72,
219, 616, 617
Sulfacetamide sodium, 399, 400, 402
Sulfadiazine, 587
Sulfo-N-hydroxysuccinimidyl biotin, 82
Sulfonic stilbenes, 83
Sulphonimide, 591
Superficial punctate keratitis, 596, 599
Superoxide, 475, 477
Supradyn$^{TM}$, 286, 287
Symblepharon, 478, 481
Sympathetic innervation, 2-4, 7, 13, 67, 68, 393,
396, 397

α-Sympathicomimetic agents, topical, 385, 386
Synaptophysin, 615
Systemic lupus erythematosus, 481, 496, 586,
610, 618, 619, 623, 655, 681, 682,
684, 687
Systemic sclerosis, 481, 610

T cell, 11, 17, 18, 163, 166, 181, 223, 624, 625,
626, 628, 647, 679, 682, 685, 687
autoreactive, 583, 589
CD4$^+$, 609, 611, 613, 614, 615, 625, 626,
628, 638, 640, 641, 651-654
CD8$^+$, 613, 618, 625, 628, 641, 644
CD45$^+$, 613
cytotoxic, 642, 644
"double negative", 626
helper, 587, 680, 685
suppressor, 587, 588
suppressor/cytotoxic, 685
Tear film
advances in ocular tribology, 275-283
aqueous layer, 1, 3-7, 231, 233, 234, 239,
240, 242-247, 277, 279-281, 432,
489, 491, 492
boundary lubrication, 243
breakup time, 232, 233, 236, 250, 367, 378,
473, 474, 476, 481, 483, 489, 491,
492, 496-498, 500, 501, 506, 507,
509, 525-527, 539, 541, 557, 601,
602, 667, 668, 669, 670, 692, 693
non-invasive, 476, 479, 480, 541
closed eye, 432
albumin, 428
complement activation, 428, 429
conjunctival epithelial cells, 428
plasmin activity, 428, 429
polymorphonuclear leukocytes, 428, 429
secretory IgA, 428
composition and biophysical properties, 231-
238
composition and function, 1, 335
contact lens interactions, 403-410
evaporation, 235, 242, 243, 491
effect of age, 271-274
influence of gender, 272, 274
lipid layer, 231, 232, 235, 239, 240, 242-244,
277, 279, 281, 293-297, 432
distribution and spreading, 491
effect on tear evaporation, 271, 274
structure and abnormalities, 489-493
mucous layer, 231-234, 236, 239-247, 255,
260, 277, 278, 280, 282, 359, 432
rheological properties, 234, 236, 267-270

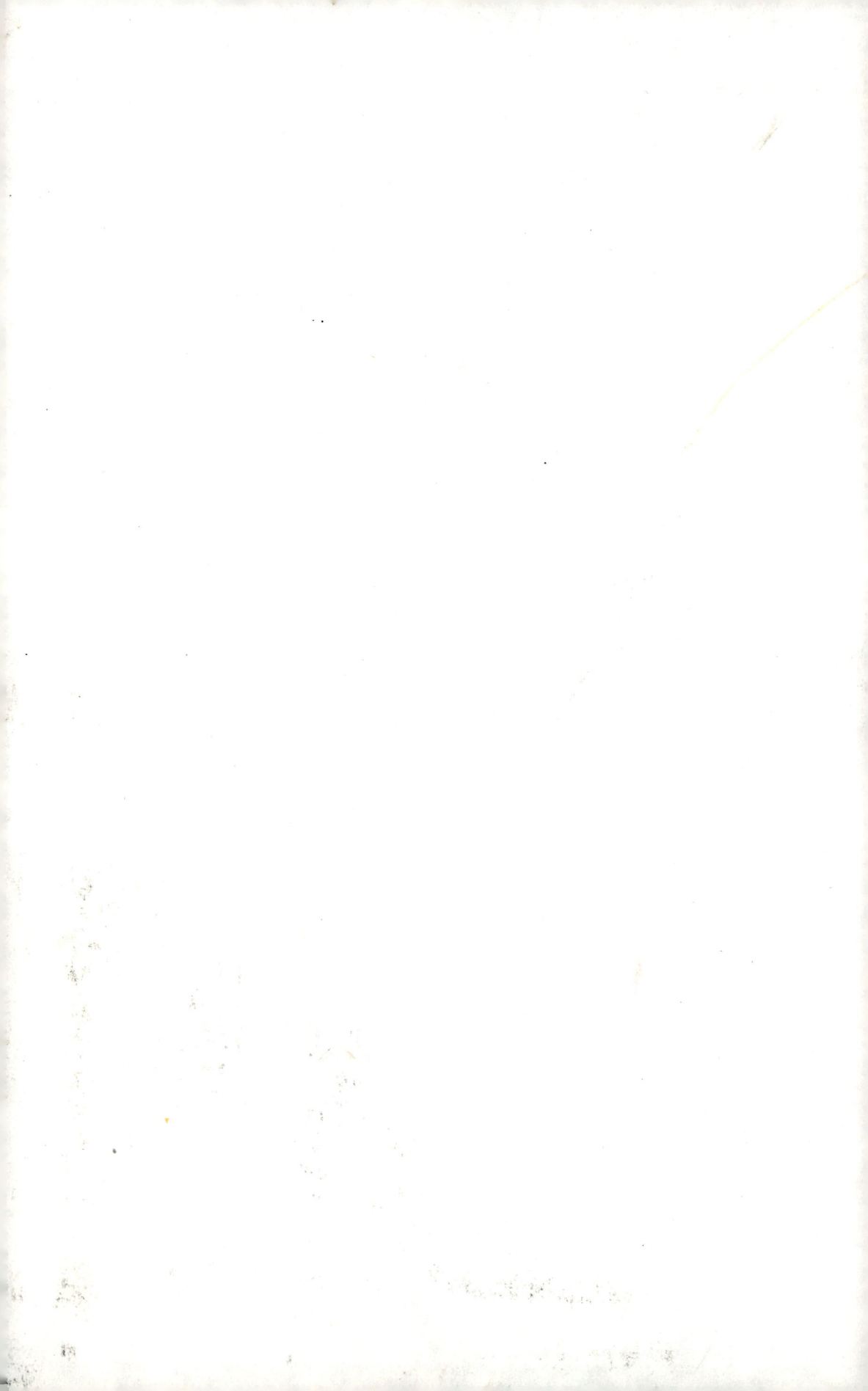